High-Brightness Accelerators

NATO ASI Series

Advanced Science Institutes Series

A series presenting the results of activities sponsored by the NATO Science Committee, which aims at the dissemination of advanced scientific and technological knowledge, with a view to strengthening links between scientific communities.

The series is published by an international board of publishers in conjunction with the NATO Scientific Affairs Division

A	**Life Sciences**	Plenum Publishing Corporation
B	**Physics**	New York and London
C	**Mathematical and Physical Sciences**	Kluwer Academic Publishers Dordrecht, Boston, and London
D	**Behavioral and Social Sciences**	
E	**Applied Sciences**	
F	**Computer and Systems Sciences**	Springer-Verlag
G	**Ecological Sciences**	Berlin, Heidelberg, New York, London,
H	**Cell Biology**	Paris, and Tokyo

Recent Volumes in this Series

Series B: Physics

High-Brightness Accelerators

Edited by

Anthony K. Hyder and
M. Franklin Rose

Auburn University
Auburn, Alabama

and

Arthur H. Guenther

Air Force Weapons Laboratory
Kirtland AFB, New Mexico

Plenum Press
New York and London
Published in cooperation with NATO Scientific Affairs Division

Proceedings of a NATO Advanced Study Institute on
High-Brightness Accelerators,
held July 13–25, 1986,
in Pitlochry, Scotland

Library of Congress Cataloging in Publication Data

NATO Advanced Study Institute on High-Brightness Accelerators (1986:
Pitlochry)
    High-brightness accelerators / edited by Anthony K. Hyder, M. Franklin
Rose and Arthur H. Guenther.
        p.   cm.—(NATO ASI series. Series B, Physics; v. 178)
    "Proceedings of a NATO Advanced Study Institute on High-Brightness
Accelerators, held July 13–25, 1986, in Pitlochry, Scotland"—T.p. verso.
    "Published in cooperation with NATO Scientific Affairs Division."
    Includes bibliographical references and index.
    ISBN-13: 978-1-4684-5510-6   e-ISBN-13: 978-1-4684-5508-3
    DOI: 10.1007/978-1-4684-5508-3
    1. High-brightness accelerators—Congresses. 2. Particle beams—Con-
gresses. I. Hyder, A. K. (Anthony K.) II. Rose, M. F. (M. Franklin) III. Guenther,
Arthur Henry, 1931–   . IV. North Atlantic Treaty Organization. Scientific
Affairs Division. V. Title. VI. Series.
QC787.H53N38   1986   88-17662
539.7'3—dc19   CIP

© 1988 Plenum Press, New York
Softcover reprint of the hardcover 1st edition 1988

A Division of Plenum Publishing Corporation
233 Spring Street, New York, N.Y. 10013

PREFACE

A NATO Advanced Study Institute (ASI) on High-Brightness Accelerators was held at the Atholl Palace Hotel, Pitlochry, Perthshire, Scotland, from July 13 through July 25, 1986. This publication is the Proceedings of the Institute.

This ASI emphasized the basic physics and engineering of the relatively new and fast-emerging field of high-brightness particle accelerators. These machines are high- to very-high-current (amperes to hundreds of kiloamperes), modest-voltage (megavolt to tens of megavolts) devices, and as such are opposed to those historically used for high-energy physics studies (i.e., gigavolt and higher energies and rather low currents).

The primary focus of the Institute was on the physics of the accelerator and the beam, including the dynamics, equilibria, and instabilities of high-current beams near the space-charge limit; accelerator engineering techniques; and the applications of high-brightness beams in areas such as free-electron lasers, synchrotron-radiation sources, food processing, and heavy- and light-ion fusion.

The Institute concentrated on bringing together several diverse but related communities which, we hope, benefited from this opportunity to interact: the North American activity in machine technology, engineering, and diagnostics with the strong European theoretical community; the basic beam physicists with the engineering technologists.

The theme of the Institute was the enhancement of scientific communication and exchange among academic, industrial, and government laboratory groups having a common interest in high-brightness accelerators and their applications. In line with the focus of the Institute, the program was organized into five major sessions dealing with the current and historical concepts of brightness and perveance, the physics of high-brightness accelerators, the physics of high-brightness beams, the engineering of high-brightness accelerators, and applications of high-brightness beams in science, industry, and defense.

The initial pace of the Institute allowed ample time for informal discussion sessions to be organized and scheduled by the participants and lecturers with the encouragement and assistance of the Directors. As the Institute progressed, the interest and demand for these additional sessions grew and they consumed much of the unscheduled time.

A departure from the normal format of an ASI which greatly enhanced this Institute was the opportunity for participants to contribute poster papers in two evening sessions. The posters remained in place during virtually the entire ASI and served as a catalyst for technical interaction among the participants during all of the breaks. Eleven of these poster papers have been selected by the editors for inclusion in the proceedings and are presented in the appendix.

The Institute was attended by one hundred participants and lecturers representing Belgium, Canada, the Federal Republic of Germany, France, Italy, Turkey, Spain, the United Kingdom, the United States, Israel, Japan, and Switzerland.

A distinguished faculty of lecturers was assembled and the technical program organized with the generous and very capable assistance of an Advisory Committee composed of Dr. Henri Doucet (France), Dr. Terry Godlove (USA), LtCol Richard Gullickson (USA), Dr. Bruce Miller (USA), Dr. John Nation (USA), Dr. Jean Pierre Rager (Belgium), Dr. Charles Roberson (USA), Dr. Winfried Schmidt (FRG), Dr. Alan Toepfer (USA), and Mr. Richard Verga (USA). The Institute was organized and directed by Dr. Anthony K. Hyder (USA), Dr. M. Franklin Rose (USA), and Dr. Arthur H. Guenther (USA).

The value to be gained from any ASI depends on the faculty - the lecturers who devote so much of their time and talents to make an Institute successful. As the reader of these proceedings can see, this ASI was particularly honored with an exceptional group of lecturers to whom the organizers and participants offer their deep appreciation.

We are grateful to a number of organizations for providing the financial assistance that made the Institute possible. Foremost is the NATO Scientific Affairs Division which provided not only important financial support for the Institute, but equally important organizational and moral support. In addition, the following US sources made significant contributions: Auburn University, Naval Surface Weapons Center, Air Force Office of Scientific Research, Office of Naval Research, Office of the Secretary of Defense, Science Applications International, Department of Energy, Los Alamos National Laboratory, Livermore National Laboratory, Sandia National Laboratory.

We would also like to thank Dougal Spaven and his staff at The Atholl Palace Hotel for a truly enjoyable and memorable two weeks in the Scottish Highlands. The accommodations, food, service and meeting facilities were superb. And to the Tenth Duke of Atholl, our thanks for the use of Blair Castle for a magnificent banquet on the evening of the Royal Wedding.

Our appreciation goes also to Pat Whited for her assistance as Secretary to the Institute, to Kate Hyder for work on the graphics in the book, to the EG&G Washington Analytical Service Center in Dahlgren, Virginia, which undertook the task of centrally retyping the lecturers' manuscripts and producing a camera-ready document to Plenum Press, and to Susie M. Anderson of EG&G for producing an exceptional product.

An finally to the people of Perthshire, Scotland, who certainly displayed, in every way, 'Ceud Mile Failte' (Gaelic for 'A Hundred Thousand Welcomes').

Anthony K. Hyder
M. Franklin Rose
Auburn, Alabama

Arthur H. Guenther
Albuquerque, New Mexico

December, 1987

CONTENTS

INTRODUCTION TO HIGH BRIGHTNESS

ACCELERATOR PHYSICS

BEAM PHYSICS

ACCELERATOR ENGINEERING

APPLICATIONS OF HIGH-BRIGHTNESS BEAMS

SUMMARY

Appendix A: Poster Papers

INTRODUCTION

HIGH-BRIGHTNESS ACCELERATORS

This volume contains the proceedings of the 1986 "High Brightness
Accelerators" NATO Advanced Study Institute. The object of the Institute
was to address issues concerned with the generation and applications of
low emittance, high current beams. The scope of the workshop was very
broad. It addressed the fundamentals of accelerator design both from the
point of view of the needs of the high energy physicist and also from the
perspective of more general applications. The range of accelerator
designs and concepts reviewed varied from the TeV, low current devices to
the multi-megampere, MeV accelerators used for the study of inertial
confinement fusion. In almost all applications the beam brightness, the
current per unit area per unit solid angle, plays an important role.

A recent study of the adequacy of U.S. accelerator technology
carried out for the Department of Energy[1] concluded that there is a need
to expand the funding commitment to this technology since there appears
to be major payoff potential in several areas of national need. Areas
singled out in the list of potential applications included:

Energy Source Department	Nuclear Waste Management
Acid Rain Control	Food and Materials Processing
Medical Technology	Microelectronics
Nuclear Reactor Reliability	

The report addresses these issues from the point of the view of the U.S.
National needs. It is apparent however that the applications have a far
broader interest than those of any one country.

Accelerator applications require a spectrum of beam energies ranging
from tens of kilo-electron volts for semiconductor processing to energies

---

[1] Department of Energy Report ER-0176, "Assessment of the Adequacy
of U.S. Accelerator Technology for Department of Energy Missions," E. T.
Gerry and S. A. Mani, September 1983.

in excess of 10^{12} eV for high energy physics. Similarly the instantaneous beam current requirements span more than thirteen orders of magnitude. To illustrate the variety of beam requirements we consider the
wealth of potential industrial applications which require beams having
energies in the range of 10^7 eV. These include waste transmutation,
materials testing, light ion fusion, nuclear medicine, radiative processing of foods, fossil fuel smoke stack cleanup, and laser drivers. From
this range of applications we note that food processing requires beams of
about 100 mA in about 10 microsecond pulses. In comparison, light ion
fusion requires beam currents of order 10 MA in short (~20 nsec) duration
pulses. Other applications, such as the flue gas cleanup, require more
modest (~1 MeV, 3-30 A) steady state DC beams.

A useful parameter to categorize the beam requirements is the beam
brightness. A very bright beam has a low transverse energy spread. The
beam quality, which is typically limited by thermal velocities at the
source, aberrations in the beam transport system and other imperfections,
is expressed quantitatively by the normalized emittance, a measure of the
area occupied by the beam in the transverse xx' (or yy') space. In this
expression x represents the particle position and x' the divergence of
the particle trajectories measured along their equilibrium orbits.
Physically the transverse emittance determines the focal spot size for a
beam. The beam brightness is closely related to the transverse emittance
and is proportional to the beam current divided by the product of the x
and y transverse emittances. The concept of beam brightness has its
origins in electron optics and is closely related to the beam brightness
used in the study of optical systems. During the course of the institute
the utility of the so called four dimensional brightness was questioned
in relation to a six dimensional brightness or ideally the particle
distribution function. Nonetheless the concept has wide use and is at
minimum a useful criterion to specify the beam quality.

In several of the applications listed earlier the beam brightness is
of critical importance. The viability of a particular accelerator design
for high energy physics or heavy ion fusion, for example, is determined
by the value of the beam brightness with a high premium being placed on
maximizing this measure of beam quality. The free electron laser is
another example where the physics of the device places strict requirements on the beam brightness. One might note that the six dimensional
brightness is also of interest in this latter application since the
number of particles interacting to generate a coherent radiation output
also depends on the longitudinal beam energy spread. In other applications such as waste transmutation or smoke stack clean-up the emittance
(brightness) is of only limited importance.

An interesting feature of recent work is the attempt to exploit a mechanism which leads to a growth in the beam emittance in an accelerator, and hence limits the accelerator performance, to serve as a new and potentially attractive driver for an accelerator. In a practical accelerator the beam emittance typically grows as the particle traverses the accelerator as a result, for example, of the feedback of the beam electromagnetic fields, through the accelerator cavities, on the particles in the beam. A proposal has been made recently to use this 'wake field' effect usefully as a driver for a new type of accelerator. This is one of the new approaches to accelerator design which was to be reviewed during the meeting.

In the proceedings of the Advanced Study Institute the concepts of accelerator design are systematically developed. A total of 30 invited papers are presented. In addition about 24 contributed poster papers were presented; 11 of these are also included in the proceedings. The papers are printed in the order they were presented at the meeting. The first session was devoted to the lessons to be learned from the historical development of accelerators with special attention paid to the concept of beam brightness. The next major topic addressed dealt with the principles underlying the various types of accelerator. This session not only addressed the classical accelerator concepts but also reviewed novel accelerator ideas currently being explored. A major portion of the meeting was devoted to the study of beam physics. At the start of this section the program focused on the physics of the particle codes, which today play a central role in the design of any accelerator. This is followed by a study of the beam dynamics in linear and circular accelerators and includes a discussion of the limits on the performance of accelerators due to beam instabilities which might, for example, place an upper bound on the beam current. In this section of the program some of the constraints placed by specific applications on the brightness of the beam are developed and show the interplay between accelerator design and the proposed accelerator application. The next section of the program addressed specific engineering aspects of accelerator design ranging from power conditioning to superconducting cavity design and included a discussion of focusing structures and field gradient limitations. The final section which spanned three days of the ASI deals with applications of accelerators. In view of the current high interest in free electron lasers there is a special emphasis on this application. The proceedings concludes with an overview of the applications of high brightness accelerators.

The ASI was planned to take a non specialist in accelerator design from the basic concepts of accelerator physics through to obtaining at least a superficial understanding of the current status and problems of contemporary accelerator design. As a result this volume should serve as a useful and timely reference text for those interested in exploring this field.

BRIGHTNESS, EMITTANCE AND TEMPERATURE

J. D. Lawson

Rutherford Appleton Laboratory
Chilton
Oxon, United Kingdom

INTRODUCTION

During the past fifty years there has been a spectacular growth in the number of applications for particle beams. The various practical requirements scan orders of magnitude in energy, current, and particle mass, from electrons to uranium ions. The behaviour of some of these beams, especially when self-fields are large and interaction with walls or background plasma is significant, can be extremely complex. Conditions can vary markedly along the length of the beam, and change rapidly with time.

In order to discern some overall pattern, it is helpful to look for relations between the parameters of the beam which enable one to classify the different types of behaviour, in the form of scaling laws, and suitably defined invariants and figures of merit.

In the past, emphasis has very often been on achieving high energies and high current, though for many applications, such as electron microscopy and electron beam lithography, a small spot size with adequate current density has always been the goal. More recently the need for very precisely focused beams with high quality has become increasingly evident in other fields. An important example, looking towards the future, is the electron-position linear collider for high energy physics. Because of the enormous energy loss from synchrotron radiation when electrons are bent into circular orbits, linear machines must be used at energies exceeding about 100 GeV. In order to utilize the available kinetic energy of the beams, head-on collisions are required; in linear as opposed to orbital machines the beams can only collide once, and this implies extremely small bunches of particles if an adequate collision rate is to be obtained with affordable power.

Another topical device requiring a high quality beam is the free-electron laser, an important source of intense radiation at infrared and optical wavelengths. In order to obtain adequate coherence of the radiation a beam is required which is well localized at the centre of the undulator, with very small spread of particle energies.

Quantitatively, beam quality is determined by the related concepts of emittance and brightness. Two components, transverse and longitudinal, can be distinguished, the latter being equivalent to energy spread. This means that particles with different energies travel in different trajectories through lenses and deflecting magnets, giving rise to dispersion, often referred to by the optical term 'chromatic aberration'.

In this talk the meaning of brightness will be examined, and related to the emittance concept developed by the accelerator community. Also related is the beam temperature, sometimes a useful concept, reflected in the term 'cold' for a good quality beam. Before attempting formal definitions, it is useful to look qualitatively at different kinds of beam. A good pictorial understanding of some simple, idealized beams forms a useful conceptual framework for understanding more complicated ones.

THE CONCEPT OF A BEAM

A good, universal, definition of a beam is not easy to provide. Roughly one might say that it is an ensemble of particles moving in almost the same direction, and having almost the same energy. The transverse dimensions are generally small compared with the length. It may be rectilinear in shape, or bent round in a curve as in synchrotron accelerators and storage rings. Very often an 'axis' or 'equilibrium orbit' can be defined, which represents the path of a particle with particular position and energy. For rectilinear systems this is often conveniently taken as the symmetry axis, which is independent of particle energy.

In many situations the self-interaction between the particles in the beam is small, and can be considered as a perturbation. Nevertheless, even if small it can have important consequences. This is the case in most conventional accelerators, where the self-fields ultimately become destructive and limit the beam current or degrade the beam quality. In 'intense relativistic electron beams', where extremely high currents flow for a very short time, these self-fields determine the beam behaviour and must be taken into account ab initio. In this case the beam may be regarded as a rather special form of plasma, and the concepts of plasma physics are useful in describing its behaviour.

6

It is important here to emphasize a property of beams of low or moderate intensity that often makes them much simpler to deal with than conventional plasma. To first approximation the transverse and longitudinal motion can be decoupled and treated separately. Indeed, in very many situations the two transverse degrees of freedom can be decoupled also, and we later take this very simple situation as a starting point.

If one moves 'with the beam', at a velocity corresponding to the mean of all the particles in the neighbourhood, then the environment is a charged gas; this is in general confined transversely by a focusing system, often an array of lenses of some sort which as seen by the particle appears as a potential well that varies in time. Typically this is provided by an array of quadrupoles, giving a focusing force of rapidly alternating sign, positive in one plane and negative in the other, resulting in overall dynamic focusing. At high current this focusing force is opposed by the outward self-field of the beam. The limits to the current that can be transported in such systems, and also the degradation of beam quality, have recently been the topic of much study in connection with fusion driven by heavy ions. Details will be presented later during this meeting.

In conventional synchrotrons and storage rings there are bending magnets in addition to focusing quadrupoles, and both in linear accelerators and synchrotrons there is a longitudinal accelerating field, which can be considered locally as a travelling wave with a component of electric field along the orbit, with phase velocity equal to the velocity of the beam. Although in practice such a field may be provided by a succession of gaps, it can be fourier analysed into a set of travelling harmonic components. This provides a longitudinal potential well, of the form shown schematically in Fig. 1. Particles can either be trapped, and carried upwards in energy, or they can 'phase slip' continuously, without net gain of energy.

To a rough but useful approximation, known as 'the smooth approximation' the varying transverse potential that contains the particles can be replaced by an equivalent steady potential well that varies slowly. The particles can be considered as a gas contained in a 'soft walled' box. For small oscillations in all directions the motion is simple harmonic, the transverse and longitudinal frequencies are known as betatron and synchrotron frequencies, ω_b and ω_s. These may vary during the acceleration. If the changes are adiabatic the action, equal to energy/frequency (observed in the moving frame) remains constant, the particle 'gas' heats or cools according to the classical gas laws.

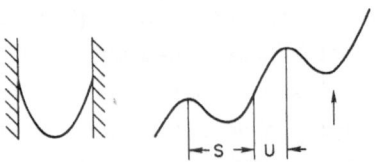

Fig. 1. Transverse and longitudinal potential wells seen by
a particle confined by a focusing system and accelerated in a travelling wave. In the former, which
is in general time-dependent, motion is limited by
the walls of the apparatus; in the latter there is
a finite range of phases over which stable motion
is possible.

The subject is rather more complicated in detail, especially for
orbital machines. First, the appropriate relativistic transformation for
fields and frequencies from the laboratory to the moving frame must be
made and second, the effective mass of the particles must be adjusted.
It is convenient to consider motion with respect to a reference particle
sitting at the bottom of the potential wells. Rectilinear motion, as in
a linear accelerator, is considered first. For transverse motion, the
mass of the particle is simply its rest mass m_o in the moving frame, or
γm_o in the laboratory frame. In the longitudinal direction, the mass
observed in the laboratory frame is $\gamma^3 m_o$; this is readily seen from the
equation of motion in the laboratory frame

$$eE = d(\beta\gamma\, m_o c)/dt = \gamma^3 m_o d(\beta c)/dt \; . \tag{1}$$

The situation is more complicated for orbital machines. If a particle originally stationary at the bottom of the longitudinal potential
well is given additional acceleration, it will travel on an orbit of
greater radius, and not move forward as rapidly as in a linear machine.
This makes it appear heavier. Indeed, the mass becomes infinite when the
effect of increasing speed is just compensated by the longer path, so
that the time for one revolution of the machine is not altered, and
beyond this it appears negative. This mass can be written

$$m^\star = \frac{\gamma\, m_o}{\dfrac{1}{\gamma^2} - \dfrac{1}{\gamma_T^2}} \tag{2}$$

where γ_T is known as the 'transition energy' where the mass changes sign.
For a synchrotron accelerator with strong focusing this can occur during
the accelerating cycle. As γ increases, the change of velocity has less
effect than the change of orbit radius. Its value depends on how tightly
the particle is focused transversely. For a uniform field, as in a
cyclotron $\gamma_T = 1$, and $m^\star = -\,\gamma\, m_o/\beta^2$. This negative mass effect has

8

interesting consequences; the gas has an effective longitudinal temperature that is negative, this is important when collisions (discussed
later) couple the degrees of freedom. Also, self-fields become effectively attractive rather than repulsive in the longitudinal direction,
and if the potential well is weak, (or absent, as in a betatron) this can
cause 'condensation' as in star clusters. In accelerators and plasmas
this is the 'negative mass instability', responsible for the mechanism
made use of in gyrotrons. With the mention of collisions, negative mass
instability, and gyrotrons, the subject of interaction between particles
in the accelerator has been introduced. Such interaction has many
important consequences, and limits the current that can be accelerated in
conventional machines. It is discussed qualitatively in the next
section.

THE EFFECTS OF INTERACTIONS BETWEEN PARTICLES IN THE BEAMS

Analysis of motion in classical particle accelerators is now well
understood. Hamiltonian dynamics provides a basis for calculating the
motion of single particles. Ensembles, without particle-particle interaction require simple statistical mechanics, as we see later. Before
proceeding, we look at a general way what happens when interaction
between the particles is included.

We distinguish two types of interaction that become important as the
current is increased; first, a particle 'becomes aware' of the smooth
'collective' field of the other particles. For example, in a uniform
cylindrical beam, a radial electric field develops. This can be found
from Gauss' theorem, by treating the beam as charged cylinder. If this
radial force exceeds the focusing force, the beam is no longer confined.
For relativistic beams there is also an inwardly directed magnetic force,
equal to β^2 times the electric force.

These self-fields can disrupt the beam by perturbing the focusing.
In circular accelerators they impose a severe limit to the current. If
the number of betatron oscillations Q per revolution is integral or half-
integral, magnet errors drive resonances which make the beam particles
hit the wall. Practical machines, therefore, operate with non integral
values of 2Q. Self-fields weaken the focusing and give rise to
'Q-shifts' ΔQ. These must, then, be less than about 1/4Q. For large
machines Q is typically between 10 and 100 so that 1/4Q is small. The
self-fields can also give rise to collective behaviour of the beam in the
form of longitudinal or transverse waves that can propagate along it.
When these waves have negative energy, that is, there is less energy in
the beam in the presence of waves than when they are absent, they can
interact with the resistive and reactive components of the walls, giving

rise to various types of unstable growth. Similarly, beams propagated through plasma couple with plasma waves to give, for example, longitudinal two stream instability. (It is interesting that ions trapped in electron beams can also give rise to transverse two stream instability.)

Instabilities, though interesting, are not the topic of this paper, and we now note the second type of interaction observed in beams. There is a fundamental difference between these, as we describe later. This second type of interaction occurs through individual particle-particle collisions. It can be important even where there are rather few electrons in a beam. For example, in a very low diameter 'cross-over' where an electron microscope beam is focused through a very small diameter waist, the electron spacing is often greater than the beam diameter. Under these circumstances two electrons close together can interact in such a way that the front one is accelerated and the rear one decelerated. This induces enough energy spread to spoil the final focus. This is the 'Boersch effect' (1954) that is the limiting physical effect in beam shaping for electron beam lithography. It occurs as 'Touschek effect' in low energy electron storage rings, and ultimately disperses the beam in proton storage rings. (Note the 'hot' negative temperature in the longitudinal direction, which prohibits equipartition of the three degrees of freedom!)

We return later to a discussion of these two classes of interaction.

INTERACTION OF BEAM WITH GASES AND PLASMA: NEUTRALIZATION

As noted already, the interaction of beams and plasma can be very complex. This is particularly true if ionization processes are initiated by the beam. This topic is not pursued here, but we note that positive ions can reduce the radial electric field in a high current beam. Since the electric and magnetic fields are in the ratio 1 to β^2, the net force for neutralization fraction f is proportional to $1 - \beta^2 - f$; this becomes inward if $f > 1 - \beta^2 = 1/\gamma^2$, and the beam then forms a 'pinch'. When the behaviour is time dependent, inductive effects can cause 'magnetic neutralization', a reversed current flowing in the plasma, that reduces the pinching. Another way of explaining this is to say that magnetic fields cannot be quickly created in the high conductivity plasma.

A number of phenomena have been touched on in a descriptive way in this introductory section. Many of these will be amplified and explained more fully by other speakers. Others will be taken up again in later sections.

THE PARAXIAL RAY EQUATION FOR A BEAM WITH NON-INTERACTING PARTICLES, AND
THE EMITTANCE ELLIPSE

A good starting point for studying beams is to take the simplest
possible situation, and add the complications as we proceed. We start,
then, with a monoenergetic beam of non-interacting particles focused
about the z-axis by an array of lenses, (or a continuous focusing
channel). Angles that the particles make with the axis are small, and
the focusing elements provide forces towards the axis that are proportional to the particle displacement. The axis may be gently curved, but
in a plane, (to include bending magnetic fields in orbital accelerators).
The x and y directions are perpendicular to the orbit in and perpendicular to the orbit plane respectively, and z is the direction measured
along the axis. (For curved axes this is often denoted by s, but we use
z here.) Equations of motion in the x and y directions are uncoupled and
of similar form; it is only necessary, therefore, to consider one of
them.

If the force towards the axis is αx, then the equation of motion in
the x-direction is

$$\gamma m_o \ddot{x} + \alpha(z)x = 0 \ . \tag{3}$$

This may be written as a trajectory equation by transferring the independent variable to z rather than t

$$\overset{''}{x} + \kappa_o(z)x = 0 \ . \tag{4}$$

Evidently for a uniform focusing channel, with κ_o independent of z,
the particle moves in a sinusoidal orbit with $\kappa_o = (2\pi/\lambda_b)^2$, where λ_b is
the 'betatron wavelength'. This is illustrated in Fig. 2(a).

Equation (2) is a second order equation with two independent solutions. These can be chosen as $x = c(z)$ and $x = s(z)$, with initial conditions (1, 0) and (0,1). These are known as the 'cosine-like' and 'sine-
like' solutions. Any solution $x(z)$ be expressed as the sum $a_1 c(z) +$
$a_2 s(z)$ with gradient $a_1 c'(z) + a_2 s'(z)$. If c and s are known, the value
of x and x' at $z = z_1$ can be related to the values at $z = z_o$ by the
matrix

$$\begin{pmatrix} x_1 \\ x'_1 \end{pmatrix} = \begin{pmatrix} c(z_1) & s(z_1) \\ c'(z_1) & s'(z_1) \end{pmatrix} \begin{pmatrix} x_o \\ x_o' \end{pmatrix} \ . \tag{5}$$

Associated with this matrix is the Wronskian determinant $W = cs' -$
$c's$. Differentiating and substituting in Eq. (4) yields $W' = 0$. The

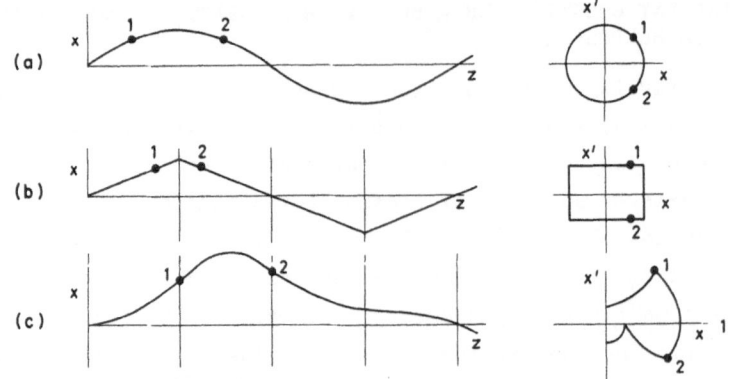

Fig. 2. Paraxial trajectories in a) a channel with uniform
transverse focusing, b) an array of equal focusing
lenses and c) alternating focusing and defocusing
sections of equal strength, corresponding to a
quadrupole array. Trajectories in xx' space are
also shown.

determinant of the matrix is therefore invariant, equal to its initial
value cs' - c's = 1.

It is interesting now to consider the mapping in the xx' plane of
the position of a particle as it moves along the z axis. If the focusing
is uniform, κ_o = constant, and the trajectory is clearly an ellipse. If
the particle is focused by a succession of thin lenses, then the trajectory consists of lines parallel to the x axis (with constant x') which
'jump' to another value of x' as the particles cross the lens. This is
shown in Fig. 2.

If, now, we consider a short length Δz of the beam surrounding one
of the beam particles rather than a single particle, the xx' plane
contains an ensemble of points, rather than just one. We enquire how
such an ensemble develops as the beam passes down the axis. This can be
done algebraically, by making a change to phase-amplitude variables. It
is straightforward to show that:

1) A distribution that is originally a centered ellipse of uniform
    density in xx' space remains elliptical and uniform.

2) The eccentricity of the ellipse, and the orientation of its axes
    in general vary, but the <u>area</u> <u>is</u> <u>invariant</u>.

If the area of this ellipse is A, then the quantity A/π may be
called the 'emittance' of the beam. It is further readily shown that:

3) The equation for the <u>envelope</u> of the beam is given by

$$a'' + \kappa_o a - \frac{\varepsilon^2}{a^3} = 0 \ . \qquad\qquad (6)$$

12

This is the same as the single particle equation with x = a and the addition of the term $\varepsilon^2/a^3 = 0$.

4) If the particles are accelerated in the direction of motion, ε is no longer constant. If instead of x we choose as independent variable the 'reduced' coordinate

$$X = (\beta\gamma)^{1/2}x \tag{7}$$

then the 'normalized emittance' $\beta\gamma\varepsilon$ is conserved, and the envelope equation becomes

$$A'' + \frac{\gamma^2(\gamma^2+2)}{4\beta^4\gamma^4} A - \frac{\varepsilon_n^2}{A^3} = 0 . \tag{8}$$

Although all these facts follow fairly directly from the theory of linear second order differential equations, the invariance of ε_n can be seem more directly as a consequence of Liouville's theorem. This can be stated in several different ways. Here we use a form appropriate to the development in time of the distribution of an ensemble of non-interacting particles in an external potential. The theorem follows directly from Hamilton's equations and the equation of continuity, and states that in phase-space the density of points in the neighborhood of a particular point remains invariant. The phase space has twice the dimensionality of ordinary space, the coordinates being the positions of the particles and their momenta.

In the example chosen the canonical coordinates are x, and the transverse coordinate of mechanical momentum $p_x = \beta\gamma m_0 c x'$. Since the density of points remains constant, the area in xp_x space remains invariant. This area is just $m_0 c$ times the invariant normalized emittance described above. Strictly it is the density in the six dimensional $xyzp_xp_yp_z$ space that is conserved. To paraxial approximation with monoenergetic particles, however, the distribution has no thickness to first order in the p_z direction, since p_x is small, and $\Delta p_z/p_z = \frac{1}{2}(p_x/p_z)^2$ is a second order quantity. Further because of the linearity of the x and y motion the emittances ε_x and ε_y are separately conserved.

The distribution represented by a uniformly populated ellipse with sharp edges is not physically realistic, but it is a simple and useful idealization for approximate calculations. Such a distribution represents the projection of a uniformly populated hyper-ellipsoidal three dimensional shell in the four dimensional phase-space. It is known as a K-V distribution after the Soviet physicists Kapchinskij and Vladimirskij who described it in 1959. For the special case of an upright ellipse,

13

corresponding to a point in the beam where a' and b' are both zero, (a maximum or minimum in the envelope), this has the simple form

$$\frac{x^2}{a^2} + \frac{y^2}{b^2} + \frac{x'^2 a^2}{\varepsilon_x^2} + \frac{y'^2 b^2}{\varepsilon_y^2} = 1 \tag{9}$$

with axes parallel to the coordinate axes. Projections of a uniform hollow shell from four to two dimensions results in a uniform distribution of points. This may be compared with the theorem of Archimedes; a spherical shell in three dimensions projected on to an axis (one dimension) also gives uniform distribution with a sharp edge.

A simple beam, consisting of sinusoidal particle trajectories in a uniform focusing system, is illustrated in Fig. 3. In this case the orbits correspond to points on the circumference of the ellipse, and the whole ellipse rotates as the distance z along the axis is increased. Where the beam is narrow, x is small and x' large; where it is wide on the other hand, x' is small and x large. Roughly, the two dimensional 'temperature' is large when the 'volume' is small and vice versa. The gas law $TV^{\gamma-1}$ = constant is easily verified.

If all phases of oscillation of the beam particles were present, the envelope would be uniform. The beam would then be 'matched', and the ellipse would be stationary. (With the scale chosen in Fig. 3 it would be a circle.) In such a beam a" = 0, so that from Eq. (4)

$$\varepsilon = a^2 \kappa_o^{1/2} \sim a^2/\lambdabar \tag{10}$$

the emittance represents, therefore, the product of the beam radius a and $a/\lambdabar$, a typical angle that the beam makes with the axis.

Another interesting case is when there is no focusing; particles move in straight lines with envelope, again from Eq. (4)

$$a" = \varepsilon^2/a^3 . \tag{11}$$

This represents a hyperbola; if the waist has radius a_o, then the asymptotic half angle is $\theta = \varepsilon/a_o$.

FURTHER DISCUSSION OF THE MEANING OF EMITTANCE

The definition of emittance in the previous section was a very restricted one, appropriate to a beam governed by linear forces, and having a special and unnatural configuration. In this section we discuss how this can be extended.

A more realistic distribution might be hyperellipsoidal in phase space, with density decreasing from the centre as a gaussian in all four directions x, y, x', y'. The projections in this case are all gaussian

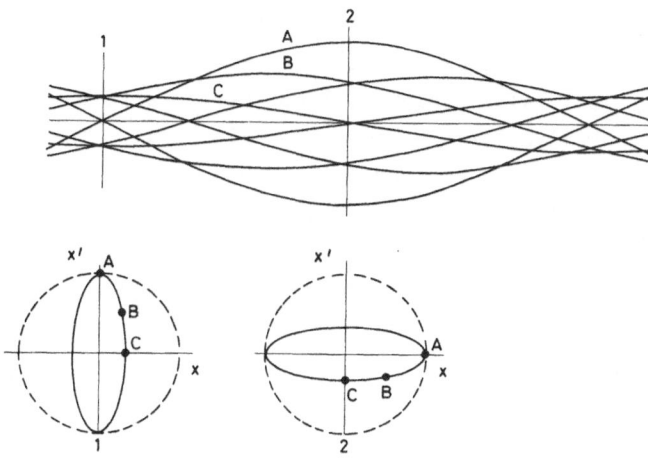

Fig. 3. Trajectories of particles forming a mis-matched
beam in a uniform focusing channel. At any point
in the beam x and x' for the trajectories illustrated lie on an ellipse, which rotates as the
particles move along the beam. Points on the
ellipses corresponding to these typical trajectories are shown.

also. The real space part just represents a gaussian density distribution; a gaussian in x' space corresponds to a Maxwellian distribution of
velocities, and hence a transverse temperature.

In order to make a formal connection between emittance and temperature, it is necessary to extend the definition to distributions that are
not uniform, elliptical, and sharp edged. It is required to represent a
distribution with a great deal of structure by a single number. There
are several ways of doing this, appropriate to different situations. A
convenient convention is to take the r.m.s. emittance ε, defined as

$$\bar{\varepsilon} = 4(\langle x^2 \rangle \langle x'^2 \rangle - \langle xx' \rangle^2)^{1/2} \tag{12}$$

where $\langle x^2 \rangle$ is the mean squared value of x of all the points. The
normalizing factor 4 is introduced to provide compatibility with the
previous definition.

For an elliptical gaussian distribution the second term is zero, and
$\bar{\varepsilon}$ is readily evaluated. For example, a cylindrical beam of r.m.s. radius
a (with no z-variation), and transverse temperature $kT_{\perp} = 1/2(\gamma m_o \beta^2 c^2 x'^2)$
has r.m.s. emittance

$$\bar{\varepsilon} = 4a(kT/\gamma m_o \beta^2 c^2)^{1/2} . \tag{13}$$

It is not difficult to prove that for a linear focusing system, and
non-interacting particles, the r.m.s. emittance is conserved. This is
not in general true if either of these conditions is violated.

The term 'acceptance' or 'admittance', closely related to emittance, is often used for a focusing channel, or other piece of hardware. Such a device has physical walls, which will not accept particles too far from the axis, or travelling at too large an angle to it. These limitations can often be expressed as contours on the xx' and yy' planes at the point of entry. The smallest centred elliptical area that is within this contour (which may not be elliptical) is known as the acceptance, α. Clearly a K-V beam with $\varepsilon < \alpha$ will be passed without loss. Sometimes with realistic distributions the emittance is defined in terms of the ellipse containing some fraction, say 95%, of the particles. A beam with $\varepsilon = \alpha$ would then lose 5% to the walls.

THE LONGITUDINAL EMITTANCE ε_z

So far a monochromatic beam has been assumed, with zero extent in zz' space. In considering the transverse emittance, particles in a thin slice of length Δz were considered. In a long beam with no structure in the z direction, the longitudinal emittance cannot usefully be defined. If, however, there is an accelerating wave that provides an effective potential well of periodic length λ_0, then a momentum Δp_z can be defined equal to $p_z - p_0$, where $p_0 = \gamma m_0 \beta_w c$ and $\beta_w c$ is the phase velocity of the wave. The position Δz is measured from the bottom of the potential well shown in Fig. 1. The coordinates of the emittance diagram are Δz, and $\Delta p/m_0 c$. If there is finite transverse emittance also, then for small oscillations the distribution is xyz space may be roughly ellipsoidal. There is, however, no simple six dimensional distribution corresponding to the K-V distribution.

NON-LINEAR FOCUSING SYSTEMS

So far, only linear systems obeying Gaussian optics, have been considered. If aberrations are included, then third order terms must be included in the trajectory equation. In general there are very many such terms, including cross terms containing products of x, y and their derivatives.

In a non-linear system the density in the four dimensional xx'yy' space is conserved. If, however, the distribution starts off as an ellipsoid, the shape can become distorted and 'filamented' in such a way that the effective volume is increased. This effect is illustrated in Fig. 4 for a hypothetical one dimensional lens with spherical aberration. The _actual_ area is constant, but the _effective_ area is increased. This phenomenon of 'mixing' is well known in statistical mechanics, and is illustrated by the classical example of dropping ink into water. When the mixing is so thorough such that the individual black and colorless

16

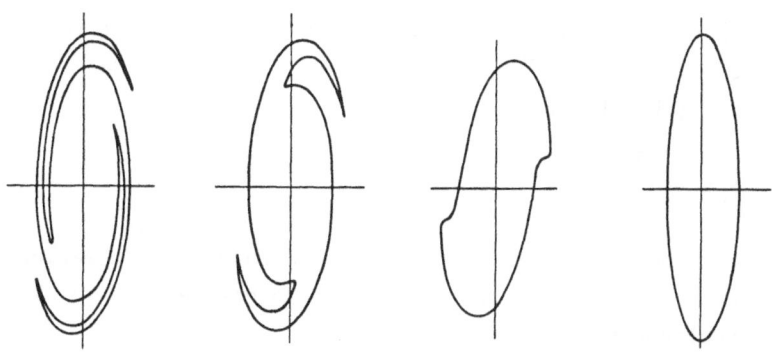

Fig. 4. Progressive distortion of a phase-space ellipse at points in a long focusing channel, consisting of an array of one dimensional lenses with small 'spherical' aberration. The true area is invariant, but the effective area is increased by phase-space filamentation.

streaks can only be perceived as grey, the density is truly diluted. Similarly if the fine structure of the emittance diagram can no longer be perceived, the emittance has 'effectively' grown. This comparison can be put on a formal footing by noting that the entropy corresponds to the logarithm of the emittance. The 'cell size' corresponds to the resolution of the apparatus for actually measuring the emittance.

It is important also to note in non-linear systems that the transverse emittances, defined as projections, are no longer independent. Although the density in xx'yy' space is conserved it is not true that the projected densities in xx' space and yy' space are also conserved, even for mild distortions, where filamentation has not developed. Growth of r.m.s. emittance (or, less commonly, decrease) can occur.

EFFECT OF COLLECTIVE FORCES

When the intensity of an initially dilute beam is increased, space-charge forces appear, which weaken the external focusing. Only when the density of the beam is uniform with radius do these additional forces increase linearly with x or y, and thus maintain linear paraxial optics. This is only possible with the unrealistic K-V distribution. Nevertheless, this is sometimes a useful approximation. Space-charge forces can readily be incorporated into Eqs. (6) and (8) by adding an additional term $-K/a$ or $-K/A$, where K is the beam perveance, written in terms of the line density of charges in the beam N and their classical radius $r_c = Nq^2/4\pi\varepsilon_o m_o c^2$, or alternatively the ratio of the current I to the Alfven current $I_A^o = (4\pi\varepsilon_o m_o c^3/q)\beta\gamma$:

$$K = \frac{2Nr_c}{\beta^2\gamma^3} = \frac{2}{\beta^2\gamma^2}\frac{I}{I_A} . \tag{14}$$

For non relativistic beams this can be expressed in terms of the familiar relation $k = I/V^{3/2}$ amps/(volt)$^{3/2}$ as $K = 15,000k$. Equation 6 thus becomes, (when b = a)

$$a'' + K_o a - \frac{\epsilon^2}{a^3} - \frac{K}{a} = 0 . \tag{15}$$

(The familiar 'beam spreading' curve for a zero emittance laminar beam is found by equating the first and last terms to give $aa'' = K$.) When b ≠ a, a must be replaced by (a + b)/2.

It was shown by Sacherer in 1971 that Eq. (15) is in fact true even for non-linear distributions, if r.m.s. values are taken for a and ϵ. Unfortunately, however, $\bar{\epsilon}$ is no longer invariant, and it varies in an undetermined way. Further insights into the nature of emittance growth in space-charge dominated beams have recently been obtained, but a discussion of these is outside the scope of the present review; the topic will be discussed later at this meeting.

Incidentally, the condition that a beam should be 'space-charge dominated' or 'emittance dominated' can easily be found from Eq. (15) by comparing the ratio of the third and fourth terms, $Ka^2/\epsilon^2 >> 1$ or $<< 1$ respectively.

It is interesting to note that Eq. (15) also has a hydrodynamic interpretation. It can be considered as applying to a small volume element of the charged fluid. The inward and outward forces arising from the focusing and space-charge are straightforward; the emittance terms may be interpreted as a further outward force arising from the radial pressure gradient in the beam. For Maxwellian beams the pressure is simply related to the temperature; in general it can be defined formally in terms of the pressure tensor.

In the presence of smooth space-charge forces, it may be shown Liouville's theorem still applies. Formal proof of this fact, from plasma kinetic theory, is rather lengthy. Situations in which the theorem does not apply are described in the next section.

NON-LIOUVILLIAN EFFECTS IN BEAMS

The validity of Liouville's theorem for 6-dimensional phase-space requires that there be no correlation between particle position and inter-particle interaction, except through the smoothed out collective force. In situations where inter-particle scattering, or scattering in a background gas occurs, Liouville's theorem is not valid; the more general Boltzmann equation must be used, and emittance is not conserved. Scattering tends to disperse the beam, and increases the emittance.

It is possible, however, to decrease the emittance by making use of smooth dissipative forces. Since these are not describable by a Hamiltonian, again Liouville's theorem does not apply. The transverse and longitudinal motion of the beam electrons were described earlier in terms of oscillations in potential wells. If smooth dissipation is present the oscillators damp and the emittance decreases. This process occurs in high energy electron synchrotrons and storage rings in which the electrons radiate synchrotron radiation. Indeed, this process is made use of in the 'damping rings' designed to reduce the emittance of the electron and positron beams in the Stanford Linear Collider. Such rings are likely to be a feature of future machines of this class.

The efficacy of the scheme is limited, however, because in fact the radiation is emitted not smoothly, but in pulses corresponding to a succession of quanta of finite energy. This acts as a source of noise that maintains the amplitude of the oscillation at an equilibrium level. At high energies the damping is fast, but the quantum fluctuations are large. At low energies the damping is so slow that it is counteracted by non-Liouvillian inter-particle scattering.

BRIGHTNESS

We now come to the central theme of this meeting, 'brightness'. The idea is intuitively well understood, and has been used in light optics and electron optics for many years. Unlike emittance, it is a function both of position within the beam and direction. It may simply be defined as the current per unit area per unit solid angle. This is evidently directly proportional to the four dimensions phase-space density at the point. It has the same invariance properties as emittance; and in light optics these are expressed in terms of the 'Helmholtz-Lagrange invariant'. It is, of course, possible to define averaged brightness along the axis, as the current divided by the beam area and by the solid angle occupied by the particles. In this case

$$B = \eta I / \pi^2 \varepsilon_x \varepsilon_y \tag{16}$$

where η is a form factor of order unity that depends on the form of the distribution function.

It is interesting to note that for the singular K-V distribution described earlier the phase-space distribution is a hollow shell, so that the brightness along the axis is zero. In this distribution there are no particles moving along the axis, all particles crossing the axis lie on the surface of the cone, as indeed they do at all points in the beam. The angle of this cone decreases with radius, and is zero at the edge. For a matched cylindrical beam of radius a the cone angle is simply $(\varepsilon/a)(1 - (x^2 + y^2)/a^2)^{1/2}$, as can be seen from Eq. (9).

For the more realistic gaussian phase-space distribution the brightness also decreases as a gaussian. Another distribution often used as a
model of a realistic beam is the 'waterbag', in which the phase-space
density is uniform with a sharp edge, so that the brightness is every-
where the same. In the absence of space-charge a matched waterbag
distribution with hyper-ellipsoidal boundary projects to a parabolic
density distribution in real space. It will feature in later lectures.

BEAMS FOCUSED TO A SMALL SPOT

In many applications high brightness is required in order to produce
a finely focused spot. In a monochromatic, paraxial beam, the spot size
can be reduced by progressively increasing the strength of the focusing
lens, and hence the convergence angle. In practice there are well
defined limits arising from chromatic aberration, spherical aberration,
space-charge, and for very small spots, diffraction. Whereas spherical
aberration becomes more serious at large convergence angles, the diffraction and space-charge effects are worse when the angles are smaller. All
this is well understood, and the nature of the 'trade-offs' is clear.

In the newer fields of linear colliders, ion beam driven inertial
fusion, and free electron lasers on the other hand, the situation is not
so clear. More work is needed not only to understand clearly how the
limitations arise, but also how best to design beams to overcome them.
Two problems are strongly interconnected, the first is to find how to
produce high quality beams in the first place, the second is to learn how
to propagate and manipulate them in such a way that the quality is not
degraded. In the presence of space-charge forces, and the inevitable
non-linearity associated with them, and also of interactions with the
walls, which are particularly complicated when the beam is in the form of
bunches small compared with the diameter of the vacuum vessel, this
produces many challenging problems, some of which we shall hear about
during the next two weeks.

CONCLUSION

In this talk some general properties of beams in accelerators have
been outlined in a qualitative way, leading up to a discussion of emittance, and finally a definition of brightness. Although an attempt has
been made to keep the treatment simple, it must be emphasized that when
self-fields are present the behavior of real beams can be very far from
simple. We are still a long way from understanding all the subtleties of
beam behavior in accelerators, and in particular have much to learn about
what brightness can be achieved in the energy and current ranges of interest in the important new developments to be discussed at this meeting.

HISTORICAL OVERVIEW OF HIGH-BRIGHTNESS ACCELERATORS

F. T. Cole

Fermi National Accelerator Laboratory
Batavia, Illinois 60510 USA

INTRODUCTION

The history of high-brightness accelerators cannot be separated from that of accelerators as a whole. Advances in accelerators have been almost as closely connected with intensity as with energy, particularly in fields other than particle physics, in which energy is at a premium. Even here, the relatively new (or perhaps reawakened) interest in colliding beams has been a strong incentive to work on raising intensity and brightness.

The concept of brightness is taken from electron optics; it is defined as the current per unit area and per unit solid angle. Brightness is closely related conceptually to emittance. The emittance θ is the phase-space area in each transverse dimension occupied by particles. (It is conventionally divided by π and by the total momentum p. The units are usually meter-radians and the usual emittance is of the order of a few μm-rad in conventional accelerators. Note that there are different conventions about the factor π. Here we follow Lawson (1977). By Liouville's theorem, the total six-dimensional emittance is conserved and in many cases, the invariant emittance $\varepsilon_N = \beta\gamma\varepsilon$ (where β is the particle speed in units of c and γ is the Lorentz factor) is constant for each transverse dimension throughout acceleration. The factor $\beta\gamma$ is responsible for what is usually called adiabatic damping, the shrinking of transverse momentum and amplitude as the particle's total momentum is increased.

There are two general ways to achieve high brightness, with large current and moderate cross-sectional area or with moderate current and very small cross-sectional area. There are applications in which the goal is high luminosity (event rate per unit interaction cross section)

and high current may not be relevant. For example, the SLAC Linear
Collider now being built will have peak currents within individual
bunches of less than a kiloamp and beams in the crossing region of linear
dimensions a few hundred Angstroms. The first way, using intense
relativistic electron beams accelerating tens or hundreds of kiloamps is
perhaps more directly relevant to this study group, but the second should
not be overlooked.

THE UNDERSTANDING OF SINGLE-PARTICLE MOTION

There was little concern with orbit stability in the early 1930's.
In high-voltage accelerators, the major concern was with voltage holding.
This concern constrained high-voltage accelerators until the development
of gas-insulated systems (Herb et al, 1935). The first cyclic accelerator, in which particles passed through a low accelerating voltage more
than once, was Rolf Wideroe's linac (1928). Wideroe, a seminal figure in
accelerator development, is shown in Fig. 1 as a young man. His success
inspired Ernest Lawrence to invent the cyclotron (Lawrence and Edelfsen,
1930). Lawrence is shown in Fig. 2 with a major piece of equipment he
had charmed out of a radio company. Lawrence and his student M. S.
Livingston soon discovered that they needed to make the theoretical
uniform magnetic guide field taper slightly with radius to observe
accelerated beam and understood qualitatively that this decrease was
necessary for vertical focusing. A series of cyclotrons were built by
Lawrence and his collaborators in the 1930's with largely empirical
understanding of particle orbits. In 1938, Robert Wilson (1938) made the
first quantitative study of cyclotron orbits, in which electric focusing
by the accelerating voltage plays an important part at low energy. This
was one of the few studies of orbit properties to come from Lawrence's
laboratory in the 1930's.

The Lawrence cyclotron was limited to non-relativistic energies (of
the order of 15 MeV for protons) by the onset of the relativistic increase of mass, which disturbed the isochronicity of orbits of different
energy. There were many efforts to invent ways to overcome or circumvent
this relativistic difficulty. In one of these, L. H. Thomas wrote a
paper (1938) widely considered to be mysterious at the time, but later
understood in the context of alternating-gradient focusing, in which he
proposed a solution of the problem of cyclotron motion in the relativistic regime by introducing azimuthal variation of the guide field. This
was in a sense a precursor of the alternating-gradient principle, but it
also had alternating values of the guide field itself. It was also the
first proposal for use of edge focusing in an accelerator.

22

Fig. 1.   Rolf Wideroe in the Laboratory (c. 1929).

Fig. 2.   Ernest Lawrence and the magnet of his second cyclotron.
It had been part of a spark radio transmitter.

Independently, E. T. S. Walton was one of the many inventors of the
induction accelerator, another way to avoid the difficulty of relativity,
and had written down the radial and vertical equations of motion (Walton,
1929) long before anyone else understood that there was even a relevant
problem. He did not pursue the betatron beyond this paper, but turned
with Cockcroft to the voltage-multiplier system, shown in Fig. 3, with
which they observed the first accelerator-induced nuclear reaction
(Cockcroft and Walton, 1932).

The first successful induction accelerator was built by Kerst
(1940). He christened it the betatron. Kerst and his betatron appear in
Fig. 4. What made Kerst's effort successful after several others had

Fig. 3. The first Cockroft—Walton voltage multiplier in the Cavendish Laboratory.

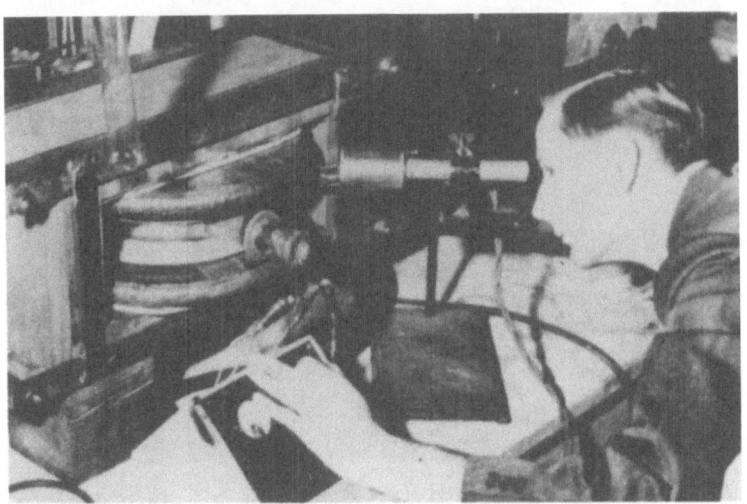

Fig. 4. Donald Kerst and his first betatron.

failed was his understanding of the focusing process and of the injection process (Kerst, 1941). He and Serber wrote a classic paper on focusing (Kerst and Serber, 1941) that affected thinking in the field for many years. Unlike Thomas' work, Kerst's was understood by others. Unlike Walton's work, there was experimental verification in the form of accelerated beam, although not in the form of confirmation of orbit predictions, because experimental methods were not available for such detailed work. There was no confirmation of orbit theory until two accelerator generations later, when the Cosmotron aperture was large enough to allow observation of successive turns at injection with simple fluorescent screens. This first experimental confirmation had a profound effect on the development of accelerators.

The betatron itself has had a long, useful life (in fact, some are still used for medical and industrial radiography), but it was not the road to indefinitely high energy because the weight of the transformer core needed increased so rapidly with energy. The largest betatron built was Kerst's 300-MeV accelerator in the form of the modified betatron, which is discussed below.

World War II interrupted almost all work on accelerators. At the end, when the pressure of war work was easing off, there were two independent inventions of the synchrotron principle, by Veksler in the USSR (1944) and by McMillan in the US (1945). This principle, which is the basis of synchrotrons, synchrocyclotrons, and rf linear accelerators for both electrons and heavier ions, shows that there are stable oscillations in phase and energy about an equilibrium particle being accelerated by a radiofrequency electric field, even when the frequencies of revolution and acceleration are modulated as the energy increases. The validity of the principle was quickly demonstrated by the conversion of a cyclotron to a synchrocyclotron at Berkeley. Then a whole generation of 300-MeV electron synchrotrons was built and used to explore the new world of the pi meson. Figure 5 shows McMillan beside the first 300-MeV synchrotron.

Without explicit knowledge of the synchrotron principle, M. L. Oliphant and his coworkers had proposed (Oliphant et al, 1947) and built a 1-GeV proton synchrotron at Birmingham, which came into operation in 1953. Its usefulness was always hampered by lack of money and space for experiments. In the United States, two proton synchrotrons, the 3-GeV Cosmotron at Brookhaven and the 6-GeV Bevatron at Berkeley, were constructed and put into operation in the early 1950's. Both had long, successful lives as physics-research accelerators. Strange particles, which had been seen in cosmic rays, were identified and studied in detail in Cosmotron experiments. The antiproton was discovered in Bevatron experiments, as well as many other new particles.

The back legs of the Cosmotron magnets interfered with secondary beams coming from internal production targets and M. S. Livingston asked E. D. Courant if reversing some of the magnets, so that their gradients were reversed, would seriously disturb the focusing of the ring. Courant soon found that certain periodic arrangements would in fact improve the focusing. In this way, they, together with H. S. Snyder, invented the alternating-gradient or strong-focusing principle (Courant, Livingston and Snyder, 1952). After strong focusing was announced, Nicholas Christofilos pointed out that he had previously (1950) suggested the same

Fig. 5. Edwin McMillan standing by the first 300-MeV synchrotron.

principle in an unpublished paper sent to the Radiation Laboratory at
Berkeley. Christofilos and a fragment of a newspaper story are shown in
Fig. 6. He was invited to join the Brookhaven group and went on to a
career full of imagination and innovation. Blewett (1952) showed that
alternating-gradient focusing could also be applied to drift-tube linear
accelerators, which had previously utilized grids or solenoids for
focusing. With alternating-gradient focusing, linear accelerators became
high-intensity sources of particles, both for direct use in experiments
and for use as injected beam into synchrotrons. In addition, in the next
few years, Courant and Snyder (with contributions from others) laid the
foundation for an elegant, far-reaching theory of single-particle motion
in accelerators (Courant and Snyder, 1958).

In the early days of alternating-gradient focusing, there was wide-
spread concern about the existence of betatron-oscillation resonances
which could steer the beam out of the vacuum chamber and about the transition energy, the energy at which longitudinal focusing changes from one
side of the voltage wave to the other. When alternating-gradient synchrotrons were built and tried, both of these were real effects, but
neither was as catastrophic as some had feared and they could be managed
easily in practice.

The initial promise of strong focusing was to reduce accelerator
aperture and thus cost. This promise has been richly realized. The cost
of synchrotrons per electron-volt has decreased by an order of magnitude
with strong focusing. But at the same time there have been other
advances that were striking, even though less anticipated, arising from
strong focusing.

26

Fig. 6. Nicholas Christofilos and a part of a New York Times
story about the beginning of strong focusing.

Beam intensity has risen by two orders of magnitude. The Cosmotron
and Bevatron struggled to reach intensities of 10^{11} per pulse while the
Fermilab Tevatron and CERN SPS accelerate more than 3×10^{13} per pulse.
Research programs of more than a dozen simultaneous experiments are run
at these laboratories. But the higher intensity also brings a host of
problems of many-particle effects, which we shall discuss later.

Strong focusing also stimulated great interest in orbit theory.
Collins showed (1961) that it was possible to interrupt the periodic
lattice with long straight sections without disturbing the focusing
properties. All modern synchrotrons have some form of long straight
sections for injection, acceleration, beam extraction, beam manipulation
and observation, and colliding-beam experiments. It is now possible to
design synchrotron rings with widely varying, almost arbitrary properties
of orbit separation and dispersion of different momenta at various
azimuths around the ring.

It was shown by Symon and Sessler (1956) that it is possible in a
ring with guide field constant in time to store successively accelerated
circulating beams. This beam stacking led Kerst (Kerst et al, 1956) to
reinvent colliding beams as a method of reaching higher energy. Many
people had thought of colliding beams as a way to avoid dissipating a
large fraction of the energy in forward motion when a fixed target is
struck by an accelerated particle. Wideroe had even patented the idea in
1943. But the density of particles in a beam used as a target is too

27

small by many orders of magnitude to produce collisions at a useful rate. Kerst pointed out that beam stacking can produce a useful circulating particle density even within the limits imposed by Liouville's theorem. Symon, Sessler, and Kerst conceived of stacking taking place in a fixed-field alternating-gradient accelerator (an outgrowth of Thomas' cyclotron), but it was quickly realized that acceleration and stacking can be done in different rings. Later events showed that a simple synchrotron ring with fixed field, a storage ring, is an efficient way to carry out colliding beams. Colliding beams took some years to come to fruition, but it is now recognized almost universally that any future steps to higher energy will be made by colliding beams. Long straight sections are used in storage rings for collisions, because the detectors are very large in size and require considerable longitudinal space. It may also be noted that it was in this work that Kerst made important advances in the application of the concepts of phase space and emittance to accelerators.

Most single-particle theory has been concerned only with linearized motion about equilibrium. But the time during which a beam circulates and the number of periods of focusing through which it passes in a storage ring are long enough that nonlinear restoring forces arising from imperfections in the system and from kinematic effects can disturb stability. It is well known that there are many nonlinear resonances that can give rise to unbounded motion. It is also well known that there are stochastic effects that can give rise to chaotic motion. These motions have been active topics of research in the accelerator field for many years. Nonlinear resonances have also found a beneficial application in accelerators: they are used to extract beams slowly over many revolutions (Hammer and Laslett, 1961). The combination of nonlinear extraction and thin septa to separate extraction orbits from circulating orbits gives relatively smooth extracted beams over periods of many seconds with beam losses of the order of 1% averaged over months of operation. Beam loss is not only a waste of particles, but also creates residual radioactivity that hinders maintenance.

In the last decade, a number of accelerators have been built with superconducting magnets or acceleration systems. Superconductivity offers no new accelerator principle, nor does it appear to offer smaller magnet-construction cost, but it is important in large systems in reducing the length and construction cost of the accelerator enclosure and in reducing the later cost of operation to somewhat more manageable levels. Without superconducting magnets, it would be economically impossible to build the Supercollider now being considered. In electron rings, even though synchrotron radiation makes the high fields of superconducting

magnets uninteresting, superconducting radiofrequency systems are needed
to reach the full potential of such a large ring as LEP.

Linear accelerators were developed in parallel with circular accelerators. Even before Blewett's suggestion of strong focusing in proton
linacs, a very high intensity proton linear accelerator had been built
and operated successfully (Livdahl, 1982) for breeding of fissile
material (although unfortunately for that accelerator, the shortage of
fissile material went away while it was being built). It was later
dismantled, but there has been discussion of building a similar linear
accelerator for testing fusion materials. Many proton linear accelerators now act as injectors for synchrotrons, producing adequately large
currents. Others act as high-intensity accelerators in their own right.
A superconducting drift-tube linear accelerator, ATLAS at Argonne
(Bollinger, 1983) has also been built.

As discussed below, proton linear accelerators cannot utilize very
high frequencies, because of the need to provide space for focusing.
They therefore usually have frequencies of 200 MHz or less. Electron
linear accelerators have usually used much higher frequencies (approximately 3 GHz). At relativistic energies, there is no need to provide
room for focusing within the rf structure and a simple disc- loaded
waveguide is used. New technological ground was broken in the construction of the Stanford Linear Accelerator (SLAC) in the simplified
fabrication of waveguides by brazing and in the development of high-power
klystrons. With this technology, electron linear accelerators for
medicine and industry have completely swept other kinds of electron
accelerators from that market. Considerable progress has been made in
recent years in superconducting waveguide accelerators, starting from the
first work at the High Energy Physics Laboratory at Stanford.

Thomas focusing was reinvented by many people and enlarged by
Kerst's invention of spiral-sector focusing (alternating edge focusing).
many cyclotrons have been built for nuclear-physics research. It has
also been possible to separate the sectors of cyclotrons, improving the
focusing and making it possible to inject beam from one cyclotron into a
higher-energy one. The first superconducting cyclotron has been built
(Blosser, 1979).

MANY-PARTICLE EFFECTS IN CONVENTIONAL ACCELERATORS

Static Effects

In the 1920's, Langmuir and Tonks (1929) founded plasma physics in
studies of oscillations in electrical discharges, deriving the plasma
frequency (which was in fact known to Lorentz many years earlier in the
electron theory of metals). Later, in the 1930's, self-field effects

were known and understood semi-quantitatively in the development of high-power amplifier triodes and the traveling-wave and klystron amplifiers. These applications were all non-relativistic at that time. The first consideration of self-fields in accelerators was by Kerst (1941), who calculated the poloidal magnetic field of a particle beam and demonstrated the cancellation of electrostatic space-charge forces by this magnetic field. The magnetic field gives rise to a factor $1/\gamma^2$ in the space-charge force (there is another factor $1/\gamma$ from the relativistic increase of mass). Kerst showed that the space-charge force decreases the restoring force of the transverse focusing, reducing the oscillation frequency, a tune shift in modern terms, and gives a space-charge limit when the tune has shifted to zero. In a strong-focusing ring, the tune need shift only to the nearest resonance.

Kerst observed the space-charge limit in the early betatrons. The tune shifts were later measured directly in experiments at MURA. Another effect was also seen in these MURA experiments: an electron beam provides a potential well for positive ions generated by the beam in collisions with the residual gas and these ions can be trapped in the well. The effect of ion trapping is to cancel the electrostatic force in the beam, but not the magnetic force, so neutralization reduces the decrease in tune. The transverse distributions of electrons and positive ions are are not identical and electrons of different amplitude therefore have different cancellations, so there is a tune spread. Trapping is a much larger effect with electron beams than with positive beams, because free electrons are energetic enough that they are not usually trapped and negative ions are not made as easily as positive ions.

Accelerator beams do not circulate in free space, but inside a vacuum chamber. The vacuum chamber walls are conducting in order to avoid charge buildup that would deflect the beam. There are therefore electrostatic and time-varying (eddy-current) magnetic images of the beam induced in the chamber walls and dc magnetic images induced in the magnetic structure. These image forces make important contributions to space-charge forces at extreme relativistic energies ($\gamma \gg 1$), but then the space-charge forces are small. They were first discussed in the literature by Laslett (1963).

Nowadays, static space-charge tune shifts of 0.1 to 0.2 are considered common and acceptable. Space-charge effects are most important at injection, because of the factors of γ. Of course, much larger tune shifts are encountered in the transport of high-power beams, which pass through the system only once and therefore do not feel the full force of resonances. Dynamic effects, which we shall discuss next, are usually more important in limiting intensity, with the exception that ion

30

trapping in electron storage rings is an important limit in practice. In synchrotron radiation rings, which have average circulating currents of hundreds of mA, clearing electrodes, systems to provide transverse electric fields of the order of 100 V, are needed to reach design currents. Many synchrotron radiation rings of the next generation are planned for the storage of positrons to avoid trapping.

It might be thought that SLC would experience large space-charge forces because the beam is so small in size. But the energy is so high ($\gamma \approx 10^5$) that these forces are entirely negligible and do not broaden the focus appreciably.

Dynamic Effects

Breaking up of the beam at higher intensity and intensity limitation from it were first observed explicitly in the Cosmotron, using the longitudinal position electrodes installed to operate the rf feedback system as an observation system. More detailed experiments showed that the cause was a longitudinal instability, the negative mass instability (Nielsen, Sessler, and Symon, 1959). The Cosmotron is a weak focusing ring and therefore above transition energy, so that a proton near the head of a bunch responds to the repulsive space-charge force of the bunch with increasing energy and therefore decreasing revolution frequency, as if it had a negative mass. This increases the bunching of the beam and leads to instability. This is a coherent effect; it only looks like a random breakup because many harmonics are involved. Because it is coherent, the instability can be ameliorated by external forces, which affect all the particles the same way.

Soon after, Laslett, Neil, and Sessler (1965) showed that this is one of many such effects arising from the environment of the beam. The beam current I excites electromagnetic fields E in the environment and these fields affect the beam. The fields can be described by an impedance $Z = E/I$ per unit length. There are separate longitudinal and transverse impedances. Out-of phase components excited by the finite resistance of the chamber walls (or some other part of the environment) can generate coherent resistive instabilities. Above an intensity threshold, the coherent amplitude has a growth rate that continues until the amplitude is large enough that Landau damping stops the growth. Because they are coherent, the instabilities can be quelled by feedback systems, just as in the negative-mass case. These systems need large bandwidth to deal with many harmonics of the revolution frequency. Modern large accelerators and storage rings have elaborate damping systems that are crucial to reaching high intensity. Stronger fields are excited by discontinuities in the environment and scrupulous attention is therefore paid to minimizing impedance during the design of a ring.

31

There are conceptually similar effects in linear accelerators. SLAC was troubled in its early days by a regenerative beam breakup along the linac and accompanying intensity limitation. These effects were explained (Panofsky and Bander, 1968) as transverse electromagnetic dipole fields generated by off-axis beam in the waveguide that excite oscillations in subsequent bunches. These oscillations are curable by careful alignment and choice of operating point. Proton linear accelerators have increases of transverse emittance that have recently been understood as exchange of longitudinal and transverse emittance caused by fields excited by the beam in the environment. Any emittance increase persists throughout the remainder of the acceleration process and has deleterious effects on beam extraction or colliding-beam luminosity.

The understanding and mastering of coherent instabilities was largely work of the 1960's and 1970's (although there are still people who work at calculating impedances of complicated systems). Quite separately, during these same years instabilities in plasmas were being studied and understood in depth by a number of people. The close relationship between the two fields was slowly grasped by many of us until Lawson clarified and extended the relation in a number of papers, culminating in his book (1977).

There is a different class of phenomena in accelerator beams that are basically diffusion effects. The first such effect uncovered was the Touschek Effect, which is the loss of particles from an accelerating bucket by single Coulomb scattering from other particles in the bucket. Later, multiple Coulomb scattering has been studied (Piwinski, 1974). This slow loss mechanism, called intrabeam scattering, is important for colliding beams, because of the long storage times.

All these are effects that occur in one accelerator beam. There can be analogous effects between two colliding beams. It was recognized early that the cancellation between electric and magnetic fields that occur in a single beam is replaced by summation of the fields for two oppositely directed beams of the same charge. There are also beam-beam dynamic effects. In a linear collider, the luminosity can be depressed by the blowup of the beams during collision by their electromagnetic forces, a phenomenon called disruption. There have been many calculations of disruption; the first experimental evidence will come from SLC when it begins operation.

Beam Cooling is a new feature of heavy-particle accelerators in the last decade. There had been, of course, cooling by synchrotron radiation in electron rings for many years and it had been the basis for the great success of electron storage rings. Cooling for heavy particles comes in two flavors, electron cooling (Budker, 1967) and stochastic cooling (van der Meer, 1972). Electron cooling is the temperature equilibrium of an electron gas and a heavy-particle gas interacting through multiple Coulomb scattering in their center-of-mass system. The two beams move along at the same speed and energy of longitudinal and transverse oscillations is transferred from the heavy particles to the electron gas as they move. Because of the relativistic increase of the masses and the Lorentz contraction of the fields, the cooling rate of electron cooling varies with energy as γ^{-5}, so that it is necessary in many applications to decelerate the heavy-particle beam to cool it. Stochastic cooling, on the other hand, is a very specialized many-particle effect. The position (longitudinal or transverse) of each particle is measured by a pickup electrode, then amplified and sent to a kicker that corrects the particle's motion. The amplified signals from all other particles appear to a given particle as noise (Schottky noise) that gives rise to diffusion. Stochastic cooling depends on the fact that there are a finite number of particles in the beam. Thus, as cooling proceeds and particles move toward the same position, the signal from an individual particle is overwhelmed by the signals of other particles (signal suppression) and there is a limit to cooling. As the number of particles being cooled increases, say, by stacking, there is also a limit to cooling.

Thus stochastic and electron cooling have disjoint regimes of applicability. Electron cooling has been demonstrated in many experiments (see the review by Cole and Mills, 1981). It will also be used in a 200-MeV proton storage ring being built for nuclear physics research at the Indiana University Cyclotron. Stochastic cooling has been used very successfully in the CERN Antiproton Accumulator, forming the basis for spectacular discoveries in high-energy physics, and later in the Fermilab Antiproton Source. There is a long-range plan to combine their best features by using a 4-MeV electron beam from an electrostatic generator to cool the 8-GeV antiproton stack in the Fermilab Antiproton Source. This cooling, even though it is slow, does not get slower as the stack is increased and therefore a stack of higher ultimate density and luminosity can be achieved by this combination of methods.

Beam cooling represents a new concept in conventional accelerators, whereas the other major new development of the last decade, the application of superconductivity to magnetic and acceleration systems, which was discussed above, can be considered more of a new technological development. Even if both these ideas are utilized to their fullest, a major step in energy represents a financial investment that is noticeable even on national terms and at this time we are seeing some hesitation on the part of governments to begin on major new projects. The LEP ring being built at CERN may represent a technological limit for electron storage rings in that it becomes prohibitively expensive to supply enough rf power to larger rings. Presently proposed proton storage rings such as the Superconducting Super Collider (SSC) will be well within this technological limit, but perhaps may represent an economic limit. Although these limits are soft, awareness of them has stimulated interest in linear colliders for later steps in high-energy physics. Linear colliders have in turn stimulated interest in brightness and accelerating fields beyond those that have been achieved in conventional synchrotrons and storage rings.

HIGH BRIGHTNESS IN CONVENTIONAL ACCELERATORS

Synchrotron-Radiation Sources

Synchrotron radiation in electron rings was considered a nuisance for many years. Mills and Rowe of the University of Wisconsin were the first to propose utilizing this radiation for experiments and the first such ring, TANTALUS, was built at Wisconsin in the later 1960's. Synchrotron radiation has since developed into a major tool of material science. Many electron storage rings have been adapted for this work and now several rings designed specifically for synchrotron-radiation research are in use. Ion-trapping effects were discussed above. The techniques of wigglers and undulators have been taken over from free-electron lasers and the rings now being designed for the next generation make use almost exclusively of radiation from these devices mounted in straight sections.

Intense Neutron Sources

These accelerators are competitors to fission reactors as sources of intense neutron beams of relatively low energy for research. The first intense neutron accelerator was built at Argonne (Rauchas et al, 1979) in the 1970's initially utilizing a rapid-cycling magnet system left over from a 500-MeV proton synchrotron that had been an injector for a high-energy physics accelerator. Other intense neutron-source devices have since been built at the Los Alamos and Rutherford-Appleton Laboratories. These devices accelerate more than 10^{14} protons per second. Most have used rapid cycling to produce high intensity, although there have been proposals to utilize fixed-field beam-stacking rings.

Radio-Frequency Quadrupoles

The frequency in a drift-tube linear accelerator must be chosen low to give space inside the drift tubes for the focusing quadrupoles. This is particularly difficult at low energies and particle speeds, because the drift tubes are short. To provide this space, accelerating frequencies used for proton linear accelerators have been approximately 200 MHz or lower. This difficulty has now been circumvented by the invention of the radiofrequency quadrupole (RFQ) (Kapchinski and Teplyakov, 1970), where the accelerating rf field is shaped by artfully placed conductors to provide both focusing and acceleration. RFQ's have completely taken over the field of linear accelerators, making it possible to inject at low energy (say 50 keV) and accelerate high enough intensity to make them interesting for high-brightness applications.

Linear Colliders

The energy limits for electron rings apparently imposed by synchrotron radiation have led people to propose single-pass colliders, in which beams from two linear accelerators cross once at an interaction region. The luminosity of particle collisions depends on particle density not on the total number of particles and an acceptable interaction rate can be achieved if the beam size can be made small enough that the density is very large. Thus linear colliders are high-brightness systems. A major test of the linear-collider concept is being built at the SLAC Linear Collider (SLC), in which one linear accelerator is used with a damping ring for positrons and two 180-deg arcs to bring the beams together. If it is as successful as its designers hope, it will be a competitor to LEP. SLC will begin commissioning in the near future.

Modified Betatron

The addition of a toroidal (longitudinal) field to a conventional betatron increases greatly the momentum spread that can be contained and thus the intensity that can be accelerated. The idea appears to have occurred to Kerst and to others, but was never successfully implemented experimentally until much later (see Roberson et al, 1985 for discussion and references on the history). Self-fields are very important in the modified betatron, but its basic operation is by external fields and it is in this sense a conventional accelerator. The modified betatron has become an active competitor of other high-brightness devices and is the subject of active research (see the paper of Rostoker at this meeting).

New Kinds of Accelerators

There has been a small, low-key, but dedicated effort over many years to study new methods of acceleration, but interest has expanded greatly over the last few years as more and more people have begun to seek new ways to reach desirable new parameter ranges and to build devices for new applications that have arisen. Recent work has included a strikingly diverse range of new concepts and it is difficult in a finite space to do more than characterize each with a brief sketch. We begin with a discussion of the history of intense relativistic electron beams because they are basic in many new accelerator concepts.

Intense Relativistic Electron Beams

A remarkable feature of the history of pulsed-power technology and intense relativistic electron beams (IREB) is its large measure of independence from the development of other accelerators. Not until the mid-1970's, when heavy-ion fusion began to be studied, did people who worked in pulsed-power technology and people who worked in conventional accelerators begin to interact to any significant degree. This problem of cultural isolation has eased a great deal since then, in part because of workshops and institutes like this one.

There is almost universal agreement that J. C. Martin founded the field of pulsed-power technology in 1962 with his combination of the Marx generator, Blumlein transmission line and high-power diode. These were not all new elements; Marx generators were widely used in testing electrical equipment for high-voltage transmission lines and had been made into an early accelerator (Brasch and Lange, 1930). Blumlein (1948) had demonstrated his transmission line some years before. Martin put all these elements together to make the basis for an entirely new field of accelerator technology. There is an excellent review of the technology by Nation (1979).

Accelerators for High Energy

Collective Acceleration. Graybill and Uglum (1970) first observed collective acceleration of ions when an IREB strikes a neutral gas. The details of the acceleration are complicated (see Olson, 1979 for discussion) but the basic physics is clear. Large electric fields (of the order of 100 MV/m) exist at the head of the column because of the self-fields of the beam and they ionize gas atoms and accelerate them. The trick is to control the phase velocity of the moving field at the column front and to maintain it moving with the ions over some distance in order to accelerate them to high energy. One example of a way to do this is the Ionization Front Accelerator of Olson (1979), in which a staged pulsed laser beam is used at intervals along the acceleration path to ionize gas and therefore neutralize the virtual cathode at the head of the beam.

Wave Acceleration. Another approach to acceleration has been to generalize a conventional travelling-wave linear accelerator by generating some kind of waves in a moving non-neutral plasma that is an IREB. Fainberg and his colleagues in Kharkov have worked in this field for many years (see Fainberg, 1976). Proposals to utilize cyclotron waves (Sloan and Drummond, 1973) or plasma waves (Sprangle, Drobot, and Manheimer, 1976) have had some preliminary success in tests. The difficulty is again to control the phase velocity over a long acceleration distance. Even if this problem is overcome, the accelerating fields achievable in the linear field regime appear to be limited to the order of 1 GeV/m. To achieve higher fields requires operation in non-linear regimes.

Wake-Field Acceleration. It had been well known from the study of coherent instabilities that a particle beam creates an electromagnetic wake field in a conducting enclosure. Voss and Weiland (1982) suggested using these wake fields to accelerate particles by arranging a wave-guide geometry to amplify the wake field of a hollow intense electron beam at the axis of the guide, where the beam to be accelerated moves. It is in this sense an application of IREB. In many geometries, there is a limit to the total acceleration achievable imposed by conservation of energy. There are also geometries for which these limits do not apply. A number of variants using different geometries, other particles, or plasmas have been suggested. At this time, several comprehensive experiments are being constructed at various laboratories.

Laser Acceleration. Another proposed method of accelerating particles to very high energy utilizes a laser beam rather than pulsed power. The electric field in a laser beam can be very large, but unfortunately it is transverse and cannot be used directly to accelerate particles. A method of adapting laser beams to acceleration was put forward by Trajima and Dawson (1974). They propose to excite a longitudinal plasma beat wave in a plasma by beating two frequencies of a laser beam and to use this field to accelerate particles. Experiments have demonstrated fields of at least several hundred MeV/m and it appears that much higher fields are possible. It is not yet clear that these fields can be sustained over the distances needed to accelerate particles to high energy. A continuing series of workshops has been held on laser acceleration, branching out into many other varieties of acceleration methods. The proceedings of these workshops have been published by the American Institute of Physics.

Accelerators for High Brightness

Linear Induction Accelerator. The linear induction accelerator is a conceptual outgrowth of the betatron and can be thought of in some ways as a betatron of infinite radius. it was apparently first proposed as early as 1923 by Bouwers and later (1939) discussed by him in more detail. The real development was begun when Christofilos built the first one as an injector for a controlled fusion device, the Astron. There have been many others constructed, both in the US and in the USSR.

Induction linear accelerators have unmatched capability of accelerating short (nanoseconds) pulses of high currents (kiloamps) of electrons from an IREB to moderate energies, 50 MeV in the largest, ATA at Livermore. They have the additional feature of pulse compression or current amplification by shaping the primary voltage pulse. This compression is important in heavy-ion fusion applications. For these applications, it is also necessary to accelerate multiple beams, then combine them to make a large instantaneous current. It appears to be possible to accomplish this in a single induction linear accelerator. (See Keefe, 1986 and Wangler, 1986 for reviews of the current status of the field).

The beam-breakup phenomena that occur in rf linacs also occur with a vengeance in linear induction accelerators and have been the subject of extensive work (see the paper of Prono at this meeting).

High-Current Beam Transport. The needs of heavy-ion fusion, particle acceleration with IRE, and other applications give new interest in the transport of very high intensity beams. In contrast to the tune shifts of 0.1 to 0.2 in circular rings, much larger tune depressions are encountered and can be withstood in beam transport. There are a number of theoretical studies of such transport (see, for example, Reiser (1976), Struckmeyer and Reiser (1984) and Struckmeyer, Klabunde and Reiser (1984) and references therein). There is a reasonable amount of information and understanding from computational studies and somewhat less understanding from analytical work (see Hofmann, Laslett, Smith and Haber, 1983, for example) because of incomplete knowledge of the distribution function of particles in phase space.

There have also been proposals to make use of ion columns to transport and focus particle beams and experiments have been carried out to investigate ion focusing (see the paper of Miller at this meeting).

Applications of High-Brightness Accelerators

Free-Electron Laser. Coherent, efficient, and tunable high-power radiofrequency sources in the microwave and millimeter regions have many uses in radar and other direct microwave applications. They also have application in accelerator technology, because the Kilpatrick sparking limit in rf cavities increases with frequency and higher accelerating gradients can therefore be achieved at higher frequency. The free-electron laser (FEL) couples the considerable energy of an electron beam to the radiation field by means of a space-modulated magnetic guide field. The Ubitron of Phillips (1960) was the first such device to produce radiation. The FEL provides a radiofrequency source of considerable power (see Miller, 1982 for a good discussion and Morton, 1983 for a discussion of the underlying physics in accelerator terms). Experimental verification has proven the principles (Elias et al, 1976). The FEL is now being utilized as the power source for the two-beam accelerator (Sessler, 1982)

Heavy-Ion Fusion. Many people have suggested compression of a deuterium-tritium pellet by a high-energy (say 10 GeV) heavy-ion beam as a method of initiating a thermonuclear reaction. Several devices have been proposed for acceleration of these ions, synchrotrons (Arnold and Margin), radiofrequency linear accelerators (Maschke) and linear induction accelerators (Keefe). Over some years of work, the synchrotron appears to have lost out as a competitor because of its lower intensity. Many workshops were held to start this work and a longer-term effort is now being carried out (see the paper of Keefe at this meeting).

CONCLUSION

I have sketched the history of the development of accelerators in the pages above with a broad brush. I have emphasized the intellectual development of understanding and concepts, not the history of large projects. I should note that there are entire worlds of particle accelerators that I have not touched in this survey. For example, there are many different uses of accelerators in medicine and industry. These are important, but the accelerators designed for these purposes emphasize reliability and ease of operation rather than innovation.

The development of particle accelerators has been spectacular. For example, in high-energy physics, where energy is the primary parameter, the accompanying graph, Fig. 7 a Livingston plot, shows that particle energy has increased by a factor 10 every 7 years since 1930. Other fields of application of accelerators show similar spectacular increases. For example, in free-electron lasers, heavy-ion fusion and other applications where intensity and brightness are the primary parameters, the last few years have brought significant advances. The wiggler techniques of FEL's has been taken over to make much brighter synchrotron-radiation sources.

The initial hopes that self-fields could be used to provide much-higher accelerating fields has not been borne out so far. Many of the most successful high-brightness accelerators discussed above utilize conventional external fields for basic acceleration, not self-fields. But a large measure of understanding of the effects of self-fields has been won and many of these high-brightness devices take self-fields into account as part of their basic containment systems. In this sense, the devices are hybrids between conventional external-field accelerators and self-field-dominated accelerators.

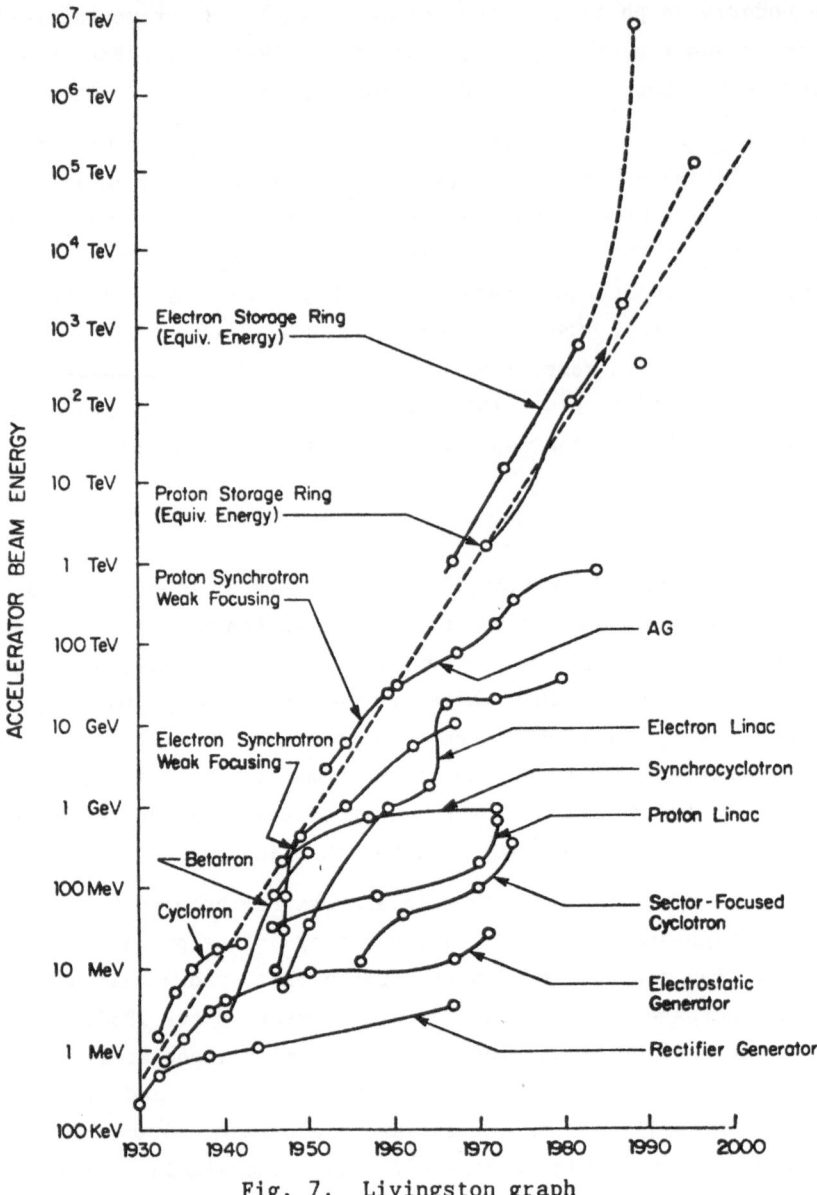

Fig. 7. Livingston graph

Are there lessons for use to learn from the history of accelerators? It seems to me that the history of accelerators is a story of individuals and small bands of brothers (and sisters) striving to reach new heights, motivated by their overwhelming curiosity and by their desire to go further than those before them, not just to set a record, but to use for some scientific or technologic end. It is the drive of the people that is important, not the amount of money poured in.

The field of particle accelerators is in a state of splendid motion and confusion just now, going off in many different directions and doing many different things. If my story has any moral, it is probably that this state of disarray is a constant of the motion.

REFERENCES

Blewett, J. P., 1952, Phys. Rev. 88:1197.
Blosser, H., 1979, IEEE Trans. Nucl. Sci, NS-26:3653.
Blumlein, A. D., 1948, US Patent 2, 465, 84.
Bollinger, L. M., 1983, IEEE Trans. Nucl. Sci. NS-30, 2065.
Bouwers, A., 1939, Elektrische Hochspannung Springer, Berlin.
Brasch, A. and Lange, F., 1930, Naturwiss, 18:769, 1931, Z. Physik 70:10.
Budker, G. I., 1967, Proc. 1966 Symp. on Electron and Positron Storage
    Rings, Saclay, Atomnaya Engergiya 22:346.
Christophilos, N. C., 1950, unpublished letter.
Cockcroft, J. D., and Walton, E. T. S., 1932, Proc. Roy. Soc. (London)
    A136:619.
Cole, F. T. and Mills, F. E., 1981, Increasing the Phase-Space Density of
    High Energy Particle Beams, Ann. Rev. Nucl. & Part. Sci. 31:295.
Collins, T. L., 1961, Cambridge Electron Accelerator Report CEA-86,
    (unpublished).
Courant, E. D., Livingston, M. S., and Snyder, M. S., 1952, Phys. Rev.
    88:1190.
Courant, E. D. and Snyder, H. S., 1958, Annals of Physics 3:1.
Elias, L. R., Fairbank, W. M. Madey, J. M. J., Schwettman, H. A., and
    Smith, T. I., 1976, Phys. Rev. Ltrs. 36:717.
Fainberg, Y. B., 1976, Part. Accel. 6:95.
Graybill, S. E. and Uglum, J. R., 1970, J. Appl. Phys. 41:236.
Hammer, C. L. and Laslett, L. J., 1961, Rev. Sci. Instr. 32:144.
Hofmann, I., Laslett, L. J., Smith, L. and Haber, I., 1983, Part. Accel
    13:145.
Herb, R. G., Parkinson, D. B., and Kerst, D. W., 1935, Rev Sci. Instr.
    6:261.
Kapchinski, I. M. and Teplyakov, V. A., 1970, Pribory i Teknika Eksp.
    2:19.
Keefe, D., 1986, Experiments and Prospects for Induction Linac Drivers,
    Proc. Int Symp. on Heavy-Ion Fusion, Washington, DC.
Kerst, D. W., 1940, Phys. Rev 58:841.
Kerst, D. W., 1941, Phys. Rev. 60:47.
Kerst, D. W., and Serber, R., 1941, Phys. Rev. 60:53.
Kerst, D. W., and Serber, R., 1941, Phys. Rev. 60:53.
Kerst, D. W., Cole, F. T., Crane, H. R., Jones, L. W., Laslett, L. J.
    Ohkawa, T., Sessler, A. M. Symon, K. R., Terwilliger, K. M. and
    Vogt-Nilsen, N., 1956, Phys. Rev. 102:590.
Langmuir, I. and Tonks, L., 1929, Phys. Rev. 33:195.
Laslett, L. J., Neil, V. K., and Sessler, A. M., 1965, Rev. Sci. Instr.
    36:429 and Rev. Sci. Instr. 36:436.
Laslett, L. J., 1963, Proc. 1963 Summer Study on Storage Rings,
    Accelerators, and Experimentation at Super-High Energies, Brookhaven
    National Laboratory, Upton, N.Y., p. 324.
Lawrence, E. O., and Edlefsen, N. E., 1930, Science 72:376.
Lawson, J. D., 1977, The Physics of Charged-Particle Beams, Clarendon
    Press Oxford.
Livdahl, P. V., 1982, Proc. of the 1981 Linear Accelerator Conf., Los
    Alamos, N.M., p. 5.
McMillan, E. M., 1945, Phys. Rev. 69:143.
Miller, R. B., 1982, An Introduction to the Physics of Intense Charged
    Particle Beams, Plenum Press, New York and London.
Morton, P. L., 1983, Proc. of the 12th Int Conf. on High-Energy Acc.,
    Fermilab, p. 477.
Nation, J. A., 1979, Part. Accel. 10:1.

Nielsen, C. E., Sessler, A. M. and Symon, K. R., 1959, _Proceedings of the 1959 International Conference on High-Energy Accelerators_, CERN, Geneva, p. 239.

Oliphant, M. L., Gooden, J. S., and Hide, G. S., 1947, _Proc. Roy. Soc._ (London) 59:666.

Olson, C. L., 1979, _Collective Ion Acceleration_, Springer-Verlag, Berlin and New York.

Panofsky, W. K. H. and Bander, M., 1968, _Rev. Sci. Instr._ 39:206.

Phillips, R. N., 1960, _IRE Trans Electron Devices_ ED-7:231.

Piwinski, A., 1974, _Proc. 9th Int. Conf. on High Energy Accelerators_, Stanford, Cal. p. 347.

Rauchas, A. V., Brumwell, F. R., and Volk, G. J., 1981, _IEEE Trans. Nucl. Sci._ NS-26:3006.

Reiser, M., 1976, _Part. Accel._ 8:167.

Roberson, C. W., Mondelli, A., and Chernin, C., 1985, _Part. Accel_, 17:79.

Sessler, A. M., 1982, _Workshop on Laser Acceleration of Particles_, AIP Conf. Proc. 91, AIP, NY, p. 154.

Sprangle, P., Drobot, A., and Manheimer, W., 1976, _Phys. Rev. Ltrs._ 36:272.

Struckmeyer, J. and Reiser, M., 1984, _Part. Accel._ 14:227.

Struckmeyer, J., Klabunde, J. and Reiser, M., 1984, _Part. Accel._ 15:47.

Symon, K. R. and Sessler, A. M., 1956, _Proc. 1956 International Conference on High-Energy Accelerators_, CERN, Geneva, p. 44.

Tajima, T. and Dawson J. M., 1974, _Phys. Rev. Ltrs._ 43:267.

Thomas, L. H., 1938, _Phys. Rev._ 54:580.

van der Meer, S., 1972, _Stochastic Damping of Betatron Oscillations in the ISR_. CERN/ISR-PO/72-31 (unpublished).

Veksler, V. I., 1944, _Doklady USSR_ 43:444, 44:393, _and J. Phys. (USSR)_, 9:153, 1945.

Voss, G. A., and Weiland, T., 1982, _Particle Acceleration by Wake Fields_, DESY Report M-82-10, (unpublished).

Walton, E. T. S., 1929, _Proc. Camb. phil Soc._ 25:469.

Wangler, T. P., 1986, _High Current, High Brightness, and High Duty Factor Ion Injectors_, AIP Conference Proc. 139, AIP, NY, p. 173.

Wideroe, R., 1928, Arch. Elektrotech, 21:387.

Wilson, R. R., 1938, _Phys. Rev._ 53:408.

HIGH-INTENSITY CIRCULAR PROTON ACCELERATORS

M. K. Craddock

Physics Department, University of British Columbia
and TRIUMF, 4004 Wesbrook Mall
Vancouver, B. C., Canada V6T 2A3

ABSTRACT

Circular machines suitable for the acceleration of high intensity proton beams include cyclotrons, FFAG accelerators, and strong-focusing synchrotrons. This paper discusses considerations affecting the design of such machines for high intensity, especially space charge effects and the role of beam brightness in multistage accelerators. Current plans for building a new generation of high intensity "kaon factories" are reviewed.

INTRODUCTION

The concept of beam brightness had come to play an important role in the design of multistage high intensity circular proton accelerators through the space charge defocusing effects associated with high brightness. In this paper we first review the different types of circular proton accelerators and how focusing is provided; then discuss the effects of space charge forces, including the role of beam brightness in determining the number of accelerator stages required; and finally describe some current proposals for high intensity machines.

Good general introductions to circular accelerators are given in the standard texts by Livingood (1961), Livingston and Blewett (1962) and Bruck (1966). Valuable review articles are also available on sector-focusing cyclotrons (Richardson, 1965), and high intensity accelerators (Courant, 1968). More recent developments are covered in the proceedings of the summer schools on high energy accelerators which have been held annually for the past few years in the U.S. (1981 -) and in Europe; the schools held at Erice (CERN-ISPA, 1977) and in Paris (CERN, 1985) offer particularly good introductory articles. Finally, an excellent short

survey of modern proton synchrotron design for beginners is given by
E. J. N. Wilson (1977).

By "circular" accelerators we shall mean any machine in which the particles are bent into closed orbits, even though these may not be perfect circles but include segments that are straight or of different curvatures. This "recirculation" principle allows the beam to be passed repeatedly through the same electric accelerating fields, vastly reducing the complexity and cost of the accelerating equipment; the cost of the magnetic bending system is modest by comparison. The net result is a very high effective accelerating field gradient - as much as 150 MV/m for the Fermilab Tevatron, or 240 MV/m for the proposed SSC.

THE CYCLOTRON RESONANCE PRINCIPLE

For a proton of mass m and charge e moving with velocity v normal to magnetic induction **B**, the Lorentz force e**v** x **B** provides the centripetal acceleration to bend the trajectory with radius of curvature ρ:

$$evB = \frac{mv^2}{\rho} . \tag{1}$$

From this we can derive equations for the radius

$$\rho = \frac{mv}{eB} \tag{2}$$

which increases in proportion to momentum $p \equiv mv$, and for the angular frequency of rotation

$$\omega = \frac{eB}{m} . \tag{3}$$

Notice that whereas the radius ρ increases in proportion to momentum p for fixed B and m, the angular frequency ω and orbit time are constant, independent of velocity. Lawrence (1930) realized that this property of "isochronism" greatly simplified the problem of designing a circular accelerator, as a fixed frequency rf accelerating field would remain in synchronism with the ions as they were accelerated to greater momenta and radii (Fig. 1). Not only was the hardware design simplified but ions could be accelerated on every rf cycle in a quasi-continuous (cw) stream. The fruit of this "cyclotron resonance principle" was a successful series of fixed frequency cyclotrons of increasing size built by Lawrence and his students in the 1930s.

For protons of kinetic energy $T \simeq 20$ MeV, however, a limit was reached. The relativistic increase in mass

$$\frac{m}{m_0} = \frac{E}{E_0} = 1 + \frac{T}{E_0} \equiv \gamma = \frac{1}{\sqrt{1-\beta^2}} \tag{4}$$

44

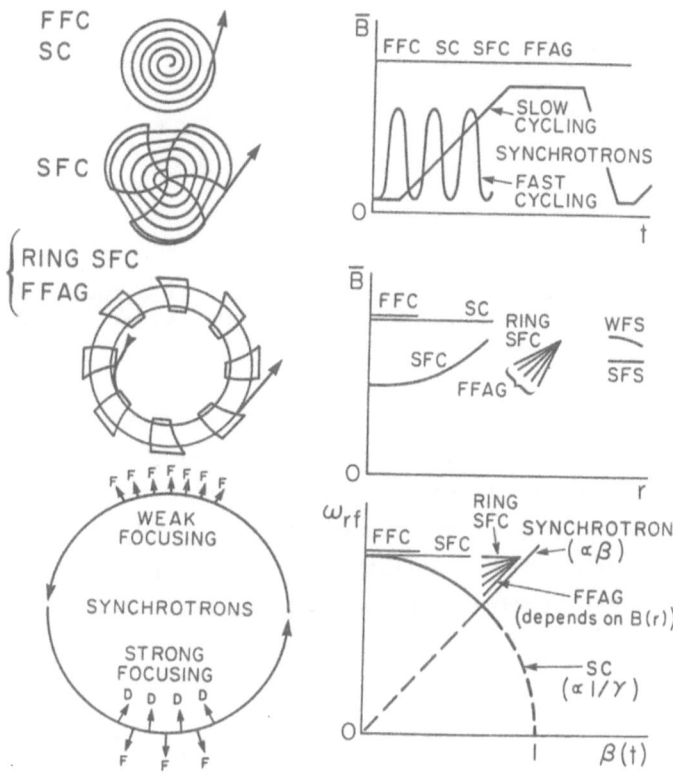

Fig. 1.   Beginner's guide to circular proton accelerators – schematic
plans, magnetic field dependence on time and radius, and
radiofrequency variation with velocity.

FFC – fixed frequency
          cyclotron
SC – synchrocyclotron
SFC – sector-focused
          cyclotron

WFS – weak focusing
          synchrotron
SFS – strong focusing
          synchrotron
FFAG – fixed field
          alternating gradient

of about 2% disturbed the constancy of ω enough that over many turns the
protons lost synchronism with the rf voltage peak and were no longer
accelerated.   (In the above notation $E_0 = m_0 c^2$ denotes the "rest energy"
and $\beta \equiv v/c$.)

WEAK FOCUSING

This problem was enhanced by the natural drop-off with radius of the
magnetic field strength B between flat pole-faces — a drop-off that was
found, however, to be essential to provide vertical focusing of the
particles.   (In accelerator usage "focusing" means containment of the
particle beam within a not-too-large diameter, rather than formation of
an image spot.)   Figure 2 shows how a field decreasing with radius r is

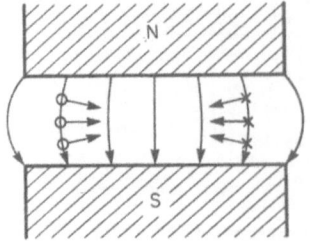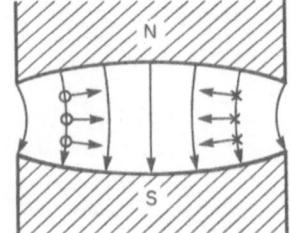

Fig. 2. Vertical focusing (defocusing) effect of a radially decreasing (increasing) magnetic field. Protons travel in counterclockwise orbits seen from above.

associated with a convex field pattern which provides restoring forces towards the median plane, while a field increasing with radius does the opposite. The horizontal and vertical focusing strengths are conventionally described in terms of the "tunes" ν_r and ν_z (Q_r and Q_z in Europe) — the numbers of "betatron" oscillations performed per turn about the stable orbit. The vertical component of the Lorentz force is given by

$$F_z = evB_r \simeq ev \frac{\partial B_r}{\partial z} z \ . \tag{5}$$

But for stable betatron oscillations $F_z = -m\omega^2 \nu_z^2 z$. Using Maxwell's equations and writing the logarithmic field gradient

$$k \equiv \frac{r}{B_z} \frac{\partial B_z}{\partial r} \tag{6}$$

the student will find (Exercise 1) that

$$\nu_z = \sqrt{-k} \ . \tag{7}$$

Similarly, for horizontal motion (Exercise 2)

$$\nu_r = \sqrt{1+k} \ . \tag{8}$$

Clearly k must be negative for vertical stability, but not too strongly so, or horizontal stability is lost. In this situation we require

$$-1 < k < 0 \tag{9}$$

but achieve only "weak" focusing

$$0 < \nu < 1 \ . \tag{10}$$

SYNCHROCYCLOTRONS AND SYNCHROTRONS

Historically, the first technique used to overcome the energy limit posed by rising mass was modulation of the rf frequency in sympathy with the orbital frequency given by Eq. (3). For this to be practical for a real beam, spread out in time and momentum around the synchronous

46

condition, there must be stable motion about this point in longitudinal phase space. The existence of longitudinal or "phase" focusing was proved independently by Veksler (1944) and McMillan (1945) and first demonstrated experimentally by Richardson et al. (1946).

In synchrocyclotrons the magnetic field is kept constant while the radiofrequency is modulated. The orbit radius increases with energy, just as in fixed frequency cyclotrons. The energy attainable is set by the cost of providing a large area of magnetic field; the highest energy synchrocyclotron, at Gatchina in the USSR, has a pole diameter of 6.85 m, uses 7800 tons of steel and reaches 1000 MeV:

$$B \simeq \text{constant}$$

$$\omega_{rf} \propto 1/\gamma = \sqrt{1-\beta^2} \tag{11}$$

$$\rho \propto \beta\gamma .$$

In synchrotrons higher energies are brought within economic reach by keeping the orbit radius constant, thus shrinking the vacuum chamber from a hollow disc to a hollow ring and drastically reducing the area of magnetic field required for a given energy. To achieve this, the magnetic field strength must be modulated as well as the radio frequency:

$$\rho = \text{constant}$$

$$B \propto \beta\gamma \tag{12}$$

$$\omega \propto \beta .$$

The $B(t)$, $B(r)$ and $\omega_{rf}(\beta)$ dependence of these machines is illustrated in Fig. 1. In fast-cycling synchrotrons ($\gtrsim 3$ Hz) the magnetic field is modulated harmonically; in slow cycling synchrotrons ($\lesssim 3$ Hz) it follows a linear ramp, often with a short flat bottom for multi-turn injection and a long flat top for slow extraction. The latter is an essential feature for many counter-based experiments, which depend on coincidence techniques to identify the numerous reaction products and cannot tolerate a beam which is sharply pulsed.

This highlights one of the major characteristics of all frequency-modulated machines – they are operated in a pulsed mode where one group of particles has to go through a complete cycle of capture, acceleration and extraction before another can begin. As a result the beam intensities achievable are much lower (<1 μA on average) than for fixed frequency cyclotrons, which can be operated cw at milliampere currents.

Lower intensity is the price paid for reaching higher energies. In principle there is no upper limit to the particle energy which can be achieved in a synchrotron. In practice the limit is set by cost (mostly for the magnet). The largest weak focusing synchrotron using constant field gradient focusing is the 10 GeV "Synchrophasotron" at Dubna; 36000 tons of steel were used in the magnet, whose circumference is 175 m and beam aperture 1.50 m x 0.36 m. The enormous - and expensive - apertures of such machines are directly related to their large radii and weak focusing. To see this quantitatively, suppose the maximum amplitude betatron oscillation is described by

$$y = a \sin(\nu\theta + \delta) , \qquad (13)$$

where y stands for transverse displacement, ν is the tune and θ the azimuthal angle. Then it is easy to show (Exercise 3) that if the maximum divergence angle is α, then

$$a = \alpha r/\nu , \qquad (14)$$

so that small ν and big r lead to large a. To maintain constant a and α, ν must increase with r, as is obvious if one considers betatron oscillations of fixed wavelength in machines of different radii. Apparently stronger focusing is required to achieve higher energies economically.

SECTOR FOCUSING

An alternative source of vertical focusing had been proposed by Thomas (1938) in a scheme to allow higher energy cyclotrons to be built at fixed frequency. He proposed to maintain a constant orbit frequency Eq. (3) as the mass increased with energy and radius ($m = \gamma m_0$) by increasing the magnetic field strength commensurately (Fig. 1):

$$B = \gamma B_c = \frac{B_c}{\sqrt{1-(r/r_c)^2}} . \qquad (15)$$

Here $B_c \equiv \omega m_0/e$ is the "central" field and $r_c \equiv c/\omega$ is the radius at which $v \to c$. Such a field profile of course gives a vertically defocusing field gradient (Exercise 4)

$$\nu_z^2 = - \beta^2 \gamma^2 . \qquad (16)$$

To counteract this Thomas proposed contouring the magnet pole faces to provide a sinusoidal azimuthal variation in field strength. In practice this can be achieved by dividing the poles into N symmetrical sectors, each consisting of a "hill" with small gap and high field B_h and a "valley" with large gap and low field B_v (Fig. 3). The different orbit curvatures in hill and valley lead to a scalloped closed orbit oscillating around a perfect circle. This implies a radial velocity component v_r, strongest at the hill-valley "edges" where the hill fringing field provides B_θ components away from the median plane. The result is a vertical Lorentz force component F_z which is focusing at every edge.

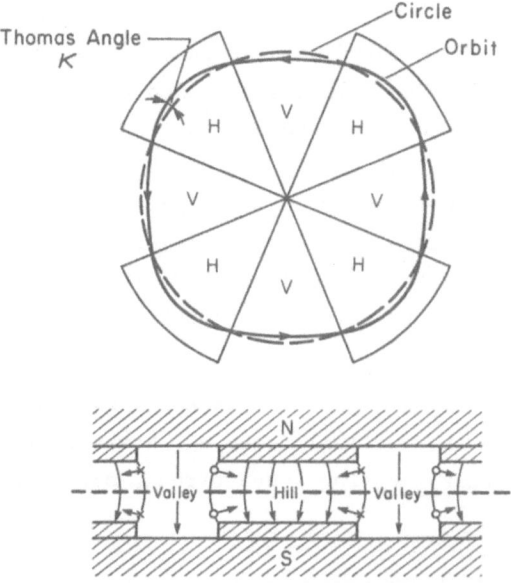

Fig. 3.   Vertical (Thomas) focusing effect of an azimuthally varying
magnetic field created by radial magnet sectors; the radial
velocity components associated with orbit scallopping provide
vertical forces at the edge of each sector.

This is in fact the edge focusing effect familiar in spectrometers and
other magnets where a particle crosses a field change ΔB at an angle κ to
the normal: the edge acts like a thin lens whose focal length f may be
obtained (Exercise 5) by integrating along the orbit to obtain the total
impulse and then using Stokes' theorem:

$$\frac{1}{f} = \frac{e\,\Delta B}{mv}\,\tan\,\kappa\;.\tag{17}$$

Some wrestling with the geometry of the scalloped orbit in this "hard
edge field" approximation (Exercise 6) will convince the persevering
reader that the "Thomas angle" κ is given by

$$\kappa = \frac{\pi}{N}\,\frac{(B_h - \bar{B})(\bar{B} - B_v)}{(B_h - B_v)\bar{B}}\;,\tag{18}$$

where $\bar{B}$ is the azimuthal average of B_h and B_v [and has to obey (15)].
Normally κ is small enough to make small angle approximations valid.
Exercise 13 below offers an opportunity to demonstrate that the lenses
described by (17) make a net contribution to the vertical focusing equal
to the field "flutter" F^2:

$$\Delta v_z^2 = F^2 \equiv \frac{\overline{(B - \bar{B})^2}}{\bar{B}^2} = \frac{(B_h - \bar{B})(\bar{B} - B_v)}{\bar{B}^2}\;.\tag{19}$$

Sector focusing was first demonstrated experimentally in 1950-53 in model cyclotrons built by Richardson's group (Kelly et al., 1956) for electrons up to $\beta \simeq 0.5$. The principle has led to the construction of a large number of high current "isochronous" fixed frequency cyclotrons. By removing all steel from the valley, making $B_v = 0$, and using only narrow hills, the "flutter" factor on the right-hand side of Eq. (18) may in principle be made as large as desired. In practice the availability of strong spiral focusing (see below) has made the use of purely radial sector focusing uncommon for protons above 50 MeV.

Edge focusing was also used in the Argonne 12.5 GeV Zero Gradient Synchrotron, where the magnet was built in 8 separate sectors with angled ends. The strength of the vertical focusing, however, was no greater than in conventional weak focusing synchrotrons, since the drift spaces were rather short.

STRONG FOCUSING

It was the discovery of the strong focusing principle by Courant, Livingston and Snyder (1952), and independently by Christofilos, that provided the mechanism allowing both synchrotrons and cyclotrons to be built to even higher energies. They observed that a succession of focusing and defocusing lenses of equal strength have an overall focusing effect (provided their spacing is not large enough to allow cross-over). On average the displacement is greater at the F than at the D lenses (Fig. 4) and so the deflexions towards the axis are greater than those away from it. Since the F-D combination is focusing in both transverse planes, much stronger lenses can be used than allowed by Eq. (9), and true values $\nu \gg 1$ achieved. A formal treatment is given below.

For synchrotrons the use of magnets with alternating (focusing and defocusing) gradients made it economically feasible to build machines such as the Bookhaven AGS (33 GeV) and CERN PS (28 GeV) in which the tune values were raised to 6-9, the magnet apertures reduced to 6 cm x 3 cm, and the circumference increased to 600-800 m.

While these early designs used "combined function" gradient magnets which both bent and focused the beam, since about 1970 it has been usual to employ "separate function" magnets - quadrupoles for focusing and zero-gradient dipoles for bending.

With the help of stronger fields from superconducting magnets, proton synchrotron energies have recently been pushed towards 2000 GeV with the 7 km long Fermilab Tevatron. Future plans call for the construction of the 20 TeV Superconducting Super Collider (SSC); improvements in magnet design keep the circumference to 83 km, while the tune grows to

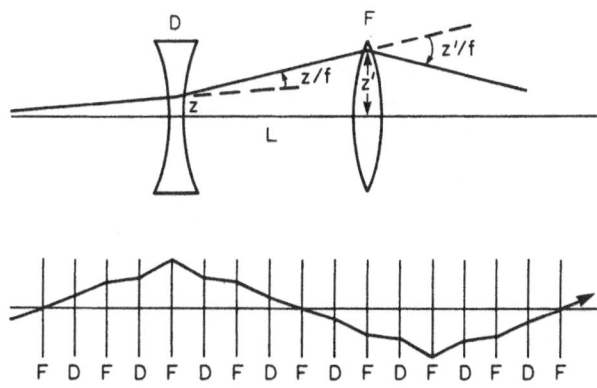

Fig. 4. The strong focusing principle – net focusing effect of alternating focusing and defocusing lenses of equal length f. Note that the effect is the same in both directions, i.e., for FD and DF arrangements.

78.3 and the magnet aperture stays at 3 cm x 3 cm. These very high energy accelerators consist of a number of stages and we shall see below that the concept of beam brightness plays an important part in matching these stages together.

Fixed field alternating focusing accelerators have traditionally been known as FFAG (fixed field alternating gradient) accelerators, after the first electron models built at MURA, although later designs have tended to use edge rather than gradient focusing. In particular Kerst (Symon et al., 1956) suggested setting the hill edges at large angles ε to a radius to form spiral ridges or sectors, obtaining much stronger edge focusing (Fig. 5). Alternate edges would be focusing and defocusing, with focal powers given by

$$\frac{1}{f} = \pm \frac{e\,\Delta B}{mv}\,\tan(\varepsilon \pm \kappa) \tag{17'}$$

but the net effect would of course be strongly focusing.

FFAG accelerators, with fixed magnetic field but modulated rf, are the strong focusing analogues of synchrocyclotrons; however, they are generally designed with a rising rather than flat field profile B(r), in order to narrow the range of orbit radii, and hence the magnet aperture, giving a less costly ring-shaped machine (Fig. 1). No FM proton FFAGs have ever been built, although there have recently been proposals for 1–3 GeV versions as spallation neutron sources (Khoe and Kustom, 1983; Meads and Wüstefeld. 1985).

Fixed frequency alternating focusing accelerators are however ubiquitous, in the form of isochronous cyclotrons with spiral sectors. These may be regarded as an extreme form of FFAG accelerator in which

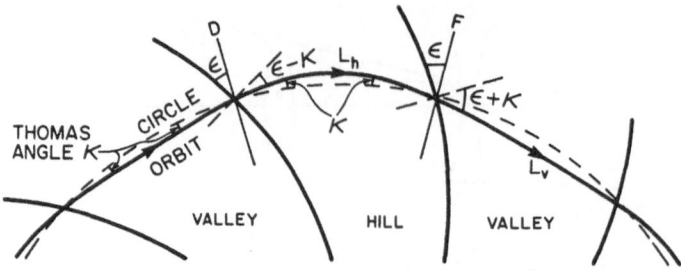

Fig. 5. Geometry of scallopped orbits in a cyclotron with spiral
sectors.

B(r) follows Eq. (15) to keep the orbital frequency constant, and the rf
frequency modulation is consequently reduced to zero. The additional
vertical focusing provided by the spiral edges allows the defocusing
[Eq. (16)] associated with isochronism to be compensated to much higher
energies, while cw operation permits high currents to be accelerated.
Thus proton beams exceeding 200 µA have been extracted from the 500 MeV
TRIUMF cyclotron (Zach, Dutto et al., 1985) and the 590 MeV SIN
cyclotron. With a new injector cyclotron (Joho et al., 1985) SIN expects
to raise the beam current above 1 mA. Botman et al., (1983) at TRIUMF
have studied spiral ring cyclotron designs using superconducting magnets
that would accelerate protons above 10 GeV.

TRANSVERSE MOTION IN PERIODIC LATTICES

Suppose a curvilinear co-ordinate system is taken, based on the
equilibrium closed orbit, with s tangential, x normal horizontal out-
wards, and z vertical upwards (Fig. 6). The equations of motion for
transverse (betatron) oscillations in either x- or z-plane then take the
form of the Hill equation

$$\frac{d^2 y}{ds^2} + k(s)y = 0 \qquad\qquad (20)$$

where y stands for x or z, second and higher order terms have been
omitted, and momentum-dependent effects are ignored. The function k(s)
describes the variation in focusing strength along the orbit. In any
accelerator k(s) will be periodic over the complete circumference C; in
addition in strong focusing synchrotrons and cyclotrons where the machine
is composed of a number (N) of identical superperiods, cells or sectors,
k(s) will be periodic over their length (L). The solutions to (20) may
be shown to be of the quasi-periodic form

$$y(s) = \sqrt{\varepsilon\beta(s)}\, \cos[\psi(s) + \delta] \; . \qquad\qquad (21)$$

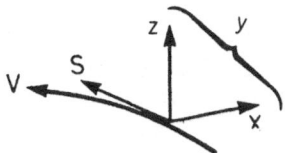

Fig. 6. Orbit co-ordinate system.

Here the cosine term represents the betatron oscillation, whose phase is determined partly by the "phase function" $\psi(s)$, which varies around the orbit, and partly by the arbitrary angle δ. The amplitude of the oscillation is given by $\sqrt{\varepsilon\beta(s)}$ – the product of a position-independent factor ε and the "amplitude function" or "beta function" $\beta(s)$, which varies with position within a cell, but is the same for each identical cell.

Differentiating Eq. (21) we obtain an expression for the divergence angle

$$y'(s) \equiv \frac{dy}{ds} = -\sqrt{\varepsilon\beta}\,\frac{d\psi}{ds}\,\sin(\psi + \delta) + \frac{1}{2}\sqrt{\frac{\varepsilon}{\beta}}\,\frac{d\beta}{ds}\,\cos(\psi + \delta). \qquad (22)$$

Equations (21) and (22) will be recognized as the parametric equations for a tilted ellipse in y-y' phase space (Fig. 7). As a particle moves around the machine $\beta(s)$ will change, altering the shape of the ellipse, and so will $\psi(s)$, moving the representative point around the ellipse once for each betatron oscillation and ν times for each orbit. At a given point of the orbit, different values of δ (ranging from 0 to 2π) describe different locations around the ellipse, while different values of ε represent ellipses of different sizes (but the same shape). A beam of particles with all phases of oscillation, and all amplitudes up to some maximum, can therefore be described by a set of representative points entirely filling an ellipse in phase space defined by the largest ε and the local $\beta(s)$.

Eliminating the phase angles $(\psi + \delta)$ between Eqs. (21) and (22) (Exercise 7) gives the Cartesian equation of the ellipse

$$\gamma(s)y^2 + 2\alpha(s)yy' + \beta(s)(y')^2 = \varepsilon \qquad (23)$$

where the "Twiss parameters" are defined by

$$\alpha(s) \equiv -\frac{1}{2}\frac{d\beta}{ds} \qquad (24)$$

$$\gamma(s) \equiv \frac{1 + \alpha^2}{\beta(s)}. \qquad (25)$$

The parameter $\alpha(s)$ describes the tilt of the ellipse ($\alpha = 0$ for one that is upright), while $\gamma(s)$ defines the maximum divergence $y'_{max} = \sqrt{\varepsilon\gamma}$ [just as $\beta(s)$ defines the maximum displacement $y_{max} = \sqrt{\varepsilon\beta}$].

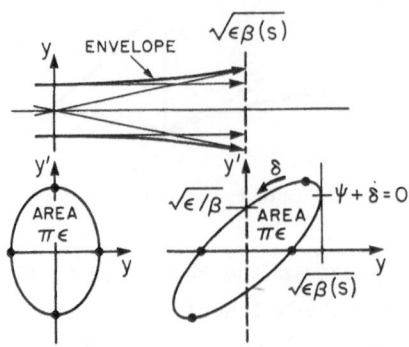

Fig. 7. Elliptical envelopes of a beam in the y–y' phase plane:
left, at a waist; right, in a divergent region.

The area of the ellipse given by Eq. (23) represents the "emittance" of a beam of particles whose representative points are enclosed by it:

$$E = \pi\varepsilon \ . \tag{26}$$

Thus the phase space area is constant, as required by Liouville's theorem for all processes describable by a Hamiltonian function, such as charged particle motion in electric and magnetic fields. But in deriving Eq. (23) it is only possible to make the area constant, independent of s, by requiring

$$\frac{d\psi}{ds} = \frac{1}{\beta(s)} \ . \tag{27}$$

Thus there is only one independent function of s – usually taken to be $\beta(s)$. Not only $\alpha(s)$ and $\gamma(s)$, but also the phase angle $\psi(s)$, can be derived from it:

$$\psi(s) = \int \frac{ds}{\beta(s)} \ . \tag{28}$$

Integrating Eq. (28) around a complete orbit of circumference $2\pi R$ we see that the average

$$\left\langle \frac{1}{\beta(s)} \right\rangle = \frac{\nu}{R} = \frac{2\pi}{\lambda} \tag{29}$$

where λ represents the wavelength of a betatron oscillation.

A differential equation for the function $\beta(s)$ may be obtained (Exercise 8) by substituting Eq. (21) into Eq. (20):

$$\frac{1}{2}\beta \frac{d^2\beta}{ds^2} - \frac{1}{4}\left(\frac{d\beta}{ds}\right)^2 + k(s)\beta^2 = 1 \ . \tag{30}$$

This may readily be converted into a differential equation for the "envelope function" $y_m(s) = \sqrt{\varepsilon\beta}$ – just Eq. (20) modified by a cubic term:

$$\frac{d^2 y_m}{ds^2} + k(s)y_m - \frac{\varepsilon^2}{y_m^3} = 0 \ . \tag{31}$$

Figure 8 illustrates the envelope function and several individual orbits for a FODO lattice of regularly spaced quadrupoles. Notice that the envelope is periodic over the cell length while the betatron oscillations have a much longer wavelength.

EVALUATION OF THE FOCUSING STRENGTH

It is a property of the Hill equation (Eq. 20) that solutions $y(s)$ at point s can be expressed in terms of the displacement $y_0 \equiv y(s_0)$ and divergence $y_0' \equiv y'(s_0)$ at some point s_0 upstream by means of a linear superposition of "cosine-like" and "sine-like" solutions $C(s)$ and $S(s)$, where $C(s_0)=1$ and $S(s_0) = 0$:

$$y(s) = C(s)y_0 + S(s)y_0' \ . \tag{32}$$

Consequently it is possible to express the relation between the co-ordinates at s and s_0 by a matrix equation

$$\begin{bmatrix} y(s) \\ y'(s) \end{bmatrix} = \begin{bmatrix} M_{11} & M_{12} \\ M_{21} & M_{22} \end{bmatrix} \begin{bmatrix} y_0 \\ y_0' \end{bmatrix} \tag{33}$$

where the elements of the "transfer matrix" M depend on both s and s_0. Now suppose that the matrix M describes one complete cell of the lattice, so that

$$s = s_0 + L$$

and

$$\psi(s) = \psi(s_0) + \mu \tag{34}$$

where μ denotes the "phase advance". Then the Twiss parameters, α, β, γ take the same values at s and s_0, and if we use Eq. (34) to expand the expressions (21) for $y(s)$ and (22) for $y'(s)$ in terms of y_0 and y_0' we find (Exercise 9)

$$M = \begin{bmatrix} \cos\mu + \alpha \sin\mu & \beta \sin\mu \\ -\gamma \sin\mu & \cos\mu - \alpha \sin\mu \end{bmatrix} \ . \tag{35}$$

If M can be constructed independently of Eq. (35) by multiplying together known matrices for each element of the cell, then the resultant matrix elements M_{ij} will yield solutions for α, β, and γ at the ends of

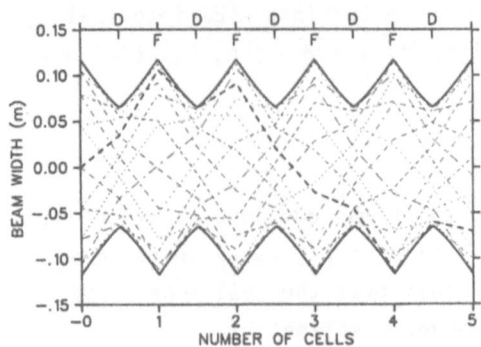

Fig. 8. Selected orbits and their envelope function
for a regular FODO thin lens lattice.

the cell, and for

$$\cos \frac{2\pi\nu}{N} = \cos\mu = \frac{1}{2}(M_{11} + M_{22}) \; , \tag{36}$$

yielding the value of the betatron tune ν.

For a _separated function synchrotron_ we suppose that the quadrupole magnets are arranged in a regularly-spaced FODO lattice (Fig. 8) and that they can be treated as thin lenses. Any focusing action by the bending magnets will be neglected; [even with zero field gradient, there may be small edge focusing or weak focusing effects (cf. Fig. (8) above)]. Then a complete cell can be thought of as consisting of four elements - a focusing lens of focal length f, a drift space L/2, a defocusing lens of focal length -f and a second drift space L/2. [For a quadrupole of length and field gradient g, $f \simeq (B\rho)/\ell g$.] The transfer matrix for the whole cell can be formed by multiplying together the matrix operators for each element in sequence:

$$
\begin{aligned}
M &= M_0 M_D M_0 M_F \\[4pt]
&= \begin{bmatrix} 1 & L/2 \\ 0 & 1 \end{bmatrix} \begin{bmatrix} 1 & 0 \\ 1/f & 1 \end{bmatrix} \begin{bmatrix} 1 & L/2 \\ 0 & 1 \end{bmatrix} \begin{bmatrix} 1 & 0 \\ -1/f & 1 \end{bmatrix} .
\end{aligned} \tag{37}
$$

Readers not familiar with matrix optics should check (_Exercise 10_) that the forms given for M_0, M_F and M_D do have the expected effects on parallel and divergent incident rays (represented by vectors $\binom{1}{0}$ and $\binom{0}{1}$).

Multiplying out the matrix M and comparing it with Eq. (35) above (_Exercise 11_) yields the following expressions for the phase advance μ and the β-value at the F quadrupole (or at the D quadrupole by changing the sign of f)

$$\sin \frac{\mu}{2} = \frac{L}{4f} \tag{38}$$

$$\beta_{\pm} = \frac{L}{\sin \mu} (1 \pm \sin \frac{\mu}{2}) \ . \tag{39}$$

The maximum and minimum β-values β_+ and β_-, occurring at the F and D quadrupoles respectively, are plotted in Fig. 9 as a function of μ. There is a rather flat minimum in β_+ for μ ≃ 78°; for economy in magnet apertures μ is usually chosen in this region, leading to tune values ν ≃ N/4.

The usual scaling law with machine radius R is to increase the number of cells N ∝ $\sqrt{R}$ (Reich, Schindl & Schönauer, 1983). The reader may confirm (Exercise 12) that this choice results in several other quantities [tune ν, cell length L, beta function β and quadrupole strength (∝ p/f)] having to grow only at the same modest rate ∝ $\sqrt{R}$. Thus the SSC and the Brookhaven AGS, whose circumferences are in the ratio 83 km/0.81 km = 103, have ratios of 444/60 = 7.4 in N and 78.3/8.75 = 9.0 in ν, close to $\sqrt{103}$ = 10.1.

For a <u>sector focusing cyclotron</u> we will consider just vertical focusing, which is the most crucial because of the defocusing associated with isochronism (Eq. 16) and the proximity of the magnet poles. We suppose that the hills are much wider than the pole gap so that the hill and valley fields B_h and B_v are uniform and hard-edged. Each sector thus consists of two edge-focusing/defocusing thin lenses separated by drift spaces of length L_h and L_v (Fig. 5). The lenses have focal powers $1/f = \pm G \tan(\epsilon_{\pm} \kappa)$ [see. Eq. (17′)] where G can be expressed in terms of the radii of curvature, ρ_h, ρ_v:

$$G \equiv \frac{e \Delta B}{mv} = \frac{B_h - B_v}{\bar{B} R} = \frac{1}{\rho_h} - \frac{1}{\rho_v} \ . \tag{40}$$

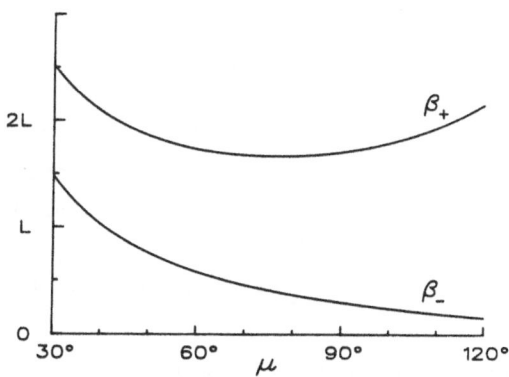

Fig. 9. Variation of maximum and minimum β-values (β_+ and β_- at F and D quadrupoles respectively) with phase advance μ for a regular FODO thin lens lattice.

If the spiral angle ε > κ, the Thomas angle, we have a FODO cell, but a more complicated one than evaluated for the synchrotron above, since the lenses are of different strengths and the drift spaces of different lengths. (For very small spiral angles, ε < κ, the cell is reduced to FOFO form.) The transfer matrix is given by

$$M = M_0' M_D M_0 M_F$$

$$= \begin{bmatrix} 1 & L_v \\ 0 & 1 \end{bmatrix} \begin{bmatrix} 1 & 0 \\ G\tan(\epsilon-\kappa) & 1 \end{bmatrix} \begin{bmatrix} 1 & L_h \\ 0 & 1 \end{bmatrix} \begin{bmatrix} 1 & 0 \\ -G\tan(\epsilon+\kappa) & 1 \end{bmatrix} \quad (41)$$

Multiplying this out and comparing it to Eq. (35) (Exercise 13) we can obtain expressions for the phase advance μ per cell and the β-values at the focusing and (changing the sign of ε) defocusing edges

$$\sin \frac{\mu}{2} = \frac{1}{2} \sqrt{2GL\kappa(1 + \tan^2\epsilon) + G^2 L_h L_v \tan^2\epsilon} \quad (42)$$

$$\beta_{\pm} = \frac{L}{\sin \mu} [1 \pm 2\kappa \tan\epsilon - \kappa^2(2 + 3\tan^2\epsilon)] . \quad (43)$$

where small angle approximations have been used for κ. Note that the smallness of the Thomas angle also limits the oscillations in the β-function, which are generally not as marked as in synchrotrons. The phase advance is also usually small enough to allow use of the small angle approximation for μ/2 in Eq. (42), so that, using Eqs. (18) and (40), and including the isochronous defocusing term (16), the reader will find the following expression for the overall vertical tune

$$\nu_z^2 = -\beta^2\gamma^2 + F^2(1 + 2\tan^2\epsilon) . \quad (44)$$

Spiraling the sectors clearly provides a strong alternating focusing enhancement to the weak Thomas flutter focusing. For zero spiral angle Eq. (44) agrees with Eq. (29) quoted above.

TRANSVERSE SPACE CHARGE DEFOCUSING

High beam intensity and brightness affect many aspects of accelerator design and operation, but particularly beam dynamics (defocusing and instabilities), the rf accelerating system (beam loading), shielding and safety. The principal beam dynamic effect is transverse defocusing, due to mutual repulsion of the electrically charged particles making up the beam.

To evaluate this, we assume for simplicity that the beam is of circular cross-section (area A) with uniform charge density ρ, and that these quantities do not vary around the orbit (radius R, charge Ne), so that $\rho = \frac{Ne}{2\pi RA}$. Also we use polar co-ordinates (r, φ) centred on the beam axis (Fig. 10), with r << R. Then the self-fields produced by the electric charge and associated current can be written

$$E_r = \frac{\rho}{2\varepsilon_0} r \qquad (45)$$

$$B_\phi = \frac{\rho}{2\varepsilon_0} \frac{v}{c^2} r \qquad (46)$$

where $v = \beta c$ is the particle velocity. These fields exert a Lorentz force $e(\mathbf{E} + \mathbf{v \times B})$ on individual protons, where the two terms tend to cancel as $v \to c$; the net force is easily seen to be of the entirely radial form

$$\Delta F_r = \frac{e}{2\varepsilon_0} \frac{\rho}{\gamma^2} r \ . \qquad (47)$$

(Confirmation of this and the following derivations constitutes <u>Exercise 14</u>.) Note that the defocusing force is linear in r just like the basic focusing force F_r which determines the tune ν:

$$F_r = -m(\omega\nu)^2 r = -\frac{m_0 c^2}{R^2} (\beta^2\gamma) \nu^2 r \ . \qquad (48)$$

The net effect of ΔF_r is thus to decrease the tune by an amount $\Delta\nu$, the "tune shift," given by

$$2\nu\Delta\nu \simeq \Delta\nu^2 = -\frac{r_p R}{A} \frac{N}{\beta^2\gamma^3} \qquad (49)$$

where a number of constants have been collected together in $r_p = e^2/4\pi\varepsilon_0 m_0 c^2 = 1.5347 \times 10^{-18}$ m, the classical radius of the proton. Besides its not unexpected dependence on N and A, the tune shift depends very strongly on energy, but inversely, so that $\Delta\nu$ is greatest at the lowest energy – at injection. Since there is an upper limit to $\Delta\nu$ if serious betatron resonances are not to be crossed, this defines the lowest injection energy which can be used for a given R, N and A. Note, however, that the full $\beta^2\gamma^3$ dependence only applies for beams of <u>fixed</u>

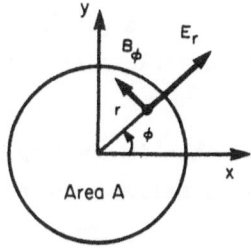

Fig. 10. Cross-section and self-fields of an axially symmetric beam.

area A, independent of energy, such as those defined by a magnet aperture. Thus, increasing the Brookhaven linac injector energy from 50 MeV to 200 MeV potentially increases the charge accelerable in the AGS for a fixed tune shift by the $\beta^2\gamma^3$ factor of 5.0, but to achieve it the magnet aperture must be filled to the same extent at the two injection energies.

A given beam, however, will shrink in transverse dimensions as it is accelerated. This "adiabatic shrinking" is essentially due to the longitudinal momentum of the particles $p_s \simeq p$ being increased while their transverse momentum components remain unchanged. Liouville's theorem requires the area occupied by the beam in y-p_y phase space to be conserved. Areas in y-y' space, which we have been considering, will change with forward momentum p. To recover the invariant property we multiply the emittance by a factor proportional to momentum to obtain the "normalized emittance"

$$\varepsilon^* = \beta\gamma\varepsilon \ . \tag{50}$$

Since the aperture and tune define ε independent of energy, the normalized emittance of the beam injected at 200 MeV in the example above will be larger by the momentum factor 2.1 than that injected at 50 MeV, and its physical diameter will be larger by a factor $\sqrt{2.1}$ at any given energy.

If, on the other hand, we consider a given beam with a fixed normalized emittance ε^* then its cross-sectional area will decrease inversely as $\beta\gamma$:

$$A = \pi\varepsilon\beta_y = \pi \frac{\varepsilon^*}{\beta\gamma} \frac{R}{\nu} \tag{51}$$

where we have used Eq. (29) for the transverse beta-function β_y. In this case Eq. (49) becomes

$$\Delta\nu = - \frac{r_p}{2\pi} \frac{N}{\varepsilon^*} \frac{1}{\beta\gamma^2} \tag{52}$$

showing a reduced $\beta\gamma^2$ energy dependence. This shows how the tune shift varies as a given beam is accelerated, or how the accelerable charge Ne increases with injection energy for a given emittance and tune shift (a factor 2.4 for the Brookhaven example if no effort is made to maintain A by transverse stacking).

Equation (52) is also notable in showing that the tune shift is directly proportional to the normalized one-dimensional beam brightness

$$B_1^* = \frac{e\omega}{2\pi} \frac{N}{\varepsilon^*} \ . \tag{53}$$

Equation (52) has been derived for the simplest conditions - a continuous "coasting" beam of uniform charge density and circular cross-section which has no interaction with its environment. For more complicated conditions the same basic dependencies remain valid but additional factors are required (Reich et al., 1983):

$$\Delta\nu = -\frac{r_p}{\pi}\frac{N}{\epsilon^*}\frac{1}{\beta\gamma^2}\frac{F\,G\,H}{B_f}\,.\qquad(54)$$

The "bunching factor" $B_f \equiv \bar{I}/I_{max}$ describes the extent to which the beam is bunched longitudinally by the rf accelerating field. For a given average current $\bar{I}$, the greater charge density in a bunched beam increases $\Delta\nu$ proportionately.

The factor F describes the effect of image forces from the vacuum chamber and magnet poles. It was first derived by Laslett (1963) and indeed the whole expression (54) is often referred to as the Laslett incoherent tune shift. F has a complicated dependence on the energy, the bunching factor and the heights and widths of the beam, the vacuum chamber and the magnet poles. At low energies the image terms are negligible and $F \simeq 1$. For $\Delta\nu_z$ the electric image terms become important for $\gamma >$ $g/b\sqrt{B_f}$ where h, g and b are the half-heights of the vacuum chamber, magnet poles and beam respectively (Fig. 11).

The factor G describes the transverse density distribution. G equals 1 for a uniform and 2 for a parabolic distribution; in practice G usually lies between 1 and 2. The factor H takes into account the aspect ratio (width/height = a/b) of non-circular beams. Locally $H_z = 1/(1 + a/b)$ and averaging over the orbit $\bar{H}_z = 1/(1 + \sqrt{\epsilon_x\nu_z/\epsilon_z\nu_x})$. Thus widening the beam can be effective in reducing H_z below the "circular" value 1/2, at the expense of increasing H_x.

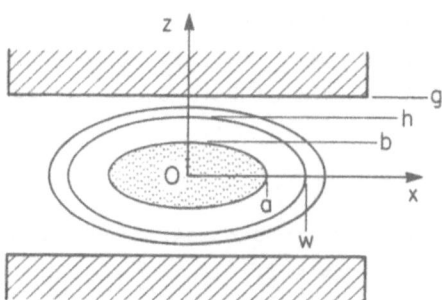

Fig. 11. Cross-section of the beam aperture, showing magnet poles, vacuum chamber and beam.

While Eq. (54) is valid for all types of circular accelerator, for the same average current its effects are of course much more pronounced for pulsed machines (synchrocyclotrons, sychrotrons, FFAG) than for cw ones (isochronous cyclotrons). For slow cycling sychrotrons the duty factor is of order 10^{-6}, giving a million-fold enhancement and often making tune shift the crucial intensity-limiting factor. Synchrotrons designed for the highest intensities therefore tend to be fast-cycling, with lower charges Ne per pulse.

OFF-MOMENTUM ORBITS

As the momentum varies, so will the average radius R, the period τ and angular frequency ω of an orbit. For cyclotrons and FFAG machines with fixed magnetic field there exists an equilibrium (closed) orbit for each momentum, with a radius which generally increases with momentum.

For synchrotrons, where the magnetic field is raised in proportion to the momentum, a central orbit (C.O.) is defined by the geometry (e.g. by the axes of the quadrupole magnets). At any given field level this central orbit is closed for one particular momentum value p_0. For a slightly different momentum $p_0 + \delta p$ there will be a slightly different closed orbit, deviating from the C.O. by

$$x(s) = \eta_x(s)\ \delta p/p_0 \tag{55}$$

where this equation defines the horizontal dispersion function $\eta_x(s)$. For weak focusing synchrotrons it is easy to show (<u>Exercise 15</u>) that

$$\eta_x(s) = \frac{R}{1 + k} = \frac{R}{\nu_x^2} \tag{56}$$

so that with $\nu_x^2 < 1$ the dispersion is quite large and uniform around the machine.

For a strong focusing synchrotron with a regular FODO lattice $\eta_x(s)$ will be periodic over each cell, oscillating in phase with $\beta_x(s)$; by symmetry its slope will be zero at both focusing and defocusing quadrupoles

$$\eta'_+ = \eta'_- = 0 \ . \tag{57}$$

Adding a third component to our matrix representation to describe the momentum deviation we may write

$$\begin{bmatrix} \eta_- \\ 0 \\ 1 \end{bmatrix} = M \begin{bmatrix} \eta_+ \\ 0' \\ 1 \end{bmatrix} \tag{58}$$

where the matrix M describes the effect of the half-cell between the centres of an F and a D quadrupole:

$$M = M_{D/2} \; M_{\theta/2} \; M_{F/2}$$

$$= \begin{bmatrix} 1 & 0 & 0 \\ 1/2f & 1 & 0 \\ 0 & 0 & 1 \end{bmatrix} \begin{bmatrix} 1 & L/2 & L\theta/8 \\ 0 & 1 & \theta/2 \\ 0 & 0 & 1 \end{bmatrix} \begin{bmatrix} 1 & 0 & 0 \\ -1/2f & 1 & 0 \\ 0 & 0 & 1 \end{bmatrix} \tag{59}$$

where the angle of bend $\theta/2$ is assumed small enough for the small angle approximation to hold. [A comparison with Eq. (37) above will show the equivalence of the 2x2 x–x' sub-matrices to those used previously.] Evaluating Eq. (59) (<u>Exercise 16</u>) shows that

$$\eta_{\pm} = R \left(\frac{\pi/N}{\sin \mu/2} \right)^2 [1 \pm \tfrac{1}{2} \sin \mu/2] \tag{60}$$

so that on average

$$\overline{\eta}_x \simeq \frac{R}{\nu_x^2} . \tag{61}$$

This is the same as Eq. (56) for a weak focusing synchrotron, although the stronger tune here drastically reduces the magnitude of the dispersion. It should be noted, however, that relation Eq. (61) can break down for more complicated lattices, such as those with superperiodicity.

What of the momentum-induced changes in orbital period τ and angular frequency ω? Here we have the differential relations

$$\frac{\delta\omega}{\omega} = - \frac{\delta\tau}{\tau} = \frac{\delta v}{v} - \frac{\delta R}{R} . \tag{62}$$

At low energy it is possible for the increase in velocity to be greater than that in radius, and a higher momentum particle will orbit faster. At higher energies, however, as $v \to c$, the increase in radius will dominate and extra momentum will produce a slower orbit. Equation (62) can be rewritten (<u>Exercise 17</u>) in terms of the parameter η (not to be confused with η_x)

$$\eta \equiv \frac{\delta\omega/\omega}{\delta p/p} = \frac{1}{\gamma^2} - \frac{\overline{\eta}_x}{R} . \tag{63}$$

This parameter plays a crucial role in the theory of longitudinal motion in synchrotrons. At low enough energies η is positive, but it decreases as the energy rises (Fig. 12), approaching $-\overline{\eta}_x/R$ asymptotically. (Readers are warned that η is sometimes defined with the opposite sign.) The critical energy at which η changes sign and a high momentum orbit changes

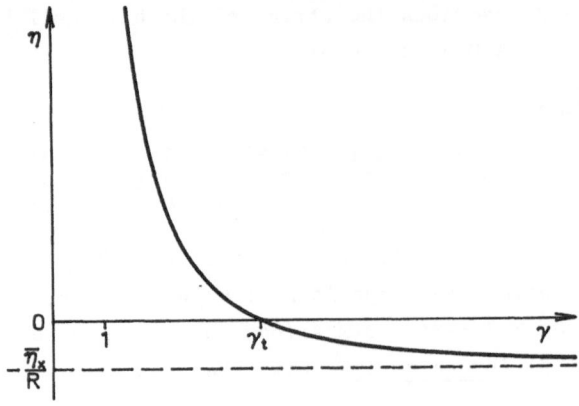

Fig. 12. Dependence of η on energy.

from being faster to slower is known as the <u>transition energy</u> γ_t:

$$\gamma_t = \sqrt{\frac{R}{\eta_x}} \simeq \nu_x \ . \tag{64}$$

Weak focusing machines have $\nu_x < 1$ and therefore always operate above transition. In strong focusing machines with regular lattices the horizontal tune is usually such that γ_t lies within the acceleration range. At this energy the machine becomes isochronous for all momenta and phase focusing disappears, as we shall see below, creating a potential trouble spot.

ACCELERATION AND PHASE STABILITY

At each instant t the magnetic field strength B(t) defines the momentum $p_0(t)$ for the central orbit, and the corresponding orbital frequency $\omega_0(t)$. The rf voltage at an accelerating gap should therefore oscillate at an integer multiple frequency $h\omega_0$ given by the "harmonic number" h:

$$V = V_0 \sin h\omega_0 t \ . \tag{65}$$

In discussing the longitudinal motion the rf phase angle $\phi = h\omega_0 t$ at the moment of an ion's crossing the accelerating gap is the conventional choice for co-ordinate. (Note that ϕ is also a measure of the longitudinal position of different ions at a given instant.) As time progresses the field B will rise, and with it the C.O. momentum p_0. Assuming V_0 is big enough there will be some "synchronous phase" ϕ_s for which the energy gain $eV_0 \sin\phi_s$ provides just the right momentum gain to keep the "synchronous particle" on the C.O. and at that same phase on later turns (<u>Exercise 18</u>):

$$V_0 \sin\phi_s = 2\pi R^2 \frac{dB}{dt} \ . \tag{66}$$

(Here we have assumed that the orbit encloses zero magnetic flux, so that there is no inductive "betatron" acceleration.) In practice B is increased either sinusoidally or linearly, depending on whether the cycling rate is fast or slow; in the latter case $V_0 \sin\phi_s$ = constant.

Since the rf voltage oscillates h times during one orbit period, there will be h places around the orbit where a synchronous particle can be found. As we shall now see, neighbouring particles undergo stable oscillations about these places, if their amplitudes are not too large, defining stable "buckets" in longitudinal phase space.

To understand the motion of these not-quite-synchronous particles we must investigate the rates of change of their energy and phase (<u>Exercise 19</u>). Comparing a particle at phase ϕ, for which the energy gain per turn is

$$\frac{2\pi}{\omega} \frac{dE}{dt} = eV_0 \sin\phi \ , \tag{67}$$

with the synchronous particle, it may be shown that the energy difference ΔE between the two obeys

$$\frac{d}{dt} \left(\frac{\Delta E}{\omega_0} \right) = \frac{eV_0}{2\pi} (\sin\phi - \sin\phi_s) \ . \tag{68}$$

(ΔE, Δp, $\Delta\omega$ etc., will all denote differences from the synchronous values.) The rate of change of phase may be derived from Eq. (63):

$$\frac{d\phi}{dt} = \frac{-h\eta}{m_0 R_0^2} \left(\frac{\Delta E}{\omega_0} \right) \ . \tag{69}$$

Noting that the "momentum" co-ordinate canonically conjugate to the "position" co-ordinate ϕ is

$$W \equiv \frac{\Delta E}{\omega_0} = R_0 \Delta p \tag{70}$$

we see (<u>Exercise 20</u>) that Eqs. (68) and (69) have the form of Hamilton's equations of motion for a system defined by the Hamiltonian function

$$H(\phi,W) = \frac{h\eta}{2m_0 R_0^2 \gamma} W^2 + \frac{eV_0}{2\pi} [\cos\phi_s - \cos\phi - (\phi-\phi_s)\sin\phi_s] \ . \tag{71}$$

Curves of H = constant in the longitudinal ϕ–W phase plane represent the particle trajectories (Fig. 13). We see that the curves are closed for small deviations from the synchronous condition (ϕ_s,0), indicating stable <u>synchrotron oscillations</u> about this point. In fact there are two distinct situations, depending on whether the energy is below or above transition.

Below transition $\gamma < \gamma_t$, the parameter $\eta > 0$, and the synchronous phase occurs on the rising side of the rf voltage wave ($0 \leq \phi_s \leq \pi/2$).

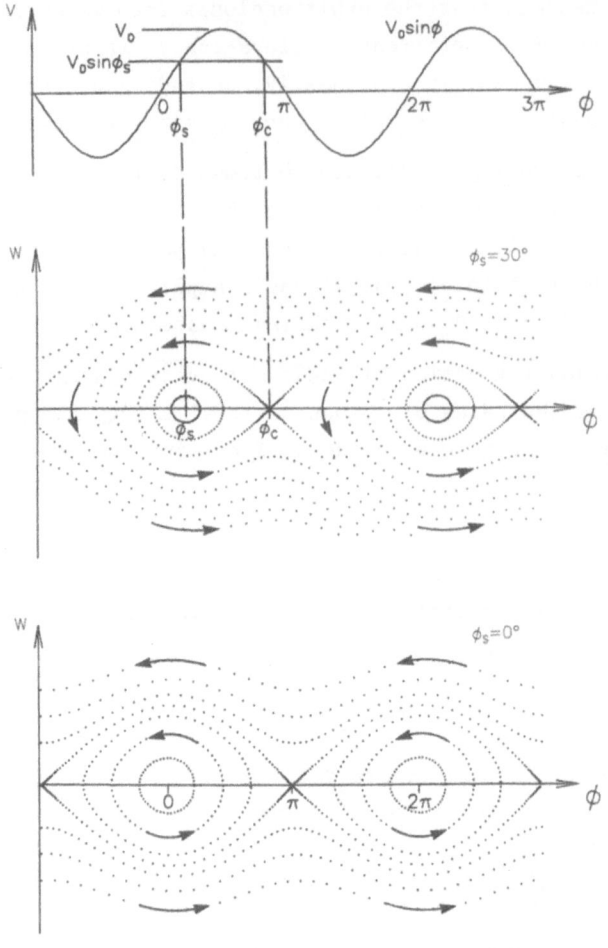

Fig. 13. Stroboscopic view of particle trajectories in the W-ϕ phase plane; the dots are plotted at equal time intervals.

The reason for this becomes clear if the motion is followed in detail in the phase plane. For a particle lagging behind in phase ($\phi > \phi_s$) the accelerating voltage will be higher on the rising side (since $\sin\phi > \sin\phi_s$ there) and hence by Eq. (68) the energy difference ΔE will grow more positive. But according to Eq. (69) $\Delta E > 0$ will lead to a faster orbit ($d\phi/dt < 0$) and negative phase change (a succession identifiable as the first quadrant of a counterclockwise orbit in the phase plane) - <u>provided</u> $\eta > 0$, that is, below transition.

Above transition, where η is negative, slower orbits and negative phase change are contrariwise produced by negative ΔE. In this case stable orbits require $\sin\phi < \sin\phi_s$ (for $\phi > \phi_s$) and hence the synchronous phase occurs on the falling side of the voltage wave ($\pi/2 \leq \phi_s \leq \pi$). The trajectories in the phase plane are mirror images of those for $\gamma < \gamma_t$.

The differential equation for the phase oscillation may be obtained by eliminating $\Delta E/\omega_0$ between Eqs. (68) and (69):

$$\frac{d^2\phi}{dt^2} = -\frac{heV_0}{2\pi R_0^2 m_0}\frac{\eta}{\gamma}(\sin\phi - \sin\phi_s) . \tag{72}$$

For large deviations from ϕ_s, the restoring force is clearly non-linear, due to the sinusoidal voltage variation. For small deviations, however, the force is linear in $(\phi - \phi_s)$, resulting in a simple harmonic oscillation at a frequency $\nu_s\omega_0$, where ν_s is the "synchrotron tune," given by (Exercise 21)

$$\nu_s = \sqrt{\frac{h}{2\pi}\cdot\frac{eV_0\cos\phi_s}{m_0c^2}\cdot\frac{|\eta|}{\beta^2\gamma}} . \tag{73}$$

Because $eV_0 \ll m_0c^2$, the tune $\nu_s \ll 1$, so that particles take tens or hundreds of turns to complete a synchrotron oscillation. (Even so the oscillations are much faster than the rates at which ω_0, p_0, η, V_0 and ϕ_s change, so that our tacit assumption of their constancy has been reasonable.) Note that the tune varies strongly with energy, reaching zero at transition, where the oscillations come to a standstill.

For large deviations from the synchronous condition the trajectories do not close, and the motion becomes unstable, W becoming more and more negative as the particle is left behind the accelerating bucket. A pear-shaped separatrix divides the unstable region from the stable "bucket," whose area gives the longitudinal acceptance of the machine.

The width of the bucket can be shown to depend on ϕ_s alone. The phase of the cusp ϕ_c is simple to evaluate. Like the minima of the unstable trajectories it occurs where dW/dt = 0. From Eq. (68) this implies that

$$\phi_c = \pi - \phi_s \tag{74}$$

so a phase oscillation may extend over the top of the voltage wave to the point where V has dropped to $V(\phi_s)$. Inserting the coordinates of the cusp $(\phi,W) = (\pi-\phi_s,0)$ in (71) we may obtain (Exercise 22) the value of the constant H for the separatrix and hence the height of the bucket:

$$\hat{W} = W(\phi_s) = \sqrt{\frac{m_0R_0^2\,eV_0\gamma}{\pi h\eta}[2\cos\phi_s - (\pi-2\phi_s)\sin\phi_s]} . \tag{75}$$

For a given value of ϕ_s the bucket height and area may be seen from Eq. (75) to be proportional to $\sqrt{V_0\gamma/|\eta|}$. The equation for the separatrix may also be used to obtain the extreme phase ϕ_e at the opposite end of the bucket to the cusp, by setting W = 0 and solving the resulting transcendental equation

$$\cos\phi_e + \phi_e\sin\phi_s = (\pi-\phi_s)\sin\phi_s - \cos\phi_s \; . \tag{76}$$

Neither ϕ_e nor the area of the bucket can be expressed in terms of simple functions of ϕ_s, but tabulated numerical values are available (see e.g. Bovet et al., 1970).

Below transition the range of synchronous phases extends down to $\phi_s = 0$. This extreme value is symmetrically placed on the voltage wave, and consequently the bucket for this case is mirror-symmetric (Fig. 13). One cusp occurs at $\phi_c = \pi$ and another at $-\pi$ (otherwise regardable as the $+\pi$ cusp of the neighbouring bucket). The net effect is that the stable region extends over the complete 2π (at $\Delta E = 0$) - at the price of there being zero energy gain, averaged over a complete synchrotron oscillation. This is not an academic curiosity, but a useful mode for accumulation, capture or storage of a beam at constant energy. Above transition the corresponding condition occurs for $\phi_s = \pi$.

As ϕ_s is increased from 0 (or decreased from π) towards $\pi/2$, the width, height and area of the stable bucket all decrease towards zero, for given values of γ, η and V_0. Indeed (74) shows that ϕ_c and ϕ_s become coincident for $\phi_s = \pi/2$. Thus the condition for maximum energy gain completely removes phase stability! To achieve adequate stable bucket area a lower energy gain per turn must be accepted. From a beam dynamics point of view the optimum values for synchronous phase $\phi_s(t)$ and rf voltage $V_0(t)$ at each point in the cycle can be determined uniquely from the required energy gain per turn Eq. (66) and bucket area. In practice other considerations, particularly rf engineering ones, may prevail. A value of ϕ_s around 30° (150°) is often used.

For most existing proton synchrotrons the transition energy lies within the acceleration range. In approaching γ_t at constant ϕ_s and V_0 the width of the bucket remains constant but its height and area increase as $\sqrt{\gamma/|\eta|}$, i.e. towards infinity. For a bunch of particles the emittance area will remain unchanged, but the aspect ratio will follow that of the bucket, at least at first; as a result the bunch length will shrink as $(|\eta|/\gamma)^{1/4}$ and its momentum spread will increase as $(\gamma/|\eta|)^{1/4}$. These effects reach a natural limit as the rate of rise of the bucket becomes too fast for the bunch to follow adiabatically, the synchrotron motion becoming increasingly sluggish as the tune $\nu_s \to 0$. Nevertheless both effects are sufficiently large to be undesirable. The bunching enhances longitudinal space charge effects (see below), firstly a change in bucket area across transition, causing mismatch, and secondly microwave instabilities. The increased momentum spread requires tighter tolerances on chromaticity and stopband width. In order to minimize these effects it has been usual to programme γ_t with a sudden jump so that no time is spent close to the transition condition. At the same moment the rf phase must be shifted by $\pi - 2\phi_s$ so that the bunch finds itself on the falling side of the rf wave, the stable side above γ_t.

Isochronous cyclotrons operate right on transition at all energies, in the sense that $\eta = 0$ and the orbital frequency is independent of momentum and phase. (No transitions are necessary, of course, so the terminology is redundant.) Under these conditions the analysis above breaks down. There is no special synchronous phase ϕ_s (all phases are equally synchronous) and no stable bucket. In a W-ϕ phase plane defined for some reference energy, the representative points are stationary. In an E-ϕ phase plane the points move up in energy at fixed phase according to Eq. (67). After $n = \omega_0 t/2\pi$ orbits starting from injection energy E_i the ion energy is given by

$$E = E_i + neV_0 \sin\phi . \qquad (77)$$

While Eq. (69) suggests that $d\phi/dt = 0$, Joho (1974) has pointed out that another effect must be considered if the accelerating voltage varies with radius so that $V_0 = V_0 (R)$. In this case the longitudinal rf electric field will be accompanied by a vertical rf magnetic field $B_0(R)$ which will modify the orbital frequency and lead to a variation of phase with radius and energy $\phi(E)$. In place of Eq. (71) above we have the Hamiltonian

$$H(\phi,E) = -\frac{eV_0}{2\pi} \cos\phi , \qquad (78)$$

which defines the flowlines in the longitudinal phase plane. These are illustrated in Fig. 14, which shows how a decreasing $V_0(R)$ produces a "phase expansion" effect. Beam emittance is conserved since the stretching in phase is balanced by compression in energy. The example is taken from Joho's proposal (1984) for the ASTOR 2 GeV isochronous cyclotron, to be used as an intermediate stage between the SIN 590 MeV cyclotron and a 20 GeV high-intensity proton synchrotron. ASTOR would compress 250-turn packets into a small enough radial interval for efficient extraction, while leaving them with a suitable phase-energy distribution to match and partly fill the synchrotron buckets.

LONGITUDINAL SPACE CHARGE EFFECTS

Any variation in density along the orbit (i.e. bunching) will produce longitudinal electric fields which will modify the accelerating voltage $V_0 \sin\phi$ provided by the rf cavities, changing the phase focusing strength and bucket area, and possibly producing instabilities. The longitudinal component of electric field derives both from the direct

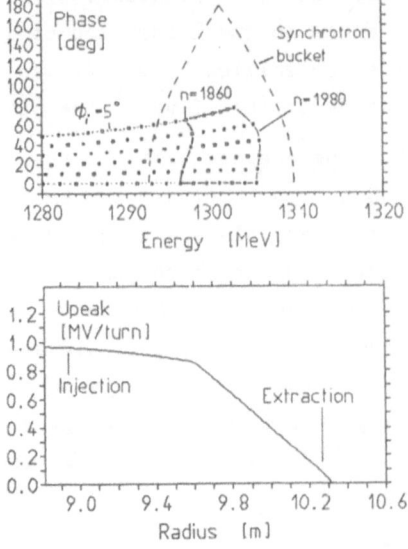

Fig. 14. Phase expansion and energy compression (above) with a radially
          decreasing accelerating voltage (below). The phase width of
          the beam increases from ±5° at injection to ±75° at extraction.

space charge of the beam and from the (opposite) charge induced on the
surrounding surfaces. For simplicity we assume that both the beam and
vacuum chamber are of circular cross-section, with radii a and w respectively (Fig. 15). The transverse fields within the beam are given by
Eqs. (45) and (46), those outside the beam by

$$E_r = \frac{e\lambda}{2\pi\varepsilon_0}\frac{1}{r} \tag{79}$$

$$B_\phi = \frac{\mu_0 e\lambda v}{2\pi}\frac{1}{r} \tag{80}$$

where $\lambda(s) = dN/ds = \pi a^2 \rho/e$ represents the numerical line density. In
order to evaluate the electric field component E_s along the axis (Exercise 23) we use the integral form of Faraday's Law of Magnetic Induction

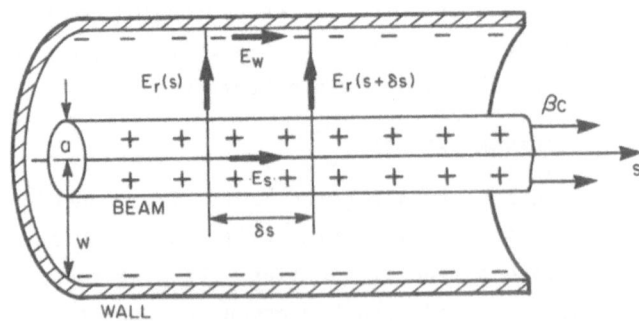

Fig. 15. Longitudinal space charge fields.

over a rectangular path (w x δs) from axis to wall and back, obtaining

$$E_s = - \frac{eg_0}{4\pi\varepsilon_0\gamma^2} \frac{d\lambda}{ds} + E_w , \qquad (81)$$

where $g_0 \equiv 1 + 2\ln \frac{w}{a}$ $\qquad\qquad\qquad (82)$

and E_w is the longitudinal field at the wall. This drives the wall current I_w, which is equal in magnitude but opposite in sign to the a.c. component of the beam current. In most accelerators the reactive wall impedance is inductive at low and medium frequencies. Denoting the total inductance around the machine by L, the wall field

$$E_w = \frac{L}{2\pi R} \frac{dI_w}{dt} = \frac{ev^2 L}{2\pi R} \frac{d\lambda}{ds} . \qquad (83)$$

Thus both space charge and wall contributions to the longitudinal field E_s experienced by a particle on the axis are directly proportional to the charge density gradient in its vicinity, so that overall

$$E_s = - \frac{e}{4\pi R} \left[\frac{g_0 R}{e_0 \gamma^2} - \beta^2 c^2 L \right] \frac{d\lambda}{ds} . \qquad (84)$$

The two effects act in opposition, the wall being inductive and dominant at high energy, the space charge capacitive and dominant at low energy. Over one complete orbit (for which the synchrotron motion is negligible) the particle will experience an effective voltage

$$\Delta V(\phi) = 2\pi R E_s = e\beta c \frac{d\lambda}{d\phi} h \left[\frac{g_0 Z_0}{2\beta\gamma^2} - \omega_0 L \right] \qquad (85)$$

where $Z_0 = 1/c\varepsilon_0 = 377$ Ω. Note that, for better comparison with cavity voltages, the density gradient has been expressed in terms of its time variation at an accelerating gap rather than its spatial variation at an instant; the sign change occurs because more positive s corresponds to more negative ϕ (leading particles arrive early).

The expression in square brackets in Eq. (85) represents the effective impedance at the orbit frequency ω_0; the factor h describes the increased impedance at the bunch frequency $h\omega_0$. To maintain a formula applicable to any harmonic frequency $n\omega_0$ it is usual to define

$$\frac{Z_e}{n} \equiv j \left[\omega_0 L - \frac{g_0 Z_0}{2\beta\gamma^2} \right] . \qquad (86)$$

At low energies the capacitive space charge term predominates. The induced voltage $\Delta V(\phi)$ therefore swings from positive to negative, just like $d\lambda/d\phi$, as ϕ passes through the phase for which λ is a maximum – ϕ_s for any single-peaked time-invariant particle distribution (Fig. 16). If

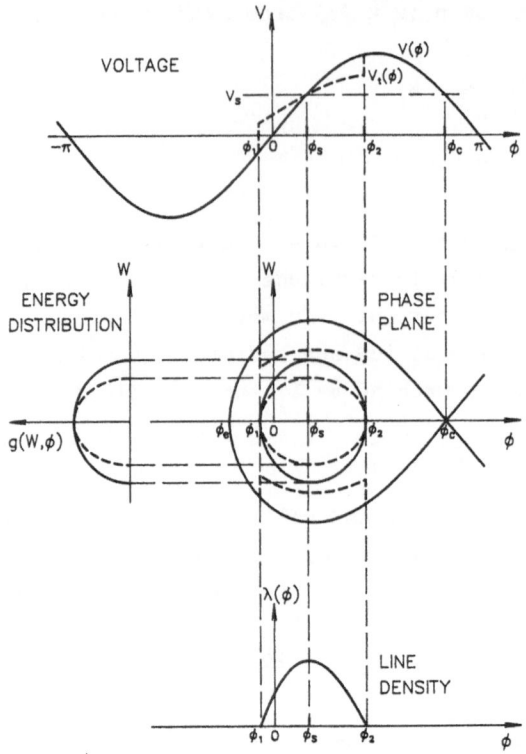

Fig. 16. Effect of space charge in a beam bunch on the rf voltage
and bucket: with (– – –) and without (————)space charge.

the energy is also low enough to be below transition ($\gamma < \gamma_t$), ϕ_s will
lie on the rising side of the rf voltage wave $V(\phi)$, and the net effect of
$\Delta V(\phi)$ will be to decrease the slope of the rise around ϕ_s. But from Eqs.
(72) and (73) we see that the synchrotron restoring force is proportional
to this slope, and the synchrotron tune ν_s to its square root. The space
charge $\Delta V(\phi)$ will thus reduce them both. The stable bucket and the bunch
contour will also shrink in height and area, as V_0 is in effect
decreased.

To be more specific we follow Hofmann and Pedersen (1979) in considering (Exercise 24) a particle distribution which is elliptic in
energy

$$g(W, \phi) \equiv \frac{d^2N}{dWd\phi} \propto \sqrt{W_b^2(\phi) - W^2} \qquad (87)$$

where $W = W_b$ on the bunch boundary in phase space. From Eq. (71) we see
that the density $g(W, \phi) = g(H)$ remains invariant in time as the particles
flow along lines of constant H in phase space. Integrating over W to
obtain the line density

$$\lambda(\phi) = \frac{h}{R} \frac{dN}{d\phi} \propto W_b^2(\phi) \qquad (88)$$

we find the same phase dependence as in Eq. (71). The **density gradient** therefore has the same shape as the applied voltage (relative to $V_s \equiv V(\phi_s)$ - see Fig. 15)

$$\frac{d\lambda}{d\phi} \propto V_0 \sin\phi - V_0 \sin\phi_s \; . \tag{89}$$

From Eq. (85) so also does the space-charge induced voltage $\Delta V(\phi)$, **and as** a result the total voltage V_t can be written

$$V_t(\phi) \equiv V_s + k_t[V(\phi) - V_s] \tag{90}$$

where
$$k_t \equiv 1 + 2\pi h \, \frac{\overline{I}}{V_0} \, \frac{\mathrm{Im}(Z_e/n)}{f(\phi_1, \phi_2)} \tag{91}$$

and
$$f(\phi_1, \phi_2) \equiv \sin\phi_2 - \sin\phi_1 - \frac{1}{2}(\phi_2 - \phi_1)(\cos\phi_1 + \cos\phi_2) \; . \tag{92}$$

The magnitude of the voltage change, described by the factor k_t, is seen to depend directly on the time-averaged beam current $\overline{I}$ and the effective impedance $h(Z_e/n)$, and inversely on the applied voltage V_0 and the function $f(\phi_1, \phi_2)$, which depends on the phase values at the ends of the bunch ($\phi_1 \leq \phi \leq \phi_2$) and grows with its length $\phi_\ell = \phi_2 - \phi_1$. For very short bunches

$$f(\phi_1, \phi_2) = \frac{1}{12} \, \phi_\ell^3 \, \cos\phi_s \; . \tag{93}$$

At low energies the (capacitive) impedance is negative and so $k_t < 1$ and $dV/d\phi$ is reduced as described above. As a result the heights of the bucket and of the other contours of constant H, including the bunch boundary, are reduced proportional to $\sqrt{k_t}$ over the length of the bunch. The same is true for the area occupied by the bunch:

$$\frac{A_t}{A} = \frac{W_t(\phi)}{W(\phi)} = \sqrt{k_t} \; . \tag{94}$$

Beyond the ends of the bunch the voltage, the bucket shape and the H contours are unaffected. The change is not completely abrupt at the ends of the bunch, although it appears to be so in our simple model in which fringing fields are neglected. Of course, if the bunch fills the whole bucket then Eq. (94) applies to its area too.

From Eq. (91) it would appear that for a sufficiently high beam current the bucket would be completely suppressed. In practice high-frequency __instabilities__ will develop within the bunch before the space charge limit is reached. Instabilities are treated in a later lecture by J.-L. Laclare (1986). For present purposes it will be sufficient to quote the Keil-Schnell criterion (1969) for the stability of coasting (uniform) beams, as applied by Boussard (1975) to bunched beams:

$$I(\phi) \leq F' \frac{|\eta|}{e\beta^2 E} \frac{[\Delta E(\phi)]^2}{\left|Z_e'/n\right|} . \tag{95}$$

Here F' is a form factor, Z_e'/n is the microwave coupling impedance, and $\Delta E(\phi)$ refers to the full width of the energy distribution at half height (FWHH). The criterion sets a current threshold for microwave instability, which in general will vary along the bunch. For the elliptic energy distribution, however, we know from Eq. (88) that the local charge density is proportional to the square of the local energy spread, so that in this case the criterion is independent of phase. Evaluating the microwave threshold current for the reduced voltage given by Eq. (90), and substituting it back in Eq. (91) to get the corresponding κ_t, Hofmann and Pedersen find (Exercise 25) that most of the physical quantities cancel out, leaving an almost purely numerical expression

$$k_t = \left[1 + \frac{3F'}{\pi} \frac{Im(Z_e/n)}{\left|Z_e'/N\right|}\right]^{-1} . \tag{96}$$

For not too short bunches and well-damped resonance the impedance at high frequency and reactance at low frequency may be assumed roughly equal, leading to $k_t \approx 0.6$ for $F \approx 0.7$, i.e. the microwave instability will break in when the space charge induced voltage rises to about 40% of the applied voltage.

The numerical nature of Eq. (96) is a reflection of the similar forms of the instability criterion Eq. (95) and the standard relation Eq. (75) between bucket height and rf voltage. Although dispersion relations are needed to derive Eq. (95), its similarity to Eq. (75) is suggested by some elementary considerations. The criterion must express the balance between driving and damping mechanisms, the former described by the induced voltage IZ_e, the latter by the spread in orbital frequencies $\Delta\omega$ which inhibits coherent motion through Landau damping. The frequency spread is provided by the momentum spread ($\Delta\omega/\omega = \eta\Delta p/p$) and hence the criterion is of bucket height-voltage form.

So far we have considered only energies low enough to be below transition ($\gamma < \gamma_t$) and to give capacitive coupling impedance. For these conditions $k_t < 1$ and the rf voltage and bucket height are reduced. But in fact the same effects result at high energies – provided they are high enough to be simultaneously above transition ($\gamma > \gamma_t$) and to give indicative impedance. While the impedance change reverses the sign of $\Delta V(\phi)$, the shift of the synchronous phase to the falling side of the rf voltage wave again produces a reduction in the slope of the voltage.

74

At intermediate energies, however, there is the possibility of the opposite situation. Either below γ_t with inductive Z_e, or above γ_t with capacitive Z_e, the slope of the voltage is increased, $k_t > 1$ and the bucket grows bigger. The energy for impedance reversal is determined by the inductance and the pipe/beam diameter ratio w/a:

$$\beta\gamma = \sqrt{\frac{g_0 Z_0}{2\omega_c L}} \tag{97}$$

where $\omega_c = c/R_0$. With $\omega_c L \simeq 10\ \Omega$, typical of well engineered machines, and w/a = e (g_0 = 3), we obtain $\beta\gamma = 7.5$. This is close to γ_t for 30 GeV machines like the CERN PS and Brookhaven AGS. For higher energy machines, however, γ_t is higher, and there exists a significant energy range over which the bucket will increase rather than decrease in height. (Efforts to reduce $\omega_c L$, for instance to < 1 Ω, will of course narrow this range). As the bucket grows taller the beam bunch will tend to follow its shape and grow narrower; this narrowing of the bunch as a result of space charge represents a "negative mass" effect.

We are now in a position to explain the perils of crossing transition, alluded to above, in more detail. Firstly, the switch in synchronous phase from the rising to the falling side of the rf wave will produce an abrupt change of sign in $f(\phi_1, \phi_2)$ and hence in $k_t - 1$; the resulting sudden jump in bucket height will leave the bunch mismatched. Secondly, the bucket height decrease/increase is greatest near transition, making the beam more susceptible to microwave instabilities. Recent designs for very high intensity proton synchrotrons (kaon factories) have tended to avoid crossing transition altogether by driving $\gamma_t \gg \nu_x$ (or $\ll \nu_x$). In some cases this results in enhanced rf voltage and bucket height over a large energy range.

BEAM BRIGHTNESS AND MULTISTAGE ACCELERATORS

Modern high energy proton accelerators consist of several synchrotron stages following an injector, normally a linac. Table 1 lists the energies of the various stages for existing and (below the dashed line) proposed machines.

There are a number of reasons for breaking these accelerators into stages, each with a restricted energy range. One is to avoid operating the magnets in a very low field region where field quality and control is poor. A second, especially important for rapid-cycling machines, is to separate the lower energies (say below 3 GeV, where $\beta \simeq 0.97$) where a large swing in radiofrequency is required, from the main energy range, where a high rf voltage must be provided. A third, of particular concern in high intensity machines, and therefore worthy of elaboration here,

Table 1. Multistage Proton Accelerator Energies

Institute	Injector	Stage 1	Stage 2	Stage 3	Stage 4
Brookhaven	200 MeV	1.5/2.5 GeV	33 GeV		
CERN	50 MeV	0.8 GeV	10/28 GeV	450 GeV	
Fermilab	200 MeV	8 GeV	150/500 GeV	1000 GeV	
KEK	40 MeV	0.5 GeV	12 GeV		
Serpukhov	100 MeV	1.5 GeV	76 GeV	600 GeV	3 TeV
DESY	50 MeV	7.5 GeV	40 GeV	820 GeV	
SSC	600 MeV	7 GeV	100 GeV	1000 GeV	20 TeV
TRIUMF	450/520 MeV	3 GeV	30 GeV		
LAMPF	800 MeV	6 GeV	45/60 GeV		
EHF	1200 MeV	9 GeV	30 Gev		

stems from the space-charge induced shift $\Delta\nu$ in betatron tune, given by Eq. (52) or (54). The sharp rise in $\Delta\nu$ with decreasing energy ($\Delta\nu \propto N/\varepsilon^*\beta\gamma^2$) and the need to keep its magnitude below about 0.2 to avoid crossing fourth or lower order resonances, determine the minimum injection energy for a machine designed for charge Ne and emittance ε^*. If this is above the injector energy, a lower energy synchrotron stage is required, and so on....

For a chain of synchrotrons it is helpful to express the tune shift equation as far as possible in terms of quantities that are invariant through the chain. The normalized emittance ε^* is one such quantity, assuming no deterioration in beam quality during acceleration or transfer between the stages. The number of particles N varies from ring to ring, but is related to another invariant, the average current passing through the system, $\bar{I} = eN_1/\tau_1 = eN_2/\tau_2 = \ldots$, where τ_i is the cycle time for the ith ring, and beam losses are assumed negligible. Equation (54) then shows that the tune shift for each ring is directly proportional to the time-averaged one-dimensional <u>beam brightness</u> $\bar{B}_1^* = \bar{I}/\varepsilon^*$ (invariant through the system), but depends also on terms peculiar to that ring:

$$\Delta\nu_i = - \frac{r_p \bar{B}_1^*}{\pi e}\left(\frac{FGH}{B_f} \cdot \frac{\tau}{\beta\gamma^2}\right)_i . \qquad (98)$$

[Note that the instantaneous brightness B_1^* defined by Eq. (53) is not the appropriate quantity for a chain of synchrotrons, since it is not invariant; being based on the circulating current, it increases in direct proportion to particle velocity].

An alternative form for $\Delta\nu_i$ is obtained by writing $N = 2\pi R \bar{\lambda}$, where $\bar{\lambda}$ is the spatially averaged line density. This is also invariant from machine to machine, provided pulses from one ring are simply stacked end-to-end, box car fashion, in the next ring. The resulting expression is

$$\Delta\nu_i = - 2r_p \cdot \frac{\bar{\lambda}}{\varepsilon^*} \left(\frac{FGH}{B_f} \cdot \frac{R}{\beta\gamma^2} \right)_i . \tag{99}$$

If the tune-shift at injection is limited to the same value at each stage, the term in parentheses will also take a constant value. Thus for each ring in a given chain

$$\left(\frac{B_f \beta\gamma^2}{FGH} \right)_{inj} = \frac{2r_p}{|\Delta\nu|} \frac{\bar{\lambda}}{\varepsilon^*} R \propto R . \tag{100}$$

There is a fixed relation between conditions at injection and the average radius R. Since this is a parameter determined by the top energy, Eq. (100) defines the energy range for each stage. In terms of the top momentum

$$(\beta\gamma^2)_{inj} = \frac{2m_0 c r_p}{e(\rho/R)} \frac{\bar{\lambda}}{\varepsilon^*} \left(\frac{FGH}{B_f |\Delta\nu|} \right)_{inj} \left(\frac{\beta\gamma}{B} \right)_{max} \tag{101}$$

where ρ/R is the dipole packing fraction and B is the dipole magnetic field strength. In so far as we can neglect variations in the various parameters from ring to ring, we may write

$$(\beta\gamma^2)_{inj} = k(\beta\gamma)_{max} . \tag{102}$$

In practice we can only expect k to be very roughly constant from stage to stage. Even if $\Delta\nu$ is the same for each ring, the packing fraction will vary with the lattice, B_{max} will depend on the cycling rate, and the bunching factor B_f will probably be larger for the first stage than for subsequent ones.

Values of k for a number of multistage machines are plotted in Fig. 17. For most lower energy stages k is constant within a factor 2. Higher intensity machines tend to have "shorter" stages and higher k values, presumably reflecting the use of brighter beams or more conservative $\Delta\nu$. Large changes in k show where $\Delta\nu$ was not critical in choosing the staging energies – where tunnels already existed (DESY-HERA), to avoid depolarization (EHF) and for very high energies. The case of the proposed SSC is an interesting one. Here considerations of tune shift and beam brightness (Ankenbrandt, 1984) led to a decision to add a third Low Energy Booster (LEB) to the original two booster stages. With the

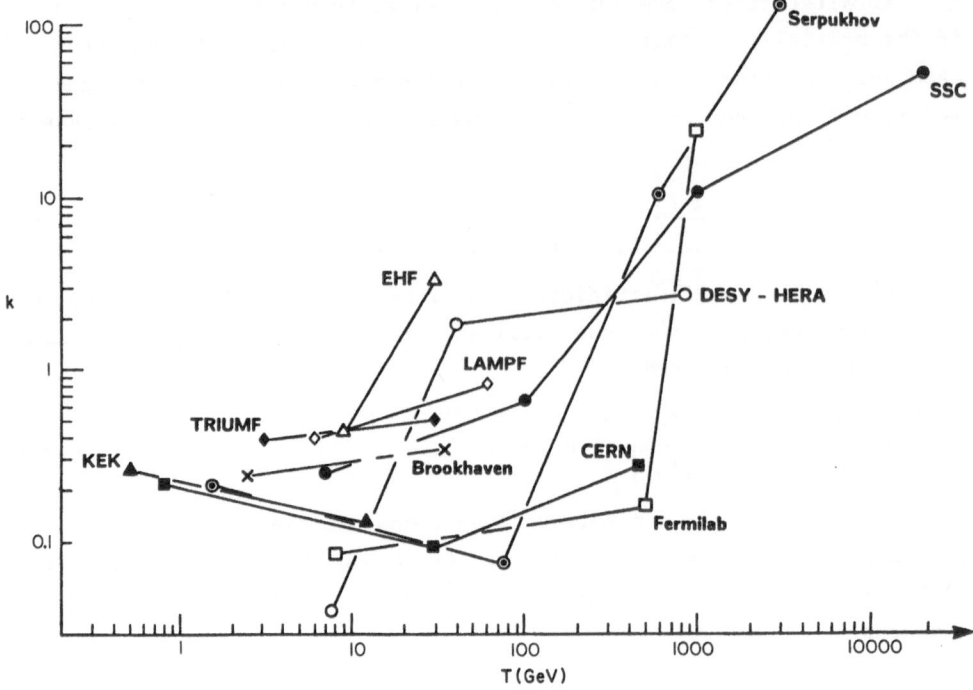

Fig. 17. Values of the parameter $k = (\beta\gamma^2)_{inj}/(\beta\gamma)_{max}$, plotted against energy for each stage of various multistage machines.

energies of the original stages more or less fixed, however, the tune shift is only critical in the LEB, and k increases strongly with energy.

HIGH INTENSITY PROTON SYNCHROTRONS

Ever since the first proton synchrotrons were built 30 years ago there has been a steady effort to increase primary beam intensity and hence the fluxes of secondary particles (kaons, antiprotons, neutrinos, etc.) and the capability of studying rarer and rarer particle inter-actions. The most recent initiative has been the proposal of a new generation of fast-cycling synchrotrons – the so-called kaon factories – machines with beam intensities a hundred times greater than those available at present.

In considering the maximum intensity which can be accelerated in a synchrotron two parameters are of particular importance, the number of particles per pulse N and the circulating current I. N is critical because it determines the incoherent space charge tune shift $\Delta\nu$, given by

Eq. (54), the decrease in the betatron tune due to the defocusing effects of space charge. Of course, a drop in tune by itself could be compensated by adjusting the quadrupole magnets: the problem here is that the shift can vary across and also along the bunch, so that $\Delta\nu$ also represents a spread in tunes. In order to avoid coming to close to lower-order resonances it is generally agreed that $\Delta\nu$ should be kept below 0.2. The $\beta\gamma^2$ factor makes this condition most critical near injection. It was to take advantage of this energy dependence that the injection energies of the Brookhaven AGS and CERN PS were raised from the original 50 MeV to 200 MeV and 800 MeV, respectively. The crucial role $\Delta\nu$ plays in the design of multistage accelerators was discussed in the previous section.

The circulating current I is important through its involvement in longitudinal space charge effects and beam stability. We saw above how the current appears directly in Eq. (91) for k_t, the factor describing the effect of space charge on bucket height, and in Eq. (95), the Keil-Schnell-Boussard criterion for microwave stability.

The energy and intensity parameters, including N and $\bar{I}$, are listed in Table 2 for existing and proposed high energy proton synchrotrons. The existing higher energy machines achieve average beam currents of ~1 μA. These currents are limited both by the slow-cycling rate (<1 Hz) and by their low injection energies ($\leq$200 MeV into their first synchrotron stages) which restrict N to ~2×10^{13}. The circulating current $I \approx 1$ A.

Higher intensities have been achieved in machines using faster cycling rates (10-50 Hz). A record current (for a synchrotron) of 40μA was recently achieved at the Rutherford ISIS spallation neutron source, and this will be raised to 200 μA when commissioning is completed. The number of protons per pulse N will then be 2.5×10^{13}, only a little more than in the slow-cycling machines, but the circulating current $\bar{I}$ will rise to 6A.

The proposed kaon factories aim at energies in the 25-45 GeV range with proton currents of 30-100 μA. Proposals have come from all three existing pion factories at LAMPF, SIN and TRIUMF, these laboratories being unique in already possessing operating machines with adequate energy and current to act as injectors, and also from a European consortium and from Japan. All the proposals also involve intermediate booster synchrotrons with energies in the 3-9 GeV range. These have a dual purpose. In the first place they raise the injection energy into the main ring, and therefore through Eq. (54) the charge per pulse that

Table 2. High-Intensity proton synchrotrons

	Energy (GeV)	Average current (μA)	Rep. rate (Hz)	Protons/ pulse N (x 10^{13})	Circulating current $\bar{I}$ (A)
Slow Cycling[a]					
KEK PS	12	0.32	0.6	0.4	0.6
CERN PS	26	1.2	0.38	2	1.5
Brookhaven AGS	28.5	0.9	0.38	1.6	0.9
– with Booster		(3)		(5)	(3)
Fast Cycling[b]					
Argonne IPNS	0.5	8	30	0.17	2.3
Rutherford ISIS	0.55(0.8)	40(200)	50	(2.5)	(6.1)
Fermilab Booster	8	7	15	0.3	0.3
AGS Booster	(1.5)	(20)	(10)	(1.25)	(3)
Proposed Boosters[b]					
TRIUMF	3	100	50	1.2	2.7
European HF	9	100	25	2.5	2.5
LAMPF	6	144	60	1.5	2.2
KEK Booster	1-3	100	15	4	8
Kaon Factories[a]					
TRIUMF	30	100	10	6	2.8
European HF	30	100	12.5	5	2.5
LAMPF	45	32	3.33	6	2.2
Japan – Kyoto	25	50	30	1	0.5
– KEK	30	30	1	20	7

[a]Slow extraction
[b]Fast extraction

can be accelerated, to N ~ 6×10^{13} ppp. This enables the desired current
of 100 μA to be achieved with only moderately fast cycling rates ~10Hz.
The second reason for using a booster synchrotron concerns the radio-
frequency acceleration requirements. In a fast-cycling machine the much
more rapid acceleration requires a much higher rf voltage than has been
conventional at slower cycling rates – about 2 MV for a 10 Hz, 30 GeV
machine. At the same time a large frequency swing (20-30%) is required
when starting from pion factory energies of 500-800 MeV. The use of a
booster enables these demands to be handled separately. Almost the
entire frequency swing can be provided in the booster at relatively low
rf voltage, while the main ring provides the 2 MV with only a few percent

frequency swing. Being smaller, the booster must cycle faster (15-60 Hz) in order to fill the circumference of the main ring. The charge per pulse would be $N \sim 10^{13}$, comparable to existing machines injected in the same energy range. The circulating current in both booster and main ring would be $\bar{I} < 3$ A, a level which is not expected to present any problems.

Many of the design features required for these high energy high intensity machines are common to all the proposals; to avoid repetition and because I am most familiar with the TRIUMF design, I will describe its rationale in some detail, reporting only distinctive features in the other cases. Before doing this, however, it will be appropriate to discuss the project already under construction at Brookhaven to enhance the AGS performance by the addition of a booster synchrotron.

BROOKHAVEN AGS BOOSTER

Funding for the booster began in October 1985 and the project is expected to be complete by the end of 1989. This is in fact a multi-purpose project aimed at the acceleration of heavy ions as well as polarized and unpolarized protons (Brookhaven, 1984; Lee, 1985). The booster ring is one-quarter the circumference of the AGS and is located in the angle between the linac tunnel and the AGS ring. The modes of operation for various particles are illustrated in Fig. 18, where the time scale covers the 2.8 s of a single slow extracted pulse. For unpolarized protons four booster pulses would be injected at a 10 Hz repetition rate within a 300 ms flat bottom, enabling the present 1.6×10^{13} ppp to be increased to 5×10^{13} ppp. Initially protons would be accelerated to 1.5 GeV although the bending capability provided for heavy ions would eventually allow protons to be accelerated to 2.5 GeV. For heavy ions a slower acceleration time is required in the booster, and only one pulse would be injected into the main ring. For polarized protons there is the option of stacking up to 28 pulses in the booster ring before injecting them into the AGS. Further improvements beyond this program include the possibility of adding a 30 GeV stretcher ring and of making modifications to the AGS (rf, etc.) to accommodate $>5 \times 10^{13}$ ppp and increase the beam intensity to as much as 2×10^{14} p/s (32 μA).

TRIUMF KAON FACTORY

The TRIUMF proposal (1985) is based on a 30 GeV main "Driver" synchrotron 1072 m in circumference accelerating 10 μC pulses at a 10 Hz repetition rate to provide an average beam current of 100 μA. For the reasons explained above a Booster synchrotron is used to accelerate protons from the TRIUMF cyclotron at 450 MeV to 3 GeV: this machine is

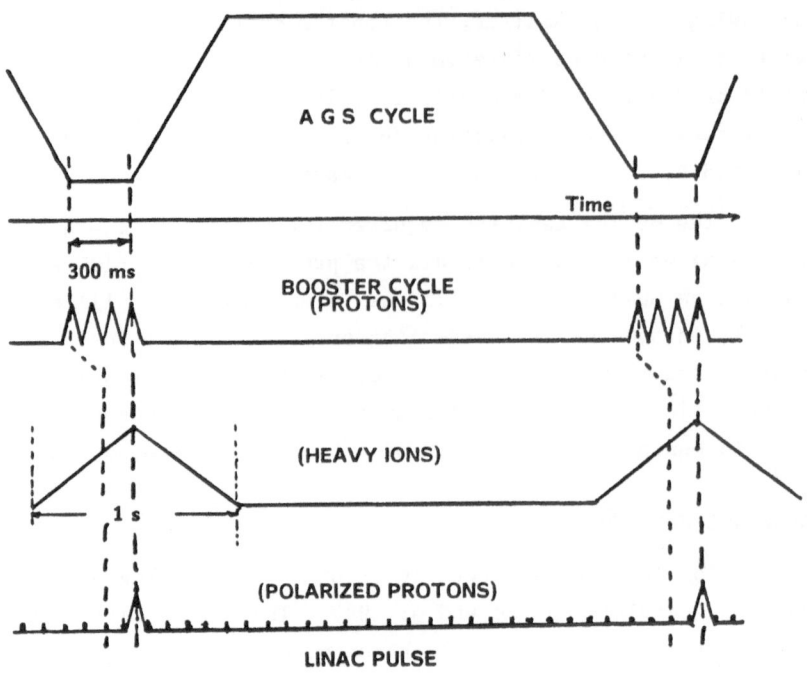

Fig. 18.   Injection programs for the AGS.

1/5 the radius of the main ring (Fig. 19) but cycles five times faster at 50 Hz. The Booster energy is chosen to minimize the total cost of the project. This depends mainly on magnet costs, and in particular on the magnet apertures. The minimum cost condition occurs when the emittances set by the space charge tune shift formula [Eq. (54)] are the same for both machines (Wienands and Craddock, 1986).

Each of the three accelerators is followed by a dc storage ring to provide time-matching and finally a slow extracted beam for coincidence experiments. These are relatively inexpensive, accounting for only 25% of the total cost. Thus the TRIUMF cyclotron would be followed by a chain of five rings, as follows:

A	Accumulator:	accumulates cw 450 MeV beam from the cyclotron over 20 ms periods
B	Booster:	50 Hz synchrotron; accelerates beam to 3 GeV
C	Collector:	collects 5 booster pulses and manipulates the beam longitudinal emittance
D	Driver:	main 10 Hz synchrotron; accelerates beam to 30 GeV;
E	Extender:	30 GeV storage ring for slow extraction

As can be seen from the energy-time plot (Fig. 20) this arrangement allows the cyclotron output to be accepted without a break, and the B and D rings to run continuous acceleration cycles without wasting time on

82

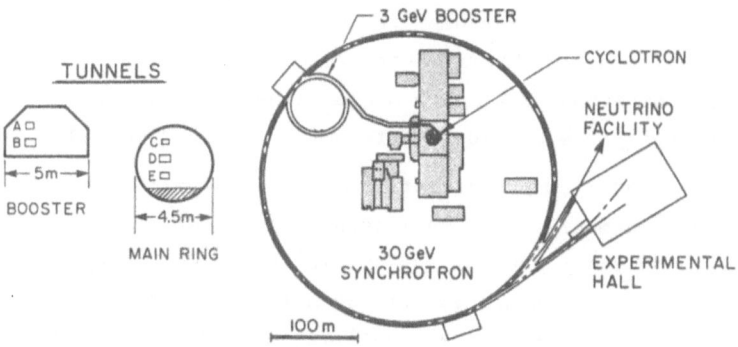

Fig. 19. Proposed layout of the TRIUMF KAON Factory accelerators and cross sections through the tunnels.

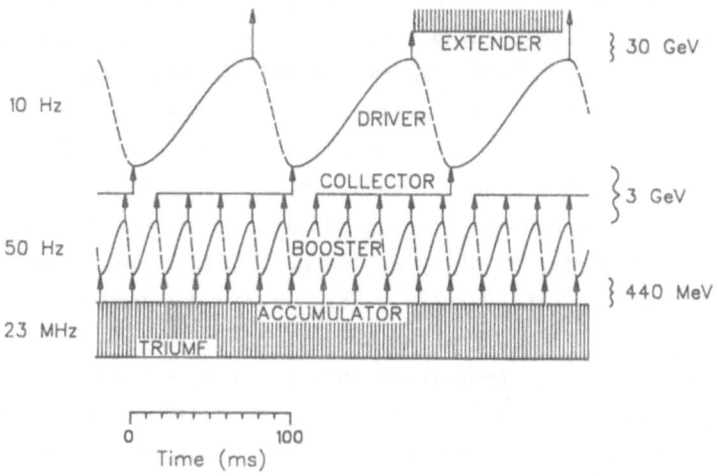

Fig. 20. Energy-time plot showing the progress of the beam through the five rings.

flat bottoms or flat tops; as a result the full 100 µA from the cyclotron can be accelerated to 30 GeV for either fast or slow extraction. Figure 20 also illustrates the asymmetric magnet cycles used in both synchrotrons. The rise time is three times longer than the fall, reducing the rf voltage required by 1/3, and the number of cavities in proportion. Full-scale power supplies providing such a cycle have been developed at Argonne (Praeg, 1983) with the encouragement of Los Alamos.

Figure 19 shows the location of the Accumulator directly above the Booster in the small tunnel, and of the Collector and Extender rings above and below the Driver in the main tunnel. Identical lattices and tunes are used for the rings in each tunnel. This is a natural choice providing structural simplicity, similar magnet apertures and straightforward matching for beam transfer. Multi-ring designs are now conventional at the high-energy accelerator laboratories, and the use of bucket-to-bucket transfer at each stage rather than the traditional coasting and recapture should keep transfer spills to a minimum.

Separated function magnet lattices are used with a regular FODO quadrupole arrangement, but with missing dipoles arranged to give super-periodicity 6 in the A and B rings and 12 in the C, D and E rings. This automatically provides space for rf, beam transfer and spin rotation equipment. It also enables the transition energy to be driven above top energy in both synchrotrons, avoiding the beam losses usually associated with crossing transition in fast-cycling machines and difficulties antic-ipated in making that jump under high beam-loading conditions. It will be recalled from Eq. (64) that the transition energy may be expressed roughly in terms of the momentum dispersion $\eta_x = \Delta x/(\Delta p/p)$ by $\gamma_t \simeq \sqrt{R/\langle \eta_x \rangle}$, where R is the machine radius. In conventional alternating gradient proton synchrotrons with regular dipole lattices, $\eta_x \simeq$ constant and $\gamma_t \simeq \nu_x$. In a missing dipole lattice the dispersion function will oscillate (see Fig. 21) and its average value $\langle \eta_x \rangle$ can be driven down towards zero ($\gamma_t \to \infty$) or even to negative values (making γ_t imaginary). This effect can be enhanced, without perturbing the other lattice functions too strongly, by bringing the horizontal tune value ν_x towards, but not too close to, the integer superperiodic resonance. Values of $\nu_x \simeq 5.2$ for the S = 6 Booster and $\nu_x \approx 11.2$ for the S = 12 Driver prove to be quite convenient. Associated choices of $\nu_y = 7.23$ and 13.22, respectively, keep the working points away from structural resonances and allow room for the anticipated space charge tune spreads, $\Delta \nu_y = 0.18$ and 0.11.

In fast-cycling machines the synchrotron tune ν_s is relatively large and care must be taken to avoid synchro-betatron resonances of the form

$$\ell \nu_x + m \nu_y + k \nu_s = n . \tag{103}$$

Here k, ℓ and m are integers whose sum gives the order of the resonance, while n is an integer giving the Fourier harmonic of the field component driving it. Only the lowest order of these resonances, satellites of the integer betatron resonances, are of serious concern, but in the Booster ν_s is as large as 0.04, so that these could significantly reduce the working area available for the tune spread. Fortunately it is possible to minimize their ill effects by suppressing the driving terms for the nearby integer resonances. This can be achieved by placing the rf accelerating cavities symmetrically with the magnet superperiodicity S = 6. The synchro-betatron resonances near $\nu_x = 5$ and $\nu_y = 7$ are then

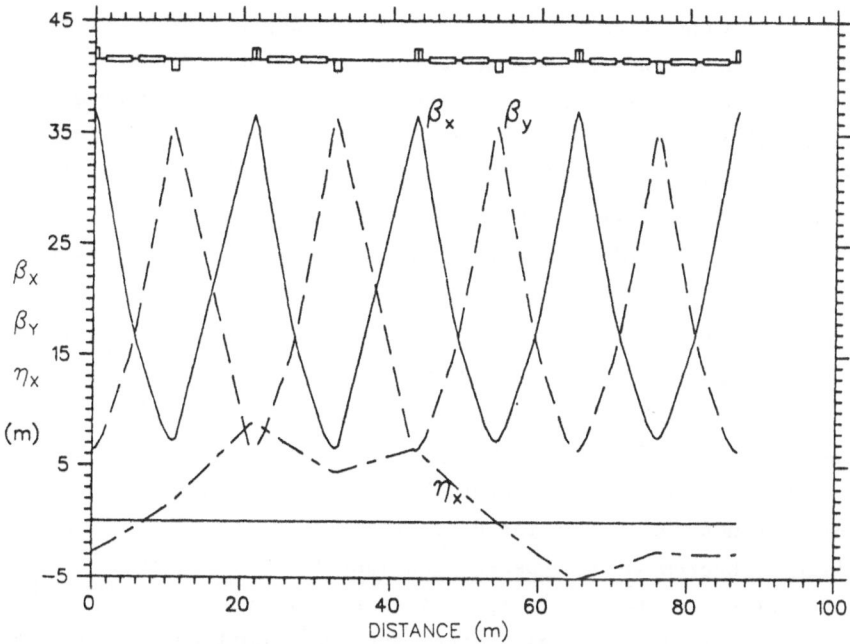

Fig. 21. Magnet arrangement and lattice functions for the
TRIUMF KAON Factory Driver synchrotron.

eliminated except for fifth harmonic variations in the cavity voltages
and seventh harmonic errors in vertical dispersion.

The superperiodicity of the magnet lattice also drives depolarizing
resonances. Compared to the AGS the superperiodicity is stronger but
passage through the resonances is more rapid, so that the overall depolarizing effects are comparable in magnitude. At the higher energies it
becomes impractical to built pulsed quadrupoles fast enough and strong
enough to jump these resonances. Instead it is proposed to use helical
"snakes" as proposed by E.D. Courant (1986). These would fit into the
10 m long drift spaces and cause a closed orbit distortion of <4 cm at
3 GeV. Pulsed quadrupoles would be practical for the lower momenta
resonances in the Booster.

At injection into the A and B rings the rf accelerating system will
operate at 46.1 MHz, twice the radio frequency of the TRIUMF cyclotron.
So that every other rf bucket is not missed the radius of these rings is
made a half-integer multiple (4.5) of that of the last orbit in the
cyclotron. The Booster cavities are based on the double gap cavities
used in the Fermilab booster. They will develop a voltage of 26 kV at
each gap for a frequency swing of 46-61 MHz. Twelve cavities will
develop the required voltage of 580 kV. The cavities for the other rings
are based on the single gap cavities designed for the Fermilab main ring.

For the main synchrotron 18 cavities each developing 140 kV give a total of 2520 kV with a frequency swing from 61.1–62.9 MHz. To keep the rf power requirements within a factor two of the 3 MW beam power and to provide stability under high beam loading conditions, a powerful rf control system has been designed, based on experience at CERN, including fast feedback around the power amplifiers, and phase, amplitude, tuning, radial and synchronization feedback loops. One-turn-delay feedforward is used to control transient loading effects.

The Accumulator ring is designed to provide the matching between the small emittance cw beam from the TRIUMF cyclotron and the large emittance pulsed beam required by the Booster. The Accumulator will stack a continuous stream of pulses from the cyclotron over a complete 20 ms Booster cycle. Injection and storage over this 20,000 turn long cycle is only possible through the use of the H^- stripping process. This enables Liouville's theorem to be bypassed and many turns to be injected into the same area of phase space. In fact it is not necessary or desirable to inject every turn into the same area; the small emittance beam from the cyclotron [$(2\pi \text{ mm·mrad})^2$ x 0.0014 eV-s] must be painted over the much larger three-dimensional emittance 31 x 93 (π mm·mrad)^2 x 0.048 eV-s needed in the Booster to limit the space charge tune shift [Eq. (54)]. Painting also enables the optimum density profile to be obtained and the number of passages through the stripping foil to be limited. The stripping foil lifetime is estimated to be > 1 day. The painting is achieved by a combination of magnetic field ramping (Fig. 22), vertical steering and energy modulation using additional rf cavities in the injection line. A group of 5 rf buckets (out of 45) will be left empty to provide ~100 ns for kicker magnet rise and fall times during subsequent beam transfers.

At present, of course, the H^- ions are stripped in the process of extracting them from the cyclotron. To extract them whole from the cyclotron a new extraction system will be required, and elements of this have been under test for the last year. The first element is an rf deflector operating at half the fundamental frequency: it deflects alternate bunches in opposite directions. Because the cyclotron operates on an odd harmonic (h=5) and the ν_x = 3/2 resonance is nearby, this produces a coherent radial betatron oscillation. Although successive turns are not completely separated a radial modulation in beam density is produced. An electrostatic deflector is then placed so that the septum is at a density minimum. In fact the septum is protected by a narrow stripping foil upstream so that no particles hit it; instead they are safely deflected out of the machine. The particles entering the electrostatic deflector eventually pass into magnetic channels and then into the

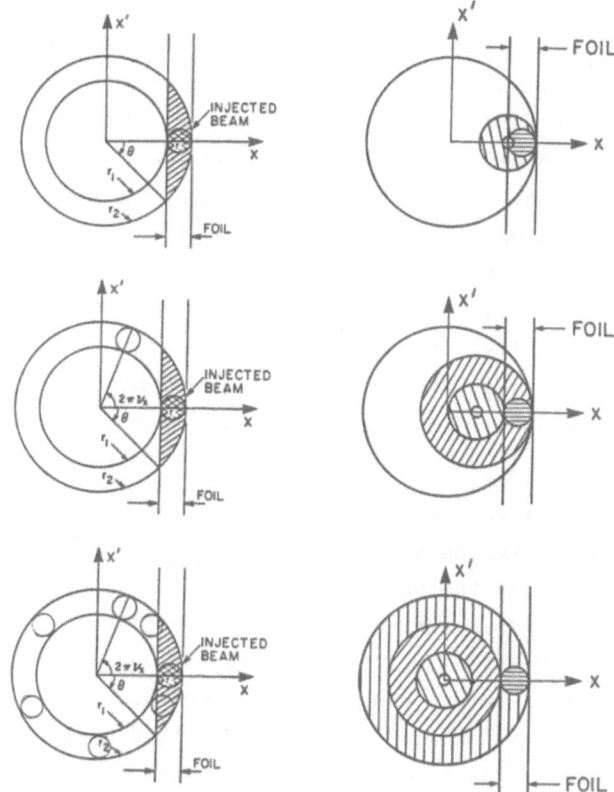

Fig. 22. Painting the beam in the horizontal phase plane. Left: over many turns betatron oscillations will fill an annulus. Right: the central orbit is initially shifted to the edge of the stripping foil (and incoming beam spot) using bump magnets (top); their gradual deexcitation moves the C.O. (and stored beam) off the foil and back to the machine axis (bottom).

Accumulator injection line. The magnetic channels have only just become available but tests with the first two elements have demonstrated 85% extraction efficiency.

Successful operation of a high-intensity accelerator depends cru-cially on minimizing beam losses and the activity they produce. Where some loss is expected, near injection and extraction elements, collima-tors and absorbers will be provided and equipment will be designed for remote handling. The beam current and any spill will be carefully monitored, and in case of the beam becoming unstable at any time through component failure or power excursions, each of the five rings is equipped with a fast-abort system which would dump the entire beam safely within one turn.

Several processes which give rise to losses in existing machines have been avoided entirely in this design. The use of H⁻ ions for

injection into the Accumulator ring will almost entirely eliminate
injection spill. The use of bucket-to-bucket transfer between the rings
will avoid the losses inherent in capturing coasting beams. The buckets
will not be filled to more than 60% of capacity; this should avoid beam
losses while providing a sufficient bunching factor to minimize the space
charge tune spread at injection and sufficient spread in synchrotron tune
to give effective Landau damping. Magnet lattices are designed to place
transition above top energy in all the rings, thus avoiding the instabilities and losses associated with that passage. Moreover, with the beam
always below transition, it is no longer advantageous to correct the
natural chromaticity, so that sextupole magnets are needed only for error
correction, and geometric aberrations in the beam are essentially reduced
to zero.

Beam instabilities will be suppressed or carefully controlled.
Although all five rings have large circulating currents, the rapid
cycling times give the instabilities little time to grow to dangerous
levels. Coupled-bunch modes, driven by parasitic resonances in the rf
cavities and by the resistive wall effect, will be damped using the
standard techniques (Landau damping by octopoles, bunch-to-bunch population spread and active damping by electronic feedback). The longitudinal
microwave instability is a separate case because of its rapid growth
rate. It will be avoided by making the longitudinal emittance sufficiently large at every point in the cycle and by minimizing the high
frequency impedance in the vacuum chamber as seen by the beam. At this
stage of the design it is not possible to make accurate estimates of beam
blow-up due to instabilities or non-linear resonances, but to be safe,
the magnet apertures have been designed to accommodate a 50% growth in
the horizontal, and 100% growth in the vertical beam emittance.

The proposal was submitted to TRIUMF's funding agencies, the
National Research Council and the Natural Sciences and Engineering
Research Council, in September 1985. The total cost, including salaries
and E,D&I, but not contingency, is estimated to be $427M (1985 Canadian
dollars). Two review committees have been set up. The first, an
international "Technical Panel" of particle, nuclear and accelerator
physicists, met in February and has produced a very favourable report.
The second "Review Committee" consisting of Canadian industrialists and
scientists from various disciplines met recently and is expected to
finalize its report later this summer.

LAMPF II

The Los Alamos proposal (1984, 1986) for a 45 GeV, 32 µA facility
aims at significantly higher energies but lower currents than the other

88

schemes. It is argued that higher energy protons will be more suitable
for Drell-Yan studies of quark confinement in the nucleus, besides providing higher momentum secondary beams. In fact it is aimed to achieve
60 GeV by running the main ring bending magnets at 2.0 T. Figure 23
shows the proposed layout, with the existing 800 MeV linac injecting
every other H^- pulse into a 6 GeV circular booster synchrotron of circumference 330.8 m, operating at 60 Hz and providing an average current of
144 µA. Out of every 18 pulses, 14 (112 µA) are directed to Area N for
neutrino studies, while 4 are transferred to the main ring of circumference 1333.2 m for acceleration to 45 GeV. With the magnets flat-topped
for 50% of the cycle for slow spill to Area A, and a 3.33 Hz repetition
rate, the average proton current is 32 µA. There are two modes of slow
spill, either fully debunched with 100% microscopic duty factor, or with
rf on giving 1.75 ns pulses (FWHM) every 16.7 ns. Fast extraction
without the flat-tops would provide 64 µA currents. Storage rings have
not been included in the initial proposal, but space is available for
later installation of collector and stretcher rings. This will enable
the main ring cycle rate to be increased to 10 Hz and the average fast-
or slow-extracted current to be increased to 96 µA.

In the latest design both synchrotrons use separated-function
magnets. The booster lattice has a superperiodicity S=6, created by
omitting the dipoles from every other focusing cell and modulating the
lengths of the cells. The choice of working point (ν_x = 5.22, ν_y = 4.28)
drives transition above the top energy (γ_t = 14.5). The six short empty
cells are then used to accommodate 18 rf cavities, while the long empty
cells are available for beam transfer elements.

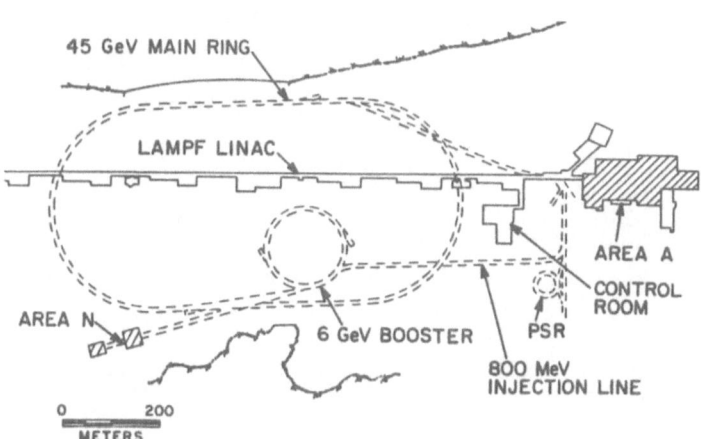

Fig. 23. Site plan of the proposed LAMPF II accelerators and
experimental areas.

The main ring race-track lattice consists of several functionally different sections — two 144° bending arcs, four 18° missing magnet dispersion suppressors, four short matching sections, and two 90 m long straight sections to accommodate rf cavities, beam transfer elements, and Siberian snakes. The 28 m long high-β section will reduce the beam spill on the extraction septum. The working point (ν_x = 7.45, ν_y = 6.45) was chosen close to a half-integer resonance for slow extraction. The high average dispersion in the bending arcs ($\eta_x \simeq 6$ m) keeps the transition energy below the acceleration range (γ_t = 6.4, 5.06 GeV).

The Los Alamos group is pioneering a number of potentially important technical developments. First among these is the use of an asymmetric magnet cycle (Praeg, 1983), (see above) with slow rise and fast fall to reduce the rf voltage requirements. In the latest design this scheme is applied only to the booster synchrotron. For the main ring a linear ramp is proposed with equal 50 ms rise and fall times. Instead of a resonant power supply system this would use SCR bridge rectifiers operating from three phase 60 Hz, supplied by a generator-flywheel set.

To achieve tunable high-power rf cavities economically, the Los Alamos group are proposing to bias the ferrite with magnetic field perpendicular to that of the rf, rather than parallel, as has been conventional. Figure 24 shows how with TDK G26 ferrite this results in cavities with much higher Q values. Tests of a full-scale cavity under high-power conditions are beginning in the laboratory, and will be continued in an existing machine such as the Los Alamos Proton Storage Ring under high beam-loading conditions.

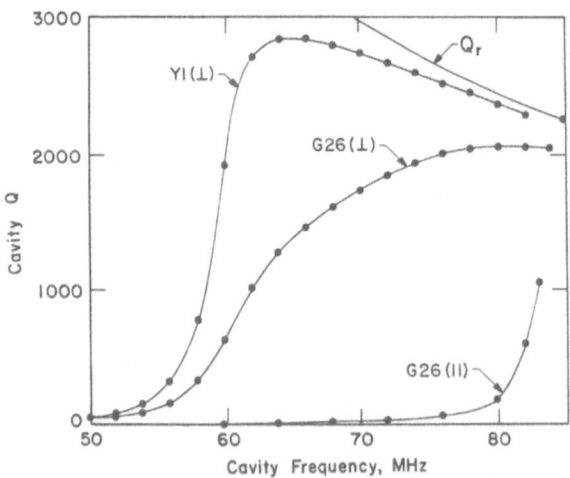

Fig. 24. Frequency variation of the Q-factor of test cavities for perpendicular- and parallel-biased ferrite.

The third development concerns the design of suitable vacuum chambers for the fast-cycling magnets. There must be a conducting surface on the inside of the chamber to prevent the build-up of electrostatic charge and also to provide a low-impedance path for the high-frequency image currents involved in maintaining beam stability. On the other hand eddy currents must be suppressed to minimize heating and magnetic field distortion. The only present example of such a system is at the Rutherford ISIS synchrotron (Bennett and Elsey, 1981), where a ceramic vacuum chamber is used, fitted with an internal cage of longitudinal wires to provide an rf shield (Fig. 25). The Los Alamos group is building test sections of a ceramic vacuum chamber with the conducting surfaces provided by an internal 1 μm copper coating and external layers of copper strips (Fig. 26). The system offers a smaller magnet aperture but will require careful design of the end connections.

The initial proposal appeared in December 1984, the second edition in May 1986. The cost has been reduced somewhat by eliminating one experimental area. This has been done without cutting down the number of secondary channels by using the MAXIM scheme of C. Tschalär (1986) to produce beams to the left and right of each target. The total cost, including ED&I but not contingency, is estimated to be $328M (FY 1985 US dollars).

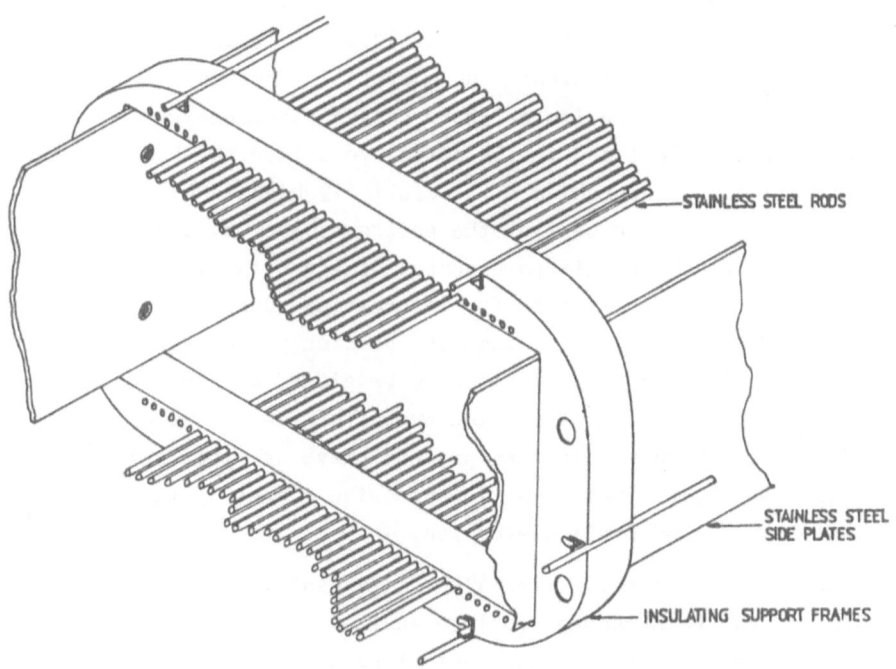

Fig. 25.   Ceramic vacuum chamber and wire rf shield
used in the Rutherford ISIS synchrotron.

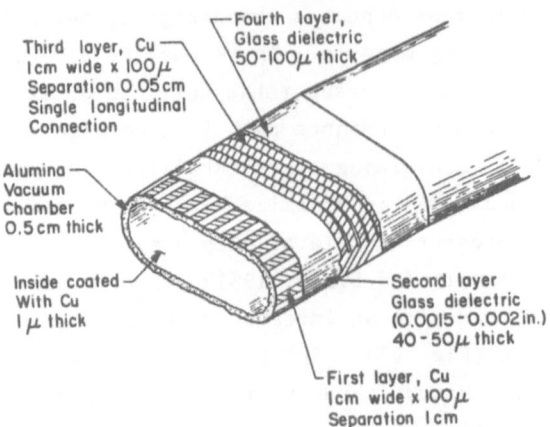

Fourth layer,
Glass dielectric
50-100μ thick

Third layer, Cu
1cm wide x 100μ
Separation 0.05cm
Single longitudinal
Connection

Alumina
Vacuum
Chamber
0.5cm thick

Inside coated
With Cu
1μ thick

Second layer
Glass dielectric
(0.0015 - 0.002 in.)
40 - 50μ thick

First layer, Cu
1cm wide x 100μ
Separation 1cm

Fig. 26. Los Alamos design for booster dipole magnet vacuum chamber.

EUROPEAN HADRON FACILITY

The first European scheme for a kaon factory came from SIN. This
was for a 20 GeV 80 µA synchrotron fed from the 590 MeV SIN cyclotron and
the proposed ASTOR isochronous storage ring (Joho, 1984). Rapidly
growing support led to the formation of an international study group.
The conceptual design has been discussed at a number of workshops during
the last nine months, and the reference design was agreed on in March
(EHF Study Group, 1986).

The layout of the proposed EHF is shown in Fig. 27. To accelerate
100 µA to 30 GeV the main ring cycles at 12.5 Hz; it is fed by a 9 GeV
booster of half the circumference and a specially designed 1.2 GeV linac,
both cycling at 25 Hz. Collector (here called "accumulator") and
stretcher rings are included in the design to follow the booster and main
ring, respectively. The choice of a relatively high 9 GeV for the
booster energy raises the cost of the entire project by about 10% over
the minimum at ~4 GeV but offers a number of advantages. Not only does
it provide greater opportunities for interesting physics at an intermediate construction stage but it brings the booster radius to half that of
the main ring, allowing the collector to be placed in the booster tunnel,
halving its length and equalizing the number of rings in each tunnel.
The desire to accelerate polarized protons plays an important role in
this design, and with 9 GeV injection Siberian snakes in the main ring
will cause less closed orbit distortion.

Although the design is officially "siteless" the circumference of
the main ring has been chosen equal to that of the CERN ISR for reference
purposes. The advantages of tunnel-stuffing are clearly as obvious in
Europe as they have been in the USA. Besides the tunnel, CERN's site

92

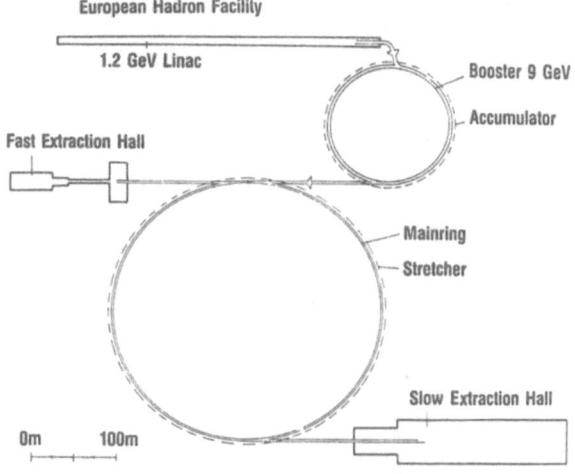

Fig. 27. Schematic layout of the European Hadron Facility.

offers other advantages — the existence of interim injectors, the availability of the West Hall after the experimental program has moved to LEP, and the infrastructure (even if only as a model for an independent EHF laboratory). The most obvious advantage to CERN would be the provision of a back-up PS. But these are mere speculations and at the moment it seems more likely that the EHF would be located elsewhere, say in Italy or at SIN.

All the rings use separated-function magnets and superperiodic lattices with transition energy outside of the acceleration range. The booster uses an interesting doublet lattice with S = 6 and nine cells in each superperiod. Dipole magnets are omitted from three cells, two of which can then be made dispersionless and therefore ideal for the location of rf cavities. The quadrupole doublets give low β-functions and high tunes (ν_x = 13.4, ν_y = 10.2); the transition energy γ_t = 12.7. The main ring has eight superperiods with seven regular FODO cells in each; dipoles are omitted in four of the half-cells, bringing the dispersion close to zero in three of them grouped together. These provide 25 m long straight sections, four of which are used for rf cavities, two for Siberian snakes and two for beam transfer.

The 1200 MeV linear injector consists of a series of linacs of different types (Fig. 28). Following the H$^-$ ion source and dc accelera- tion to 30 keV, two RFQs operating at 50 MHz and 400 MHz take the beam to 200 keV and 2 MeV, respectively. A 400 MHz drift tube linac then accelerates the ions to 150 MeV and finally a 1200 MHz side-coupled linac to 1200 MeV. Figure 28 also illustrates the time structure, showing how only two out of 24 SCL buckets are filled. These are then painted over

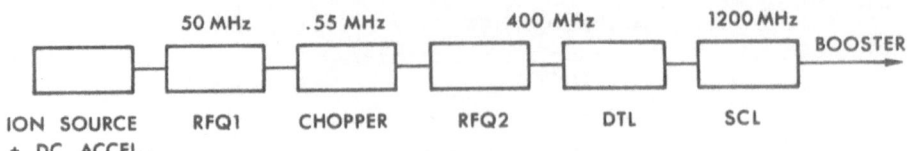

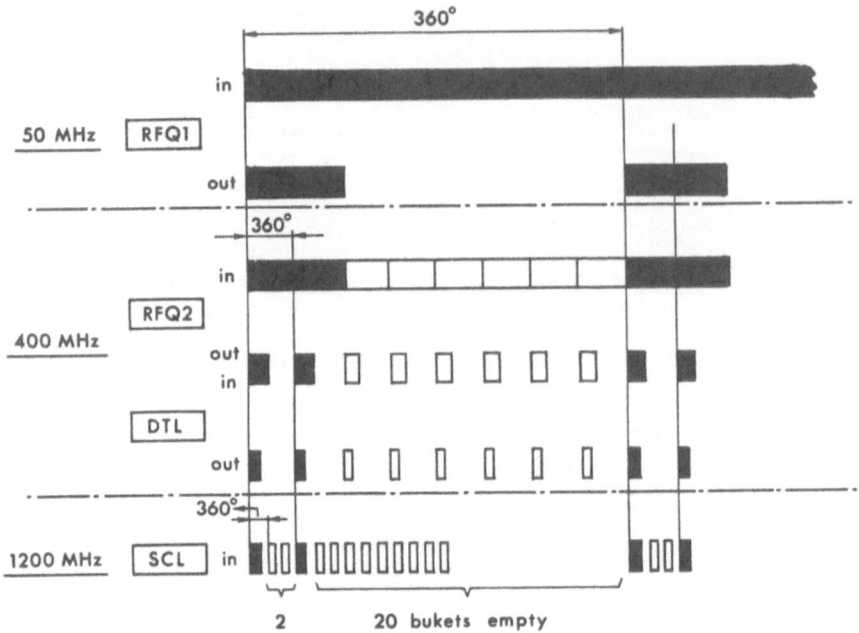

Fig. 28. Schematic diagram of the E.H.F. linear accelerators and of the phase-compression of the dc beam from the ion source into the 2:8 bucket scheme for optimal injection into the Booster.

the central 50% of the 50 MHz booster bucket to provide the desired density distribution.

Work is continuing with the aim of producing a formal proposal within about a year. Some temporary office space has been made available at CERN and the study group plans to make this their headquarters. The total cost is estimated to be MDM867, not including controls, contingency or inflation.

JAPANESE PROPOSALS

There is considerable interest in kaon factories in Japan, with schemes being promoted by Kyoto University and jointly by KEK and the University of Tokyo.

The most fully developed scheme is that from Kyoto, where for some time there has been a proposal for an 800 MeV linac pion factory. The proposed kaon factory complex (Imai et al., 1984) would add an 800 MeV compressor ring, a 25 GeV fast-cycling synchrotron, a stretcher ring, and antiproton accumulation and storage rings. The linac would operate at 60 Hz, feeding every other pulse to the synchrotron, operating at 30 Hz with an average current of 50 μA. The magnet lattice has a superperiodicity S = 4 with regular FODO cells and dispersion-free straight sections. The betatron tunes are ν_x = 11.2, ν_y = 10.2; the transition γ_t = 7.9 GeV.

At KEK and the University of Tokyo there has been longstanding interest (Yamazaki, 1985) in increasing the intensity of the KEK 12 GeV PS by using a higher energy booster (1-3 GeV). There is also the independent GEMINI project, (Saski et el., 1984) a spallation neutron source and pulsed muon facility, based on an 800 MeV synchrotron operating at 50 Hz to produce 500 μA proton beams. There is now a move to combine these schemes into a single coherent project which might also act as the injector of a kaon factory. One scheme which is being considered would involve a 30 GeV, 30 μA proton synchrotron cycling at 1 Hz fed by a 3 GeV, 100 μA booster cycling at 15 Hz. The booster might be a new fast-cycling synchrotron or it could be a combination of GEMINI with a 3 GeV FFAG after-burner. The booster would form the core of a multi-purpose laboratory, feeding not only the main ring but also a storage ring for spallation neutrons and pulsed muons. The acceleration of heavy ions is also being considered.

A meeting is expected to be held shortly by the proponents of all these schemes to settle the parameters of an accelerator complex - referred to as the "large hadron project" - satisfying their various needs. A possible arrangement of the booster stages is shown in Fig. 29. Here a heavy ion linac (8 MeV/nucleon) and a proton linac (1 GeV) are shown feeding individual booster and cooler rings; these in turn supply the existing 12 GeV synchrotron, or in a later phase, a new 30 GeV synchrotron.

CONCLUSIONS

There is active interest in a number of fields in building circular proton accelerators with the highest possible beam intensity. At low energies isotope production and cancer therapy provide the rationale for high current cyclotrons. At higher energies a variety of important questions in both particle and nuclear physics could be attacked using the more intense and/or clean beams of kaons, antiprotons, neutrinos and other particles that high-intensity proton synchrotrons could supply. In response to these rich possibilities a booster is under construction at the Brookhaven AGS and proposals for kaon factories have been completed at LAMPF and TRIUMF and are in preparation in Europe and Japan. The

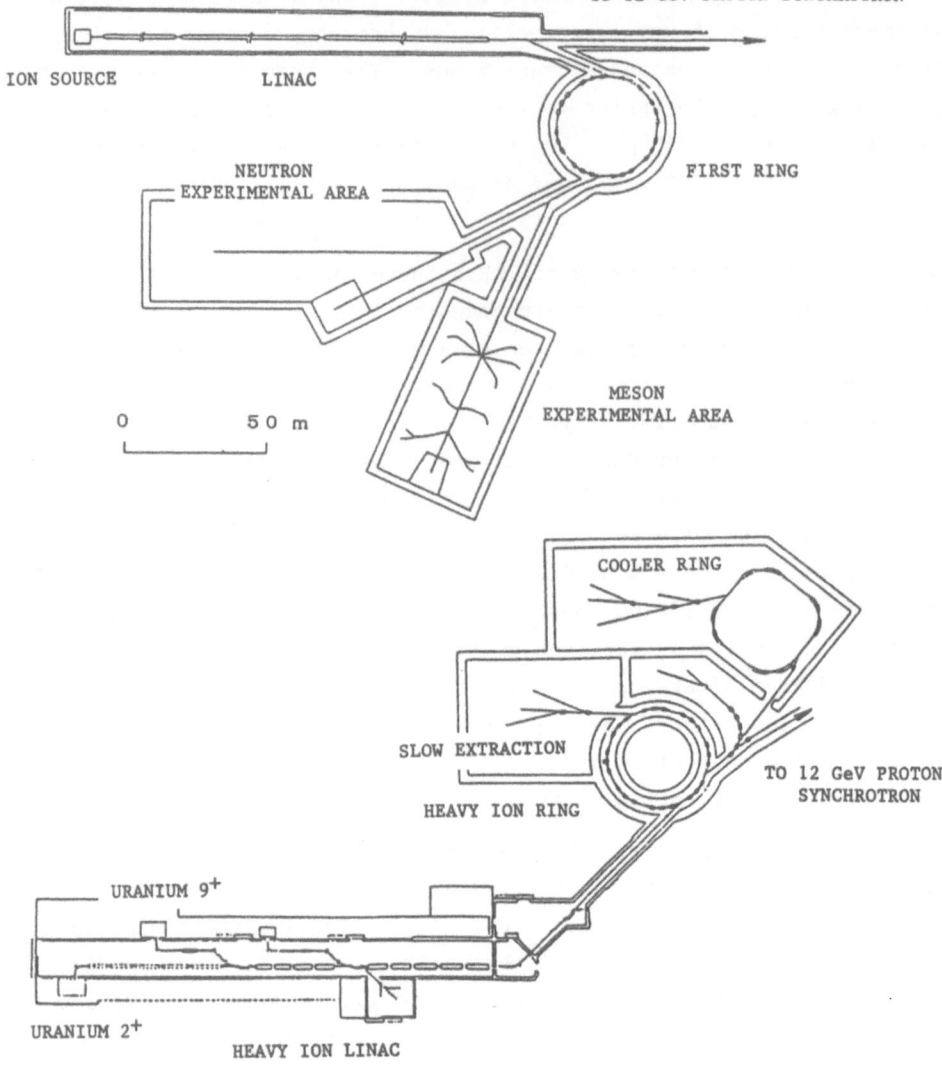

Fig. 29. Booster accelerator complex for the
Japanese Large Hadron Project.

concept of beam brightness has played a major role in determining the number of stages in these machines and in the Superconducting Super Collider.

ACKNOWLEDGEMENTS

The author is delighted to acknowledge the information and material that have been generously supplied to him by Drs. F. Bradamante, M. Inoue, T. Ludlam, A. Masaike, H. Sasaki, T. Suzuki, H.A. Thiessen, and T. Yamazaki. He has been heavily dependent on previous authors covering this ground and hopes that imitation of their treatments will be treated as flattery rather than plagiarism. He is also very grateful for the comments of colleagues at TRIUMF on the manuscript, especially to Richard Baartman. Finally it is a pleasure to acknowledge Denise Dale's and Lorraine King's meticulous typing, Anna Gelbart's and Terry Bowyer's excellent drawings and Fred Jones' and Alistair Martin's impressive computer graphics.

REFERENCES

Ankenbrandt, C., 1984
>The case for a small SSC booster, Proc. 1984 Summer Study on the Design and Utilization of the Superconducting Super Collider, Snowmass, ed. R. Donaldson, J.G. Morfin, (Division of Particles and Fields of the American Physical Society), p. 315.

Bennett, J.R.J., and Elsey, R.J., 1981
>Glass-jointed alumina vacuum chambers, IEEE Trans. Nucl. Sci., NS-28:3336.

Botman, J.I.M., Craddock, M.K., Kost, C.J. and Richardson, J.R., 1983,
>Magnet sector design for a 15 GeV superconducting cyclotron, IEEE Trans. Nucl. Sci., NS-30:2007.

Boussard, D., 1975
>Observation of microwave longitudinal instabilities in the CPS, CERN-Lab:II/RF/Int./75-2.

Bovet, C., Gouiran, R., Gumowski, I, and Reich, K.H., 1970
>A selection of formulae and data useful for the design of A.G. synchrotrons, CERN/MPS-SI/Int. DL/70/4.

Bradamante, F., 1986
>Conceptual design of the EHF, INFN/AE-86/7.

Brookhaven National Laboratory, 1984
>AGS booster conceptual design report, BNL-34989.

Bruck, H., 1966
>"Accélérateurs Circulaires de Particules", Presses Universitaires, Paris. [Available in an English translation as Los Alamos report LA-TR-72-10].

CERN - International School on Particle Accelerators, 1977
>Theoretical aspects of the behaviour of beams in accelerators and storage rings, Erice, 1977, ed. M.H. Blewett, CERN 77-13.

CERN Accelerator School, 1985
>General accelerator physics, ed. P. Bryant and S. Turner, CERN 85-19.

Courant, E.D., Livingston, M.S., and Snyder, H.S., 1952
>The strong focusing synchrotron - a new high energy accelerator, Phys. Rev., 88:1190.

Courant, E.D., 1968
>Accelerators for high intensities and high energies, Ann. Rev. Nucl. Sci., 18:435.

Courant, E.D., 1986
>private communication.

EHF Study Group, 1986
>Feasibility study for a European Hadron Facility, ed. F. Bradamante, J. Crawford and P. Blüm, EHF-86-33.

Hofmann, A., and Pedersen, F., 1979
>Bunches with local elliptic energy distributions, IEEE Trans. Nucl. Sci., NS-26:3526

Imai, K., Masaike, A., Matsuki, S., and Takekoshi, H., 1984,
>A proposal of high intensity accelerator complex for meson science in Kyoto University, Proc. 5th Symposium on Accelerator Science and Technology, KEK, p. 396.

Joho, W., 1974
Application of the phase compression – phase expansion effect for isochronous storage rings, Part. Accel., 6:41.

Joho, W., 1984
Interfacing the SIN ring cyclotron to a rapid cycling synchrotron with an Acceleration and STOrage Ring ASTOR, Proc. 10th Int. Conf. on Cyclotrons and their Applications, East Lansing, ed. F. Marti, IEEE, New York, p. 611.

Joho, W., Adam, S., Berkes, B., Blumer, T., Humbel, M. Irminger, G., Lanz, P., Markovits, C., Mezger, A., Olivo, M., Rezzonico, L., Schryber, U., and Sigg, P., 1985
Commissioning of the new high intensity 72 MeV injector II for the SIN ring cyclotron, IEEE Trans. Nucl. Sci., NS-32:2666.

Keil, E., and Schnell, W., 1969
Concerning longitudinal stability in the ISR, CERN-ISR-TH-RF/69-48.

Kelly, E.L., Pyle, R.V., Thornton, R.L., Richardson, J.R., and Wright, B.T., 1956
Two electron models of a constant frequency relativistic cyclotron, Rev. Sci. Instr., 27:493.

Khoe, T.K., and Kustom, R.L., 1983
ASPUN, Design for an Argonne Super intense PUlsed Neutron source, IEEE Trans. Nucl. Sci., NS-30:2086.

Laclare, J.-L., 1986
Beam current limits in circular accelerators and storage ring longitudinal coasting beam instabilities (these proceedings).

Laslett, J.L., 1963
On intensity limitations imposed by transverse space-charge effects in circular particle accelerators, Proc. Brookhaven Summer Study on Storage Rings, BNL-7534, p. 324.

Lawrence, E.O., and Edlefsen, N.F., 1930
On the production of high speed protons, Science, 72:376.

Lee, Y.Y., 1985
The AGS improvement program, IEEE Trans. Nucl. Sci., NS-30:1607.

Livingood, J.J., 1961
"Cyclic Particle Accelerators", van Nostrand, Princeton.

Livingston, M.S. and Blewett, J.P., 1962
"Particle Accelerators", McGraw-Hill, New York.

Los Alamos National Laboratory, 1984
A proposal to extend the intensity frontier of nuclear and particle physics to 45 GeV (LAMPF II), LA-UR-84-3982.

Los Alamos National Laboratory, 1986
The physics and a plan for a 45 GeV facility that extends the high-intensity capability in nuclear and particle physics, LA-10720-MS.

McMillan, E.M., 1945
The synchrotron – a proposed high energy accelerator, Phys. Rev., 68:143.

Meads Jr., P.F., and Wustefeld, G., 1985
An FFAG compressor and accelerator ring studied for the German spallation neutron source, IEEE Trans. Nucl. Sci., NS-32:2697.

Praeg, W.F., 1983
Dual frequency ring magnet power supply with flat-bottom, IEEE Trans. Nucl. Sci., NS-30:2873.

Reich, K.H., Schindl, K., and Schonauer, H., 1983
An approach to the design of space charge limited high intensity synchrotrons, Proc. 12th Int. Conf. on High Energy Accelerators, ed. F.T. Cole, R. Donaldson, Fermilab, p. 438.

Richardson, J.R., MacKenzie, K.R., Lofgren, E.J., and Wright, B.T., 1946
Frequency modulated cyclotron, Phys. Rev., 69:669.

Richardson, J.R., 1965
Sector focusing cyclotrons, Prog. Nucl. Techniques & Instrumentation 1:1.

Sasaki, H., and GEMINI Study Group, 1984
    Accelerator project GEMINI for intense pulsed neutron and meson
    source at KEK, Proc. Int. Collaboration on Advanced Neutron Sources
    VII, Chalk River, September 1983, AECL-8488, p.50.
Symon, K.R., Kerst, D.W., Jones, L.W., Laslett, L.J., and Terwilliger,
K.M., 1956
    Fixed field alternating gradient particle accelerators, Phys. Rev.,
    103:1837.
Thomas, L.H., 1938
    The paths of ions in the cyclotron, Phys. Rev., 54:580.
TRIUMF, 1985
    KAON factory proposal, TRIUMF, Vancouver.
Tschalar, C., (in press)
    Multiple achromatic extraction system, Nucl. Instrum. Methods. (in
    press)
U.S. Summer Schools on High Energy Particle Accelerators, 1981-
    Physics of high energy particle accelerators, A.I.P. Conf. Proc.,
    vols. 87, 92, 105, 127, 153.
Veksler, V.I., 1944
    A new method for acceleration of relativistic particles, Doklady,
    43:329; J. Physics USSR, 9:153.
Wienands, U., and Craddock, M.K., 1986
    Variation of cost of KAON factory accelerators with beam energy and
    intensity, TRIUMF internal report TRI-DN-86-7.
Wilson, E.J.N., 1977
    Proton synchrotron accelerator theory, CERN 77-07.
Yamazaki, T., 1985
    Toward high intensity proton facilities - a promising future of the
    KEK accelerator complex, Proc. KEK Workshop on Future Plans for High
    Energy Physics, preprint UTMSL-116.
Zach, M., Dutto, G., Laxdal, R.E., Mackenzie, G.H., Richardson, J.R.,
Trellé, R., and Worsham, R., 1985
    The H⁻ high intensity beam extraction system at TRIUMF, IEEE Trans.
    Nucl. Sci., NS-32:3042.

HIGH-BRIGHTNESS CIRCULAR ACCELERATORS

Norman Rostoker

Department of Physics
University of California
Irvine, California 92717   USA

PARTICLE DYNAMICS

In plasma physics the magnetic fields are usually sufficiently large that the gyro-radius of particles is small compared to any characteristic dimension so that the adiabatic or drift approximation is valid. This is illustrated in Fig. 1 where the particles follow the helical field lines except for slow drifts of the guiding center. The small oscillations shown are the cyclotron motion which is averaged out, i.e., it does not appear in the approximate equations of motion.

In conventional accelerator physics the gyro-radius of particles is about the same as the size of the machine - the particles do not follow field lines or drift. This case is illustrated in Fig. 2. If the particles followed the field lines they would simply collide with the wall! A different approximation called the paraxial approximation is employed. This means that the orbit is circular with small oscillations about the circle. The tangential velocity is much larger than any perpendicular component of velocity and is nearly constant. The small oscillations illustrated do appear in the equations of motion.

A conventional betatron as illustrated in Fig. 2 would have magnetic fields that can be described as

$$B_y \simeq B_{yo} \left[1 + (sx/R) \right] \tag{1}$$

$$B_x \simeq (sy/R) \, B_{yo}$$

In these equations s is the field index and R is the major radius. The coordinate systems used to describe particle dynamics in a torus are illustrated in Fig. 3. For a constant energy beam of electrons the

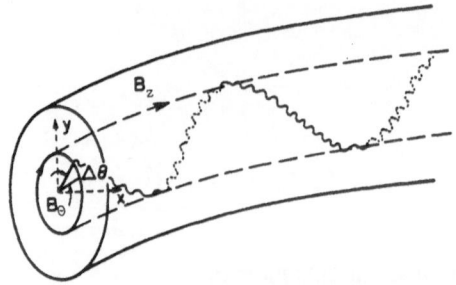

Fig. 1.   Plasma Physics – adiabatic particle dynamics.

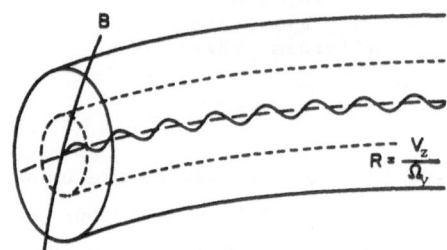

Fig. 2.   Accelerator Physics – non adiabatic dynamics.

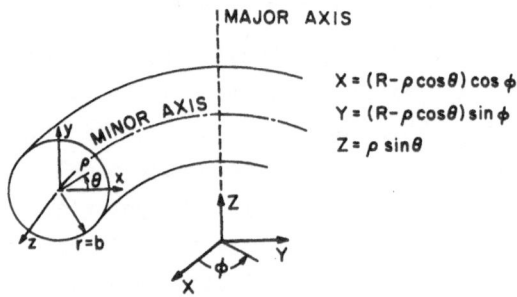

Fig. 3.   Fixed and local coordinate systems
for toroidal geometry.

equations of motion in the local coordinate system are

$$m\gamma \left(\dot{v}_x + \frac{v_z^2}{R-x} \right) = e \frac{v_z}{c} B_y \tag{2}$$

$$m\gamma \, \dot{v}_y = - e \frac{v_z}{c} B_x$$

and

$$v_z \simeq V_z = \text{constant}.$$

We expand the denominator in Eq. (2) to first order in x/R and substitute Eq. (1) for B_x, B_y. Assuming $m\gamma V_z^2/R = eV_z B_{yo}/c$ the standard equations for betatron oscillations are

$$\ddot{x} + \Omega_y^2(1-s)\, x = 0 \tag{3}$$

$$\ddot{y} + \Omega_y^2 sy = 0$$

where $\Omega_y = eB_{yo}/\gamma mc$, $\dot{v}_x = \ddot{x}$ and $\dot{v}_y = \ddot{y}$. The condition $\gamma m V_z = eB_{yo} R/c$ will be called momentum matching. It determines the average beam radius. For a standard betatron one must use the paraxial approximation and not the adiabatic approximation.

When we consider a modified betatron, a large toroidal magnetic field is added. Then the adiabatic approximation is applicable.

$$B_z = B_{zo}\,[1 + (x/R)], \tag{4}$$

If $v_z \gg v_x, v_y, |v_z - V_z|$, the paraxial approximation may also be applicable. For example the momentum matching condition obtains in the adiabatic approximation although the physical interpretation seems to be different from the paraxial approximation. The toroidal field B_z and the vertical field B_y constitute a spiral field line so that in following the field lines the electron must have a velocity in the y-direction $(B_{yo}/B_{zo})V_z$ assuming that $B_{yo} \ll B_{zo}$ and $V_z \gg v_x, v_y$. In addition there is a toroidal drift because the toroidal field is not constant. It is in the opposite direction and of magnitude $[V_z^2 + (V_\perp^2/2)]/R\Omega_z$ where $\Omega_z = eB_{zo}/\gamma mc$ and $V_\perp^2 = v_x^2 + v_y^2$. Containment requires that

$$V_y = (B_{yo}/B_{zo})\, V_z = [V_z^2 + (V_\perp^2/2)]/R\Omega_z = 0. \tag{5}$$

If $V_\perp^2$ is negligible compared to V_z^2 the condition that $V_y = 0$ is

$$\gamma m V_z = eB_{yo} R/c . \tag{6}$$

In the paraxial approximation this momentum balance condition is simply the balance of the Lorentz force and the centripetal acceleration term. In the adiabatic approximation it expresses the balance of two guiding center drifts. It should be noted that the adiabatic approximation can also treat the case where $V_\perp \sim V_z$. Then the momentum balance condition is

$$\gamma m\, [V_z + (V_\perp^2/2\, V_z)] = (eB_{yo}\, R/c) . \tag{7}$$

In the UCI Betatron experiments with inductive charging $V_\perp > V_z$ initially is expected; thus $B_{yo} > (\gamma m V_z c/eR)$. If acceleration takes place eventually $V_z \gg V_\perp$ and the usual momentum balance condition Eq. (7) must be satisfied.

In order to treat a beam rather than single particles it is necessary to consider the self-fields of the beam which are neglected in conventional accelerators, but are important for high brightness accelerators. The Vlasov-Maxwell formalism provides a self-consistent treatment, but is mathematically complex. To avoid this, some assumptions are made about the beam, for example, it has a minor radius "z" and a constant density n. The form is fixed but not the location. Thus the self-fields of the beam within the beam are

$$\underset{\sim}{E}_s = -2\pi ne\left[(x-x_o)\hat{x} + (y-y_o)\hat{y}\right] \tag{8}$$

$$\underset{\sim}{B}_s = 2\pi ne(v_z/c)\left[(x-x_o)\hat{y} - (y-y_o)\hat{x}\right]$$

$\hat{x}$ and $\hat{y}$ are unit vectors; (x_o, y_o) locates the center of the beam. The beam is usually in a vacuum region bounded by a conductor at radius $\rho = b$. Then there are image fields from the wall charge and current. If we expand the image fields to the lowest order in b/R and $(x_o^2 + y_o^2)/b^2$, they are

$$\underset{\sim}{E}_i \simeq -\frac{2\,Ne}{b^2}\left[x_o\hat{x} + y_o\hat{y}\right] \tag{9}$$

$$\underset{\sim}{B}_i \simeq -\frac{2\,Ne}{b^2}\left(\frac{v_z}{c}\right)\left[x_o\hat{y} - y_o\hat{x}\right] .$$

$N = n\pi a^2$ is the beam line density.

If the conducting wall at $\rho = b$ is slotted, the image magnetic field can be neglected.

The adiabatic equations of motion for a particle including the fields Eq. (8) and Eq. (9) are

$$\dot{x} = \frac{-\omega_o}{\gamma_o^2}(y-y_o) + \frac{V_z\,B_{yo}}{B_{zo}}\frac{sy}{R} - \frac{2\,Nec}{b^2 B_{zo}}\,y \tag{10}$$

$$\dot{y} = \frac{\omega_o}{\gamma_o^2}(x-x_o) + \frac{V_z\,B_{yo}}{B_{zo}}\left[1 + \frac{(s-1)}{R}x\right] - \frac{[V_z^2 + (1/2)\,V_\perp^2]}{R\Omega_z} + \frac{2\,Nec}{b^2 B_{zo}}\,x_o$$

In these equations $\omega_0 = \omega_p^2/2\Omega_z$, $\omega_p^2 = 4\pi ne^2/\gamma_0 m$, $\gamma_0 = [1 - (V_z/c)^2]^{-1/2}$. To identify the center of the beam (x_0, y_0), Eqs. (10) are averaged over all of the beam particles assuming $x_0 = \langle x \rangle$, $y_0 = \langle y \rangle$, $\langle \dot{x} \rangle = \dot{x}_0$, $\langle V_z x \rangle = \langle V_z \rangle x_0$ etc. The eqs. for the beam center are referred to simply as the beam equations:

$$\dot{x}_0 = \frac{\langle V_z \rangle B_{yo}}{B_{zo}} \frac{sy_0}{R} - \frac{2\,Nec}{b^2 B_{zo}} y \qquad (11)$$

$$\dot{y}_0 = \frac{\langle V_z \rangle B_{yo}}{B_{zo}} [1 + (s-1)\frac{x_0}{R}] + \frac{2\,Nec}{b^2 B_{zo}} x_0 - \frac{[\langle V_z^2 \rangle + (1/2)\langle V_\perp^2 \rangle]}{R\Omega_z}$$

The solution of these equations is of the form

$$x_0(t) = \bar{x}_0 + C_x \sin(\Omega t + \alpha_x) \qquad (12)$$

$$y_0(t) = C_y \sin(\Omega t + \alpha_y)$$

where C_x, C_y; α_x, α_y are constants determined by initial conditions and

$$\Omega^2 = (\frac{2\,Nec}{b^2 B_{zo}} - \frac{s\langle V_z \rangle B_{yo}}{B_{zo} R}) \left(\frac{2\,Nec}{b^2 B_{zo}} - \frac{(1-s)\langle V_z \rangle B_{yo}}{B_{zo} R} \right) \qquad (13)$$

If $s = 1/2$, $\Omega^2 > 0$ and the beam is stable. If $s \neq 1/2$ the beam will become unstable for a sufficiently large value of B_{yo}. The equilibrium position is the x-position where all drifts cancel, i.e.

$$\frac{\langle V_z^2 \rangle + (1/2)\langle V_\perp^2 \rangle}{R\Omega_z} - \frac{B_{yo}}{B_{zo}} \langle V_z \rangle [1 + \frac{(s-1)}{R}\bar{x}_0] - \frac{2\,Nec}{b^2 B_{zo}} \bar{x}_0 = 0 \qquad (14)$$

The first term is the toroidal drift. The second term is the drift caused by particles following the field lines and the last term is the drift caused by the electric image. (The magnetic image has been neglected; to include it, divide the last term by γ^2).

The paraxial equations (Eq. (2)) can similarly be generalized to include a toroidal magnetic field, the self-fields of the beam and the image fields. Then the particle equations can be averaged over all particles to give beam equations which are

$$\ddot{x}_0 + \Omega_z \dot{y}_0 + \omega_x^2 x_0 = 0 \qquad (15)$$

$$\ddot{y}_0 - \Omega_z \dot{x}_0 + \omega_y^2 y_0 = 0$$

The general solution can be expressed as a sum of two oscillations with frequencies

$$\Omega^2_{1,2} = 1/2 \; \{(\Omega_z^2 + \omega_x^2 + \omega_y^2) \pm [(\Omega_z^2 + \omega_x^2 + \omega_y^2)^2 - 4\omega_x^2\omega_y^2]^{1/2}\}$$

(16)

In the above equations

$$\omega_x^2 = (1-s) \; \Omega_y^2 - \Omega_b^2$$

$$\omega_y^2 = s\Omega_y^2 - \Omega_b^2$$

and

$$\Omega_b^2 = (a/b)^2 \; (\omega_p^2/2)$$

When $\Omega_z^2 \gg \omega_x^2, \; \omega_y^2$ there are two basic frequencies,

$$\Omega_1^2 \simeq \Omega_z^2$$

(17)

and

$$\Omega_2^2 \simeq \omega_x^2 \; \omega_y^2/\Omega_z^2$$

The result for Ω_2^2 agrees with Eq. (13). The additional high frequency oscillation would not be expected in the adiabatic approximation since the cyclotron motion is "averaged out". We conclude that as long as there is a large toroidal magnetic field, the adiabatic approximation describes the low frequency behavior of the beam correctly whether or not the paraxial approximation is valid.

To the modified Betatron, stellarator windings may be added as in Fig. 4 (Roberson, Mondelli and Chernin, 1983) where $\ell=2$ (quadrupole) helical windings are indicated. The "Stellatron" has in addition to betatron and toroidal fields the following

$$B_{xs} = \frac{\epsilon N_F B_{zo}}{R} \; [- x \sin 2\alpha z - y \cos 2\alpha z]$$

(18)

$$B_{ys} = \frac{\epsilon N_F B_{zo}}{R} \; [- x \cos 2\alpha z + y \sin 2\alpha z]$$

$\alpha = 2\pi/L$ where L is the pitch length of the helix, $N_F = 2\pi R/L$, and $\epsilon N_F B_{zo} = B_s$, the stellarator field amplitude. The rotational transform produced by the stellarator and toroidal magnetic field is

$$\lambda = \iota/2\pi = \epsilon^2 N_F/2$$

(19)

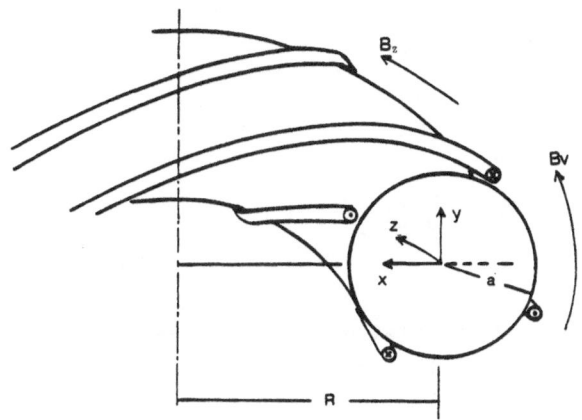

Fig. 4.   The Stellatron configuration.

For $s = 1/2$, Eqs. (11) can be written

$$\dot{x}_0 = \omega\, y_0 \tag{20}$$

$$\dot{y}_0 = -\,\omega\, x_0 + \langle V_D \rangle \; ,$$

where

$$\omega = \frac{\langle V_z \rangle}{2R}\,\frac{B_{yo}}{B_{zo}} - \frac{2\,Nec}{b^2 B_{zo}} \tag{21}$$

and

$$\langle V_D \rangle = \langle V_z \rangle\, \frac{B_{yo}}{B_{zo}} - \frac{\langle V_z^2 + (V_\perp^2/2) \rangle}{R\Omega_z}$$

To include the stellarator windings it is only necessary to redefine ω as

$$\omega^* = \omega + (\lambda \langle V_z \rangle / R) \tag{22}$$

The beam deflection is

$$\bar{x}_0 = \langle V_D \rangle / \omega^* \tag{23}$$

By adding stellarator windings it is thus possible to reduce the deflection of the center of the beam. The stellarator windings also increase the tolerance of confinement to momentum mismatch. Consider the case where $V_\perp \ll V_z$. The momentum matching condition is $p = eB_{yo}R/c$; the allowable mismatch for a conventional betatron is

$$\frac{\Delta p}{p} = \frac{\Delta R}{R} < \frac{b-a}{R} \; . \tag{24}$$

The orbits have betatron oscillations about a circle of radius R. The amplitude of oscillations is of the order of the beam radius a. The center of a particle orbit is from Eqs. (10).

$$\bar{x} = V_D / \omega_p^* \tag{25}$$

107

where

$$V_D \cong V_z \frac{B_{yo}}{B_{zo}} - \frac{V_z^2}{R\Omega_z}$$

$$\omega_p^* = \frac{V_z B_{yo}}{2R B_{zo}} - \frac{\omega_o}{\gamma_o^2} + \lambda \frac{V_z}{R} \; .$$

Assuming the amplitude of betatron oscillations to be about a, $\Delta\bar{x} \lesssim b-a$ for confinement which can be expressed as

$$\frac{\Delta p}{p} \lesssim \frac{b-a}{R} \frac{B_{zo}}{B_{yo}} \frac{\omega_p^*}{\Omega_y} \cong \frac{(b-a)}{R} \frac{B_{zo}}{B_{yo}} \lambda \; . \tag{26}$$

Since $B_{zo} \gg B_{yo}$ a large increase in momentum mismatch is possible. We assumed that for matching, $V_D = 0$; therefore $\Delta V_D = -\Delta V_z B_{yo}/B_{zo}$ and $\bar{x}_o = 0$.

INJECTION AND TAPPING

Inductive Charging and Runaway Electrons

Experiments at AVCO by (Daugherty, Eninger and Janes, 1971) and a similar device at Maxwell Laboratories (W. Clark et al. 1976) used a method of injection called inductive charging. Electrons are thermionically emitted near the torus wall and the emitter is held at 10-20 kV below the potential of the conducting torus wall. The injector is illustrated in Fig. 5 and the experiment in Fig. 6. A fast rising toroidal magnetic field is required which carries electrons away from the injector into the center of the torus. In the UCI experiments, the toroidal magnetic field increased to a peak of 10-15 kG in about 90 μsec. In the experiments at AVCO and Maxwell Laboratories sufficient charge was trapped by this method to produce a ring current of about 10 kA, if the electrons were accelerated to a velocity c. In the UCI experiments with a modified betatron this method was successful. A peak current of about 200 A was trapped and accelerated to about 1 Mev. The current increased linearly with injector voltage (Ishizuka et al., 1984). The highest beam current attained was about 250 A, limited apparently by injector voltage.

When helical windings were added to the modified betatron as illustrated in Fig. 7, similar results were obtained with inductive charging. However, the beam current did not increase with the injector voltage as in the modified betatron (Mandelbaum, 1985) for voltages from 20-40 kV. In order to increase the beam current some plasma was injected into the torus. This was accomplished by pulsing a coaxial transmission line with an open end. (See Fig. 6). Several such sources were

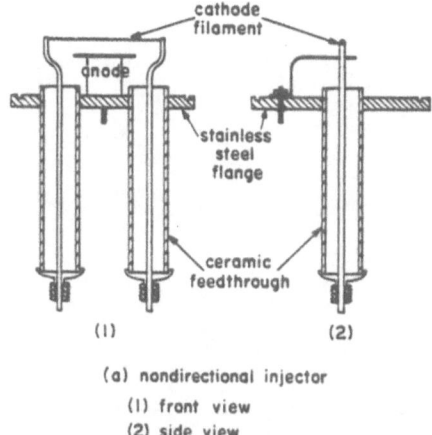

(a) nondirectional injector
(1) front view
(2) side view

Fig. 5.   Injector for inductive charging.

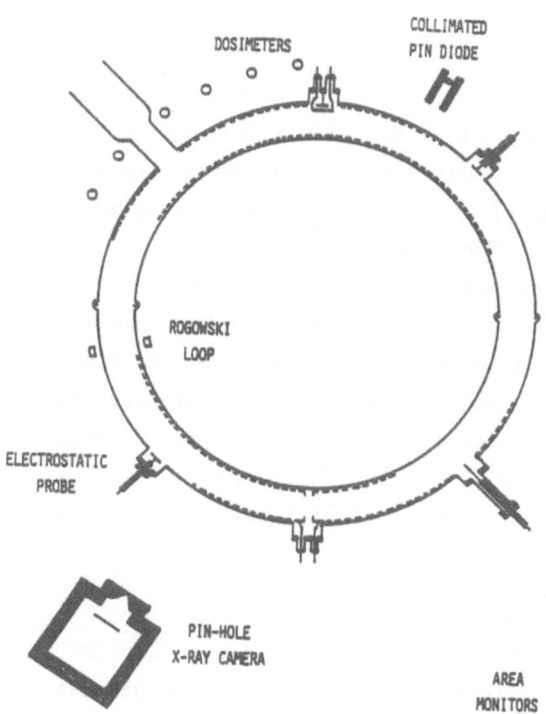

Fig. 6.   Schematic diagram of inductive
charging experiment.

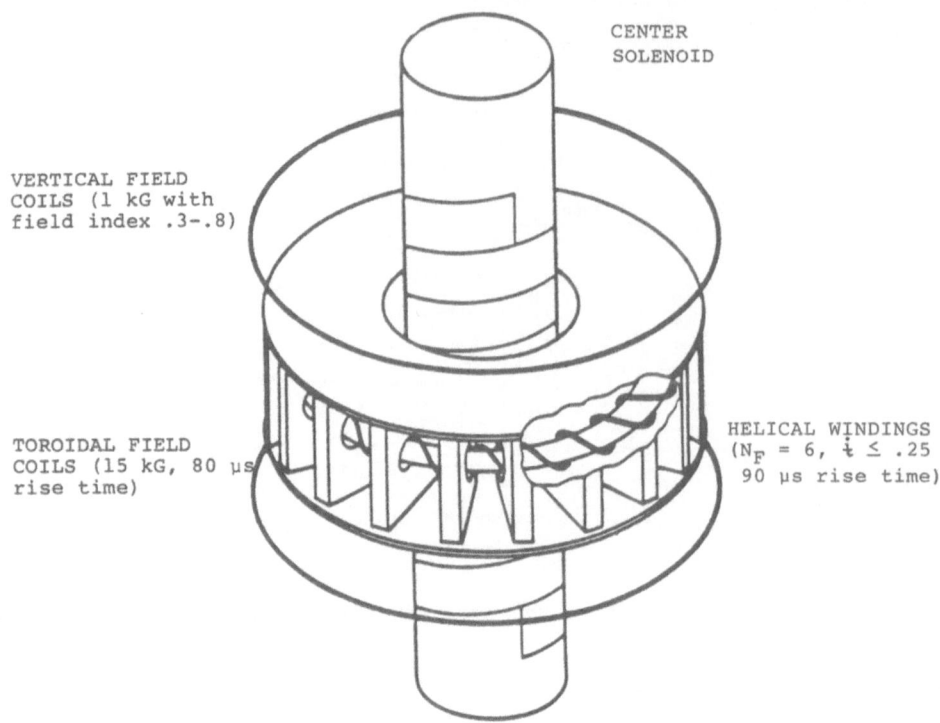

CENTER
SOLENOID

VERTICAL FIELD
COILS (1 kG with
field index .3-.8)

TOROIDAL FIELD
COILS (15 kG, 80 μs
rise time)

HELICAL WINDINGS
(N_F = 6, $\ell \leq$ .25
90 μs rise time)

Fig. 7.   UCI Stellatron experiment (Glass Torus
with R = 41 cm and b = 4 cm).

employed. The plasma was trapped and compressed in the rising fields and
spread around the torus. The plasma injector was first used together
with the electron injector to neutralize the space charge of the electron
beam. Advantages of a tenuous background have been discussed by
Manheimer (1985). The result showed that the trapped beam current was
increased substantially. It was noted that there was no observable
difference with or without the electron injector i.e., the electron beam
formed from runaway electrons such that it was indistinguishable from an
injected beam. Subsequent experiments used only the plasma injector with
which beam currents of about 1 kA have been accelerated to 10 Mev. Beam
currents higher than 1 kA were obtained, but the beam current is reduced
by various instabilities and so far no more than 1 kA survived to 10 Mev.
(H. Ishizuka et al., 1986).

Injection Into the NRL Modified Betatron

    The NRL Torus has a 1 m major radius and a 15.3 cm minor radius. An
electron beam is injected tangentially. The beam is formed by a pulse
line and diode that produces a 1.5-4 kA beam with an energy of 1-3 Mev
and a pulse length of about 30 nsec. The toroidal magnetic field has a
risetime of 1.6 msec to 2-5 kG. When the beam is injected at the torus

110

midplane near the conducting wall there is a drift in the poloidal plane due to focusing fields and image fields. In one toroidal transit (about 20 nsec) it must drift enough to miss the injector. In one poloidal drift time (several hundred nsec) external fields can be changed to reduce the minor radius of the drift trajectory sufficiently so that it continues to miss the injector. On a longer time scale wall resistivity causes the beam to drift inward, if the net force on the beam is inward. Recent experiments (J. Golden et al, 1986) confirm the azimuthal drift with various measurements including open shutter photography of the beam interacting with a thin polycarbonate film target that spans the minor cross section of the torus. The drift trajectory radius is reduced due to parasitic fields associated with late pulses in the pulse-line-diode circuit (self trapping). The beam lasts several microseconds or about 100 transits. The termination of confinement is not understood.

Equation (11) describe the motion of a beam. The azimuthal drifts result in an angular velocity

$$\omega = \frac{2\ Nec}{(b^2-\rho^2)B_{zo}} \left(\frac{1}{\gamma^2} - f\right) - \frac{B_{yo}}{2\ B_{zo}} \frac{\langle V_z \rangle}{R} \ .$$

(27)

The first term is produced by the electric and magnetic fields of the images. In the NRL experiment the current image is not eliminated, which accounts for the factor $1/\gamma^2$ that was not included in Eq. (11). The factor f has been included to describe the possibility of ionization of background gas; f is the neutral fraction. The denominator $b^2-\rho^2$ applies for all values of ρ; it was previously approximated by b^2 for a beam near the minor axis. The second term in Eq. (27) is due to the focusing fields. We assume the following parameters appropriate to the NRL experiment B_{zo} = 2 kG, R = 1 m, b = 15 cm, a = 2 cm, γ = 2.6, beam current 1.5 kA and f = 0. The angular velocity ω = 0 when

$$B_{yo}^* = (\frac{1}{5} \frac{I}{a})\ (\frac{R}{\gamma^2 b}) \stackrel{\sim}{=} 148\ \text{gauss} \ .$$

(28)

When $B_{yo} < B_{yo}^*$ as it generally was in the experiments, drift due to the images dominates. However, as the beam drifts inward the image drift can decrease by a factor of $b/2a \approx 3.75$ so that the frequency ω can pass through zero. (Manheimer, 1985). In that case, the toroidal drift would carry the beam to the wall unless momentum matching is perfect, i.e., unless

$$\frac{\langle V_z \rangle\ B_{yo}}{B_{zo}} = \frac{\langle V_z^2 \rangle + 1/2\ \langle V_z^2 \rangle}{R\Omega_z} \ .$$

(29)

A similar problem arises with the particle orbits. According to Eq. (10)

$$\omega = \frac{\omega_p^2}{2 \, \gamma_o^2 \Omega_z} - \frac{B_{yo} V_z}{2 \, B_{zo} R} \quad .$$ (30)

The first term is diamagnetic and the second term is paramagnetic. If the beam is accelerated sufficiently the particles undergo a diamagnetic to paramagnetic transition [Manheimer, (1985)] which would probably cause a disruption. If there is some background ionization or plasma

$$\frac{1}{\gamma_o^2} \rightarrow \frac{1}{\gamma_o^2} - f = 1 - \left(\frac{V_z}{c}\right)^2 - f$$ in Eq. (30). For $f \simeq 1$ the beam will always

be diamagnetic. However, another problem arises when $\omega R/V_z = n$ where $n = \pm 1, \pm 2 \ldots$ etc. Then the rotational transform is $2\pi n$ which is the Krushal-Shafranov limit, familiar for tokamaks. $\omega R/V_z = -1$ when

$$I_o = \frac{5a^2}{R} (B_{zo} - \frac{B_{yo}}{2}) \cong 400 \text{ A}$$

for the parameters of the NRL modified betatron. For the parameters of the UCI experiment with plasma injection, the limiting current is $I_o = 1.5$-6 kA assuming $R = 40$ cm, $B_{zo} = 12$ kG and $a = 1$-2 cm.

INSTABILITIES

The present treatment will be mainly directed at the interpretation of current experiments and anticipation of future experiments.

Beam Instability When $s \ne 1/2$

The beam equations (Eq. (13)) have an instability whenever

$$\omega_x^2 = (1-s) \, \Omega_y^2 - \Omega_b^2$$

$$\omega_y^2 = s\Omega_y^2 - \Omega_b^2 = s(V_z/R)^2 - (4 \, Ne^2/\gamma m \, b^2)$$

have opposite signs, because according to Eq. (17), one solution is $\Omega_2^2 \cong \omega_x^2 \, \omega_y^2 / \Omega_z^2$. Assume that $s \ne 1/2$ and Ω_b^2 dominates initially corresponding to a high current beam. Then initially $\omega_x^2 < 0$, $\omega_y^2 < 0$. As the beam is accelerated Ω_b^2 must decrease like $(1/\gamma)$. If $s < 1/2$, ω_x^2 must change sign before ω_y^2 as Ω_b^2 decreases. This will take place when

$$\gamma\beta^2 = 4 \, N \, r_o \left(\frac{R}{b}\right)^2 (1-s)$$ (31)

$r_o = e^2/mc^2$ is the free electron radius and $\beta^2 = 1 - (1/\gamma^2)$. If γ continues to increase ω_y^2 changes sign at

$$\gamma \beta^2 = 4 N r_o \left(\frac{R}{b}\right)^2 s \tag{32}$$

Between these two values of γ, $\Omega_2^2 < 0$ and the beam is unstable. The results are similar for $s > 1/2$. For example, if $N = 10^{11}$ cm^{-1}, $R = 40$ cm, $b = 1-2$ cm, $\gamma = 90 - 22.5$ assuming $s \simeq 1/2$. These are the parameters of the UCI experiment. For the parameters of the NRL experiment this instability would appear at $\gamma \tilde{=} 7.5$.

Orbital Resonance Instabilities

In a conventional betatron described by Eq. (3) the frequency of oscillations is $\omega_\beta = \sqrt{s}\ \Omega_y$ or $\sqrt{1-s}\ \Omega_y$. Field errors make Ω_y a periodic function of time with period $2\pi R/V_z$; in terms of perturbation theory field errors produce periodic force terms for Eq. (3). There are resonances when

$$\omega_\beta = \frac{n}{2}\ \Omega_b \tag{33}$$

where $n = \ldots 2, -1, 1, 2,$ etc. and $\Omega_o = V_z/R$. At resonance, the amplitude x or y grows secularly with time. Resonances can be avoided; for example if $s = 1/2$, $\omega_\beta = \Omega_y/\sqrt{2} = \Omega_o/\sqrt{2}$. Resonances are at $\Omega_o/2$, Ω_o so that $\Omega_o/2 < \omega_\beta < \Omega_o$, i.e. there will be no resonance. Since $\Omega_y = eB_y/\gamma mc$ and B_y and γ increase proportionately during acceleration, resonances can be avoided. In the modified betatron the resonances are at

$$\{\Omega_z^2 + 2\ \Omega_y^2 - 4\ \Omega^2\}^{1/2} = n\Omega_o \tag{34}$$

for $s = 1/2$. $\Omega^2 = (a/b)^2(\omega_p^2/2)$ for beam equations and $\Omega^2 = \omega_p^2/2\gamma^2$ for particle equations. For $s \neq 1/2$ there are additional resonances (Barak and Rostoker 1983). It is not possible to accelerate electrons without passing through resonances. For the parameters of the UCI modified betatron experiment (Ishizuka et al., 1984), Eq. (34) simplifies to a good approximation to

$$\Omega_z \tilde{=} n\Omega_o \tag{35}$$

Assuming an initial energy of 20 kev and a final energy of 1 Mev, resonances must be passed during acceleration from $n = 342$ initially to $n = 100$ finally. The growth in amplitude from crossing a resonance is proportional to n^{-4} (Barak and Rostoker, 1983) so that only low n-resonances are expected to be important.

In an $\ell = 2$ stellatron there are four betatron modes or tunes. They are (Roberson et al., 1985)

$$\omega_\beta = \left(\frac{k}{2} \pm \nu_\pm\right) \Omega_y \ .$$

where

$$\nu_{\pm}^2 = \hat{n} + (1/4) \ \hat{k}^2 \ \pm \ (\hat{n}\hat{k}^2 + \mu^2)^{1/2} \tag{36}$$

$$\hat{n} = 1/2 - [\omega_p^2/2\gamma_0^2\Omega_y^2] + (1/4) \ d^2 \ \simeq \ \frac{d^2}{4}$$

$$\hat{k} = k + d$$

$$d = B_{zo}/B_{yo} \ , \ k = 2 \ N_F = 8 \ , \ \mu = B_s/B_{yo} = \varepsilon N_F d$$

The resonance condition is $\omega_\beta = n\Omega_o$. For the parameters of the UCI stellatron k = 8; d $\simeq$ 15 and $\mu \simeq$ 19 at the highest energy of about 10 Mev. Resonances would be at n $\simeq$ 26, 16, -18 and -8. To reach this energy it is necessary to pass through some low integer resonances that occur when d > 15. It is in principle possible to avoid all of the resonances by keeping d constant during acceleration at a non-resonant value. It is necessary that $d_2 \gg \omega_p^2/2 \ \gamma_0^2\Omega_y^2$ because this quantity must change during acceleration. Similar remarks apply to the modified betatron.

Negative Mass Instability

Sufficient conditions for stability can be derived simply from particle orbit considerations (Roberts and Rostoker, 1986). Consider a clump of particles as illustrated in Fig. 8. The fields from the clump will perturb the toroidal velocity $v_z = (R-x) \ \dot{\Phi}$ (see Fig. 3).

$$\delta v_z = \delta \ [(R-x) \ \dot{\Phi}] \tag{37}$$

$$= R \ \delta\dot{\Phi} - \delta \ x \ \dot{\Phi} \ .$$

If $\delta\dot{\Phi} \gtrless 0$ when $\delta v_z \gtrless 0$, the particle behaves as though the mass is positive. If $\delta\dot{\Phi} \gtrless 0$ when $\delta v_z \lessgtr 0$ the particle behaves as though the mass is negative. The condition for positive mass is

$$\frac{\delta\dot{\Phi}}{\delta v_z} = (\dot{\Phi} \ \frac{\delta x}{\delta v_z} + 1) \ \frac{1}{R} > 0 \ \ . \tag{38}$$

This condition can be evaluated with the particle orbit equations (Eqs. (10)) and the equation for the time averaged beam centroid $\bar{x}_o$ given by Eq. (14). The time average of Eqs. (10) is

$$\bar{x} = \frac{(V_z^2/R) - (\Omega_y V_z) + [(\omega_p^2/2 \ \gamma^2) - (\omega_p^2/2)(\frac{a}{b})^2] \ \bar{x}_o}{[(\omega_p^2/2\gamma^2) + (s-1/r) \ \Omega_y V_z]} \tag{39}$$

where

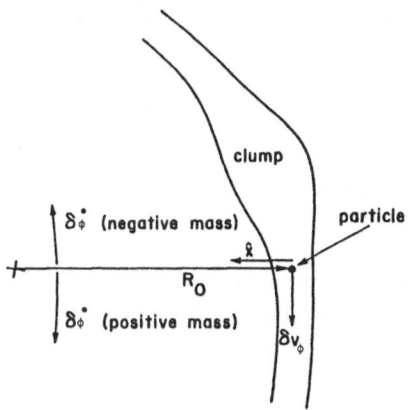

Fig. 8. Particle orbits illustrating positive and negative mass.

$$\bar{x}_o = [(<V_z^2/R>) - \mathcal{Q}_y V_z]/[(\omega_p^2 a^2/2b^2) + (s-1/R)\, \mathcal{Q}_y V_z]$$

according to Eq. (14). We have assumed $V_\perp = 0$ and $V_z = <V_z> = R\dot{\Phi}$. Let $\bar{x} \to \bar{x} + \delta x$ and $V_z \to V_z + \delta v_z$. After calculating $\delta x/\delta v_z$ and substituting into Eq. (38) the result is

$$\frac{\gamma^2 \mathcal{Q}_y^2}{(\omega_p^2/2)\,(a/b)^2 - (1-s)\,\mathcal{Q}_y^2} + 1 > 0 \tag{40}$$

For a conventional betatron with $\omega_p^2 \approx 0$ the condition is $\gamma^2 < 1-s$ which is never satisfied. (R. Landau and V. Neil, 1966). According to Eq. (40) the toroidal field does not change the condition, but the finite beam density is stabilizing. A sufficient condition for stability of the simple or modified betatron is $\omega_p^2 (a/b)^2 > \mathcal{Q}_y^2$ or

$$\gamma < 4\nu\,(R/b)^2 \tag{41}$$

where $\nu = Ne^2/mc^2$ is the Budker parameter. If the current image is included γ should be replaced by γ^3 in Eq. (41).

For the stellatron, the condition is

$$\frac{\gamma^2 \mathcal{Q}_y^2}{\mathcal{Q}^2 - \{\mathcal{Q}_s^2/[(2N_F)^2 + 2N_F d + (\mathcal{Q}/\mathcal{Q}_y)^2]\}} + 1 > 0 \tag{42}$$

where

$$\mathcal{Q}^2 = (\omega_p^2/2)\,(a/b)^2 + (s-1)\,\mathcal{Q}_y^2$$

and

$$\mathcal{Q}_s = \frac{e}{mc}\,\varepsilon\, N_F\, B_{zo}\ .$$

Even if Eq. (41) is not satisfied, the stellarator fields can provide stability providing that

$$\gamma^2 < B_s^{\ 2}/\ 2N_F\ B_y\ B_z \neq \lambda d \quad . \tag{43}$$

For example, if $\lambda = .2$ and $d = 50$, which are typical values for the UCI stellatron, $\gamma_c \cong 3.2$.

The negative mass instability has been the subject of a great deal of investigation (Landau and Neil, 1966; Landau, 1968; Godfrey and Hughes, 1985; Sprangle and Chernin, 1984; Davidson and Uhm, 1982). The sufficient conditions are, at best, a guide when to expect the instability. The more detailed analysis that gives necessary conditions and growth rates has not yet been useful in interpreting experiments. For example, instability below the threshold energy of Eqs. (41) or (42) has been predicted (Godfrey and Hughes, 1985) but not observed in the UCI experiments. Instead the current problem seems to be explaining the absence of the negative mass instability up to $\gamma \cong 9$, which probably involves analysis of distributions with a finite spread in γ. (Landau and Neil, 1966).

EXPERIMENTS ON THE MODIFIED BETATRON AND THE STELLATRON

Modified Betatron

The experimental results of interest are as follows:

- $p \neq eB_{yo}R/c$; $eB_{yo}R/pc$ is initially 3-5 and after the beam has been accelerated it is about 2.

- departures from the betatron condition $\langle B_y \rangle / B_{yo} = 2$ are observed.

- the position of the beam centroid is not correctly given by Eq. (14).

- electrostatic probes show two frequencies.

With inductive charging $V_D = \langle V_z \rangle\ B_{yo}/B_{zo} - \langle V_z^{\ 2} + (V_\perp^{\ 2}/2) \rangle R\Omega_z$. For beam formation $V_D \simeq 0$ and $\langle V_\perp^{\ 2} \rangle \gg \langle V_z^{\ 2} \rangle$ so that

$$p = m\langle V_z \rangle \simeq 2\ e\ \frac{B_{yo}R}{c}\ \frac{\langle V_z \rangle^2}{\langle V_\perp^{\ 2} \rangle} \quad . \tag{44}$$

Initially $eB_{yo}R/pc = \langle V_\perp^{\ 2} \rangle / 2\ \langle V_z^{\ 2} \rangle > 1$ simply because there is initially a large value of $\langle V_\perp^{\ 2} \rangle$ from magnetic compression. After the beam has been accelerated $\langle V_z \rangle^2 \simeq \langle V_z^{\ 2} \rangle \gg \langle V_\perp^{\ 2} \rangle$ so that

$$V_D \simeq - \frac{\langle V_z^{\ 2} \rangle}{R\Omega_z}\ [1 - \frac{eB_{yo}R}{pc}] \quad . \tag{45}$$

Initially $(eB_{yo}R/pc) \sim 3-5$ and after acceleration it is about 2. Therefore the particle acquires a drift as it accelerates and a displacement
of the form

$$\bar{x}_o = V_D/\omega \tag{46}$$

where

$$\omega = \{\frac{(1-s)}{R} \, \Omega_y \, \langle V_z \rangle - \frac{\omega_p^2}{2} \, (\frac{b}{a})^2\}/\Omega_z \; .$$

Assuming $\omega < 0$, $\bar{x}_o < 0$ because $V_D > 0$ according to Eq. (45) so that
the beam expands. In fact it was observed to hit the outer wall for a
final energy of 500 kev and a current of 150 A. $|\bar{x}_o|$ can be reduced by
using an auxiliary coil to reduce B_{yo} after beam formation. This means
that $\langle B_y \rangle/B_{yo} > 2$. By means of the auxiliary coil the expansion of the
beam was reduced and the energy increased to 1 Mev.

$\bar{x}_o$ can be calculated from experimentally determined quantities. For
example, $(\gamma-1) \, mc^2 = 500$ kev determines $\langle V_z \rangle = 2.6 \times 10^{10}$ cm/sec. Then
$I = 150$ A determines $N = I/e\langle V_z \rangle = 3.6 \times 10^{10}$/cm. With $b = 5$ cm, $R = 40$
cm, $B_{yo} = 130$ G and $s = .8$, it follows that $\bar{x}_o = -76$ cm. For this and
several other reasons we have assumed that there is a group of quasi-
confined electrons that are not accelerated by the toroidal electric
field. The evidence is as follows

- electrons can be trapped between mirrors and will not accelerate
  around the torus if $\mu \partial B/\partial z > e \, E_z$; $\tilde{=} .8$ volts/cm and
  $\mu \tilde{=} (1/2) \, mV_\perp^2/B_z$ is the electron magnetic moment. If B_z changes
  by about 1% over $L = 10$ cm, electrons will not accelerate if

  $$(1/2) \, mV_\perp^2/e > E_z \, L \, \frac{B}{\Delta B} \sim 10^3 \text{ volts} \; . \tag{47}$$

  The method of inductive charging involves an initial energy of
  10^4 ev and compression so that most of the electrons would not be
  in the loss cone, and would satisfy Eq. (47).

- the injector has a current of about 10 A initially which drops to
  2 A after trapping. If the injector is turned off the beam
  crashes in a short time. The 2 A current presumably supplies
  losses to the "trapped" electrons for which the lifetime is
  estimated to be about 10 μsec.

- trapped electrons are well known in tokamaks where typically 10%
  of electrons are trapped between mirrors. They were previously
  assumed to interpret HIPAC experiments (W. Clark et al., 1976),
  where more than 90% of electrons were trapped.

- diochotron oscillations have a frequency that is proportional to N, the beam line density. Electrostatic probes indicate two frequencies at about 10^7 sec^{-1} and 10^8 sec^{-1}. Similar frequencies are observed with X-ray signals.

To include the quasi-confined electrons, ω in Eq. (46) is replaced by

$$\omega = - \{\frac{\omega_p^2 a^2}{2 b^2} + \frac{(\omega_p')^2}{2} + \frac{(s-1)}{R} \Omega_y <V_z> \}/\Omega_z \qquad (48)$$

where $(\omega_p')^2 = 4\pi n' e^2/\gamma m$ and n' is the density of trapped electrons. $N = n\pi a^2 = 3.6 \times 10^{10}$/cm in the example previously discussed. Assuming $N' = n'\pi b^2 = 4.2 \times 10^{11}$/cm in the predictions for beam deflection and oscillation frequencies are consistent with the experiment (Roberts and Rostoker, 1985). The sufficient condition for the negative mass instability [Eq. (40)] is similarly altered. Instead of Eq. (41) the energy threshold becomes

$$\gamma < (4\pi n' e^2/m) (R/c)^2 \simeq 100 . \qquad (49)$$

Stellatron

With injection by inductive charging a maximum current of 250 A was trapped. The ring expansion was delayed compared to the modified betatron and the final energy reached a maximum of about 4 Mev (B. Mandelbaum, 1985). The current reached a peak and then declined slowly until the beam was finally dumped in a location of a few cm dimensions according to X-ray measurements. In contrast with the modified betatron experiments the injector was turned off immediately following beam trapping; otherwise the beam would disrupt. The slow loss of current was due to skimming by the injector. The localized loss was due to a field error involving high resistance of cables connecting several toroidal coils. When this was corrected almost all of the beam loss was at the injector.

With the limited voltage possible for the electron injector (20-40 kev), the trapped current could not be increased beyond 250 A. With the addition of plasma, the current was increased up to about 4 kA. For currents above 2 kA, a disruption took place after which there was no current. This was evidently the Krushal-Shafranov limit. For lower current there was also a disruption, but a finite current survived and was accelerated to as much as 10 Mev. Typical experimental measurements are illustrated in Fig. 9.

There was microwave activity during injection/trapping and at the disruption. The spectrum was peaked at harmonics of 115 MZ which is

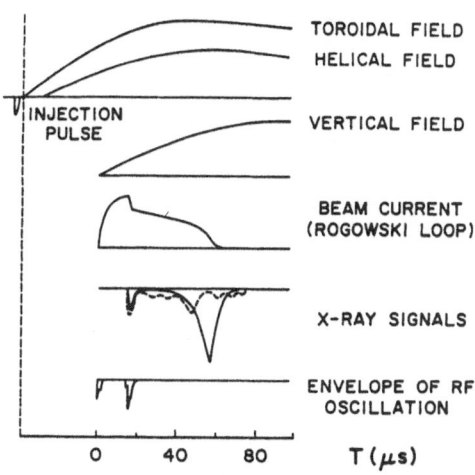

Fig. 9. Typical experimental data for the UCI Stellatron.

$V_z/2\pi R$. After the disruption, the current decayed slowly during which
time there was no microwave activity, but there were X-ray signals until
finally the beam dumped indicated by a large X-ray pulse.

The disruption must be a collective instability and a leading
candidate is the negative mass instability. Detailed calculations have
not been done for a neutralized beam, but threshold calculations such as
Eq. (43) require little modification. If the rotational transform from
the current dominates $\Lambda = B_\theta R/a\, B_z$ and Eq. (43) becomes

$$\gamma^3 < (R/a)^2\,(I/7.5) \tag{50}$$

where the current I is in kA. The threshold energy is 1 Mev (1.9 Mev)
for I = 500 A and a = 2 cm (I = 1 kA and r = 1.4 cm). These numbers are
consistent with the experiment and involve $\Lambda = .1 - .4 < 1$. After disruption the absence of microwave activity during current decay indicates
that orbital resonance instabilities are being crossed with the final
disruption when one of the resonances with low n is reached as suggested
by Eq. (36). The maximum current to survive to 10 Mev is about 1 kA.
Further improvement will require a stronger field B_z and reduction of
field errors which are very large in the current experiment.

Beam Extraction

Extraction in a conventional betatron has been done with a magnetic
peeler as illustrated in Fig. 10 (Kerst et al., 1949/1950). After acceleration the beam can be expanded by changing $B_y < 2\,\bar{B}_y$. When it reaches
a field free region it escapes tangentially in the form of a beam. The
addition of a toroidal magnetic field and stellarator fields certainly
complicates the extraction. However, in all experiments to date the beam

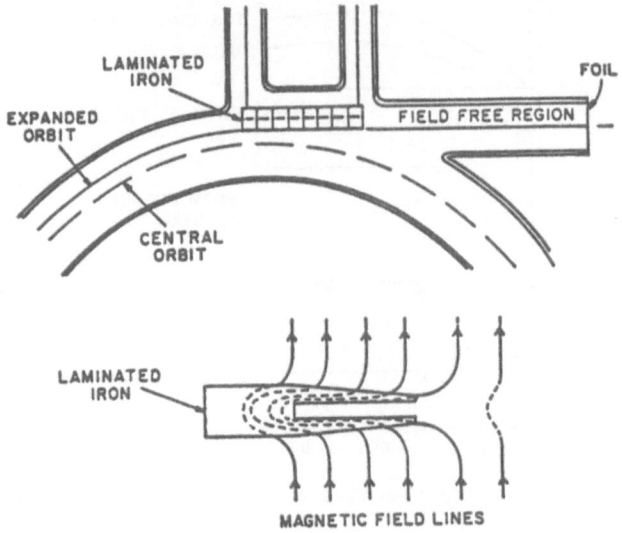

Fig. 10. Magnetic peeler for electron beam extraction.

has expanded, indeed the problem has been to slow down the expansion.
The betatron condition $p = eB_{yo}R/c$ survives the additional magnetic
fields so that the beam radius can be controlled as in a conventional
betatron. An additional possibility is to use resonance instability
selected by controlled field errors to move the beam into a magnetic
peeler. That this is possible has been demonstrated accidentally –
because of a field error in one of the toroidal field coils, the beam was
observed to dump on the wall in a space of a few centimeters. The next
phase of the UCI program will include a design with sufficient space
between coils to permit extraction by a magnetic peeler.

MODIFIED ELONGATED BETATRON (MEBA)

The MEBA is a descendant of the Astron. It is illustrated in Fig.
11. The Astron program did not achieve field reversal and was eventually
terminated for this reason. However, it did succeed in confining several
kA of electrons in a magnetic containment geometry that is similar to
Fig. 11. The relative ease with which injection and trapping were
accomplished motivated the attempt to make an accelerator (Blaugrund et
al., 1985). It is also expected that extraction will be easier in MEBA
geometry.

Referring to Fig. 11 the beam of radius 6 cm, thickness 0.5 cm and
length 80 cm was confined in an annular space of radii 4.5 cm and 9.5 cm.
Injection was tangential with an electron gun that emitted 70 kev electrons. Approximately 100 nc of electrons were trapped in initial
magnetic fields $B_p \simeq 90$ G $\simeq B_t$. By increasing the poloidal magnetic

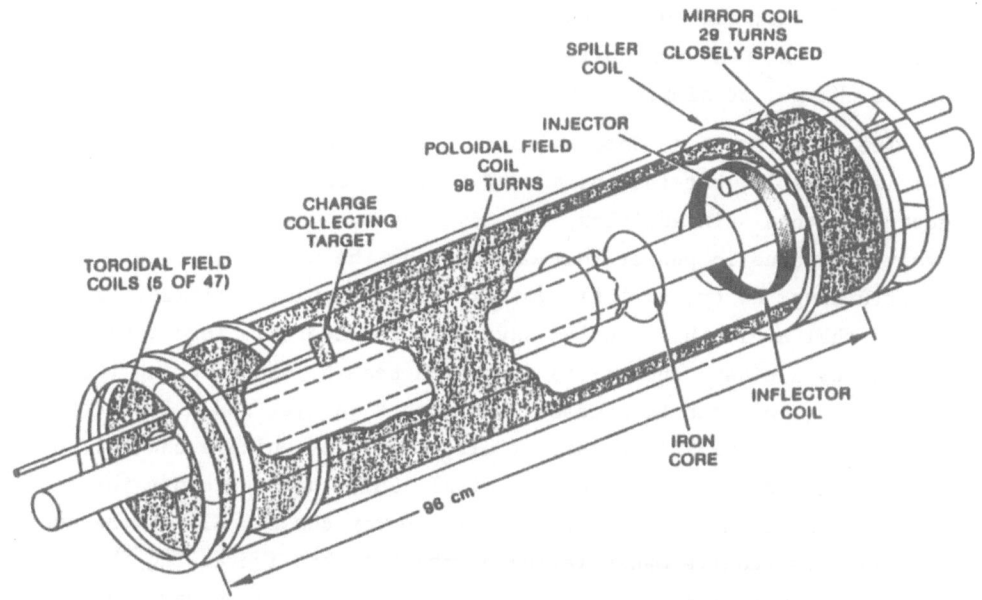

Fig. 11. Modified elongated betatron experiment.

field to about 1.5 k=Gauss electrons were accelerated to about 2 Mev.
The betatron flux condition $\bar{B}_p = 2\ B_p$ was maintained by means of a
magnetic core of radius 2.7 cm filled with powdered iron of effective
permeability 5.

The poloidal· magnetic field coil was in series with the toroidal
field coil so that $B_p \simeq B_t$. Rapid acquisition of data was facilitated by
a .5 Hz repetition rate. Extraction in the form of a beam was accomplished but not without degrading the accelerator performance because the
extraction port requires a radical modification of one of the mirror
coils; the present accelerator was not designed for this and the resultant field perturbation substantially affects the accelerator performance. However, significant progress was made in extracting the beam with
cylindrical symmetry maintained. A spiller coil was used to weaken the
downstream mirror and the entire E-layer was dumped onto the glass wall.
The resultant Bremstrahlung emissions were observed through a 3 mm wide
vertical slit collimator and a lead shielded plastic scintillation
detector. Approximately 27,000 shots (2 days of running) of the accelerator were made and the X-rays were scanned along the betatron axis.
The width of the circular band emitting Bremstrahlung was less than 7 mm.
(The width of the distribution was almost entirely due to the resolution
of the collimator which was checked using a point source of Cs^{137}).

Internal extraction was done routinely to make beam measurements;
i.e., the beam was released from mirror confinement by means of ejector

121

coils which increase the mirror ratio at one end and decrease it at the other end. The beam was collected on a target which was a Faraday cup to collect charge, or simply an X-ray target.

The measurement to date indicate that extraction in the form of a beam is possible, but the coils must be replaced with spaced coils to accomplish this. In addition, the most obvious method to extract a beam involves a magnetic peeler that may degrade the accelerator performance by perturbing the cylindrical symmetry of the magnetic fields. Several methods that avoid this problem have been considered. They involve creation of a plasma channel to guide the beam by ion focusing (Martin et al., 1985). This can be accomplished with a gas jet and a laser (B. Hui and Y. Y. Lau., 1985; A. Fisher et al., 1985) to ionize a channel, or a neutralized ion beam (S. Robertson et al 1981) that can cross the magnetic fields of the MEBA if it is sufficiently dense. All of the above possibilities require manipulation of the beam and increasing the beam radius. A study of these methods of extraction is being carried out at UCI.

The MEBA is physically quite different from the modified betatron or the stellatron. Although there is a toroidal magnetic field, its only purpose is to stabilize the precessional mode and as such it is small enough that the electron gyro-radius is the same as the beam radius. Therefore, adiabatic dynamics is inapplicable. Particles do not follow field lines; this makes the analysis more difficult but there are compensations such as the absence of the Kruskal-Shafranov limit. The toroidal field does not increase the space charge limit very much; instead the beam charge is increased by the length of the beam. After acceleration the beam can be contracted since the space charge problem reduces as γ increases.

The cold beam energy threshold for the negative mass instability is $\gamma = 1 + (B_t/B_p)^2 \stackrel{\sim}{=} 2$ which is already exceeded in the present experiments. A relativistic Vlasov-Maxwell treatment of the MEBA has been developed (Cavenago and Rostoker, 1986).

The procedure is to solve self-consistently the relativistic Liouville equation

$$\frac{\partial f}{\partial X^\mu} \frac{\partial h}{\partial P_\mu} - \frac{\partial h}{\partial X^\mu} \frac{\partial f}{\partial P_\mu} = [f,h] = 0 \tag{51}$$

and Maxwell's equations

$$\Box A^\mu - \frac{4\pi e^2 J^\mu}{mc^2} \tag{52}$$

(X^μ, P_μ) are the position and momentum four-vectors. In Eq. (51), $f(X^\mu, P_\mu)$ is the distribution function; h is the Hamiltonian, defined as

$$h = -\frac{1}{2} g^{\mu\nu} (P_\mu + A_\mu)(P_\nu + A_\nu) = -\frac{1}{2} \tag{53}$$

$g^{\mu\nu}$ is the metric tensor and A_μ is the vector potential.

The only constant of the motion is P_Θ. For the radial and axial coordinates r and z we assume that all quantities are slowly varying so that

$$I_z = \oint P_z \, dz \tag{54}$$

$$I_r = \oint P_r \, dr$$

are adiabatic invariants. Any function of I_r, I_z and P_Θ is a solution of Eq. (54). We select a suitable function and calculate

$$J^\mu = \int d^4P \, (P^\mu + A^\mu) \, f \tag{55}$$

Some examples of the distribution of I_r are illustrated in Fig. 12 along with the resultant current distribution which is proportional to rg(X) where $X = r - r_A$ and r_A is the orbit center. The end result of the calculation is the average radial oscillation frequency

$$\omega_{rA}^2 = (\partial^2 h / \partial r^2)_{r=r_A}$$

$$= \left\{ -\left(\frac{\partial A_4}{\partial r}\right)^2 + \left(\frac{\partial A_z}{\partial r}\right)^2 + \frac{(P_\Theta + A_\Theta)^2}{r^4} + \left[\frac{P_\Theta}{r^2} - \frac{\partial}{\partial r}\left(\frac{A_\Theta}{r}\right)\right]^2 \right\}_A$$

$$- \frac{\nu}{\gamma_A r_A} \left(\frac{\omega_{rA}}{I}\right)^{1/2} . \tag{56}$$

$\nu = Ne^2/mc^2$ is the Budker parameter.

$$N = \int_{r_A - s}^{r_A + s} n(r) \, 2\pi r \, dr \tag{57}$$

is the line density, n(r) is the electron density, r_A is the beam average radius and 2s is the beam thickness. For an equilibrium to exist ω_{rA}^2 must be real which sets a limit to ν or to the circulating current. This is the space charge limit.

Stability is investigated by means of a linear perturbation expansion about the above equilibria. The details of the calculations are available in UCI reports (available on request).

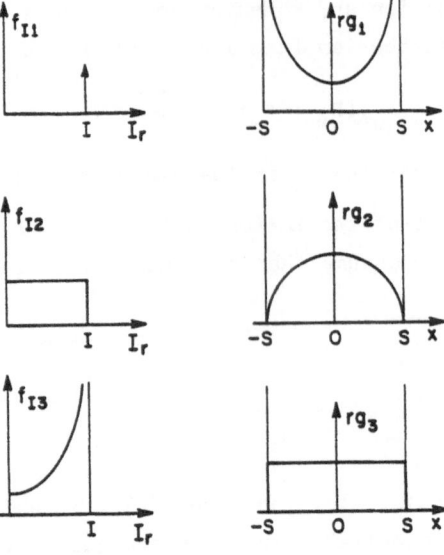

Fig. 12. Some examples of electron distribution functions based on the radial invariant I_r.

Results

The space charge limit is given by the formula

$$\gamma = \frac{\nu [2 + \dfrac{r_A}{s(\gamma^2 - 1)}]}{1 + [\dfrac{B_\Theta(r_A)}{B_z}]^2 - \dfrac{1}{\gamma^2-1} (\dfrac{Ir_A}{s^2})^2}$$ (58)

$B_\Theta(r_A)$ is the toroidal magnetic field at the average radius, B_z is the poloidal magnetic field and $\gamma = \gamma_A$ is average relativistic energy factor required at injection to obtain the charge represented by ν. I is the maximum value of I_r. Initially we neglect $I \simeq 0$. For $\gamma \simeq 1$, $[r_A/s (\gamma^2 - 1)] \gg 2$ and the formula reduces to the usual scaling for a modified betatron. For $\gamma \gg 1$, ν scales linearly with γ rather than with γ^3.

The other new feature is that finite values of I reduce the space charge limit.

The stability analysis shows that a cold beam would always have a range of γ for which the beam has a negative mass instability. A spread in P_Θ and I_z in the equilibrium distribution reduces the growth rate, but the beam remains unstable for sufficiently large γ. Complete stability can be obtained only with a finite value of the radial action I. This in turn will modify the space charge limit of Eq. (58). With the present theory it is possible to determine the injection γ and the quantity I corresponding to stable confinement of the charge ν. From I the emittance can be determined.

There is another space charge limit due to action of the space charge fields in the axial direction. Confinement requires that $e\Delta\Phi < \mu\Delta B_z$ where $\mu = P_\perp^2/2\gamma\, mB_z$ is the magnetic moment of the electron, ΔB_z is the change in magnetic field over half the length $L/2$ of the beam. $\Delta\Phi$ is corresponding change in electric potential Φ. According to this criterion the minimum injection γ is

$$\gamma = \frac{4\nu}{M}\,\frac{\ln(r_o/r_A)}{[1 - (1/\gamma^2)]} \tag{59}$$

where $M = 1 + \dfrac{\Delta B_z}{B_z}$ is the mirror ratio.

In Table 1, the predictions of Eqs. (58) and (59) are compared with experimental results. Predictions of the γ required for injection are made for a large MEBA.

Table 1. Comparisons of experiment with predictions of Eqs. (58) and (59)

PARAMETER		EXPERIMENT	LARGE MEBA
Average Radius of Beam	r_A cm	6	100
Length of Beam	L cm	80	600
Outside Wall Radius	r_o cm	9.5	102.5
Inside Wall Radius	r_i cm	4.5	97.5
Beam Thickness	$2s$ cm	.5	2
Magnetic Field Ratio	$B_\theta(r_A)/B_z$	1	2
Mirror Ratio	M	.05	.15
Budker Parameter	ν	0.41	14
Total Charge	Q_r Radial Limit nC	1850	5×10^6
	Q_1 Longitudinal nC	350	5×10^6
	Q nC Observed	100	
Total Circulating	I Observed A	100	
Current	I Required kA		240
Injection	γ_r Radial		7.8
	γ_ℓ Longitudinal		5.4
	γ Observed	1.1	

Some revision of these calculations will be required because of the radial action I, necessary for stability.

ACKNOWLEDGMENTS

The research on the modified betatron and the stellatron was supported by ONR. The research on MEBA was supported by DOE. This paper is mainly a review of the work of the past four years by myself and my colleagues E. Blaugrund, G. Barak, A. Fisher, M. Cavenago, H. Ishizuka, B. Mandelbaum, R. Prohaska and G. Roberts.

REFERENCES

Barak, G., and Rostoker, N., 1983, Orbital Stability of the High Current Betatron, Phys. Fluids, 26:856.

Blaugrund, A. E., Fisher, A., Prohaska, R., and Rostoker, N., 1985, J. Appl. Phys., 57:2474.

Cavenago, M., and Rostoker, N., 1986, Modified Elongated Betatron Accelerator, "Proc. 6th Int. Conf. on High Power Particle Beams", June 9-12, 1986, Kobe, Japan.

Clark, W., Korn, P., Mondelli, A., and Rostoker, N., 1976, Experiments on Electron Injection into a Toroidal Magnetic Field, Phys. Rev. Lett., 37:592.

Daugherty, J. D., Eninger, J., and Janes, G. S., 1971, "AVCO Everett Research Report 375" (unpublished).

Davidson, R., and Uhm, H., 1982, Stability Properties of an Intense Relativistic Non-neutral Electron Ring in a Modified Betatron Accelerator, Phys. Fluids, 25:2089.

Fisher, A., Rostoker, N., Pearlman, J., and Whitham, K., 1985, Laser Extraction of an Electron Beam from MEBA, "Particle Accelerator Conference, Vancouver, B.C.", May 13-15.

Godfrey, B. B., and Hughes, T. P., 1985, Long Wavelength Negative Mass Instabilities in High Current Betatrons, Phys. Fluids, 28:669.

Golden, J., Mako, F., Floyd, L., McDonald, K., Smith, T., Dialetis, D., Marsh, S. J., and Kapetanakos, C. A., 1986, Progress in the Development of the NRL Modified Betatron Accelerator, "Proc. 6th Int. Conf. on High Power Particle Beams", June 9-12, 1986, Kobe, Japan.

Hui, B., and Lau, Y. Y., 1984, Injection and Extraction of a Relativistic Electron Beam in a Modified Betatron, Phys. Rev. Lett., 53:2024.

Ishizuka, H., Lindley, G., Mandelbaum, B., Fisher, A., and Rostoker, N., 1984, Formation of a High Current Electron Beam in Modified Betatron Fields, Phys. Rev. Lett., 53:266.

Ishizuka, H., Saul, J., Fisher, A., and Rostoker, N., 1986, Beam Acceleration in the UCI Stellarator, "Proc. 6th Int. Conf. on High Power Particle Beams", June 9-12, 1986, Kobe, Japan.

Kerst, D. W., Adams, G. D., Koch, H. W., and Robinson, C. S., 1949/50, An 80 Mev Model of a 300 Mev Betatron and a 300 Mev Betatron. Phys. Rev. Lett., 75:330; 78:297. Rev. Sci. Inst., 21:462.

Kapetanakos, C. A., Sprangle, D., Chernin, D. P., Marsh, S. J., and Haber, I., 1983, Equilibrium of a High Current Electron Ring in a Modified Betatron Accelerator, Phys. Fluids, 26:1634.

Landau, R., 1968, Negative Mass Instability with B_θ Field, Phys. Fluids, 11:205.

Landau, R., and Neil, V., 1966, Negative Mass Instability, Phys. Fluids, 9:2412.

Mandelbaum, B., 1985, A Study of a Modified Betatron with Stellarator Windings, Ph.D. thesis, University of California, Irvine.

Manheimer, W., 1985, The Plasma Assisted Modified Betatron, <u>Particle Accel.</u>

Martin, W. E., Caporaso, C. J. Fawley, W. M., Prosnitz, D., and Cole, A. G., 1985, Electron Beam Guiding and Phase Mix Damping by a Laser-Ionized Channel, <u>Phys. Rev. Lett.</u>, 54:685.

Roberson, C. W., Mondelli, A., and Chernin, D., 1983, High-current Betatron with Stellarator Fields, <u>Phys. Rev. Lett.</u>, 50:507.

Roberson, C. W., Mondelli, A., and Chernin, D., 1985, The Stellatron Accelerator, <u>Particle Accel.</u>, 17:29.

Roberts, G. A., and Rostoker, N., 1985, Adiabatic Beam Dynamics in a Modified Betatron, <u>Phys. Fluids</u>, 28:1968.

Roberts, G., and Rostoker, N., 1986, Effect of Quasi-Confined Particles and l = 2 Stellarator Fields on the Negative Mass Instability in a Modified Betatron; <u>Phys. Fluids</u>, 29:333.

Robertson, S., Ishizuka, H., Peter, W., and Rostoker, N., 1981, Propagation of an Intense Ion Beam Transverse to a Magnetic Field, <u>Phys. Rev. Lett.</u>, 47:508.

Sprangle, D., and Chernin, D., 1984, Beam Current Limitation Due to Instabilities in Modified and Conventional Betatrons, <u>Particle Accel.</u>, 15:2089.

WAKEFIELD ACCELERATION: CONCEPTS AND MACHINES*

Perry B. Wilson

Stanford Linear Accelerator Center
Stanford, California 94305

INTRODUCTION

An introduction of the basic concepts of wakefields and wake potentials is presented in Richard K. Cooper's paper "Wake Fields: Limitations and Possibilities" in these proceedings. In this lecture we explore the possibility of using wakefield concepts to accelerate electrons and positrons to the high energies required by the next generation of linear colliders.

Brightness in Linear Colliders

In a linear collider, electron and positron bunches accelerated in opposing linacs are directed toward a collision point between the two linacs. The intensity of the collision is measured by the luminosity, defined as

$$\mathsf{L} = \frac{N^2 f_r}{4\pi\sigma^2_\perp} \tag{1}$$

Here N is the number of particles in the electron and positron bunches (assumed to be equal), f_r is the bunch repetition frequency and $\sigma_\perp$ is the transverse dimension of the bunch, assumed for simplicity to be Gaussian and round. We note that the units of L are cm^{-2} sec^{-1}. Thus the luminosity multiplied by the cross section in cm^2 for the occurrence of a particular event in the beam-beam collision gives the number of events produced per second. The transverse bunch dimension in Eq. (1) can be expressed in terms of the invariant emittance ε_n, the beam energy γ normalized to the electron rest energy, and the beta function at the

---

* Work supported by the Department of Energy, Contract DE-AC03-76SF00515.

beam collision point $\beta*$ as

$$\sigma_\perp = (\epsilon_n \beta*/\gamma)^{1/2}$$

Using this expression together with the average beam current $I_{ave} = eNf_r$ and the peak beam current $I_p = eNc/(2\pi)^{1/2}\sigma_z$ in equation (1) we obtain

$$L = \frac{1}{2(2\pi)^{1/2}e^2c}\left[\frac{I_{ave}\,I_p}{\epsilon_n}\right]\left[\frac{\gamma\sigma_z}{\beta*}\right] \tag{2}$$

The bunch length σ_z must be less than $\beta*$ in order to avoid loss of luminosity due to variation of the transverse bunch dimensions during the beam-beam collision. Furthermore, it can be shown under reasonable assumptions that $\beta*$ scales with energy as $\gamma^{1/2}$ and that at least approximately $\sigma_z \sim \lambda_{rf} \sim \gamma^{1/2}$ in order to keep the total AC wall plug power for a collider with reasonable bounds. Thus the last factor in Eq. (2) is roughly invariant, and the luminosity then depends only on the "brightness" $B \equiv I_{ave}\,I_p/\epsilon_n$ given by the second factor.

The concept of brightness in a linear collider can be carried further by introducing the relative energy spread in the bunch $\delta \equiv \Delta E/E$. The $\beta*$ that can be achieved for a given δ varies as δ^2. Thus an extended definition of collider brightness would be $B \equiv I_{ave}\,I_p/\epsilon_n\delta^2$.

Acceleration Concepts

A number of mechanisms have been proposed for the acceleration of particles to the 300 GeV - 5 TeV energy range desired for future linear colliders. (An efficient and economical single bunch acceleration method could also be useful for an injector into storage rings for particle physics, synchrotron radiation and FEL's.) The concepts that have been suggested range from the more-or-less conventional, such as a traditional disk-loaded accelerator structure powered by microwave tubes, to more exotic schemes, such as laser and plasma accelerators. Although the principal topic of this paper is the wakefield acceleration mechanism, it is useful to place the wakefield scheme in context by comparing it with other acceleration concepts proposed for high energy linear colliders. First, however, we will discuss some simple but basic concepts which are common to any acceleration method.

Figure 1 gives a conceptual diagram of an accelerator. In general an accelerator consists of some sort of driver which produces electromagnetic energy (not necessarily rf), which is, in turn, converted into an accelerating field in some kind of structure. A figure of merit for the driver is the efficiency with which it converts average input AC power into the electromagnetic power delivered to the structure. A figure of merit for the structure is the effective elastance per unit

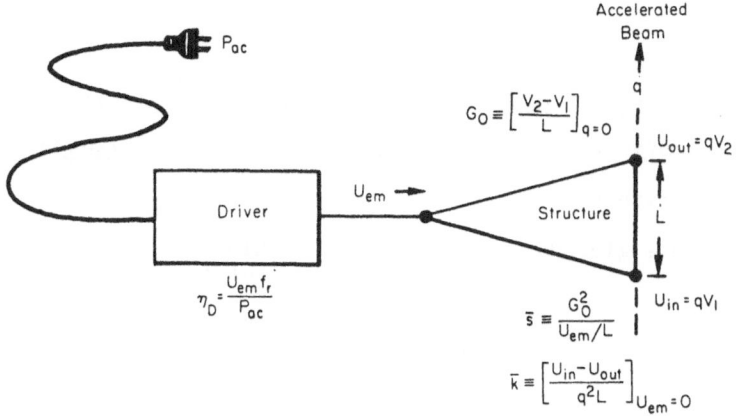

$$\text{Accelerated Beam}$$

$$G_0 \equiv \left[\frac{V_2 - V_1}{L}\right]_{q=0}$$

$$U_{out} = qV_2$$

$$\eta_D = \frac{U_{em} f_r}{P_{ac}}$$

$$\bar{s} \equiv \frac{G_0^2}{U_{em}/L}$$

$$U_{in} = qV_1$$

$$\bar{k} \equiv \left[\frac{U_{in} - U_{out}}{q^2 L}\right]_{U_{em}=0}$$

Fig. 1. Conceptual diagram of an accelerator.

length, defined as the square of the unloaded accelerating gradient divided by the input electromagnetic energy per unit length. The accelerating structure must, of course, be appropriate for each particular driver. Some proposed drivers and structure combinations are listed in Table 1.

In the first three concepts in Table 1, the electromagnetic energy is produced external to the accelerating structure. Concept number three is the so-called two beam accelerator. The high current external beam, running parallel to the accelerator, can be accelerated either by induction linac units or by superconducting rf cavities. Rf energy can be extracted either by an FEL interaction or by bunches interacting with longitudinal fields in a transfer cavity. In both cases the driving beam is simply an external source of rf energy. Concepts 4 through 8 involve the production of electromagnetic energy internal to the structure.

Concepts 4 and 5 are the standard wakefield accelerator mechanisms in which a high-charge driving bunch is injected into an appropriate metallic structure. Electromagnetic wakefields set up behind the driving bunch can in turn be used to accelerate a trailing lower-charge bunch. Figure 2 shows the details of concept 4, in which a ring driving beam is used. In concept 5 both beams are injected on the axis of a more or less conventional periodic accelerating structure. Both of these wakefield schemes will be discussed in detail in the following sections. In concept 7 a driving electron bunch is injected into a plasma, setting up intense plasma oscillations and associated electromagnetic fields behind the bunch. This scheme, which is also a type of wakefield acceleration, will be discussed briefly in a later section.

Table 1. Some Driver and Structure Combinations

Driver	Structure
A. External Drivers	
1. Discrete microwave tubes	Disk-loaded or other periodic metallic structures.
2. Laser	Metallic grating or other periodic open resonator.
3. Low energy, high current parallel external beam	Disk-loaded or other periodic metallic structure.
B. Internal Drivers	
4. Coaxial ring driving bunch	Radial-line wakefield transformer with annular and axial beam apertures.
5. Collinear driving bunch	Disk-loaded or other periodic structure with axial beam aperture.
6. Laser-switched photodiode	Radial line transformer.
7. Driving electron bunch	Plasma
8. Laser	Plasma

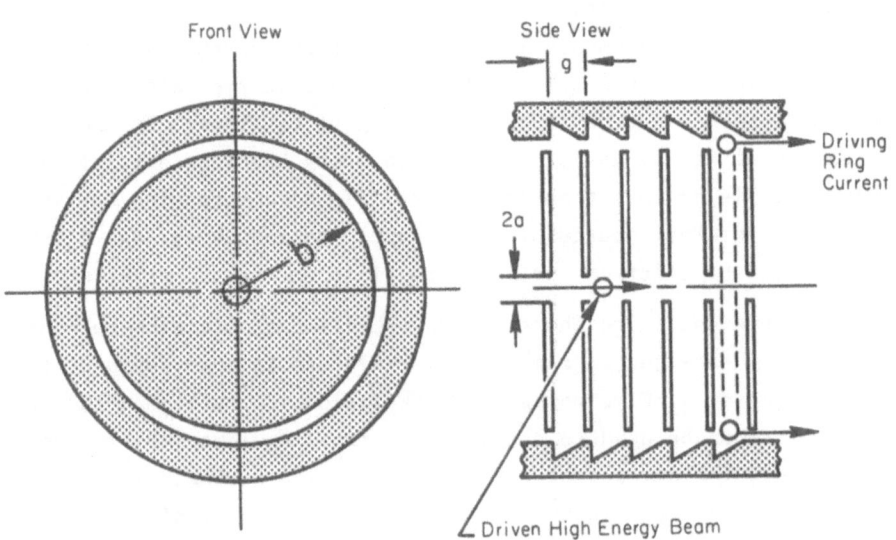

Fig. 2. Wakefield accelerator with a ring driving beam.

Concept 6, sometimes called switched-power acceleration, is illustrated in Figure 3. In this method, photocathodes arranged in a ring at the outer perimeter of a radial transmission line are charged from a DC power supply. Short laser pulses trigger the photocathodes in synchronism with the accelerated beam traveling on the axis of the structure, but with an appropriate fixed time advance. The intense current emitted from the photocathode crosses the gap of the radial transmission line and induces an electromagnetic pulse which travels toward the axis of the structure. As the radius of the ring-shaped pulse decreases, the volume occupied by the electromagnetic fields also decreases and from conservation of energy the field strength must increase. This "transformer action" produces a large ratio between the longitudinal field strength on the axis of the structure and the field in the neighborhood of the photodiode (which in turn is approximately equal to the DC charging voltage divided by the longitudinal gap g). This transformer ratio is given approximately by

$$R = 2 \left[\frac{2b}{g + c\,\tau_r} \right]^{1/2} \tag{3}$$

where τ_r is the rise time of the electromagnetic pulse and b is the radius at the photocathodes (Villa 1975). It is assumed that both g and $c\tau_r$ are approximately equal to the axial hole radius $\underline{a}$. It is seen that the switched power concept is closely related to the ring-beam wakefield accelerator scheme shown in Figure 2.

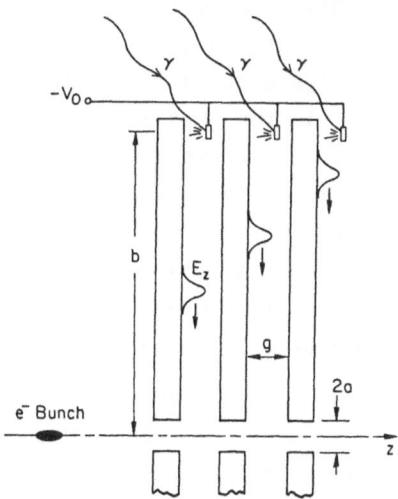

Fig. 3. Switched power accelerator concept.

Concept 8 is the laser beatwave acceleration mechanism, in which two laser beams having a frequency difference equal to the plasma frequency drive intense plasma oscillations. The fields associated with these oscillations can in turn be used to accelerate a trailing electron bunch. Only the wakefield acceleration schemes will be discussed in more detail in this paper. For a more complete description of the other acceleration mechanisms, and for a review of the current status of analytic and experimental work in this area, the reader is referred to Joshi and Kastsouleas (1985) and the Symposium on Advanced Accelerator Concepts.

Comparison of Acceleration Concepts

It is reasonable to ask what physical limitations there are on the structure figure of merit, the elastance. Suppose a uniform longitudinal accelerating field E_z is present inside a cylindrical region of radius a_{eff} and is zero outside this radius. Since magnetic fields are also necessary for propagation of electromagnetic energy, assume also that an equal amount of stored magnetic energy is present. The stored energy per unit length is then $u_{em} = \pi a_{eff}^2 \epsilon_0 E_z^2$, and the elastance per unit length is

$$S = \frac{E_z^2}{u_{em}} = \frac{1}{\pi \epsilon \, a_{eff}^2} = \frac{3.6 \times 10^{10}}{a_{eff}^2} \qquad \frac{V}{C-m} \qquad (4)$$

We see that the elastance increases inversely as the square of the energy confinement radius a_{eff}. As an example of the application of the concept of effective radius for energy confinement, consider the SLAC disk-loaded structure, with an elastance of 77 V/pC-m. The effective energy radius from Eq. (4) is 2.17 cm, while the actual beam hole radius is 1.17 cm. Since energy stored outside of the beam aperture region is of no use for accelerating particles, a figure of merit for the structure is $a_{eff}/a_{geo} = 1.85$ where a_{geo} is the physical, or geometric, hole radius. From the definition of elastance it is seen that to obtain a high value for S it is only necessary to go to a structure with small transverse dimensions. For the case of an rf structure, this implies a short operating wavelength for a large elastance, giving $S \sim \lambda^{-2}$. The ratio a_{eff}/a_{geo} tells how effective structures with similar physical dimensions are in confining electromagnetic energy near the beam axis.

In Table 2 some structures are compared for the various acceleration concepts that have been discussed. A rough estimate of the driver efficiency is also included. We see that the rf, wakefield and switched power acceleration schemes all have values of a_{eff}/a_{geo} in the range 1-2. The rf acceleration method has the advantage of high efficiency, although

conventional microwave tubes with very high peak output power do not
exist at very short wavelengths where the elastance is highest. The
two-beam and wakefield acceleration schemes were devised basically to get
around this wavelength limitation on the peak power obtainable from
conventional sources.

Table 2. Comparison of Acceleration Schemes

Scheme	a_{eff}	a_{eff}/a_{geo}	a_{eff}/λ	η_D
RF (SLAC structure	2.17 cm	1.85	0.21	0.5
Ring beam wakefield[1]	2.7 mm	1.34	—	0.1
Switched power[2]	2.7 mm	$\simeq 2$	--	few %/pulse
Plasma wakefield[3]	0.2-0.6 mm	1.0	1-2	0.1
Plasma beatwave[3] ·	0.3-3 mm	3.4	3-6	< 0.1

Notes: 1) The Voss-Weiland proposal (Weiland and Willeke, 1983).
       2) Based on calculations by I. Stumer. The efficiency can be
          improved using multiple laser pulses to the photocathodes.
       3) Based on data of Chen (1986).

In concluding the discussion in this section, we note that all
internal beam accelerating schemes (under B in Table 1) suffer from a
fundamental difficulty. the accelerated beam "sees" the transverse
wakefields produced by the driving beam, in addition to the longitudinal
component. In effect, transverse deflecting fields are produced which
are of the same order as the accelerating field. Thus the driving beam
must be oriented with extreme precision in the transverse direction, and
in addition must be azimuthally homogeneous in order not to produce
deflecting fields which are unacceptably large. In an external beam
scheme, the rf is generated outside of the accelerating structure and
only the accelerating component of the wakefields produced by the high
current beam is propagated into the structure.

Delta-Function Wake Potentials

In this section we consider wake potentials for closed cavities and periodic structures. The delta-function wake potentials for such structures are an important concept. These wake potentials are the longitudinal and transverse potentials experienced by a test charge moving at a fixed distance s behind a unit point driving charge passing through the structure. The test charge is assumed to move along a path which is parallel to that of the driving charge, and both test and driving charges are assumed to move with a velocity $v \approx c$. For $v \neq c$ the expressions for the wake potentials are in general much more complicated and the wake potential concept is less useful. The geometry of the problem is illustrated in Fig. 4.

Expressions for the longitudinal and transverse delta-function wake potentials for the general case are introduced by Cooper (1988). These potentials are simply the integrated effect of the longitudinal and transverse forces acting on a test charge as it passes through a structure (e.g., from z_1 to z_2 for the test charge T in Fig. 4) following behind a unit point driving charge.

Under certain conditions, which will be spelled out in detail later, it can be shown (Bane, et al., 1983; Bane, 1980) that the longitudinal and transverse wake potentials can be written in terms of the properties

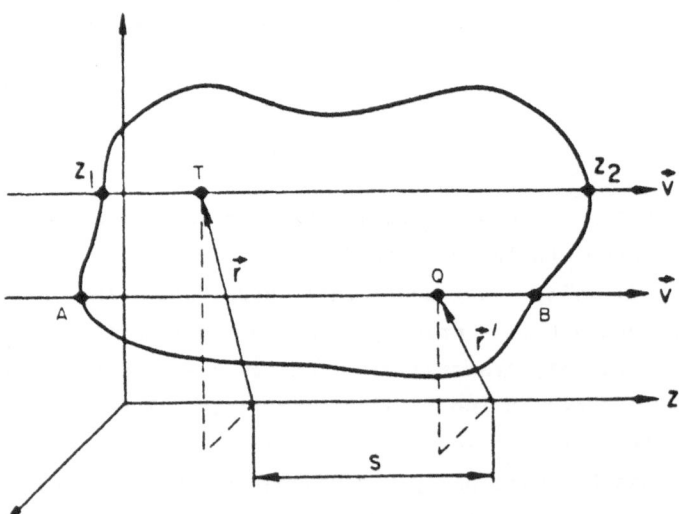

Fig. 4. A driving charge Q, moving at constant velocity V parallel to the z-axis, enters a closed cavity at A (r', $z = 0$) at $t = 0$ and leaves at B(r',$z = L$). A non-perturbing test particle T also moves at the same velocity v, but at transverse position r and at longitudinal distance s behind Q.

of the normal modes of a charge-free cavity in a relatively simple way:

$$W_z(r',r,s) = 2H(s) \sum_n k_n(r',r) \cos \frac{\omega_n s}{c} \tag{5a}$$

$$W_\perp(r',r,s) = 2H(s) \sum_n k_{n\perp}(r',r) \sin \frac{\omega_n s}{c} \tag{5b}$$

where

$$H(s) \equiv \begin{cases} 0 & s < 0 \\ 1/2 & s = 0 \\ 1 & s > 0 \end{cases}$$

and

$$k_n(r',r) = \frac{V_n^*(r') \, v_n(r)}{4U_n} \tag{6a}$$

$$k_{n\perp}(r',r) = \frac{V_n^*(r') \, \nabla_\perp V_n(r)}{4U_n \omega_n/c} \tag{6b}$$

Here ω_n is the angular frequency the nth mode, and $V_n(r)$ is the voltage that would be gained by a nonperturbing test particle crossing the cavity in which energy U_n is stored in the nth mode. Assuming the electric field for the nth mode varies with time as $\exp(i\omega t)$ and the position of the test particle is given by $z = ct$, this voltage is

$$V_n(r) = \int_{z_1}^{z_2} dz \, E_z(r,z) \exp\left(\frac{i\omega_n z}{c}\right) \tag{7}$$

The conditions under which eqs. (5) are valid for the longitudinal and transverse wake functions are discussed in detail in Refs. 8 and 9, and are summarized in Table 3. We see that if the driving charge and test particle follow different paths in a closed cavity of arbitrary shape, neither Eq. (5a) nor (5b) give a valid description of the wake potentials. If the particles follow the same path in a closed cavity of arbitrary shape, Eq. (5a) is valid for the longitudinal wake potential but Eq. (5b) does not correctly describe the transverse wake potential. Formal expressions can indeed be written down for the non-valid cases, but the integrals are much more complicated, and the wake potentials for a given mode do not separate neatly into a product of an s-dependent factor and a factor which depends only on r.

Note that Eqs. (5a) and (5b) are related by

$$\frac{\partial W_\perp}{\partial s} = \nabla_\perp W_z. \tag{8}$$

137

This relation between the longitudinal and transverse wakes is sometimes termed the Panofsky-Wenzel theorem (Panofsky and Wenzel, 1956). It was originally derived to calculate the transverse momentum kick received by a nonperturbing charge traversing a cavity excited in a single rf mode.

Table 3. Cases for which Eqs. (5a) and (5b) give the wake potentials in the limit $v \approx c$

	Case	Eq. (5a) Valid for W_z	Eq. (5b) Valid for W
(a)	Test charge and driving charge follow different paths in a closed cavity of arbitrary shape.	No	No
(b)	Test charge and driving charge follow the same path in cavity of arbitrary shape.	Yes	No
(c)	Velocity v is in the direction of symmetry of a right cylinder of arbitrary cross section.	Yes	Yes
(d)	Both driving charge and test charge move in the beam tube region of an infinite repeating structure of arbitrary cross section.	Yes	Yes
(e)	Both particles move near the axis of any cylindrically symmetric cavity.	Yes	Yes

The wake potential formalism, using properties of the charge-free cavity modes, makes it possible to calculate useful quantities for the charge-driven cavity. An important example is the longitudinal wake potential for the case in which the test charge and driving charge follow the same path. Equations (5a) and (6a) reduce then to

$$W_z(r,s) = \sum_n k_n(r) \cos \frac{\omega_n s}{c} \times \begin{cases} 0 & s < 0 \\ 1 & s = 0 \\ 2 & s > 0 \end{cases} \tag{9}$$

$$k_n(r) = \frac{|V_n(r)|^2}{4U_n} \; .$$

The total potential seen by the charge itself is

$$V(r,0) = -q \, W_z(r,0) = -q \sum_n k_n(r), \tag{10a}$$

while the component in the nth mode acting on the charge is

$$V_n(r,0) \equiv V_n(0) = -q \, k_n. \tag{10b}$$

The energy left behind in the nth mode after the driving charge has left the cavity is

$$U_n = -q \, V_n(0) = q^2 \, k_n. \tag{11}$$

The parameter k_n is the constant of proportionality between the energy lost to the nth mode and the square of the driving charge, hence the name "loss parameter" or "loss factor".

Note from Eq. (9) that an infinitesimal distance behind a driving point charge the potential is retarding for the nth mode with magnitude

$$V_n(0^+) = 2V_n(0) = -2q \, k_n. \tag{12a}$$

As a function of distance s behind the driving charge, the potential for the nth mode varies as

$$V_n(s) = V_n(0^+) \cos \frac{\omega_n s}{c} = -2q \, k_n \cos \frac{\omega_n s}{c} \; . \tag{12b}$$

Equation (12a) expresses what is sometimes termed the fundamental theorem of beam loading (Wilson, 1981): the voltage induced in a normal mode by a point charge is exactly twice the retarding voltage seen by the charge itself.

The case of a periodically repeating structure is of obvious importance in accelerator design. Although real periodic structures are of course never infinite, practical structures at least a few periods in length seem to fulfill condition (d) of Table 3. Thus the wake potentials can be computed by a summation over the normal modes of the infinite structure. For the case of a cylindrically symmetric structure, all modes depend on the azimuthal angle ϕ as $e^{im\phi}$. The wake potentials can then be written (Bane, et al., 1983) for $s > 0$,

$$W_{zm} = 2 \left(\frac{r'}{a}\right)^m \left(\frac{r}{a}\right)^m \cos m\phi \sum_n k_{mn}^{(a)} \cos \frac{\omega_{mn} s}{c} \tag{13a}$$

$$W_{\perp m} = 2m \left(\frac{r'}{a}\right)^m \left(\frac{r}{a}\right)^{m-1} (\hat{r} \cos m\phi - \hat{\phi} \sin m\phi)$$

$$x \sum_n \frac{k_{mn}^{(a)}}{\omega_{mn} a/c} \sin \frac{\omega_{mn} s}{c} . \tag{13b}$$

Here $\hat{r}$ and $\hat{\phi}$ are unit vectors and $k_{mn}^{(a)}$ is the loss factor per unit length calculated at $r = a$, where a is the radius of the beam tube region. That is

$$k_n^{(a)} \equiv \frac{\left[E_{zn}(r = a)\right]^2}{4u_n}$$

where E_{zn} is the synchronous axial field component for the nth mode and u_n is the energy per unit length. The longitudinal cosine-like wake potential per period for the SLAC structure is shown in Fig. 5. Note the very rapid fall-off in the wake immediately behind the driving charge, from a peak wake of 8 V/pC per period at time $t = s/c = 0^+$. The wake seen by a point charge would be just one half of this wake, or 4 V/pC per cell. The sine-like transverse dipole ($m = 1$) wake potential for the SLAC structure is shown in Fig. 6. This figure illustrates the fact that the total wake potential is obtained by summing a finite number of modes that can be obtained using a reasonable computation time, and then adding on a so-called analytic extension to take into account the contribution

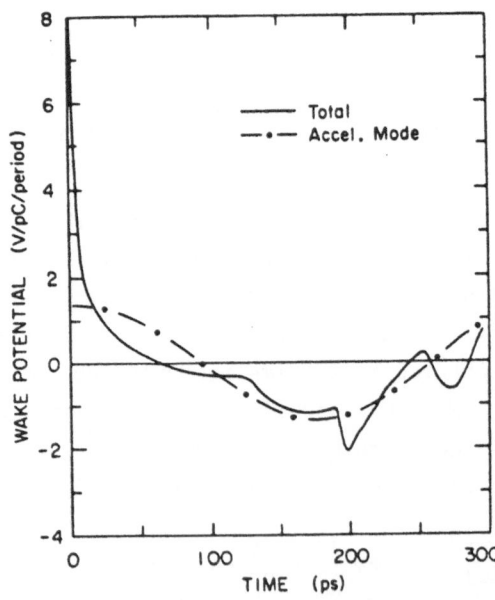

Fig. 5. A longitudinal delta-function wake potential per cell for the slac disk-loaded accelerator structure at time s/c behind a point driving charge.

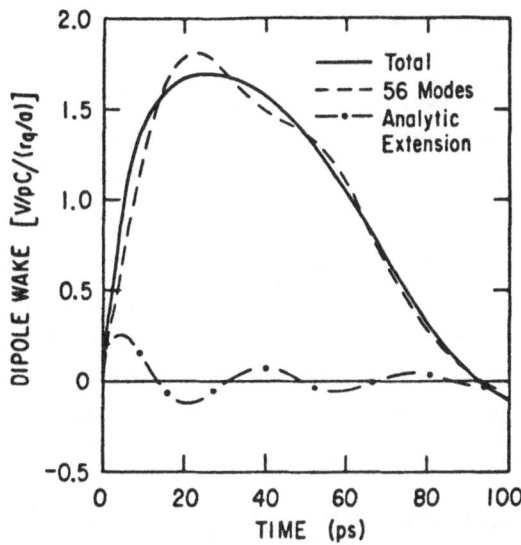

Fig. 6. Dipole delta-function wake potential per cell for the SLAC structure.

from very high frequency modes. Details are discussed by Bane and Wilson (1980).

Transformer Ratio for a Point Driving Charge

The transformer ratio is defined in this case as the maximum accelerating gradient anywhere behind the driving charge divided by the retarding gradient experienced by the driving charge itself. It is seen from Eq. (12) that for a single mode the transformer ratio R is

$$R = \frac{\text{Max } [V_n(s)]}{V_n(0)} = \frac{V_n(0^+)}{V_n(0)} = 2 \; . \tag{13c}$$

It is readily shown that this factor of two also follows directly from conservation of energy (Wilson, 1981). We will show later that this restriction on the transformer ratio for a point charge does not necessarily apply to an extended driving charge distribution.

A physical wake for a real cavity is a summation over many modes. Perhaps the modes might add up to produce a transformer ratio greater than two, even for a point charge. We note, however, that the wake for each mode varies with s as $W_n = 2k_n \cos(\omega_n s/c)$. At $s = 0^+$ the wakes all add in phase, and the sum of the wakes for all the modes gives a retarding potential which is exactly twice the retarding potential seen by the driving charge itself at $s = 0$. At any value of s where the net wake is accelerating, the cosine wakes for the individual modes can never do better than add exactly together in phase, as they do at $s = 0^+$. Thus

$$|W(s)| \leq \sum_n W_n(s = 0^+) = 2 \sum_n W_n(s = 0) \; , \tag{14}$$

141

and the transformer ratio for a real cavity with many modes, driven by a point charge, is equal to or less than two. In practice it will be considerably less than two, since the modes will never come close to adding in phase anywhere except at $s = 0^+$, where the net wake is retarding. It is easy to show that Eq. (14) also follows from conservation of energy (Wilson, 1986).

If the driving charge and the accelerated charge follow different paths through the cavity, the situation becomes more complicated. We first note from Eqs. (5a) and (6a) that the longitudinal wake potential is unchanged if the paths of the driving charge and the test charge are interchanged. If we now apply conservation of energy to two charges q_1 and q_2 following different paths, we can show (Wilson, 1986) that

$$\left| W_{12}(s) \right| = \left| W_{21}(s) \right| \leq 2 \left[W_1(0) \, W_2(0) \right]^{1/2} ,$$

where $W_{12}(s)$ is the wake along path 2 produced by a charge traveling on path 1, and so forth. If we define a transformer ratio R_{12} by

$$R_{12}(s) \equiv \frac{\left| W_{12}(s) \right|}{W_1(0)}$$

and similarly for R_{21}, then for any value of s

$$R_{12} \leq 2 \left[\frac{W_2(0)}{W_1(0)} \right]^{1/2} = 2 \left[\frac{\sum_n k_n(r_2)}{\sum_n k_n(r_1)} \right]^{1/2} \tag{16a}$$

$$R_{21} \leq 2 \left[\frac{W_1(0)}{W_2(0)} \right]^{1/2} = 2 \left[\frac{\sum_n k_n(r_1)}{\sum_n k_n(r_2)} \right]^{1/2} \tag{16a}$$

and

$$R_{12} \, R_{21} \leq 4. \tag{17}$$

Wake Potential in a Charge Distribution

Once the wake potential for a unit point charge is known, the potential at any point within or behind an arbitrary charge distribution with line density $\rho(s) = I(t)/c$ can be computed by

$$V(s) = - \int_s^\infty W_z(s'-s) \, \rho(s') \, ds' \tag{18a}$$

142

or

$$V(t) = - \int_{-\infty}^{t} W_z(t-t') \, I(t') dt'. \qquad (18b)$$

Here $W_z(\tau) = W_z(s/c)$, where τ is the time of delay after passage of the point driving charge. Similar expressions hold for the transverse potential within a charge distribution. Figure 7 shows examples of the net longitudinal potential, using the delta-function wake potential of Fig. 5 in Eq. (18a), for a Gaussian bunch of unit charge in the SLAC structure. Note the reduction in amplitude of the net wake potential, and the suppression of higher modes, as the bunch length increases.

The maximum accelerating gradient, E_a, behind a driving charge distribution is often of interest. It is useful to define three loss parameters for the distribution as follows:

$$k_\ell \equiv \frac{E_a^{\,2}}{4u} \qquad (19a)$$

$$k_a \equiv \frac{E_a}{2q} \qquad (19b)$$

$$k_u \equiv \frac{u}{q^2} \; . \qquad (19c)$$

Here u is the total electromagnetic energy per unit length deposited by the driving charge. The three loss parameters are related by $k_a^2 = k_\ell k_u$. For a single mode k_ℓ does not depend on the charge distribution, and $k_\ell = k_n$. For a Gaussian bunch interacting with a single mode, the loss parameters k_a and k_u are given in terms of $k_\ell = k_n$ by

$$k_u = k_n \, e^{-\omega_n^2 \sigma_t^2} = k_n \, e^{-4\pi^2 \sigma_z^2/\lambda_n^2} \qquad (20a)$$

$$k_a = k_n e^{-\omega^2 \sigma_t^2/2} = k_n e^{-2\pi\sigma_z^2/\lambda_n^2} \qquad (20b)$$

for each mode. Thus as the bunch length increases, coupling to higher modes is rapidly suppressed by the exponential factor. For the SLAC structure, $k_\ell p = 0.70/V/pC/cell$ for the fundamental mode, where p is the cell length. The amplitude of the accelerating mode voltage per cell excited by a Gaussian bunch with total charge q is therefore

$$\frac{V_1}{q} = 2k_a p = 1.40 \, e^{-2\pi^2 \sigma^2/\lambda_0^2} \qquad V/pC/cell$$

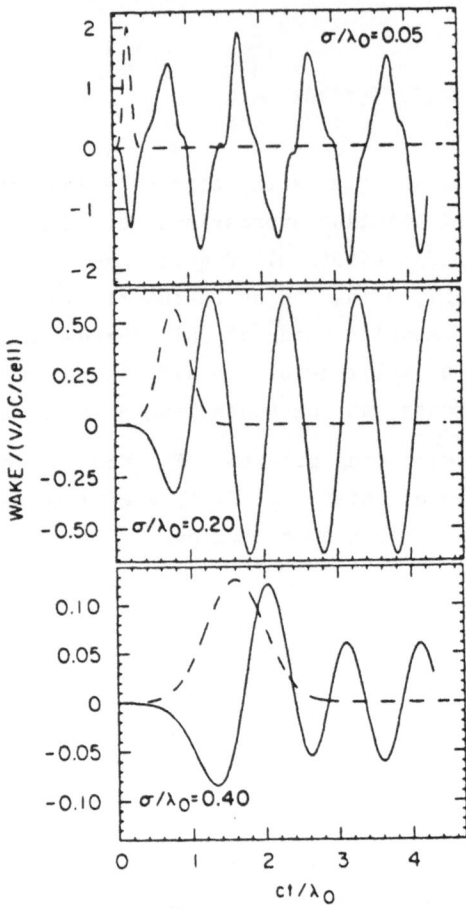

Fig. 7. Potential in and behind a Gaussian bunch interacting with the longitudinal modes of the SLAC structure for several values of $\sigma = \sigma_z$.

For σ/λ_0 = 0.05, 0.20, and 0.40, this gives V_1/q = 1.33, 0.64 and 0.06 V/pC/cell. These values can be compared with the computer calculation shown in Fig. 7, which takes into account all modes.

The plot for σ/λ_0 = 0.4 in Fig. 7 also illustrates the phenomenon of auto-acceleration, in which fields induced by particles at the front of the bunch can accelerate particles at the tail of the same bunch.

Transformer Ratio and Efficiency for a Charge Distribution

The transformer ratio for a charge distribution is

$$R = \frac{E_a}{E_m^-} \ , \tag{21}$$

where E_a is the maximum accelerating gradient behind the bunch and E_m^- the maximum retarding gradient within the bunch. It is useful also to define an efficiency for the transfer of energy from the bunch to the electromagnetic energy per unit length, u, in the wakefield,

$$\eta = \frac{u}{q \, E_m^-} \, . \tag{22}$$

Equations (21) and (22) can be combined with Eqs. (19) to obtain

$$k_a = \frac{2\eta \, k_\ell}{R} \tag{23a}$$

$$q = \frac{E_a R}{4\eta \, k_\ell} \, . \tag{23b}$$

COLLINEAR WAKEFIELD ACCELERATION

Using the results of the previous sections, we now examine the possibility of collinear wakefield acceleration; that is, the case in which the driving charge and accelerated charge follow the same path through the cavity or structure. Specifically, we are interested in the transformer ratio for various charge distributions.

For a point charge we found previously that

$$V(t) = -2q \sum_n k_n \cos \omega_n t.$$

If such a charge having initial energy qV_0 is just brought to rest by the retarding wake potential at t = 0, then $V_0 = q \sum_n k_n$ and

$$V(t) = - \frac{2V_0 \sum_n k_n \cos \omega_n t}{\sum_n k_n} \, . \tag{24}$$

If the structure supports only a single mode, then $V(t) = 2V_0 \cos \omega_n t$. However, a physical bunch, even a very short bunch consists of a large number of individual charges which are not rigidly connected. Thus the leading charge in such a physically real bunch will experience no deceleration, while the trailing charge will experience the full induced voltage, or twice the average retarding voltage per particle (assuming the bunch length is short compared to the wavelengths of all modes with appreciable values of k_n). The wake potential for a short charge distribution extending from t = 0 to t = T, interacting with a single mode, is illustrated in Fig. 8a. Within such a bunch the wakefield is constant ($\omega t \ll 1$) and the potential is given by

$$V(t) = -\frac{2V_0}{q} \int_0^t I(t') \, dt' , \qquad (25)$$

where V_0 is the average energy loss per particle in the distribution. This can be seen by substituting Eq. (25) in

$$\bar{V}(t) = \frac{1}{q} \int_0^T V(t) \, I(t) \, dt = V_0,$$

and working out the double integral. Note from Eq. (25) that for $t = T$ at the end of the distribution $V(T) = -2V_0$. Therefore $V_m^+ = 2V_0$, $V_m^- = |-2V_0| = V_m^+$ and the transformer ratio is $R = V_m^+/V_m^- = 1$.

The potential in and behind a long charge distribution is shown schematically in Fig. 8b. We consider first the case for a single mode. From Eq. (18b) with $W_z(t) = 2k_n \cos \omega_n t$,

$$V_n(t) = -2k_n \int_{-\infty}^t I(t') \cos \omega_n(t-t') \, dt' \qquad (26)$$

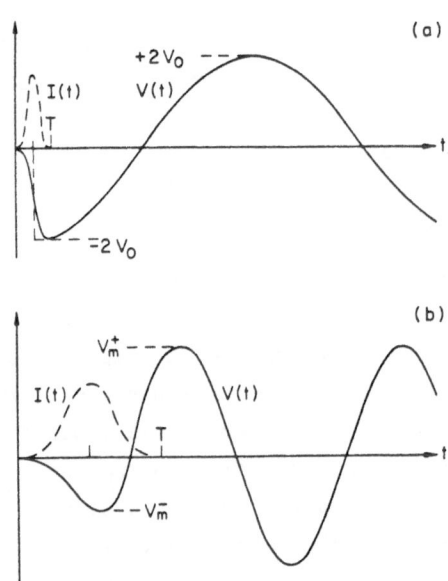

Fig. 8.   Potential in and behind a charge distribution interacting with a single mode for (a) a short bunch, and (b) a long bunch.

Assume now that the bunch extends in time from −T to +T. Within the bunch (−T < t < T) the retarding potential is

$$V_n^-(t) = -2k_n \left[\cos \omega_n t \int_{-T}^{t} I(t') \cos \omega_n t' \, dt' + \right.$$

$$\left. \sin \omega_n t \int_{-T}^{t} I(t') \sin \omega_n t' \, dt' \right]. \tag{27}$$

Following the bunch (t > T) the accelerating potential is

$$V_n^+(t) = 2k_n \left[\cos \omega_n t \int_{-T}^{T} I(t') \cos \omega_n t' \, dt' + \right.$$

$$\left. \sin \omega_n t \int_{-T}^{T} I(t') \sin \omega_n t' \, dt' \right]. \tag{28}$$

If the bunch is symmetric about t = 0, the second integral in Eq. (28) vanishes, and $V^+(t)$ reaches a maximum value given by

$$V_m^+ = 2k_n \int_{-T}^{T} I(t') \cos \omega_n t' \, dt' \quad . \tag{29}$$

The retarding potential at the center of such a symmetric bunch is given by Eq. (27) for t = 0,

$$V^-(0) = -2k_n \int_{-T}^{0} I(t') \cos \omega_n t' \, dt' = -\frac{1}{2} V_m^+ \; . \tag{30}$$

If $V^-(0)$ happens also to be the maximum (absolute) value of the retarding potential, then $|V^-(0)| = V_m^- = \frac{1}{2} V_m^+$ and the transformer ratio is R = 2. If $V^-(0)$ is not the peak of the retarding potential, then $V_m^- > |V^-(0)|$ and R < 2. Thus for symmetric bunches interacting with a single mode, the transformer ratio cannot exceed two. This upper limit is reached only if the maximum retarding potential is reached at the center of symmetry of the distribution. Otherwise, the transformer ratio is less than two. If the bunch is not symmetric, the preceding argument does not apply. The transformer ratio can then in principle be arbitrarily large, as we will see shortly.

The limitation R ≤ 2 tends to apply for symmetric bunches even in the case of a physical structure with many modes. For example, in Fig. 7, showing Gaussian bunches in the SLAC structure, the transformer ratios for σ/λ_o = 0.05, 0.20, and 0.40 are seen to be 1.4, 1.9 and 1.4 respectively. It is possible in principle to imagine a structure in which several modes cooperate to produce R > 2 for symmetric bunches (Bane, et al., 1985). However it is not probable that the limitation R = 2 can be exceeded by a significant amount in any physically realizable structure.

Asymmetric Driving Bunches

Let us now turn to the case of an asymmetric driving bunch. Take as an example a triangular current ramp in a single mode cavity with frequency ω. Let $I(t) = I\omega t$ for $0 < t < T$ and $I(t) = 0$ otherwise. For simplicity let the bunch length be $T = 2\pi N/\omega$, where N is an integer. Then within the bunch

$$V^-(t) = 2kI\omega \int_0^t t'\cos \omega(t - t')\, dt' = -\frac{2kI}{\omega}(1 - \cos \omega t) \quad , \qquad (31a)$$

whereas behind the bunch

$$V^+(t) = 2kI\omega \int_0^T t'\cos \omega(t - t')\, dt' = 2kIT \sin \omega t \quad . \qquad (31b)$$

Thus $V_m^- = 4kI/\omega$, $V_m^+ = 2kIT = 4\pi kIN/\omega$ and

$$\qquad\qquad\qquad\qquad\qquad\qquad\qquad\qquad\qquad\qquad\qquad\qquad (32)$$

$$R = \frac{V_m^+}{V_m^-} = \pi N \qquad \left\{ \begin{array}{l} \text{current ramp,} \\ \text{single mode} \end{array} \right\}$$

The wake potential for a current ramp of length N = 2 interacting with a single mode are shown in Fig. 9a.

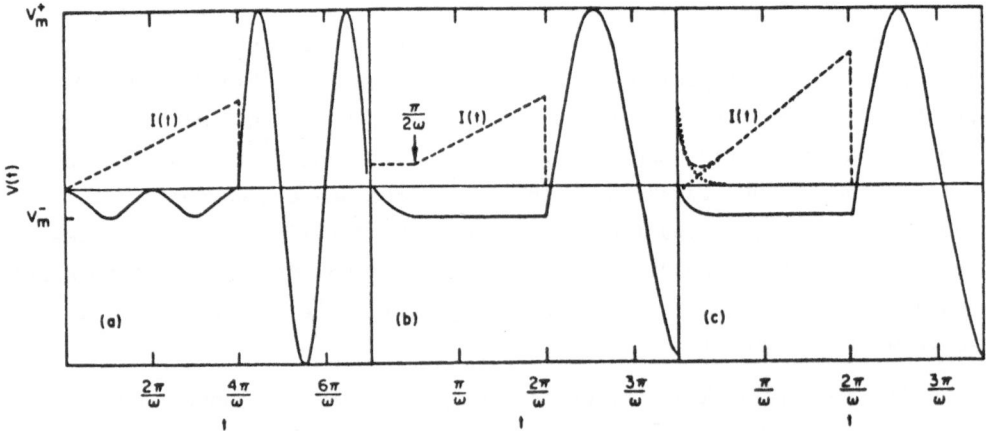

Fig. 9. The voltage induced by three different asymmetric current distributions interacting with a single mode.

In a real structure with many modes, one might expect that the transformer ratio will be less than that given by Eq. (32). The potential excited in the SLAC structure by a current ramp with N = 2 is shown in Fig. 10. Within the bunch the retarding potential has a behavior close to the single mode calculation, $V(t) \sim 1 - \cos \omega t$. However, some energy goes into higher modes, as is evident by ripples on the cosine wave behind the bunch. This causes a degradation of the transformer ratio to R = 4.86 from the single mode prediction R = 2π. The degradation worsens as the bunch gets longer, as can be seen in Fig. 11.

The efficiency for energy extraction from a driving bunch extending from t = 0 to t = T in which all of the electrons have the same energy $eV_0 = eV_m^-$ is

$$\eta = - \frac{1}{qV_m^-} \int_0^T I(t) \, V^-(t) \, dt \; .$$ (33)

For a linear current ramp interacting with a single mode, substitution of Eq. (31a) together with appropriate expressions for I(t), V_m^- and q into Eq. (33) gives an efficiency of 0.5 if $\omega T = 2\pi N$. A higher efficiency and a higher transformer ratio could be obtained if the retarding potential could be made as flat as possible across the current distribution. In the limit $V^-(t) = V_m^- = $ constant, Eq. (33) gives an efficiency of 100%. In Ref. 14 it is proven that the potential can be exactly flat only for a current distribution which consists of a delta function followed by a linear current ramp, where the proper relation exists between the value of the delta function and the slope of the current ramp. In this limit the transformer ratio is given by (Bane, et al., 1985)

$$R = \left[1 + (2\pi N)^2 \right]^{1/2} \qquad \left\{ \begin{array}{c} \text{Delta function plus} \\ \text{current ramp,} \\ \text{single mode} \end{array} \right\} \; .$$ (34)

Here $N = \omega T/2\pi = cT/\lambda$, and N can now take non-integer values. For large N the transformer ratio approaches $R \approx 2\pi N$ and the efficiency approaches 100%. The transformer ratio for the delta function alone ($N \to 0$) is R = 1, as we know is the case for all short bunches, and the efficiency is 0.5. An approximation to this distribution, in which the wake potential is driving negative by an exponentially decaying spike and then held constant by a rising current ramp, is illustrated in Fig. 9c.

A third distribution of interest is a linear current ramp preceded by a quarter wavelength rectangular pulse. The response to this distribution is shown in Fig. 9b. The transformer ratio in the case of this "doorstep" distribution is (Bane, et al., 1985)

$$R = \left[1 + \left(1 - \frac{\pi}{2} + 2\pi N\right)^2\right]^{1/2} \qquad \left.\begin{array}{l} \text{Doorstep plus} \\ \text{current ramp,} \\ \text{single mode} \end{array}\right\} \qquad (35)$$

In the limit of large N, the transformer ratio again approaches $R \approx 2\pi N$. For long bunches the transformer ratio and the efficiency are again approximately twice that for the linear current ramp alone. Except for particles in the first quarter wavelength of the bunch, all particles experience the same retarding potential. At the end of the doorstep (N = 1/4), $R = \sqrt{2}$ and $\eta = 2/\pi$.

As a numerical example, consider an accelerator operating at $\lambda = 1$ cm with a desired gradient of 200 mV/m. A SLAC-type structure at this wavelength would have a loss parameter on the order 2×10^{15} V/C-m. With a transformer ratio of 20, driving bunches with an energy of 100 MeV

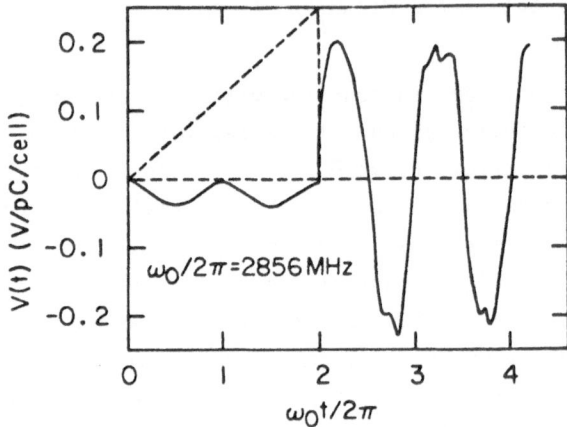

Fig. 10. The potential induced by a linear current ramp interacting with the modes in the SLAC structure.

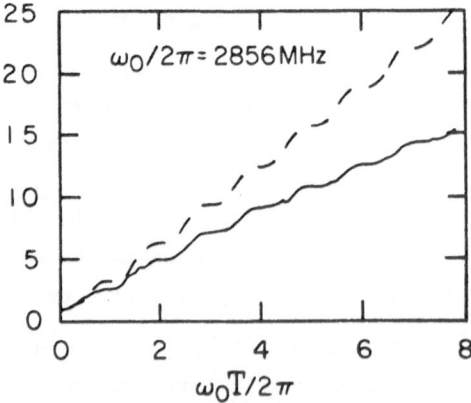

Fig. 11. The transformer ratio for a linear current ramp in the SLAC structure as a function of bunch length. The dashed line gives single mode results.

150

would need to be injected every ten meters. The charge per bunch as given by Eq. (23b) is

$$q = \frac{E_a R}{4k\eta} \approx 0.5 \ \mu C \ ,$$

assuming that most of the energy goes into a single mode and that the efficiency is close to 100%. The bunch length is approximately $R\lambda/2\pi = 3.2$ cm or 100 ps, and the peak current at the end of the bunch is 10 kA. Many practical questions must be addressed, such as feasibility of creating properly shaped bunches with very high peak currents. The deflecting fields induced if the driving bunch wanders off the axis of the structure are also a serious problem.

The Plasma Wakefield Accelerator

The plasma wakefield accelerator is another type of collinear acceleration scheme in which the metallic rf structure is replaced by a plasma medium. If the plasma is cold, one expects that only a single mode, the oscillation at the plasma frequency, will be excited. Thus a plasma is, in essence, a single mode structure in which the axis of symmetry is defined by the driving beam.

Figure 12 shows the result of a simulation (Chen, et al., 1985) in which a triangular bunch one wavelength long is injected into a plasma. The transformer ratio as measured from the figure is $R \approx \pi$, in agreement with the theoretical prediction for a triangular bunch given by Eq. (32).

RING BEAMS

Transformer Ratio for a Ring Beam

Consider first a circular bunch of radius $\underline{a}$ passing between two parallel metallic planes spaced apart by a distance $g \approx a$. The energy deposited between the two plates, initially contained in a volume $V_1 \approx \pi a^2 g$, will be distributed at some later time t over a spreading ring-shaped region of radius b = ct, thickness $\approx 2a$ and volume $V_2 \approx 4\pi abg$. Thus the ratio of field strengths will be

$$R_1 \sim \left(\frac{V_1}{V_2}\right)^{1/2} = \left(\frac{a}{4b}\right)^{1/2} \ .$$

This also gives the approximate transformer ratio at transverse distance b from a single bunch passing between the two plates. The transformer ratio for a ring-shaped beam of radium b can now be obtained by considering that the ring is made up of $n = A_2/A_1 = 4b/a$ beamlets, giving for the net transformer ratio $R = n \ R_1 = 2(b/a)^{1/2}$.

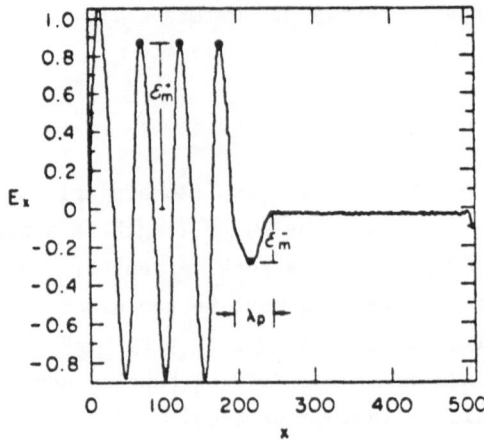

Fig. 12.   Wakefield of a triangular bunch in a plasma
(Chen, et al., 1985).

If now an outer cylindrical wall is added, as shown in Fig. 13, the
energy that would propagate outward away from the axis is directed toward
the axis. This increase R by a factor of $\sqrt{2}$, giving

$$R = 2 \left(\frac{2b}{a}\right)^{1/2} . \tag{36}$$

This is in agreement with the transformer ratio given by Eq. (3) for the
switched radial transmission line with the assumption $a \approx g$.

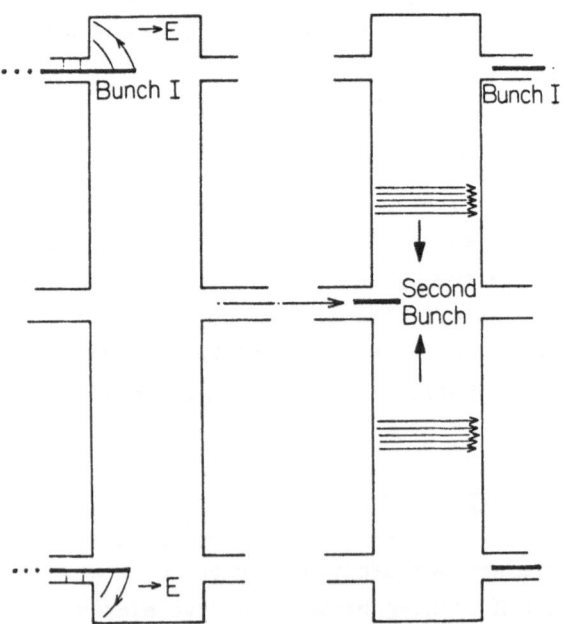

Fig. 13.   Qualitative picture of the field induced by a ring
bunch passing through a pillbox cavity.

The Wakefield Transformer

The wakefield transformer, illustrated in Fig. 2, consists essentially of a series of pillbox cavities, of the type shown in Fig. 13,
having a ring gap near the outer radius for the driving beam and a hole
on the axis for the accelerated beam. A wakefield transformer of this
type was originally proposed by Voss and Weiland (1982). In Ref. 4 the
transformer ratio is computed for a wakefield transformer of this kind
with a = 2 mm, b = 26 mm and g = 1.5 mm, using a particle tracking code.
The peak energy gain on the axis is calculated to be 20 times the average
energy loss in the ring driving bunch. However, the energy gain on the
axis divided by the <u>peak</u> energy loss in the driving bunch (our definition
of transformer ratio) is about 12. Putting the transformer dimensions
into Eq. (36) we obtain R = 10, in substantial agreement with the results
from the tracking code.

An experiment is underway at DESY (Deutsches Elektronen Synchrotron
in Hamburg) to test this type of coaxial wakefield transformer principle
as a means to obtain high accelerating gradients for future linear
colliders. The eventual goal is a collider in the 1 TeV energy range
operating at a gradient on the order of 200 MV/m. In the prototype
experiment presently in progress, a ring-shaped bunch with a charge of
1 μC will be injected at 8 MeV into a wakefield transformer to accelerate
a second bunch from 8 MeV to about 80 MeV at a gradient of 100 MV/m or
greater. Details of the experiment, and the current status, are given
elsewhere (Weiland and Willke (1982), Weiland (1985)).

PROTON WAKEFIELD ACCELERATION AND TEST FACILITIES

In the last section it was noted that the transformer ratio $\bar{R}$,
defined as the (unloaded) energy gain in the accelerated bunch divided by
the <u>average</u> energy loss in the driving bunch, is typically larger than
the transformer ratio R based on the <u>peak</u> energy floss in the driving
bunch. For relativistic electrons the ratio R would normally apply,
since a driving bunch of electrons injected with uniform energy would
degrade rapidly once electrons in the region of peak decelerating fields
have been brought to rest. For non-relativistic particles (e.g., few
hundred MeV protons) the situation is different. The particles can move
back and forth within the bunch, with the lead particles and trailing
particles continuously changing places. By this process of "mixing", it
is possible for all of the particles to experience the average
decelerating gradient. This is the basis for the proton wakefield
accelerator (WAKEATRON) proposed by A. Ruggiero (1986).

For a Gaussian bunch interacting with a single mode with loss parameter k_o, it is seen from Eqs. (19b) and (20b) that the accelerating gradient behind the bunch is

$$E_a = 2 \, q \, k_o e^{-\omega^2 \sigma_t^2 / 2} \tag{37a}$$

The average loss, on the other hand, is seen to be from Eqs. (19c) and (20a)

$$\overline{E^-} = u/q = q \, k_o e^{-\omega^2 \sigma_t^2} \tag{37b}$$

Therefore

$$\overline{R} \equiv \frac{E_a}{\overline{E^-}} = 2 \, e^{\omega^2 \sigma_t^2 / 2} \; . \tag{38}$$

For a Gaussian or any symmetric bunch it was shown that the transformer ratio $R \equiv E_a / E_m^- \leq 2$, while from Eq. (38) we see that the transformer ratio $\overline{R}$, based on the average energy loss in the driving bunch, can in principle increase without limit as the bunch length increases. On the other hand, the charge required in the driving bunch for a given accelerating gradient must also increase as $\overline{R}$ increases. From Eq. (37a)

$$q = \frac{E_a}{2k_o} e^{\omega^2 \sigma_t^2 / 2} = \frac{E_a \, \overline{R}}{4k_o} \; . \tag{39}$$

A facility (the Advanced Accelerator Test Facility (Simpson et al., 1985)) has just been completed at Argonne National Laboratory to test some of the wakefield acceleration techniques that have been described here. A 22 MeV electron linac can produce driving bunches of $1 - 2 \times 10^{11}$ electrons with pulse lengths of 5 - 150 ps and an emittance of 7 π-mm-mr. A second bunch (witness beam) can be injected to probe the longitudinal and transverse wake potentials in the range 0 - 2.4 ns behind the driving bunch. Initial experiments to test the WAKETRON and plasma wakefield accelerator techniques are currently in the planning stage, and tests of other new acceleration concepts have been proposed for the future (see article in Ref. 3).

REFERENCES

Bane, K., CERN/ISR-TH/8-47, 1980.
Bane, K. and Wilson, P. B., <u>Proceedings of the 11th International Conference on High-Energy Accelerators</u>, CERN, July 1980 (Birkhauser Verlag, Basel, 1980), p. 592.

Bane, K. L. F., Wilson, P. B., and Weiland, T., in Physics of High Energy Particle Accelerators, BNL/SUNY Summer School, 1983; M. Month,P. Dahl and M. Dienes, eds., AIP Conference Proceedings No. 127. (American Institute of Physics, New York, 1985), pp. 875-928. Also available as SLAC-PUB-3528, December 1984).

Bane, K. L. F., Chen, Pisin and Wilson, P. B., IEEE Trans. Nucl. Sci. NS-32, No. 5, 3524 (1985). Also available as SLAC-PUB-3662, April 1985.

Chen, P., Su, J. J., Dawson, J. M., Wilson, P. B. and Bane, K. L., UCLA Report No. PPG-851, 1985.

Chen, Pisin, to be published in the Proceedings of the 1986 Linear Accelerator Conference, Stanford, California, June 2-6, 1986. Also available as SLAC-PUB-3970, May 1986.

Cooper, Richard K., "Wake Fields: Limitations and Possibilities". These Proceedings.

Joshi, C. and Kastsouleas, T., eds. Laser Acceleration of Particles, AIP Conference Proceedings No. 130 (American Institute of Physics, New York, 1985).

Panofsky, W. K. H. and Wenzel, W. A., Rev. Sci. Inst. 27, 967 (1956).

Ruggiero, A. G., Argonne National Laboratory Report ANL-HEP-CP-86-51 (1986).

See articles by the Wakefield Accelerator Study Group and by T. Weiland and F. Willeke in Proceedings of the 12th International Conference on High Energy Accelerators, (Fermi National Accelerator Laboratory, Batavia, Illinois, August 1983), pp. 454-459.

Simpson, J., et al., IEEE Trans. Nucl. Sci. NS-32, 3492 (1985).

Stumer, I., private communication.

Symposium on Advanced Accelerator Concepts, Madison, Wisconsin, August 21-27, 1986. To be published by American Institute of Physics, AIP Conference Proceedings.

Voss, G. and Weiland, T., DESY Report 82-074 (1982).

Villa, F., SLAC-PUB-3875, Stanford Linear Accelerator Center, Stanford, California.

Weiland, T., IEEE Trans. Nucl. Sci. NS-32, 3471 (1985).

Wilson, P.B. in Proceedings of the Thirteenth SLAC Summer Institute on Particle Physics, SLAC Report No. 296 (Stanford Linear Accelerator Center, Stanford, California), pp. 273-295. Also available as SLAC-PUB-3891, February 1986.

Wilson, P. B. in Physics of High Energy Particle Accelerators, Fermilab Summer School, 1981; R. A. Carrigan, F. R. Huson and M. Month, eds., AIP Conference Proceedings No. 87 (American Institute of Physics, New York, 1982), Sec. 6.1. Also available as SLAC-PUB-2884, February 1982.

WAKE FIELDS: LIMITATIONS AND POSSIBILITIES

Richard K. Cooper

Los Alamos National Laboratory
Los Alamos, NM 87545 USA

INTRODUCTION

If we consider the electric and magnetic fields surrounding a relativistic point charge, we see that they are nearly transverse to the direction of motion, having an angular spread of the order of $1/\gamma$, as shown in Fig. 1. A conducting obstacle to one side of the charge a distance b, say, will intercept a portion of these fields and scatter them so that a second charge, following the first on the same path a distance s behind, will experience the scattered fields after the time it takes the speed of light to go from the scatterer to the axis of travel.

Exercise for the student. Show that this time is $(b^2 + s^2)/2sc$. For $ct \gg b$ show that the distance s behind the first point charge at which the effects of the scattered fields will first be felt is $s = b^2/2ct$.

These scattered fields can result in both longitudinal (in the direction of motion) and transverse forces on the second charge. These scattered fields are known as "wake fields" and are clearly proportional in intensity to the magnitude of the incident (driving or exciting) charge. The longitudinal and transverse forces on charge following the driving charge are functions of position and time; we will find it useful to define the integrals of these forces on a given (test) particle as "wake potentials". In the case of particles moving in periodic structures, we will define (in more detail in the next section) the longitudinal wake potential as the integral of the longitudinal electric field per unit exciting charge **following the test charge** over one period of the structure, with a similar definition for the transverse force (produced by both electric and magnetic fields). For bunches passing obstacles such as cavities and beam tube discontinuities, the integrals are taken over the entire length

157

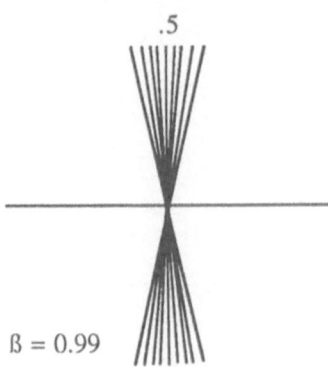

.5

β = 0.99

Fig. 1. Electric field lines of a point charge with β = v/c = 0.99.

of the interaction. In this paper I will point out how these wake fields can be troublesome, and also how we might use them to advantage.

WAKE POTENTIALS

The literature on wake fields has burgeoned in recent years, although be strict definition one could say that the subject is practically as old as electromagnetic scattering theory itself. A good selection of references can be found in a recent paper by P. B. Wilson (1986). In most accelerator applications, i.e., those for which the transverse motion of the charges can be neglected over the period in which wake field forces are operating, it is the integrated effect of these forces that matters, for it is these integrated forces that determine transverse momentum changes and longitudinal energy spread.

Thus we define the longitudinal wake potential per unit exciting charge (q) at a longitudinal distance behind the exciting charge as the integral

$$W_z(r',r,s) = -\frac{1}{q} \int dz \; [E_z(r,z,t)]_{t=z'+s)/c'} \tag{1}$$

where r is the transverse location at which the integrated wake field is desired (i.e., r is a vector in the plane perpendicular to the motion) and r' is the transverse location of the driving charge, which has longitudinal position z'(t). The units of W_z are clearly volts per coulomb. Similarly we define the transverse wake potential (a vector)

$$W_\perp(r',r,s) = \frac{1}{q} \int dz \; [E_\perp(r,z,t) + (v \times B(r,z,t)_\perp)]_{t=(z'+s)/c'} \tag{2}$$

Analytic solutions for wake potentials are very few in number. K. Bane (1984) has used the expressions given by Chao and Morton (1975) for the time-dependent electric fields of a point charge passing between two

158

parallel conducting plates to obtain the wake-potential expression

$$2\pi\varepsilon_o W_z(o,o,s) = 2\delta(s)\ln\frac{g}{s} - 2\sum_{n=1}^{\infty} \delta(2ng - s)\ln\left(\frac{s^2}{(s+g)(s-g)}\right)$$

$$-\frac{1}{2}\left[\frac{1}{[s/2g]_{IP} + s/2g} - \frac{1}{[s/2g]_{IP} + s/2g + 1}\right]. \tag{3}$$

In this expression g is the distance between the plates and IP means the integer part of the term within the brackets. This function is shown in Fig. 2. As with so many Green's function (delta-function) expressions, this one is full of divergences which must be integrated over for realistic distributions. Note that the integrated wake field is accelerating ($W_z < 0$) for all $z > 0$.

Another analytic solution which has been obtained is that for a bellows; that is, a periodically varying cylindrically symmetric wall (beam tube) surrounding the flight path (Cooper, Krinsky, and Morton, 1982). If the radius of the wall is given by the (periodic, with period L) function

$$a(z) = a_o\left[1 + \sum_{p=-\infty}^{\infty} \tilde{C}_p e^{j2\pi pz/L}\right] \qquad p \neq 0, \tag{4}$$

the a perturbation solution of Maxwell's equations allows one to calculate the longitudinal wake potential on axis as

$$W_z(o,o,s) = \frac{8\pi}{cL}\sum_{p=1}^{\infty} p\left|\tilde{C}_p\right|^2 \sum_{s=1}^{\infty} \omega_{os} e^{j\omega_{os}\tau}, \tag{5}$$

where x_{os} is the sth root of $J_o(x) = 0$ and $v(\approx o)$ is the velocity of the The angular frequency ω_{os} is given by, for each p value,

$$\omega_{os} = \frac{\pi p}{L} + \frac{Lx_{os}^2}{4\pi pa_o^2}. \tag{6}$$

As in the case of the parallel-plate wake function, this function has a divergent behavior. When integrated over a Gaussian distribution as the exciting distribution, however,

$$\rho(z) = \frac{Ne}{\sqrt{2\pi}\sigma}e^{-z^2/2\sigma^2}, \tag{7}$$

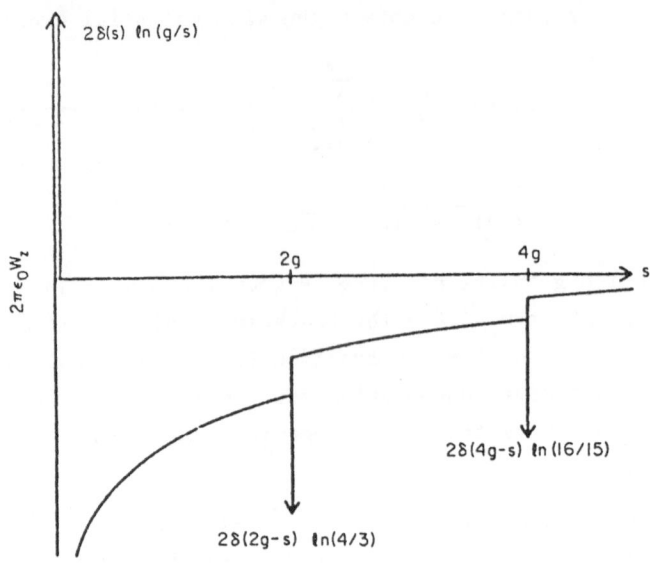

Fig. 2.　Longitudinal wake potential of a point charge passing between two parallel conducting plates.

we obtain for the wake potential

$$W_z(0,0,s) = \frac{2\pi}{cL} \sum_{p=1}^{\infty} p \left|\tilde{C}_p\right|^2 \sum_{s=1}^{\infty} \omega_{0s} \mathrm{Re} \left[\frac{1}{2} e^{-s^2/2\sigma^2} \; w \left(\frac{\omega_{0s}\sigma}{\sqrt{2}v} - j \frac{s}{\sqrt{2}\sigma} \right) \right], \quad (8)$$

where w is the complex error function (Abramowitz and Stegun)

$$w(z) = e^{-z^2} \left(1 + \frac{2j}{\sqrt{\pi}} \int_0^z e^{t^2} \, dt \right) = e^{-z^2} \mathrm{erfc}(-jz). \quad (9)$$

Figure 3 shows plots of the average force per period for a beam pipe of radius 5 cm with a sinusoidal modulation of period 5 cm for two different bunch lengths. Figure 3a shows the results for a bunch σ of 5 mm, while Fig. 3b shows results for a bunch σ of $5\sqrt{10}$ mm. The first (negative) peak occurs within the bunch itself and is a decelerating force. The following peaks represent the force on a test particle following the bunch. Whether these peaks represent an accelerating or decelerating force obviously depends on the sign of the test charge.

If we consider only cylindrically symmetric cavities or periodic structures, the fields can be decomposed into a Fourier series in the azimuthal variable ϕ, as can then also the wake potentials. In this case it can be shown that if the driving charge moves at radius r' at $\phi' = 0$, then the mth term in the series for the longitudinal wake potential is

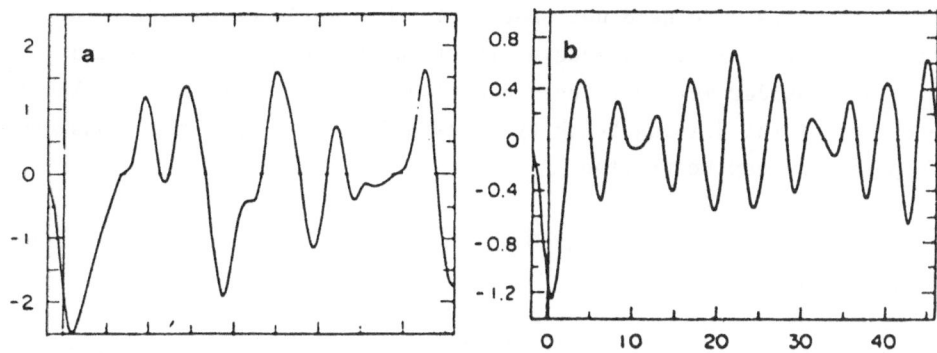

Fig. 3. Plots of the wake force for L = 5 cm, b = 5 cm and
a) σ = 5 mm   b) σ = 18.8 mm.

given by (Bane, Wilson, and Weiland, 1983)

$$W_{zm} = 2 \left(\frac{r'}{a}\right)^m \left(\frac{r}{a}\right)^m \cos m\phi \sum_{n=1}^{\infty} k_{mn}^{(a)} \cos\omega_{mn}s/c, \qquad (10)$$

$$W_{\perp m} = 2m \left(\frac{r'}{a}\right)^m \left(\frac{r}{a}\right)^{m-1} (\hat{r} \cos m\phi - \hat{\phi} \sin m\phi) \sum_{n=1}^{\infty} \frac{k_{mn}^{(a)}}{\omega_{mn}a/c} \sin \omega_{mn}s/c. \quad (11)$$

In these expressions the quantity $k_{mn}^{(a)}$ is the "loss factor" calculated at r = a, where a is the radius of the beam tube region, i.e., there is no conducting material inside r = a. This loss factor is given by

$$k_{mn}^{(a)} = \frac{\left| \int E_{zmn}(r = a,z)e^{j\omega_{mn}z/c} dz \right|^2}{4U_{mn}}, \qquad (12)$$

where U_{mn} is the energy stored in the mn mode. This quantity $k_{mn}^{(a)}$ is a function only of the geometry. Note that the transverse dipole wake potential is not a function of r, i.e., the deflecting field is uniform.

If we consider the effect of a Gaussian bunch such as in Eq. (7), we find that the loss factor $k_{mn}^{(a)}$ in Eq. (12) must be multiplied by the factor $\exp(-\omega_{mn}^2 \sigma^2/c^2)$. Thus for long enough bunches it may be sufficient to consider only a single mode in evaluating the long-range wake effects. This single transverse mode effect is considered in the next section.

LONG-RANGE WAKE EFFECT: BEAM BREAKUP

We want to consider here a long train of bunches interacting with a series of cavities. We are concerned with the long-range dipole wake (m = 1) which can deflect bunches, possibly to the point where they

161

strike the walls. We need not specify which $m = 1$ mode we consider; we will simply refer to it as mode n with frequency ω_{1n} and loss factor $k_{1n}^{(a)}$. We consider motion in only one plane (the x - z plane) and label the offset of the driving bunch by ξ instead of r'. Thus the transverse wake potential can be written

$$W_\perp = 2\frac{\xi c}{\omega_{1n}a^2} k_{1n}^{(a)} \sin(\omega_{1n}s/c) \, \hat{x}, \tag{13}$$

where $\hat{x}$ is the unit vector in the x direction. A bunch following a distance $c\tau$ behind the first bunch will have its transverse momentum changed by an amount, if both bunches contain $N_p e$ coulombs,

$$\Delta p_{x1} = \frac{2N_p e^2 \xi}{\omega_{1n}a^2} k_{1n}^{(a)} \sin \omega_{1n}\tau. \tag{14}$$

A third bunch following a distance τ behind the second bunch will be deflected by the wake fields of the two preceding bunches. Applying superposition, labeling the displacements of the first two bunches by ξ_0 and ξ_1, respectively, and allowing the cavity mode to have a finite Q gives us the following expression for the momentum change of the third bunch:

$$\Delta p_{x2} = \frac{2N_p e^2 k_{1n}^{(a)}}{\omega_{1n}a^2} \left[\xi_0 e^{-2\omega_{1n}\tau/2Q} \sin 2\omega_{1n}\tau + \xi_1 e^{-\omega_{1n}\tau/2Q} \sin \omega_{1n}\tau \right]. \tag{15}$$

The extension to bunch number M is obvious. So far we have considered only the first cavity. If the cavities are separated by a distance L, then in the absence of focusing, the parameter ξ for the third bunch, say, when that bunch gets to the next cavity, will be

$$\xi_{2\text{second cavity}} = \xi_{2\text{first cavity}}$$

$$+ L(p_{x2\text{first cavity}} + \Delta p_{x2})/p_{z\text{after first cavity}}, \tag{16}$$

and in general, the displacement of bunch number M at cavity number $N + 1$ can be written in terms of the values at cavity N:

$$\xi_M^{N+1} = \xi_M^N + L\frac{\left(p_{xM}^N + \Delta p_{xM}\right)}{p_z^N}, \tag{17}$$

where p_z^N is the longitudinal momentum of the particles in the bunch (assumed to be the same for all bunches) **after** the interaction with cavity number N. Physically what can happen is that the first cavity produces a modulation on the beam which then drives the second cavity at its resonant frequency. This cavity further modulates the beam and so on, leading to exponential growth with distance in the beam displacement.

The last equation, coupled with Eq. (15), has been solved by Gluckstern, Cooper, and Channell (1985) for a variety of initial conditions and including the effect of focusing. Figure 4 shows the displacement ξ vs. bunch number M after a beam has interacted with cavities, when all bunches were initially offset by 1 mm. There are three regions to observe on this plot: the exponential growth region, the maximum displacement region, and the steady-state region. The first two regions can be avoided by turning the beam on slowly; the steady-state region is characterized by a growth factor

$$\left| \frac{\xi^N}{\xi_0} \right| = \exp \left[3^{3/4} \left(\frac{z I_0 k_{1n}^{(a)} Q}{c L V' a^2} \right)^{1/2} \right], \tag{18}$$

where V' is the voltage gain per unit length (assumed constant in z), z is the distance from the first cavity, and I_0 is the average beam current. Thus for high-current accelerators care must be taken in the design of cavities to minimize the loss factor $k_{1n}^{(a)}$ and/or to minimize the Q of the deflecting mode.

SHORT-RANGE EFFECT: BUNCH BLOWUP

The previous section focused on the long-range wake effects that produce bunch-to-bunch coupling. In that analysis the bunches were treated in effect as point charges. If the amount of charge in a single bunch is high enough, and/or the interaction long enough, the tail of the bunch can be affected by the wake fields produced by the head of the bunch. This effect has been studied in connection with the beam dynamics of the Stanford Linear Collider (Bane, 1985).

The longitudinal and transverse point charge wake potentials of the SLAC accelerating structure have been calculated and are shown in Fig. 5; the wake potentials are given by the solid lines. In operation for the Stanford Linear Collider, each accelerated electron bunch will contain 5×10^{10} electrons, and the effect of the longitudinal wake potential would be to induce an energy difference from the head of the bunch to the tail of 2 GeV (out of 50 GeV total). By placing the bunch ahead of the crest of the rf wave the energy gain of the head of the bunch can be reduced slightly so as to correspond more closely to the total gain of the middle of the bunch, which suffers an energy loss due to the longitudinal wake fields. By this means the final energy spread in the bunch can be reduced by a factor of 3 over what it would be if the bunch center were to ride at the peak of the rf wave.

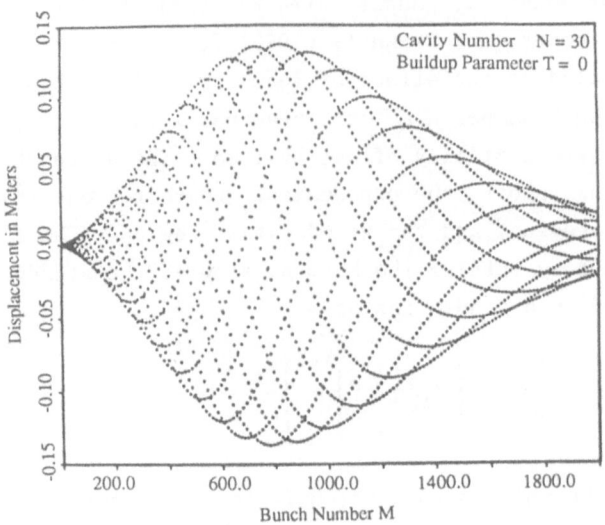

Fig. 4. Displacement of individual bunches (labeled by bunch number M) after passing through 30 cavities.

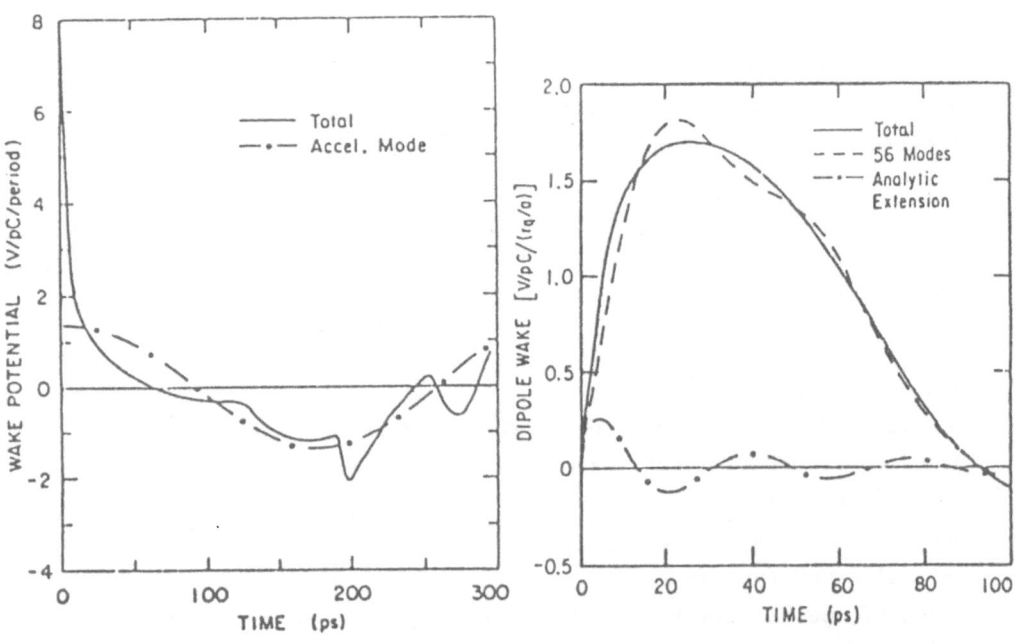

Fig. 5. Longitudinal and transverse point-charge wake potentials for the SLAC disk-loaded accelerating structure.

Over the approximately 13 picosecond (ps) length of the bunch in the SLAC linac the transverse (dipole) wake varies nearly with distance from the driving charge. If one were to assume that the bunch density were uniform from head to tail, a straight-forward (integro-differential) equation could be written for the transverse motion of any longitudinal slice of the bunch. Such an analysis has been carried out by Chao, Richter, and Yao (1980) for a bunch which is assumed to have no energy spread. A simpler analysis which gives insight into the consequences of the transverse wakefield and at the same time gives order-of-magnitude estimates of the effect of energy spread is the two-particle model described by Wilson (1982) and Chao (1983). In this model the bunch is considered to consist of just two equal point charges of charge $Q/2$ separated by a longitudinal distance z. The two particles represent the "head" and the "tail" of the bunch, and have (assumed constant) energies E and $E + \Delta E$, respectively. The "head" particle feels no transverse wake force and thus simply performs the normal transverse oscillations satisfying the equation

$$x_1'' + k^2 x_1 = 0, \tag{19}$$

where $k = 2\pi/\lambda$ is the (also assumed constant) wave number for the transverse oscillations due to the transverse focusing forces.

An individual electron in the "tail" particle not only is focused with wave number $k + \Delta k$ corresponding to its different energy, but it experiences the additional force due to the transverse wake $F_x = eQW_x(z)x_1/2$. Thus the "tail" particle satisfies the differential equation for transverse motion

$$x_2'' + (k + \Delta k)^2 x_2 = Cx_1, \tag{20}$$

where $C = eQW_x(z)/2(E + \Delta E)$, and the independent variable is distance s down the machine.

Exercise for the student. Show that if $\Delta k = 0$, i.e., the two particles have the same energy and hence the same number of transverse oscillations per unit length, then if both particles have the same initial conditions, then

$$x_2 - x_1 = \mathrm{Re}\left[\frac{iCs\hat{x}}{2k}\, e^{iks}\right], \tag{21}$$

where $\hat{x}$ is a complex constant determined by the initial conditions. Thus the first particle simply oscillates transversely, while the second particle oscillates with ever growing amplitude.

Under conditions in which the second particle and the first oscillate with the same wavenumber, the amplitude of the transverse oscillation grows without bound. In a real bunch this would mean that the tail of the bunch would, if it propagated the entire length of the accelerator, be displaced transversely with respect to the head of the bunch, thus lowering the effectiveness of the bunch as it collides head on with another bunch (also probably distorted head to tail).

Exercise for the student. Show that if $\Delta k \neq 0$, but for $|\Delta k / k| \ll 1$,

$$x_2 - x_1 = \mathrm{Re}\left[\hat{x}\left(1 - \frac{C}{2k\Delta k}\right)2i\,\sin(\Delta ks/2)e^{i(k+\Delta k/2)s}\right]. \qquad (22)$$

For Δk not equal to zero we have two possibilities for minimizing the influence of the transverse wake field, namely to choose ΔE, and hence Δk, so that the quantity $1 - C/2k\Delta k$ vanishes, and the second particle simply tracks the first through the linac. The second possibility is to choose Δk to make $\Delta ks/2$ be some multiple of π when s is equal to the length of the accelerator. For real bunches with charge distributed throughout the bunch, these simple considerations clearly will not be sufficiently quantitative. Bane (1985) has used a simulation code to study the effects of a deliberate energy spread in the bunch and finds that the beam behavior can be greatly improved thereby, so that the alignment tolerances of the machine can be relaxed by an order of magnitude. The cost of this deliberately induced spread in energy is a somewhat lower final energy of the bunch.

A WAKE-FIELD ACCELERATOR EXPERIMENT

Having seen what havoc can be wrought by wake fields, the question arises, can they do any good? We have seen in the longitudinal wake function of the bellows, for instance, that at certain positions behind the driving bunch the wake field is an accelerating field. One could hope to find a structure that would maximize the accelerating field. One could hope to find a structure that would maximize the accelerating field relative to the decelerating field experienced by the driving charge. Then one could use a high-intensity, low energy beam to create high accelerating fields to accelerate a lower intensity beam. Figure 6 shows a schematic longitudinal potential for this application. Analyses indicate that for bunches with symmetric charge distributions, the **transformer ratio** $R = V_m^+/V_m^-$ is of the order of 2 for bunches which follow the same path through the structure (Wilson, 1986). Recent work by Bane, Chen, and Wilson (1985), indicates that by shaping the charge distribution in the bunch the transformer ratio can be significantly increased.

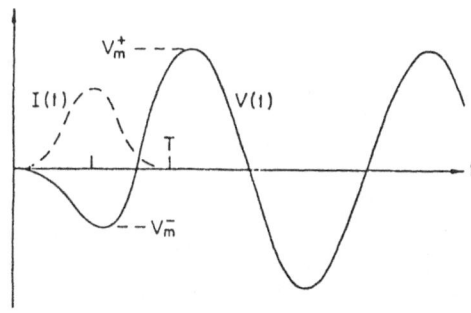

Fig. 6. Wake potential illustrating decelerating potential V_m^-
and accelerating potential V_m^+.

Another approach to improving the transformer ration involves decelerating an intense bunch in one region of the structure and using the
fields in another region. For example, if a beam were decelerated near
the periphery of a pillbox cavity and the resulting fields on the axis
were used for acceleration, then an analysis based on the field generated
in a closed cavity indicates that the transformer ratio for such a device
would be given by $2/J_0(2.405r_0/b)$ where b is the radius of the cavity and
r_0 is the radius at which the driving beam is decelerated (Wilson, 1986).

At DESY (the Deutsches Elektronen Synchrotron) Voss and Weiland (The
Wake Field Study Group, 1983) have proposed an experiment to provide high
accelerating gradients through wake fields, generated by decelerating an
intense ring of charge near the outer radius of s structure that then
acts as a series of radial transmission lines conducting the fields to
the center $r = 0$ where these fields can be used for acceleration.
Figure 7 shows the concept. The transformer ratio for the experiment is
calculated to be about 8. The experiment is well under way, details are
provided in (Weiland, 1985).

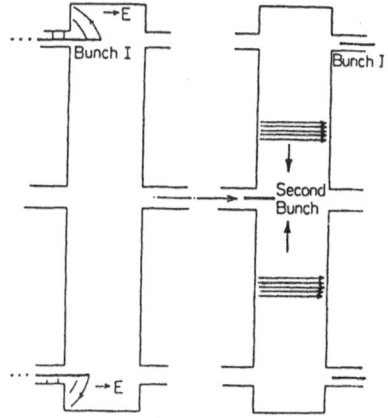

Fig. 7. The wake field transformer experiment concept.

REFERENCES

Abramowitz, M. and Stegun, I. A., eds., "Handbook of Mathematical Functions", NBS AMS-55, p. 297.

Bane, K. L. F., Private Communication, 1984.

Bane, K. L. F., Landau Damping in the SLAC Linac, IEEE Trans. Nucl. Sci. NS-32, 1985, pp. 2389-2891.

Bane, K. L. F., On Collinear Wake Field Acceleration, IEEE Trans. Nucl. Sci. NS-32, 1985, pp. 3524-3526.

Bane, K. L. F., Wilson, P. B., and Weiland, T., "Physics of High Energy Particle Accelerators", BNL/SUNY Summer School, 1983, M. Month, P. Dahl, and M. Dienes, eds., AIP Conference Proceedings No. 127 (American Institute of Physics, New York, 1985), pp. 875-928.

Chao, A. and Morton, P. L., PEP-Note-105, SLAC, 1975.

Chao, A., Coherent Instabilities of a Relativistic Bunched Beam, Physics of High Energy Particle Accelerators, AIP Conf. Proc. No. 105, (Am. Inst. of Physics, New York, 1983).

Cooper, R. K., Krinsky, S., and Morton, P. L., Transverse Wake Force in Periodically Varying Waveguide, Particle Accelerators 12, 1979, pp. 1-12.

Gluckstern, R. L., Cooper, R. K., and Channell, P. J., Cumulative Beam Breakup in RF Linacs, Particle Accelerators 16, 1985, pp. 125-153.

Wake Field Study Group, The, A Wake Field Transformer Experiment, Proc. 12th International Conference on High Energy Accelerators, Fermilab, 1983, pp. 454-456.

Weiland, T., Wake Field Work at DESY, IEEE Trans. on Nucl. Sci. NS-32, 1985, pp. 3471-3475.

Wilson, P. B., Physics of High Energy Accelerators, AIP Conference Proceedings No. 87 (Am. Inst. of Physics, New York, 1982).

Wilson, P. B., Wake Field Accelerators, SLAC PUB-3891, February, 1986.

HIGH-BRIGHTNESS RF LINEAR ACCELERATORS*

Robert A. Jameson

Accelerator Technology Division, MS H811
Los Alamos National Laboratory
Los Alamos, New Mexico 87545 USA

INTRODUCTION

Soon after electrons and ions were discovered, production of practical generators of particle beams began, and a succession of machines were invented that could produce more energetic and more intense beams. Progress on the energy frontier is often charted from the 1930s in the form of the Livingston Chart, Fig. 1, showing that particle accelerator energy has increased by a factor of about 25 every 10 years. The corresponding cost per million electron volts has decreased by about a factor of 16 per decade (Lawson, 1982). The physics principles on which all of these devices work were deduced long ago; the energy increase were possible because of cost reductions from thorough exploitation of parameters, engineering perfection, systems integration, and advanced manufacturing methods (Voss, 1982).

At the same time, the development of more intense sources proceeded. Linear accelerators (linacs) are suited to intense sources because the beam can easily exit the machine.

Although technology to increase energy and intensity tended to be pursued separately in the past, recent applications have had to consider both, along with the ability to keep the beam very precisely confined, aimed, or focused. The figure of merit used is called brightness, defined (variously) as the beam power (sometimes only the beam current divided by the phase space appropriate to the problem at hand. Phase space for the beam as a whole is six-dimensional, describing the physical

---

*Work supported by the U.S. Department of Energy.

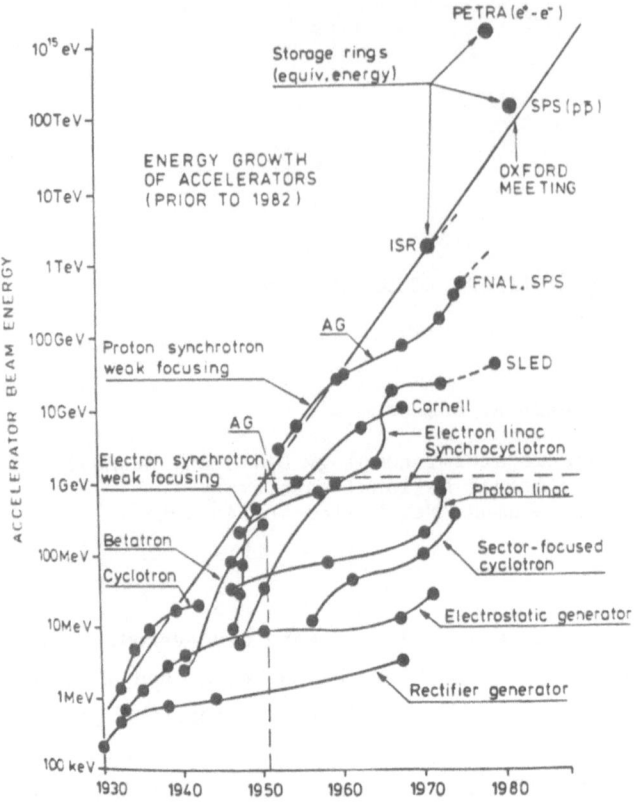

Fig. 1. The Livingston chart, showing the evolution of various types of accelerators with time.

size of the beam and the change in size with time or distance; the area projected on one plane is called emittance.

This discussion will concentrate on a particular kind of linear particle accelerator--the kind whose driving energy is provided by radio-frequency fields--that is well suited to producing high-brightness electron or ion beams. We will concentrate on the issue of high brightness and its ramifications.

CONTEXT OF LINEAR ACCELERATORS

It is useful to place the rf linac in context with other accelerators, using a classification proposed by Lawson (Lawson, 1982) to illustrate the physical principles used in various accelerator types. Figure 2 shows a division between machines where the accelerating field at a point various harmonically and those in which it does not. These categories are then divided, depending on whether the particles move in free space or in a medium, which could be a plasma or an intense beam of a different kind of particle. The free-space category is subdivided,

170

	CATEGORY 1		CATEGORY 2
	ACCELERATED PARTICLES IN FREE SPACE		ACCELERATED PARTICLES IN A MEDIUM
	CATEGORY 1.A NO FREE CHARGES IN SYSTEM	CATEGORY 1.B FREE CHARGES IN SYSTEM	
ACCELERATING FIELDS — HARMONIC	• Linacs • Synchrotrons • Inverse Free-Electron Laser	• Linac plus rf Drive System	• Inverse Cherenkov • Beam-Wave • Laser Beat-Wave
ACCELERATING FIELDS — NONHARMONIC	• Betatron • Induction Linac • Electrostatic Accelerator	• Ion-Drag Accelerator • Wake-Field Accelerator	• Ionization Front • Electron Ring

Fig. 2. Classification of accelerators used by Lawson.

depending on whether the charges that produce the accelerating and focusing fields are all bound in metals or dielectrics or if they are free parts of a plasma or particle beam. In a generic sense, most applied accelerator systems today are in Category 1 and are based on classical electromagnetic (EM) physical principles. Category 2 basically involves plasma physics, which is much less tractable and has not led to significant practical applications in accelerator technology.

As the beam brightness is raised, the particles cease to be acted upon by the EM fields independently but begin to feel the repulsive force of the other, like-charged, particles, and the total EM interaction becomes the collective effect of the whole ensemble of particles and fields. The limit at which the particle self-fields cancel the externally applied fields, called the space-charge limit, is where control is lost of the acceleration and/or focusing process, a condition obviously deleterious to brightness. Another basic limiting phenomenon, called beam breakup (BBU), occurs when the intense beam interacts electromagnetically with its surroundings, creating waves that interact back on the beam, causing it to be diverted or diluted.

For the Category 1 high-brightness linacs we are now building, we must consider these collective effects in the accelerated beam, but do not rely on them for acceleration, as would be the case in Category 2. As the need for brighter beams grows, we continue to explore the collec- tive-effect boundary, requiring better understanding of plasma effects in the beam itself or as an efficient acceleration mechanism; therefore, an understanding of plasma physics is becoming a prerequisite for workers in this field.

Before delving into some of the details of linear accelerators, let us look at some current applications that are strongly driving progress in rf linear accelerators.

APPLICATIONS STIMULATING RF LINAC DEVELOPMENT

Physics Research

Nuclear and particle physics, and the increasing blurred interface between these traditional fields, continue to stimulate linac development. The Los Alamos Meson Physics Facility (LAMPF) is the most intense operational proton linac in the world, producing a 1-mA average current at 800 MeV. LAMPF was recently upgraded to produce a bright H^- beam for injection into the new Proton Storage Ring. Both the Superconducting Super Collider (SSC) and the heavy-ion-collider facilities that will likely be the next generation of large ion accelerators probably will have linac-based injectors that use new techniques involving higher frequencies, higher accelerating gradients, radio-frequency quadrupole (RFQ) preaccelerators, and other advanced accelerator structures. Figure 3 shows such a machine and outlines the innovations that influence most new ion-linac initiatives now in progress.

In electron machines, a new US Continuous Electron Beam Accelerator Facility (CEBAF), Newport News, Virginia, is proposed, based on very recent advances in superconducting linac technology, to provide a high-intensity cw 200-μA, 4-GeV electron beam. Very high energy physics (HEP) machines now use colliding beams to reach the highest center-of-mass energies, and construction of an important proof of principle is nearing completion at SLAC's Linear Collider (SLC). Here, two intense beams will be collided at a spot about 1μm in diameter. The brightness figure of merit for these machines is called luminosity--a combination of bright ness and the event rate and data collection characteristics of the physics experiment. In the long term, luminosity goals of 10^{33-34} $cm^{-2}s^{-1}$ at energies in the 3-TeV range are sought for electron colliders, compared to the design goal of 6×10^{30} cm^{-2} sec^{-1} at 50-GeV energy for the SLC. Control of BBU is important in the linac drivers for these colliders. Similar considerations influence the design of microtron electron accelerators (Fig. 4) and free-electron lasers (FELs) (Fig. 5). The microtron application also stresses development of room-temperature cw accelerating structures; the Los Alamos/NBS program produced an advanced 2400-MHz cw side-coupled structure capable of 2-MeV/m accelerating gradient.

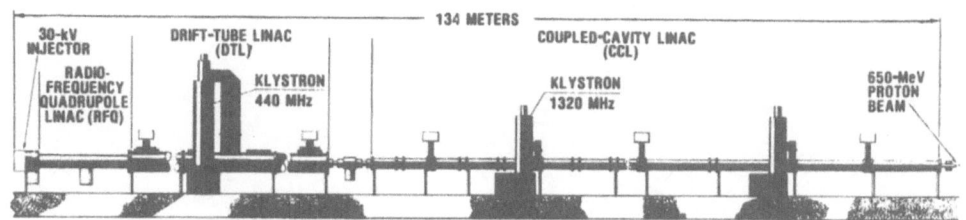

MAJOR TECHNICAL INNOVATIONS	PROTON BEAM PARAMETERS	
HIGHER FREQUENCIES	INJECTION ENERGY	30 keV
HIGHER GRADIENTS	RFQ/DTL TRANSITION ENERGY	2.5 MeV
LOWER INJECTION ENERGY	DTL/CCL TRANSITION ENERGY	125 MeV
RFQ LINAC STRUCTURE	FINAL ENERGY	650 MeV
POST-COUPLED DTL STRUCTURE	PEAK BEAM CURRENT	28 mA
PERMANENT-MAGNET QUADRUPOLE LENSES	PULSE LENGTH	60 μs
DISK-AND-WASHER CCL STRUCTURE	REPETITION RATE	60 Hz
COAXIAL BRIDGE COUPLERS	AVERAGE BEAM CURRENT	100 μA
DISTRIBUTED MICRO-PROCESSOR CONTROL		

PROTON LINAC PARAMETERS

	FREQUENCY	KYLSTRONS	GRADIENT
RFQ & DTL SECTION	440 MHz	1	6 MV/m
CCL SECTION	1320 MHZ	6	8MV/m

Fig. 3.   PIGMI program.

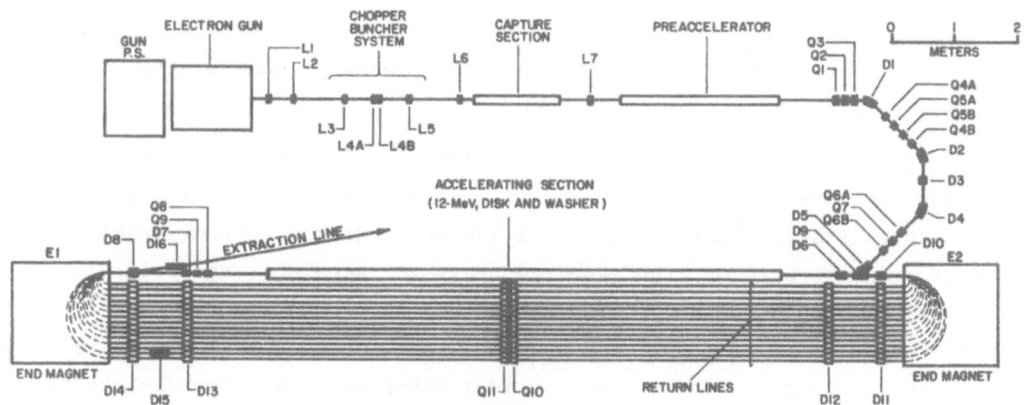

L = SOLENOID LENS
Q = QUADRUPOLE MAGNET
D = DIPOLE MAGNET

Fig. 4.   The NBS–LASL Racetrack Microtron.

173

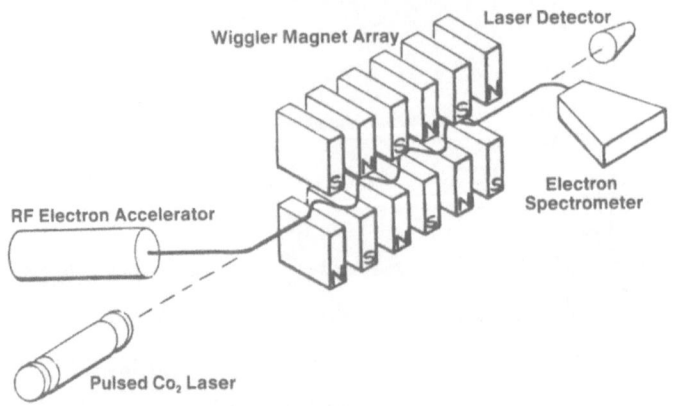

Fig. 5. The free-electron laser.

Fusion

The Fusion Materials Irradiation Test (FMIT) project, now in abey-
ance, was to test materials in a neutron flux produced by a cw, 100-mA,
35-MeV linac accelerating a deuteron beam that would hit a molten lithium
target, as outlined in Fig. 6. The cw, very high intensity nature of
this linac presented great challenges in two major areas. Efficiency
required operation near the space-charge limit while, at the same time,
residual beam losses that would cause radioactivity build-up in the
machine had to be minimized so that machine maintenance problems would
not be too severe. The development work for this program contributed
much to our present understanding of the dynamics of high-intensity
linacs (Jameson, 1983; Jameson, 1982; and Hofmann, 1983) and a new
accelerator type, the RFQ (Stokes et al., 1981). The other challenge was
the engineering requirements of such a high power, cw system that must
run with very high availability. A 2-MeV prototype accelerator was
operated cw at 50 mA this spring.

In inertial confinement fusion, high-energy heavy ions might
interact classically with the target, avoiding problems that have
prevented laser, electron, or light-ion beams from achieving practical
performance. However, the required heavy-ion accelerators would still be
large and complex devices. Two approaches have been studied--the
rf-linac/storage ring approach and the induction linac. The primary
system requirement is for a very bright 6-D phase space because the beam
must deposit its energy in a short time on a small target. Thus, the
heavy-ion fusion (HIF) program has been a primary motivation toward
understanding space charge and instability limits in both types of
machines (Hoffman, 1983; Darmstadt report, 1982; Tokyo report, 1984; and
Washington, D.C., report, 1986) and toward practical techniques for
phase-space manipulation and control that will not spoil the brightness.

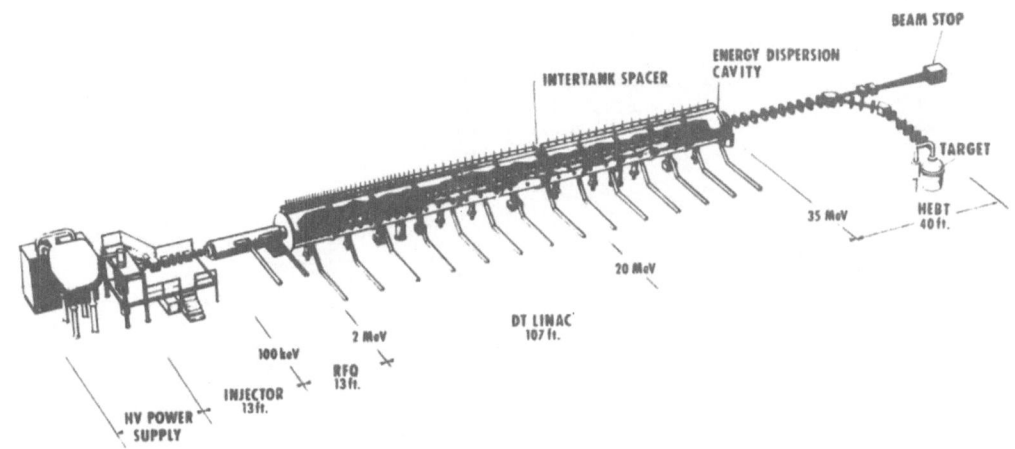

Fig. 6. The FMIT accelerator.

Industrial/Medical

As indicated in Fig. 3, high-brightness ion-linac technology is
being applied to cancer treatment or radioisotope production. Advanced
electron linacs are being considered for radiography, free-electron
lasers, and food processing. FELs present challenging demands on
electron linac performance; considerably more intense beams with better
emittance, compared with existing machines, are required, and this makes
understanding and control of BBU phenomena essential for both beam
acceleration and the energy-recovery beam-deceleration scheme now being
tested at Los Alamos (Watson, 1985) (Fig. 7). Applications of rf-linac
based FELs to infrared on ultraviolet light sources, process chemistry,
and other industrial uses are under study.

Strategic Defense

The possibility of using particle beams for defense against nuclear
weapons has resulted in increased attention to rf linac development
(Jameson, SLAC report to be published). Neutral particle beams, which
would be undeflected by the earth's electromagnetic fields, and FELs are
being studied. Such systems require exceedingly bright beams and present
overall system challenges of a new scope--in particular, the prospect for
accelerator of substantial size operating in space. Beyond any defense
application, this environment would afford many scientific and practical
initiatives, and the techniques developed will influence all linac
construction.

BASICS OF LINAC BRIGHTNESS

So linac devices for a variety of applications have similar
challenges, problems, and approaches to solutions--the basic problems of

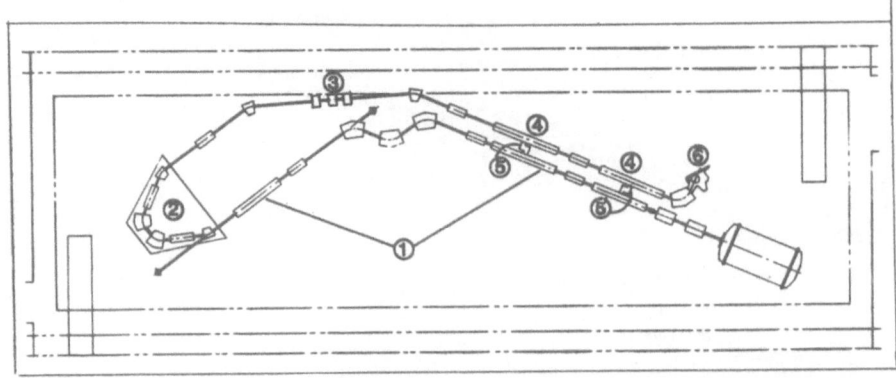

Fig. 7. Los Alamos rf-linac-driven FEL energy-recovery experiment
layout: (1) accelerator and FEL operating at two-fold
increase in peak current; (2) isochronous 180° bend on
translation table; (3) isochronous 60° bend; (4) two 1.8-m
decelerator (energy-recovery) sections; (5) two variable
rf bridge couplers; (6) beam dump for 2- to 3-MeV beam;
(7) 20-MeV diagnostics.

attacking the numerator or the denominator of the brightness equation.
The numerator can be raised by brute force, but the large power
requirements and engineering problems are formidable, and better system
efficiency is a desirable research goal. Power scale-up may tend to
spoil the beam quality because of intensity-related phenomena. Emittance
preservation in each case also requires that aberration effects in the
beam transport optics be avoided. Thus, a long-term development program
in advanced linac-based drivers is required. We turn to a short outline
of how linacs work (Humphries, 1986) and further development of the
high-brightness theme.

As we have noted, rf linacs accelerate particles in a beam through a
resonant interaction with external charge distributions and the coupling
EM fields that transfer energy to the beam particles. The applied EM
fields exert forces on the beam that we will vectorize as acting longitudinally along the beam direction to accelerate or decelerate it and,
transversely, to confine the beam. At high enough beam currents, the
self-consistent solution of the equations describing a particle's motion
must account for the total field generated by the external charges and by
fields generated by other particles. This a nonlinear problem and can be
handled in detail only by computer programs, using successive iteration.
However, much of the useful design information comes from smoothed
approximations of the detailed motion, dealing in particular with the rms
properties.

In the resonant rf linac, the resonance properties of the circuit are used to obtain voltage amplification, and the time-varying fields are used specifically to influence the particle motion. In particular, particles must be at the right place at the right time to see an accelerating field, and this synchronism must be maintained over a long distance to produce high energy. In rf linacs, the energizing field is expressed as a sum of traveling waves, and one wave is made to travel near the average velocity of the particles. The truly synchronous particle would travel exactly along the axis with the wave, whereas particles with different phases or energies would oscillate around the synchronous particle, or if too far from synchronism, would be only jostled as the wave passed. As indicated in Fig. 8, (nonrelativistic) particles arriving earlier than the synchronous particle see a lower voltage, are accelerated less, tend to converge on the synchronous phase point ϕ_s, and so on; thus, there is a region of phase stability over some phases of the wave, known as the rf bucket, in which the particles oscillate around the synchronous particle. An approximate equation describing this phase oscillation is that of a nonlinear oscillator:

$$d^2\phi/dt^2 = -K_\ell(\sin \phi - \sin \phi_s).$$

The solution for small-amplitude oscillations is harmonic, characterized by the phase advance σ of the oscillation over an accelerator system period. The particle with synchronous phase clearly also has a synchronous energy as well, and off-synchronous particles define oscillation trajectories around the synchronous particle in energy and phase. A plot of the displacement from synchronism is called the longitudinal phase space. The boundary within which particles oscillate stably is called the acceptance, as indicated in Fig. 9.

If the distribution of the beam particles, called the emittance, is congruent with the acceptance, as in A, the beam is said to be matched, with the particles moving on congruent orbits in the linear approximation. Particles injected in the shape B would sweep out a larger area in phase space, like C, and if nonlinear forces are present, the particles would eventually disperse to fill the area C (or worse). Thus, we have introduced the concepts of matching and emittance growth and have suggested that matching is to be desired and emittance growth avoided. With relativistic particles, assuming the wave travels at the speed of light c, acceleration occurs if the phase is less than π, with a slippage until the particle reaches π because the particle is not quite at c. However, if the acceleration voltage is high enough, acceleration takes place so fast that the particles can be trapped in a bucket and carried to arbitrarily high energy. The solution is not oscillatory;

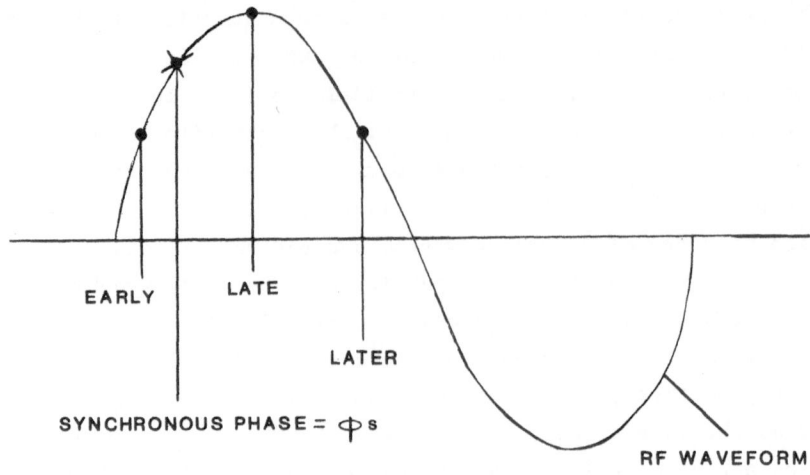

Fig. 8. Time or equivalent position along rf wave.

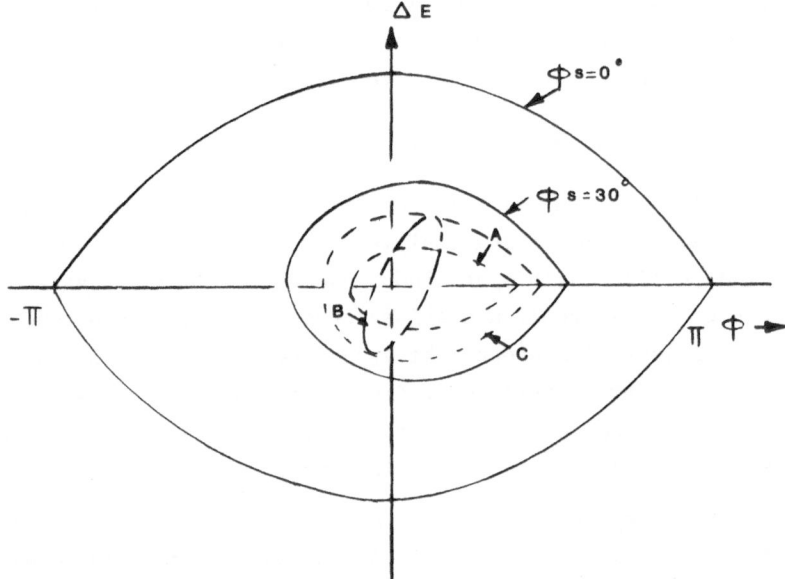

Fig. 9. Longitudinal acceptance diagrams for different ϕ_s.

because time dilation dominates, the particles monotonically approach a constant phase. We will have to differentiate between relativistic (electron) machines and nonrelativistic (ion) aspects as we go along.

The traditional physical structures for setting up the desired fields are basically configured to produce the longitudinal field. For short pulses, the electron linac structure is simple pipe with periodic loading in the form of iris disks with a beam hole at tie center, spaced to slow a traveling wave to match the beam velocity, as in Fig. 10. For

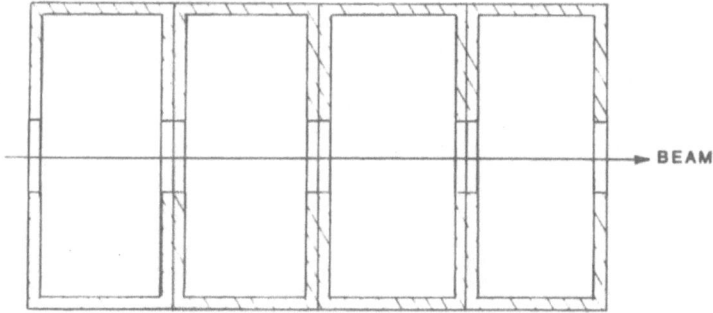

Fig. 10. Iris-loaded waveguide for traveling-wave electron linac.

longer pulses, a standing-wave structure is more efficient; in the
structure of Fig. 11, the field can again be explained using the sum of
traveling waves, one of which matches the beam, or by considering that
the beam is only exposed to the field during the proper interval of the
rf wave that results in acceleration. During the rest of the rf cycle,
the beam is in the tunnel, or drift-tube, between cells, and does not
feel the field.

   In ion linacs, the traditional structure for energies above an MeV
or so is called the Alvarez or drift-tube linac, Fig. 12; it operates as
a standing-wave structure. Above 100 MeV or so, the drift tubes become
long and efficiency arguments require a transition to a higher frequency
and a structure like that of Fig. 11.

   In the transverse plane, the unavoidable existence in cylindrically
symmetric rf linacs of the slow traveling wave's radial field components
results in orbits that are transversely unstable when the longitudinal
orbits are stable; therefore, ion linacs must use arrays of focusing
elements to contain the beam. This focusing requirement introduces a
host of new considerations but, for now, the essential point is that the
added transverse focusing also has a periodic property. In electron rf
linacs, the beam is traveling so fast that the radial defocusing is less
apparent, and added transverse focusing elements are needed much less
frequently. A new type of structure, the RFQ, has azimuthally asymmetric
transverse fields and is basically a transverse focusing structure that
is perturbed to set up a longitudinal accelerating component (Fig. 13).
This structure uses the electrostatic focusing from the rf fields to
provide both the focusing and acceleration and has great advantages for
low-velocity ions in the tens of keV to few MeV range. The RFQ is also
well described by the smoothed oscillator equations.

   Off-axis particles oscillate in a transverse phase space characterized by position with respect to the axis and angle with (or velocity

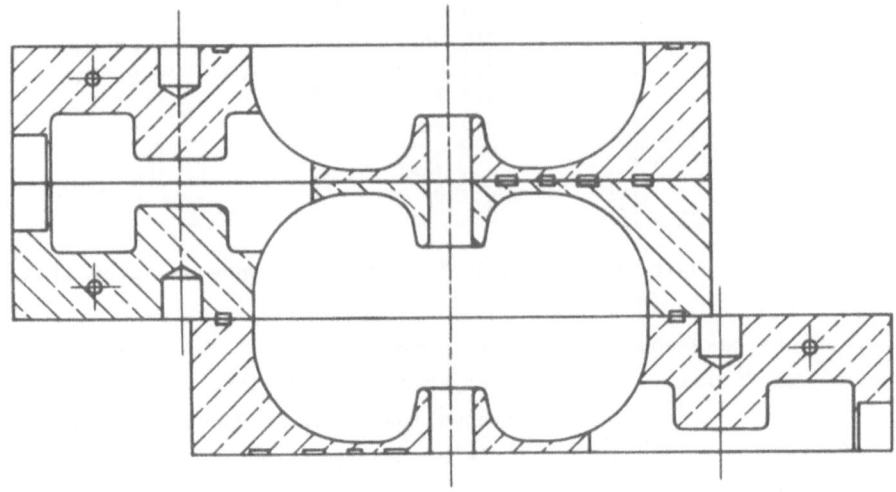

Fig. 11.   Standing-wave structure for electron linac.

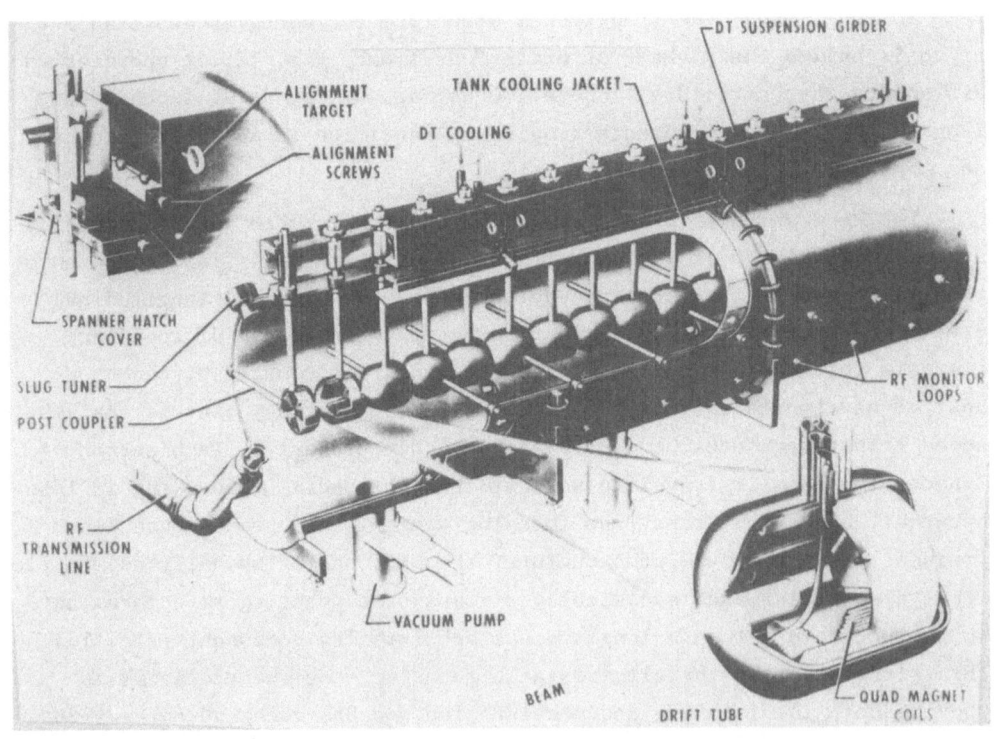

Fig. 12.   Alvarez or drift-tube linac.

Fig. 13.   The radio-frequency quadrupole (RFQ) accelerator.

away from) the axis.   Their motion can be described by a nonlinear
oscillator equation similar to that of the longitudinal space, with a
phase advance per acceleration transverse period of σ^t.   As with the
longitudinal phase space, it is important to match the shape of the beam
distribution to the shape of the transverse acceptance to avoid emittance
growth.   Such growth leads to loss of brightness or (eventually) even to
loss of particles, which not only reduces the transmitted current
(further brightness loss) but also causes unwanted problems with heat
dissipation or radioactivity build-up in the accelerator walls.

The characteristic phase advances σ^t and σ are not independent because of coupling between the transverse and longitudinal external fields and because the fields of the beam itself are also coupled between longitudinal and transverse. The equations for the phase advances have the forms

$$\sigma = \cos^{-1}[\cos \sigma_0 + f(\text{beam current, beam size a and b})],$$

where a is the average transverse rms beam radius; 2b is the physical rms bunch length; and σ_0 is the zero current phase advance, which in turn is a function of the structure geometry and the external fields, including the transverse and longitudinal couplings.

The rms space-charge forces in the beam directly cancel the rms external restoring forces, so the phase advances tend to zero as the current rises. We define the relation between space charge and external forces in terms of the phase advances as

$$\mu_t = \left[1 - \left(\sigma^t/\sigma_0^t\right)^2\right] \text{ and } \mu_\ell = \left[1 - \left(\sigma^\ell/\sigma_0^\ell\right)^2\right].$$

Next we need to look at the idea of emittance a little more closely. Figure 14 indicates a collection of particles in transverse phase space that has some particle density distribution. The rms emittance* ε_{rms} of the distribution is given by

$$\varepsilon_{rms} = \left(\overline{x^2}\ \overline{x'^2}\ \overline{-xx'}^2\right)^{1/2},$$

which is the equation of an ellipse. The rms beam size is $a = \sqrt{\overline{x^2}}$ and the rms beam divergence is $a' = \sqrt{\overline{x'^2}}$. In a linear periodic system, the ellipse would have the same shape at similar points of each period; one convenient point is where the correlation terms vanish and the ellipse is upright. If an ellipse of the rms shape is fit through each particle in the distribution, an effective ε_{total} is found. While the actual area occupied by particles in phase space is conserved, the effective area is of more practical import because nonlinear effects tend to push particles out in diffusion or filamentation processes until they occupy the larger area. Because brightness is a key figure of merit, we see that it would be desirable to keep the maximum number of particles in the smallest rms area, or perhaps in the smallest total area.

$$\text{Brightness } \alpha \ \frac{\text{current}}{\varepsilon_x \varepsilon_y \varepsilon_z}.$$

---

*This is true rms emittance, without the factor of 4 used by some.

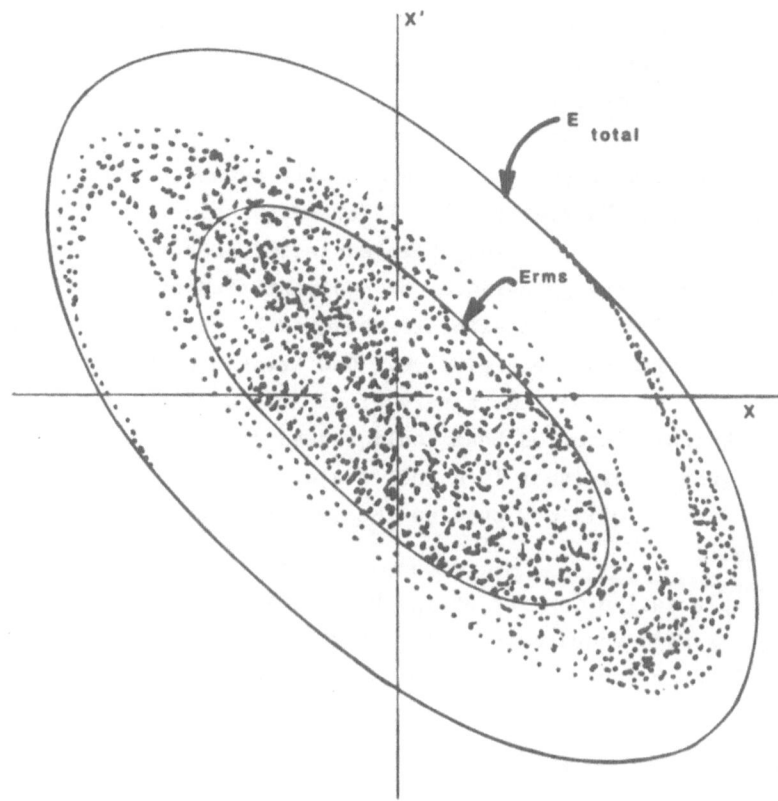

Fig. 14. A particle distribution in x-x' transverse phase space.

The accelerated beam can be no brighter than the input beam, but there are a number of processes that can make it worse. The study of these processes and attempts to control them have absorbed accelerator designers for many years. The difficulties of creating the particle beam and delivering it to the beginning of the rf linac are another whole story, fraught with poorly understood phenomena involving material properties and partially neutralized plasma effects. In the rf linac, the EM fields ensure that the beam plasma is nonneutralized.

The nonlinearities and couplings in the external fields have been thoroughly studied, and their control is a large part of the art of making practical accelerating structures. At present, configuration of the external field is done well enough that most particles are confined near the center of the phase-space diagrams where the motion is nearly linear, on average. It is important that the rms shape properties of the emittance be matched to the acceptance. The beam must be kept centered on the transverse axis and straight along it, and the longitudinal centroid must be at the synchronous phase and energy. Also, the rms-emittance shape and orientation of the longitudinal distribution must be right.

However, it was long a mystery that, even though the ellipse properties were matched in each plane, transverse rms emittance growths of a factor of 2-3 occurred in the low-velocity section of rf ion linacs with intense beams. The cause has been the subject of an intense search by a few people, with some real progress recently (Wangler and Guy, 1986; Guy and Wangler, 1986; Wangler et al., 1986; and Wangler, this advanced Study Institute report). It had long been suspected (Lapostolle, et al., 1968) that some kind of energy balance, or equipartitioning, between the degrees of freedom would ameliorate the growth, but the way to characterize the physics was very elusive. One breakthrough occurred (Jameson, 1981) when it was shown that a very simple rms equipartitioning requirement on a bunched injected beam could indeed produce remarkably small emittance growth, at least in a full-scale, nonlinear computer simulation of the linac. We will now outline the set of equations leading to these conditions.

A simple derivation for energy balance in a weakly coupled harmonic oscillator system requires equality of the average kinetic and potential energies in each degree of freedom;

$$\langle 1/2 \ mv_i^2 \rangle \ = \ \langle 1/2 \ k_i x_i^2 \rangle,$$

where k_i is the appropriate force constant. If we characterize the motion in terms of the oscillation's phase advance σ, over an accelerator system period $N\beta\lambda$, we can write the mean-square velocity as

$$\langle v_i^2 \rangle \ = \ \sigma_i^2 \langle x_i^2 \rangle / (N\beta\lambda)^2.$$

At a location where the correlation $\langle x_i v_i \rangle$ is zero, rms emittance is defined as

$$\varepsilon_i \ = \ \langle x_i^2 \rangle^{1/2} \langle v_i^2 \rangle^{1/2},$$

and the two equations that describe the motion of the particle envelopes follow directly:

$$\varepsilon_t \ = \ \sigma^t a^2 / N\beta\lambda \ \text{ and } \ \varepsilon_\ell \ = \ \sigma^\ell b^2 / N\beta\lambda.$$

It can be shown rigorously that these are the matched envelope equations ($\ddot{a} = \ddot{b} = 0$) for the rms envelope behavior of particle distributions in linearized periodic systems. The simultaneous solution of these two equations was a necessary condition for preventing emittance growth, but was not enough. If we require equal average energy in each of the coupled degrees of freedom, by equating

$$\langle v_i^2 \rangle \ = \ \langle v_j^2 \rangle \ \text{ and } \ \sigma_i^2 \langle x_i^2 \rangle / (N\beta\lambda)^2 \ = \ \sigma_j^2 \langle x_j^2 \rangle / (N\beta\lambda)^2,$$

we find

$$\frac{\varepsilon_\ell}{\varepsilon_t} = \frac{\sigma^t}{\sigma^\ell} = \frac{b}{a} \, .$$

Systems satisfying this equation and the envelope equations
simultaneously will be both matched and equipartitioned. We have
observed, in simulation studies of completely described accelerating
channels, emittance growths of only about 20% over a large number of
cells for linacs quite near the space-charge limit ($\mu_t = 0.9$) (Fig. 15).

The problem with earlier drift-tube linacs (DTLs) was that the
injected longitudinal emittance was typically four to five times larger
than the injected transverse emittance because of the way the particles
came to be bunched around the synchronous phase. Ion beams are generated
from sources with extraction voltages typically less than 50 kV; thus the
ions are traveling very slowly, around $\beta = 0.001$, and are very nonrelativistic and susceptible to emittance growth phenomena. Large initial
acceleration forces, although convenient in terms of length, are disruptive. It was long thought desirable to accelerate ion beams adiabatically, but this required excessive length. The compromise was to use a
crude buncher system consisting of one or two rf cavities, separated by
the proper drift lengths between the buncher cavities and the linac input
to cluster as much of the initially dc beam as possible near the synchronous phase at the first linac cell. The bunching process itself leads to
the too-large longitudinal emittance and also to transverse emittance
growth. The RFQ accelerator mentioned is now the preferred method in
most applications for converting the initially dc particle distribution
into an appropriately bunched beam for further acceleration in a DTL. A
fundamental advantage of the RFQ is that it focuses at all energies.
When perturbations are added to produce longitudinal fields, short cell
lengths with precisely controllable properties result, allowing many
cells in a reasonable length and the capability to longitudinally bunch
and accelerate the beam gently, keeping the emittance from growing very
much. The RFQ can thus be used to capture the beam at the typical ion-
source extraction voltage and deliver the beam to a drift-tube-type
accelerator at about 2 MeV. Before the RFQ was invented, ion sources had
to inject directly into drift-tube linacs in which the required magnetic
focusing strength varies inversely with particle energy. Restrictions of
space inside the drift tubes and on magnetic strength make it difficult
to build drift tubes below about 700 keV; therefore, the ion-source beam
was accelerated in a dc Cockcroft-Walton up to that energy. The RFQ
system is dramatically smaller and much better in terms of beam dynamics
properties.

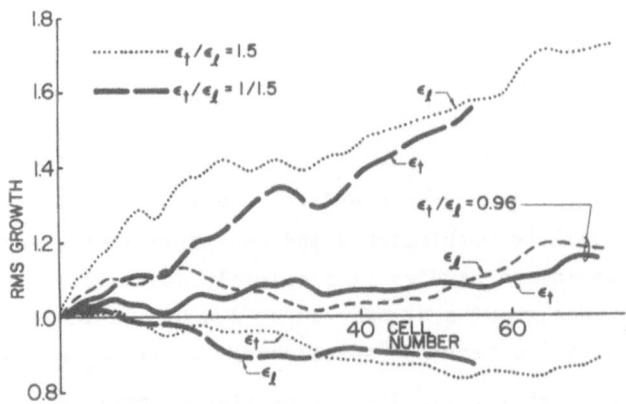

Fig. 15.   Rms emittance growth in 72-cell, constant μ_t = 0.9,
constant E_0 linac. Initial $\mu \sim 0.8$. ε_t and ε are
transverse and longitudinal rms emittances; ratios
are initial conditions at injection into linac.

In turns out, however, that we still cannot prepare beams as we
would like at various points in the system without some unwanted side
effect, for example, an overly long RFQ. Thus, we have had to continue
looking for a more complete understanding of the detailed physical
processes that lead to emittance growth. At this point in our discussion, we have noted that emittance growth can occur from the following.

- Nonlinear external forces: these forces particularly affect the
  longitudinal phase-space, but include important longitudinal/
  transverse coupling effects, and may be a dominant factor in
  processes that cause growth of the total emittance through the
  formation of halos at the beam edge.

- Mismatching: certainly the rms beam properties must be matched to
  the channel properties. At a deeper level, minimum emittance
  growth would require **all** beam properties, including those of the
  density distribution, to be perfectly balanced against the channel
  properties and to repeat each period--we do not understand how to
  do this yet.

- Missteering: the beam must start and stay on-axis and the longitudinal centroid must be at the synchronous energy and phase.

- Energy unbalance, or nonequipartitioning.

The fact that using rms equations to set up the injected beam and
initial linac parameters resulted in low emittance growth held some
important clues: the main ones being that

- these equations cover a large fraction of the problem if the
  conditions could be met in practice, and

- the equations involve only linear forces.

Linear space-charge forces result from uniform particle distribution in the beam. In the space-charge limit, the beam behaves like a plasma and arranges itself to shield the external field from the interior of the beam; inside the shielding layer, which is of an equivalent Debye-length thickness, the particle distribution is uniform. It was also often observed in computer simulations that, at high currents, the beam tended to homogenize. So it made sense to look at the differences between typical beam density distributions and uniform beams to look further at processes causing emittance growth. A differential equation was discovered (Wangler and Guy, 1986; Guy and Wangler, 1986; Wangler et al., 1986; and Wangler, this Advanced Study Institute report) that relates the rate of change of rms emittance and the rate of change of nonlinear field energy. The nonlinear

$$\frac{d\varepsilon^2}{dz} = - \ (\text{Constant}) \ \frac{dU}{dz}$$

field energy corresponds to a residual field energy, available for emittance growth, of beams with a nonuniform charge density distribution, depending only on the shape of the distribution. The quantity U is the difference between the self-electric-field energies of the actual beam and the equivalent uniform beam with the same rms properties as the actual beam and the equivalent uniform beam with the same rms properties as the actual beam. Using the property that a matched beam near the space-charge limit stays the same size (laminar flow) and assuming the tendency to uniformity of the final charge density, the equation can be integrated and predictions made of the final emittance. Two effects are described: in the first, an adjustment of the beam's charge distribution occurs to match the external focusing forces. The redistribution occurs very quickly, within about one-fourth a plasma period, and results in transfer of the nonlinear field energy to particle kinetic energy and an approximately uniform beam distribution with a tail of about the Debye length at the edge of the beam. On a slower timescale, any unbalance in the kinetic energy from one coordinate direction to another also equilibrates, resulting in partial or complete kinetic energy equipartitioning. The degree of final equipartitioning is somewhat difficult to predict at this stage of the theory development; simulations show three distinct regions of behavior for a given initial distribution, departure from equipartitioning, and number of plasma periods.

1) Above a threshold in σ/σ_o, the (emittance-dominated) beam is stable and no kinetic energy transfer occurs for a uniform beam. For nonuniform beam above this threshold, only charge redistribution occurs.

2) At high current, with σ/σ_0 far below the threshold, the kinetic energy becomes completely equipartitioned for all initial charge distributions after a few plasma periods.

3) Between these regions is a transition region where the beam moves more slowly toward equipartitioning. An empirical equation for the equipartitioning rate has been derived, but more theoretical work is needed.

Wangler's lecture at this ASI (Wangler, this ASI report) will develop these topics in detail. Another cause of emittance growth is coherent modes that can be excited. Thresholds for these modes have been derived (Hofmann, 1981) and checked for periodic transport systems; the thresholds appear to be approximately correct for a variety of charge distributions and also for accelerator systems where the rms envelope approximations are valid (Jameson, 1982). In the recent simulation work (Guy and Wangler, 1986) described above, rather complicated behavior is seen near initial tune depressions corresponding to the coherent mode thresholds. In some cases, the beam seems to be attracted to integer or half-integer ratios of x and y tunes, which may result in partial kinetic energy change, in more transfer to the lower initial energy plane (over-partitioning), or even in kinetic-energy transfer from the lower energy plane to the higher. These effects require further study.

The equations can be used to estimate the minimum final emittance, corresponding to the initial conditions of the extreme space-charge limit where the initial emittances are zero. These estimates predict that the minimum final emittances depend on the initial nonlinear field energy but not on the degree of departure from initial equipartitioning. For bunched beams, the final emittance are predicted to scale as the beam current to the two-thirds power.

Simulation studies (Guy and Wangler, 1986) also suggest that uniform charge-density distributions are best for controlling the total beam emittance of the surrounding halo beam. We do not yet understand the mechanisms involved, but observe that the charge redistribution process, creating a shielded inner core and Debye-thickness shield, produces a halo that can extend to many standard deviations beyond the rms core. We have simulated nonstationary laminar beams (at the extreme space-charge limit) for which the initial total field is nonlinear, and have observed that particles initially at large radii in an initially Gaussian profile do not remain laminar but move into a halo.

Summarizing, the effort to understand single-channel current limits and emittance behavior has recently been successful in separating and elucidating several effects that can cause emittance growth. This work explains why a uniform beam distribution, kinetic energy balancing, and careful matching are desired and indicates that we must know how to avoid coherent instabilities. (The latter is easier in a linac where the parameters change as energy increases.)

We thus are returned, with greater confidence, directly to the challenge--how to produce such beams at the source and maintain them through the accelerator.

Equivalent problems govern the peak current and brightness that can be achieved in electron linacs. Electrons are very light and become essentially relativistic at around 1 MeV. Because electron sources typically start at lower voltages, it is still necessary to use great care, at the beginning of acceleration, in shaping the accelerating and focusing fields. However it is impractical to use many cells; thus, the strategy at this time is to accelerate very fast to the relativistic regime. Producing a cleanly bunched beam without longitudinal tails is important, and instead of an RFQ, which would be too long, photocathode sources such as the one shown in Fig. 16 are being developed. The cathode is driven by a laser synchronized to the rf waveform; electrons are emitted only during the desired time and are extracted directly by the rf wave in the cavity gap. The fields in the first few gaps are optimized to produce very rapid acceleration, typically 1 MeV per gap, but with as little brightness degradation as possible. Tailored solenoidal magnetic fields provide transverse focusing.

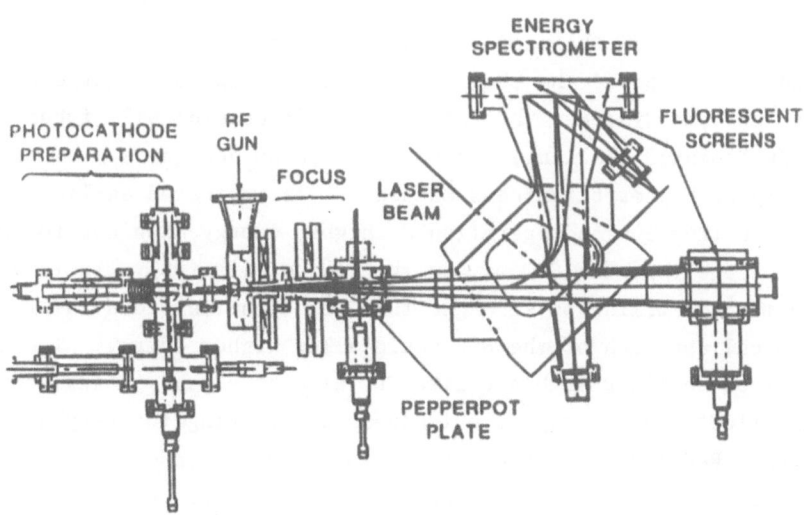

Fig. 16. Laser-driven photocathode and rf-cavity injector-development experiment at Los Alamos.

At higher energies, beam breakup effects (Gluckstern, 1986; Gluckstern, 1985; and Wilson, 1986) limit the available brightness in the electron* linac or electron transport lines, for example, in an FEL. The intense bunch shock excites fields at discontinuities in the channel as it travels along, generating transverse and longitudinal EM "wake fields." These can affect the particle distribution within the bunch; for example, the fields excited by the head of the bunch can contain components that steer the tail of the bunch away from the axis or cause the tail to gain or lose energy relative to the head. At the cavity gaps, the wake fields can build up resonantly and affect subsequent bunches, and the effects can be aggravated if the bunches are arriving at intervals corresponding to structure resonances. In recirculating devices such as the microtron, the resonances associated with the circulation time add another, severe, constraint to the achievable current. The avoidance of beam breakup depends on a detailed knowledge of the modes that can be excited in the system and on development of techniques to suppress their excitation. These techniques include smooth walls wherever possible, perturbations to break the symmetry of unwanted nodes, special consideration to allow bad modes to quickly dissipate their energy through Q-spoiling or propagation to external loads, and other techniques. The design is complicated by the 3-D nature of the problem and the difficulty of analysis. There have been considerable advances in the past few years in the understanding and theoretical treatment of the problem, and the advent of 3-D cavity codes (Weiland, 1986) will make detailed design more tractable.

HIGH BRIGHTNESS ECONOMICS

Having discussed the need for and characteristics of high-brightness beams, and having posed the challenge of needing uniform charge distributions in ion-beam systems, the rest of this discussion will focus on another fundamental challenge facing accelerator designers--the economic feasibility of higher brightness machines. As indicated earlier, the cost for construction and operation of higher energy machines for HEP has become so large that the SSC may be the last of its type. In this case, the economic constraint is stronger than the technical constraint because the technical approach of the SSC would allow higher energy. The electron-positron collider machines like SLC also have heavy power demands, spurring efforts to get higher brightness by striving for very small emittances. Reducing emittance is necessary, but at some point the cost for further reduction will rise and require a balance to be struck with other system costs. The need for more efficient machines has become a major consideration.

Efficiency could have several aspects. The most common requirement is to achieve more beam energy, or current, or brightness, or power, per dollar cost. However, in some applications, more beam power per unit weight or unit volume might be more important than the cost. In all cases, higher conversion efficiencies from the prime power source to beam power are needed. It may be difficult to make a good estimate of the ultimate system efficiency of a new scheme until after prototypes have been tested, but the need for efficiency should always be kept in mind. As an example a discussion of the rf linac is useful.

Tigner shows in Fig. 17 (Tigner, 1982) an evolution of a conventional linac circuit that guides us from today's separate linac structure and microwave tubes to a coupled source and accelerator structure and, finally, to a fully integrated system in which the transformer action between a low-voltage/high-current driving beam is integrally coupled to a high-voltage/low current accelerated beam. The idea is to force consideration of the overall system efficiency, beam power divided by prime power, as the product of power conversions through the system. In Fig. 17a., typical present-day efficiencies would be about

$\geq$ 97%, energy-reservoir to rf-source dc input;

50-70%, rf source, dc to rf;

~ 90%, coupling-network, rf source to accelerator structure;

60-80%, accelerator-structure losses;

1-10%, structure to beam;

0.1-10% overall system efficiency.

Accelerator structure losses result from field multiplication in the structure; more multiplication causes more dissipation and lower efficiency; the converse requires high peak input power and gives higher group velocity. Structures today with group velocities of 0.1-0.2 c and have around 80% efficiency, but would take several times more peak power to achieve the same accelerating gradient (for example, several gigawatts for 100 MeV/m).

The structure-to-beam efficiency is the ratio of energy gained to energy stored, per unit length of structure. This efficiency is limited from ~ 1 to 10% by beam breakup and phase-space dilution effects.

Multiplying yields overall efficiencies from less than a percent to 6-10%, with the larger numbers requiring development. Further inefficiencies usually result from the beam to the desired output, for example, in a HEP event rate, or a loss in stripping to a different charge state. The progression from Fig. 17.a. to 17.c. is to suggest that a more tightly coupled system might raise or eliminate some of the serial efficiencies. Practical schemes with significantly better efficiency are

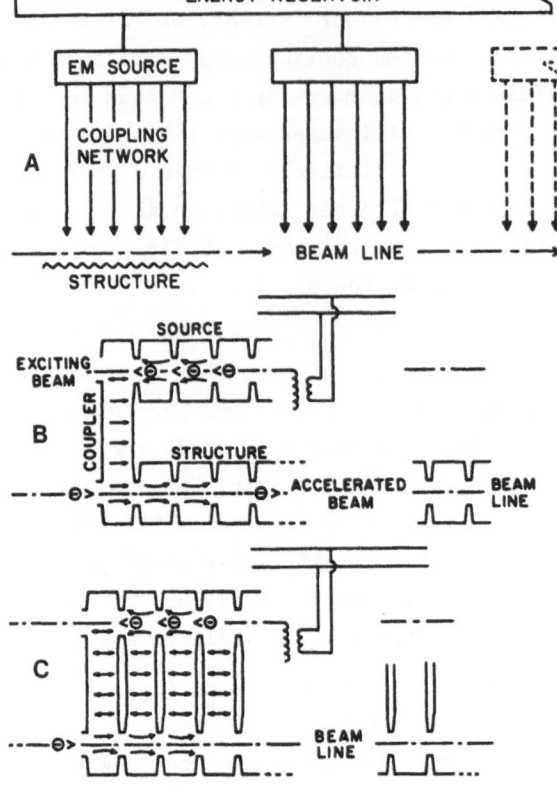

Fig. 17.   An evolution of a conventional linac circuit
to a two-beam system.

as yet elusive. It is useful to consider some of the basic tradeoffs in
the conventional system to establish a frame of reference.

First, however, a basic note of caution on what are sometimes called
tradeoff studies, or scaling studies, or system studies. A practical aid
to keeping a fresh outlook is to emphasize that the solution to a given,
specific problem does not need to be a generic solution; in fact, the
solution sought will be determined as much by the constraints imposed as
by the basic principles. A classic example occurred some years ago in
the HIF studies: in trying to determine how much power could be transported through a focusing channel, one analysis showed beam emittance
entering in the numerator and another in the denominator! A scholarly
and well-written explanation and resolution was written by M. Reiser
(Reiser, 1978), who clearly showed how the choice of constraints and
fixed or variable parameters could so drastically shape the result. The
danger, of course, is that an improper statement of the problem prevents
the needed insight. Reiser's article should be regarded as part of the
"Art of War" (Tzu, 1971) of the accelerator designer. Another example

was the revelation (Jameson, 1981) that high-intensity rf linac beams, which also required small emittance, should use higher rf frequency rather than moving to lower frequencies as was commonly supposed.

RF Power and Accelerator Structure Tradeoffs

We can use a simple linac costing relationship to elaborate the relative influence of the rf power and accelerator-structure subsystem efficiencies mentioned above and development directions that should be taken.

Basic linac costs are given by

$$\text{Cost} = R(P_{cu} + P_b) + SL + AC(\hat{P}_{cu} + \hat{P}_b),$$

where R   = cost/peak rf watt;

P_{cu} = accelerator-structure peak power that is due to losses;

P_b = beam peak power;

S   = structure cost/unit length;

L   = accelerator length;

AC = ac unit power cost; and

$\hat{P}_{cu}$ and $\hat{P}_b$ are structure and beam average power, equal to peak power times duty factor.

The first two terms represent capital investment; the last term adds in the operating cost over the expected life.

$$P_{cu} \propto (E_o L)^2 / ZL \propto (\Delta W)^2 / ZL,$$

where E_o is accelerating gradient/unit length;

Z is effective structure shunt impedance/unit length (includes transit time and synchronous phase-angle factors);

ΔW is the desired, fixed, particle-energy gain of the linac; and

$P_b = (\Delta W)(\text{beam current})$.

Substitution shows the structure power cost varies inversely with length, whereas the structure cost varies directly with length. Therefore, there is a strong tradeoff between accelerating gradient and length, and choice of the maximum achievable accelerating gradient is not a priori desirable. Ignoring the operating cost, differentiation with respect to length yields the optimum length, and thus gradient, for lowest cost:

$$E_{o\ opt} = (SZ/R)^{1/2}, \text{ independent of } \Delta W;$$

$$L_{opt} = \Delta W (R/SZ)^{1/2};$$

$$C_{opt} = \Delta W[2(SR/Z)^{1/2} + RI], \text{ linear in } \Delta W.$$

At the optimum, $RP_{cu} = SL$. Folding in operating cost will push the optimum E_o down and optimum L up.

We need to examine the cost equation further to see more of the influencing factors. It is reasonable to expect that we would want to exploit the accelerator structure to some physical limit, even though the cost relation warns us to be careful. The applicable physical limit will depend on the application and could be, for example, removal of average waste power, voltage breakdown, surface damage that is due to high peak power, magnetic field limitations, space charge limit on current, and so on. Typical proton rf linacs today might be designed at around 440 MHz for the RFQ/DTL, and around 1320 MHz (X3) for the high-beta stage. In this frequency range, a limiting factor comes from the electric-field sparking limit as defined by the Kilpatrick Limit (KL), a frequency scaling for allowable peak surface field based on an ion-multipactoring model and empirical determination of constants known as the Kilpatrick Criterion (Kilpatrick, 1957):

$$f = 1.643 \ E^2 \ \exp\text{-}(8.5/E).$$

The field E_{KP} thus found is multiplied by a "bravery factor" to determine the actual allowed peak surface field by accounting for the influence of modern techniques in raising the sparking limit; E_{KP} is 20 MV/m at 440 MHz and 32 MV/m at 1320 MHz.

The experience factor $K = E/E_{KP}$, by which E_{KP} may be multiplied for modern structures, appears to be as high as 2.5 to 3.0 for RFQs, and up to 2.0 for DTL and SSC structures. Thus, for our 1320-MHz coupled-cavity linac (CCL), we can consider peak surface fields of up to about 64 MV/m.

All the peak surface field, however, cannot be used for acceleration--geometry factors in practical structures reduce the effective gradient on-axis by some factor. This factor can be minimized but usually at some cost, say in shunt impedance Z or transit-time factor, which would directly offset the increased accelerating gradient E_o. For example, one structure with many desirable properties is called the disk-and-washer (DAW) type (Fig. 18). The addition of noses around the beam hole increases the transit-time factor, at some loss in shunt impedance, and increases the peak-surface-field to accelerating-field ratio (E/E_o) from 1.94 with no nose to 5.37 with full nose. The Vaguine structures has a somewhat better efficiency in using peak surface field as accelerating field, with the Chalk River structure intermediate.

Vaguine Structure; $E/E_0 = 1.70$

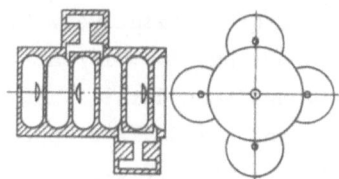

Chalk River on-axis
Coupled Structure; $E/E_0 = 3.95$

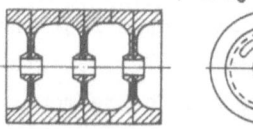

Disk-And-Washer E/E_0 w/o nose = 1.94
Structure; E/E_0 w nose = 5.37

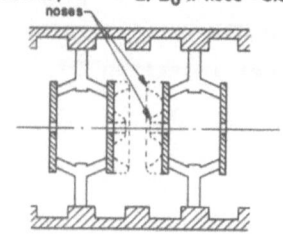

Fig. 18.   Cross sections of four CCL types: the DAW with and
without nose, the Chalk River on-axis coupled structure
(McKeown and Schriber, 1981) and the Vaguine structure
(Vaguine, 1977). E/E_0 is the ratio of peak surface
field to accelerating gradient.

The fabrication cost/unit length S of all these high-β structures is
roughly the same, \$50–100 K/m. The tradeoff among shunt impedance
(~50–100 MΩ/m), transit time (0.8–0.92), and other detailed factors are
also not dramatic. Therefore, the gradient versus length-cost tradeoff
must dominate the choice of optimum gradient. Figure 19 illustrates this
result, showing the cost curves for a linac that was designed as an
injector for the proposed SSC, and relating E, E_{KP}, and E_0 for the four
structures. The cost minima are all at about \$20 M and require an
accelerating gradient of ~ 20 MeV/m. The available E_0 (30–40 MeV/m) at
K = 2 of the more efficient structures cannot be used economically, but
the 20 MV/m E_0 giving the cost minimum is available below the sparking
limit. The less efficient structures cannot reach the cost minimum
without sparking, although this is not too serious because the cost
minima are broad. Another look at the cost equation shows the optimum
$E_0 \propto (SZ/R)^{1/2}$; thus, we could use a higher accelerating gradient if we
could get the effective structure shunt impedance up or the unit rf power
cost down.

A great deal of rf accelerating structure development has occurred
at frequencies $\leq$ 3 GHz, and it is unlikely that major increases in shunt

impedance can be achieved. Also, as with structure cost, the cost per peak rf watt at low duty factor is relatively independent of frequency in this frequency range, at about $0.01-0.015/watt. The cost of rf power is beginning to be looked at, and some details are given in a later lecture at this school. Both tube and solid-state approaches are pushing toward higher conversion efficiency at better weight and volume ratios.

Figure 19 indicates that a point design accelerating gradient of only 8 MeV/m was selected. We had already concluded for economic reasons that only half of the 40-MV/m available from the Vaguine structure could be used; why did we limit the design by more than another factor of 2? The answer is in the emittance growth arguments of the preceding discussion. In this particular study, we used a conventional DTL that incorporated no special provisions for preparing a beam that would stay equipartitioned across the DTL/CCL interface. We studied the emittance growth resulting from direct injection into the CCL as a function of the

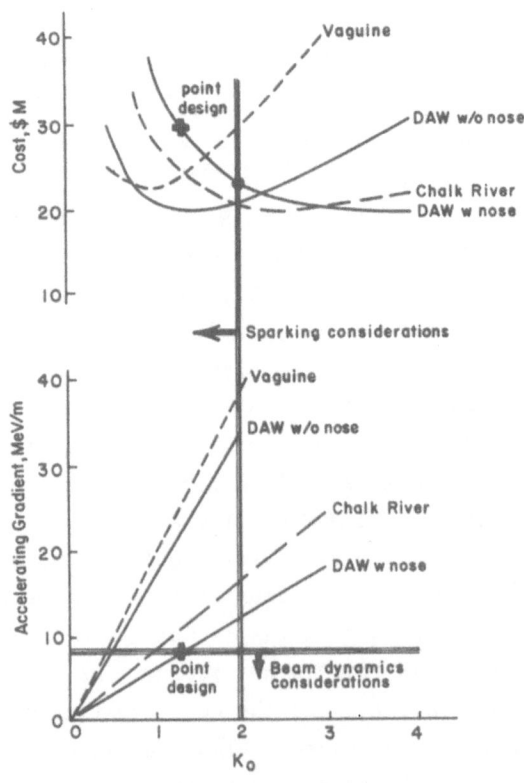

Fig. 19.   Cost estimate for the SSC 2.5-GeV injector linac as a function of K, the ratio of peak-surface-field E in the CCL accelerating gradient E_o as a function of K. Curves for the four CCL geometries of Fig. 18 are plotted.

CCL accelerating gradient, with that gradient held constant along the CCL. Unacceptable growth occurred above 8 MeV/m. The cost impact of operating at this nonoptimum gradient would be significant. More recently, we have devised workable recipes for injecting at a low gradient with approximate equipartitioning, then raising the gradient to the cost optimum level at a controlled rate that causes little emittance growth, which can be done at either an RFQ/DTL or a DTL/CCL interface and can result in a cost and/or length advantage. (If length were more important than cost, appropriate weights could be assigned.)

Full optimization is seen to be a complicated nonlinear optimization with many constraints. More and more attention is being given these days to finding advanced methods of acceleration that would be more efficient and cost less, particularly in the field of HEP because of the cost barrier to even higher energy machines. Ideas involving laser drives have so far come up short because lasers are considerably less efficient than rf sources on an average power basis. At this point, it appears that scaling rf linac technology might be the best bet until some kind of collective-effect accelerator is mastered.

The most important physical limits on accelerating gradient in an rf linac as a function of wavelength are indicated on Fig. 20. A frequency around 30 GHz may be at about the point of diminishing returns, and there a gradient of a few hundred MeV/m may be possible, assuming that other constraints do not intervene.

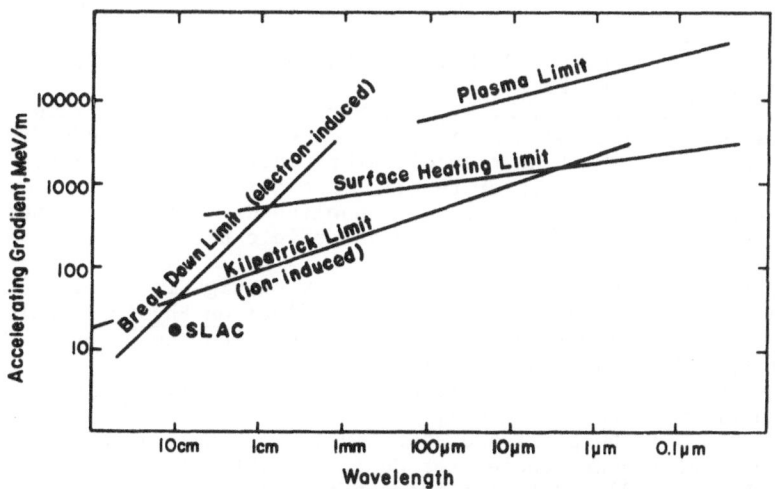

Fig 20. Approximate limits on accelerating gradient, for structures with assumed ratio of peak surface field to accelerating gradient equal to 2, vs wavelength. Kilpatrick-Limit line also assumes peak surface field of twice KL.

The linac structure at these frequencies would be only a centimeter or so in diameter, and the rf power requirements at a few hundred MeV/m would be a few hundred MW/m; thus only very short pulse machines might be considered, but this might be all right for HEP requirements. An innovative research program (Sessler, 1986) is under way to see if a system will work that uses distributed induction-linac modules and single-pass FEL amplifiers to generate rf for the linac structures, prototypes of which have been fabricated. If the system operates successfully, and if one assumes that the linac structure cost is still in the range of $50-100K/m, then the rf cost per peak watt would have to be reduced to around 5×10^{-4} $/rf watt to make optimal use of a 200-MeV/m accelerating gradient. It appears that such costs are not out of the question.

CONCLUSION AND ACKNOWLEDGMENTS

This field of particle accelerators, and linacs in particular, is rich in challenge and subtlety. This has been only an overview in simple terms, describing a few key issues on the basic requirements for bright beams and some of the economic impacts of brightness on accelerator design. It is hoped that the treatment might reveal something of the thrill of addressing an unresolved problem, the satisfaction of solving it, and the posing of future work that might be addressed by newcomers. It is my privilege to discuss here the knowledge accumulated by many colleagues; the many discussions are gratefully acknowledged.

REFERENCES

Darmstadt report, 1982, Proc. Symp. Accelerator Aspects of Heavy Ion Fusion, GSI-82-8.
Gluckstern, R. L., Cooper, R. K., and Channell, P. J., 1985, Particle Accelerators, 16:125.
Gluckstern, R. L., Cooper, R. K., and Neri, F., to be published, "Cumulative Beam Breakup with a Distribution of Deflecting Mode Frequencies," Proc. 1986 Linear Accelerator Conf., 2-6 June 1986, Stanford Linear Accelerator Center report, Palo Alto, CA.
Guy, F. W., and Wangler, T. P., to be published, "Numerical Studies of Emittance Exchange in 2-D Charged Particle Beams," Proc. 1986 Linear Accelerator Conf., op. cit.
Hofmann, I. 1983, "Transport and discussing of High-Intensity Unneutralized Beams," Adv. Electron, Electron Physics, Supl. 13C:49.
Hofmann, I., 1981, "Emittance Growth of Beams Close to be Space-Charge Limit," IEEE Trans. Nucl. Sci., 28(3):2399.
Humphries, S., Jr. 1986, "Principles of Charged Particle Acceleration," John Wiley & Sons, New York.
Jameson, R. A., to be published, "RF Linacs for Esoteric Applications," Proc. 1986 Linear Accelerator Conf. op. cit.
Jameson, R. A., 1983, "Beam Intensity Limitations in Linear Accelerators," IEEE Trans. Nucl. Sci., 30(4):2408.
Jameson, R. A., 1982, "Equipartitioning in Linear Accelerators," Los Alamos National Laboratory report LA-9234-C:125.
Jameson, R. A., 1981, "Beam Intensity Limitations in Linear Accelerators," IEEE Trans. Nucl. Sci., 28(3):2408.

Kilpatrick, W. D., 1957, "Criterion for Vacuum Sparking Designed to Include both rf and dc," Rev. Sci. Instr., 28:824.

Lapostolle, P., Taylor, C., Tetu, P., and Thorndahl, L., 1968, "Intensity Dependent Effects and Space-Charge Limit Investigations on CERN Linear Injector and Synchrotron," CERN report 68:35.

Lawson, J. D., 1982, "Physics of Particle Accelerators," Proc. ECFA-RAL Topical Meeting, Oxford, UK, ECFA 82/63:29.

McKeown, J., and Schriber S. O., 1981, IEEE Trans. Nucl. Sci., 28 (3):2755.

Reiser, M., 1978, "Periodic Focusing of Intense Beams," Particle Accelerators, 8 (3):167.

Sessler, A. M., to be published, "The Two-Beam Accelerator," Proc. 1986 Linear Accelerator Conference, op. cit.

Stokes, R. H., Wangler, T. P., and Crandall, K. R., 1981, "The Radio-Frequency Quadrupole--A New Linear Accelerator," IEEE Trans. Nucl. Sci., 28 (3):1999.

Tigner, M., 1982, "Near Field Linear Accelerators," Proc. ECFA-RAL Topical Meeting, Oxford, UK, ECFA 83/63:229.

Tokyo report, 1984, Proc. Int. Symp. on Heavy Ion Accelerators and Applications to Inertial Fusion, INS.

Tzu, Sun, 1971, "Art of War," Translated and Introduced by Samuel B. Griffith, Oxford University Press, Oxford, UK.

Vaguine, V. A., 1977, IEEE Trans. Nucl. Sci. 24 (3):1084.

Voss, G. A., "Limits of Conventional Techniques," Proc. ECFA-RAL Topical Meeting, Oxford, UK, ECFA 83/63:45.

Wangler, T. P., to be published, "Brightness Limits in Ion Accelerators," this Advanced Study Institute.

Wangler, T. P., Crandall, K. R., and Mills, R. s., to be published, "Emittance Growth from Charge Density Profile Changes in High Current Beams," Proc. 1986 Int. Symp. on Heavy-Ion Fusion.

Wangler, T. P., and Guy, F. W., to be published, "The Influence of Equipartitioning on the Emittance of Intense Charged Particle Beams," Proc. 1986 Linear Accelerator Conf., op. cit.

Washington, D.C., report, to be published, Proc. 1986 Int. Symp. on Heavy Ion Fusion, May 27-29, 1986, sponsored by U.S. Department of Energy, op. cit.

Watson, J. M., 1985, "The Los Alamos Free-Electron Laser," IEEE Trans. Nucl. Sci., 32 (5):3363.

Weiland, T., to be published, "RF Cavity Design and Codes," Proc. 1986 Linear Accelerator Conference, op. cit.

Wilson, P. B., to be published, "Future e^+e^- Linear Colliders and Beam-Beam Effects," Proc. 1986 Linear Accelerator Conference, op. cit.

INDUCTION LINACS*

Denis Keefe

Lawrence Berkeley Laboratory
University of California
Berkeley, California 94720 USA

INTRODUCTION

Induction acceleration -- in a circle as in a betatron, or in a straight line as in the induction linac -- has a venerable history. The idea of using the electric field produced by a time-varying magnetic field to accelerate particles (exclusively electrons, until recently) were first actively explored in the 1920's. The relationship between the electric and magnetic fields is

$$\oint \underline{E} \cdot d\underline{\ell} = - \frac{1}{c} \oint \frac{d\underline{B}}{dt} \cdot d\underline{S} \tag{1}$$

where the line integral is taken around the circular orbit in the betatron or along the core axis in an induction linac. The surface integral is over the orbital area in the first case, and over the core cross-section in the second. Betatron acceleration of electrons was first suggested by Slepian in 1922, and the famous "two-to-one" betatron condition for an orbit of constant radius was discovered independently by Wideroe in 1928 and Walton in 1929. Nonetheless, development of a working betatron -- despite many experimental efforts in the meantime -- had to wait over a decade more, until the classic analysis and experiments by Kerst and Serber (1941).

Bouwers (1939) discusses clearly the principle of linear induction acceleration. He points to the fact that in a betatron the time-varying magnetic field is in the poloidal direction leading to a toroidal electric field. In the induction linac case the time-varying magnetic

---

*This work was supported by the Office of Energy Research, Office of Basic Energy Sciences, U.S. Department of Energy, under Contract No. DE-AC03-76SF00098.

field is toroidal around the axis and the electric field is poloidal. He claims to have proposed this method in 1923, referring to it as "reversal of the transformer method". (Why he chose these words escapes me.) Again decades were to elapse before Christofilos and his coworkers (1964) were to demonstrate a multi-gap working induction linac.

PRINCIPLES OF THE LINEAR INDUCTION ACCELERATION UNIT

In a familiar low-frequency analogy, an induction accelerating unit can be thought of as a torus of ferromagnetic material which acts as a transformer core surrounding the beam, a primary (pulsed) power supply providing excitation by means of a one-turn primary winding looping the core, and the beam acting as the secondary.

More insight can be obtained by examination of the transmission line analogy pictured in Fig. 1(a). (Keefe, 1981). This shows a bent coaxial line, with a hollow inner-conductor, which is driven from the side and shorted at the end. Acceleration occurs across the gap (two holes allow passage of the beam) and will continue from the time the start of the voltage pulse, V, arrives at the gap until its inverted reflection arrives back at the gap and cancels the field. Thus the accelerating pulse lasts only as long as the double transit time from the gap to the short-circuit. If V were 1 MV and the length of the axial part of the line were 1 meter, the gradient would be 1 MV/m and the pulse duration 6 nanoseconds. To provide for this "transit-time isolation" for longer pulses would involve increasing the line length, and hence would soon lead to a serious reduction in gradient.

One way of avoiding this problem is to load the line with either ferromagnetic (μ) or dielectric (σ) material to slow the speed of propagation. Figure 1(b) illustrates how the physical size, axially, can be reduced by this means. Ferromagnetic material, in the form of tape-wound cores to allow for rapid field penetration, is preferred over dielectric because it presents a higher electrical impedance to the driving source. Enough cross-sectional area in the core is required so that saturation does not occur before the end of the desired pulse duration. See Humphries (1986) for details.

A second way around the gradient problem for moderately short pulses is to flare the coaxial line to form a radial line as in the Radlac (see below); this leads to a transversely bulky structure if carried too far.

Several cores can be driven in parallel to provide increased gap voltage. They may be stacked axially [Fig. 1(c)] or radially [Fig. 1(d)]; the latter was the choice for the NBS 2 μ-sec induction linac built by Leiss et al. (1980).

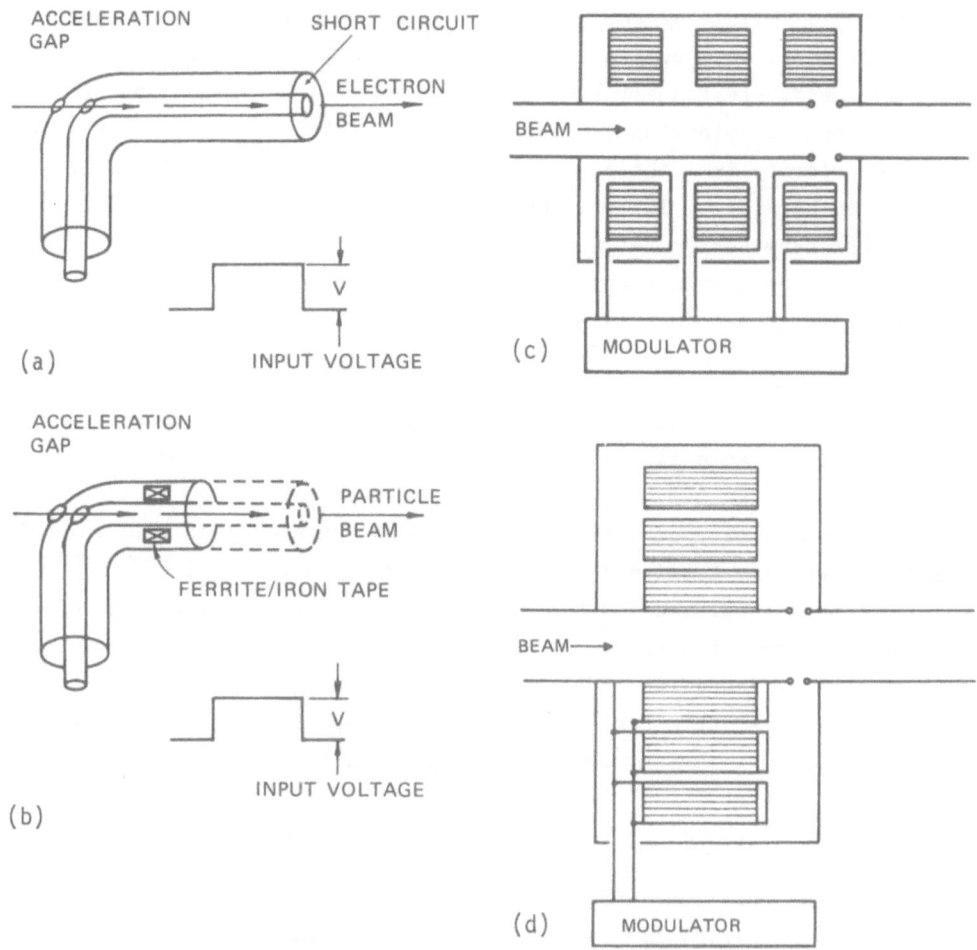

Fig. 1. Evolution of the induction linac geometry. See text.

PRACTICAL CONFIGURATIONS FOR INDUCTION LINACS; EXAMPLES

A practical induction linac is made up of a succession of small
pulse-power modules each timed to give an energy increment to the
particles at the moment of passage of the beam. Pulse-power devices have
the special advantage that the peak power capacity can exceed the average
power rating by a factor of 10^4 to 10^6. (By contrast the size and cost
of a pulsed radio-frequency system designed for a certain peak power
turns out to be roughly the same as one that could deliver about one-
tenth as much power on a cw basis. Since, therefore, it is reasonable to
supply power to each module at the gigawatt level and above -- usually at
a voltage level in the 100 kV range -- the induction linac is ideally
suited (and efficient) for accelerating very large beam currents (100-
100,000 Amps.). Most often, a Marx generator is used to charge a pulse-
forming network or transmission line, the geometry of the line being so

arranged that voltage of only one polarity (accelerating) is seen by the beam. For short beam pulses a vacuum or dielectric line can be used provided the double-transit time is adequately long; for long pulses a high impedance termination (ferromagnetic toroid) is used to exclude the unwanted polarity from the beam for as long as it takes the magnetic material to saturate. Table 1 shows a listing of a number of induction linacs that have been constructed or proposed. Figure 2 shows a schematic of the system (top) and a practical configuration used in the LBL ERA injector and in ETA, and ATA.

Advanced Technology Accelerator (ATA)

At present under construction at Lawrence Livermore Laboratory, this 50 MeV ferrite-loaded linac is intended to deliver 10,000 Amps of electrons in 50 nsec pulses. The average repetition rate is 50 Hz with a burst-mode capability of 1 kHz for 10 pulses. Water-filled Blumleins are used for the pulse-forming lines. The 2.5 MeV gun of the 5-MeV injector (ETA) has been completed and has so far delivered the design current at the desired repetition rate. About 10 kA of beam has been accelerated through further induction module stages to an energy of 4.5 MeV. The development of reliable high-voltage (250 kV) spark-gap switches to operate at 1,000 times per second was a significant technological advance.

Long Pulse Induction Linac

For pulse durations much longer than 100 nsec large volumes of ferromagnetic material are needed and ferrite becomes unduly expensive. The National Bureau of Standards had a program (now discontinued) to address the problem of using thin (1-mil) inexpensive iron sheet, insulated layer to layer, as a core material suitable for a pulse duration of 2 μsec (Leiss et al., 1980). In addition, this design included the novel feature of stacking several (n) nested ferromagnetic toroids of successively larger radii. These can all be driven in parallel from a single pulse line (of voltage V) so that an accelerating voltage nV can be developed across a single gap. Units with n = 4 and 5 were successfully built. Such an arrangement leads to a reduction in the overall length of the accelerator at the expense of a more bulky transverse dimension. [See Fig. 1(d).]

The NBS machine was operated at 0.8 MeV and 1000 Amps electron beam current. Experiments with this beam gave a striking demonstration that a gas-focused beam can propagate for long distances in low pressure gas (1 to 30 Torr) and can even be bent through 360° with dipole magnets only (no additional focusing lenses are needed) for recirculation through the accelerating cavity.

Table 1. High current linear induction accelerators.

ACCELERATOR	ASTRON INJECTOR (ORIG.)	ASTRON INJECTOR (UPGRADE)	ERA INJECTOR	ERA INJECTOR "SILUND"	NEP2 INJECTOR	PROPOSED	ATA	LIVERMORE FXR	PAVLOVSKI	HIF REQUIREMENTS
LOCATION / YEAR BUILT, PROPOSED, OR PUBLISHED	LIVERMORE 1963	LIVERMORE 1968	BERKELEY 1971	DUBNA 1968	DUBNA 1971	NBS** 1971	LIVERMORE 1982	LIVERMORE 1981	USSR 1975	BERKELEY 1976
PARTICLE	e	e	e	e	e	e	e	e	e	HEAVY ION, A > 100
KINETIC ENERGY	3.7 MEV	6 MEV	4 MEV	2.4 MEV	30 MEV	100 MEV	50 MEV	22 MEV	10-12 MEV	10 GEV
BEAM CURRENT ON TARGET	350 AMPS	800 A	900 A	700 A	250 A	2000 A	10,000 A	200 A	100,000 A	20,000 A
PULSE DURATION	300 NS	300 NS	2-45 NS	20 NS	500 NS	2 μS	50 NS	60 NS	20-40 NS	10 NS / 20 μSEC AT INPUT, DECREASING TO 50 NS
PULSE CHARGE / PULSE ENERGY	100 μC / 390 J	240 μC / 1 KJ	30 μC / 100 J	14 mC / 5 J	125 μC / 3.75 KJ	4 MC / 400 KJ	500 μC / 25.0 KJ	120 μC / 2.5 KJ	2-4 MC / 20-50 KJ	300 μC / > 1 MJ
REP RATE, PPS	0-60 / 1400 BURST	0-60 <5> / 800 BURST	0-5		50	1	5 / 1000 BURST	1		1-10
MOMENTUM			<10⁻³	<10⁻³						<10⁻²
NUMBER OF SWITCH MODULES	300	496 (~550 BY 1975)	17	160 ?	1500 ?	250	200	62	24	~ 10,000
CORE TYPE	NI-FE TAPE	NI-FE TAPE	FERRITE	FERRITE	NI-FE TAPE	STEEL TAPE	FERRITE	FERRITE	WATER	TAPE AND FERRITE
SWITCH	THYRATRON	THYRATRON	SPARK GAP	THYRATRON	THYRATRON	SPARK GAP	SPARK GAP	SPARK GAP	SPARK GAP	SPARK GAP
MODULE VOLT. / CORE VOLT.	250 KV / (12.5 KV)	250 KV / (12.5 KV)	250 KV	180 KV / 15 KV	250 KV / 22 KV	400 KV	250 KV	300 KV	500 KV	20-500 KV
ACCELERATOR LENGTH	~ 10 M	30 M	14 M	~ 10 M	210 M	~ 250 M	53 M	40 M		~ 5 KM

* UNDER CONSTRUCTION
** PROPOSED, PROTOTYPE MODULES BUILT

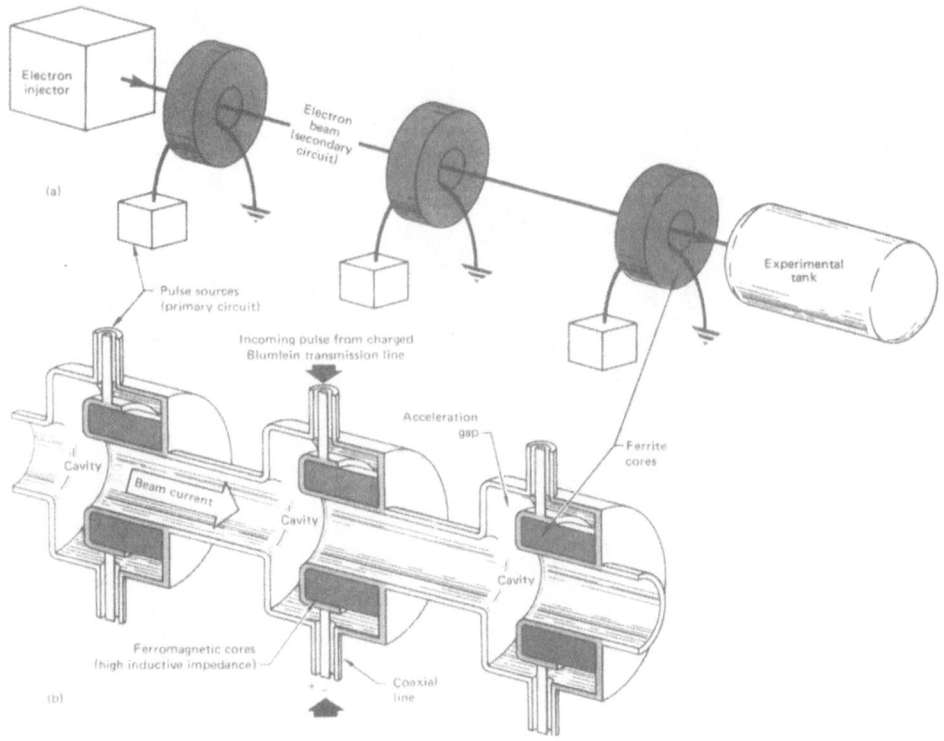

Fig. 2. Two diagrams of the accelerator modules in the ATA. (a)
         Generalized representation of the relationship of the
         two circuits in the accelerator, the electron beam and
         and the pulsed power that drives it through the ferrite
         cores. (b) A more specific drawing (longitudinal section)
         of the accelerator module. This structure is essentially
         a long metal tube consisting of a series of chambers, or
         cavities, containing ferrite rings that prevent the current
         n the coaxial line from shorting. Blumlein transmission
         lines deliver a high-voltage pulse to each cavity just as
         one of the electron clusters that make up the beam reaches
         the cavity. The electron clusters thus pass from one
         cavity to the next, increasing in momentum each time.

Radial Line Accelerator (RADLAC)

For short pulses (~ 20 nsec) the pulse-forming line can be a radial
transmission line closed at the outer radius and, if one wishes to minimize the transverse size, it can be filled with dielectric. Such a
device was first assembled in the USSR by Pavlovskij (1975). RADLAC-1
consisted of 4 such radial lines each supplying 2 MV across a 2-in. gap
(Miller et al., 1981). With the use of a 2-MV pulse-power relativistic
electron-beam (REB) diode as an injector, the final performance was
intended to be acceleration to 10 MeV of a 50-kA annular electron beam
with a pulse length of 15 nsec.

If one analyzes a transmission line initially charged to voltage V, which is suddenly shorted by a fast switch at one end, one finds the following voltage behavior at the open-circuit end. The voltage remains at the value V for a single transit-time τ after switch closure; the wave reflects at the open circuit end, with the voltage doubling to -2V. The resultant voltage amplitude is (V-2V) = -V which persists for a double-transit time 2τ, by which time the pulse has returned from the shorted end, and is now inverted to +V. Thus it can be seen that, in the absence of losses, the output voltage will be a train of square pulses each 2τ long and alternating in amplitude from +V to -V. The only exceptional pulse is the first one, which is only τ in length. For acceleration one can choose to use either the first pulse or, if the longer pulse length is desired, one of the later pulses.

Figure 3 shows how the radial lines are arranged in the RADLAC. Each consists of a flat inner conductor flanked on either side by slightly conical outer conductors to form a tapered line of constant impedance (~ 10 ohms). It is a "folded" geometry with both the switch and gap located at the inner radius. The oil-filled cavities are 3 m in diameter and of fairly simple sheet-metal construction. A circular hole in the center is surrounded by a graded insulator (which provides the oil-vacuum envelope) and allows for passage of the beam. Arranged symmetrically around the cylindrical insulator are eight self-closing oil switches that fire in synchronism as the potential on the inner conductor is brought up rapidly. During the passage of the beam, no field is present on the (shorted) switch side of the line, while the other side acts as an accelerating gap. Solenoid lenses provide magnetic focusing. The injector and the four cavities have operated as to produce a 25-kA beam at 9 MeV, with current losses of only 10%. The average accelerating gradient is 3 MV/m.

Magnetically-Insulated Electron-Focused Ion Linac (PULSELAC)

Results to date from this program are very promising. The basic acceleration scheme is a conventional one using pulsed drift tubes to accelerate a long slug of ions. Ions are accelerated into a drift tube and when the head of the beam reaches the downstream end the voltage is removed from the drift tube and the succeeding one switched on. Instead of using conventional focusing, Humphries et al. have arranged to inject electrons into the drift tubes to provide transverse focusing of the ion beam; a convenient arrangement is an array of field emission points. The key feature of the scheme, however, is to prevent the electrons from crossing the accelerating gap between successive drift tubes so that they do not constitute an inordinate current drain on the power supply. This is accomplished by magnetic insulation whereby a magnetic field is

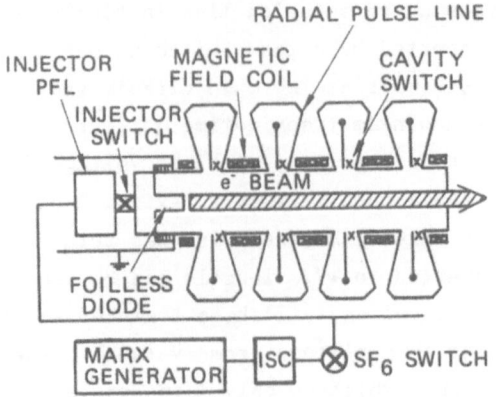

Fig. 3. The RADLAC comprises four folded radial pulse lines that are filled with oil. The taper is chosen to provide constant characteristic impedance in the radial direction. The 2 MeV injector is an E-beam generator and diode.

applied in such a direction that the electrons perform magnetron orbits (with an E x B drift) but can never cross the gap and so drain the voltage generator. Obviously, fresh electrons must be injected into successive drift tubes.

Creating such a situation requires the drift tube to consist of two concentric tubes with an annular ion-beam contained between them (Fig. 4b). Conductors wound around the outer radius at the tips of the outer tube, and around the inner radius of the inner tube can provide a magnetic field to meet the requirements of magnetic insulation. A useful feature of this arrangement is that the E x B drift can carry the electrons around the axis again and again; thus charge-accumulation, which can be troublesome in other geometries, is avoided.

A set of plasma guns arranged in an annulus supplies about 3,000 to 4,000 Amps of carbon ions for injection; a 5-gap pulsed drift-tube system now in operation produced at its exit an impressive 3,000 Amps of carbon ions at an energy of 600 keV, with good emittance. These results seem to indicate that the mobile electron species can adjust its distribution in a benign way to provide focusing that is both strong and, as far as one can judge, reasonably linear.

A RELATED CONCEPT: THE AUTO ACCELERATOR

Auto-Accelerator

This program is an ingenious effort to exploit the high-current electron beam technology that has been developed in the sub-10 MeV region to produce electron-beams at very much higher energies, perhaps in the

208

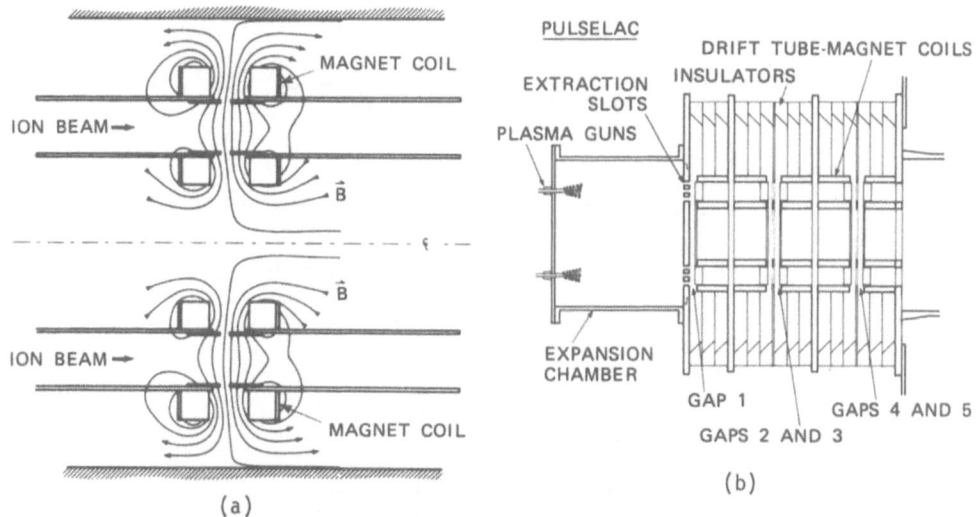

Fig. 4. (a) The arrangement of the four field coils to produce the
desired magnetic field in an accelerating gap of the
PULSELAC. Note that the ions form a hollow cylindrical
situated in the space between the two coaxial conductors
that make up a drift tube. (b) A schematic of the
PULSELAC that shows the three tubes and the annular carbon
ion source.

range 100–1000 MeV. In contrast to the RADLAC geometry, the cavities
have their long dimension (~ 1 m) in the axial, not the radial, direction
(Fig. 5). Each cavity acts as a transmission line with a double transit
time, $2\tau = 6$ nsec. The mode of operation is highly novel; an intense
electron beam passing through the pipe is arranged to charge the cavities
with magnetic energy on a slow time scale and this energy is later
extracted quickly, in a double-transit time, to accelerate a 6-nsec pulse
of electrons near the tail of the beam pulse.

The relativistic electron "charging" beam rises linearly from zero
to $I = 30$ kA in a time of 800 nsec. The beam current $i(t)$ acts as a
current source for the transmission line and instantaneously contributes
a voltage, $Z_o i(t)$, at the gap. If one follows how each such signal
increment reflects back and forth along the line with inversion at the
shorted end, a doubling at its first return to the open-circuit end and,
in the absence of losses, repeated reflections of alternating sign thereafter, one can synthesize the voltage waveform developed across the
gap. This turns out to be a linear rise to a value $Z_o i(2\tau)$, followed by
a linear fall to zero at $t = 4\tau$ and a repetition of this triangular form
as long as the current rise continues. Thus, the average value of the
gap voltage is $1/2\ Z_o i(2\tau)$. Bearing in mind that $Z_o = (L/C)^{1/2}$ and
$c = 1/(LC)^{1/2}$, we find that this voltage is equal to $(Lc\tau)\ dI/dt$, where
$Lc\tau$ is the lumped-element (long time-scale) inductance of the coaxial

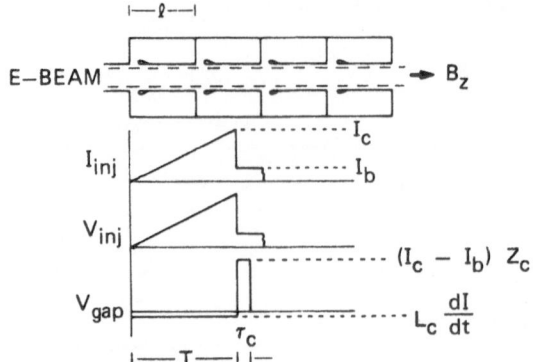

$$T = \text{CURRENT RISETIME} = 800 \text{ ns}$$
$$\tau_c = \frac{2\ell}{c} = 6 \text{ ns}$$
$$Z_c = \text{CAVITY IMPEDANCE} = 70\Omega$$
$$L_c = \text{CAVITY INDUCTANCE} = 0.23 \ \mu H$$

Fig. 5. The NRL auto-accelerator concept. From top to bottom
the figures show: Cavity structure; injected current
i(t); voltage developed across each gap showing the
time averaged retarding voltage L_c di/dt during the
current rise, and the accelerating voltage during the
current drop.

cavity. The sign of this voltage is such as to provide a slight
deceleration of the electron beam.

If the current rise is halted at i(t) = I and the electron-beam
current returned to zero, the destructive reflections that keep the gap
voltage at this low value are suddenly removed and it can quickly be
verified that a large accelerating voltage, $Z_o I$, appears on the gap for a
time 2τ. In the NRL auto-accelerator the electron-beam current is
switched not to zero but to I/5, so that the accelerating voltage per gap
is 0.8 $Z_o I$. (See Fig. 5.)

What is distinctive about this device is that it circumvents two of
the major problems of pulse-power accelerators--the switches and the
insulators. Since the magnetic energy release from a cavity begins just
when the downward step in beam-current occurs at the gap location, the
accelerator is automatically self-synchronized from gap to gap; jitter is
eliminated because switches are not needed. Insulators at the acceleration gaps are also not required; for short pulses, very high voltages
(~ 3 MV) can be achieved across just a few centimeters in vacuum.
Finally, the accelerator can be designed to have relatively high
efficiency from the wall plug to the beam, perhaps in the region of 30%.

In the experiments, the injector is an E-beam generator, with a
transmission line for pulse forming, which produces a 30 kA, 1.5 MeV
hollow beam from a foiless diode. This beam is transported in a uniform-
field solenoid magnet (15 kG) through a sequence of coaxial cavities.

Six cavities were planned for the proof-of-principle experiment but only two installed. Electrons were accelerated from 0.3 MV to 3 MV with 4 kA beam current. Some "cross-talk" was encountered between the two cavities, but it was eliminated by reducing the Q-factor of the cavities.

It is tempting to call this device a collective accelerator in which electrons are used to accelerate other electrons, but in fact the electromagnetic field occurs as an intermediary between the action of one set of electrons and the reaction of the other set. (Note, for comparison, that in a conventional rf accelerator electron-beam tubes create rf fields that are coupled via wires or waveguides to cavities and thence to the beam. The Two Beam Accelerator shares similar properties.)

INDUCTION LINACS FOR IONS

The very intense (20 kA) short-pulses (10 nsec) of heavy ion beams needed for a heavy-ion driver for inertial fusion led to the proposal that an induction linac could provide a suitable solution (Keefe, 1976). The non-relativistic nature of ions beams makes for considerable difficulties with transverse focusing, but also allows for the interesting strategy of current amplification. The special application to fusion drivers is discussed in a later talk in this series.

Another application of a current-amplifying ion induction linac was pointed out by Keefe and Hoyer (1982) namely the acceleration of protons to the 50-250 MeV range to generate an intense short burst of spallation neutrons.

An interesting result of their studies was that when the beam energy was fixed at 10 kJ per pulse it was advantageous, from both capital and operating cost considerations, to choose low kinetic energy and high beam charge (125 MeV, 80 μC) rather than high kinetic energy and low beam charge (2 GeV, 5 μC). The initial pulse length was typically 5 μsec and the final pulse length 20 nsec.

EFFICIENCY

When the beam accelerated in an induction linac is in the range of a few hundred amperes to many kiloamperes, the energy transfer efficiency can be very high. This is illustrated by a somewhat simplified analysis presented by Faltens and Keefe (1977). In constructing an equivalent circuit for an induction linac, Faltens has repeatedly stressed that the beam should be treated as a current source not an impedance. The analysis uses the concept of "gap impedance", Z_g, which for high frequencies (Faltens and Keefe, 1981) can be written as:

$$Z_g = 60 \; g/a \; \text{ohms}, \tag{1}$$

where g is the gap length and a is the gap radius. The voltage across the gap will result from three terms: the incident voltage from the feed line V_o^+, the voltage reflected from the gap, $V_r^- = V_o^+(Z_g - Z_o)/Z_g + Z_o)$, and the voltage wave generated by the beam current $V_b^- = I_b(Z_g Z_o)/(Z_g+Z_o)$ since the beam "sees" the gap impedance paralleled by the transmission-line impedance. The efficiency of the induction-linac module can be defined as the ratio of the power delivered to the beam to the power delivered to a matched load. This efficiency is

$$\eta = \frac{V_{gap}I_b}{\left(V_o^+\right)/Z_o} = \frac{\left(V_o^+ + V_r^- + V_b^-\right)I_b}{V_o^{+2}/Z_o} = \frac{\left[V_o^+\left(\dfrac{2Z_g}{Z_g+Z_o}\right) - I_b\dfrac{Z_gZ_o}{Z_g+Z_o}\right]I_b}{V_o^{+2}/Z_o}$$

$$= \frac{Z_g}{Z_g+Z_o}\left(2 - \frac{I_bZ_o}{V_o^+}\right)\left(\frac{I_bZ_o}{V_o^+}\right). \tag{2}$$

When maximized, this efficiency is

$$\eta_{max} = \frac{Z_g}{Z_g+Z_o}, \tag{3}$$

which can be made close to unity by choosing $Z_o \ll Z_g$. The achievement of such a high efficiency, however, would require a precise matching of the voltage and current waveform, and a specific design value of the beam current.

CONCLUSION

The induction linac represents well-established technology for efficiency acceleration of beams with currents in the range of hundreds and thousands of amperes. The lifetime can be long and the repetition rate high, because the linac is engineered as a multigap device with low energy per stage (a few kilojoules) but very high power per stage (gigawatts).

REFERENCES

Bouwers, A., 1939, Electrische Hochstpannung, Berlin, Chapter 1, p. 80.
Christofilos, N. C., Hester, R. E., Lamb, W. A. S., Reagan, D. D.,
    Sherwood, W. A. and Wright, R. E., 1964, Rev. Sci. Inst., 35:866.
Faltens, A., Keefe, D., 1977, Proc. Xth Int. Conf. on High Energy Accel.,
    (Protvino), Vol. I, p. 358.
Faltens, A. and Keefe, D., 1981, Proc. 1981 Linac Conf., (Los Alamos),
    LANL Report LA-9234-C, p. 205.
Humphries, S. J. et al., 1979, IEEE Trans. Nuc. Sci., NS-26:4220.

Humphries, S. J., Jr., 1986, "Principles of Charged Particle
    Acceleration", Wiley-Interscience, New York.
Keefe, D., 1976, Proc. ERDA Summer Study of Heavy Ions for Inertial
    Fusion, LBL-5543, p. 21.
Keefe, D., 1981, Particle Accelerators, 11:187.
Keefe, D. and Hoyer, E., 1981, Proc. Workshop on High Intensity
    Accelerators and Compressor Rings, (ed. M. Kuntze), Karsruhe, June
    1981, Kernforschungszentrum Karlsruhe Report KFK 3228, p. 64.
Kerst, D. W. and Serber, R., 1941, Phys. Rev., 60:53.
Leiss, J. E., Norris, N. J., and Wilson, M. A., Particle Accelerators,
    10:223.
Lockner, T. R. and Friedman, M., 1979, IEEE Trans. Nuc. Sci., NS-26:3036.
Miller, R. B., Prestwich, K. R., Poukey, J. W. Epstein, B. G., Freeman,
    J. R., Sharpe, A. W., Tucker, W. R. and Slope, S. L., 1981, Journ.
    Appl. Phys.
Pavlovski, A. I., 1975, Sov. Phys. Dokl., 20:441.

ADVANCED CONCEPTS FOR ACCELERATION*

Denis Keefe

Lawrence Berkeley Laboratory
University of California
Berkeley, CA 94720 USA

INTRODUCTION

The study of advanced methods of acceleration has been largely
concentrated on two quite different areas of applicability -- first, the
acceleration of electrons and positrons to ultra-relativistic energies
(> 1 TeV) and second, the acceleration of protons and ions from rest or
very low energies to a speed that is a significant fraction of the speed
of light. Both applications share the common goal of seeking to achieve
accelerating gradients far in excess of those obtainable with conventional accelerating structures, e.g., E >> 10 MV/m.

The energy frontier in high-energy physics has advanced exponentially with time simply because new inventions have been successfully
developed that have kept the accelerator costs within bounds. The largest accelerator under serious consideration today is the Superconducting
Super-Collider (SSC) which will accelerate protons by the traditional
synchrotron method to 20 TeV, and then act in the storage ring mode to
collide two such beams together. It seems likely that the next step in
energy beyond the SSC will employ the Linear Collider technology now
being developed at SLAC, and will use electron-positron collisions in the
multi-TeV range ("Equivalent" physics can be accomplished with electrons
at about one-fifth the energy of protons). Electron linacs at such
energies if based on today's technology would be prohibitive in cost,
size and power consumption. If the message of the Livingstone Chart of
exponential growth continues to hold, new accelerating concepts must be
developed to enable us to advance in center of mass energy.

*This work was supported by the Office of Energy Research, Office of
Basic Energy Sciences, U.S. Department of Energy, under Contract No.
DE-AC03-76SF00098.

In contrast to electrons, which are relativistic ($\gamma > 2$) when they leave a typical electron gun and can be accelerated with high gradient, ions gain velocity at a much lower rate. After the 500 m of linac in AMPF, the protons have attained only a γ of 1.8. Hence much work has centered on collective methods of ion acceleration in which the charge density of an intense electron-beam (IREB) is modulated to produce very high fields within the beam that can then be used to accelerate ions embedded in the beam. These fields must be programmed to have a phase-velocity that varies in synchronism with the changing speed of the ions; the problems of creating slow-moving modulations and controlling their speed are indeed severe. The IREB sources are of two types, pulse-power single-gap diodes driven by a Marx generator and pulse-forming-line (PFL), or multigap induction linacs. The latter have the advantage of offering high repetition rate, high efficiency, and excellent beam quality.

Advanced concepts have been the subject of several recent conferences and workshops [Corbett et al. (1983), Bryant and Mulvey (1985), Channell (1982), Joshi & Katsouleas (1985)] in the hope that some might provide promising breakthroughs in technology, decades in the future. It has become customary to categorize advanced concepts (as shown in Table 1).

For relativistic electrons, the c.w. Near-Field devices offer the opportunity of moving continuously up in frequency from the conventional linac structures for which a highly sophisticated data-base of experience exists; calculations of expected beam quality and behavior, and parameters such as projected efficiency, can be on more or less firm ground depending on how far one extrapolates downward in wavelength. For other entries in the Table, answers to such questions are much more speculative; laboratory demonstrations of the physics concepts do not, indeed, exist in many areas.

The accelerator examples listed in the Table offer, to varying degrees, the promise of controlled or staged acceleration in which the phase velocity of the field can be kept in synchronism (or nearly so) with the beam. (Acceleration by uncontrolled virtual cathode formation is not listed, for example).

Only selected examples will be reviewed in this paper; other reports at this meeting by Wilson, Cooper, Jameson and Neil will cover some of the topics in greater detail.

Table 1. A classification of accelerator schemes.

Medium	Comment	Example Accelerators
Vacuum (Near Field)	Tuned EM Wave	Advanced Linac Structures powered by microwave tube, free-electron laser, or optical laser
Vacuum (Near Field)	EM Pulse	Wakefield Transformer, Autoaccelerator, Switched power
Vacuum (Far Field)		Inverse Free Electron Laser (IFEL)
E (Electron)-beam		Auto-resonant (ARA) Ionization Front (IFA) Waves (e.g., space-charge) *Electron Ring (ERA)
Gas		Inverse Cerenkov (ICA)
Plasma		Laser Beam Wave (LBW) Plasma Wake Field (PWF)

*ERA is fundamentally different from other E-beam devices in that the electrons providing the field remain in the frame of the ions rather than flowing through the region of charge modulation; the electron ring behaves more as a vehicle than a medium.

METHODS FOR RELATIVISTIC ELECTRONS/POSITRONS

Approach

Obtaining high gradients in near-field vacuum structures is limited by the surface breakdown field. This limit is increased by using higher frequencies than today (SLAC operates at 3 GHz), hence the interest in pushing to shorter wavelength microwaves and, ultimately, lasers. Unconventional linac structures for very short wavelength excitation are discussed by Palmer (1985). See Fig. 1. We address below advanced concepts that seek to achieve fields beyond those attainable in microwave structures by use of two approaches:

i) Develop the fields within a plasma either neutral or non-neutral; such fields can be local and need never appear near the structure walls.

ii) Take advantage of the fact that the surface breakdown field in vacuum can be very large for extremely short pulses [V_b scaling roughly

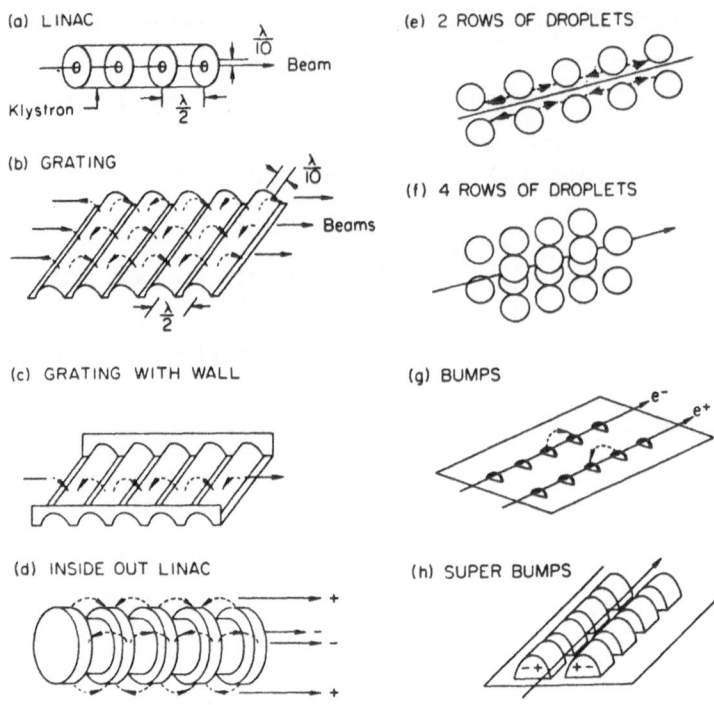

Fig. 1. Some unusual RF near-field structures discussed by
Palmer (1985).

as $(\tau)^{-1/4}$]. Very large fields on axis can be produced by radial line
transformers.

Plasma Accelerators

The last few years have seen a rapid expansion of interest in
devising ways of controlling large-amplitude waves in plasmas to make
them useful for the acceleration of electrons to high energies. Some
twenty-five years ago, the invention of the laser caused much speculation
on whether the very large coherent electric fields it can produce might
be useful for acceleration. Unfortunately, the laser fields are transversely polarized and so could not be used for acceleration over long
distances in vacuum. In a neutral plasma, however, the intense radiation
from a laser can modulate the plasma to produce plasma waves (Langmuir
waves) with longitudinal electric fields of enormous magnitude. These
fields can accelerate particles - as witnessed in laser fusion experiments, where the effect is undesired - and, if the mechanism is controllable over significant distances, might offer for the future a compact
way of producing very high energy electron beams. The interest in the
use of plasma waves that followed the Tajima-Dawson (1979) proposal of
the Beat-Wave Accelerator has led to several other suggestions for plasma

accelerators, including the substitution of a particle beam for the laser to excite the waves. Several typical schemes are summarized briefly below.

Plasma Beat Wave Accelerator (PBWA)

Two coherent laser beams, with frequencies ω_1 and ω_2, are injected into an underdense plasma with plasma frequency, ω_p. If the difference frequency $(\omega_1-\omega_2)$ matches the plasma frequency, ω_p, forward Raman scattering of the higher frequency beam (ω_1) gives rise to a photon of the lower frequency beam (ω_2) and a forward-going plasmon (ω_p). A very large electric field due to the temporary charge separation in the plasma can be built up resonantly and can propagate forward with a phase velocity $v_p = c(1-\omega_p^2/\omega^2)^{1/2}$ where $\omega \approx \omega_1 \approx \omega_2 \gg \omega_p$. (See Fig. 2.) The group velocity of the light pulse is also equal to v_p; thus the laser pulse and the wake of excited plasma waves move forward as a unit into the as-yet undisturbed plasma downstream.

For electrons moving in synchronism with the plasma wave the accelerating gradient (proportional to ω_p) can be very large, i.e.,

$$E(\text{volts per cm}) \approx \left[n_o \ (\text{particles/cm}^3) \right]^{1/2}$$

or, for a plasma density of 10^{18}cm^{-3}, some 100 GeV/m. For the accelerated particle to remain synchronous, it should travel at a speed, $v = v_p$; v_p is, however, less than c, and the accelerated particle will move forward out of synchronism at some limiting kinetic energy which is proportional to $1/\omega_p^2$, or $1/n_o$. thus there is a competition between the desires for high gradient (large n_o) and long accelerating distance (small n_o).

The surfatron concept is a proposal to circumvent this difficulty by application of a magnetic field at right angles to the direction of propagation of the beat wave. The accelerated particles thereby acquire

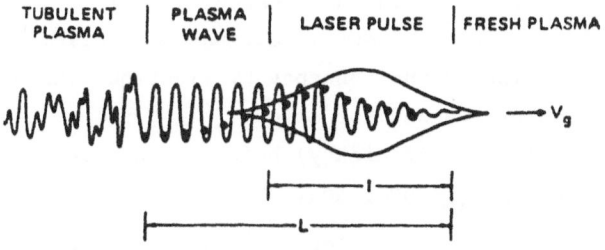

Fig. 2. Build up of accelerating field which moves forward in beat-wave excitation method.

a small transverse component of velocity with respect to the direction of light propagation and can thus keep in step with the wave even when their speed exceeds v_p. (See Fig. 3.) This technique can be shown to have the desirable property of phase-stability. The laser beams in a surfatron must, of course, have significant lateral extent to accommodate the sideways motion of the accelerated particles. An alternative scheme proposed by Katsouleas et al. (1983) called "optical mixing" results in significantly less width; here the two laser beams are tilted at a slight angle to each other and the plasma wavefronts move at an angle to the direction of the accelerated beam.

While extensive work with computer simulation has taken place, much more needs to be done to sort our the complicated physics of wave-wave and wave-particle interactions when a high-power laser interacts with a plasma. Instabilities can occur and ultimately, very short laser pulses (10-50 psec) must be employed to avoid such unwanted waves. Initial experiments on exciting plasma waves have taken place at UCLA, RAL, and LANL. The most successful of these, by Joshi et al. (1985) at UCLA, using two laser beams at $\lambda = 10.6$ μm and 9.6 μ, has demonstrated the generation of the fast plasma beat-wave in a plasma of the right density ($10^{17} cm^{-3}$) for resonance. The scale length of the wave region was about 1 mm and a field of some 500 MV/m has been inferred.

The beat wave accelerator is unique among advanced concepts in the magnitude of the accelerating fields it could perhaps achieve. A vast amount of theory and simulation is still needed to sort out the extraordinarily complex physics of the interaction of intense laser beams with plasma. In addition, a succession of scaled experiments, beginning with a demonstration of modest acceleration, will be needed over many years to establish whether a viable accelerator system is possible and, if so, what its limitations are. Developments in high power laser technology will also be needed.

Plasma Wakefield Accelerator (PWFA)

Excitation of intense Langmuir waves in a plasma can alternatively be achieved by use of small, intense bunches of relativistic electrons rather than by laser light. There is already a substantial technology base for the production of such bunches by electron linacs; by contrast, technology advances are needed for lasers (e.g., pulse length ~ 10 picoseconds, higher wall-plug efficiency) before an attractive laser-plasma accelerator system might be proposed.

In the plasma wakefield accelerator (PWFA) proposed by P. Chen et al. (1984), the driving stream of electron bunches, with dimensions less than a plasma wavelength, excite plasma waves by virtue of a two-stream

220

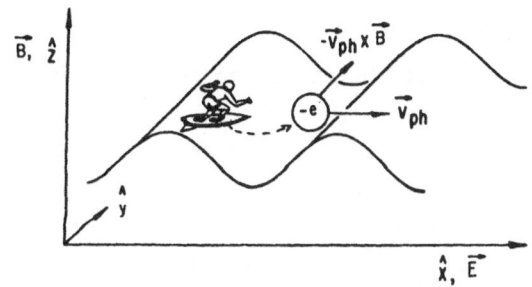

Fig. 3. A magnetic field gives sideways motion to keep the particles in phase with the wave.

instability -- as opposed to the ponderomotive force in the laser case. The plasma waves growing in the wake of each driving bunch to large amplitude (and eventually becoming turbulent) can produce a very high gradient and so accelerate a second driven bunch traveling in phase with the wave. The attainable gradients, and the phase-slip problem are the same as for the PBWA.

The energetics governing the acceleration of the driven bunches at the expense of the acceleration of the driving bunches (assumed to be much higher in intensity) was quickly perceived for the PWFA as related to the classic beam-loading problem in conventional accelerators, namely, the maximum kinetic energy gain per driven particle will not exceed twice the kinetic energy of a driving particle. Thus, a driving beam of 1 GeV electrons could add nearly 2 GeV to the high-energy particles, meaning that acceleration to the TeV range would require many stages. (In principle, if the charge profile within the already tiny drive bunch could be tapered, the "transformer ratio" could be increased above two; this seems, however, difficult to achieve in practice.)

Use of an electron beam rather than laser light to excite accelerating plasma waves seems to offer some practical advantages. As mentioned, the technology is largely in hand, and the wall-plug efficiency could be significantly greater. Unless laser self-focusing is found to be realizable in a controlled and reproducible way, the laser beam divergence sets a limit to the length of the acceleration region (Rayleigh length) obtainable with a laser. Wave excitation by an electron beam can be maintained over a longer distance, first because of the very small emittance and, second, because the electrons can be confined by magnetic quadrupoles (i.e., non-material lenses) exterior to the plasma.

An interesting proof-of-principle experiment with a 22 MeV short-pulse linac (now 30 psec; being modified for 5 psec operation) is in the planning stage at ANL, by Rosenzweig et al. (1985).

Plasma "Grating" Accelerator

Simulation studies of the surfatron indicate that particle energy gain is limited by wave dynamics rather than by particle dynamics. While most of the laser energy is transferred to plasma wave energy, the plasma wave group velocity is very small, and the energy that can be transferred to the driven beam diminishes with distance. Important consequences of this pump energy depletion are reflected in limitations on the acceleration distance and in the overall electrical efficiency.

In an effort to circumvent pump depletion, Katsouleas et al. (1985) have explored the possibility of injecting laser light from the side, along the length of the accelerating region. In this case, therefore, one can choose the laser beam (one frequency only) to be polarized parallel to the direction of acceleration. Creating a periodic density modulation in the same direction along the plasma then leads to a beat-wave between the laser light-wave and the zero-frequency density wave, which is analogous to the two-laser beat-wave in the PBWA. Thus a propagating accelerating plasma wave can be set up with analogous properties but in a different -- and presumably more favorable -- geometry. Work on this approach has consisted, so far, of some simulation studies.

Pulsed E-Beam, Pulsed Laser Accelerator

Following the very successful demonstration that an intense relativistic electron beam (IREB) to 10 kA, 3-50 MeV, can be guided in a straight line over a long length (50 m) by a laser beam, Briggs (1985) proposed that such a beam might be helpful in creating the rapid acceleration of a small bunch of very high energy electrons following several centimeters behind the tail of the drive beam. While transient low density plasmas are involved, extended plasma wave phenomena, which can often lead to uncontrolled and undesired effects, are not invoked. Two lasers are used (one long pulse, one short pulse) to control the sequence of events, but these are used basically to cause ionization in gas and are of modest intensity. High-gradient acceleration of the driven beam is envisaged to take place as a result of four steps (see Fig. 4).

i) A uv laser photo-ionizes a narrow column of gas along the axis of the drive-beam accelerator. The gas (benzene) density needed is in the range of 10^{13}-10^{14}cm^{-3} -- orders of magnitude less than the plasma densities referred to in earlier sections.

ii) The electron beam is injected on axis. As it propagates, it rapidly expels the cold electrons created during Step (i) leaving an ion column about a millimeter across on axis, and furthermore, creates additional ionization thus enhancing the density of positive ions on axis. (Note that Step (i), in fact, is not absolutely necessary as the ion-beam

222

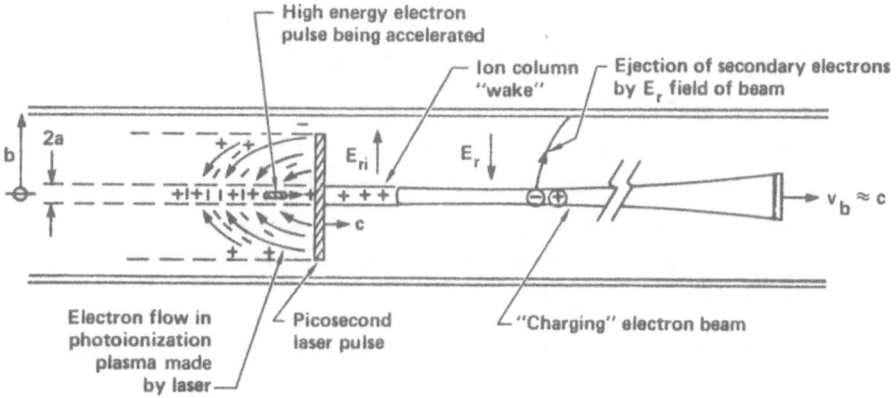

Fig. 4. Schematic of the electron beam/laser accelerator.

will propagate stably down the tube by virtue of creation of the pulses, the ion column remains on axis for some nanoseconds before dispersing. Steps (i) and (ii) have been demonstrated at the Advanced Technology Accelerator at LLNL, with a 10 kA 50 nsec long beam.

iii) Next, a one picosecond laser pulse (a few Joules) is injected in coincidence with the high-energy beam bunch to be accelerated. In time it should follow the driving beam tail by about one nanosecond, creating a broad plasma column whose head follows closely behind the drive pulse tail. Transversely, the light pulse should be larger than either the driving or driven beam.

iv) Finally, cold electrons created off axis by the picosecond laser pulse rush radially inward towards the residual positive ion column on axis. Briggs (1985) has shown that this inward flow of electrons can create a large electric field on axis that could accelerate the high energy bunch with a field of perhaps 250 MV/m or more. Also, transverse focusing of the driven beam is taken care of automatically by the presence of the on-axis positive-ion column.

Special attention has to be paid to generating and maintaining a sharp edge on the tail of the drive beam; this requirement, together with the dispersion of the laser light and the energy depletion of the charging beam, will result in the need for multiple stages to reach the TeV region. The length per stage, however, could be substantially greater than in the PBWA and PWFA. It is especially noteworthy that this concept is not a wakefield accelerator, the function of the leading beam simply being to create enhanced ionization on axis and to eject the cold electrons (in the process, of course, acting as an energy source). Thus, the system is not subject to the transformer ratio limit of two referred to in the aforementioned section, Plasma Wakefield Accelerator (PWFA), for the PWFA; instead, the ratio of kinetic energy gain (driven) to kinetic energy loss (drive) can be of order 10,000.

The applicability of the scheme to the acceleration of positrons has not been considered in any detail. Both an accelerating and a focusing field suitable for positrons will occur some distance behind the optimum location for electrons, corresponding to about a half a plasma oscillation period. Simulation studies are needed to determine the quality and reproducibility of this field.

Because of the multi-stage nature of the process envisioned, a sequence of experiments will probably be needed to establish whether the concept works or not, and to determine how reliable and reproducible it could be. An encouraging technological aspect is that an induction linac driven by magnetic modulators based on new ferromagnetic materials has recently been operated at LLNL at a repetition rate of 10 kHz; extension of this pulsing technique well into the megahertz range seems to be straightforward.

Inverse Cerenkov Accelerator

A relativistic particle ($\beta \approx 1$) traveling through a refractive medium produces Cerenkov radiation continuously along its path at a cone angle θ, where

$$\text{Cos } \theta = 1/\beta n,$$

and $\underline{n}$ is the refractive index of the medium. In the limit of large γ and small angles, this can be rewritten:

$$2(n-1) = \theta^2 + \frac{1}{\gamma^2}$$

so that $\theta \approx$ constant if $\theta \gg 1/\gamma$. In the inverse process, coherent laser light is brought in with a corresponding conical geometry and a particle on axis sees a continuous accelerating field $E \sin\theta \cos\psi$, where E and ψ are the laser field and phase. Such an interaction over a distance of a few centimeters has been experimentally observed by Fontana and Pantell (1983) at Stanford. An interesting feature of this mechanism is that the slipping out of phase of the particle with the accelerating field frequently noted earlier need not occur. If ψ lies between $\pi/2$ and π, particles are both focused to the axis and accelerated; furthermore, the motion is phase-stable.

To use this mechanism to accelerate electrons (or positrons) to high energy requires an index of refraction corresponding to about 1 atmosphere of hydrogen gas. To minimize multiple scattering, one would like to choose as low a gas pressure as possible; this, however, would decrease the accelerating component of the electric field by decreasing θ.

224

Conical illumination over significant distances can be accomplished by containing the laser light within a reflecting cylinder.

In this scheme, the usable laser intensity is limited by ionization breakdown in the gas. More damaging, however, for high-energy applications is multiple scattering in the hydrogen gas which produces an intolerable increase in beam emittance for acceleration to energies in excess of a few tens of GeV.

Inverse Free-Electron Laser

This concept falls within the far-field vacuum category in that no medium is used and the laser accelerating field is many wavelengths away from neighboring surfaces. Neil will later discuss the free-electron laser in which a fast-electron beam is wiggled in a spatially reciprocating magnetic field and emits synchrotron radiation in the forward direction with a wavelength $\lambda \approx L/2\gamma^2$ where L is the magnetic-field periodicity. The electrons lose energy while the photon field is amplified by coherent lasing action.

In the inverse device, the electrons are injected into a wiggler field and irradiated in the direction of motion with an intense laser beam, which can give energy to the electrons in a resonant way, by the inverse process. Since the laser light is of fixed wavelength, λ, a transverse magnetic field is added to maintain the condition $\lambda = L/2\gamma^2$ as acceleration proceeds and γ increases.

Pellegrini (1982) has presented a design for an IFEL for 300 GeV electrons. The synchrotron radiation loss (which increases as $B^2\gamma^2$ -- just as in a storage ring) becomes intolerable at higher energies.

Switched Radial-Line Accelerators

The field that can be sustained without breakdown by an accelerating gap in vacuum depends upon the pulse duration for pulses shorter than, say, one microsecond. For nanosecond and subnanosecond durations the field can be many times the static breakdown field. This concept relies on supplying a very short voltage pulse to the periphery of a pair of parallel discs which have a small hole on-axis through which the accelerated beam passes. As the pulse travels inwards to the axis of the radial transmission line it becomes amplified exactly in the fashion Wilson has described earlier for the wakefield accelerator (WFA). Indeed, the disc structures when stacked together very much resemble the WFA structure except for details at the periphery, where the drive electron beam is replaced by a sequence of suitably phased, independent switches.

This method was proposed in 1968 for the acceleration of electron-rings (containing ions; see Section 3) from low energies to high energies. Hartwig (1968) has reported results for a copper disc structure fed at 16 points on the periphery (to simulate an approach to ideal symmetry) and indeed he observed the expected voltage amplification (about x2.5) for pulses in the 1-3 ns range. The transformer ratio was not large for this application because the hole in the plates had to be several centimeters in diameter to allow for passage of the electron rings.

Interest in the method was renewed when Willis (1985) pointed out the possible advantages for ultra-high energy electron linacs. Here the pulse-length could be vastly shorter (~ 10 psec) and the hole size much smaller (~ 1 mm) leading to a much higher breakdown limit and a larger transformer ratio. He estimated that accelerating gradients of 1 GV/m might be achieved. The need for switching times of order 1 psec has stimulated studies of laser-switched gallium-arsenide switches. The amount of laser energy needed to accomplish the switching at a kilohertz rate may be very large, however.

Two-Beam Accelerator

Hopkins et al. (1984) have proposed using the immense power (~ 1000 MW) available in the 30 GHz range from free electron lasers to power an ultra-short-wave linac structure ($\lambda \approx 1$ cm). The concept is shown in Fig. 5. The low energy beam -- a few MeV -- has thousands of amperes and as it loses energy to the microwave radiation in the wigglers it is periodically rejuvenated by further induction acceleration modules. Gradients of 200 MeV/m and more have been observed in test structures. The Livermore FEL work is presented in this volume by Neil.

ADVANCED CONCEPTS FOR ION ACCELERATION

In the past twenty years, several avenues have been explored in an attempt to harness the collective fields of intense electron beams to accelerate ions from low to high energies. Some of these methods could also be used, in principle, to accelerate high energy electrons or positrons. To date, only three methods have been successful in accelerating ions in a moderately orderly way, and then only to modest energies of the order of tens of MeV/amu.

Early observations of collective ion acceleration were with disorderly processes in which an intense electron beam formed a virtual cathode which halted the propagation of the beam by the large space charge potential at its head. When ions were formed -- either from a background low-pressure gas or surface flashover of an insulator (Luce diode) -- the virtual cathode became neutralized and the beam front propagated forward. In the process of dynamic relaxation of the virtual

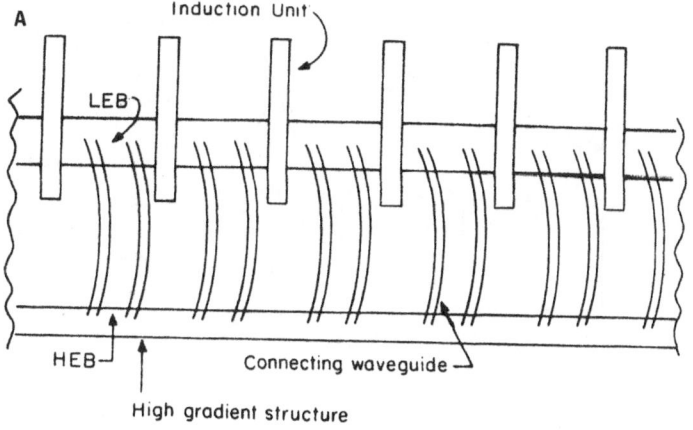

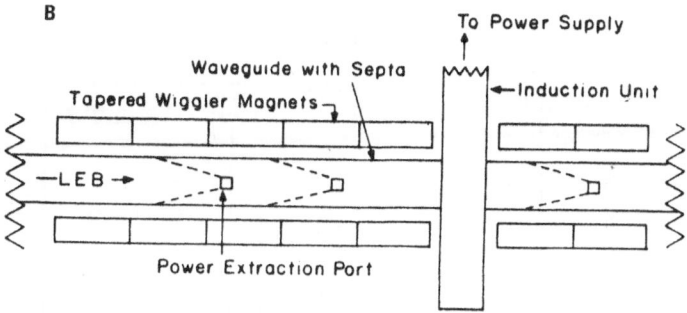

Fig. 5. The two-beam accelerator concept showing (a) the power generating (LEB) and high gradient structures (HEB) side-by-side and (b) the low energy beam in the wiggler structure.

cathode, ions were accelerated with a broad energy spectrum centered about two or three times the electron kinetic energy. While these results were interesting, this so-called "natural" acceleration process is basically uncontrollable and incapable of significant extension.

A degree of control proposed by Olson (1974) was to use a laser beam to produce a spatially narrow region of ionization of the background gas just in front of the virtual cathode; by programming the motion forward of the laser beam, the virtual cathode (and the large accelerating field at its head) could be dragged along in synchronism with the accelerating ions. (See Fig. 6.) This method has recently been shown to work (Olson 1985); a modest number of protons and deuterons were accelerated to an energy of 10 MeV over a distance of 0.3 m.

A unique method proposed by Veksler (1968) -- the electron ring accelerator (ERA) -- uses acceleration of a ring of relativistic electrons seeded with a small number of positive ions. The detailed principles of the method are discussed by Keefe (1970). Multigap acceleration of rings holding xenon ions has been achieved by Sarantsev

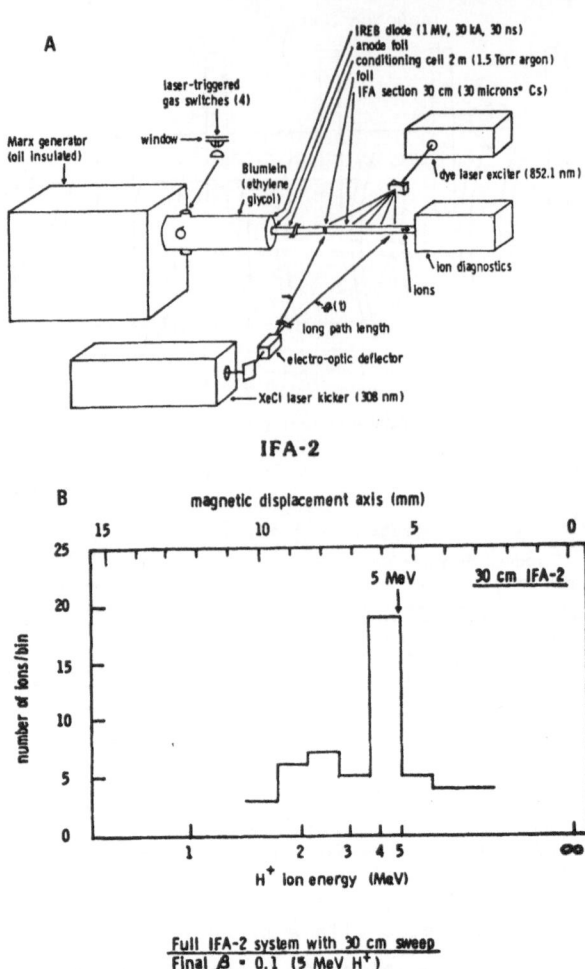

A

laser-triggered
gas switches (4)

IREB diode (1 MV, 30 kA, 30 ns)
anode foil
conditioning cell 2 m (1.5 Torr argon)
foil
IFA section 30 cm (30 microns° Cs)

Marx generator
(oil insulated)

window

Blumlein
(ethylene
glycol)

dye laser exciter (852.1 nm)

ion diagnostics

ions

θ(t)

long path length

electro-optic deflector

XeCl laser kicker (308 nm)

IFA-2

B   magnetic displacement axis (mm)

5 MeV   30 cm IFA-2

number of ions/bin

H^+ ion energy (MeV)

Full IFA-2 system with 30 cm sweep
Final β = 0.1 (5 MeV H^+)
Ion source: 50 microns hydrogen
Bin size: 0.1 mm x 0.1 mm

Fig. 6.   Ionization Front Accelerator

et al. (1986) through a sequence of induction acceleration units at
Dubna.   Experiments on formation of suitable rings (major radius $\approx$ 30 mm,
minor radius $\approx$ 2 mm, number of electrons $\geq 10^{13}$) have been also carried
out at LBL, Garching and Maryland but these efforts were not continued
long enough to achieve controlled acceleration of ions.   Figure 7 is a
picture of the ring-forming apparatus at LBL.   A 4-MeV electron beam of
hundreds of amperes from an induction linac injector is injected on an
orbit near the outside radius.   Pulsed coils apply a magnetic field
transversely to the ring and cause simultaneous acceleration (to 20 MeV)
and compression of the ring (to R = 30 mm).   If ions are added in an
amount of about 1% of the electrons, they provide enough focusing to
cancel the residual electrostatic self-defocusing and render the ring

Fig. 7. The ring-forming chamber at LBL (1970). The rapidly pulsed
inflection field can be seen at the outer radius. The
injected beam enters from the top left.

stable. When accelerated in an axial field, the ring electrons respond
with a transverse mass $m_e \gamma \approx 40\ m_e$, i.e., about 50 times lighter than a
proton. Thus the ions trapped in the ring experience an accelerating
gradient very much larger than the externally applied field.

Excitation of waves on intense electron beams (i.e., dense non-
neutral plasmas) has been explored in an effort to develop an orderly
method of ion acceleration with a tailored velocity increase with distance [see Keefe (1981)]. The charge density modulations in the wave act
as accelerating buckets. Efforts to grow cyclotron waves (E-beam in
axial magnetic field were never successful, but growth of a slow space-
charge wave (E-beam in a disc-loaded wave guide has been achieved at
Cornell by Nation (1985). Reproducibility of the space-charge wave at
low phase-velocity, however, is poor.

The most promising approach to wave acceleration on electron beams
would seem to involve combining two waves, one of which is usually a
zero-frequency wave provided, for example, by a rippled magnetic solenoid
field along the axis. Adjustments can be made easily to the periodicity
of the magnetic field or to the ripple amplitude to allow one a number of
external degrees of control. In experiments at NRL, Friedman (1979) has
used a sequence of electron rings (formed by longitudinal chopping of a
hollow beam) to inject a high-speed space charge wave of a somewhat

unusual character. Space-charge waves of the conventional kind grown in a disc-loaded structure have been used by Nation at Cornell and have delivered some modest energy gain to low energy protons.

While such two-wave schemes are still in an infant stage of development, it is to be noted that, besides slow waves (of interest initially for slow ion acceleration), these systems can support fast waves with v_{phase} = c and end hence, in the future, offer (in principle at least) the possibility of accelerating relativistic electrons or positrons in a collider.

In summary, most of the research on these other concepts has been with the aim of accelerating ions to intermediate energies and some, such as ERA, are of interest exclusively for ions. The physical mechanisms involved are very complicated and take years of study to understand. With the exception of the ERA, where acceleration (of ions) has been successfully extended to a multi-gap structure, the research has been most valuable in weeding out the less promising candidates and in identifying others (and there is no shortage of them) that seem worthy of pursuit. Major questions of interest such as available flux and beam quality (emittance, energy spread) remain to be addressed.

REFERENCES

Briggs, R. J., 1985, Phys. Rev. Lett., 54, 2588.
Bryant, P. and Mulvey, J. H. (eds.), 1984, "The Generation of High Fields for Particle Acceleration to Very High Energies" in Proceedings of the CAS-ECFA-INFN Workshop, Report No. ECFA 85/91, CERN 85/07 Frascati.
Channell, P., 1982, Proc. 1st Workshop on Laser Acceleration of Particles, AIP Conference Proceedings No. 91, Los Alamos.
Chen, P., Dawson, J. M., Huff, R. W. and Katsouleas, T., 1985, Phys. Rev Lett., 54, 693.
Corbett, I., Lawson, J. D. and Amaldi, U. (eds.), 1982, "Challenge of Ultra-high Energies" in: Proceedings of the ECFA-RAL Meeting, ECFA Report No. ECFA 83/68, Oxford.
Fontana, J. R. and Pantell, R. H., 1983, J. Appl. Phys., 54, 4285.
Friedman, M., 1979, IEEE Trans. Nuc. Sci., 26, 4186.
Hartwig, E. C., 1968, Proc. ERA Symposium, Lawrence Berkeley Laboratory Report UCRL-16830.
Hopkins, D. B., Sessler, A. M. and Wurtele, J. S., 1984, LBL-17800.
Joshi, C., Clayton, C. E., Darrow, C. and Umstedter, D., 1985, Proc. 2nd Workshop on Laser Acceleration of Particles, AIP Conference Proceedings No. 130, p. 99, Malibu.
Joshi, C. and Katsouleas, T., 1985, Proc. 2nd Workshop on Laser Acceleration of Particles, AIP Conference Proceedings No. 130, Malibu.
Katsouleas, T., Dawson, J. M., Sultana, D., and Yan, Y. T., 1985, IEEE Trans. Nuc. Sci., 32, 3554.
Katsouleas, T., Joshi, C., Mori, W., Dawson, J. M., and Chen, F. F., 1983, Proc. 12th Int. Conf. on High Energy Accel. (Fermilab).
Keefe, D., 1970, Particle Accelerators 1, 1.
Keefe, D., 1981, Particle Accelerators, 11, 187.
Nation, J., 1984, in Proceedings of the CAS-ECFA-INFN Workshop, Report No. ECFA 85/91, CERN 85/07 Frascati, p. 115.

Olson, C. L., 1974, <u>Proc. 9th Int. Conf. High Energy Accel. (SLAC)</u>, 272.

Palmer, R. B., 1985, See Joshi & Katsouleas (1985), p. 234.

Rosenzweig, J. B., Cline, D. B., Dexter, R. N., Larson, D. J., Leonard, A. W., Mengelt, K. R., Sprott, J. C., Mills, F. E., and Cole, F. J., 1985, <u>Proc. 2nd Workshop on Laser Acceleration of Particles</u>, AIP Conference Proceedings No. 130, p. 266, Malibu.

Sarantsev, V. P. et al., 1986, <u>Proc. of 1986 Linac Conference</u>, Stanford, June 1986, (in press).

Tajima, T. and Dawson, J. M., 1979, <u>Phys. Rev. Lett.</u>, 43, 267.

Veksler, V. I. et al., 1968, <u>Atomnaya Energiya</u>, 24, 317.

Willis, W., 1985, See Joshi & Katsouleas (1985), p. 421.

THE PHYSICS OF CODES

Richard K. Cooper and Michael E. Jones

Los Alamos National Laboratory
Los Alamos, NM 87545 USA

INTRODUCTION

The title given this paper is a bit presumptuous, since one can hardly expect to cover the physics incorporated into all the codes already written and currently being written. We will focus on those codes which have been found to be particularly useful in the analysis and design of linacs. At that we will be a bit parochial and discuss primarily those codes used for the design of radio-frequency (rf) linacs, although the discussions of TRANSPORT and MARYLIE have little to do with the time structures of the beams being analyzed. We may also be a bit parochial in our choice of codes to discuss; this is a personal choice which fits within the confines of the time for oral presentation and the (self-imposed) limits of space in the proceedings. If we omit mention of somebody's favorite code it is possibly due to ignorance, but hopefully just due to the need to be selective and not overly detailed in the presentation of this tutorial material.

The plan of this paper is first to describe rather simply the concepts of emittance and brightness, then to describe rather briefly each of the codes TRANSPORT, PARMTEQ, TBCI, MARYLIE, and ISIS, indicating what physics is and is not included in each of them. At the outset we will state that since we are dealing with linear machines we will not mention synchrotron radiation. We expect that the vast majority of what we cover will apply equally well to protons and electrons (and other particles). This material is intended to be tutorial in nature and can in no way be expected to be exhaustive.

In case the concepts of emittance and brightness have not been discussed yet, let me present a quick introduction to these concepts, indicating why they are useful. We want to consider a beam of particles, by which we mean a collection of particles with one component (the z component, say) of their velocity vectors very much greater than the other two (the x and y components). For the moment, in fact, let us take the z components to be all the same, so that particles do not overtake one another. Real beams have finite extent in the x and y directions (referred to often as the transverse direction) as well as some distribution in position and velocity in those directions. The distribution of particles in the z direction (often called the longitudinal direction) may be characterized by dimensions of the same order as the transverse dimensions, or may be many orders of magnitude greater. The electron beam in an rf linac, for example, may be a millimeter or so in diameter and a centimeter or thereabouts in length, while in a high-current induction linac the diameter may be a centimeter while the length is tens of meters. At any rate, let us consider a beam cut up longitudinally into slices of thickness δz, and focus our attention on one of the slices, containing, say, N particles. We define the root mean square (rms) size of the beam slice in the x and y directions R_x and R_y by the equations[1]

$$R_x^2 = \frac{4}{N} \sum_{i=1}^{N} x_i^2 \tag{1}$$

$$R_y^2 = \frac{4}{N} \sum_{i=1}^{N} y_i^2 \ . \tag{2}$$

As the beam slice moves in the z direction the individual particles follow trajectories appropriate to the forces acting on them:

$$\frac{dp_i}{dt} = F_i \ . \tag{3}$$

The transverse motion of the particles gives rise to a change in R_x and R_y:

$$R_x \dot{R}_x = \frac{4}{N} \sum_{i=1}^{N} x_i \dot{x}_i \tag{4}$$

---

[1]This definition is in fact a factor of 2 times the true rms value; this factor of 2 is included to make R_x and R_y equal to the actual beam radius for a beam of uniform density.

$$R_y \dot{R}_y = \frac{4}{N} \sum_{i=1}^{N} y_i \dot{y}_i \tag{5}$$

Concentrating on just the x component and solving for $\dot{R}_x$ we have

$$\dot{R}_x = \frac{4}{NR_x} \sum_{i=1}^{N} x_i \dot{x}_i \tag{6}$$

which, when differentiated a second time, yields, after some rearrangement,

$$\ddot{R}_x - \frac{E_x^2}{R_x^3} = \frac{4}{NR_x} \sum_{i=1}^{N} x_i \ddot{x}_i \tag{7}$$

where the quantity E_x^2 is given by

$$E_x^2 = \left(\frac{4}{N}\right)^2 \left(\Sigma x_i^2 \, \Sigma \dot{x}_i^2 - (\Sigma x_i \dot{x}_i)^2 \right) . \tag{8}$$

Exercise for the student. Show that if $\ddot{x}_i = Kx_i$, i.e., the individual particles are subject to forces linear in their displacement, then $dE_x/dt = 0$. Note that K may depend on z.

Exercise for the student. Show that if the rms quantities are defined relative to the center of mass values, e.g.,

$$R_x^2 = \frac{4}{N} \sum_{i=1}^{N} (x_i - x_{cm})^2 \text{ etc.}$$

then $dE_x/dt = 0$ if $\ddot{x}_i = K(x_i - x_{cm}) + B$.

When the quantity E_x^2 is divided by v_z^2 and it is observed that $\dot{x}_i / v_z = x'_i = dx_i/dt$, that is, the slope of the trajectory of the ith particle, then we have defined the rms emittance,

$$\varepsilon_x \equiv \frac{4}{N} \left(\Sigma x_i^2 \, \Sigma x'^2_i - (\Sigma x_i x'_i)^2 \right)^{1/2} . \tag{9}$$

Exercise for the student. Show that if the distribution of particles in a beam can be characterized by a uniform density in x-x' space bounded by the ellipse $x^2/x_0^2 + x'^2/x'^2_0 = 1$, then the rms emittance is given by $\varepsilon_x = x_0 x'_0$, which is the area of the ellipse divided by π.

Exercise for the student. Show that if the distribution of particles in a beam can be characterized by a uniform density in

235

x-x' space bounded by the ellipse

$$\hat{\gamma} x^2 + 2\hat{\alpha} xx' + \hat{\beta} x'^2 = W \tag{10}$$

then the rms emittance is given by $\varepsilon_x = W/(\hat{\beta}\hat{\gamma} - \hat{\alpha}^2)^{1/2}$, which is the area of the ellipse divided by π. (In a periodic system, $\hat{\alpha}$, $\hat{\beta}$, and $\hat{\gamma}$ are known as the Courant-Snyder (1958) parameters and satisfy the relation $\hat{\beta}\hat{\gamma} - \hat{\alpha}^2 = 1$).

Exercise for the student. Using the fact that $p_x = \gamma\beta\, m_0 cx'$ and Eq. (3), show that for relativistic beams the quantity $\gamma\beta\varepsilon_x$ is constant if the transverse force on the ith particle is given by $F_{xi} = Kx_i$.

If the beam has x-x' and y-y' phase space distributions which are ellipses with uniform densities, the space-charge forces will be linear, and Eq. (7) can be written [the Kapchinsky-Vladimirsky (1959) envelope equation], assuming linear focusing forces (Lapostolle, 1971), (Sacherer, 1971)

$$\ddot{R}_x - \frac{E_x^2}{R_x^3} - \frac{qI}{2\pi\varepsilon_0 mv_z(R_x + R_y)} + \kappa(z)R_x = 0 \tag{11}$$

where I is the beam current and k(z) represents the (spatially varying) external focusing. In the limit of zero current the second term controls the minimum size to which the beam can be focused. Such beam are said to be emittance dominated. For high-current beams the third term controls the behavior of the beam envelope; such beams are called space-charge dominated.

The quantity $\gamma\beta\varepsilon_x$, shown by the last exercise to be invariant under the influence of linear forces, is referred to as the normalized emittance:

$$\varepsilon_{nx} = \gamma\beta\varepsilon_x \ . \tag{12}$$

The brightness of a beam can be variously defined. For applications in which the time-averaged beam properties are of principal interest, we define the normalized brightness (van Steenbergen, 1965):

$$B_n \equiv \frac{I}{\varepsilon_{nx}\varepsilon_{ny}} \tag{13}$$

where I is the beam current. Thus brightness is a measure of the density of points in four-dimensional phase space, and if there is no coupling of the longitudinal and transverse phase spaces, Liouville's theorem states that for Hamiltonian systems this density is a constant. In point of fact, Liouville's theorem applies to the density of points in the phase space actually occupied by the particle distribution, whereas the rms

236

definition of emittance we use here may not be a precise measure of the area actually occupied, since the distribution may look like an ellipse. Thus even for Hamiltonian systems the rms emittance is not necessarily a constant of the motion. Indeed, recent interesting work by Wangler, Crandall, Mills, and Reiser (1985) has related the emittance growth observed in simulations of intense beams to the distribution of energy in the space-charge field distribution. Under the influence only of linear forces, however, as shown in the exercises, the rms emittance is constant and thus the brightness will also remain constant.

In recent years circular accelerators have been able to "beat" Liouville's theorem and increase beam brightness by multiturn injection using the non-conservative stripping interaction which converts H^- ions to H^+ ions, thereby creating points in the phase space of the proton beam.

TRANSPORTING THE BEAM

If we consider a beam characterized at some point by the upright ellipse of the exercise following Eq. (9) and then imagine a free drift of this beam, each point in x-x' phase space moves in time (or, equivalently, with longitudinal distance) in a very simple fashion. By free drift we mean under the influence of no forces, so that the slope of an individual trajectory x'_i is constant, and thus the position x_i increases linearly with time (or distance). After a drift of length L we can write for the coordinates of the i_{th} particle

$$x_{iL} = x_{i0} + Lx'_{i0} \tag{14}$$

$$x'_{iL} = x'_{i0} . \tag{15}$$

Thus points with x' > 0 will move to the right linearly with time (and linearly proportional to the value of x'), while points with x' < 0 will move to the left. That is, the ellipse will shear into the first and third quadrants, preserving its area. The laboratory observer will see the beam expanding. After some distance, if the experimenter wishes to restore the beam to its original size, he must use a lens. Recall from optics that a converging lens has the property of bending rays through an angle proportional to their distance from the optic axis. If we could produce a y-directed magnetic field in a region of length l, say, along the z direction

$$B_y = B'x \tag{16}$$

then, if we neglect the change in transverse position of a particle as it moves through this magnetic field, it will have its transverse momentum changed according to

$$\Delta p_x = -qv_z B_y \Delta t = -qv_z B'x\frac{\ell}{v_z} \tag{17}$$

so that the change in slope is given by

$$\Delta x' = \frac{\Delta p_x}{p} = -\frac{q}{p} B'\ell x . \tag{18}$$

That is, the change in slope is proportional to the distance from the axis. The quantity p/q is the magnetic rigidity of the particle (in Tesla-metres) and is often written as $[B\rho]$. Thus the linearly varying magnetic field acts as a focusing lens with focal length

$$f = \frac{[B\rho]}{B'\ell} . \tag{19}$$

The fly in the ointment here is that the magnetic field must satisfy, in free space (and with no time dependence),

$$\nabla \times B = 0 , \tag{20}$$

one component of which gives

$$\frac{\partial B_y}{\partial x} = \frac{\partial B_x}{\partial y} . \tag{21}$$

Thus if we have produced the postulated $B_y = B'x$, we must also have $B_x = B'y$, which then has the action of a diverging lens of focal length $f = [B\rho]/B'\ell$ on motion in the y direction. Thus any attempt to focus a beam in the x direction will cause a defocusing in the y direction and vice versa.

> **Exercise for the student.** Show that the fields B_x and B_y given above result from a vector potential having only a z-component and which satisfies Laplace's equation with $\cos 2\theta$ dependence (cylindrical coordinates).

All is not hopeless for focusing with these (quadrupole) fields, however, as it is well known that the combination of a focusing lens and a defocusing lens can be overall focusing. The principle is shown in Fig. 1. The essential point is that if one has equal strength lenses, then a particle receiving a converging "kick" at the first lens receives a weaker diverging impulse at the second lens, since it is nearer the axis, and vice versa. These observations can be made quantitative and at the same time matrix notation can be introduced as follows. The equations for the action of a thin quadrupole lens as introduced above are:

$$x_{out} = x_{in} \tag{22}$$

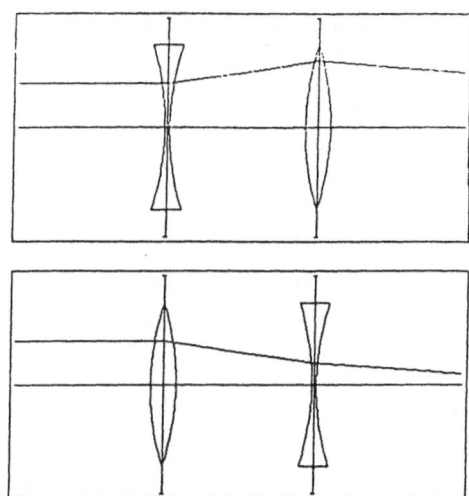

Fig. 1. Illustrating the principle of focusing with lenses of equal but opposite strength.

$$x'_{out} = x'_{in} - \frac{x_{in}}{f} \tag{23}$$

$$y_{out} = y_{in} \tag{24}$$

$$y'_{out} = y'_{in} + \frac{y_{in}}{f} \tag{25}$$

where the subscripts in and out refer to values of a particle's coordinates entering and leaving the lens, respectively. Concentrating on the x motion for the moment, we can write a matrix equation

$$\begin{pmatrix} x \\ x' \end{pmatrix}_{out} = \begin{pmatrix} 1 & 0 \\ -\frac{1}{f} & 1 \end{pmatrix} \begin{pmatrix} x \\ x' \end{pmatrix}_{in} . \tag{26}$$

From the equations of drifting motion, Eqs. (14) and (15), we can write a matrix equation for a drift of length L:

$$\begin{pmatrix} x \\ x' \end{pmatrix}_{L} = \begin{pmatrix} 1 & L \\ 0 & 1 \end{pmatrix} \begin{pmatrix} x \\ x' \end{pmatrix}_{0} . \tag{27}$$

Thus, if a particle is sent through a converging lens of focal length f, followed by a drift of length L, and then a diverging lens of focal length -f, we have for the coordinates of ith particle upon exiting the second lens, in terms of its coordinates upon entering the first lens:

$$\begin{pmatrix} x \\ x' \end{pmatrix}_{out} = \begin{pmatrix} 1 & 0 \\ \frac{1}{f} & 1 \end{pmatrix} \begin{pmatrix} 1 & L \\ 0 & 1 \end{pmatrix} \begin{pmatrix} 1 & 0 \\ -\frac{1}{f} & 1 \end{pmatrix} \begin{pmatrix} x \\ x' \end{pmatrix}_{in} . \tag{28}$$

Performing the indicated matrix multiplications gives

$$\begin{pmatrix} x \\ x' \end{pmatrix}_{out} = \begin{pmatrix} 1 - \dfrac{L}{f} & L \\ -\dfrac{L}{f^2} & 1 + \dfrac{L}{f} \end{pmatrix} \begin{pmatrix} x \\ x' \end{pmatrix}_{in} . \tag{29}$$

Exercise for the student. Show that all particles initially traveling parallel to the z axis (i.e., $x'_i = 0$) will be focused at a distance $f(f/L - 1)$ from the center of the second lens. Notice that if $L > f$ this is a virtual focus.

Exercise for the student. Suppose that one wants to transport a beam a long distance using many focus/defocus lens combinations as above. If the lenses are all separated by distance ℓ and all have equal focal lengths $\pm f$:

1. Show that the transfer matrix from focusing lens to focusing lens is given by

$$M = \begin{pmatrix} 1 - \dfrac{L}{f} - \dfrac{L^2}{f^2} & L(2 + \dfrac{L}{f}) \\ -\dfrac{L}{f^2} & 1 + \dfrac{L}{f} \end{pmatrix} . \tag{30}$$

2. The transfer matrix for N periods of this periodic transport system is $M_N = M^N$, i.e., the matrix M raised to the Nth power. For the particle displacement not to grow with N, it is necessary to insure that the eigenvalues of M have absolute values less than 1. Show that this requires $f > L/2$.

In designing transport systems to move a beam from one place to another, from ion source to accelerator, say, or from accelerator to target, use is made of computer codes which allow one to fit automatically certain desired parameters. Dipole (bending) magnets are arranged to bend the beam through the various angles required for the end use and then the focusing is designed to manipulate the beam size as required. Probably the most commonly used program for these calculations is the program TRANSPORT (Brown, Carey, Iselin, and Rothacker, 1973). This code uses the six-dimensional description of the particle motion:

$$X = \begin{pmatrix} x_1 \\ x_2 \\ x_3 \\ x_4 \\ x_5 \\ x_6 \end{pmatrix} = \begin{pmatrix} x \\ x' \\ y \\ y' \\ 1 \\ \delta \end{pmatrix} \tag{31}$$

where x and y are transverse displacements from the design trajectory, δ is the fractional momentum difference, i.e., the difference in momentum between the particle being traced through the system and the momentum for which the design trajectory is valid divided by the design momentum. That is,

$$\delta = \frac{p - p_0}{p_0} \ .$$

(32)

The variable ℓ represents the longitudinal separation between the particle being traced and the reference particle, when the two particles start at the beginning of the system at the same time.

After passing through an element or group of elements, the particle coordinates can be expressed as a Taylor-series expansion in the input coordinates. That is, one can write the pth component of a given particle

$$x_p = \sum_{j=1}^{6} R_{pj} x_j(0) + \sum_{j=1}^{6} \sum_{k=1}^{6} T_{pjk} x_j(0) x_k(0) +$$

$$\sum_{j=1}^{6} \sum_{k=1}^{6} \sum_{\ell=1}^{6} W_{pjk\ell} x_j(0) x_k(0) x_\ell(0) + \dots$$

(33)

where p takes on values from 1 to 6. If one retains only the first summation on the right-hand side, one is dealing with linear charged-particle optics. For a thin quadrupole lens focusing in the x direction the matrix R has the form

$$R = \begin{pmatrix} 1 & 0 & 0 & 0 & 0 & 0 \\ -\frac{1}{f} & 1 & 0 & 0 & 0 & 0 \\ 0 & 0 & 1 & 0 & 0 & 0 \\ 0 & 0 & \frac{1}{f} & 1 & 0 & 0 \\ 0 & 0 & 0 & 0 & 1 & 0 \\ 0 & 0 & 0 & 0 & 0 & 1 \end{pmatrix}$$

(34)

The TRANSPORT code calculates matrices for bending magnets, drifts, and quadrupoles, and the linear parts of sextupoles, solenoids, etc. If also calculates the second order elements T_{pjk} and recently third-order elements $W_{pjk\ell}$ have been included for some elements.

Note that in principle there are 6^3 = 216 elements T_{pjk} and there are 6^4 = 1296 elements of $W_{pjk\ell}$. These elements are not all independent, of course, but there are a very large number of them. For Hamiltonian systems the condition that the Hamiltonian equations of motion should be preserved in any physical transformation is known as the symplectic condition. Truncation of the Taylor-series expansion at any order automatically violates the symplectic condition to some order, and thus gives in principle a non-physical description of the particle motion. In the last decade a new approach to the description of nonlinear effects has been developed and incorporated into a code called MARYLIE (Dragt, et al., 1985). In this approach one defines the coordinates of a particle as the canonically conjugate ones:

$$z = (x \quad p_x \quad y \quad p_y \quad \tau \quad p_\tau) \tag{35}$$

where τ plays the role of the time variable in the Hamiltonian. We also write $q = (x,y,\tau)$ and $p = (p_x, p_y, p_\tau)$. Motion through a physical system produces a mapping M of the initial coordinates into the final coordinates:

$$z^{fin} = M z^{in} . \tag{36}$$

From the mapping one can form the Jacobian matrix M, of which the ab element is

$$M_{ab}(z^{in}) = \frac{\partial z_a^{fin}}{\partial z_b^{in}} . \tag{37}$$

From the invariance of the Poisson brackets it follows that the matrix M satisfies the condition

$$\tilde{M}(z^{in})JM(z^{in}) = J \text{ for all } z^{in} \tag{38}$$

where the matrix J is given by

$$J = \begin{pmatrix} 0 & 1 & 0 & 0 & 0 & 0 \\ -1 & 0 & 0 & 0 & 0 & 0 \\ 0 & 0 & 0 & 1 & 0 & 0 \\ 0 & 0 & -1 & 0 & 0 & 0 \\ 0 & 0 & 0 & 0 & 0 & 1 \\ 0 & 0 & 0 & 0 & -1 & 0 \end{pmatrix} . \tag{39}$$

A matrix that satisfies this next to the last equation is said to be a **symplectic** matrix; the map M is a symplectic map. This symplectic condition puts restrictions on the Taylor-series expansion coefficients R, T, and W. This approach taken to insure that the mappings produced by such elements as dipole magnets, quadrupoles, sextupoles, etc., are symplectic involves the use of Lie algebraic techniques.

To give the reader a flavor of the Lie algebraic method, we introduce here the concept of the Lie operator associated with a function $f(z)$. This (differential) operator is denoted by $: f :$ and is defined to be, for our problems having 3 spatial dimensions,

$$: f : \equiv \sum_{i=1}^{3} \frac{\partial f}{\partial q_i} \frac{\partial}{\partial p_i} - \frac{\partial f}{\partial p_i} \frac{\partial}{\partial q_i} . \tag{40}$$

If we let $h(z)$ be any other function, then the result of operating on h by $: f :$ is

$$: f : h = \sum_{i=1}^{3} \frac{\partial f}{\partial q_i} \frac{\partial h}{\partial p_i} - \frac{\partial f}{\partial q_i} \frac{\partial h}{\partial p_i} = [f, h] \tag{41}$$

that is, the operator $: f :$ acting on a function h produces the Poisson bracket f and h.

Exercise for the student. Show that the Lie operator corresponding to the communicator of two functions f and g, i.e.,
$: f :: g : - : g :: f :$ is the Lie operator associated with the Poisson bracket $[f,g]$.

We can define powers of a Lie operator as repeated use of the operator

$$: f :^2 h =: f :: f : h =: f : [f,h] = [f,[f,h]] \tag{42}$$

$$: f :^3 h =: f :: f :^2 h =: f : [f,[f,h]] = [f,[f,[f,h]]] \tag{43}$$

et cetera, which, along with the definition,

$$: f :^0 h = h \tag{44}$$

allows one to define power series of the operator: f:. In particular we can define the exponential operator

$$\exp(: f :) \equiv \sum_{n=0}^{\infty} : f :^n /n! \tag{45}$$

the operation of which is called a **Lie transformation.** This transformation will also be written an $e^{: f :}$.

Exercise for the student. Recall that $[q_i, p_j] = -[p_j, q_i] = \delta_{ij}$, and show that if we consider the function $f(x, y, \tau, p_x, p_y, p_r)$ to be given by

$$f = -\frac{1}{2} a p_\tau^2 \tag{46}$$

then

$$: f : \tau = -\frac{a}{2}[p_\tau^2, \tau] = ap_\tau \tag{47}$$

$$: f : p_\tau = -\frac{a}{2}[p_\tau^2, p_\tau] = 0 \tag{48}$$

$$: f : x = -\frac{a}{2}[p_\tau^2, x] = 0 \text{ etc.} \tag{49}$$

so that the Lie transformation associated with this function is a shift in the variable τ by an amount ap_τ. In matrix notation this is

$$\exp : -\frac{a}{2} p_\tau^2 : z \leftrightarrow \begin{pmatrix} 1 & 0 & 0 & 0 & 0 & 0 \\ 0 & 1 & 0 & 0 & 0 & 0 \\ 0 & 0 & 1 & 0 & 0 & 0 \\ 0 & 0 & 0 & 1 & 0 & 0 \\ 0 & 0 & 0 & 0 & 1 & a \\ 0 & 0 & 0 & 0 & 0 & 1 \end{pmatrix}. \tag{50}$$

Exercise for the student. Show that if $f = -\frac{w}{2}(x^2 + p_x^2)$, then

$$: f : x = wp_x \tag{51}$$

$$: f : y = y, \text{ etc.} \tag{52}$$

$$: f : p_x = -wx \tag{53}$$

and hence

$$\exp : -\frac{w}{2}(x^2 + p_x^2) : x = x \cos w + p_x \sin w \tag{54}$$

and

$$\exp : -\frac{w}{2}(x^2 + p_x^2) : p_x = p_x \cos w - x \sin w . \tag{55}$$

Thus in matrix notation we would write

$$\exp : -\frac{w}{2}(x^2 + p_x^2) : z \leftrightarrow \begin{pmatrix} \cos w & \sin w & 0 & 0 & 0 & 0 \\ -\sin w & \cos w & 0 & 0 & 0 & 0 \\ 0 & 0 & 1 & 0 & 0 & 0 \\ 0 & 0 & 0 & 1 & 0 & 0 \\ 0 & 0 & 0 & 0 & 1 & a \\ 0 & 0 & 0 & 0 & 0 & 1 \end{pmatrix}. \tag{56}$$

The utility of all this has to do with the following fact. If we take an arbitrary function of the coordinates z^{in}, say $f(z^{in})$ and define the coordinates z^{fin}, of which the ath component is given by

$$z_a^{fin} = \exp(: f :)z_a^{in} \tag{57}$$

then, if we write this as a mapping

$$z_a^{fin} = M z_a^{in} \tag{58}$$

or equivalently,

$$M = \exp (: f :) \tag{59}$$

then this mapping M is guaranteed to be symplectic (Dragt, 1981). Thus assuming that we can fairly readily find functions f whose Lie transformations represent the usual dipole, quadrupole, etc. elements found in beam lines and accelerators, and further, that we can readily combine such transformations, we have a powerful tool for analyzing physical systems. Implicit in the method is inclusion of effects higher than first order.

MARYLIE 3.0 is a code that calculates transformations to third order and insures that the symplectic condition is fulfilled. MARYLIE 3.1 is a third-order code that includes the effects of misalignments of the various physical elements. Special versions of MARYLIE have been written for particular applications which require higher-order transformations, but which use only a few element types.

One of the major drawbacks of the TRANSPORT and MARYLIE codes and, indeed, most such codes, is the fact that space-charge forces are not included in them. Since high-brightness beams, at least in linacs, are space-charge dominated, this is a serious drawback indeed. There exist some TRANSPORT-like codes that include space-charge effects [e.g., GIOS (Wollnik, 1986)], but these lack the sophistication and versatility of TRANSPORT and MARYLIE. The problem of the proper calculation of space-charge forces is a formidable one, as it must include non-relativistic and relativistic regimes, longitudinal-to-transverse size ratios varying over orders of magnitude, and sometimes rather complicated boundary conditions.

ACCELERATING THE BEAM

Accelerating a beam from low-energy (typically 100 kV or so) ion sources is complicated by the fact that the particle velocities are slow (we are speaking here of particles more massive than electrons) so that focusing by means of magnetic fields becomes difficult. The past decade has seen the practical development of the "radio-frequency quadrupole", which provides both transverse focusing and longitudinal bunching and acceleration (Kapchinsky and Teplyakov, 1970). These properties make it the ideal injector for rf linacs. Electric fields having quadrupolar symmetry are produced between vanes (see Fig. 2) embedded in an rf resonator that is essentially a waveguide operated at the cutoff frequency. The fields oscillate at a frequency much higher than the particle oscillation frequency and thereby provide an effective potential well (Landau and Lifshitz, 1951) (angle brackets indicate time average)

$$V_{effective} = \frac{1}{2m\omega^2} \langle F^2(r) \rangle \tag{60}$$

which focuses the particle of the beam.

The vanes are modulated radially, i.e., the distance between opposite vanes varies longitudinally (approximately sinusoidally), with the modulation on adjacent vanes being 180° out of phase, thus producing a longitudinal field on axis. This field can be used to bunch and accelerate the beam. The fields are given by [a is the minimum distance from the axis to the vane tips, while ma is the maximum distance (height of modulation)]:

$$E_r = -\left[\frac{XV}{a^2} r \cos 2\theta + \frac{kAV}{2} I_1(kr) \cos kz\right] \sin(\omega t + \phi) \tag{61}$$

$$E_\theta = \frac{XV}{a^2} r \sin 2\theta \sin(\omega t + \phi) \tag{62}$$

$$E_z = \frac{kAV}{2} I_0(kr) \sin kz \sin(\omega t + \phi) \tag{63}$$

where

$$A = (m^2 - 1)/(m^2 I_0(ka) + I_0(mka)) \text{ and } X = 1 - AI_0(ka) . \tag{64}$$

The quantity V is the potential difference between adjacent vanes and V A is the potential difference that exists on axis between the point of

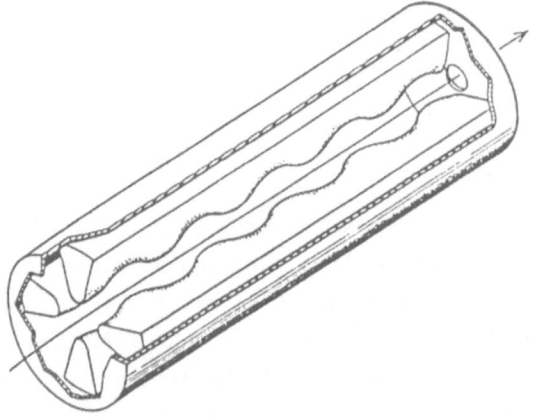

Fig. 2. Radio-frequency quadrupole geometry.

minimum radius and the point of maximum radius of a given vane. The code PARMTEQ (Crandall, Stokes, and Wangler, 1979), which stands for Phase and Radial Motion in Transverse Electric Quadrupoles, solves the equations of motion by dividing the rfq into a series of longitudinal sections and transforming the transverse coordinates from section to section using the quadrupole lens transformation, taking into account the rf phase and energy variation, and taking the radial defocusing term in Eq. (61) into account. The code is an approximate space-charge calculation that gives good results (Chidley and Desirens, 1985).

THE BEAM ACTING ON ITSELF VIA THE STRUCTURE

In addition to the externally applied forces and space-charge forces, a beam can experience forces resulting from fields it itself generates when encountering a change in its surroundings. Figure 3 shows the fields generated by a bunch of charge passing through a bellows. The first part of the bunch loses energy, while the tail of the bunch can gain some energy, and if the bunch is not quite on the center of the (assumed axially symmetric) structure, the tail can also be deflected away from the axis. These effects (called wake-field effects) can be studied by solving Maxwell's equations in the time domain, assuming that the bunch moves as a rigid body. The cumulative effects of the inter-action with the structure can then be obtained by integrating the time- and space-dependent forces on a given particle of the bunch over the flight time of travel through the structure. The code TBCI [Transverse Beam-Cavity Interaction] (Weiland, 1983) performs these calculations for cylindrically symmetric structures, and the newly-developed code T3 (Klatt and Weiland, 1986) does the same for three-dimensional.

The starting point for these calculations is the discretization of Maxwell's equations by the Finite Integration Technique (FIT) (Weiland, 1984) algorithm, which produces a first-order approximation to Maxwell's equations by replacing the line and surface integrals, appearing in Faraday's law and Ampere's law, by mean field values times path lengths and areas, respectively. Figure 4 shows the basic cell geometry used in this method. Note that the electric field components are not defined at a single point, but at the midpoints of the sides of the rectangular cells. The magnetic field components are defined in the center of the faces of the cells, and taken together form a mesh dual to that defined for the electric field. Only continuous components are involved in the assignment of field values to the mesh, and thus the cells of the mesh may be individually filled with arbitrary permittive and permeable media. Using the FIT method one can write the discrete form of Faraday's law as

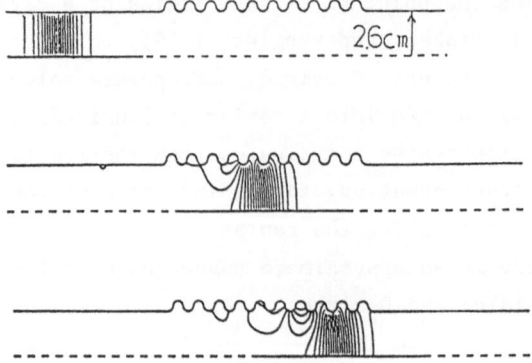

Fig. 3. Fields generated by a charged bunch passing through a bellows.

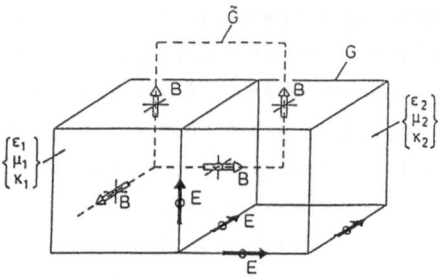

Fig. 4. Geometry and allocation of field components used in the
FIT method.

$$C\ D_S\ e = D_A\ \dot{b}$$

where e and b are vectors of length 3N where N is the number of nodes of
the mesh, C is a 3N by 3N matrix containing only the values 0, 1, and −1
and which corresponds to taking the curl of a field, D_S is a 3N by 3N
diagonal matrix containing the lengths of the sides of the mesh cells,
and D_A is a 3N by 3N diagonal matrix containing the areas of the mesh
cell surfaces. Similarly Ampere's law can be written

$$\tilde{C}\ \tilde{D}_S\ h = \tilde{D}_A\ (\dot{d} + j),$$

where the tilde indicates that the matrix corresponds to the dual mesh.
When the material distribution is taken into account and a leapfrog
integration technique in the time variable is used, one finds the
recursive algorithm for the calculation of time-dependent fields:

$$b^{n+1} = b^n - \delta t(D_A^{-1}\ C\ D_S)\ e^{n+1/2},$$

$$e^{n+3/2} = e^{n+1/2} + \delta t(D_\varepsilon^{-1}\ \tilde{D}_A^{-1}\tilde{C}\ \tilde{D}_S D_\mu^{-1})b^{n+1} - \delta t D_\varepsilon^{-1}\ j^{n+1}$$

where the superscripts refer to the time step ($t_n = n\delta t$), and D_ε and D_μ are diagonal matrices describing the filling of the mesh with permittive and permeable media. These last two equations are solved by T3; given the initial values of the fields, fields at a subsequent time require only two multiplications of a matrix with a vector, per time step.

The assumption that the bunch moves as a rigid body is not necessarily a good one, especially if the beam energy is low, or the interaction is particularly strong. In these cases the individual particle motions must be separately calculated and these motions must be fed back into Maxwell's equations as source terms. Thus a self-consistent solution of the interaction can be obtained.

Self-consistent, particle-in-cell (PIC) models were originally developed to study collective interactions in plasmas. In this method Maxwell's equations are solved on a finite difference grid with the sources (current and charge densities) determined by assigning values to the cells according to the number of particles in each cell. These particles are then allowed to move for a short time step according to their equations of motion using these fields interpolated from the grid. The fields are recomputed from the resulting change in the sources and the process is repeated. By alternately advancing the fields and particles in time this way, the self-consistent evolution of the system is obtained. In early versions of these codes the boundary conditions were simple and usually chosen to simulate a small region out of an infinite homogeneous plasma (Dawson, 1970). The fields were produced by the separation of charge and the plasma supported a number of electrostatic waves. Later, fully electromagnetic PIC models found extensive use in laser-plasma interaction studies for laser fusion (Langdon and Lasinski, 1976).

More recently, these methods have been modified to include more complex boundary conditions and more physics such as algorithms that simulate emission from conducting boundaries. Thus these codes are finding application to accelerator problems. The two codes that have been used most for accelerator studies are named MASK (Yu, Wilson, and Drobot, 1985) and ISIS (Jones and Peter, 1985). The details of the algorithms can be different but the basic ideas are the same. The details given here are for the ISIS code. Because the fields which dominate typical calculations in PIC codes come from the particles themselves, it is necessary to use many particles to simulate accurately the plasma or intense beam. The number of particles in a calculation can vary from a few thousand to over a million depending on the phenomena being studied.

However, even a million particles is far fewer than actually exist in a typical plasma. A collisionless plasma or intense charged particle beam can be modeled by a collisionless Boltzmann or Vlasov model:

$$\frac{\partial f}{\partial t} + \mathbf{v} \cdot \frac{\partial f}{\partial \mathbf{x}} + \frac{e}{mc}\left(\mathbf{E} + \frac{\mathbf{v} \times \mathbf{B}}{c}\right) \cdot \frac{\partial f}{\partial \mathbf{u}} = 0 \tag{65}$$

where f is the single particle distribution function and for relativistic particles $\mathbf{u}$ is the first three components of the four velocity divided by the speed of light, i.e., $\mathbf{u} = \gamma\mathbf{v}/c$, $\gamma^2 = 1 + \mathbf{u} \cdot \mathbf{u}$. This partial differential equation has a well known method of solution which involves integrating along the characteristics (Krall and Trivelpiece, 1973), provided the electric and magnetic fields are known. The characteristics are in fact the equations of motion for a particle

$$\frac{d\mathbf{u}}{dt} = \frac{e}{mc}\left(\mathbf{E} + \frac{\mathbf{v} \times \mathbf{B}}{c}\right) \tag{66}$$

and

$$\frac{d\mathbf{x}}{dt} = \mathbf{v} . \tag{67}$$

By alternately advancing the particles and fields in time, the PIC method in fact solves the Vlasov equation by the method of characteristics by computing a self-consistent field for up to a million characteristics. Thus rather than modeling the beam or plasma by an inadequate number of particles this method is very effective at "simulating" a plasma or beam that obeys the collisionless Vlasov equation.

Because so many particles are used in a calculation, the finite difference form of the equation of motion is chosen to give maximum speed. The time-centered difference form of the force equation is

$$\frac{\mathbf{u}^{n+1/2} - \mathbf{u}^{n-1/2}}{\Delta t} = \frac{e}{mc}\left[\mathbf{E}^n + \frac{1}{2c}\left(\mathbf{v}^{n+1/2} + \mathbf{v}^{n-1/2}\right) \times \mathbf{B}^n\right] \tag{68}$$

where the superscript n denotes the time step. The problem with solving this equation is that the $\mathbf{v}$'s and $\mathbf{u}$'s are related, making the equation implicit. ISIS uses a simple approximation to this equation which has · become almost the standard method (Langdon and Lasinski, 1976). The four velocity is first advanced 1/2 time step using only the electric field,

$$\mathbf{u}^- = \mathbf{u}^{n-1/2} + \frac{e\mathbf{E}^n\Delta t}{2mc} . \tag{69}$$

The magnetic field rotation is then performed in two steps:

$$\mathbf{u}' = \mathbf{u}^- + \mathbf{u}^- \times \mathbf{T} \tag{70}$$

$$\mathbf{u}^+ = \mathbf{u}^- + \mathbf{u}' \times \mathbf{S} \tag{71}$$

where

$$T = \frac{eB^n \Delta t}{2\gamma^n mc} \tag{72}$$

and $(\gamma^n)^2 = 1 + (u^-)^2$ and $S = 2T/(1 + T^2)$. Then the second half of the electric field acceleration is applied

$$u^{n+1/2} = u^+ + \frac{eE^n \Delta t}{2mc} . \tag{73}$$

As with all codes which solve differential equations with finite step methods, it is essential that the resulting transformations preserve the symplectic condition.

Exercise for the student. Show that the error in the angle of rotation the particle makes about the magnetic field is $T^3/3$.

The position is updated from the new velocities by a leap-frog difference. For two-dimensional, r-z, coordinates, the position is updated in a local cartesian coordinate system as follows

$$z^{n+1} = z^n + \frac{u_z^{n+1/2}}{\gamma^{n+1/2}} \Delta t \tag{74}$$

$$x = r^n + \frac{u_r^{n+1/2}}{\gamma^{n+1/2}} \Delta t \tag{75}$$

$$y = \frac{u_\theta^{n+1/2}}{\gamma^{n+1/2}} \Delta t . \tag{76}$$

The positions are then transformed back to the cylindrical coordinate system and the velocities are modified to insure conservation of angular momentum.

$$r^{n+1} = \sqrt{x^2 + y^2} \tag{77}$$

$$u_r^{n+1/2} = \frac{u_r^{n+1/2}x + u_\theta^{n+1/2}y}{r^{n+1}} \tag{78}$$

$$u_\theta^{n+1/2} = \frac{u_\theta^{n+1/2}r^n}{r^{n+1}} . \tag{79}$$

In ISIS the electric and magnetic fields are located at the same grid positions as used in TBCI. The fields are advanced by leap-frogging in time Ampere's law and Faraday's law. The location of the field components on the staggered grid insures centered differencing in space. For r–z, cylindrical coordinates, the explicit form of the difference equations are, for Ampere's law,

$$E_{z,i+1/2,j}^{n+1/2} = E_{z,i+1/2,j}^{n-1/2} - J_{z,i+1/2,j}^{n}\Delta t$$

$$+ \frac{1}{r_j}\left(r_{j+1/2}B_{\theta,i+1/2,j+1/2}^{n}\right.$$

$$\left. - r_{j-1/2}B_{\theta,i+1/2,j-1/2}^{n}\right)\frac{\Delta t}{\Delta r} \tag{80}$$

$$E_{r,i,j+1/2}^{n+1/2} = E_{r,i,j+1/2}^{n-1/2} - J_{r,i,j+1/2}^{n}\Delta t$$

$$- \left(B_{\theta,i+1/2,j+1/2}^{n} - B_{\theta,i-1/2,j+1/2}^{n}\right)\frac{\Delta t}{\Delta z} \tag{81}$$

$$E_{\theta,i,j}^{n+1/2} = E_{\theta,i,j}^{n-1/2} - J_{\theta,i,j}^{n}\Delta t + \left(B_{r,i+1/2,j}^{n} - B_{r,i-1/2,j}^{n}\right)\frac{\Delta t}{\Delta z}$$

$$- \left(B_{z,i,j+1/2}^{n} - B_{z,i,j-1/2}^{n}\right)\frac{\Delta t}{\Delta r}. \tag{82}$$

Faraday's law becomes

$$B_{z,i,j+1/2}^{n+1} = B_{z,i,j+1/2}^{n}$$

$$- \frac{1}{r_{j+1/2}}\left(r_{j+1}E_{\theta,i,j+1}^{n+1/2} - r_j E_{\theta,i,j}^{n+1/2}\right)\frac{\Delta t}{\Delta r} \tag{83}$$

$$B_{r,i+1/2,j}^{n+1} = B_{r,i+1/2,j}^{n} + \left(E_{\theta,i+1,j}^{n+1/2} - E_{\theta,i,j}^{n+1/2}\right)\frac{\Delta t}{\Delta z} \tag{84}$$

$$B_{\theta,i+1/2,j+1/2}^{n+1} = B_{\theta,i+1/2,j+1/2}^{n}$$

$$- \left(E_{r,i+1,j+1/2}^{n+1/2} - E_{r,i,j+1/2}^{n+1/2}\right)\frac{\Delta t}{\Delta z}$$

$$+ \left(E_{z,i+1/2,j+1}^{n+1/2} - E_{z,i+1/2,j}^{n+1/2}\right)\frac{\Delta t}{\Delta r} \tag{85}$$

On the row of cells just above the axis, the following formula is used

$$B_{z,i,1/2}^{n+1} = B_{z,i,1/2}^{n} - 4E_{\theta,i,1}^{n+1/2}\frac{\Delta t}{\Delta r}. \tag{86}$$

The i subscript denotes the z cell number and j denotes the r cell number.

252

Exercise for the student. Show that these field equations give second order accurate solutions to the source-free dispersion relation for light waves.

Boundary conditions are applied only to the electric field which is tangential to the boundary. Perfectly conducting boundaries require that the tangential electric field vanish.

The other two Maxwell's equations are treated as initial conditions. The method of assigning charge and current densities to the grid is not uniquely constrained. If one area weights the charge to the grid, the electric field obtained from the equations above may not satisfy Poisson's equation unless the current density is accumulated to insure the charge continuity equation is satisfied. A current accumulation method which assures charge conservation is used is ISIS. Other codes periodically adjust the electric field so that is satisfies Poisson's equation.

The necessity to use large numbers of particles for plasma problems has produced highly efficient algorithms for studying the interaction of charged particles via their electromagnetic fields. As accelerator technology moves toward more highly space-charge dominated beams, these techniques will find more and more applications.

MISCELLANY

We only have a brief space here to mention the codes most frequently used for the solution of the Poisson equation and the vector Helmholtz equation. Included in the Poisson group are the codes POISSON (Halbach, 1967) and PANDIRA, used to solve 2-D problems with source currents and magnetic materials, and SUPERFISH (Halbach and Holsinger, 1976), used to obtain resonant frequencies and fields of azimuthally symmetric electro-magnetic modes in azimuthally symmetric cavities. The group of codes denoted by URMEL (van Rienen and Weiland, 1986a) obtain the resonant frequencies and fields for azimuthally symmetric and asymmetric modes in azimuthally symmetric cavities. Both sets of codes use an irregular triangular mesh distorted to fit arbitrary boundaries and obtain the eigenmodes by various forms of direct matrix inversion. The URMEL codes have recently been extended to obtain modes in arbitrary 3-D geometries. In addition both groups of codes are being adapted to calculate disper-sion curves for periodic structures (Gluckstern and Opp, 1985) (Weiland, 1986), as well as coupling impedances for cavities in beam pipes (Gluckstern and Neri, 1986), (van Rienen and Weiland, 1986b).

CONCLUSION

We have not covered in this presentation a whole raft of important topics such as beam instabilities, beam-plasma interactions, resonances, aspects of circular machines, etc. By focusing on just a few of the codes used in accelerator physics we hope we have given an idea of the kinds of physics that goes into our codes. In simplest terms one could probably say that what we solve most in accelerator physics in $\dot{\mathbf{p}} = \mathbf{F}$ and Maxwell's equations. There's a lot of physics in those equations.

ACKNOWLEDGMENT

Dr. Robert L. Gluckstern has been extremely helpful in reviewing this work and suggesting improvements, as well as enlightening us on several points.

REFERENCES

Brown, K. L., Carey, D. C., Iselin, Ch., and Rothacker, F, TRANSPORT, A Computer Program for Designing Charged Particle Beam Transport Systems, SLAC 91, 1973 rev., NAL 91, and CERN 80-04.

Chidley, B. G. and Desirens, N. J., Beam Transmission of RFQ1 Calculated Using the Finite Element Method for Space and Image Charges, IEEE Trans. Nucl. Sci. NS-32, 1985, pp. 2459-2461.

Courant, E. D. and Snyder, H. S., Theory of the Alternating Gradient Synchrotron, Annals of Physics 3, 1958, p. 1-48.

Crandall, K. R., Stokes, R. H., and Wangler, T. P., RF Quadrupole Beam Dynamics Design Studies, Proc. 1979 Linear Accelerator Conf., Brookhaven National Laboratory, Report BNL-51134, 1979, p. 205.

Dawson, J. M., Methods Comput. Phys. 9, 1, 1970.

Dragt, A., Lectures on Nonlinear Orbit Dynamics, The Physics of High Energy Particle Accelerators, Fermi National Accelerator Laboratory, AIP Conference Proceedings No. 87, p. 147.

Dragt, A., Healy, L., Neri, F., Ryne, R., Forest, E., and Douglas, D., MARYLIE, A Program for Non Linear Analysis of Accelerator and Beam Line Lattices, IEEE Trans. Nucl. Sci. NS-32, pp. 2311-2313.

Gluckstern, R. L., and Neri, F., Longitudinal Coupling Impedance for a Beam Pipe With a Cavity, IEEE Trans. Nucl. Sci. NS-32, 1985, pp. 2403-2404.

Gluckstern, R. L., and Opp, E. N., Calculation of Dispersion Curves in Periodic Structures, IEEE MAG-21, 1985, pp. 2344-2346.

Halbach, K., A Program for Inversion of System Analysis and Its Application to the Design of Magnets, Proc. Second Conf. on Magnet Technology, Oxford, England, 1967.

Halbach, K., and Holsinger, R. F., SUPERFISH-A Computer Program for Evaluation of RF Cavities with Cylindrical Symmetry, Particle Accelerators 7, 1976, pp. 213-222.

Jones, M. E., and Peter, W., IEEE Trans. Nuc. Sci. 32:1794, 1985.

Kapchinsky, I. M., and Teplyakov, V. A., Prib. Tekh. Eksp. 2, 1970, p. 19.

Kapchinsky, I. M., and Vladimirsky, V. V., Proc. Int. Conf. on High Energy Accelerators and Instrumentation CERN, 1959, p. 274.

Klatt, R., and Weiland, T., Wake Field Calculations with Three-Dimensional BCI Code, Proceedings of the 1986 Linear Accelerator Conference SLAC, 1986.

Krall, N. A., and Trivelpiece, A. W., Principles of Plasma Physics, McGraw-Hill, New York, 1973, chap. 8.

Landau, L. D., and Lifshitz, E. M., Classical Theory of Fields, Addison-Wesley, Reading, Mass., 1951.

Langdon, A. B., and Lasinski, B. F., Methods Comput. Phys. 16 1976, p. 327.

Lapostolle, P. M., Possible Emittance Increase Through Filamentation Due to Space Charge In Continuous Beams, IEEE Trans. Nucl. Sci. NS-18, 1971, p. 1101.

Sacherer, F. J., RMS Envelope Equations With Space Charge, IEEE Trans. Nucl. Sci. NS-18, 1971, p. 1105.

van Rienen, U., and Weiland, T., Triangular Discretization Method for the Evaulation of RF Cavities with Cylindrical Symmetry, Particle Accelerators, 1986 (in press).

van Rienen, U., and Weiland, T., Impedance of Cavities with Beam Ports above Cut-Off, Proceedings of the 1986 Linear Accelerator Conference, SLAC, 1986.

van Steenbergen, A., Recent Developments in High Intensity Ion Beam Production and Preacceleration, IEEE Trans. Nucl. Sci. NS-12, 1965, pp. 746-764.

Wangler, T. P., Crandall, K. R., Mills, R. S., and Reiser, M., Relation Between Field energy and RMS Emittance in Intense Particle Beams, IEEE Trans. Nucl. Sci. NS-32, 1985, pp. 2196-2200.

Weiland, T., Transverse Beam Cavity Interaction, Part I: Short Range Forces, Nuclear Instruments and Methods (NIM) 212, 1983, pp. 13-34.

Weiland, T., On the numerical solution of Maxwell's equation and applications in the field of accelerator physics, Particle Accelerators 15, 1984, 245-292 and references therein.

Weiland, T., On the Unique Solution of Maxwellian Eigenvalue Problems in Three Dimensions, Particle Accelerators 17, 1985.

Weiland, T., Modes in Infinitely Repeating Structures of Cylindrical Symmetry, Proceedings of the 1986 Linear Accelerator Conference, SLAC, 1986.

Weiland, T., Proceedings of the URSI International Symposium on Electromagnetic Theory, Budapest, Hungary, August 1986.

Wollnik, H., Proceedings of the Second International Conference on Charged Particle Optics, Albuquerque, NM, Nuclear Instruments and Methods, 1986 (in press).

Yu, S. S., Wilson, P., and Drobot, A., Two and One-Half Dimension Particle-in-Cell Simulation of High-Power Klystrons, IEEE Trans. Nucl. Sci. NS-32, 1985, pp. 2918-2920.

HIGH-CURRENT ELECTRON-BEAM TRANSPORT IN RECIRCULATING ACCELERATORS

B. B. Godfrey and T. P. Hughes

Mission Research Corporation
Albuquerque, New Mexico   USA

INTRODUCTION

Interest in high current (i.e., > 1 kA) electron beam recirculating accelerators has grown greatly during the last several years. As evidence, this chapter's bibliography includes some 110 recent reports and is by no means complete. Applications include basic research, materials processing, food sterilization, radiography, collective ion acceleration, and free electron lasers. Some of these topics are discussed elsewhere in this book.

Present research in high electron beam recirculating accelerators has its roots in two areas, low current betatrons developed by Kerst and others in the 1940's and high current linear induction accelerators developed in the 1970's. For instance, recent proposals for recirculating accelerators driven by ferrite-loaded induction gaps can be viewed equally well as betatrons with slotted walls to let the fields in faster or as linear induction accelerators bent into closed loops. Both viewpoints are fruitful. Linear induction accelerators are reviewed in the chapters by Miller, Prestwich, and Prono in this book, and the relationship between linear and recirculating induction accelerators in the recent article by Kapetanakos and Sprangle (1985). The historical survey by Cole in this book includes early betatron developments.

This chapter consists of five principal sections. The equilibrium properties of high current beams in curved accelerator cavities are described first. Next, orbital resonances, which can cause emittance growth and current loss, are treated briefly. The negative mass instability is particularly dangerous in high current devices, and so it is presented next in considerable detail. Following the negative mass instability are three instabilities which occur also in linear accelerators: parametric, beam breakup, and resistive wall. These first four sections for the most part deal with beam transport in vacuum. The final section summarizes the advantages and complications of transport in low density ionized channels, so-called ion focused transport.

Two important topics are absent from this chapter, beam injection and extraction, and the status of ongoing experiments. Both are covered Rostoker, and beam bending experiments in ionized channels by Miller. Here, we only mention that vigorous experimental efforts are under way at the U.S. Naval Research Laboratory, the University of California at Irvine, Sandia National Laboratories, the University of New Mexico, Maxwell Laboratories, and elsewhere.

EQUILIBRIUM PROPERTIES

Trajectories of beam electrons in a conventional betatron are defined by the cyclotron orbits of the electrons in the applied magnetic field of the betatron, which is approximately uniform in the vicinity of the beam. The vertical field is increased as the electrons are accelerated in order to maintain a constant orbit radius. Gradients in the vertical magnetic field provide beam focusing and offset the net self-fields of the electrons, which would otherwise cause the beam to disperse. This focusing is, however, relatively weak, which limits conventional betatrons to about 100 A unless the beam energy is several MeV or greater. In addition, the acceptable electron energy spread in the conventional betatron is very small, because electrons of different energies follow different cyclotron orbits, which may intersect the cavity wall.

The modified betatron overcomes the current constraint by adding a strong toroidal magnetic field to confine the beam, just as an axial magnetic field confines an electron beam propagating linearly in an evacuated drift tube. Adding periodic strong focusing magnetic fields greatly increases the acceptable energy spread in a betatron. Various configurations are under investigation, including solenoidal lenses and twisted quadrupole fields.

Conventional Betatron

Figure 1 represents the basic geometry and applied fields of a conventional betatron. An electron beam of radius $\underline{a}$ circulates in a toroidal cavity of radius $\underline{b}$; the radius of curvature is $\underline{R}$. An applied magnetic field B_z, often called the vertical field, determines the curvature radius.

$$R = \gamma\, V_\theta / B_z \tag{1}$$

258

V_θ is the electron toroidal velocity, and γ is the electron relativistic energy. Here and elsewhere in this chapter the electron charge, the electron mass, and the speed of light are set equal to unity.

The vertical field must increase with the electron energy to maintain constant $\underline{R}$. In many betatron designs the changing vertical field itself inductively accelerates the electrons.

$$\dot\gamma = E_\theta V_\theta = 1/2 \ R \ V_\theta \langle \dot B_z \rangle \tag{2}$$

where E_θ is the toroidal induced electric field, and $\langle B_z \rangle$ is the average vertical field contained in the beam orbit. Consistency between Eqs. (1) and (2) then requires $B_z = 1/2 \langle B_z \rangle$ for constant $\underline{R}$ with $V_\theta \sim 1$. This is the well known 2:1 condition, which states that the average vertical field threading the torus must be twice the vertical field at the beam position. Different constraints apply when large electron poloidal velocities cause the toroidal velocities to be somewhat less than one (Roberts and Rostoker, 1983; 1985). Other acceleration methods, such as induction modules (Roberson, 1981; Mondelli and Roberson, 1983; 1984; Dialetis et al., 1986), also violate the 2:1 relation.

Electron poloidal motion and self-fields cause the beam to disassemble, if focusing is not provided (Kapetanakos, et al., 1983). The variation of centrifugal and magnetic forces across the beam provides weak radial focusing.

$$F_r = \frac{\delta}{\delta r} \left(\gamma \frac{V_\theta^2}{R} - V_\theta B_z \right) \delta r \ . \tag{3}$$

Carrying out the derivatives, introducing the betatron index,

$$n = - \frac{R}{B_z} \frac{\partial B_z}{\partial r} \tag{4}$$

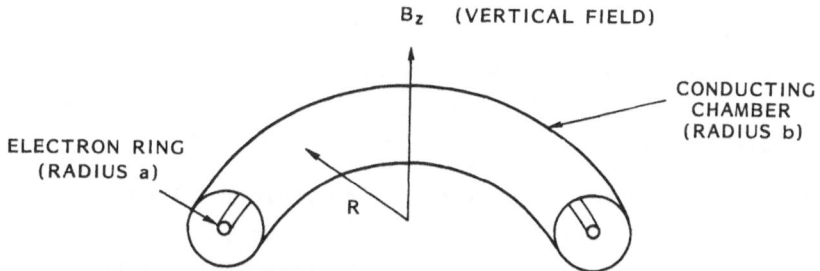

Fig. 1.   Basic betatron configuration. Cylindrical coordinates, with the z axis aligned with the vertical magnetic field, are employed.

259

and employing Eq. (1) to simplify the resulting expression yields

$$F_r = -V_\theta B_z (1 - n) \frac{\delta r}{R} .$$ (5)

Since the applied magnetic field is curl free in the vicinity of the beam, the variation in B_z requires a small B_r, which provides axial focusing.

$$F_z = -V_\theta B_r = -V_\theta B_z n \frac{\delta z}{R} .$$ (6)

The electric and magnetic fields of the beam largely cancel when acting on the beam electrons.

$$F_s \approx \gamma^{-2} \frac{2\nu}{a^2} (\delta r, \delta z)$$ (7)

where ν is the beam current normalized to 17 kA. Combining Eqs. (5) - (7) with the electron equations of motion gives the transverse oscillation frequencies,

$$\Omega_r^2 = (1 - n - n_s) \Omega_\theta^2$$ (8)

$$\Omega_z^2 = (n - n_s) \Omega_\theta^2$$ (9)

where

$$\Omega_\theta = B_z / \gamma = V_\theta / R$$ (10)

is the toroidal rotation frequency, and

$$n_s = \frac{2\nu}{\gamma^3} \frac{R^2}{a^2}$$ (11)

is the self-field index. Small toroidal self-field corrections of order $(1 + 2 \ln(b/a))\nu/\gamma$ have been ignored in Eqs. (3) - (11).

The right sides of Eqs. (8) and (9) must be positive for beam stability. Thus, $0 < n < 1$ and $n_s < \min(n, 1 - n)$ are required. Typically, $n \approx 1/2$, for which the stability criteria become

$$n_s < 1/2$$ (12)

which severely restricts the beam current at low to moderate beam energies. The γ^3 scaling of maximum current has been demonstrated experimentally (Pavlovsky, et al., 1977).

The transverse dynamics of the beam centroid is also of interest. The beam as a whole is not influenced directly by its own self-fields but instead by its image fields from the accelerator cavity wall. The image forces are smaller than the direct force by the factor $(a/b)^2$. The centroid oscillation frequencies are, therefore,

$$\omega_r^2 = \left(1 - n - n_s \frac{a^2}{b^2}\right) \Omega_\theta^2 \tag{13}$$

$$\omega_z^2 = \left(n - n_s \frac{a^2}{b^2}\right) \Omega_\theta^2. \tag{14}$$

That the right sides of Eqs. (13) and (14) must be positive for stability is less restrictive than the requirement on Eqs. (8) and (9).

Modified Betatron

The modified betatron adds a strong toroidal magnetic field B_θ to relax the beam current limit.

The toroidal field couples from the electron transverse oscillations to yield (Chernin and Sprangle, 1982)

$$\omega^2 = 1/2 \left(\Omega_c^2 + \Omega_r^2 + \Omega_z^2\right) \pm \left[1/4\left(\Omega_c^2 + \Omega_r^2 + \Omega_z^2\right)^2 - \Omega_r^2 \, \Omega_z^2\right]^{1/2} \tag{15}$$

where $\Omega_c = B_\theta/\gamma$. For Ω_c^2 much greater than Ω_r^2 and Ω_z^2, Eq. (15) reduces to a fast frequency, Ω_c, and a slow frequency,

$$\Omega_B^2 = \Omega_r^2 \, \Omega_z^2 \Big/ \Omega_c^2. \tag{16}$$

As in a straight beam confined by an axial magnetic field, electrons within the beam oscillate at the fast frequency, and the beam rotates about its axis at the slow frequency.

Two requirements for orbit stability are implied by Eq. (15).

$$\Omega_r^2 \, \Omega_z^2 > 0 \tag{17}$$

$$\Omega_c > \left(-\Omega_r^2\right)^{1/2} + \left(-\Omega_z^2\right)^{1/2}. \tag{18}$$

Eq. (17) limits the allowed range of $\underline{n}$. Eq. (18) is equivalent to

$$n_s > 1/2 + 1/4 \left(B_\theta/B_z\right)^2 \tag{19}$$

which replaces Eq. (12) as the beam current limit. Figure 2 summarizes the stable and unstable ranges of Ω_r^2 and Ω_z^2. Recall that the entire lower left quadrant is unavailable in the conventional betatron.

Also shown in Fig. 2 is the trajectory in parameter space that a beam might follow as it is accelerated. The beam is diamagnetic in the lower left quadrant and paramagnetic in the upper right quadrant. The transition between these two states, which occurs at the origin, involves a substantial change in the internal structure of the beam (Finn and Manheimer, 1983; Manheimer and Finn, 1983; Grossmann, et al., 1983; 1985). Nonetheless, limited computer simulations suggest that emittance

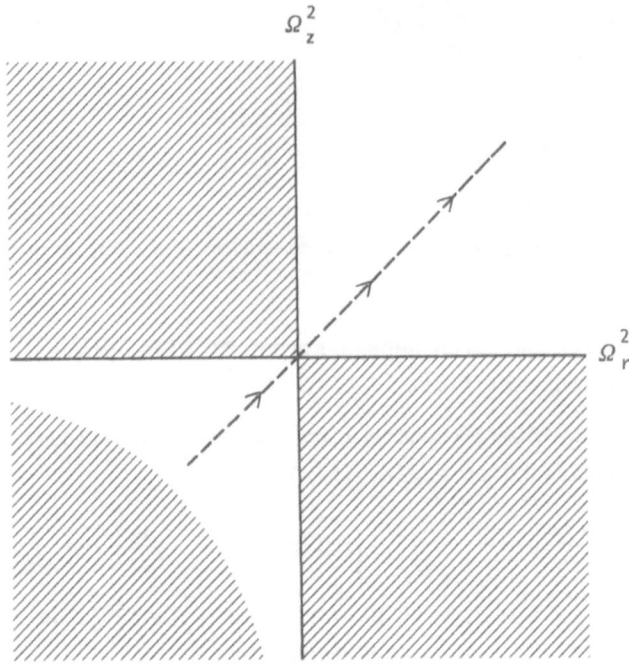

Fig. 2.   Stable (open) and unstable (cross hatched) regions
of (Ω_r^2, Ω_z^2) parameter space in the modified betatron.
The dashed line indicates the possible trajectory of
an accelerating beam.

growth is not excessive (Haber, et al., 1983), provided $\Omega_r^2 = \Omega_z^2$ is well
satisfied near the transition point. Thus n = 1/2 again is needed.

The transverse dynamics of the beam centroid in the modified
betatron are formally identical to that of the individual electrons with
Ω_r^2 and Ω_z^2 replaced by ω_r^2 and ω_z^2. Of greatest importance is the slow
poloidal oscillation frequency,

$$\omega_B^2 = \omega_r^2 \, \omega_z^2/\Omega_c^2 \ . \tag{20}$$

Stability near ω_B = 0 requires $\omega_r^2 \cdot \omega_r^2 \gtrsim 0$, so here too n = 1/2 is
needed.

Strong Focusing Betatrons

Energy bandwidth in recirculating high current electron accelerators
is desirable not only because high current beams tend naturally to have
non-negligible kinetic energy spreads but also because high gradient
acceleration gaps, if present, may introduce uncertainty in the average
beam energy. The acceptable electron energy spread in conventional and
modified betatrons is small,

$$\Delta\gamma \lesssim \left| 1 - n - n_s \right| \gamma \, b/R \ . \tag{21}$$

Periodic strong focusing, added to the modified betatron, greatly increases its tolerance to energy spread.

The most thoroughly studied of the strong focusing betatron concepts is the stellatron (Roberson, et al., 1983; 1985), which employs a continuous, twisted quadrupole field. See Fig. 3. Electron orbits are derivable analytically when n = 1/2 (Chernin, 1985; 1986).

$$\omega/\Omega_\theta = 1/2 l_o \pm \left[n_1 + 1/4 n_2^2 \pm \left(n_1 n_2^2 + \mu^2 \right)^{1/2} \right]^{1/2} \tag{22}$$

where

$$n_1 = n - n_s + 1/4 \left(B_\theta / B_z \right)^2 \tag{23}$$

$$n_2 = l_o + B_\theta / B_z \tag{24}$$

$$\mu = \frac{R}{B_z} \frac{\partial B_q}{\partial r} . \tag{25}$$

B_q is the quadrupole field strength, and l_o is the toroidal winding number. Orbits are stable if and only if

$$n_1 n_2^2 + \mu^2 > 0 \tag{26}$$

$$n_1 + 1/4 n_2^2 > 0 \tag{27}$$

$$\left(n_1 - 1/4 \, n_s^2 \right)^2 > \mu^2 . \tag{28}$$

Eqs. (26) and (27) give the beam current bound, larger than that in Eq. (18) for the modified betatron. The quadrupole field strength limit, Eq. (28), is not unduly restrictive. In particular, quadrupole field

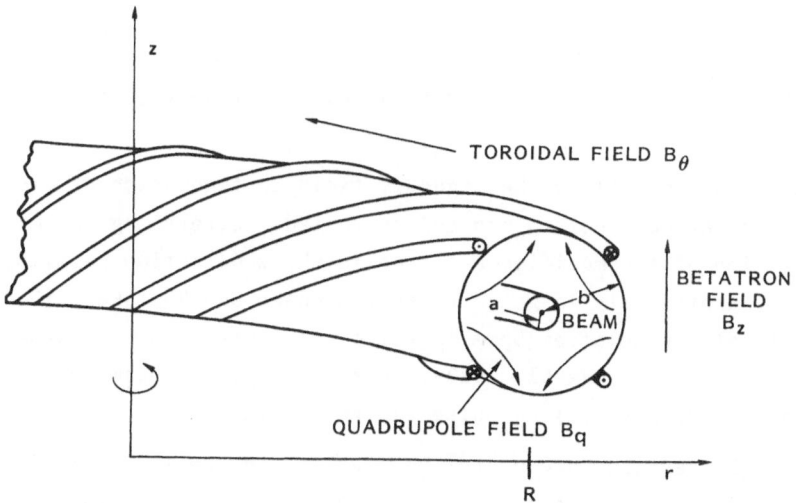

Fig. 3. Stellatron configuration with quadrupole windings shown.

strengths sufficient to eliminate the diamagnetic-paramagnetic transition of the modified betatron are permissible (Roberson, et al., 1985).

Beam centroid oscillation frequencies and constraints are obtained, as usual, by replacing n_s by $n_s(a/b)^2$. No new equilibrium issues arise.

Periodic toroidal modulation of the toroidal magnetic field provides strong focusing in the "bumpy torus" betatron (Chernin, et al., 1984; Mondelli and Chernin, 1984). For small modulation amplitudes and $n \sim 1/2$, electron orbits are stable away from

$$n_1 \sim l_o^2 \, q^2 \tag{29}$$

where q is an integer. Computational investigations show the instability bands to increase in size with the modulation.

Figure 4 (from Roberson, et al., 1985) compares the orbits of energy mismatched electrons in weak and strong focusing low current betatrons. In this example R = 100 cm, b = 10 cm, n = 1/2, and γ_o = 7. B_θ = 2 kG in all but the conventional betatron. Strong focusing fields in the l_o = 10 stellatron and the l_o = 10 bumpy torus betatron are 0.6 kG. In general, the improvement in energy bandwidth by strong focusing decreases with increasing energy.

An extreme case of B_θ modulation is the solenoidal lens betatron (Zieher, et al., 1985), shown schematically in Fig. 5. The magnetic cusps provide the strong focusing, which improves the energy bandwidth by of order $(B_\theta/B_r)^2$ relative to the modified betatron (Hughes and Godfrey, 1985). To minimize oscillations in the beam envelope, the magnitude of the toroidal field must be chosen initially to approximate the Brillouin flow condition (e.g., Davidson, 1974), here

$$n_s = 1/2 + 1/4 \, (B_\theta/B_z)^2 \; . \tag{30}$$

Thereafter, the beam radius adjusts adiabatically during acceleration to satisfy Eq. (30)

The shape as well as the magnetic field configuration of electron beam recirculating accelerators can be varied. Stretching the minor cross section of the modified betatron in the z-direction permits, in principal, arbitrarily large currents (Cavenago and Rostoker, 1985; 1986). It also simplifies beam injection and extraction (Prohaska, et al., 1983; Gisler and Faehl, 1983). Moreover, if the beam is extracted over many turns, a very long pulse can be obtained.

Straight sections are useful in high current recirculating accelerators for locating acceleration gap modules (Roberson, 1981; Mondelli and Roberson, 1983; 1984) and for simplifying injection and extraction (Lee, et al., 1983). A magnetic guide field is, of course, necessary in

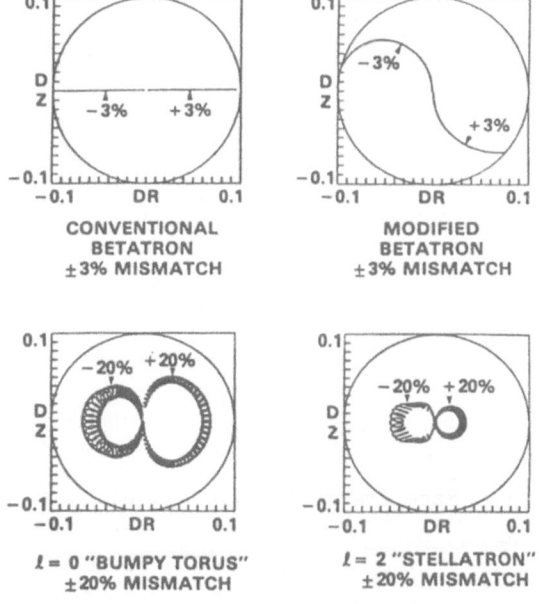

Fig. 4. Particle orbits in conventional betatron, modified betatron, bumpy torus betatron, and stellatron.

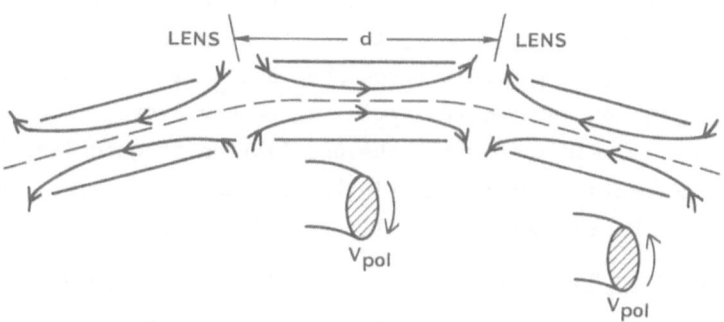

Fig. 5. Periodically reversing toroidal magnetic field and beam poloidal rotation in solenoidal lens betatron.

the straight sections, and strong focusing may be desirable in the curved. Figure 6 provides the possible layout of a race track induction accelerator with ferrite core inductive acceleration modules.

SINGLE PARTICLE RESONANCES

In simplest terms, a single particle resonance occurs whenever one of the transverse oscillation frequencies of an electron approximately

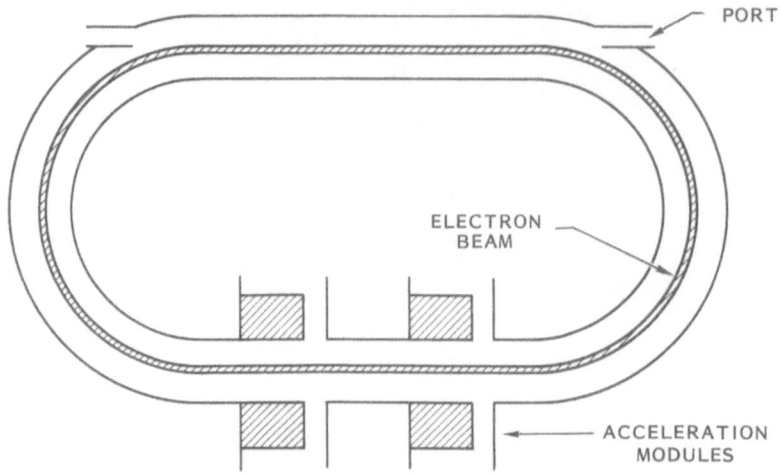

Fig. 6.   Race track induction accelerator design.

matches a harmonic of its circulation frequency around the accelerator.
This resonance condition allows the inevitable small field irregularities
in the accelerator to drive the electron transverse oscillation in phase
over several resolutions, leading to emittance growth and possible
current loss. Single particle resonances are a well known consideration
in high energy, low current recirculating accelerators. The effect of
high currents is mainly to shift the oscillation frequencies. In addition, high currents separate the single particle and centroid oscillation
frequencies, creating more possibilities for resonances.

Modified Betatron Resonances

The integer resonance condition in a modified betatron is given by

$$\Delta\Omega = 1\Omega_\Theta - \Omega_\pm \approx 0 \ . \tag{31}$$

where $\Omega_\pm$ is Ω_c or Ω_B, the fast or slow oscillation frequency defined
previously. Figure 7 illustrates the typical variation of these frequencies with γ. Ω_B/Ω_Θ is less than one, and so has no integer
resonance, except at low energies, below the diamagnetic-paramagnetic
transition. Ω_c/Ω_Θ, on the other hand, assumes several integer values.
It is assumed in Fig. 7 that B_Θ is constant in time. Increasing B_Θ with
γ is possible but requires large fields at high beam energies and may
complicate beam extraction.

If a resonance cannot be avoided, passing quickly through it may
minimize emittance growth. The condition for modest emittance growth is
(Chernin and Sprangle, 1982)

$$|\Delta\Omega| \gg 1^{-1} \, (2\,\delta B_z/B_z)^2 \qquad\qquad (32)$$

where δB_z is the toroidal Fourier component $\underline{1}$ of the errors in B_z or the equivalent force of other field errors, including image field irregularities due to openings in the cavity wall. Low $\underline{1}$ resonances are most dangerous. As an example, for the $1 = 1$ resonance in an $R = 100$ cm modified betatron accelerating an $a = 1$ cm beam from 3 MeV to 50 MeV, Eq. (32) restricts the fractional field error to one part in 10^5.

For integer resonances involving the beam centroid, Ω_B is replaced by ω_B, which for a given current has a weaker energy dependence. The fast oscillation frequency is approximately that of single electrons. A spread in Ω_Θ, as discussed subsequently in the context of collective instabilities, may damp centroid oscillation resonances (Chernin and Sprangle, 1982).

Nonuniformity of $\underline{n}$ across the beam cross section couples transverse oscillations to permit resonances involving beat oscillations

$$\Delta\Omega = 1 \, \Omega_\Theta - (\Omega_i \pm \Omega_j) \approx 0 \qquad\qquad (33)$$

where $\Omega_{i,j}$ can represent either transverse oscillation frequency. The case $i = j$ results in half-integer resonances.

$$\Delta\Omega = 1 \, \Omega_\Theta - 2\Omega_{\pm} \approx 0 \qquad\qquad (34)$$

These resonances typically are no stronger than the integer resonances. A large negative gradient in $\underline{n}$ can reduce the nonlinear growth of electron orbits due to resonances (Chernin, 1984).

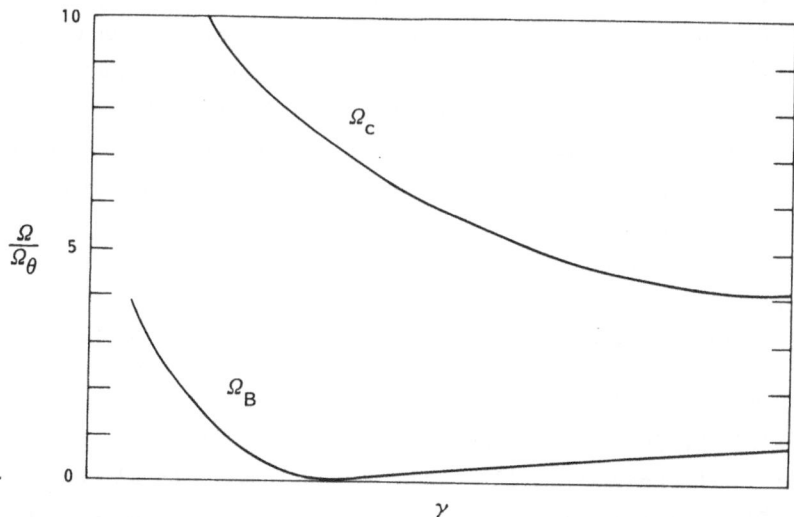

Fig. 7. Typical variation of electron transverse oscillation frequencies with energy in a modified betatron.

Stellatron Resonances

Strong focusing introduces additional transverse oscillation frequencies, creating more possibilities for resonances. The stellatron frequencies, expressed generally in Eqs. (22) – (25), often can be simplified to emphasize their energy dependence:

$$\Omega_1 \approx l_0 \Omega_\theta + \Omega_c + \frac{n - n_s}{\Omega_c} \tag{35}$$

$$\Omega_2 \approx -\Omega_c - \frac{n - n_s}{\Omega_c} \tag{36}$$

$$\Omega_3 \approx \frac{\mu^2 \Omega_\theta^2}{l_0 \Omega_c} \left(l_0 + \frac{B_\theta}{B_z} \right)^{-1} + \frac{n - n_s}{\Omega_c} \tag{37}$$

$$\Omega_r \approx l_0 \Omega_\theta - \frac{n - n_s}{\Omega_c} . \tag{38}$$

Recall that $n \sim 1/2$.

Figure 8 (from Chernin, 1986b) shows the resonance plane for an $l_0 =$ 12 stellatron. Avoiding resonances evidently is difficult unless B_θ/B_z is held constant. However, strong focusing reduces the growth of resonant orbits and may limit their nonlinear amplitudes. Experiments on the University of California, Irvine stellatron suggest that high $\underline{l}$ resonances are not serious (Ishizuka et al., 1985).

NEGATIVE MASS INSTABILITY

Although the negative mass instability is familiar from low current accelerators (Landau and Neil, 1966; Landau, 1968), its properties change somewhat at high current. It becomes as much a kink as a bunching instability and in some instances is supplanted by new instabilities which resonantly couple longitudinal and transverse beam modes. Properties of the negative mass instability also are modified by the toroidal or strong focusing fields necessary at high currents. Straight sections of race track geometry recirculating accelerators have little impact on the instability, however. Thermal spread is less effective in stabilizing high current than low current beams. On the other hand, the larger self-fields raise the negative mass threshold to higher energies.

Basic Instability Properties

In a linear device, any spontaneously occurring local excess of electrons in the beam typically tends to dissipate. Local space charge fields differentially accelerate electrons in the front of a clump and

268

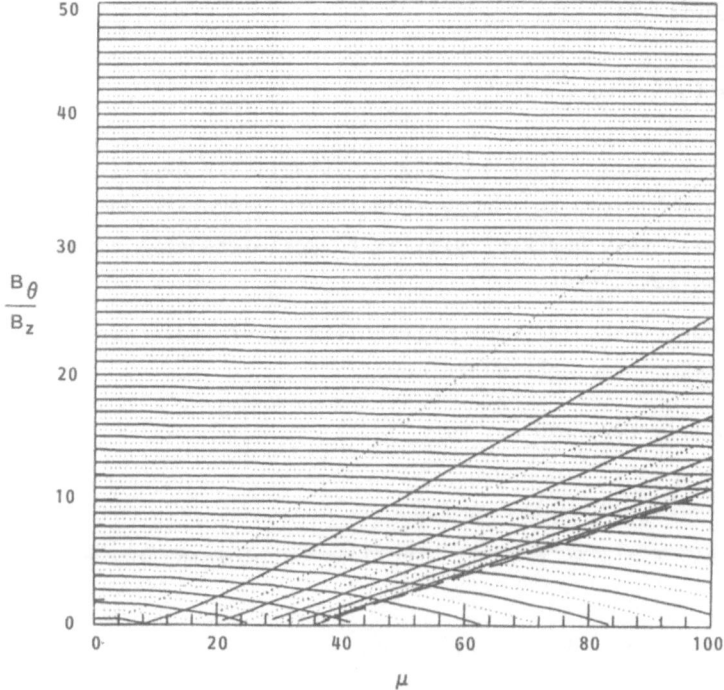

Fig. 8. Integer (solid lines) and half integer resonances (dotted lines) in l_o = 12 stellatron.

decelerate those in the back until the clump is sufficiently spread that the beam is again uniform.

The situation is different in curved sections of a recirculating device. Differentially accelerated electrons move to larger radii due to centrifugal force and so must travel further. If their increased velocity is insufficient to compensate for their increased path length, electrons in the front of a clump fall back, and the clump increases in density instead of decreasing.

The competition between increasing path length and velocity is easily quantified. A small change in the angular velocity can be expressed as

$$\frac{\delta\dot{\theta}}{\dot{\theta}} = \frac{\delta V}{V} - \frac{\delta R}{R} \tag{39}$$

where the radius of curvature $\underline{R}$ is assumed constant so that it can be related to the path length, $S = 2\pi R$. Making use of $\delta V/\delta P = \gamma^{-3}$ with $\underline{P}$ the relativistic momentum, and introducing the "momentum compaction" factor (Lawson, 1977),

$$\alpha = \frac{P}{S}\frac{\delta S}{\delta P} \tag{40}$$

we can rewrite Eq. (39) is

$$\frac{\delta\dot{\theta}}{\dot{\theta}} = (\gamma^{-2} - \alpha) \frac{\delta P}{P} . \tag{41}$$

The electron longitudinal inertia is effectively negative when $\delta\dot{\theta}/\delta P < 0$; i.e., when $\gamma^{-2} < \alpha$. We should expect such "negative mass" beams to break spontaneously into bunches, similar to Rayleigh-Jeans clumping to self-gravitating gas clouds.

The basic negative mass instability growth rate can be derived from the one dimensional cold fluid equations

$$\frac{\partial}{\partial t} \rho + \frac{\partial}{\partial S} (\rho V) = 0$$

$$\frac{\partial}{\partial t} P + V \frac{\partial}{\partial S} P = (1 - \alpha\gamma^2) E . \tag{42}$$

The electric field is given in the long wavelength, low frequency limit by

$$E = - \frac{\partial}{\partial S} (\delta\phi - V\delta A) = -\gamma^{-2} g \frac{\partial}{\partial S} \rho \tag{43}$$

because $\delta A_\theta \approx V \delta\phi$. g is a geometrical factor. Fourier transforming and then combining Eqs. (42) and (43) yields the dispersion relation

$$1 = \frac{\nu}{\gamma^3} \frac{l^2}{R^2} \frac{g}{(\omega - l\Omega_\theta)^2} (\gamma^{-2} - \alpha) . \tag{44}$$

Here, l is the toroidal mode number. One root is unstable, if $\gamma^{-2} < \alpha$.

$$\omega = l \Omega_\theta \pm i \frac{1}{R} \left[\frac{\nu}{\gamma^3} (\alpha - \gamma^{-2}) \right]^{1/2} . \tag{45}$$

The momentum compaction factor in a conventional, low current betatron is $\alpha = [1 - n]^{-1} \sim 2$. As a consequence, the conventional betatron is subject to the negative mass instability at all energies. In high current, modified, possibly strong focusing, betatrons, however, $\alpha = [1 - n - n*(\gamma)]^{-1}$, and instability occurs only above some transition energy.

Toroidal and High Current Corrections

Several more sophisticated negative mass instability linear models have been developed in the last five years to include toroidal corrections to the electromagnetic fields (e.g., Godfrey and Hughes, 1985) and poloidal displacement of the beam (Sprangle and Vomvoridis, 1981). In addition, three dimensional particle-in-cell simulations have been performed (e.g., Hughes, 1986). The disagreements, sometimes significant, among the models and between the models and the simulations

270

is attributable primarily to the delicate cancellation between ϕ and $V \cdot A$, which is evident even in the simple derivation above.

To alleviate this difficulty, Godfrey and Hughes (1986) recently incorporated exact analytical expressions for the electromagnetic fields in a torus of rectangular cross section into a negative mass instability dispersion relation. See Fig. 9. As in earlier models, the beam is treated as a flexible string of rigid disks. Thus, beam segments can move in all directions, but with no internal dynamics. Agreement between simulations and model predictions is good. Note that negative mass instability growth rates in numerical simulations are relatively insensitive to whether the torus cross section is a square or a circle, if the beam is not too near the wall.

The resulting dispersion relation can be written as

$$(\Omega^2 - \omega_z^2) \left(\Omega^2 - \omega_r^2 - \frac{X}{\Omega^2 - \epsilon} \right) = \Omega_c^2 \, \Omega^2 \tag{46}$$

with $\Omega = \omega - l\Omega_\theta$. Although the axial bounce frequency ω_z, the radial bounce frequency ω_r, the longitudinal dielectric function ϵ, and the coupling coefficient X all involve infinite sums over cavity modes, they can be approximated accurately or frequencies well below the electromagnetic cutoff at a given $\underline{l}$ by Eqs. (13), (14), and

$$\epsilon \approx \frac{\nu}{3} \left[g_1 \frac{l^2}{R^2} - g_2 \, \omega^2 \right] \tag{47}$$

$$X \approx \left[\gamma \, \Omega_\theta \Omega + \frac{\nu}{\gamma^2} \frac{l}{R^2} g_3 \right]^2 - \gamma^2 \, \Omega_\theta^2 \, (\Omega^2 - \epsilon) \; . \tag{48}$$

The three geometrical factors,

$$g_1 \approx g_2 \approx g_3 \approx 1 + 2 \, \ln(b/a) \tag{49}$$

are weak functions of ω and $\underline{l}$ and must be evaluated numerically. Small changes in g_i significantly affect ϵ and X.

The dispersion relation has been solve numerically for parameters similar to those of the Naval Research Laboratory modified betatron design: $I = 1 - 10$ kA, $B_\theta = 1$ kG, $n \sim 1/2$, $R = 1$ m, $b = 8.8$ cm, and $a = 1.76$ cm. Thus, $g \sim 3.6$. The low frequency approximation used to simplify Eqs. (47) and (48), although usually quite accurate, was not invoked. Instead, the first twenty terms of each infinite series were summed.

Figure 10(a) illustrates the variation of growth rate with beam energy for low toroidal mode numbers, in this case $l = 10$. The negative

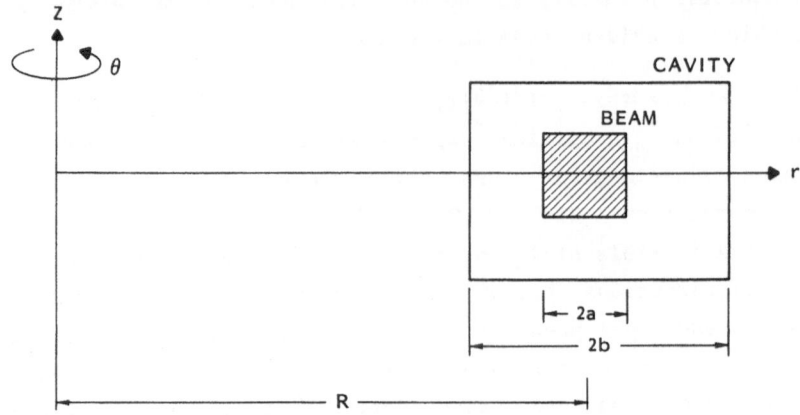

Fig. 9. Beam and cavity dimensions and cross sections
for negative mass instability dispersion
relation derivation.

mass instability occurs only above the energy at which ω_r^2 (equivalently, α^{-1}) vanishes.

$$\gamma_{t1} \approx \left[\frac{4\nu}{\Omega_\theta^2\, b^2}\right]^{1/3} \tag{50}$$

for $n = 1/2$. Its growth rate is described quite well by Eq. (45) for $\gamma > 20$. The growth rate does not, however, decrease quite as fast as $\gamma^{-3/2}$, because g_1 is slightly greater than g_2. A resonant instability coupling the $m = 0$ and $m = 1$ poloidal modes arises below γ_{t1} for I/B_θ large. Also observed in earlier models, the instability develops as a portion of the beam, oscillating poloidally at the frequency ω_B, accumulates more beam due to the change in path length and so oscillates more strongly.

For large toroidal mode numbers the inductive electric field asso-ciated with beam bunching can exceed the electrostatic field, reversing the sign of the right side of Eq. (44) and eliminating the negative mass instability at higher energies. In a sense, negative mass is canceled by negative field. From Eqs. (47) and (49), the negative mass instability ceases above a second transition energy,

$$\gamma_{t2} \approx (1 - g_1/g^2)^{-1/2} \ . \tag{51}$$

As depicted in Fig. 10(b) for $l = 30$, it usually is replaced by a second hybrid instability qualitatively resembling the one below γ_{t1}. Its growth rate for large l and moderate γ is less than that of the negative mass instability predicted by Eq. (45), however.

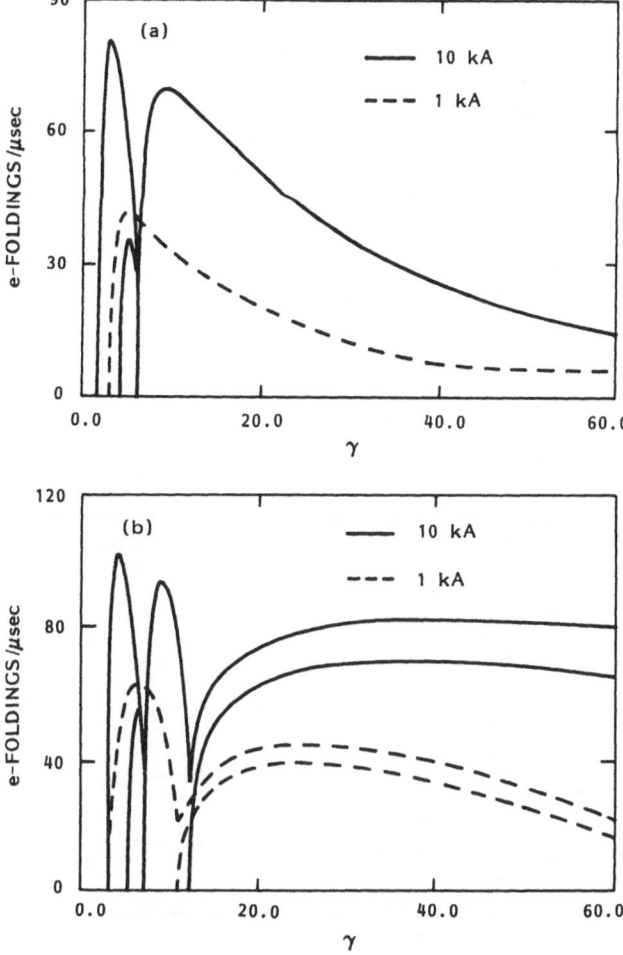

Fig. 10.   Instability growth rates versus energy for 10 kA
           and 1 kA beams in a 1 kG guide field; (a) l = 10
           and (b) l = 30.

Growth rate scaling with toroidal magnetic field is shown in Fig. 11
for l = 1 and 20.   The growth rates for B_θ = 0 are, for the most part,
indistinguishable from those for B_θ = 1 kG at those energies for which an
equilibrium exists with no toroidal guide field.   Increasing B_θ from 1 kG
to 10 kG is seen to be very effective at reducing the growth rates at low
energies, in the general vicinity of γ_{t1}, but much less so at high
energies.   Only at B_θ = 100 kG are the growth rates decreased significantly at the higher energies in Fig. 11(a), l = 1.

For completeness, we note that a very weak inductively driven
instability persists above γ_{t2} even for an infinite toroidal guide field.
From Eq. (8), its dispersion relation is $\Omega^2 = \varepsilon$.   Similar behavior has
been reported by Hughes and Godfrey (1985a) for a relativistic electron

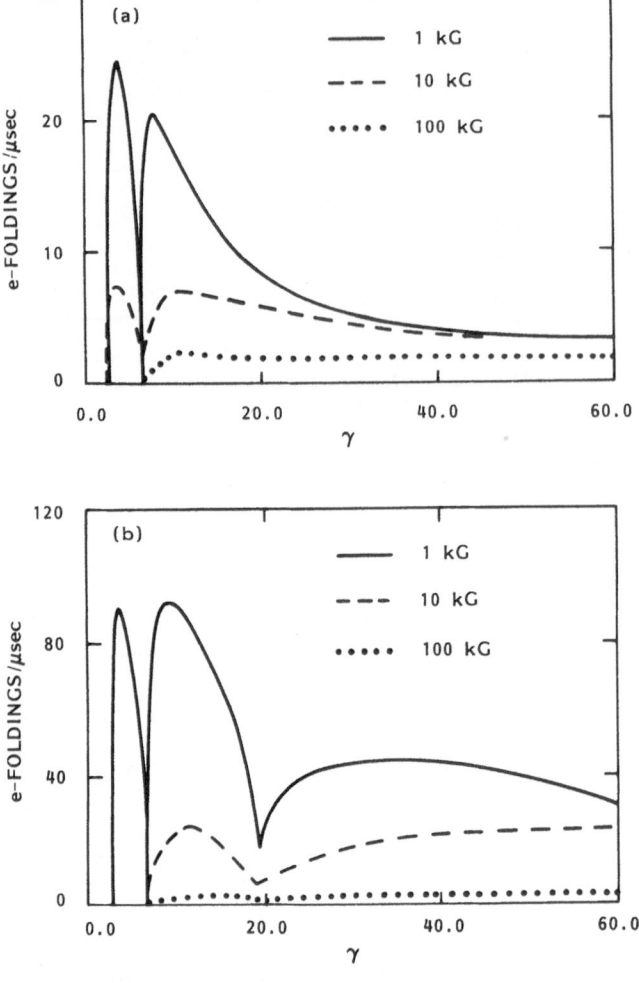

Fig. 11.　Instability growth rates versus energy for a 10 kA
beam in 100 kG, 10 kG, and 1 kG guide fields;
(a) $l = 1$ and (b) $l = 20$.

layer rotating in a strong azimuthal magnetic field between two concentric cylinders.

Approximate growth rate formulas, which reproduce fairly well the
results in Fig. 10 and 11 for $\gamma > \gamma_{t1}$, have been derived from Eqs. (46) –
(49). Several limiting cases, catalogued in Table 1, can be distinguished. Cases labeled "1" designate weak coupling between $m = 0$ and $m =$
1 beam modes, which is usually true at low $\underline{l}$, while cases labeled "2"
designate strong coupling, usually true at large $\underline{l}$.

Case 1A is applicable for $B_\theta \sim 0$, and case 1B for moderate B_θ with
$\gamma \gg \gamma_{t1}$. The second approximate solution for Ω^2 in the Table is
unstable when $\chi > 0$, usually equivalent to $\varepsilon > 0$. The growth rate,

Table 1. Approximate solutions of Eq. (46) for conventional betatrons and for modified betatrons above γ_{t1}.

B_θ	χ	$\Omega^2 =$	CASE
$\omega_B^2 \gg \omega_0^2$	$\lvert\chi\rvert \ll \left(\dfrac{\omega_r^2}{2}\right)^2$	$\omega_r^2, \ -\dfrac{\chi}{\omega_r^2}$	1A
	$\lvert\chi\rvert \gg \left(\dfrac{\omega_r^2}{2}\right)^2$	$\pm\,\chi^{1/2}$	2A
$\omega_0^2 \gg \omega_B^2 \gg \epsilon$	$\lvert\chi\rvert \ll \left(\dfrac{\omega_r \omega_B}{2}\right)^2$	$\omega_B^2, \ -\dfrac{\chi}{\omega_r^2}$	1B
	$\lvert\chi\rvert \gg \left(\dfrac{\omega_r \omega_B}{2}\right)^2$	$\pm\left(\chi\,\dfrac{\omega_B^2}{\omega_r^2}\right)^{1/2}$	2B
$\epsilon \gg \omega_B^2$	$\lvert\chi\rvert \ll \left(\dfrac{\epsilon}{2}\,\dfrac{\omega_r}{\omega_B}\right)^2$	$\epsilon, \ -\dfrac{\chi}{\epsilon}\,\dfrac{\omega_B^2}{\omega_r^2}$	1C
	$\lvert\chi\rvert \gg \left(\dfrac{\epsilon}{2}\,\dfrac{\omega_r}{\omega_B}\right)^2$	$\pm\left(\chi\,\dfrac{\omega_B^2}{\omega_r^2}\right)^{1/2}$	2C

$$\Gamma \approx \frac{1}{R}\left[\frac{\nu}{\gamma}\,\frac{1}{1-n}\,(g_3 - g_4\,v_\theta^2)\right]^{1/2} \tag{52}$$

is that of the "classical" negative mass instability, Eq. (45), for $g_1 = g_2$. Case 1C applies for very large B_θ, or for moderate B_θ with γ just larger than γ_{t1}. The second expression for Ω^2 yields

$$\Gamma \approx (1 - n)^{-1/2}\,\gamma\,\omega_B \tag{53}$$

equivalent to the formula of Landau (1968) for the negative mass growth rate in a toroidal guide field. That $\epsilon < 0$ leads to instability in case 1C has already been noted.

Case 2A applies to $B_\theta \sim 0$ betatrons at energies above γ_{t2}, which typically requires large $\underline{l}$. There is one unstable mode with a growth rate

$$\Gamma \approx \frac{l^{1/2}}{R}\left[\frac{\nu}{\gamma}\,(g_3 - g_4\,v_\theta^2)\right]^{1/4} \tag{54}$$

if χ is positive and two unstable modes with growth rates reduced from Eq. (54) by $\sqrt{2}$ otherwise. Cases 2B and 2C hold for moderate and very large B_θ, again at energies above γ_{t2}. The one or two unstable modes have growth rates reduced from those of case 2A by the factor $(\omega_z/\Omega_c)^{1/2}$.

Applying these results to the Naval Research Laboratory betatron in its present state (Golden et al., 1986), we find that the 1 MeV beam is at or just below γ_{t1} for $I = 2 - 3$ kA. Growth rates of the low energy hybrid instability at low $\underline{l}$ are up to 6 e-foldings/μ sec, sufficient to explain the observed current loss. There are, of course, other possible interpretations of the experimental results.

Many particle-in-cell three-dimensional numerical simulations have been performed (e.g., Hughes, 1986), which corroborate the linear growth rate model. Figure 12 presents comparisons for a 10 kA beam in a 1 kG guide field for $l = 1$ and 20. The low growth rate in the $l = 1$, $\gamma = 3.8$ simulation probably is attributable to temperature effects.

Simulations show that negative mass instabilities, at least up to $l = 20$, grow until the beam strikes the wall, although growth rates usually decline as instability amplitudes becomes nonlinear. Evolution of the $l = 1$ negative mass instability for a 10 kA, 5 MeV beam in a 1 kG guide field is indicated in Fig. 13. (Larger l values were suppressed numerically.) The instability, which grows from noise, was still linear at 175 nsec. By 260 nsec it had become quite nonlinear in amplitude. The beam radius had not expanded much, however. The beam struck the wall in less than 1 µsec (not shown) with significant current loss.

The nonlinear development of the high energy hybrid mode at $l = 20$ is strikingly different. Growing transverse oscillations were accompanied by comparable increased beam radii. Simulation of a 10 kA, 25 MeV beam showed instability saturation after 300 nsec, when the beam radius had a bit more than doubled.

Strong Focusing Effects

One expects intuitively that strong focusing in the stellatron (Roberson et al., 1983a; 1983b; 1985) and bumpy torus betatron (Chernin et al., 1984) should reduce the negative mass instability growth rate. Chernin (1986) demonstrated that growth rates of low frequency, long wavelength modes in the stellatron can be obtained from modified betatron formulas simply by replacing the momentum compaction factor of the modified betatron,

$$\alpha^{-1} = 1 - n - n_s \frac{a^2}{b^2} \tag{55}$$

by that of the stellatron,

$$\alpha^{-1} = 1 - n - n_s \frac{a^2}{b^2} + \mu^2 \left(l_o^2 + l_o \frac{B_\theta}{B_z} \right)^{-1} . \tag{56}$$

The quadrupole field index μ in Eq. (56) is given by Eq. (25).

From Eq. (41) the negative mass transition energy is given by $\gamma_{t1} = \alpha^{-1/2}$, or approximately

276

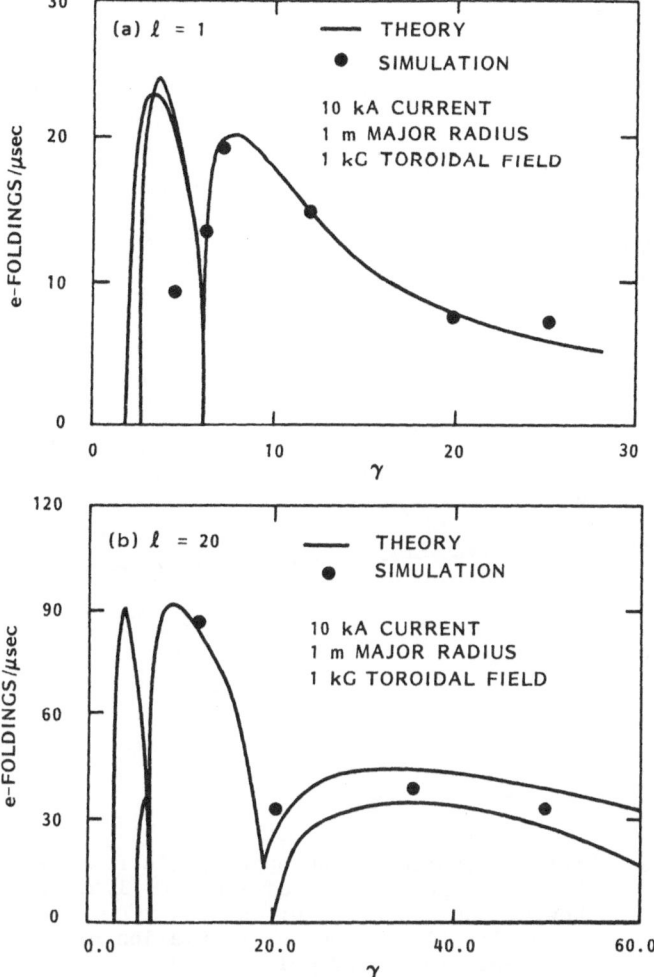

Fig. 12. Instability growth rates from theory and simulation
for a 10 kA beam in a 1 kG guide field;
(a) l = 1 and (b) l = 20.

$$\gamma_{t1} \approx \mu \left(l_o^2 + l_o \frac{B_\theta}{B_z}\right)^{-1/2} \tag{57}$$

independent of beam current for μ sufficiently large. Furthermore, based
on Eq. (45) the negative mass growth rate in the stellatron at large
energies in reduced from that of the modified betatron by the factor
$\sqrt{2/\gamma_{t1}}$ for $n \sim 1/2$.

Computer simulations (Hughes and Godfrey, 1984b) confirm this
behavior. A 10 kA, 5.5 MeV beam in an $l_o = 8$, $B_\theta = 1$ kG stellatron was
treated. For a modest quadrupole field of $\mu = 7$, corresponding to
$\gamma_{t1} = 2.3$, the development of the l = 8 negative mass instability was

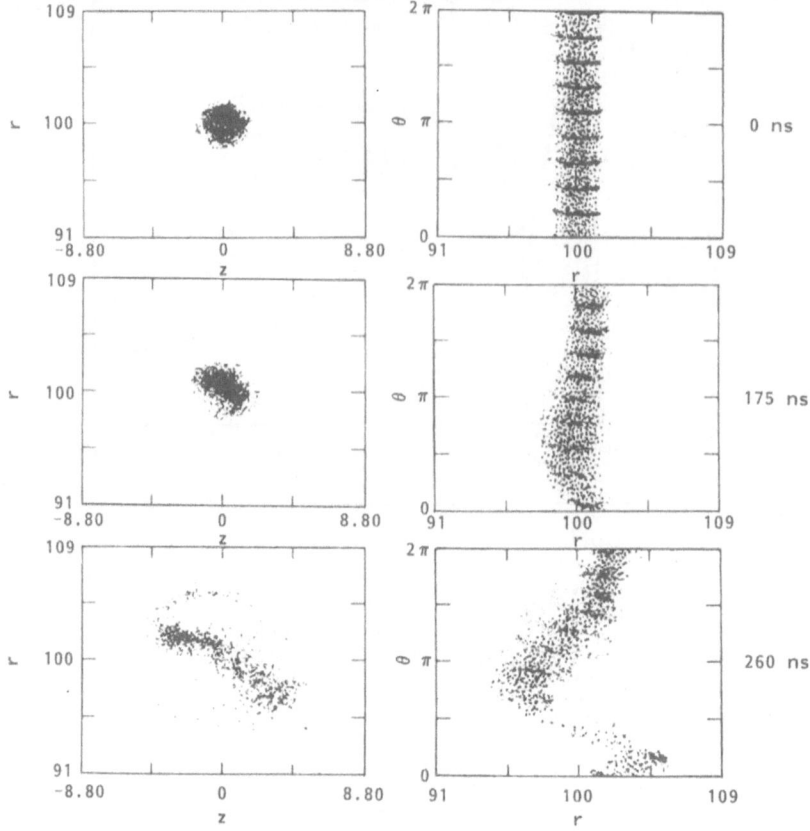

Fig. 13. Particle configuration space plots r–z (left column)
and θ–r (right column) at times 0, 175, and 260 nsec
showing the evolution of the $l = 1$ negative mass
instability in a computer simulation of a 10 kA,
5 MeV beam in a 1 kG guide field.

scarcely changed by the strong focusing. Eventually, about 75% of the
beam was lost to the wall. Results were quite different when the
quadrupole field was increased by a factor of four, $\mu = 28$ and $\gamma_{t1} = 5.2$.
The growth rate fell from 60 e-foldings/μsec to about 22 e-foldings/μsec.
Moreover, while the beam became strongly bunched in the toroidal
direction, much less current loss occurred. Figure 14 depicts the
instability near saturation.

Other simulations suggest that quadrupole focusing can, at least in
some cases, suppress the low energy hybrid instability observed in the
modified betatron. However, at large toroidal mode numbers a fast
growing electromagnetic instability sometimes was encountered. This new
effect is discussed later.

Similar stability behavior is expected in the bumpy torus betatron.

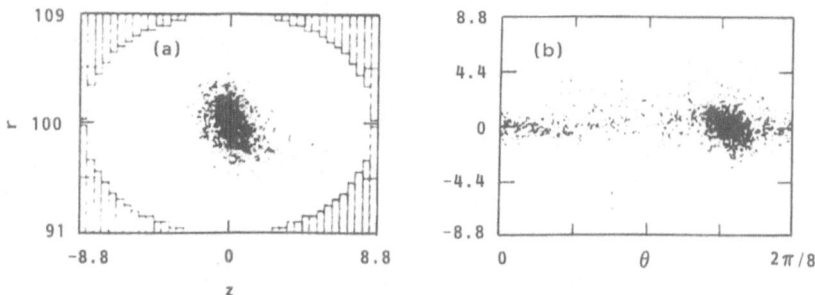

Fig. 14.   Particle configuration space plots (a) r-z and
(b) θ-r of the l = 8 negative mass instability
near saturation in a computer simulation of a
10 kA, 5.5 MeV beam in a μ = 28, B_θ = 1 kG
stellatron.

Stability properties of the solenoidal lens betatron have been
investigated to a limited extent by Hughes and Godfrey (1985b). The
momentum compaction factor at long wavelength, $l < l_o$, is

$$\alpha^{-1} = 1 - n - n_s \frac{a^2}{b^2} + \frac{1}{4} \frac{B_\theta^2}{B_z^2} \, . \tag{58}$$

Since the beam equilibrium in this betatron is determined by the
Brillouin flow condition, Eq. (30), the last term in Eq. (23) is simply
n_s. The transition energy for $n_s > 1/2$ is, therefore,

$$\gamma_{t1} \approx \left(2\nu \frac{R^2}{a^2} \right)^{1/5} \, . \tag{59}$$

Computer simulations of the 100 A, 1 MeV beam of the University of
New Mexico l_o = 15 solenoidal lens betatron showed no instability growth
for l = 1 but 66 e-folding/μ sec for l = 15. This rapid growth at large
l is more consistent with growth rates for a conventional betatron than
with the predictions of Eq. (45) with α specified by Eq. (58). The
instability saturated with a small loss of beam current and a large
increase in beam radius.

Simulations of a 10 kA beam at various energies in an l_o = 20
solenoidal lens betatron were performed to explore further the unexpect-
edly high growth rates at large toroidal mode numbers. The solenoidal
field was chosen to keep the beam radius constant at 2 cm; R = 100 cm.
Table 2 compares the l = 20 instability growth rates measured in the
simulations (Γ_a) against the predictions of three models: Eq. (45) with
the solenoidal lens betatron long wavelength expression, Eq. (58), for
α (Γ_b); the modified betatron dispersion relation in Eq. (46) with B_θ as

Table 2. Different $l = 20$ growth rate predictions (in e-foldings/ μsec), described in the text, for a 10 kA beam in an $l_o = 20$ solenoidal lens betatron.

γ	Γ_a	Γ_b	Γ_c	Γ_d
7	210	77	70	300
12	140	74	105	160
17	40	63	66	70

given in the table (Γ_c); and the Eq. (46) dispersion relation with $B_\theta = 0$, a conventional betatron (Γ_d). The simulation growth rates agree best with those of the conventional betatron and are much greater than those of a modified betatron with equal solenoidal field strength. Apparently, the effects of the alternating solenoidal fields on stability average to zero for $\underline{l}$ large.

Modified Self-Field Effects

Modifying the effective equilibrium self-fields of the beam by introducing a background plasma can change the momentum compaction factor significantly, possibly reducing negative mass instability growth. For instance, a low energy electron cloud in a modified betatron shifts α^{-1} by $-\gamma^2 n_s'$ (Roberts and Rostoker, 1986), which increases the transition energy to

$$\gamma_{t1} \approx 4\nu' \frac{R^2}{a'^2} . \tag{60}$$

Here, $\underline{a}'$ is the minor radius of the cloud, ν' is its line density, and n_s' is the self-field index with ν replaced by ν'.

Although the negative mass instability is eliminated at low energies by increasing the transition energy, it may be replaced by the hybrid instability discussed earlier. A brief analysis based on Eqs. (46) – (49) indicates the hybrid instability also is avoided, at least for low $\underline{l}$, when

$$\gamma > \left[\nu g\, l^2 \left(\frac{-\omega_r^2}{\Omega_\theta^2} \right)^{-2} \left(\frac{-\omega_z^2}{\Omega_\theta^2} \right)^{-1} \right]^{-1/5} . \tag{61}$$

Therefore, the toroidal magnetic field should not be too large. It must, however, be large enough to confine radially both the beam and the electron cloud.

As a second example of the effect of plasmas on betatron stability, consider an ionized toroidal channel of line density just less than that of the beam (Manheimer, 1984). If B_θ is not too large, the channel electrons are expelled by the beam space charge, and the beam is tightly confined by the electric field of the remaining ions. In terms of the simple negative mass analysis, α^{-1} is shifted by $+\gamma^2 n_s'$, where now n_s' is determined by the ion line density and radius. The transition energy becomes approximately

$$\gamma_{t1} = \left(4\nu' \, \frac{R^2}{a_1^2} \right)^{1/3} \tag{62}$$

which significantly exceeds the transition energy of a modified betatron when $b/a' \gg (\nu/\nu')^{1/2}$. Due to the strong attractive force of the ions, it seems unlikely that a hybrid instability can occur below the transition energy. The ions also should exert a stabilizing influence at nonlinear amplitudes when linear instabilities develop.

Two-stream instabilities between the beam and the plasma are a potentially serious disadvantage to using plasmas to suppress the negative mass instability in betatrons. Streaming instabilities are discussed briefly in the last section of this chapter.

Self-fields can be modified without using plasmas by, for instance, slotting the betatron cavity wall to impede toroidal (but not poloidal) return currents, thereby eliminating the image current restoring force. If the slots are sufficiently closely spaced, the wall can be treated as a continuous boundary conducting only in the poloidal direction, and α^{-1} is shifted by a term $-\gamma^2 n_s a^2/b^2$. Note that this term is similar to that of the electron cloud treated above, and its consequences are qualitatively the same. The transition energy increases to

$$\gamma_{t1} \approx 4\nu' \, \frac{R^2}{b^2} \tag{63}$$

above which the negative mass instability occurs and below which a hybrid instability may occur.

The stabilizing influence of modified equilibrium fields is easily quantified by solving numerically the Eq. (46) dispersion relation with the additional equilibrium forces included. In general, the lowest $\underline{l}$ modes are found to be stable over moderately large parameter ranges, but the higher modes over increasingly narrow ranges. To be specific, with $B_\theta = 1.5$ kG, $I = 1 - 10$ kA, and other parameters as in Fig. 10 - 12, the negative mass and hybrid instabilities both are eliminated for the $l = 1$ mode below $\gamma = 50$ in the presence of an electron column with $\nu' = \nu$ and a' = b. However, the hybrid mode eventually arises as $\underline{l}$ increases, especially at the higher currents. For $l = 10$, the 1 kA beam is stable only for $5 < \gamma < 46$, and the 10 kA beam is as unstable as it is without the electron cloud. Our study of this topic has not been exhaustive, and it may be that more favorable parameters can be identified. In any event, stabilizing low toroidal modes in this way even without stabilizing high modes represents progress, because low modes are most difficult to suppress by thermal effects, the next subject of discussion.

Instability Damping by Electron Ω_θ Spread

Since the negative mass instability in a conventional betatron is, in a first approximation, a bunching of the beam in the toroidal direction, spread in the toroidal rotation frequency, Ω_θ, of the beam electrons should tend to reduce the instability growth rate. The rotation frequency spread can be incorporated into the basic instability derivation, Eqs. (39) – (44), by employing the Vlasov equation instead of the cold fluid equations. There results (Landau and Neil, 1966)

$$1 = \frac{S}{\gamma^2 R} \int \frac{\frac{\partial f}{\partial P} \, dP}{\left(\Omega_\theta - \frac{\omega}{l}\right) - \frac{P}{\gamma R}\left(\alpha - \frac{1}{\gamma^2}\right)} \tag{64}$$

where $f(\Omega_\theta)$ is the electron distribution function averaged over the minor cross section of the beam.

For a relativistic beam, in which all electrons move at approximately the speed of light, any spread in toroidal rotation frequency arises solely from a spread in path lengths around the torus. For a given beam spatial distribution, the maximum spread in path lengths is achieved when each electron moves in a circle centered on the major axis,

$$\Delta\Omega_\theta / \Omega_\theta \approx a/R \qquad \text{(half width)} \,. \tag{65}$$

As depicted in Fig. 15, poloidal motion causes electrons to sample a spread in major radii, reducing $\Delta\Omega_\theta$. Fat, cold beams exhibit strongest Landau damping.

A realistic upper bound on instability suppression by the spread in Ω_θ is obtainable, therefore, by evaluating Eq. (64) for a uniform density, circular cross section, approximately monoenergetic beam with no internal poloidal motion. The resulting dispersion relation

$$1 = \frac{2\nu \, g}{\gamma^3 a^2 \Omega_\theta^2}\left(\alpha - \frac{1}{\gamma^2}\right)\left[1 - \frac{\Omega_\theta - \omega/l}{\left((\Omega_\theta - \omega/l)^2 - a^2 \Omega_\theta^2 R^2\right)^{1/2}}\right] \tag{66}$$

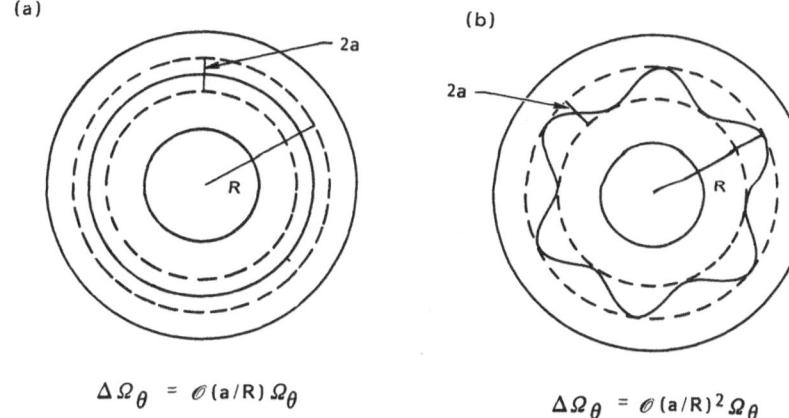

(a)   (b)

$\Delta \Omega_\theta = \mathcal{O}(a/R)\,\Omega_\theta$ $\Delta \Omega_\theta = \mathcal{O}(a/R)^2\,\Omega_\theta$

Fig. 15. The toroidal rotation frequency spread decreases sharply
          as the electron poloidal oscillation amplitudes increase
          from (a) zero to (b) the beam radius. (Beam boundaries
          are represented by dashed lines.)

reduces to Eq. (440, as it should, when a → 0. Expanding Eq. (66) about
Γ_0, the Eq. (45) zero-spread growth rate, gives the stability criterion

$$1\ \Omega_\theta\ a/R > \Gamma_0 \ . \tag{67}$$

This result has been derived, more generally but less rigorously, for the
modified betatron as well (Sprangle and Chernin, 1983; 1984).

Electron poloidal motion in a conventional betatron can, in principle, be reduced to zero by injecting a laminar beam with a slight
energy spread of half width $\Delta\gamma$,

$$\frac{a}{R} \approx \frac{\Delta\gamma/\gamma}{1 - n - n_s} \ . \tag{68}$$

Eliminating poloidal thermal motion may be difficult in practice.

In the modified betatron even a laminar beam slowly rotates at the
poloidal rotation frequency, Ω_B, which vanishes only at the paramagnetic-
diamagnetic transition. It can be shown that poloidal rotation can be
ignored in estimating instability damping when $\Omega_B < 1\ \Omega_\theta\ a/R$. That
slower poloidal rotation improves stability is illustrated in Fig. 16 for
a 10 kA, 5.5 MeV beam in a 1 kG guide field. It compares $l = 20$ negative
mass instability simulations which differ only in the beam initial
radius. In Fig. 16(a), the initial radius was 1.5 cm, leading to $\Omega_B = 1.5 \cdot 10^8\ \text{sec}^{-1}$, nearly twice $1\ \Omega_\theta\ a/R$. The instability developed at the
cold beam growth rate, and the beam soon struck the cavity wall. In
Fig. 16(b), the initial rate was doubled, reducing the poloidal rotation
frequency by an order of magnitude. The instability grew much more

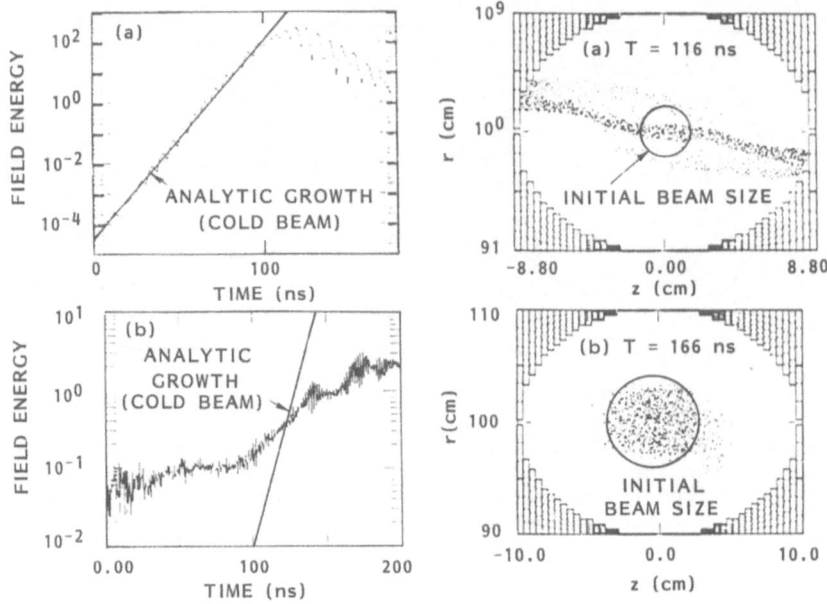

Fig. 16. The l = 20 field energy history and final beam cross
section of unstable 10 kA, 5.5 MeV beams with
(a) 1.5 cm and (b) 3.0 cm initial radii. R = 100 cm
and B_θ = 1 kG.

slowly and saturated at an innocuous level. Incidentally, the slight
expansion of the beam radius during the second simulation reduced the
poloidal rotation frequency still further.

Poloidal rotation occurs as well in the stellatron and bumpy torus
betatron. However, the average poloidal rotation frequency in the solenoidal lens betatron is zero, because the direction of beam rotation
reverses with that of B_θ. It is uncertain whether instability damping by
the spread in Ω_θ played any significant role in the few solenoidal lens
betatron simulations presented earlier.

Effect of Straight Sections

The negative mass effect vanishes in straight sections of a recirculating accelerator. Indeed, beam clumps may tend to smooth out there due
to electrostatic repulsion. Simple numerical investigations (Hughes,
unpublished) suggest, however, that straight drift tube sections have
only a small, although favorable, effect on overall negative mass instability growth per pass around the accelerator.

OTHER COLLECTIVE INSTABILITIES

Instabilities encountered in electron linear accelerators also may
arise in recirculating accelerators. Generally, they are less serious

than the negative mass instability but nonetheless should not be ignored. Two of the more important modes, the beam breakup and resistive wall instability, are considered here. Emphasis is given to features unique to recirculating devices.

Discussed first is a new instability discovered in simulations of the stellatron, although it can equally well occur in periodic, strong focused, high current linear accelerators. Its growth rate can be very large.

Parametric Instabilities

If matching conditions are satisfied, the periodic transverse magnetic fields of the stellatron or other strong focusing device parametrically couple beam modes to electromagnetic modes of the accelerator cavity (Hughes and Godfrey, 1986):

Pump mode $\qquad \omega_1 = 0 \qquad\qquad\qquad\qquad k_1 = l_0/R$

TE mode $\qquad \omega_2 = \left(\omega_0^2 + k_2^2\right)^{1/2} \qquad\qquad k_2 = 1/R$ $\qquad$ (69)

Beam mode $\qquad \omega_3 = (1 + l_0)\omega_\theta - \omega_B \qquad k_3 = (1 + l_0)/R$

where ω_0 is the TE_{11} electromagnetic cutoff frequency of the betatron cavity. The beam mode toroidal mode number is chosen in Eq. (69) such that $k_1 + k_2 = k_3$. Instability then occurs when $\omega_1 + \omega_2 \approx \omega_3$.

A dispersion relation for the stellatron can be derived in closed form, if nonresonant terms are ignored. The resulting peak growth rate is approximately

$$\Gamma^2 = \frac{\nu}{\gamma} \frac{(\omega_3 - \omega_2)\,(\omega_2 - l\Omega_\theta)\,\omega_0^2}{\omega_2(\omega_2 - l\Omega_\theta + \Omega_c)\,(\omega_0^2 b^2 - 1)\,J_1^2(\omega_0 b)} \qquad (70)$$

and the bandwidth in $\underline{l}$ of unstable modes scales as $\nu^{1/2}$. (J_1 is a Bessel function.) Figure 17 illustrates the coupling among modes and gives growth rates for an $l_0 = 20$, $\mu = 326$, $B_\theta = 5$ kG stellatron. The beam current and energy are 10 kA and 3 MeV, and other parameters are as in Figs. 10 - 12. Note the characteristic symmetry in $\underline{l}$ about $-l_0/2$.

Predicted instability frequencies and growth rates have been corroborated by computer simulations. Figure 18(a) shows analytical and computational growth rates versus the quadrupole periodicity l_0 for the $l = -22$ beam mode with the same beam and accelerator parameters used in Fig. 17. Agreement is good. A comparison of the results of rigid disk and discrete particle simulations for $l_0 = 22$ suggests that a spread in electron orbital frequencies reduces instability growth. Other

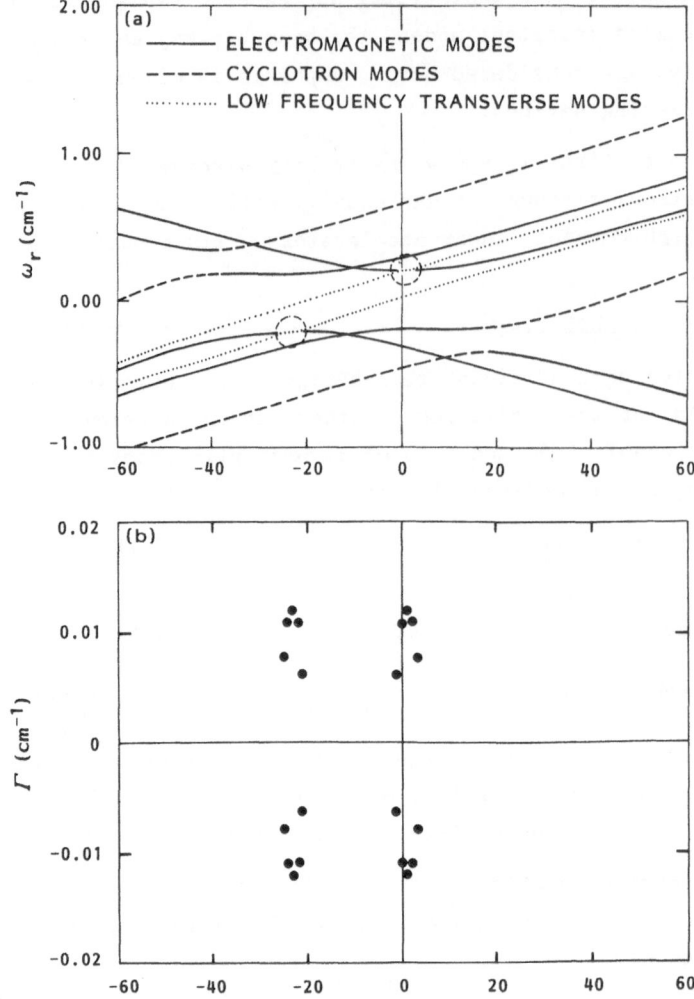

Figure 17. Parametric instability (a) frequencies and (b) growth rates for a 10 kA, 3 MeV beam in a B_θ = 5 kG, l_o = 20, μ = 326 stellatron.

simulations confirmed the predicted scaling of growth rate with beam current and with cavity minor radius.

In simulations the parametric instability saturated only when the beam struck the accelerator wall; current loss was great. Figure 18(b) illustrates the severe beam kinking which arose after one pass around the stellatron for the parameters of Fig. 18(a).

Because the bandwidth of unstable toroidal modes scales roughly as $\nu^{1/2}$, it may be possible to choose conditions such that the instability falls between modes in a low current stellatron. Reduced growth rates at low currents, coupled with thermal damping, should help as well.

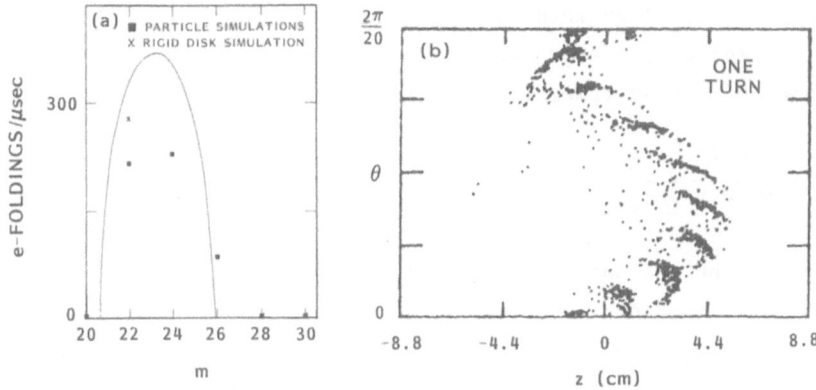

Figure 18.   (a) Growth rates from Eq. (70) and from computer simulations for $1 = -22$ and various 1_o; (b) nonlinear beam profile after one pass around stellatron from simulation with $1_o = 22$. Parameters are as in Fig. 17.

Beam Breakup Instability

The beam breakup instability has long been recognized as a serious problem in high energy accelerators (Lawson, 1978). It arises from magnetic coupling between beam transverse oscillations and a periodic array of resonant structures in the accelerator. Gaps in the accelerator wall, perhaps for applying an accelerating voltage, are a likely source of beam breakup instability, as discussed elsewhere in this book in the context of linear induction accelerators. Periodicity is automatic in a recirculating accelerator.

The dispersion relation in Eq. (46) is readily generalized to include the approximate effect of a localized transverse force upon the beam (Godfrey and Hughes, 1983).

$$\left[2\pi R - F_r \sum_1 \frac{\Omega_1^2 - \omega_z^2}{D_1} \right] \left[2\pi R - F_z \sum_1 \frac{\Omega_1^2 - \omega_r^2 - \chi_1/(\Omega_1^2 - \varepsilon_1)}{D_1} \right]$$
$$- \Omega_c^2 F_r F_z \left[\sum_1 \frac{\Omega_1}{D_1} \right]^2 \tag{71}$$

with

$$D_1 = (\omega_1^2 - \omega_z^2) \left(\Omega_1^2 - \omega_r^2 - \frac{\chi}{\Omega_1 - \varepsilon_1} \right) - \Omega_c^2 \Omega_1^2 \tag{72}$$

An equation including multiple localized transverse forces is similar in structure but longer. The radial and axial force constants, F_r and F_z, could represent, for example, the impulse beam electrons experience when moving between straight and curved portions of a race track geometry (mentioned earlier) or when passing by an interruption in the wall image current (image displacement instability). The force constant for a localized TM_{1no} electromagnetic field is

$$F = \frac{\nu}{\gamma} \frac{\omega_0 Z}{Q} \frac{\omega_0^2}{\omega^2 + i\omega\omega_0/Q - \omega_0^2} \; . \tag{73}$$

Here, ω_0 is the resonant frequency, Q is the quality factor, and Z/Q is the coupling impedance.

A numerical study of Eq. (71) – (73) with a highly simplified expression for χ has shown the coupling between the beam breakup and negative mass instabilities to be small (Godfrey and Hughes, 1983). Moreover, a single resonant toroidal mode, $1 \approx \omega/\Omega_\theta$, typically provides the dominant contribution to Eq. (71). Under these conditions, the beam breakup instability growth rate can be approximated by

$$\Gamma \approx \omega_0 Q \frac{\nu Z/Q}{2\pi R\gamma\Omega_c} \tag{74}$$

when $1/2 \, \Omega_c \gg \omega_r$, and

$$\Gamma \approx \omega_0 Q \frac{\nu Z/Q}{4\pi R\gamma\omega_z} \tag{75}$$

when $1/2 \, \Omega_c \ll \omega_r$. Eq. (74) is familiar for linear accelerators. It appears that the beam breakup instability does not occur in recirculating devices at beam energies below γ_{t1}.

Beam breakup growth has been estimated numerically for a race track induction accelerator design provided by C. Roberson. See Table 3. Total growth is predicted to be $5 \cdot (Q/6)$ e-foldings, which is acceptable for $Q \approx 6$. In comparison, total negative mass instability growth in the resonant toroidal mode, $1 = 13$, is about 20 e-foldings. The spread in Ω_θ may reduce both values a bit.

Resistive Wall Instability

The resistive wall instability is due to a resistively induced phase lag between motion of the beam and its image current in the accelerator wall. For the most part, resistive wall instability behavior is the same in betatrons as in linear accelerators, described by

$$\omega^{1/2}(\omega - \omega_0) = \frac{\nu}{\gamma} \frac{4f_m}{\sigma^{1/2} \Omega_c b^3} e^{i\pi/4} \tag{76}$$

when $1/2 \, \Omega_c \gg \omega_r$ (Godfrey and Hughes, 1982). $1/2 \, \Omega_c$ is replaced by ω_r in the opposite limit, just as in Eq. (74) and (75). Here, ω_0 is the

Table 3. Possible parameters for a race track accelerator using ferrite induction modules

Beam Current (ν)	1 kA
Beam Energy (γ)	1 – 40 MeV
Guide Field (B_θ)	2 kG
Geometry Factor (g)	5
Curvature Radius (R)	70 cm
Number of Passes	40
Number of Gaps	4
Gap Resonance (ω_o)	880 MHz
Transverse Impedance ($Z_\perp/Q$)	15 ohms
Quality Factor (Q)	6, 60

unperturbed beam mode frequency, σ (scaled by $4\pi/c$) is the wall conductivity, and f_m is a form factor which depends strongly on the poloidal mode number $\underline{m}$. It is largest for $m = 1$, a rigid oscillation with the same helicity as that of the electrons about the guide field.

Growth rates are given approximately by

$$\Gamma \approx \frac{1}{2} \left[\frac{\nu}{\gamma} \frac{4f_m}{\sigma^{1/2} \Omega_c b^3} \right]^{2/3} . \tag{77}$$

for $\omega_o \gg 0$, and

$$\Gamma \approx \frac{\sqrt{2}}{2} \omega_o^{-1/2} \left[\frac{\nu}{\gamma} \frac{4f_m}{\sigma^{1/2} \Omega_c b^3} \right] \tag{78}$$

for $\omega_o \gg 0$. The former growth rate, although much larger than the latter, occurs over very narrow energy ranges and so is relatively unimportant. This relationship is indicated schematically in Fig. 19.

For $l = 0$ beam modes, toroidal affects do play a role in the resistive wall instability. Mode frequencies are shifted by the betatron focusing field (Kleva, Ott, and Sprangle, 1983)

$$\omega_o \approx \frac{2\nu}{\gamma^3 a^2 \Omega_c} \left[m - 1 + \left(\frac{a}{b} \right)^{2m} \right] - \frac{1}{2} \frac{\Omega_\theta^2}{\Omega_c} \tag{79}$$

so as to provide stability for sufficiently large γ, because the resistive wall instability requires $\omega_o > 0$. Note that $\omega_o = \omega_B$ for $m = 1$.

Resistive wall instability growth rate in typical modified betatron designs are only of order 0.1 e-foldings/μsec unless cavity walls are made especially resistive. A small spread in the toroidal rotation frequency should be sufficient to damp this growth. In addition, poloidal thermal motion suppresses high $\underline{m}$, flute modes.

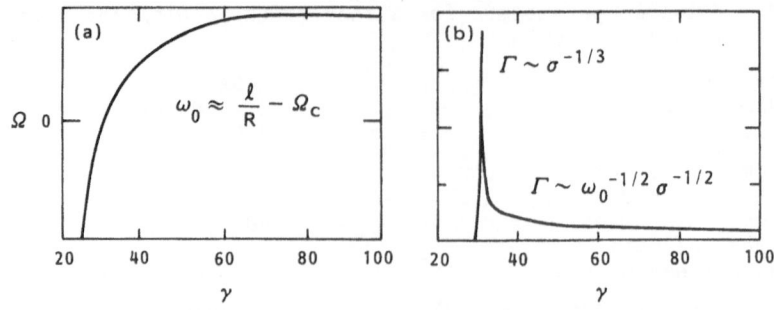

Figure 19. Typical variation with energy of (a) the frequency and (b) the growth rate of the resistive wall cyclotron mode instability for $l \neq 0$.

ION FOCUS TRANSPORT

Ion focused beam transport (IFT) offers several advantages for compact recirculating induction electron accelerators. Focusing magnets can be eliminated, requirements on bending magnets are relaxed somewhat, cavity instabilities are suppressed, and new injection and extraction possibilities are offered. Ion focusing is, however, an immature concept, and many important issues must be resolved: beam and channel equilibrium properties, beam front erosion, beam emittance growth, beam-channel instabilities, and injection and extraction.

Low energy, low current electron beams are particularly effective for channel ionization. They can be bent to form curved channels, can be used in a variety of gasses, and can be produced inexpensively (Godfrey, et al., 1985). Frost (1986), Shope (1985), and colleagues at Sandia National Laboratories have created straight and curved channels many meters in length with lower power electron beams. The chapter by Miller in this book provides additional details.

Two generic recirculating accelerating designs can be envisioned, a race track geometry in which the accelerated beam passes around a single closed drift tube many times and a helix geometry in which the beam is injected at one end and extracted at the other. The former entails complex injection and extraction procedures, while the latter is bulkier and, in particular, requires a larger acceleration module bore size. A race track concept developed by Sandia National Laboratories is shown in Fig. 20.

For purposes of discussion, we assume a relativistic electron beam propagating in an ion channel of comparable radius approximately centered in a metallic drift tube. The channel to beam line density ratio $\underline{f}$ is

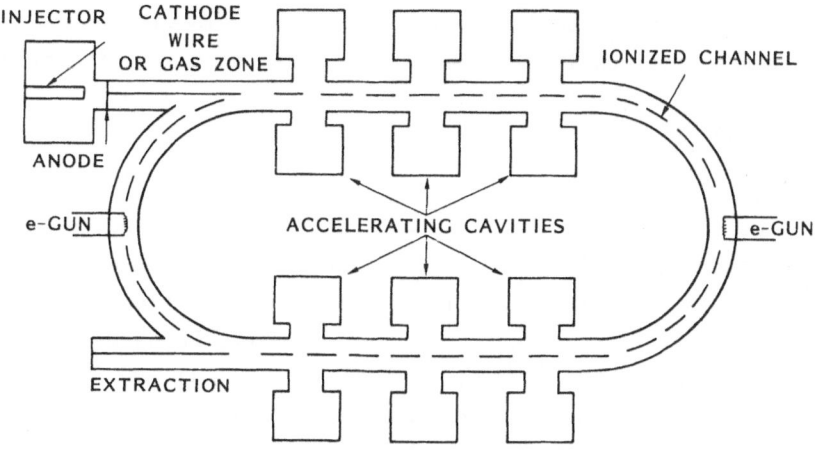

Figure 20. IFT in a recirculating induction accelerator.

taken to be less than one unless otherwise stated. For $\underline{f}$ much less than
one, the equilibrium channel radius in the presence of the beam is somewhat less than the beam radius, but this does not affect our discussion
significantly. Roughly Gaussian beam and channel radial profiles are
assumed.

Equilibrium Conditions

The usual radial force balance equation for beam electrons in the
partially charge neutralizing ion channel is

$$f \frac{\nu}{\gamma} \approx \frac{\varepsilon_n^2}{\gamma^2 a^2} \; . \tag{80}$$

where ε_n is the beam normalized emittance and may include a contribution
from beam rotation. Provided $f > \gamma^{-2}$, the beam always can find an equilibrium radius, although it may exceed the drift tube radius, in which
case current loss occurs.

Typically, as low an emittance as possible is desirable. Taking as
a lower bound the Lawson-Penner relation (Lawson, 1977), $\varepsilon_n^{LP} \approx \sqrt{\nu}$ rad-cm,
gives the simple radial equilibrium expression

$$f\gamma \, [a(cm)]^2 \approx 1 \; . \tag{81}$$

Eq. (81) is readily satisfied for even modestly relativistic injection
energies. Thus, IFT eliminates the need for magnetic focusing in the
accelerator. Care should be taken to avoid abrupt changes in the matching condition, which can cause emittance growth and current loss.

If the electron beam in the accelerator is pulsed, the ion column expands electrostatically between pulses with a doubling time of order

$$\tau_e \approx \left(\frac{a^2}{\nu f}\frac{m_i}{m_e}\right)^{1/2} . \tag{82}$$

For typical parameters $f\nu = 1/2$, $a = 1$, and singly ionized Xenon, the column radius doubles in about 20 ns. The channel contracts again with the arrival of the next pulse. The resulting oscillations increase beam emittance and may trigger instabilities. Keeping the beam interpulse spacing short is highly desirable. Alternatively, it may be possible to neutralize the ion column between beam pulses with electrons injected from the drift tube wall.

Channel Tracking

The electron beam can track a curved ion channel, if the curvature radius is not too small. Balancing electrostatic and centrifugal forces indicates that the beam in a bend displaces by an amount δ from the channel axis,

$$\frac{\delta}{a} = \frac{a}{R}\frac{\gamma}{\nu f} . \tag{83}$$

The displacement must be somewhat less than the channel radius to prevent significant loss of electrons lying near the edge of the beam emittance envelope. Computer simulations suggest $\delta/a < 0.2$, limiting the electron energy to 4 MeV or less for $R/a = 100$ and $f\nu = 1/2$. Above this energy, bending magnets are required. The channel does, however, enlarge the energy acceptance of the bending magnets by several MeV.

To the extent that the channel contributes to bending the electron beam, it feels a force and drifts sideways. A characteristic drift time is

$$\tau_d \approx \left(\frac{2fRa}{\gamma_d}\frac{m_i}{m_e}\right)^{1/2} \tag{84}$$

where γ_d is the beam energy mismatch in the turn. For the parameters used previously and a large mismatch, the transverse drift time is 70 nsec. This is about one-third the minimum circulation time in a race track geometry and may be less than the beam pulse length in a helix geometry.

Channel drift can be reduced by magnetic wall image forces, given approximately by $-2f\nu\Delta/b^2$, where Δ is the displacement of the beam and channel from the center of the accelerator cavity. Hoop stresses are negligible. Balancing the various forces gives

$$\frac{\Delta}{b} = \frac{b}{R} \frac{\gamma_d}{2\nu f}$$ (85)

or $\Delta = 0.6$ cm for $b = 5$ cm. At transitions between straight and curved drift tube sections, the channel adjusts smoothly over a large fraction of a betatron wavelength. Note that magnetic image forces decay on tens-of-μsec time scales, depending on the composition and thickness of the accelerator wall. Acceleration should be completed by that time.

Beam Front Erosion

Inductive erosion of a low current beam clearing the channel electrons in a straight drift tube is determined easily to be

$$\frac{dx}{dz} = \frac{f\nu}{\gamma} (1 + 2 \ln b/a) .$$ (86)

Most of the lost energy is carried off by expelled channel electrons. The erosion rate varies as $4.2/\gamma$ for the parameters introduced above, severely limiting the minimum practical energy at which the beam can be injected. A similar calculation shows that the beam tail gains no energy inductively, unless the pulse length is less than twice the cavity diameter. Likewise, the fronts of subsequent pulses lose no energy except to the extent that the channel has been charge-neutralized between pulses.

This simple analysis becomes questionable as ν approaches unity, because channel electrons attain mildly relativistic energies while being expelled, are trapped at a few times the beam radius by the beam magnetic field, and are swept forward with the beam front. Low energy electrons also can be accumulated at the beam head by the acceleration of channel electrons by the acceleration modules before the beam arrives. We expect the erosion rate to be unchanged to lowest order by these electrons, since $f\nu$ remains the same despite the increase in current locally.

Beam bending changes this picture. Most channel electrons moving with the beam, including those picked up in straight sections of the accelerator, are lost in the curved sections, unless their energy matches that of the main beam to within the energy bandwidth of the bend. More seriously, the limited energy bandwidth of the bends may deplete the inductively decelerated beam electrons in the beam front, enhancing inductive erosion. One might imagine as a worst case that γ should be replaced by γ_d in Eq. (86), leading to an enormous erosion rate. More realistically, only a portion of the reduced energy beam particles are lost in the turns, modestly increasing inductive erosion. The uncertainties here are large.

For completeness, we also mention what might be called centrifugal erosion. Electrons at the very front of the beam in a turn benefit not at all from the ion-generated energy bandwidth, because channel electrons are not yet expelled from the channel. A simple calculation yields a centrifugal erosion rate that is of order 0.02 for the parameters considered here.

Beam Emittance Growth

At each transition between straight and curved sections of the ion channel in race track geometry, the beam equilibrium position abruptly shifts sideways by δ. The beam itself attempts to follow its equilibrium position, overshoots, and oscillates. The oscillations damp by phase-mixing after of a few wavelengths, leaving the beam with an increased emittance. Since the initial oscillation amplitude is δ, the new emittance may be expected to be

$$\varepsilon_n \approx \varepsilon_n^o \left(1 + \frac{\delta}{a} \right) . \tag{87}$$

Rigorous calculations by Rienstra and Sloan (1985) give essentially the same result for small δ/a. After twenty cycles with four transitions per cycle, beam normalized emittance grows by an order of magnitude for the $\delta/a = 0.1$ assumed above.

That the beam relaxes in a few betatron wavelengths suggests two methods of reducing emittance growth, by varying the drift tube curvature radius over distances long compared to the betatron wavelength or by having straight sections of the drift tube short compared to the betatron wavelength. The beam betatron wavelength in the ion channel is

$$\lambda = 2 \pi a \left(\frac{\gamma}{\nu f} \right)^{1/2} \tag{88}$$

or about $6\gamma^{1/2}$ for the parameters used here. Hence, λ_β ranges between 20 cm and 85 cm for energies of 5 to 100 MeV. A roughly ellipsoidal race track geometry with straight sections about 20 cm long to accommodate an acceleration module and transition sections about 50 cm long, where 1/R is varied between 0 and 100 cm would have either long transition regions (R the local curvature radius) or short straight regions compared to the betatron wavelength for the entire 5 – 100 MeV energy range. If longer straight sections are required, as in Fig. 20, then 100 cm transition sections probably would be adequate to minimize emittance growth.

Of course, matching the bending magnetic fields precisely to the beam energy, so that $\gamma_d = 0$, also eliminates this source of emittance growth. Precisely controlling the beam energy in a device employing high voltage pulsed power acceleration modules is difficult but not impossible.

The abrupt change in beam energy at an acceleration gap also can cause emittance growth, although to a lesser degree than the effect just mentioned. Appropriately shaping the gaps (Miller, 1985) or adding solenoidal and vertical lenses to match the beam radius and position across the gap minimizes this source of emittance.

Severe emittance growth due to particle resonance effects, discussed earlier in this chapter, are unlikely in high current, ion focused, inductive gap accelerators. Ion channel focusing is nonlinear for large amplitude oscillations, saturating the resonant interaction. The number of betatron wavelengths around the race track typically is large, which narrows the resonance bandwidth. Most importantly, the rapid acceleration produced by inductively driven gaps causes the beam to pass through resonances quickly or miss them entirely.

Collective Instabilities

The accelerated electron beam may be unstable through interaction with the channel or with the accelerator. For $f > 1$, only a portion of the channel electrons are expelled by the beam, and computer simulations in this case exhibit strong instabilities between the beam and remaining channel electrons within a propagation distance of several betatron wavelengths. Analytic studies have not suggested any method of avoiding this instability except the obvious course of requiring $f < 1$.

The ion two-stream, or ion resonance, instability (Uhm and Davidson, 1982; Manheimer, 1983; 1984) may occur for $f < 1$. Its temporal growth rate at a fixed point in the accelerator is bounded by

$$\Gamma < \frac{1}{a} \left(\nu \frac{m_e}{m_i} \right)^{1/2} \tag{89}$$

resulting in an e-folding time of greater than 25 nsec. The instability can be avoided in a helix geometry simply by limiting the beam pulse length to 100 nsec or so. This option is not available in a race track geometry, and acceleration times of 100 nsec or less are unrealistic. Therefore, one must rely on effects not yet fully taken into account, such as nonlinearities, multiple ion species, or charge exchange, to reduce instability growth. The instability has been observed experimentally in only a few instances.

The negative mass instability was discussed in detail earlier. Since its e-folding time is relatively long, 100 nsec or greater for parameters of interest, it is of little concern for the rapidly accelerated beam of a recirculating inductive gap accelerator. Focusing provided by the IFT channel also should be stabilizing. The resistive wall instability is slower still and should be completely negligible. IFT is known both theoretically (Briggs, 1980) and from experiments on the ETA and ATA accelerators (Prono et al., 1985) to suppress cavity-coupled instabilities, such as the beam breakup and image displacement modes. The chapter by Prono in this book provides additional details.

Injection and Extraction

Electron beam injection and extraction are serious issues in race track geometry. As noted in the chapter by Rostoker, substantial thought has gone into this problem for high current betatrons, and proposed schemes may be practical as well for recirculating accelerators with ion channels. Here, we consider some options unique to ion focused transport.

An electron beam injected at an angle to an ion channel is captured, if the injection angle is less than $2(f\nu/\gamma)^{1/2}$, or 36° for a 5 MeV injection energy. However, by analogy with channel tracking around a curve, we expect that a much smaller angle should be used to avoid beam current loss. Only a 50 cm straight drift tube section would be needed to capture the beam injected at 6° with b = 5 cm. Conceivably, the magnetic fringing field at the transition to a curved section could be employed to shorten this distance modestly. Emittance growth could be as low as the ratio of the injection angle to the maximum injection angle, just under 20%.

An ion channel based extraction scheme is less obvious. A weak, localized, transverse magnetic field not quite sufficient to deflect the beam from the channel could be applied in a straight section of the accelerator. A laser beam fired into that region at a glancing angle to the original channel would create a new channel which the beam would follow out of the accelerator. The loss of beam quality which might result is unknown. Hui and Lau (1984) have proposed analogous schemes for injection and extraction in a modified betatron.

SUMMARY

The preceding discussions suggest that inductive erosion in curved channels, ion two-stream instabilities, and beam extraction (in race track geometry) are the most serious uncertainties confronting ion focused transport for recirculating accelerators. Understanding of streaming instabilities under realistic conditions can be improved with more detailed linear analyses. Extraction simply needs new ideas.

Beyond this, multidimensional computer simulations and parallel experiments are required.

ACKNOWLEDGMENTS

It is a pleasure to acknowledge contributions by M. Campbell, D. Chernin, C. Frost, C. Kapetanakos, J. Mack, D. Moir, A. Mondelli, B. Newberger, C. Roberson, N. Rostoker, P. Sprangle, and others. The preparation of this chapter was supported in part by the U.S. Office of Naval Research and by Sandia National Laboratories.

REFERENCES

Barak, G., Chernin, D., Fisher, A., Ishizuka, H., and Rostoker, N., 1981, "High Current Betatron," in: "High-Power Beams 81," H. J. Doucet and J. M. Buzzi, ed., Ecole Polytechnique, Palaiseau.

Barak, G., Fisher, A., Ishizuka, H., and Rostoker, N., 1981, "High Current Betatron," IEEE Trans. Nuc. Sci., NS-28:3340.

Barak, G., and Rostoker, N., 1983 "Orbital Stability of the High-Current Betatron," Phys. Fluids, 26:856.

Briggs, R. J., 1980, "Suppression of Transverse Beam Breakup Modes in an Induction Accelerator by Gas Focusing," Lawrence Livermore National Laboratory, Livermore.

Briggs, R. J., and Neil, V. K., 1967, "Negative-Mass Instability in a Cylindrical Layer of Relativistic Electrons," Plas. Phys., 9:209.

Blaugrund, A. E., Fisher, A., Prohaska, R., and Rostoker, N., 1985, "A Stretched Betatron," J. Appl. Phys., 57:2474.

Brower, D. F., Kusse, B. R., and Meixel, G. D., 1974, "Injection of Intense Electron Beams into a Toroidal Geometry," IEEE Plas. Sci., PS-2:193.

Cavenago, M., and Rostoker, N., 1985, "Modified Elongated Betatron Accelerator I - Equilibria and Space Charge Limits," University of California, Irvine.

Cavenago, M., and Rostoker, N., 1986, "Modified Elongated Betatron Accelerator - A Covariant and Systematic Description," in: "Beams '86," to be published.

Chernin, D., 1984, "Mode Coupling in a Modified Betatron," Part. Accel. 14:139.

Chernin, D., 1985, "Self-Consistent Treatment of Equilibrium Space Charge Effects in the L=2 Stellatron," IEEE Trans. Nuc. Sci., NS-32:2504.

Chernin, D., 1986, "Beam Stability in the L=2 Stellatron," Phys. Fluids, 29:556.

Chernin, D., 1986, "Orbital Resonances and Energy and Current Limits in High Current Cyclic Accelerators," Science Applications Int. Corp., Washington.

Chernin, D., and Lau, Y. Y., 1984, "Stability of Laminar Electron Layers," Phys. Fluids, 27:2319.

Chernin, D., Mondelli, A., and Roberson, C., 1984, "A Bumpy-Torus Betatron," Phys. Fluids, 27:2378.

Chernin, D., and Sprangle, P., 1982, "Transverse Beam Dynamics in a Modified Betatron," Part. Accel. 12:85.

Davidson, R. C., 1974, "Theory of Nonneutral Plasmas," Benjamin, Reading.

Davidson, R. C., and Uhm, H. S., 1982, "Stability Properties of an Intense Relativistic Nonneutral Electron Ring in a Modified Betatron Accelerator," Phys. Fluids, 25:2089.

Dialetis, D., Marsh, S. J., and Kapetanakos, C. A., 1986, "The Rebatron as a High Energy Accelerator," Naval Research Laboratory, Washington.

Felber, F. S., Mitrovich, D., Vomvoridis, J., Cooper, R. K. Fisher, A., Hughes, T. P., and Godfrey, B. B., 1983, "Relativistic Injection into a High Current Betatron," IEEE Trans. Nuc. Sci., NS-30:2781.

Finn, J. M., and Manheimer, W. M., 1983, "Self Consistent Equilibrium and Adiabatic Evolution of a High Current Electron Ring in a Modified Betatron," Phys. Fluids, 26:3400.

Frost, C. A., Shope, S. L., Ekdahl, C. A., Poukey, J. W., Freeman, J. R., Leifeste, G. T., Mazarakis, M. G., Miller, R. B., Tucker, W. K., and Godfrey, B. B., 1986, "Ion Focused Transport Experiments," Proc. 1986 Linear Accelerator Conf., to be published.

Getmanov, B. S., and Makhankov, V. G., 1977, "Study of the Longitudinal Instability of Relativistic Electron Rings," Part. Accel. 8:49.

Gisler, G. R., 1986, "PIC Simulations of Azimuthal Instabilities in Relativistic Electron Layers," submitted to Phys. Fluids.

Gisler, G., and Faehl, R., 1983, "Self-Trapping Electron Ring Accelerators," IEEE Trans. Nuc. Sci., NS-30:3204.

Godfrey, B. B., and Hughes, T. P., 1982, "Resistive Wall Instabilities in the Modified Betatron," Mission Research Corporation, Albuquerque.

Godfrey, B. B., and Hughes, T. P., 1983, "Beam Breakup Instabilities in High Current Electron Beam Racetrack Induction Accelerators," IEEE Trans. Nuc. Sci., NS-30:2531.

Godfrey, B. B., and Hughes, T. P., 1985, "Long-Wavelength Negative Mass Instability in High Current Betatrons," Phys. Fluids, 28:669.

Godfrey, B. B., and Hughes, T. P. 1985, "The Negative Mass Instability in High Current Modified Betatrons at Low Energies," IEEE Trans. Nuc. Sci., NS-32:2495.

Godfrey, B. B., and Hughes, T. P., 1986, "An Improved Negative Mass Instability Dispersion Relation for High Current Modified Betatrons," Part. Accel., to be published.

Godfrey, B. B., Newberger, B. S., Wright, L. A., and Campbell, M. M., 1985, "IFR Transport in Recirculating Accelerators," Mission Research Corporation, Albuquerque.

Golden, J., Pasour, J., Pershing, D. E., Smith, K., Mako, F., Slinker, S., Mora, F., Orrick, N., Altes, R., Fliflet, A., Chapney, P., and Kapetanakos, C. A., 1983, "Preliminary Design of the NRL Modified Betatron," IEEE Trans. Nuc. Sci., NS-30:2114.

Golden, J., Mako, F., Floyd, L., McDonald, K., Smith, T., Dialetis, D., Marsh, S. J., and Kapetanakos, C. A., 1986, "Progress in the Development of the NRL Modified Betatron Accelerator," in: "Beams '86," to be published.

Grossmann, J. M., Manheimer, W. M., and Finn, J. M., 1983, "Self-Consistent Modified Betatron Equilibria and Their Adiabatic Evolution," in: "Beams '83," R. J. Briggs and A. J. Toepfer, ed., San Francisco.

Grossmann, J. M., Finn, J. M., and Manheimer, W. M., 1985, "Acceleration of an Electron Ring in a Modified Betatron with Transverse Pressure," Phys. Fluids, 28:695.

Haber, I., Marsh, S. J., and Sprangle, P., 1983, "Emittance Growth in a Modified Betatron Crossing the Orbit-Turning-Point Transition," in: "Beams '83," R. J. Briggs and A. J. Toepfer, ed., San Francisco.

Hughes, T. P., 1985, "Estimates of Negative-Mass Instability Growth for the NRL Betatron," Mission Research Corporation, Albuquerque.

Hughes, T. P., 1986, "Theory and Simulations of High-Current Betatrons," in: "Beams '86," to be published.

Hughes, T. P., Campbell, M. M., and Godfrey, B. B., 1983, "Analytic and Numerical Studies of the Modified Betatron," IEEE Trans. Nuc. Sci., NS-30:2528.

Hughes, T. P., Campbell, M. M., and Godfrey, B. B., 1983, "Linear and Nonlinear Development of the Negative Mass Instability in a Modified Betatron Accelerator," in: "Beams '83," R. J. Briggs and A. J. Toepfer, ed., San Francisco.

Hughes, T. P., and Godfrey, B. B., 1982, "Linear Stability of the Modified Betatron," Mission Research Corporation, Albuquerque.

Hughes, T. P., and Godfrey, B. B., 1984, "Single-Particle Orbits in the Stellatron Accelerator," Mission Research Corporation, Albuquerque.

Hughes, T. P., and Godfrey, B. B., 1984, "Modified Betatron Accelerator Studies," Mission Research Corporation, Albuquerque.

Hughes, T. P., and Godfrey, B. B., 1985, "Instability in a Relativistic Electron Layer with a Strong Azimuthal Magnetic Field," Appl. Phys. Lett., 46:473.

Hughes, T. P., and Godfrey, B. B., 1985, "Equilibrium and Stability Properties of the Solenoidal Lens Betatron," IEEE Trans. Nuc. Sci., NS-32:2498.

Hughes, T. P., and Godfrey, B. B., 1986, "Electromagnetic Instability in a Quadrupole-Focusing Accelerator," Phys. Fluids, 29:1698.

Hughes, T. P., Godfrey, B. B., and Campbell, M. M., 1983, "Modified Betatron Accelerator Studies," Mission Research Corporation, Albuquerque.

Hui, B., and Lau, Y. Y., 1984, "Injection and Extraction of a Relativistic Electron Beam in a Modified Betatron," Phys. Rev. Lett., 53:2024.

Ishizuka, H., Leslie, G., Mandelbaum, B., Fisher, A., and Rostoker, N., 1985, "Injection and Capture of Electrons in the UCI Stellatron," IEEE Trans. Nuc. Sci., NS-32:2727.

Ishizuka, H., Lindley, G., Mandelbaum, B., Fisher, A., and Rostoker, N., 1984, "Formation of a High Current Electron Beam in Modified Betatron Field," Phys. Rev. Lett., 53:266.

Kapetanakos, C. A., Dialetis, D., and Marsh, S. J., 1986, "Beam Trapping in a Modified Betatron with Torsatron Windings," Part. Accel., to be published.

Kapetanakos, C. A., and Marsh, S. J., 1985, "Non-Linear Transverse Electron Beam Dynamics in a Modified Betatron Accelerator," Phys. Fluids, 28:2263.

Kapetanakos, C. A., Marsh, S. J., and Sprangle, P., 1984, "Dynamics of a High-Current Electron Ring in a Conventional Accelerator," Part. Accel., 14:261.

Kapetanakos, C. A., and Sprangle, P., 1984, "Self-Potentials of an Electron Ring in a Torus for Large Ring Displacement from the Minor Axis," Naval Research Laboratory, Washington.

Kapetanakos, C. A., and Sprangle, P., 1985, "Ultra-High-Current Electron Induction Accelerators," Phys. Today, 38(2):58.

Kapetanakos, C. A., Sprangle, P., Chernin, D. P., Marsh, S. J., and Haber, I., 1983, "Equilibrium of a High Current Electron Ring in a Modified Betatron," Phys. Fluids, 26:1634.

Kapetanakos, C. A., Sprangle, P., and Marsh, S. J., 1982, "Injection of a High-Current Beam into a Modified Betatron," Phys. Rev. Lett., 49:741.

Kapetanakos, C. A., Sprangle, P., Marsh, S. J., Dialetis, D., Agritellis, C., and Prakash, A., 1985, "Rapid Electron Beam Accelerators," Naval Research Laboratory, Washington.

Kapetanakos, C. A., Sprangle, P., Marsh, S. J., and Haber, I., 1981, "Injection into a Modified Betatron," Naval Research Laboratory, Washington.

Kerst, D. W., 1983, "Conventional and Modified Betatrons," in: "Beams '83," R. J. Briggs and A. J. Toepfer, ed., San Francisco.

Kleva, R. G., Ott, E., and Sprangle, P., 1983, "Resistive Wall Flute Stability of Magnetically Guided Relativistic Electron Beams," Phys. Fluids, 26:2689

Landau, R. W., 1968, "Negative Mass Instability with B_θ Field," Phys. Fluids, 11:205.

Landau, R. W., and Neil, V. K., 1966, "Negative Mass Instability," Phys. Fluids, 9:2412.

Lawson, J. D., 1978, "The Physics of Charged-Particle Beams," Clarendon Press, Oxford.

Lee, E. P., Faltens, A., Laslett, L. J., and Smith L., 1983, "Stabilization of Longitudinal Modes in a High Current Betatron," IEEE Trans. Nuc. Sci., NS-30:2504.

Mako, F., Golden, J., Floyd, L., McDonald, K., Smith, T., and
    Kapetanakos, C. A., 1985, "Internal Injection into the NRL Modified
    Betatron," IEEE Trans. Nuc. Sci., NS-32:3027.
Mako, F., Manheimer, W., Kapetanakos, C. A., Chernin, D., and Sandel, F.,
    1984, "External Injection into a High Current Modified Betatron
    Accelerator," Phys. Fluids, 27:1815.
Mandelbaum, B., 1985, "A Study of a Modified Betatron with Stellarator
    Windings," University of California, Irvine.
Mandelbaum, B., Ishizuka, H., Fisher, A., and Rostoker, N., 1983,
    "Behavior of Electron Beam in a High Current Betatron," in: "Beams
    '83," R. J. Briggs and A. J. Toepfer, ed., San Francisco.
Manheimer, W. M., 1983, "Electron-Ion Instabilities in a High Current
    Modified Betatron," Part. Accel., 13:209.
Manheimer, W. M., 1984, "The Plasma Assisted Modified Betatron," Naval
    Research Laboratory, Washington.
Manheimer, W. M., and Finn, J. M., 1983, "Self-Consistent Theory of
    Equilibrium and Acceleration of a High Current Electron Ring in a
    Modified Betatron," Part. Accel., 14:29.
Martin, W. E., Caporaso, G. J., Fawley, W. M., Prosnitz, D., and Cole, A.
    G., 1985, "Electron Beam Guiding and Phase-Mix Damping by a Laser-
    Ionized Channel," Phys. Rev. Lett. 54:685.
Miller, R. B., 1985, "RADLAC Technology Review," IEEE Trans. Nuc. Sci.,
    NS-32:3149.
Mondelli, A., and Chernin, D., 1984, "Envelope Stability for L=0 Focusing
    Systems," Science Applications Int. Corp., Washington.
Mondelli, A., and Chernin, D., 1985, "Plasma Focused Cyclic Accelera-
    tors," IEEE Trans. Nuc. Sci., NS-32:3521.
Mondelli, A., Chernin, D., Putnam, S. D., Schlitt, L., and Bailey, V.,
    1986, "A Strong-Focused Spiral-Line Recirculating Induction Linac,"
    in: "Beams '86," to be published.
Mondelli, A., Chernin, D., and Roberson, C. W., 1983, "The Stellatron
    Accelerator," in: "Beams '83," R. J. Briggs and A. J. Toepfer, ed.,
    San Francisco.
Mondelli, A., and Roberson, C. W., 1983, "A High-Current Race Track
    Induction Accelerator," IEEE Trans. Nuc. Sci., NS-30:3212.
Mondelli, A., and Roberson, C. W., 1984, "Energy Scaling Laws for the
    Race Track Induction Accelerator," Part. Accel., 15:221.
Mostrom, M. A., and Newberger, B. S., 1986, "Beam Energy Loss to a Back-
    ground Plasma in an Ion Focused Recirculating Accelerator," in:
    "Beams '86," to be published.
Neil, V. K., and Briggs, R. J., 1967, "Stabilization of Non-Relativistic
    Beams by Means of Inductive Walls," Plas. Phys., 9:631.
Neil, V. K., and Heckrotte, W. 1965, "Relation between Diocotron and
    Negative Mass Instability," J. Appl. Phys., 36:2761.
Newberger, B. S., and Mostrom, M. A., 1986, "Current Loss in an Ion
    Focused Recirculating Accelerator," in: "Beams '86," to be pub-
    lished.
Pavlovsky, A. I., Kuleshov, G. D., Sklizkov, G. V., et al., 1967, "High
    Current Ironless Betatrons," Sov. Phys. Dok., 10:30.
Pavlovsky, A. I., Kuleshov, G. D., Gerasimov, A. I., Klementiev, A. P.,
    Kuznetsov, V. O., Tananakin, V. A., and Tarasov, A. D., 1977,
    "Injection of an Electron Beam into a Betatron," Sov. Phys. Tech.
    Phys., 22:218.
Peter, W., Faehl, R. J., and Mako, F., 1983, "Simulation Studies of a
    Novel Betatron Injection Scheme," in: "Beams '83," R. J. Briggs and
    A. J. Toepfer, ed., San Francisco.
Peterson, J. M., 1982, "Betatrons with Kiloampere Beams," Lawrence
    Berkeley Laboratory, Berkeley.
Prakash, A., Marsh, S. J., Dialetis, D., Agritellis, C., Sprangle, P.,
    and Kapetanakos, C. A., 1985, "Recent Rebatron Studies," IEEE Trans.
    Nuc. Sci., NS-32:3265.
Prohaska, R., Blaugrund, A. E., Fisher, A., Honea, E., Schneider, J., and
    Rostoker, N., 1983, "A Stretched Betatron," in: "Beams '83," R. J.
    Briggs and A. J. Toepfer, ed., San Francisco.

Prono, D. S., and the Beam Research Group, 1985, "Recent Progress of the Advanced Test Accelerator," IEEE Trans. Nuc. Sci., NS-32:3144.

Rienstra, W. W., and Sloan, M. L., 1985, "Recirculating Accelerator Magnet Design," Science Applications Int. Corp., Albuquerque.

Roberson, C. W., 1981, "The Race Track Induction Accelerator," IEEE Trans. Nuc. Sci., NS-28:3433.

Roberson, C. W., Mondelli, a., and Chernin, D., 1983, "The Stellatron - A Strong-Focusing, High-Current Betatron," IEEE Trans. Nuc. Sci., NS-30:3162.

Roberson, C. W., Mondelli, A., and Chernin, D., 1983, "The Stellatron - A High-current Betatron with Stellarator Fields," Phys. Rev. Lett., 50:507.

Roberson, C. W., Mondelli, A., and Chernin, D., 1985, "The Stellatron Accelerator," Part. Accel., 17:79.

Roberts, G. A., and Rostoker, N., 1983, "Analysis of a Modified Betatron with Adiabatic Particle Dynamics," in: "Beams ' 83," R. J. Briggs and A. J. Toepfer, ed., San Francisco.

Roberts, G. A., and Rostoker, N., 1985, "Adiabatic Beam Dynamics in a Modified Betatron," Phys. Fluids, 28:1968.

Roberts, G. A., and Rostoker, N., 1986, "Effect of Quasi-Confined Particles and L=2 Stellarator Fields on the Negative Mass Instability in a Modified Betatron," Phys. Fluids, 29:333.

Rostoker, N., 1973, "High ν/γ Electron Beam in a Torus," Part. Accel., 5:93.

Rostoker, N., 1981, "High Current Betatron," IEEE Trans. Nuc. Sci., NS-28:3340.

Rostoker, N., 1983, "High Current Betatron Experiments and Theory," in: "Beams '83," R. J. Briggs and A. J. Toepfer, ed., San Francisco.

Shope, S. L., Frost, C. A., Leifeste, G. T., Crist, C. E., Kiekel, P. D., Poukey, J. W., and Godfrey, B. B., 1985, "Laser Generation and Transport of a Relativistic Electron Beam," IEEE Trans. Nuc. Sci., NS-32:3091.

Siambis, J. G., 1983, "Intense Beam Recirculation," IEEE Trans. Nuc. Sci., NS-30:3195.

Sprangle, P. and Chernin, D., 1983, "Current Limitations Due to Instabilities in Modified and Conventional Betatrons," in: "Beams '83," R. J. Briggs and A. J. Toepfer, ed., San Francisco.

Sprangle, P. and Chernin, D., 1984, "Beam Current Limitations Due to Instabilities in Modified and Conventional Betatrons," Part. Accel., 15:35.

Sprangle, P., and Kapetanakos, C. A., 1983, "Drag Instability in the Modified Betatron," Naval Research Laboratory, Washington.

Sprangle, P., and Kapetanakos, C. A., 1985, "Beam Trapping in High Current Cyclic Accelerators," Naval Research Laboratory, Washington.

Sprangle, P., and Kapetanakos, C. A., and Marsh, 1981, "Dynamics of an Intense Electron Ring in a Modified Betatron Field," in: "High-Power Beams 81," H. J. Doucet and J. M. Buzzi, ed., Ecole Polytechnique, Palaiseau.

Sprangle, P., and Vomvoridis, J. L., 1981, "Longitudinal and Transverse Instabilities in a High Current Modified Betatron Electron Accelerator," Naval Research Laboratory, Washington.

Taggart, D., Parker, M. R., Hopman, H., and Fleischmann, H. H., 1981, "Successful Betatron Acceleration of Kiloampere Electron Rings in RECE-Christa," IEEE Trans. Nuc. Sci., NS-30:3165.

Uhm, H. S., 1981, "Grad-B Drift of an Electron Beam in the High Current Betatron," Naval Surface Weapons Center, White Oak.

Uhm, H. S., and Davidson, R. C., 1977, "Kinetic Description of Negative-Mass Instability in an Intense Relativistic Nonneutral E Layer," Phys. Fluids, 20:771.

Uhm, H. S., and Davidson, R. C., 1978, "Influence of Axial Energy Spread on the Negative-Mass Instability in a Relativistic Nonneutral E Layer," Phys. Fluids, 21:265.

Uhm, H. S., and Davidson, R. C., 1982, "Stability Properties of an Intense Relativistic Non-Neutral Electron Ring in a Modified Betatron Accelerator," Phys. Fluids. 25:2089.

Uhm, H. S., and Davidson, R. C., 1982, "Ion-Resonance Instability in a Modified Betatron Accelerator," Phys. Fluids, 25:2334.

Uhm, H. S., and Davidson, R. C., 1985, "Influence of Electromagnetic Effects on Stabilities Properties of a High-Current Betatron Accelerator," IEEE Trans. Nuc. Sci., NS-32:2383.

Uhm, H. S., Davidson, R. C., and Petillo, J. J., 1985, "Kinetic Properties of an Intense Relativistic Electron Ring in a High-Current Betatron Accelerator," Phys. Fluids, 28:2537.

Wilson, M. A., 1981, "Recirculation Acceleration of High Current Relativistic Electron Beams - A Feasibility Study," IEEE Trans. Nuc. Sci., NS-28:3375.

Zieher, K. W., Fishbine, B., Humphries, S., and Woodall, D. M., 1985, "Beam Measurements on the Electron Injector for a High Current Betatron," IEEE Trans. Nuc. Sci., NS-32:3274.

HIGH-CURRENT ELECTRON-BEAM TRANSPORT IN LINEAR INDUCTION ACCELERATORS*

R. B. Miller

Directed Energy Research Department
Sandia National Laboratories, P.O. Box 5800
Albuquerque, New Mexico 87185 USA

INTRODUCTION

In recent years a number of advanced linear induction accelerators (LIAs) have been developed which have the capability for accelerating very high electron beam currents, provided that satisfactory answers are obtained to a number of beam transport and stability issues. Examples of this new technology include the advanced technology accelerator (Prono, 1985), the RADLAC accelerators, (Miller, et al., 1985) and the LIU accelerators of the Soviet Union (Pavlovskii, et al., 1975). In this review we analyze several important beam stability issues pertaining to the propagation of very high current electron beams in and through the accelerating structures. In particular, we develop criteria for various equilibrium configurations, injector designs for producing the equilibria, general beam stability criteria, zero-order analyses of accelerating gap designs and considerations of various multiple-gap instabilities, including transverse beam break-up and image displacement. We then show how these analyses can be practically applied by illustrating the design of the beam transport line of the RADLAC-II accelerator. Finally, we briefly describe the new transport technique of ion channel guiding.

LINEAR INDUCTION ACCELERATION PRINCIPLE

All linear induction accelerators operate by producing voltage around a circuit enclosing a time-varying magnetic flux. Mathematically, this is described by Eq. (1), where $\underline{B}$ and $\underline{E}$ represent the magnetic induction and the induced electric field, and V is the voltage induced around the circuit C which encloses the area A.

$$V = \int_C \underline{E} \cdot d\underline{l} = -\frac{1}{c}\frac{d}{dt}\int_S \underline{B} \cdot d\underline{A} \tag{1}$$

Time is represented by t, while c is the speed of light. Note that Eq. (1) allows the generation of voltage by changing either the enclosed magnetic field, or the area normal to the field lines. For example, an external circuit can be used to establish a magnetic field through a second circuit. If the current through the external circuit is time-varying, a voltage will be established at the output connections of the second circuit, i.e., this is a single-turn transformer. The output voltage can be increased by inserting a core of ferromagnetic material since the B-field enclosed is increased by the magnetic permeability of the material according to $B = \mu H$, where H represents the macroscopic magnetic field, and μ is the magnetic permeability.

As a practical embodiment of this concept consider the acceleration gap geometry of Fig. 1. A high voltage pulse is injected into an accelerating cavity containing an annular core of magnetic material. It is assumed that the magnetic induction pre-existing in the core is saturated at the value $+B_o$. Note that without the core the input pulse would be short-circuited; however, as current begins to flow around the core, the magnetic induction in the material begins to change until it reaches the value $-B_o$. During this time T, the accelerating voltage is effectively applied directly to the accelerating gap. From Eq. (1) these parameters are related according to $VT = \Delta BA = 2 B_o A$, where A is the cross-sectional area of the annular core. In other words, the product of the accelerating voltage and the pulse time is equal to the flux swing times the area of the annular core. Note that the voltage does not appear external to the accelerating gap; i.e., the exterior of the accelerator remains at ground potential and the high voltage is summed by only the beam as it propagates through the LIA structure.

An alternate description of the circuit of Fig. 1 is that the voltage pulse appears across the parallel combination of the impedance Z_b associated with the beam, the impedance Z_c associated with the flow of current around the magnetic core. When the core is saturated at the level $+B_o$, the inductance associated with the current path around the magnetic core is very high, and Z_c is very large; hence, the pulse energy is delivered directly to the beam. During the time T, however, the flow of leakage current around the core drives the core into saturation at the level $-B_o$. Z_c then drops to essentially zero, and the accelerating pulse is shorted.

Equation (1) can also be used to describe the behavior of waves in transmission lines. Consider the strip transmission line of Fig. 2. The voltage between the plates can be thought of as arising from a time-charging magnetic flux as the electromagnetic wave propagates down the line. With the current flow on the surfaces of the strip lines as shown,

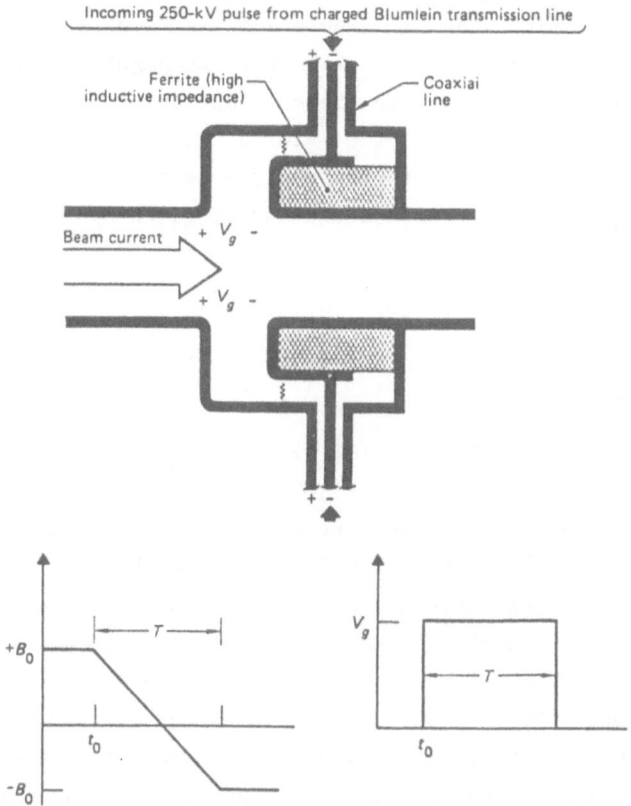

Fig. 1. A linear induction accelerator cavity which uses ferromagnetic material.

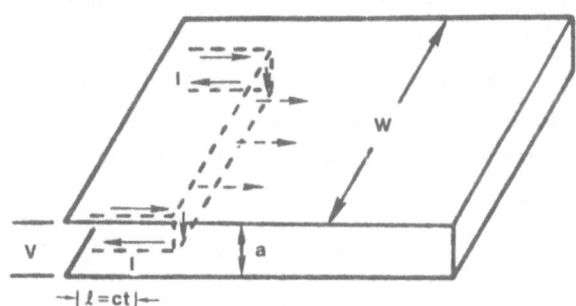

Fig. 2. A strip transmission line in which the voltage across the plates arises from a time changing magnetic flux. (pulse forming transmission line).

the magnetic induction is described by

$$B = \frac{4\pi\mu I}{cW} \tag{2}$$

where μ is the magnetic permeability of the dielectric between the electrodes, and W is the width of the lines. The area transverse to the

magnetic field is simply A = aℓ, where a is the separation of the lines
and ℓ is the linear dimension. As the wave propagates down the line the
length ℓ increases linearly with the constant of proportionality being
equal to $c/(\mu\varepsilon)^{1/2}$, where ε is the material dielectric constant. Hence,
in this case the voltage across the gap arises from the time-changing
area transverse to the magnetic field and is given by

$$V = \frac{4\pi}{c} \sqrt{\frac{\mu}{\varepsilon}} \frac{a}{W} I = Z_0 I \tag{3}$$

where we have noted that the impedance of a strip transmission line is
$Z_0 = (4\pi/c) (\mu/\varepsilon)^{1/2} a/W$.

As first noted by Pavlovskii (1975), closed transmission lines can
also be used to form the accelerating cavities of linear induction
accelerators. Consider the geometry of Fig. 3. A grounded toroidal
shield surrounds an annular disk which is pulse-charged to high negative
voltage. Since the electric fields on either side of the high voltage
disk are opposing, there is no net accelerating field across the gap.
Now suppose that an annular ring switch, internal to the cavity, is
closed. Although the line is shorted, no energy can escape since the
cavity is (almost) completely enclosed. As a result, there will be an
alternating voltage appearing across the gap of the cavity according to
the wave propagation diagrams of Fig. 4. If an impedance equal to the
characteristic impedance of the transmission line cavity is impressed
across the gap at the time τ, then the output voltage will decrease by
two, but all of the energy initially stored in the cavity will be
delivered to the load (e.g., an electron beam). Again, note that the
outside of the cavity remains at ground potential, and that only the beam
sums the voltage along the length of the accelerator. Typical high
voltage dielectric materials are transformer oil and de-mineralized water
with dielectric constants of approximately 2.1 and 81, respectively. The

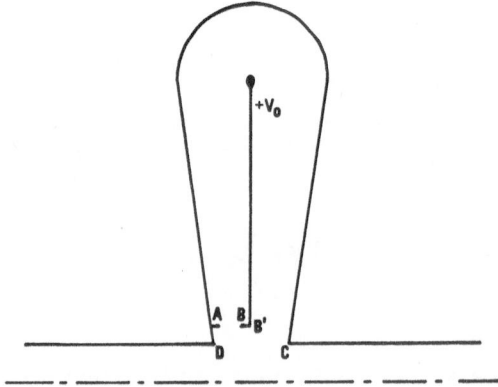

Fig. 3.   Schematic diagram of a Pavlovskii-type radial line cavity.

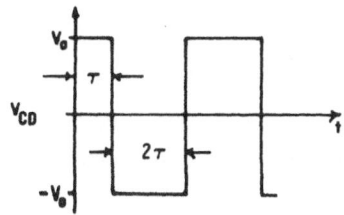

Fig. 4. Open circuit waveform of the Pavlovskii cavity.

cavity dimensions necessary to avoid high voltage insulator breakdowns
for charging pulses of typically a megavolt generally imply a characteristic cavity impedance of a few ohms for water-insulated cavities, and
a few tens of ohms for oil-insulted cavities. As a result, efficient
transfer of energy from the cavity to the beam requires that the beam
current be relatively high (of the order of 100 kA). At these current
levels, the self-fields of the beam are very large and must always be
considered in describing the beam transport through the accelerator.

BEAM EQUILIBRIA CRITERIA

Before we begin our discussion of general beam equilibrium requirements, we will first discuss several preliminary topics, including how
individual particles move in a strong magnetic field, an equation which
describes the motion of particles at the outer radial boundary of the
beam, and the limitation to beam transport posed by the potential
depression due to the beam electron space charge.

Individual Particle Orbits

Consider a cylindrically symmetric electron beam with velocity
$v = \beta c$ and density n_e propagating along a uniform, axial magnetic field
$\underline{B} = B_o \, \hat{e}_z$. From Gauss' law and Ampere's law, E_r and B_θ due to the beam
space charge and current can be found as (in Gaussian units)

$$E_r = - \frac{4\pi e}{r} \int_o^r n_e \, r \, dr \tag{4}$$

$$B_\theta = - \frac{4\pi e}{r} \int_o^r \beta \, n_e \, r \, dr \tag{5}$$

Assuming that the beam density is uniform out to a radius r_b, in
Cartesian coordinates the beam self-fields are given by

$$E_x = \frac{2Ix}{\beta c r_b^2} \quad , \; E_y = \frac{2Iy}{\beta c r_b^2} \tag{6}$$

$$B_x = -\frac{2Iy}{cr_b^2} \quad , \quad B_y = \frac{2Ix}{cr_b^2} \tag{7}$$

where the beam current $I = -\pi r_b^2 (en_o) \beta c$. The motion of a single particle of (relativistic) mass γm and charge $-e$ is described by the Lorentz force law

$$\frac{d}{dt} \underline{p} = -e \left[\underline{E} + \frac{1}{c} (\underline{V} \times \underline{B}) \right] \tag{8}$$

where $\underline{p} = \gamma m\underline{v}$ is the relativistic momentum and $\gamma = (1 - \beta^2)^{-1/2}$. Writing the transverse components explicitly gives

$$\ddot{x} = -(\Omega^2/\gamma^3) x - (\Omega_o/\gamma) \dot{y}$$

$$\ddot{y} = -(\Omega^2/\gamma^3) y + (\Omega_o/\gamma) \dot{x}$$

where $\Omega_o = eB_o/mc$ and $\Omega^2 = 2eI/m \beta cr_b^2$. Defining the parameter $s = x + iy = re^{i\theta}$, the transverse equations can be combined to yield

$$\ddot{s} = -(\Omega^2/\gamma^3) s + i (\Omega_o/\gamma) \dot{s} . \tag{9}$$

Eq. (9) has the formal solution $s(t) = Ae^{i\omega_+ t} + Be^{i\omega_- t}$, where

$$\omega_\pm = \frac{\Omega_o}{2\gamma} \left[1 \pm \left[1 + \frac{4\Omega^2}{\gamma \Omega_o^2} \right]^{1/2} \right] .$$

With the initial conditions $s(0) = s_o$, $\dot{s}(0) = 0$, in the limit of large Ω_o, the solution becomes

$$s(t) \approx s_o \left[e^{i\omega_- t} + \frac{\Omega^2}{\gamma \Omega_o^2} e^{i\omega_+ t} \right] . \tag{10}$$

Hence, the single particle motion can be considered as the sum of two rotations: a large amplitude rotation at the slow E x B rotation frequency, and a small amplitude rotation at the fast cyclotron frequency. If $\Omega^2/\gamma \Omega_o^2 \ll 1$, then the detailed fine structure of the particle orbits can be ignored. (As will be shown later, there are also fast and slow rotation modes of the electron fluid equilibrium.)

Beam Envelope Equation

It is often useful to be able to quickly predict the qualitative motion of an intense charged particle beam under various projected operating conditions. In this regard, beam envelope equations, which neglect the detailed internal state of the beam, serve a very useful function. The perpendicular components of the equation of motion, Eq. (8), are given by

$$\dot{\gamma}\, m\underline{v}_{\perp} + \gamma m\dot{\underline{v}}_{\perp} = -e\left[\underline{E}_{\perp} + \frac{1}{c}\,(\underline{v}\times\underline{B})_{\perp}\right]\tag{11}$$

where the dot notation implies total time differentiation. With the assumed cylindrical symmetry, the important field components are E_r, B_θ, and B_z. We further assume that the longitudinal magnetic field is generated by external magnetic field coils only, and is constant and uniform across the beam radial profile; i.e., $B_z = B_0$. In this case, the radial component of Eq. (11) yields

$$\ddot{r} + \frac{\dot{\gamma}\dot{r}}{\gamma} - \frac{v_\theta^2}{r} = -\frac{e}{\gamma m}\left[E_r + \frac{1}{c}\,(v_\theta B_0 - v_z B_\theta)\right]\tag{12}$$

while the azimuthal component becomes a statement of conservation of canonical angular momentum

$$\gamma r v_\theta - \frac{\Omega_0 r^2}{2} = P_\theta/m = \text{const.}\tag{13}$$

Restricting attention to a single particle at the beam edge, and assuming constant beam kinetic energy, substitution of Eq. (13) into Eq. (12) gives an equation which describes the motion of the radial envelope of the beam (Miller, 1982).

$$\ddot{r}_b + \frac{\dot{r}_b\dot{\gamma}}{\gamma}^{\,0} + \left(\frac{\Omega_0}{2\gamma}\right)^2 r_b = -\frac{e}{\gamma m}\left(E_r - \frac{v_z B_\theta}{c}\right) + \frac{(P_\theta/m)^2}{\gamma^2 r_b^3}\tag{14}$$

As an example of the use of Eq. (14) again consider the case of a beam of uniform density and energy with radius r_b propagating along a uniform longitudinal magnetic field. From Eqs. (4) and (5) the beam self-fields are given by

$$E_r = -2\pi e n_0 r_b$$

$$B_\theta = -2\pi e n_0\,\beta\,r_b = \beta E_r$$

and Eq. (14) can be written as

$$\ddot{r}_b + \left(\frac{\Omega_0}{2\gamma_0}\right)^2 r_b - \frac{\omega_e^2}{2\gamma_0^3}\,r_b = \frac{(P_\theta/m)^2}{\gamma_0^2 r_b^3}\tag{15}$$

where $\omega_e = (4\pi n_0 e^2/m)^{1/2}$ is the beam plasma frequency. Hence, the unneutralized beam self-fields cause the beam to expand, while the external magnetic field supplies a restoring force. The equilibrium radius r_{bo} ($\ddot{r}_b = \dot{r}_b = 0$) is given by

$$r_{bo}^2 = \frac{2\,(P_\theta/m)}{\left(\Omega_o^2 - 2\omega_e^2/\gamma_o\right)^{1/2}} \tag{16}$$

and there can be no equilibrium unless

$$\Omega_o^2 > 2\omega_e^2/\gamma_o \tag{17}$$

For a 1 MeV electron beam of density $10^{12}/cm^3$, the magnetic field strength required for equilibrium is only 2.6 kG; however, for a 1 MeV proton beam of the same density B_o must exceed 188 kG. Hence, while solenoidal magnetic fields can easily satisfy the equilibrium condition for transport of intense electron beams in vacuum, some charge neutralization is usually required for intense ion beam transport.

Before leaving this topic, note that the general solution of Eq. (15) is oscillatory. Eliminating ω_e^2 in favor of the total beam current magnitude, which does not vary as the beam expands or contracts, Eq. (15) becomes

$$\ddot{r}_b + \left(\frac{\Omega_o}{2\gamma_o}\right)^2 r_b - \frac{2eI}{\gamma_o^3 m\beta c r_b} = \frac{(P_\theta/m)^2}{\gamma_o^2 r_b^3}\,. \tag{18}$$

Assuming small perturbations about the equilibrium radius, $r_b = r_{bo} + \delta$, the linearization of Eq. (18) yields

$$\delta\ddot{r} + \left[\left(\frac{\Omega_o}{\gamma_o}\right)^2 - \frac{4eI}{\gamma_o^3 m\beta c r_{bo}^2}\right]\delta_r = 0$$

which has sinusoidal solutions with frequency

$$\omega = \left[\left(\frac{\Omega_o}{\gamma_o}\right)^2 - \frac{4eI}{\gamma_o^3 m\beta c r_{bo}^2}\right]^{1/2}$$

and wavelength $\lambda = 2\pi\beta c/\omega$.

Space Charge Limiting Current

Even if the confining longitudinal magnetic field is infinitely large, there exists a limit to the amount of current (space charge) that can propagate in an evacuated drift space. Consider the one-dimensional schematic in Fig. 5. A current density $j = env_z$ is injected into the space between two grounded, conducting plates. As the charge enters the cavity, a space charge potential barrier will form. However, if the current density is low, the maximum potential will occur at the mid-plane between the conducting plates and all of the current will cross the gap. Now suppose that the injected current density is raised until the

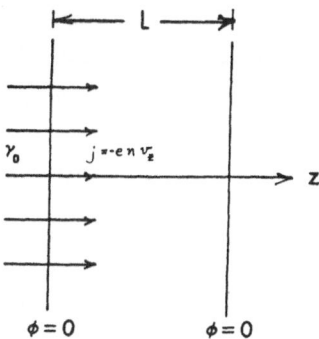

Fig. 5. Schematic geometry of a beam injected into a
one-dimensional drift space.

magnitude of the potential barrier becomes equal to the injected beam
kinetic energy. Any further increase in the beam current density will
cause an increase in the potential barrier height, resulting in a
reflection of injected beam. In this case, a virtual cathode is said to
form, while the largest current that can flow without reflection is
termed the space charge limiting current.

There are a few important special cases for which analytical solutions for the space charge limiting current can be derived. To calculate
the limiting current, we must relate the beam density and the beam velocity (through the continuity equation), the beam density and the electrostatic potential (via Poisson's equation), and the beam velocity and the
electrostatic potential (through the conservation of energy). Assuming
one-dimensional motion (implying a very large longitudinal magnetic
field), in the steady state these relations become

$$\frac{\partial}{\partial z} (nv_z) = 0$$

$$\nabla \cdot \underline{E} = -4\pi en$$

$$(\gamma - 1) mc^2 - e\phi = (\gamma_0 - 1) mc^2 .$$

When the beam density and velocity are eliminated in favor of the
electrostatic potential, the Poisson's equation becomes

$$\frac{1}{r} \frac{\partial}{\partial r} \left(r \frac{\partial \phi}{\partial r} \right) + \frac{\partial^2 \phi}{\partial z^2} = -\frac{4\pi j}{c} \left[1 - \left(\gamma_0 + \frac{e\phi}{mc^2} \right)^{-2} \right]^{-1/2} \tag{19}$$

where $E = -\nabla\phi$, $(\gamma_0 - 1) mc^2$ is the injected beam kinetic energy, and
$j = -env_z$. The solution of Eq. (19) subject to the appropriate boundary
conditions describes the space charge potential (and hence, the electron
kinetic energy) as a function of position in the cavity.

311

As an example, consider an annular beam whose thickness is much smaller than both the beam radius r_b and the distance between the beam and the chamber wall. (See, for example, the geometry of Fig. 10.) In this case, the electrostatic potential is essentially constant across the thin beam and the problem reduces to the solution of the homogeneous equation

$$\frac{1}{r} \frac{\partial}{\partial r} \left(r \frac{\partial \phi}{\partial r} \right) = 0 \tag{20}$$

subject to the boundary conditions

$$\phi(R) = 0$$

$$\left. \frac{\partial \phi}{\partial r} \right|_{r=r_b} = \frac{2I}{r_b v_b}$$

where I is the total beam current, and v_b is the longitudinal electron velocity (uniform across the thin beam). The solution is (Breizman and Ryutov, 1974)

$$\phi(r) = -2 \frac{I}{v_b} \begin{cases} \ln(R/r), & r_b < r < R \\ \ln(R/r_b), & r < r_b \end{cases} \tag{21}$$

where v_b is given by

$$v_b = c \left[1 - \left(\gamma_o + \frac{e\phi_b}{mc^2} \right)^{-2} \right]^{1/2} \tag{22}$$

Combining Eqs. (21) and (22) yields the following result for $\phi_b = \phi(r = r_b)$:

$$F(\phi_b) = \frac{-e\phi_b}{mc^2} \left[1 - \left(\gamma_o + \frac{e\phi_b}{mc^2} \right)^{-2} \right]^{1/2} = \frac{2eI}{mc^3} \ln(R/r_b) . \tag{23}$$

A plot of $F(\phi_b)$ is shown in Fig. 6. It has a maximum at $|e\phi_b| = \gamma_o^{1/3} mc^2$ given by

$$F_{max}(\phi_b) = (\gamma_o^{2/3} - 1)^{3/2} . \tag{24}$$

Since maximizing F also maximizes the current allowed to propagate, the space charge limiting current for this system is given by

$$I_\ell = \frac{mc^3}{2e} \frac{(\gamma_o^{2/3} - 1)^{3/2}}{\ln(R/r_b)} . \tag{25}$$

For currents $I < I_\ell$, there are two possible solutions for ϕ_b according to Eq. (23): (1) $0 < |e\phi_b|/mc^2 < \gamma_o - \gamma_o^{1/3}$, corresponding to particles of high velocity and low density; and (2) $\gamma_0 - \gamma_0^{1/3} < |e\phi_b|/mc^2 < \gamma_0 - 1$ corresponding to particles of lower velocity and higher density. For currents $I > I_\ell$, there is no solution to Eq. (23), and the existence of a steady state flow of current in excess of the space charge limit is not possible.

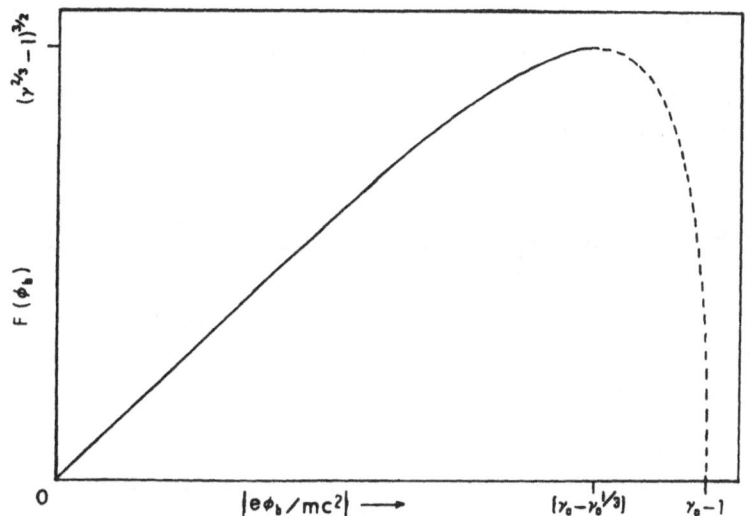

Fig. 6. The function $F(\phi_b)$ for the case of a thin hollow beam.

A calculation of the limiting current in the general case is very difficult; however, a relatively simple procedure allows a quick estimate which is generally accurate to within ten percent (Miller, 1982). To illustrate this method, consider a cylindrically symmetric, uniform density, annular electron beam of inner and outer radii r_1 and r_2 propagating in an evacuated drift tube of radius R along a very strong longitudinal magnetic field. We first use Gauss' law to show that the radial electric field is given by

$$E_r = - \frac{2\pi e n_o}{r} \times \begin{cases} 0 , & r < r_1 \\ (r^2 - r_1^2) , & r_1 < r < r_2 \\ (r_2^2 - r_1^2) , & r_2 < r < R . \end{cases}$$

A radial integration of E_r then gives the electrostatic potential in the drift tube as

$$\phi(r) = - 2\pi e n_o$$

$$
x \quad \begin{cases}
(r_2^2 - r_1^2) \ln (R/r_2) - r_1^2 \ln (r_2/r_1) + (r_2^2 - r_1^2)/2, \\
\qquad\qquad\qquad 0 < r < r_1 \\[1em]
(r_2^2 - r_1^2) \ln (R/r_2) - r_1^2 \ln (r_2/r) + (r_2^2 - r^2)/2, \\
\qquad\qquad\qquad r_1 < r < r_2 \\[1em]
(r_2^2 - r_1^2) \ln (R/r), \qquad r_2 < r < R .
\end{cases}
$$

An approximate expression for the space charge limit is then obtained by setting the magnitude of the electrostatic potential energy equal to the injected beam kinetic energy. This yields the correct expression for the extreme ultrarelativistic limit

$$
I_\ell = \beta(\gamma_0 - 1) \left(\frac{mc^3}{e}\right) \left[1 - \frac{2r_1^2}{r_2^2 - r_1^2} \ln (r_2/r_1) + 2 \ln (R/r_2) \right]^{-1} . \tag{26}
$$

To extend the range of validity to non-relativistic energies, replace the factor $\beta(\gamma_0 - 1)$ by $(\gamma_0^{2/3} - 1)^{3/2}$ obtained in Eq. (26). The final result of this ad hoc procedure is

$$
I \approx (\gamma_0^{2/3} - 1)^{3/2} (mc^3/e) \left[1 - \frac{2r_1^2}{r_2^2 - r_1^2} \ln (r_2/r_1) \right.
$$
$$
\left. + 2 \ln (R/r_2) \right]^{-1} . \tag{27}
$$

By taking appropriate limits of Eq. (27) it is easy to derive the following expressions:

Solid Beam:
(Bogdankevich and Rukhadze, 1971)
$$
I_\ell \approx \frac{(\gamma_0^{2/3} - 1)^{3/2} (mc^3/e)}{1 + 2 \ln (R/r_b)} \tag{28}
$$

Annular Beam:
$$
I_\ell \approx \frac{(\gamma_0^{2/3} - 1)^{3/2} (mc^3/e)}{[a/r_b + 2 \ln (R/r_b)]} , \tag{29}
$$
$$
a = r_2 - r_1 , \quad r_b = \frac{r_2 + r_1}{2}
$$

Annular Beam
(close to wall):
$$
I_\ell \approx (\gamma_0^{2/3} - 1)^{3/2} (mc^3/e) \left[\frac{R}{a + 2\delta}\right], \tag{30}
$$
$$
\delta = R - r_2 .
$$

Inspection of these expressions indicates that the highest beam currents can be transported if the beam is annular and located near the drift tube wall.

Laminar Flow Equilibria

Having first considered several preliminary topics, we will now develop the general equations which describe the self-consistent equilibria of unneutralized intense beams propagating in a metallic drift tube along a uniform magnetic field. The analysis is performed in cylindrical geometry (r, θ, z) with the z axis aligned with the beam and the external magnetic field. The beam is assumed to have axisymmetric equilibrium radial density and velocity profiles (no θ dependence), and is assumed to be uniform in the z direction. The beam front physics are not included, and the beam pulse is assumed to be much shorter than the time required for diffusion of the beam self-fields through the drift tube walls. The assumption of laminar flow implies that all particles (fluid elements) move on nonintersecting helical paths. Hence, the amplitude of the fast gyrational electron motion about the external field lines must be small compared with the beam radius.

The basic equations for examining the equilibrium are the fluid and Maxwell equations. For the case of an electron beam these are

$$\frac{\partial}{\partial t} n + \nabla \cdot (n\underline{v}) = 0 \tag{31}$$

$$\frac{\partial}{\partial t} \underline{p} + \underline{v} \cdot \nabla \underline{p} = -e \left(\underline{E} + \frac{1}{c} (\underline{v} \times \underline{B}) \right) \tag{32}$$

$$\nabla \times \underline{E} = -\frac{1}{c} \frac{\partial \underline{B}}{\partial t} \tag{33}$$

$$\nabla \times \underline{B} = \frac{-4\pi e n \underline{v}}{c} + \frac{1}{c} \frac{\partial \underline{E}}{\partial t} \tag{34}$$

$$\nabla \cdot \underline{E} = -4\pi e n \tag{35}$$

$$\nabla \cdot \underline{B} = 0 . \tag{36}$$

In Eqs. (31)-(36) n, $\underline{v}$, and $\underline{p}$ represent the macroscopic equilibrium electron fluid density, velocity, and momentum, while $\underline{E}$ and $\underline{B}$ represent the self-consistent equilibrium electric and magnetic fields. Following the assumptions presented in the preceding paragraph and assuming the steady state, Eq. (32) yields the radial force balance given by

$$\frac{m\gamma(r)v_\theta^2(r)}{r} - e \left\{ E_r + \frac{1}{c} \left[v_\theta \left(B_o + B_z^S \right) - v_z B_\theta \right] \right\} = 0 \tag{37}$$

where $\gamma = \left[1 - (v_\theta/c)^2 - (v_z/c)^2 \right]^{-1/2}$. The equilibrium radial electric field is determined from Poisson's equation, Eq. (35), and the equilibrium magnetic field is expressed as

$$\underline{B}(r) = (B_o + B_z^S)\, \hat{e}_z + B_\theta\, \hat{e}_\theta \tag{38}$$

where B_o is the externally applied field. The azimuthal self-field B_θ is given by Eq. (5), while the axial diamagnetic self-field B_z^S is given by

$$B_z^S(r) = -\frac{4\pi e}{c}\int_r^R dr'\, n(r')\, v_\theta(r') + B_c \,. \tag{39}$$

The constant B_c represents the uniform field due to the azimuthal image current in the drift tube wall; it is easily computed from flux conservation.

The first term in Eq. (37) is the outward centrifugal force, and the second term is the outward force of the electric field due to the beam space charge. The third and fourth terms represent the constraining forces of the magnetic fields. The third term also reflects the diamagnetic character of the beam, i.e., the reduction of the applied magnetic field strength interior to the beam. Introducing the dimensionless variables $\beta_\theta = v_\theta/c$ and $\beta_z = v_z/c$, and using Eqs. (4), (5), and (39) to eliminate E_r, B_θ, and B_z^S, Eq. (37) becomes

$$m\gamma(r)\left[\beta_\theta(r)c\right]^2 r^{-1}$$

$$+\frac{4\pi e^2}{r}\int_o^r dr'\, r'\, n(r') - \beta_\theta(r)\left[e(B_o + B_c)\right.$$

$$\left. - 4\pi e^2 \int_r^R dr'\, \beta_\theta(r')n(r')\right]$$

$$-\frac{4\pi\beta_z(r)e^2}{r}\int_o^r dr'\, r'\, \beta_z(r')n(r') = 0 \,. \tag{40}$$

We now use this radial force balance equation to derive two equilibria which have considerable practical importance.

If it is assumed that the azimuthal beam motion is very small compared to the axial motion, then the axial diamagnetic field contribution can be neglected, and Eq. (40) becomes

$$\gamma(r)\omega_\theta^2(r) + \frac{1}{r^2}\int_0^r dr' \ r' \ \omega_p^2(r') - \omega_\theta\Omega_o$$

$$- \frac{\beta_z(r)}{r^2}\int_0^r dr' \ (r') \ \beta_z(r') \ \omega_p^2(r') = 0 \tag{41}$$

where $\Omega_o = eB_o/mc$, $\omega_p^2 = 4\pi n e^2/m$, and $\omega_\theta(r) = \beta_\theta(r)c/r$.
If it is further assumed that the axial velocity profile of the electron beam is independent of radius r, i.e., $\beta_z(r)c \equiv \beta_o c = const$, then

$$\gamma(r) \simeq \left(1 - \beta_z^2(r)\right)^{-1/2} \equiv \gamma_o = const \ . \tag{42}$$

With these simplifying assumptions, Eq. (41) reduces to

$$\gamma_o \ \omega_\theta^2(r) + \frac{1}{r^2}(1 - \beta_o^2)\int_0^r dr' \ r' \ \omega_p^2(r') - \omega_\theta(r)\Omega_o = 0 \ . \tag{43}$$

Solving Eq. (43) for the angular velocity $\omega_\theta(r)$ yields

$$\omega_\theta(r) = \omega_\theta^\pm(r) = \frac{\Omega_o}{2\gamma_o}\left\{1 \pm \left[1 - \frac{4}{\gamma_o\Omega_o^2 r^2}\right.\right.$$

$$\left.\left. \times \int_0^r dr' \ (r') \ \omega_p^2(r')\right]^{1/2}\right\} \ . \tag{44}$$

For a constant density profile

$$\omega_p^2(r) = \begin{cases} \dfrac{4\pi n_o e^2}{m} = \omega_{po}^2 \ , & 0 < r < r_b \\ 0, & r > r_b \ . \end{cases} \tag{45}$$

Eq. (44) reduces to

$$\omega_\theta(r) = \omega_\theta^\pm = \frac{\Omega_o}{2\gamma_o}\left[1 \pm \left(1 - \frac{2 \ \omega_{po}^2}{\gamma_o \ \Omega_o^2}\right)^{1/2}\right] \ . \tag{46}$$

The condition for the radical to be real (the existence of an equilibrium) is

$$\Omega_o^2 > 2 \ \omega_{po}^2/\gamma_o \tag{47}$$

which is identical to Eq. (17). The beam rotates as a rigid body with two allowable rotation rates given approximately by

$$\omega_\theta^+ \simeq \frac{2\gamma_o \Omega_o^2 - \omega_{po}^2}{2\gamma_o^2 \Omega_o} \qquad (48)$$

$$\omega_\theta^- \simeq \frac{\omega_{po}^2}{2\gamma_o^2 \Omega_o} \ . \qquad (49)$$

For the slow mode, the rotation frequency is $\omega_\theta^- \approx cE(r)/r\gamma^2 B_o$, which corresponds to a beam precession in θ at the E x B drift rate with relativistic corrections (Bogema, 1971), while the fast mode rotation corresponds to the fast cyclotron motion about the axial magnetic field lines with $\omega_\theta^+ \simeq \Omega_o/\gamma_o$.

As a second application of Eq. (43), consider the case of a hollow beam with density profile given by

$$\omega_p^2(r) = \begin{cases} 0, & 0 < r < r_o \\ \dfrac{4\pi n_o e^2}{m} = \omega_{po}^2, & r_o < r < r_b \\ 0, & r > r_b \ . \end{cases} \qquad (50)$$

Substituting Eq. (50) into Eq. (43) yields the solution for the angular velocity $\omega_\theta(r)$ given by

$$\omega_\theta(r) = \omega_\theta^\pm(r) = \frac{\Omega_o}{2\gamma_o} \left\{ 1 \pm \left[1 - \frac{2\omega_{po}^2}{\gamma_o \Omega_o^2} \left(1 - r_o^2/r^2 \right) \right]^{1/2} \right\} ,$$

$$r_o < r < r_b \ . \qquad (51)$$

Since $\partial\omega_\theta/\partial r \neq 0$, the hollow beam no longer rotates as a rigid body but has angular velocity shear. This difference in rotation rate across the beam provides a source of free energy that can drive a filamentation instability.

For both the rigid rotor and hollow beam equilibria there is considerable arbitrariness; any two of the three variables $n(r)$, $v_\theta(r)$, and $v_z(r)$ could be specified. However, since the beam density is related to the beam kinetic energy through Poisson's equation, conservation of energy provides one independent relation, i.e.,

$$(\gamma - 1)mc^2 - e\phi(r) = (\gamma_o - 1)mc^2 \ . \qquad (52)$$

In addition, if the source geometry is known, then the variation of canonical angular momentum, $P_\theta(r)$, is specified as

318

$$P_\theta(r) = \gamma(r)mr\, v_\theta(r) - \frac{er}{c} A_\theta(r) = -\frac{er_c}{c} A_{\theta s}(r_c) \tag{53}$$

where $A_{\theta s}$ is given by the axial magnetic field B_z at the cathode according to

$$A_{\theta s} = \frac{1}{r} \int_0^r B_z^c(r')r'dr' . \tag{54}$$

Under the assumption of laminar flow, Eq. (53) implies that all particles at a given radius were emitted from the source at the same radius r_c and hence, have the same canonical angular momentum.

There are three source geometries of practical importance. The first is the non-immersed source for which the cathode is shielded from the axial magnetic field in the downstream region. In this case, $P_\theta = B_z^c = 0$ and Eq. (53) becomes simply

$$\gamma(r)mr\, v_\theta(r) = \frac{e}{c} \int_0^r B_z(r')r'dr' . \tag{55}$$

A second important case is that of an immersed source whose emitting area coincides with a flux surface; i.e., an annular cathode that produces a thin hollow beam. In this instance all particles have the same canonical angular momentum and Eq. (53) becomes

$$\gamma(r)mr\, v_\theta(r) - \frac{e}{c} \int_0^r B_z(r')r'dr' = -\frac{e}{2c} B_z^c(r_c)r_c^2 = P_\theta . \tag{56}$$

The final important case is that of an immersed source for which the axial magnetic field is uniform. In this instance, P_θ varies across the beam. If it is also assumed that the beam expands or contracts in self-similar fashion, then the radii of particles in the downstream region must be related to their emission radii according to $r = \alpha r_c$, where α is a constant which depends on the actual system geometry. In this case, $P_\theta(r) = -(eB_z^o/2c)(r^2/\alpha^2)$, and Eq. (53) becomes

$$\gamma(r)mrv_\theta(r) - \frac{e}{c} \int_0^r B_z(r')r'dr' = -\frac{eB_z^o}{2c}(r/\alpha)^2 . \tag{57}$$

The primary distinction between the equilibria resulting from either shielded or immersed sources is provided by a comparison of Eq. (55) with Eq. (56) or Eq. (57). For the shielded source, as the electrons leave the cathode, v_θ is initially zero, but rapidly increases as the electrons encounter the increasing external field (analogous to a magnetic mirror). The beam rotation generates an axial diamagnetic field, and equilibria for shielded sources are sometimes called diamagnetic, or rotating beam, equilibria.

For immersed sources, on the other hand, beam rotation is generated as a result of the E_r and B_z field components, and v_θ is very small compared with v_z. Hence a good first approximation is to ignore diamagnetic field contributions and set $B_z = B_0$ in Eq. (57). Solving for v_θ then yields

$$v_\theta = \frac{eB_0 r}{2\gamma mc} (1 - \alpha^{-2}) \ . \tag{58}$$

As $B_0 \rightarrow \infty$, the particles are trapped on field lines ($\alpha = r/r_c \rightarrow 1$), and it can be shown that the rotation rate drops to zero.

Equations (40), (52), and (53) form a complete set and uniquely specify the equilibrium in a self-consistent way. However, analytical solutions for this general problem are very difficult and have been obtained in only a few isolated special cases (Reiser, 1977). In practice, the usual procedure is to assume one of the simple radial force balance equilibria, such as the rigid rotor, as a first guess, and then iterate until the necessary accuracy is achieved (Godfrey, 1979a).

INJECTOR CONSIDERATIONS

The discussion of the previous section was concerned with beam equilibria in infinitely long, uniform systems. However, in a linear induction accelerator the axial symmetry is necessarily broken in the injector and accelerating gaps, and special care must be exercised in the design of these regions. In this section, we will discuss various aspects of the design of high current injectors.

Explosive Electron Emission (For a summary, see Miller, 1982)

There are several processes that can lead to the escape of electrons from the surface of a metal. The energy difference between the highest electron-filled level in the conduction band (the Fermi level) and a field-free region outside the cathode surface is called the work function, and represents a potential barrier to electron emission of typically several electron volts. In photoemission, surfaces are irradiated with photons with energy greater than the work function (visible or near ultraviolet for clean metals). In the case of thermionic emission, the metal is heated until a sufficient number of electrons acquire enough energy to escape the potential barrier. In field emission, an externally-applied electric field lowers the potential barrier to permit electron escape. While all of these processes have important uses, the dominant mechanism for generating very high beam currents is termed explosive electron emission.

On almost any surface there exists microscopic surface protrusions (whiskers) which are typically 10^{-4} cm in height with a base radius of less than 10^{-5} cm and tip radius usually much smaller than the base radius. Reported estimates of whisker concentrations have ranged from $1-10^4$ whiskers/cm^2. Application of a strong electric field across two metal electrodes will cause large field emission currents to be drawn from the micropoints. As a result, the micropoints undergo rapid resistive heating and explosively vaporize, forming localized regions of metallic plasma. The expansion and merger of these plasma blobs dramatically increase the effective electron emission area, and the expanding cathode plasma can emit an electron current which is limited only by the associated space charge cloud in the anode-cathode gap.

To illustrate this space-charge-limited emission process, assume that a plasma completely covers the cathode surface at the instant that high voltage is impressed on the diode. Initially the potential variation is simply a linear function of the distance between the electrodes. As electrons are drawn into the gap, however, a potential minimum can form outside the cathode surface and only those electrons which possess an initial energy greater than the space charge barrier height (due to electron thermal spread in the cathode plasma) can escape from the cathode and reach the anode.

In high voltage diodes, the magnitude of the (thermal) potential minimum is negligible compared with the high applied potential, and the location of the minimum is practically coincident with the cathode surface. Hence, it may be assumed that the electric field vanishes at the cathode surface and that emitted electrons have zero initial velocity. In this case, the velocity of an electron at any position between the electrodes is determined from energy conservation.

The Relativistic Planar Diode

As a simple example of space-charge-limited flow, consider the relativistic planar diode (Jory and Trivelpiece, 1969). The anode potential is assumed to be zero, while the cathode is at potential $-\phi_0$. In this case conservation of energy implies that

$$\gamma = \gamma_0 + e\phi/mc^2 \ . \tag{59}$$

For the planar geometry, Poisson's equation is

$$\frac{d^2\phi}{dz^2} = 4\pi e n_e \qquad (60)$$

where z is the distance from the cathode plasma and n_e is the electron number density. Conservation of electronic charge requires $\nabla \cdot j = 0$, or

$$j = e n_e v = \text{const} \qquad (61)$$

Combining Eqs. (59)-(61) yields

$$\frac{d^2\phi}{dz^2} = \frac{4\pi j}{v} = \frac{4\pi j}{c}(1 - \gamma^{-2})^{-1/2} \qquad (62)$$

Introducing the variable $y = d\gamma/dz$, a first integration of Eq. (62) yields

$$y^2 = 2K(\gamma^2 - 1)^{1/2} + C_1 \qquad (63)$$

where

$$K = 4\pi j \left(\frac{e}{m_o c^3}\right). \qquad (64)$$

The constant of integration C_1 is zero because $y = d\gamma/dz = d\phi/dz = 0$ at the cathode. A second integration yields

$$j = \frac{\left(m_o c^3/e\right)}{8\pi d^2} \left[F(\delta_o, \sqrt{2}/2) - 2E(\delta_o, \sqrt{2}/2) \right. $$

$$\left. + \frac{2\gamma_o \left(\gamma_o^2 - 1\right)^{1/4}}{1 + \left(\gamma_o^2 - 1\right)^{1/2}} \right]^2 \qquad (65)$$

where

$$\delta_o = \cos^{-1} \left[\frac{1 - \left(\gamma_o^2 - 1\right)^{1/2}}{1 + \left(\gamma_o^2 - 1\right)^{1/2}} \right]$$

and $F(\delta, \sqrt{2}/2)$ and $E(\delta, \sqrt{2}/2)$ are elliptic integrals of the first and second kind. d is the electrode spacing and γ_o corresponds to the applied potential $|\phi_o|$. In the nonrelativistic limit, $e\phi_o/m_o c^2 \ll 1$, Eq. (65) reduces to the familiar Child-Langmuir expression (Langmuir, 1931)

$$j = \frac{\sqrt{2}}{9\pi} \left(\frac{e}{m_o}\right)^{1/2} \frac{\phi_o^{3/2}}{d^2}. \qquad (66)$$

322

A comparison of the normalized diode impedance $\left\{ \left(Z_n = U/\xi^2, \text{ where } U = \right. \right.$

$e\phi/mc^2$ and $\xi^2 = \left[je / \left(mc^3\varepsilon_p \right) \right] d^2 \text{(MKS)} \left. \right\}$ for the relativistic and non-relativistic solutions is presented in Fig. 7. Note that the diode impedance decreases as the applied potential is increased.

The electron velocity as a function of position may be determined from Eq. (59) with the aid of Eqs. (62) and (65), and the charge density is then determined from Eq. (61). The physically impossible result that the electron density is infinite at the cathode is a consequence of the zero initial velocity assumption. In reality, the initial velocities are small but non-zero, and the charge density is large though finite.

As previously discussed, there are three injector geometries of practical importance: (1) non-immersed sources ($P_\Theta = 0$), (2) an immersed annular cathode ($P_\Theta = $ const.), and (3) an immersed source in a uniform magnetic field $\left(P_\Theta(r) \sim r^2 \right)$. For non-immersed sources a separately biased grid is often used to control the cathode emission in addition to controlling the expansion of the beam prior to its entry into the magnetic field transport system. In this case, the beam current is established by Eq. (65), where ϕ_o is the applied grid-cathode voltage, and d is the grid-cathode separation. Since the grid is an equipotential surface, the radial electric field of the beam space charge is shorted, and the beam will tend to self-pinch in the region of the grid, with subsequent expansion downstream. The profile of the magnetic field in

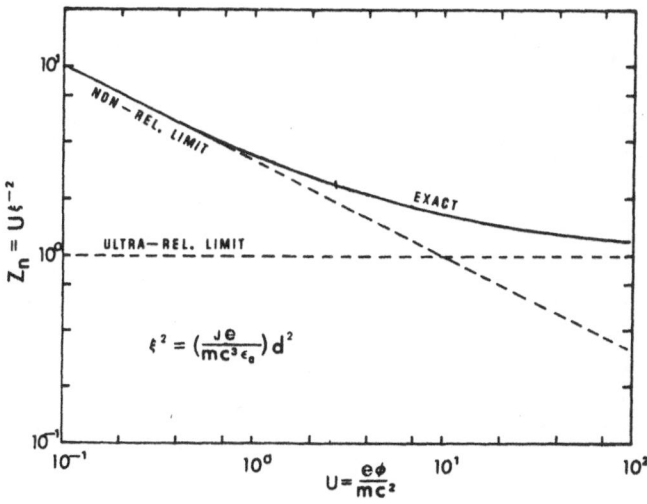

Fig. 7. A comparison of the relativistic and nonrelativistic solutions for the impedance of a one-dimensional planar diode.

the vicinity of the anode must be carefully controlled in order to optimize the beam injection efficiency. These comments are illustrated in Fig. 8 (Paul, 1981).

The Foilless Diode (Friedman and Ury, 1970)

An immersed source configuration which has proven to be extremely useful for injecting high current beams into vacuum regions is the foilless diode. In this configuration, beam expansion and transport in the diode is controlled by extending the solenoidal guide field of the accelerator into the injector diode region (Fig. 9). Theoretical analyses of Ott, et al. (1977) and Chen and Lovelace (1978) have adopted the somewhat idealized geometry of Fig. 10. A semi-infinite cathode with surface potential $-\phi_c(r)$ is joined at a right angle to a grounded anode. Electron emission is assumed to occur over a limited region of the cathode surface, and the electron motion is constrained to be one-dimensional by application of a very strong axial magnetic field. In addition, the applied voltage is assumed to be ultrarelativistic so that the narrow cathode sheath region can be ignored. .Under these conditions, a self-consistent solution for the electron current density can be found which satisfies the criterion of space-charge-limited emission. Several practically useful expressions have now been developed. For the simple limiting case of a beam which is thin compared with its distance to the anode wall δ, the diode impedance is given approximately by

$$Z = \frac{1}{c} \frac{\delta}{r_b} \ \ln\left(\frac{8\delta}{a}\right) \tag{67}$$

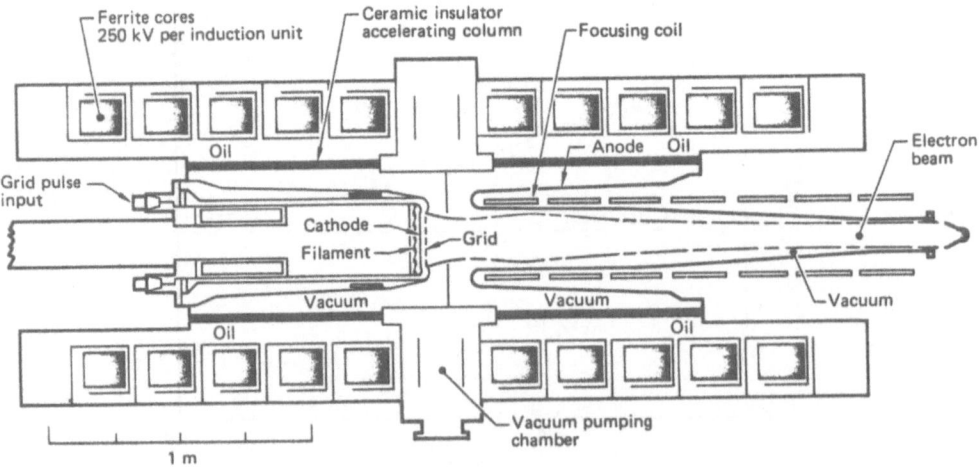

Fig. 8. Schematic diagram of the 2.5 MV electron injector for the experimental test accelerator (ETA).

324

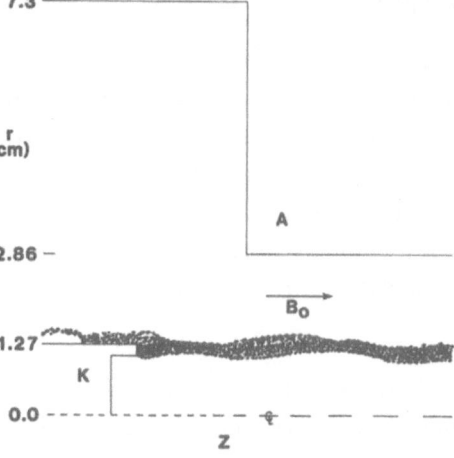

Fig. 9. A numerical simulation of a typical foilless diode geometry.

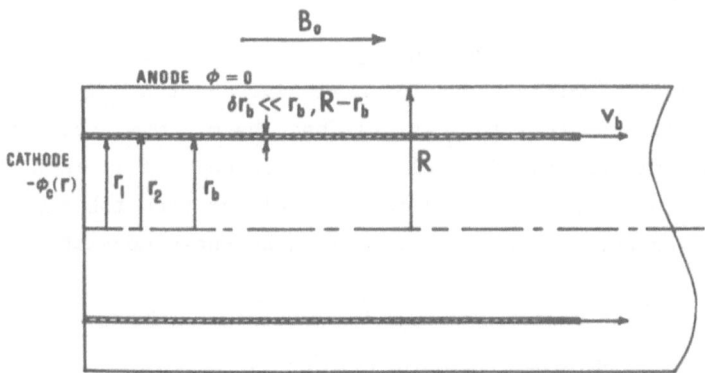

Fig. 10. Idealized geometry for the foilless diode.

and the total diode currently is approximately

$$I_t = (\gamma_0 - 1)(mc^3/e) \frac{r_b}{\delta} \left[\ln \frac{8\delta}{a} \right]^{-1}. \qquad (68)$$

An ad hoc approximation which extends the validity of Eq. (68) to non-relativistic voltages is to replace $(\gamma_0 - 1)$ by the factor $\left(\gamma_0^{2/3} - 1 \right)^{3/2}$. Relatively simple corrections to account for finite magnetic field effects (non-zero Larmor orbits and radial oscillations of the beam envelope) and the inclusion of a longitudinal anode-cathode gap to provide additional control of the diode impedance are also available in the literature (Miller, et al., 1980).

ACCELERATING GAP EFFECTS

During the course of multi-gap experiments with RADLAC I, it became apparent that the accelerating gap dimensions, if not carefully chosen, could result in serious disruption of the beam equilibrium. Consider the schematic accelerating gap geometry illustrated in Fig. 11. The zero-order gap dimensions are constrained by (1) virtual cathode formation, (2) electron emission, and (3) radial oscillations of the beam envelope. We consider these constraints in turn.

Electron Emission

As is apparent from Fig. 11, the beam space charge distorts the equipotential contours away from the cathode side of the gap; however, since this distortion typically extends radially into the gap a distance of only a gap width L, a general rule is that the gap spacing must be chosen such that

$$L \geq \phi_0/E_c \tag{69}$$

where ϕ_0 represents the accelerating potential applied to the gap and E_c is the critical electric field for explosive electron emission (Miller, et al., 1983). Eq. (69) thus establishes the minimum acceptable gap spacing for entirely avoiding electron emission. If it is necessary to violate Eq. (69), the loading effects resulting from the gap currents must be estimated. In this case, we use the one-dimensional

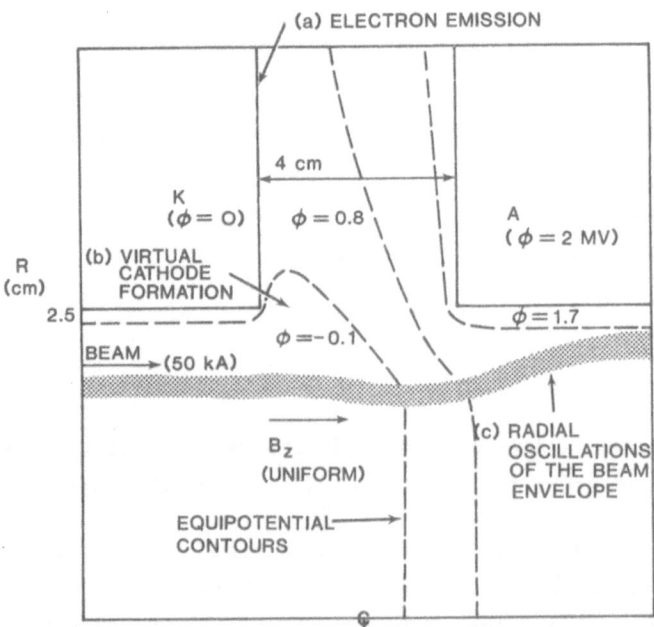

Fig. 11.   Schematic accelerating gap geometry.

space-charge-limited flow approximation developed by Jory and Trivelpiece (1969).

$$U = (\xi + 0.8471)^2 - 1 \tag{70}$$

where $U = e\phi/mc^2$ and $\xi = (2\pi e j/mc^3)^{1/2}z$. Setting $\phi = \phi_0$ and $z = L$ yields the current density flowing into the gap, and the total current can be estimated by

$$I_g \approx j \; \pi\left(R_g^{\,2} - R^2\right) . \tag{71}$$

To avoid serious loading we must require that I_g be much less than the beam current I.

Space Charge Limits

Even though accelerating voltage is applied to the gap, the equipotential contours of Fig. 11 indicate the possibility of virtual cathode formation at the gap entrance if the beam current is too high. The axial variation of the electrostatic potential at the beam radius can be modeled as

$$\phi(z) = \phi_A(z) + \phi_S(z) \tag{72}$$

where ϕ_A and ϕ_S denote the applied voltage and beam space charge contributions. We note that ϕ_A must be antisymmetric with respect to the center of the gap ($z = 0$), while ϕ_S must be symmetric. Hence, as a first approximation we use

$$\phi_A(z) \approx (\phi_0/2) \left(1 + \sin\frac{\pi z}{L\star}\right), \quad -L\star/2 < z < L\star/2 \tag{73}$$

$$\phi_S(z) \approx - \frac{2I}{c} \ln \frac{R}{r_b} \left[1 + \ln(1 + L/R) \cos\frac{\pi z}{L\star}\right],$$

$$-L\star/2 < z < L\star/2 \tag{74}$$

where R and r_b denote the drift tube and beam radii, and $L\star = L + 2(R - r_b)$. From Eqs. (72) – (74), there is a potential minimum given by

$$\phi_{min} = - \frac{2I}{c} \ln \frac{R}{r_b} + \frac{\phi_0}{2} \left\{1 - \left[1 + \left(\frac{4I}{\phi_0 c} \ln \frac{R}{r_b}\right)^2 \right. \right.$$

$$\left. \left. [\ln(1 + L/R)]^2\right]^{1/2}\right\} \tag{75}$$

which occurs at

$$Z_{min} = (L\star/\pi) \tan^{-1} \left\{ -\frac{\phi_o c}{4I} \left[\ln \frac{R}{r_b} \ln (1 + L/R) \right]^{-1} \right\} \tag{76}$$

A comparison of this simple model with an electrostatic code solution of Poisson's equation for the gap geometry of Fig. 11 indicates generally correct behavior (Fig. 12).

With Eq. (75), an approximate expression for the space-charge limiting current for the gap region can be obtained by setting

$$\left| e\phi_{min} \right| \approx \left(\gamma_o^{2/3} - 1 \right)^{3/2} mc^2 . \tag{77}$$

This procedure yields (Miller, et al., 1983)

$$I_{\ell o}/I_\ell = 1 + \frac{\phi_o c}{4I_\ell \ln R/r_b} \left\{ -1 + \left[1 + \left(\frac{4I_\ell}{\phi_o c} \ln \frac{R}{r_b} \right)^2 \right. \right.$$
$$\left. \left. \times \left[\ln(1 + L/R) \right]^2 \right]^{1/2} \right\} \tag{78}$$

where

$$I_{\ell o} = \frac{\left(\gamma_o^{2/3} - 1 \right)^{3/2} \left(mc^3/e \right)}{2 \ln R/r_b} \tag{79}$$

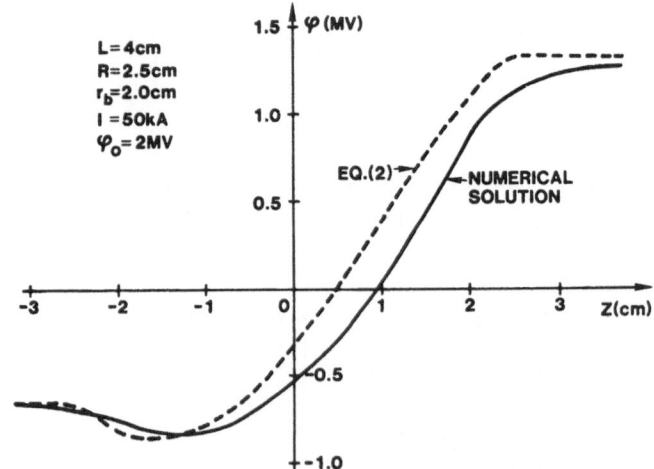

Fig. 12.   Comparison of a numerical solution of the electro-static potential energy as a function of axial position in the gap with the model solution of Eq. (75).

328

[which is Eq. (25)] denotes the limiting current for the beam in the smooth wall drift tube preceding the gap region. A graph of Eq. (78) is presented as a function of $\phi_0 c/I_{\ell 0}$ for various values of L/R in Fig. 13. Note that the limiting current can easily decrease by 50 percent at the entrance to the gap region. At the low voltage end of the accelerator it will often be the case that $4I_\ell \ln(R/r_b) \ll \phi_0 c$. In this limit

$$\frac{I_\ell}{I_{\ell 0}} \approx 1 - 2 \ln \frac{R}{r_b} \left(\frac{I_{\ell 0}}{\phi_0 c}\right)[\ln (1 + L/R)]^2 \tag{80}$$

As the applied gap voltage ϕ_0 is raised, the limiting current increases.

Radial Oscillations of the Beam Envelope

The motion of the beam envelope in the smooth drift tube sections is well described by Eq. (18). However, the presence of an accelerating gap introduces electric field variation δE_r and δE_z. In this case, the linearization of Eq. (18) yields (Miller, et al., 1983)

$$\delta \ddot{r} + \frac{4}{\gamma_b^2}\left[\left(\frac{\Omega_0}{2}\right)^2 - \frac{eI}{\gamma_b m v_z r_o^2}\right]\delta r = - \frac{2eI}{\gamma_b^4 m v_z r_o}\delta\gamma - \frac{e\,\delta E_r}{\gamma_b m} \tag{81}$$

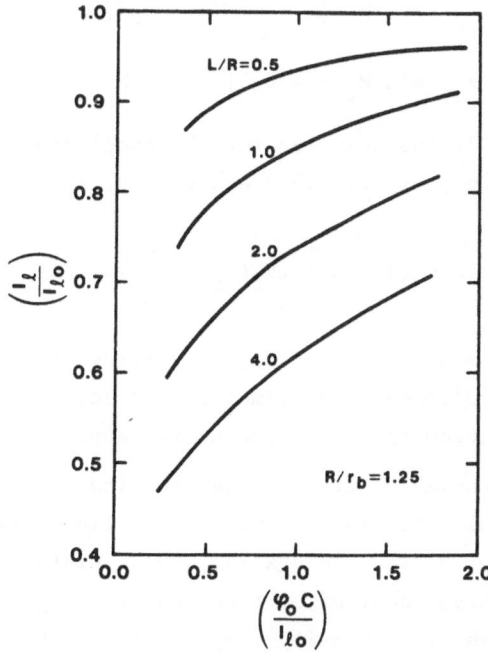

Fig. 13. Ratio of the space-charge limiting current in accelerating gap divided by limiting current in the upstream drift tube as a function of the gap width to radius ratio for various applied voltage.

$$\delta\gamma = -\frac{\left(\gamma_b^2 - 1\right)^{1/2}}{\gamma_b} \left(\frac{e}{mc}\right) \int^t \delta E_z \, dt \ . \tag{82}$$

For even moderately relativistic electron beams, the δ_γ term in Eq. (81) is negligible. Defining the quantity

$$k_c^2 = \frac{\Omega_o^2 - 4eI/\gamma_b m v_z r_o^2}{\gamma_b^2 c^2} \ . \tag{83}$$

Eq. (81) can be rewritten as

$$\delta r'' + k_c^2 \, \delta r = -\frac{e \, \delta E_r}{\gamma_b mc^2} \tag{84}$$

where the primes denote differentiation with respect to the axial coordinate. Hence, radial electric field variations in the accelerating gap region will excite radial oscillations of the beam envelope with a wave length given by $\lambda_o = 2\pi/k_c$.

By developing qualitative solutions for $\delta E_r(z)$, and assuming that γ variations are small, Eq. (83) can be solved to provide a quick estimate of the oscillation amplitudes for simple square gap geometries. Since the Green's function for Eq. (83) is given by

$$G(z|z_o) = \begin{cases} 0 \ , & z < z_o \\ (K/k_c) \sin k_c(z - z_o), & z > z_o \end{cases} \tag{85}$$

where $K = -e/(\gamma_b mc^2)$, the solution for the oscillatory beam envelope can be written as (Miller, et al., 1983 and Poukey, 1984)

$$\delta r(z) = \int_{-\infty}^{z} \frac{K}{k_c} \sin k_c(z - z_o) \, \delta E_r(z_o) \, dz_o \ . \tag{86}$$

To develop an expression for δE_r, we follow Adler (1983a) and note that $\delta E_r(z)$ can be divided into symmetric and antisymmetric contributions resulting from the beam space charge and the applied gap voltage.

To model the radial electric field contribution due to the applied gap potential, recall that the potential variation between the electrodes of a simple parallel plate accelerating gap of width L can be modeled with a negative charge sheet of surface density $- \phi_o/8\pi L$ at $z = -L/2$, and a positive charge sheet of density $+ \phi_o/8\pi L$ at $z = L/2$, where ϕ_o is the applied voltage and L is the gap width. If we insert an opening for the beam drift tube of radius R, then we must cancel the charge density interior to R; however, from Gauss' law the total charge in each half of

330

the gap cannot deviate from that of the parallel plate case. We, therefore, lump the necessary negative and positive charges in rings at the
appropriate corners of the gap. Hence, the charge density responsible
for generating radial electric fields in the region of the gap due to the
applied voltage can be modelled as

$$
\sigma_{AV} \simeq \begin{cases} \dfrac{\phi_o}{8\ \pi L}\ [\ \delta(z + L/2) - \delta(z - L/2)]\left[1 - \dfrac{R\ \delta(r-R)}{2}\right] \ , \ r \leq R \\ \\ 0 \ , \ r > R \ . \end{cases}
\tag{87}
$$

To model the δE_r contribution due to the beam space charge, recall
that the surface image charge density on the interior of a drift tube of
radius R carrying an electron beam current of magnitude I is given by
$\sigma_{SC} = I/2\pi Rc$. If we insert an accelerating gap into the drift tube, then
we must cancel the positive charge in the region $(-L/2, L/2)$. To satisfy
Gauss' law, however, we again lump the necessary positive charge into a
ring at the corner of the gap. Hence, the image charge density responsible for generating deviations from the radial electric field of a beam
in a smooth drift tube can be modelled as

$$
\sigma_{SC} \approx -\ \frac{I\delta(r-R)}{2\pi\ Rc}\ \left[1 - L/2\ [\ \delta(z-L/2) + \delta(z+L/2)]\right]\ .
\tag{88}
$$

To calculate the variation in radial electric field given by the
charge densities of Eqs. (86) and (87), we use the free space Green's
function for Poisson's equation in the form

$$
\frac{1}{\underline{x}-\underline{x}'} = \frac{2}{\pi}\ \sum_{m=-\infty}^{\infty}\ \int_o\ dk\ e^{im\Theta}\ \cos[k(z-z')]\ I_m(k\rho_<)\ K_m(k\rho_>)
\tag{89}
$$

where $\rho_{<,>} = r$ if $r <,> r'$ and $\rho_{>,<} = r'$ if $r' >,< r$. Combining Eqs.
(87) through (89), and performing the necessary integrations and
differentiations yields

$$
[\ \delta E_r\ (r,z)]_{AV} = \frac{2\phi_o R}{\pi L}\ \int_o^\infty\ dk\ \sin\ kz\ \sin(kL/2)\ I_1(kr)
$$

$$
\times\ \left[K_1(kR) + \frac{kR}{2}\ K_o(kR)\right]
\tag{90}
$$

$$
[\ \delta E_r(r,z)]_{SC} = \frac{4I}{\pi c}\ \int_o^\infty\ dk\ \cos\ kz\ K_o(kR)\ I_1(kR)
$$

$$
\times\ \left[\frac{kL}{2}\ \cos(kL/2) - \sin(kL/2)\right]\ .
\tag{91}
$$

To find the oscillatory behavior of the beam envelope, we now combine Eqs. (90) and (91), and substitute Eq. (86).

This procedure yields

$$\delta r(z) = \frac{2 \phi_0 RK}{\pi k_c L} \int_0^\infty dk \, \sin \frac{KL}{2} \, I_1(kr) \left[K_1(kR) + \frac{kR}{2} K_0(kR) \right]$$

$$\int_{-\infty}^z dz_0 \, \sin k_c(z-z_0) \, \sin k \, z_0$$

$$+ \frac{4IK}{\pi k_c c} \int_0^\infty dk \left[\frac{kL}{2} \cos \frac{kL}{2} - \sin \frac{kL}{2} \right] K_0(kR) \, I_1(kr)$$

$$\int_{-\infty}^z dz_0 \, \sin k_c(z-z_0) \, \cos k z_0 \, .$$

$$(92)$$

We will not carry out the detailed evaluation of Eq. (92) here; rather we will note two important results. First, in the asymptotic limit $z \to \infty$, the solution can be written as (Adler, 1983a)

$$\delta r(z) = C_1 \sin k_c \, z + C_2 \cos k_c \, z \qquad (93)$$

where

$$C_1 = \frac{2eI}{\gamma_b mc^3 k_c} K_0(k_c R) \, I_1(k_c r) \left[\frac{k_c L}{2} \cos \frac{k_c L}{2} - \sin \frac{k_c L}{2} \right]$$

$$(94)$$

$$C_2 = \frac{e \phi_0 R}{\gamma_b mc^2 k_c L} \sin \frac{k_c L}{2} I_1(k_c r) \left[K_1(k_c R) + \frac{k_c R}{2} K_0(k_c R) \right] \, .$$

Also, we note that after the integration over z_0, the integration over k-space of Eq. (92) will give singularities at $k_c L = \pi$. This resonant behavior indicates the physical origin of the radial oscillations, i.e., the strong excitation of zero-frequency cyclotron waves due to radial electric field variations when the gap width is approximately equal to half the beam cyclotron wavelength.

To avoid this resonant behavior, accelerating gaps must always be designed such that $L \neq \pi/k_c \approx 5.36 \, \gamma_b/B_z$ (kG) (Miller, et al., 1983). For linear induction accelerators with constantly increasing γ_b, this relation indicates that simply increasing B_z may not be sufficient to control the oscillations. However, if the injector kinetic energy is sufficiently high, the resonance can always be avoided. To be sufficiently removed from the resonance, we require

$$L \lesssim \frac{\pi}{2k_c} . \tag{95}$$

Defining an average accelerating field stress by $E_a = \phi_o L$, and assuming that $k_c \approx \Omega/\gamma_b c$, then if

$$\frac{e\phi_o}{\gamma_b mc^2} < \frac{\pi E_a}{2B_o} \tag{96}$$

the resonance condition can be entirely avoided, and the largest amplitude excitation will occur at the first accelerating gap. As an example, for a magnetic field strength of 10 kG, we must have $\Delta\gamma/\gamma_b)_1 < 0.05 \pi$ (Miller, et al., 1983), for $E_a = 300$ kV/cm.

If Eq. (96) is satisfied, then the gaps are said to be designed in the long wavelength limit, i.e., the beam cyclotron wavelength is always greater than the axial gap length.

Contoured Magnetic Fields

In principle, the radial oscillations can be completely eliminated by eliminating radial force imbalances, i.e., by adding an additional term (or terms) to the right-hand side of Eq. (82) which exactly counter balance the variations in δE_r. For example, if the axial magnetic field strength B_z were approximately varied in the gap region, an approximate radial force balance might be achieved. The envelope analysis gives the simple solution (Miller, et al., 1983)

$$B_z(z) = \frac{2mc}{e} \left\{ \frac{C^2}{r_b^4} + \frac{2eI}{\gamma_b mcr_b^2} - \frac{\gamma_b e \delta E_r}{mr_b} \right\}^{1/2} , \tag{97}$$

where C is a constant which depends on the source configuration. (If the source is immersed, $C = P_\theta/m$, where $P_\theta \approx -\left(er_c^2 B_c/2mc\right)$, with P_θ representing the canonical angular momentum, and r_c and B_c denoting the cathode radius and the applied magnetic field strength at the cathode tip.)

Equation (97) is only an approximate solution to the required magnetic field variation because it was derived assuming that the magnetic field was axially uniform; a rigorous solution of this problem is presented by Genoni, et al., (1981). Nevertheless, Eq. (97) does indicate that this force balance method is restricted in applicability to beams of relatively small kinetic energy. Note that δE_r variations are magnified by γ_b. Hence, this method will generally work provided that the beam cyclotron wavelength is typically shorter than or comparable to the axial scale length for changes in the radial force.

The correctness of this "magnetic tailoring" approach has been
verified by numerically simulating the beam transport through a single
accelerating gap and through multiple accelerating gaps (Fig. 14).
However, for beam kinetic energies in excess of nominally 10 MeV, the
required rapid variation in axial magnetic field strength cannot be
obtained with practical solenoidal field coil designs. Accordingly, we
have been investigating alternate gap geometries that substantially
reduce the electric field variations.

Radial Accelerating Gap (Miller, et al., 1983)

One alternate gap geometry that totally eliminates radial electric
field variation is shown in Fig. 15. In this structure, there is no
axial gap, and the voltage is applied across concentric cylinders. The
condition that eliminates radial electric field variations is that the
beam must not cross any equipotential contours, i.e., the beam is not
accelerated in the gap region. Far from the gap region, the radius of
the outer wall is slowly decreased such that the beam is gradually
accelerated over an axial length that is much greater than a beam
cyclotron wavelength.

Let $\phi_1 = -(2I/c) \ln (R_1/r_b)$ represent the space-charge potential
depression between the beam and the wall of radius R_1. Far from the
accelerating gap, $\phi_2 = -(2I/c) \ln (R_2/r_b)$ is the potential depression

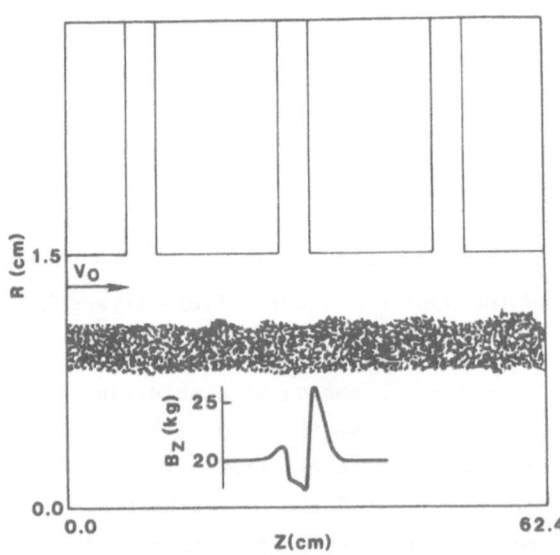

Fig. 14. Particle simulation of radial oscillation suppression
by contoured magnetic field profiles. A 50 kA beam
from a 4 MV diode traverses three gaps (each with L =
4 cm, ϕ_0 = 3 MV). In each gap, the $B_z(z)$ is as shown;
the field is uniform (20 kG) between gaps.

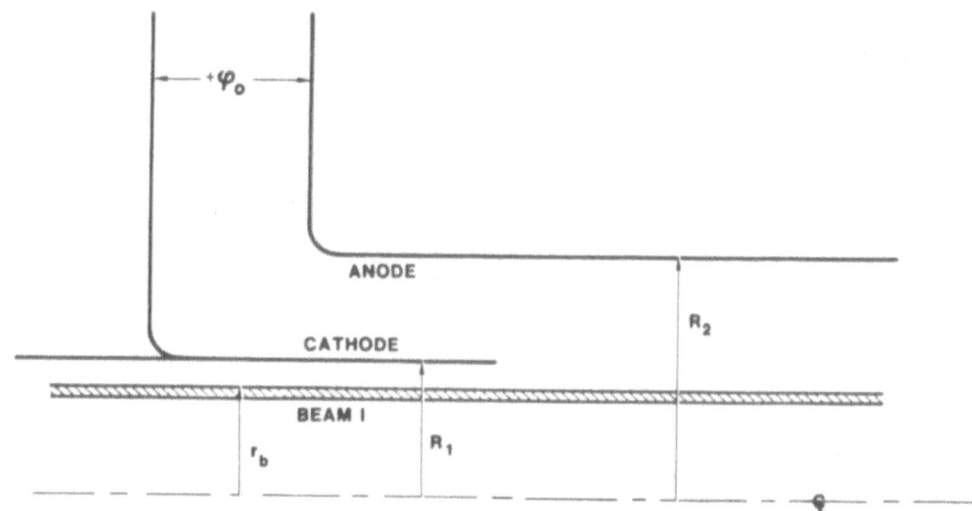

Fig. 15. Schematic geometry of a radial accelerating gap.

between the beam and the wall of radius R_2. If a voltage ϕ_o equal to

$$\phi_o = \phi_1 - \phi_2 = \frac{2I}{c} \ln \frac{R_2}{R_1} \, , \tag{98}$$

is applied, then the beam kinetic energy will not change on crossing the
gap, and there will be no radial electric field variations.

From Eq. (98), it follows that the maximum field stress in the
concentric cylinder gap region is given by

$$\left| E_{max} \right| = \frac{2I}{cR_1} \, . \tag{99}$$

Hence, for a fixed beam current, the critical electric field for electron
emission establishes the minimum radius R_1, and Eq. (97) determines the
permissible values of the applied gap potential and the outer radius R_2.
As an example, suppose that $\left| E_{max} \right|$ = 300 kV/cm = 10^3 statvolts/cm. Then

$$\phi_o(MV) = 0.06 \ I(kA) \ln \frac{5R_2(cm)}{I(kA)} \, . \tag{100}$$

Equation (99) is graphed in Fig. 16 as a function of beam current for
various values of R_2. For a 25 kA beam and R_1 = 5 cm, ϕ_o = 1.04 MV for
R_2 = 10 cm. The operating range for this gap design is quite restrictive
and requires rather large diameter drift tubes to achieve accelerating
voltages of 1 MeV or greater for beam currents in excess of ~ 10 kA. A
numerical simulation of this type of gap is presented in Fig. 17 (Poukey,
1984).

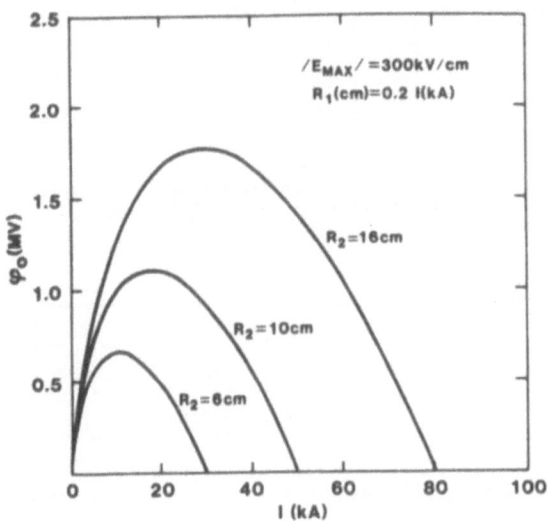

Fig. 16. Allowable voltage and current parameters for the
radial gap of Fig. 15 assuming $|E_{max}|$ = 300 kV/cm.

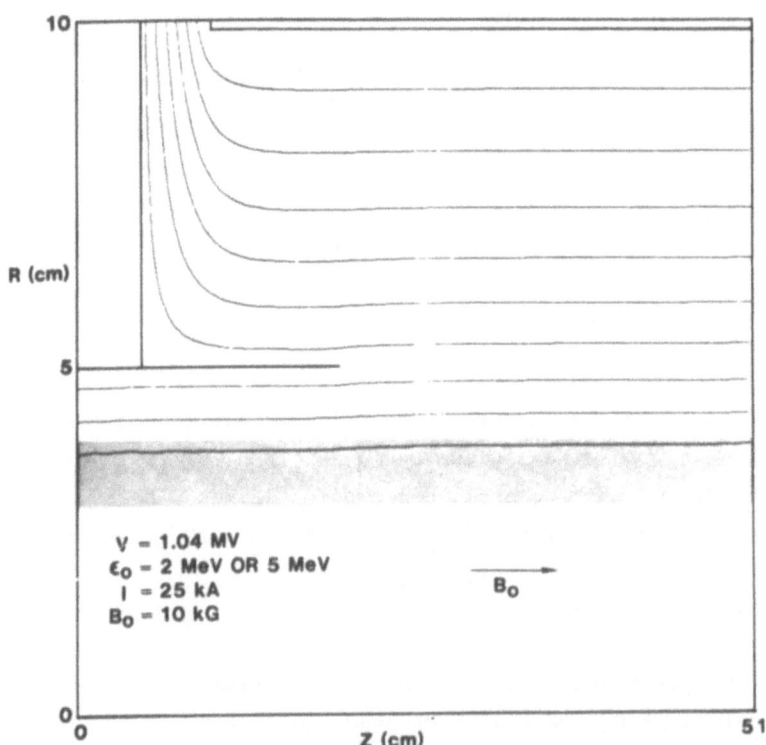

Fig. 17. A particle simulation of the radial gap geometry of
Fig. 15. The solid lines show the equipotentials
in the presence of a 25 kA beam. B_z = 10 kG, ϕ_0 =
1.04; the same result was obtained for both 2 MeV
and 5 MeV beams. (Gapless Gap Equipotentials W/Beam).

INTENSE ELECTRON BEAM INSTABILITIES

In the preceding sections the equilibria of intense relativistic beams were considered; now, attention will be focused on the stability of small amplitude oscillations about various equilibrium configurations. The mean azimuthal motion of the electron beam is assumed to be non-relativistic with

$$\frac{r^2 \omega_\theta^2(r)}{c^2} = \beta_\theta^2(r) \ll \beta_o^2 \tag{101}$$

where the unperturbed axial velocity $v_z = \beta_z c$ is assumed to be independent of radius. In this case the axial diamagnetic field contribution can be neglected and the radial force balance is approximated by

$$\gamma_o \omega_\theta^2(r) + \frac{1}{r^2} \left(1 - \beta_o^2\right) \int_0^r dr' r' \omega_p^2(r') - \omega_\theta(r)\Omega = 0 . \tag{102}$$

To determine the stability of the various equilibrium configurations, the macroscopic fluid and field quantities are expressed as the sum of their equilibrium values plus a perturbation, i.e.,

$$n(\underline{r},t) = n_o(r) + \delta n(\underline{r},t)$$

$$\underline{v}(\underline{r},t) = v_{\theta_o}(r)\hat{e}_\theta + v_{zo}\hat{e}_z + \delta\underline{v}(\underline{r},t)$$

$$\underline{E}(\underline{r},t) = E_{r_o}(r)\hat{e}_r + \delta\underline{E}(\underline{r},t)$$

$$\underline{B}(\underline{r},t) = B_o\hat{e}_z + \delta\underline{B}(\underline{r},t) .$$

In the electrostatic approximation the perturbed magnetic field, $\delta\underline{B}(\underline{r},t)$, is assumed to remain negligibly small so that the perturbed electric field, $\delta\underline{E}(\underline{r},t)$, can be expressed as the gradient of a scalar potential,

$$\delta\underline{E}(\underline{r},t) = - \nabla\phi(\underline{r},t) \tag{103}$$

A linearization of the macroscopic fluid – Poisson equations yields,

$$\frac{\partial}{\partial t} \delta n(\underline{r},t) + \nabla \cdot \left\{ n_o(r) \delta\underline{v}(\underline{r},t) + \delta n(\underline{r},t) \times \right.$$

$$\left. \left[v_{\theta_o}(r)\hat{e}_\theta + v_{zo}\hat{e}_z \right] \right\} = 0 \tag{104}$$

$$\nabla^2 \delta\phi(\underline{r},t) = 4\pi e \, \delta n(\underline{r},t) \tag{105}$$

$$\frac{\partial}{\partial t} \delta\underline{p} + \underline{v}_o \cdot \nabla\delta\underline{p} + \delta\underline{v} \cdot \nabla\underline{p}_o = -e\left(-\nabla\delta\phi + \frac{1}{c} \, \delta\underline{v} \times \underline{B}_o\right) . \tag{106}$$

337

Since the axial motion of the beam may be relativistic, the two-mass approximation is assumed for the perturbed momentum according to

$$\delta p_r = \gamma_o m \, \delta v_r$$

$$\delta p_\theta = \gamma_o m \, \delta v_\theta$$

$$\delta p_z = \gamma_o^3 m \, \delta v_z \ .$$

As a further simplification, all perturbed quantities are assumed to have harmonic dependence in time, the azimuthal coordinate, and the axial coordinate; i.e., $\delta\psi = \delta\psi(r)e^{-i\omega t}e^{i\ell\theta}e^{ik_z z}$, where ω is the characteristic frequency of oscillations, ℓ is the azimuthal harmonic number, and k_z is the axial wave number. With this assumption, the linearized equations become (Davidson, 1974)

$$-i(\omega - k_z v_{zo} - \ell\omega_{\theta_o})\,\delta n + \left[\frac{1}{r}\frac{\partial}{\partial r}(r\,\delta v_r) + \frac{i\ell\,\delta v_\theta}{r}\right.$$

$$\left. + ik_z \delta v_z\right]n_o = 0$$

$$\frac{1}{r}\frac{\partial}{\partial r}r\frac{\partial}{\partial r}\delta\phi - \frac{\ell^2}{r^2}\delta\phi - k_z^2\delta\phi = 4\pi e\,\delta n$$

$$-i(\omega - k_z v_{zo} - \ell\omega_{\theta_o})\,\delta v_r + \left(\frac{\Omega}{\gamma_o} - 2\omega_{\theta o}\right)\delta v_\theta = -\frac{e}{\gamma_o m}\frac{\partial}{\partial r}\delta\phi$$

$$-i(\omega - k_z v_{zo} - \ell\omega_{\theta_o})\,\delta v_\theta + \left[-\frac{\Omega}{\gamma_o} + \frac{1}{r}\frac{\partial}{\partial r}(r^2\omega_{\theta_o})\right]\delta v_r$$

$$= \frac{e}{\gamma_o m}\frac{i\,\delta\phi}{r}$$

$$-i(\omega - k_z v_{zo} - \ell\omega_{\theta_o})\,\delta v_z = \frac{e}{\gamma_o^3 m}ik_z\delta\phi \ .$$

In these equations $\Omega = eB_o/mc$, and the equilibrium angular velocity profile, $\omega_{\theta_o} = v_{\theta_o}/r$, is related to the equilibrium density profile by the radial force equation. Solving for the density and fluid velocity perturbations in terms of the perturbed potential, Poisson's equation can be written as

$$\frac{1}{r}\frac{\partial}{\partial r}\left[r\left(1 - \frac{\omega_{po}^2}{\gamma_o v^2}\right)\frac{\partial}{\partial r}\delta\phi\right] - \frac{\ell^2}{r^2}\left(1 - \frac{\omega_{po}^2}{\gamma_o v^2}\right)\delta\phi \tag{107}$$

$$- k_z^2 \left[1 - \frac{\omega_{po}^2 / \gamma_o^3}{(\omega - k_z v_{zo} - \ell \omega_{\theta_o})^2} \right] \delta\phi$$

$$= - \frac{\ell \, \delta\phi \ (\partial/\partial r) \left[(\omega_{po}^2 / v^2) \left(- \Omega/\gamma_o + 2\omega_{\theta_o} \right) \right]}{r \left(\omega - k_z v_{zo} - \omega_{\theta_o} \right)}$$

where

$$v^2(r) = \left(\omega - k_z v_{z_o} - \ell \omega_{\theta_o} \right)^2 - \left(-\frac{\Omega}{\gamma o} + 2\omega_{\theta_o} \right) \left[\frac{\Omega}{\gamma_o} + \frac{1}{r} \frac{\partial}{\partial r} \left(r^2 \omega_{\theta_o} \right) \right]$$

and $\omega_{po}^2 (r) = 4\pi n_o(r) e^2/m$. The procedure for using Eq. (107) is to solve for $\delta\phi$ and ω as an eigenvalue problem.

Rigid Rotor Stability

Analytical solutions for Eq. (107) are tractable for only a few simple cases. As a first example of the use of Eq. (107) consider the case of constant beam density, corresponding to the rigid rotor relativistic electron beam equilibrium. It is assumed that the beam radius is equal to the radius of a perfectly conducting wall, i.e., $[\delta\phi]_{r = r_b} = 0$.

Since the equilibrium rotation velocity $\omega_\theta (r)$, given by Eq. (51)

$$\omega_\theta (r) = \omega_\theta^\pm = \frac{\Omega}{2\gamma_o} \left[1 \pm \left(1 - \frac{2\omega_{po}^2}{\gamma_o \Omega^2} \right)^{1/2} \right], \quad 0 < r < r_b$$

is independent of radius, it follows that $v^2(r)$ is given by

$$v^2 = (\omega - k_z v_{z_o} - \ell \omega_{\theta_o})^2 - (-\Omega/\gamma_o + 2\omega_{\theta_o})^2 \tag{108}$$

which is also independent of radius. Under these conditions Poisson's equation reduces to

$$\frac{1}{r} \frac{\partial}{\partial r} r \frac{\partial}{\partial r} \delta\phi - \frac{\ell^2}{r^2} \delta\phi + k_\perp^2 \delta\phi = 0, \qquad 0 < r < r_b \tag{109}$$

where

$$k_\perp^2 = -k_z^2 \frac{\left[1 - \dfrac{\omega_{po}^2 / \gamma_o^3}{(\omega - k_z v_{zo} - \ell \omega_{\theta_o})^2} \right]}{1 - \omega_{po}^2 / \gamma_o v^2} \tag{110}$$

Equation (109) is recognized as a form of Bessel's equation. The solution which remains finite at the origin is

$$\delta\phi = AJ_\ell(k_\perp r) \tag{111}$$

where A is a constant and J_ℓ is the Bessel function of the first kind of order ℓ. Enforcing the vanishing of the perturbed potential at the conducting wall leads to the condition

$$J_\ell(k_\perp r_b) = 0 \tag{112}$$

from which it follows that

$$k_\perp^2 r_b^2 = p_{\ell m}^2; \qquad m = 1, 2, \ldots \tag{113}$$

where $p_{\ell m}$ is the mth zero of $J_\ell(x) = 0$, and $k_\perp$ is recognized as the effective perpendicular wave number quantized by the finite radial geometry.

Equation (113) is the dispersion relation for electrostatic waves in the electron-beam-filled guide tube. It relates the characteristic oscillation frequency ω to the azimuthal harmonic numbers ℓ, the axial wave vector k_z, and the properties of the rigid rotor equilibrium configuration.

Equation (113) may be expressed in the equivalent form

$$1 - \frac{k_z^2}{k^2} \frac{\omega_{po}^2/\gamma_o^3}{(\omega - k_z v_{zo} - \ell\omega_{\theta_o})^2} - \frac{k_\perp^2}{k^2} \frac{\omega_{po}^2/\gamma_o}{(\omega - k_z v_{zo} - \ell\omega_{\theta_o})^2 - \omega_v^2} = 0 \tag{114}$$

where $k^2 = k_z^2 + k_\perp^2$ and ω_v is the vortex frequency defined by

$$\omega_v = -(-\Omega_o/\gamma_o + 2\omega_{\theta_o}) . \tag{115}$$

The solution to Eq. (114) can be written as (Davidson, 1974)

$$\left(\omega - k_z v_{zo} - \ell\omega_{\theta_o}\right)^2 = \left[\omega_v^2 + \left(\omega_{po}^2/\gamma_o\right)\left[\frac{k_z^2/\gamma_o^2 + k_\perp^2}{k^2}\right]\right]$$

$$\times \left\{1 \pm \left[1 - \frac{4\left(k_z^2/k^2\right)\omega_v^2\omega_{po}^2/\gamma_o^3}{\omega_v^2 + \left(\omega_{po}^2/\gamma_o\right)\left[\frac{k_z^2/\gamma_o^2 + k_\perp^2}{k^2}\right]}\right]^{1/2}\right\} . \tag{116}$$

For the analysis to be valid, the value of the beam density must be less than the maximum density for which equilibrium solutions exist. An upper bound on the density can be determined by requiring that the radical in Eq. (116) be real; the condition that obtains (assuming the slow rotation mode) is also

$$2\omega_{po}^2/\gamma_o\Omega^2 \leq 1 \ .$$

Provided this criterion is satisfied, $\mathrm{Im}\,\omega_{\theta_o} = 0$ (i.e., ω_{θ_o} contains no imaginary component). Hence, this condition implies the existence of undamped stable oscillations for the four characteristic eigenmodes of the dispersion relation. An examination of the character of the eigenmodes is simplified by restricting the analysis of the axisymmetric ($\ell = 0$) modes and assuming the slow rotation frequency. In this case the dispersion relation can be approximately factored (Drummond, et al., 1976) and the eigenmodes can be identified as follows:

(1) Two Doppler-shifted plasma modes explicitly displaying the longitudinal electron mass effects:

$$\omega = k_z v_{zo} \pm \left(\frac{k_z^2 \omega_{po}^2}{k^2 \gamma_o^3}\right)^{1/2} \ . \tag{117}$$

(2) Two Doppler-shifted cyclotron modes:

$$\omega = k_z v_{zo} \pm \left(\frac{\Omega^2}{\gamma_o^2}\right)^{1/2} \ . \tag{118}$$

The plasma modes correspond to longitudinal bunching of the beam space charge, while the cyclotron waves are transverse modes corresponding to a traveling constriction in the beam (Fig. 18).

Before leaving this section, it should be emphasized that the relatively simple form of the dispersion relation, Eq. (115), could only

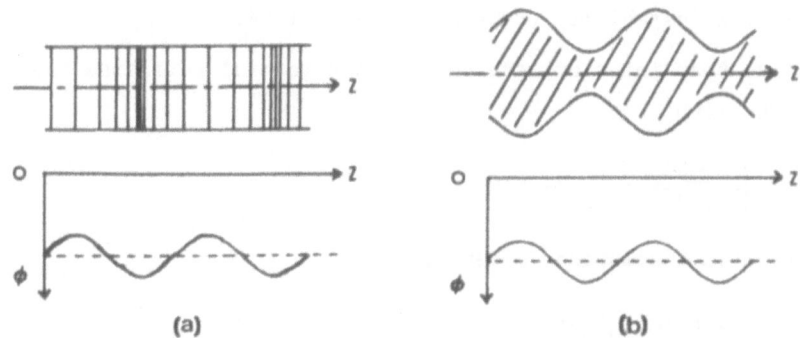

(a)   (b)

Fig. 18. Physical representation of the space charge and cyclotron waves, and attendant potential variations. (a) Space charge wave; (b) cyclotron wave.

be obtained by considering the radially homogeneous rigid rotor equilibrium. In general, radial inhomogeneities, particularly the variation of
$\gamma(r)$ across the beam, strongly influences the structure of the wave
spectra. Specific numerical solutions of the linear theory for more
realistic solid beam equilibria indicate several significant differences
(Godfrey, 1979a).

Stability of the Hollow Beam Equilibrium (Diocotron Instability)

For low-frequency perturbations of a tenuous beam in the long-wave
length limit ($k_z \rightarrow 0$), Eq. (107) can be expressed in the approximate form

$$\left(\frac{1}{r}\frac{\partial}{\partial r}\, r\, \frac{\partial}{\partial r}\right)\delta\phi - \frac{\ell^2}{r^2}\,\delta\phi = -\frac{\ell}{r}\,\frac{\delta\phi}{\left(\omega - \ell\,\omega_{\theta_o}\right)}\,\frac{\gamma_o}{\Omega}\,\frac{\partial}{\partial r}\,\omega_{po}^2 \, . \tag{119}$$

In this case, it is straightforward to show that the equilibrium
configuration is stable provided $(\partial/\partial r)(\omega_{po}^2) \leq 0$; however, for the hollow
beam equilibrium this criterion is obviously not satisfied since

$$\frac{\partial}{\partial r}\left[\omega_p^2(r)\right] = \omega_{po}^2\left[\delta(r - r_o) - \delta(r - r_b)\right] \, . \tag{120}$$

Although induced oscillations in the body of the hollow beam are
stable, with Eq. (120) the right-hand side of Eq. (119) corresponds to
perturbations in charge density on the inner and outer edges of the beam.
Because of the angular rotation shear of the hollow beam, the surface
waves propagate relative to one another, and the motion of the charge
perturbation of one wave can be modified by the electrostatic fields of
the other. Under suitable conditions, this interaction can become
synchronized so as to produce a single exponentially growing wave mode.
This type of instability has been termed "diocotron" to indicate that the
charge perturbations on the two surfaces must slip parallel to each other
to create the instability (Buneman, et al., 1966).

In the beam interior, the perturbed potential satisfies the homogeneous equation

$$\frac{1}{r}\frac{\partial}{\partial r}\left(r\,\frac{\partial}{\partial r}\right)\delta\phi - \frac{\ell^2}{r^2}\,\delta\phi = 0 \tag{121}$$

with appropriate solutions given by

$$\delta\phi_I = (A + Br_o^{-2\ell})r^\ell \, , \qquad 0 < r < r_o$$

$$\delta\phi_{II} = (Ar^\ell + Br^{-\ell}), \qquad r_o < r < r_b$$

342

$$\delta\phi_{III} = \left(Ar_b^{2\ell} + B\right)\left(\frac{R^{2\ell} - r^{2\ell}}{R^{2\ell} - r_b^{2\ell}}\right)r^{-\ell}, \quad r_b < r < R \tag{122}$$

where the condition of continuity of $\delta\phi$ across the region boundaries has been enforced. Introducing the jump conditions at the discontinuities (to eliminate the constants A and B)

$$r_o\left[\frac{\partial}{\partial r}\,\delta\phi_{II}\right]_{r\,=\,r_o} \quad -r_o\left[\frac{\partial}{\partial r}\,\delta\phi_I\right]_{r\,=\,r_o} \tag{123}$$

$$= -\ell\,[\,\delta\phi]_{r\,=\,r_o}\left\{\frac{\gamma_o\omega_{po}^2}{\Omega\left[\omega - \ell\,\omega_\Theta(r_o)\right]}\right\}$$

$$r_b\left[\frac{\partial}{\partial r}\,\delta\phi_{III}\right]_{r\,=\,r_b} \quad -r_b\left[\frac{\partial}{\partial r}\,\delta\phi_{II}\right]_{r\,=\,r_b} \tag{124}$$

$$= \ell\,\left[\,\delta\phi\right]_{r\,=\,r_b}\left[\frac{\gamma_o\omega_{po}^2}{\Omega\left[\omega - \ell\,\omega_\Theta(r_b)\right]}\right]$$

leads to an eigenvalue equation for ω given by

$$(\omega/\omega_o)^2 - b(\omega/\omega_o) + c = 0 \tag{125}$$

where

$$b = \ell\left(1 - \frac{r_o^2}{r_b^2}\right) + \left(\frac{r_b^{2\ell}}{R^{2\ell}} - \frac{r_o^{2\ell}}{R^{2\ell}}\right)$$

$$c = \ell\left(1 - \frac{r_o^2}{r_b^2}\right)\left(1 - \frac{r_o^{2\ell}}{R^{2\ell}}\right) - \left(1 - \frac{r_o^{2\ell}}{r_b^{2\ell}}\right)\left(1 - \frac{r_b^{2\ell}}{R^{2\ell}}\right)$$

$$\omega_o = \omega_{po}^2/\left(2\gamma_o^2\Omega\right).$$

If $b^2 \geq 4c$, then $\text{Im}(\omega) = 0$, and the equilibrium configuration is stable; if $b^2 < 4c$, the system is unstable with the characteristic growth rate being given by

$$\omega_i = \frac{\omega_o}{2}\,(4c - b^2)^{1/2}. \tag{126}$$

Employing these relations, the stability criterion can be explicitly expressed as

$$[-\ell(1 - \beta^2) + 2 - \alpha^{2\ell} - (\alpha\beta)^{2\ell}]^2 \geq 4\beta^{2\ell}\,(1 - \alpha^{2\ell})^2 \tag{127}$$

where $\alpha = r_b/R$ and $\beta = r_o/r_b$. It is seen by inspection that choosing $\beta = 0$ (solid beam) or $\alpha = 1$ (outer edge of the beam extending to the conducting wall) trivially satisfies the stability criterion. Physically, these cases correspond to the elimination of one of the beam surfaces which supports wave propagation.

RESISTIVE WALL INSTABILITIES

The previous section indicated that an unneutralized, solid electron beam propagating in an infinitely conducting drift tube was stable to small perturbations. Physically, the beam induces image charges and image currents of opposite sign on the surface of the conducting wall. The attractive force due to the image charge is canceled to order γ^{-2} by the magnetostatic repulsive force due to the image current, with the residual countered by the solenoidal guide field. If the drift tube is not perfectly conducting, however, the slow decay of the image currents can lead to the growth of resistive wall instabilities (Lawson, 1978). To illustrate the instability mechanism consider a step displacement of the beam from the drift tube axis. The restoring force due to the resulting asymmetric image current rises very rapidly [in a time $\sim (4\pi\sigma)^{-1}$] to the same value that it would have for a perfect conductor, and then decays, on a slower, magnetic diffusion time, $4\pi\sigma a^2/c^2$. At the end of the step displacement, a decaying wake is produced which exerts a force on subsequent particles. Depending on the wave phase velocity, sinusoidal beam displacements can grow as a result of this resistive drag. Eqs. (117) and (118) indicate that the phase velocities $(v_{ph} = \omega/k)$ associated with the space charge and cyclotron wave are given by

$$v_{ph} = v_{zo} \pm \left(\frac{\omega_{po}^2}{k^2 \gamma_o^3} \right) \qquad \text{(space charge)}$$

$$v_{ph} = v_{zo} \pm \left(\frac{\Omega^2}{k_z^2 \gamma_o^2} \right)^{1/2} \qquad \text{(cyclotron)} . \qquad (128)$$

Hence, there are both "slow" space charge and cyclotron waves in which the wave phase velocity is less than the velocity of the particles, and "fast" waves for which the phase velocity exceeds the particle velocity. As shown in Fig. 19, for a fast wave an observer moving with the beam velocity sees both wave and particles moving inward together; for a slow wave, however, the wave and particles move in opposite directions. Therefore, for a slow wave and a resistive wall, the wake fields can accelerate the particles outward while the particles are moving outward,

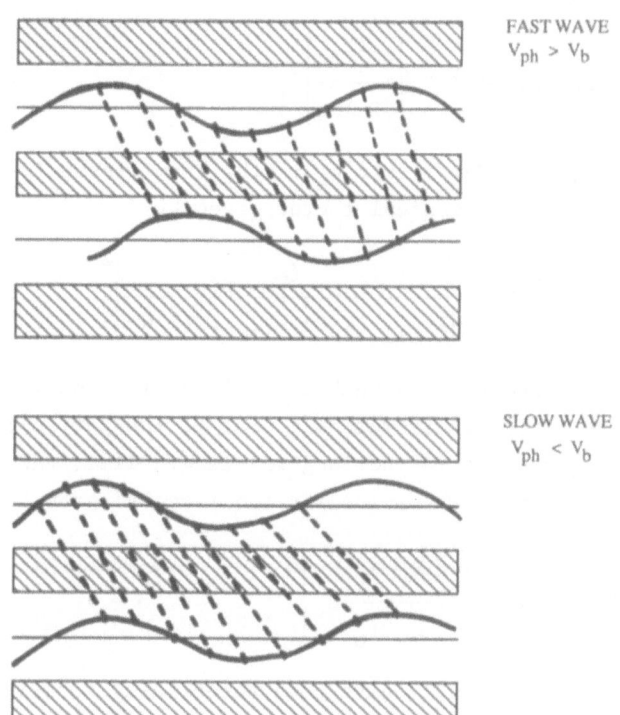

FAST WAVE
$V_{ph} > V_b$

SLOW WAVE
$V_{ph} < V_b$

Fig. 19. Diagram illustrating the motion of particles and
waves in fast and slow transverse waves. For a
fast wave, an observer moving with the beam
velocity sees both wave and particles moving
inward together. For a slow wave, the wave and
particles appear to move in opposite directions.

and the wave grows. Correspondingly, dissipation produces damping of
fast waves. Because slow waves can grow even though the system energy is
decreasing due to dissipation in the resistive walls, they are often
termed negative energy waves.

The first studies of the resistive wall instability were performed
for low current beams accelerated to high energies in circular accelerators (Laslett, et al., 1965); although the currents were low and the
characteristic growth lengths were long, the effective distance traversed
by the beam was also very long. In high current linear induction accelerators, the case is somewhat different; the effective beam transport
length in the accelerator is typically short (tens of meters), but the
high currents (tens of kiloamperes) can drastically decrease the characteristic growth length of the instability. For this case, Godfrey (1982)
has obtained the dispersion relation of the slow cyclotron instability
for an annular beam of mean radius r_b and thickness δ propagating in a
drift tube of radius R and conductivity σ as

$$\omega - \omega_0 = \pm(4\pi\sigma i \ \omega)^{-1/2} \left(\frac{2\omega_p^2 \ \delta c}{\Omega_c Rr_b}\right) f_\pm \tag{129}$$

with $k_z v_b = \Omega_c/\gamma$. In Eq. (129), ω_p is the beam plasma frequency, $\Omega_c = eB_z/mc$, and $f_\pm$ is a geometrical factor given by

$$f_\pm = \left(\ell^2 + k_z^2 \ R^2\right) \left[\frac{I_{\ell \pm 1} (k_z \ r_b)}{I_{\ell+1} (k_z R) + I_{\ell-1} (k_z R)}\right]^2 \tag{130}$$

where ℓ is the azimuthal wave number and I denotes the modified Bessel function. Waves of helicity matching that of the particles grow most rapidly. For such waves in the long wavelength limit ($kR \ll 1$) and $\omega_0 \approx 0$, an approximate expression for the temporal growth rate is

$$\Gamma_R \approx e^{i\pi/6} \ (4\pi\sigma)^{-1/3} \left\{\frac{2\omega_p^2 \ \delta c}{\Omega_c \ Rr_b} f_\pm\right\}^{2/3} \ . \tag{131}$$

The predictions of Eq. (131) have been essentially verified with the computational results of the linear theory code GRADR (Godfrey and Newberger, 1979) for several sets of parameter values relevant to high current linacs: beam kinetic energies and currents in the range of E = 4–20 MeV and I = 10–100 kA, guide fields in the range of 10–20 kG, and dimensions of order R = 2.5 cm, r_b = 2 cm, and δ = 0.2 cm. The agreement between the analytic theory and GRADR is typically better than 5%. One important result, the variation of $|\ell|$ = 1 wave growth versus beam energy is shown in Fig. 20. Also depicted in Fig. 20 is the low frequency, slow spacecharge wave resistive wall growth rate. For this mode, instability occurs at very long wavelengths, $k_z r_b \approx 0$, so helicity is not important. At low energies, this instability is strongly affected by azimuthal velocity shear and is often masked by the diocotron instability for high azimuthal wavenumbers.

IMAGE DISPLACEMENT INSTABILITY (Wood, 1970, and Neil, 1978)

In the previous section, we indicated that an uncentered beam in a uniform drift tube is attracted to the nearest point in the wall by its electrostatic image charge. Similarly, it is repelled by its image current, with the two forces canceling to order γ^{-2}. At an accelerating gap, however, the image current is "displaced" to large radius, while the image charge becomes concentrated on the gap and drift tube walls nearest the beam. The resulting loss of image force cancellation can give rise to an abrupt transverse force on the beam in the direction of its initial offset. Just as symmetric radial force imbalances can produce large radial oscillations of the beam envelope, this transverse force can excite zero-frequency transverse oscillations as the beam travels to the next gap. Depending on the oscillation phase at the next gap further growth can occur.

Assuming that each axial beam segment displaces rigidly in the transverse direction, and the displacement is small compared to the drift tube radius R, then the equation of transverse motion for the beam centroid in a highly conducting drift tube is (Adler, et al., 1983b)

$$\frac{d}{dt} \gamma \frac{d}{dt} \xi = i\Omega_c \frac{d}{dt} \xi + \frac{2eI}{\gamma^2 mcR^2} \xi \tag{132}$$

where $\xi = x + iy$ is the complex sum of transverse displacements.

Under equilibrium conditions, the magnetic focusing force dominates the residual image forces, and the spatial variation of ξ between gaps is approximately

$$\xi = \xi_1 + \xi_2 e^{ik_o z} \tag{133}$$

with $k_o = \Omega_c/\gamma c$. In words, as the beam travels from one gap to the next, the (complex amplitude of the low frequency space charge mode is essentially constant, while the amplitude of the high frequency cyclotron mode is multiplied by $\exp(i\Omega_c L/\gamma c)$, where L is the separation between gaps.

In crossing a thin accelerating gap of width $\omega \ll \gamma c/\Omega_c$ the beam will experience an impulsive force $F\xi$ which is proportional to the product of the displacement from the axis and the space charge image force, which changes the beam transverse velocity by $iF\xi/\gamma k_o c^2$, but leaves the displacement constant.

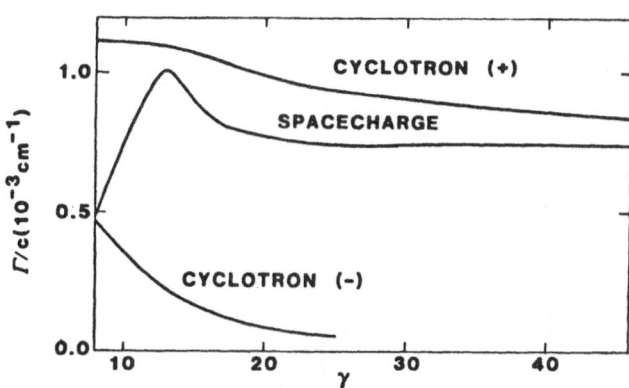

Fig. 20. Peak growth rate versus beam energy for low frequency $\ell = 1$ resistive wall instabilities.

347

The transverse motion of the beam in the drift tube after passing Through the gap can then be related to the motion prior to the gap by using Eq. (133) and the matching condition in the gap. Assuming for the moment that γ is constant, the equations for the mode amplitudes can be recast into matrix format as

$$
\begin{pmatrix} \xi_2 \\ \\ \xi_1 \end{pmatrix}_+ = e^{ik_o L/2} \begin{pmatrix} \left(1 - \dfrac{iF}{\gamma k_o c^2}\right) e^{ik_o L/2} & -\dfrac{iF}{\gamma k_o c^2} \\ \\ \dfrac{iF}{\gamma k_o c^2} & \left(1 + \dfrac{iF}{\gamma k_o c^2}\right) e^{-ik_o L/2} \end{pmatrix} \begin{pmatrix} \xi_2 \\ \\ \xi_1 \end{pmatrix}_- . \quad (134)
$$

The eigenvalues of Eq. (134) describe the growth of the gap-induced image displacement instability: these are

$$
\lambda = \psi \pm i(1 - \psi^2)^{1/2} \qquad (135)
$$

$$
\psi = \cos \frac{k_o L}{2} + \frac{F}{k_o c^2 \gamma} \sin \frac{k_o L}{2} \qquad (136)
$$

where $\dfrac{F}{\gamma k_o c^2} = \dfrac{2eI}{\gamma m^2 R^2 \Omega_c} \dfrac{fw}{}$. The image displacement force constant f is the

ratio of the transverse force in the gap to the transverse electric force in the drift tube; it is typically of order 1/2.

Returning to Eq. (135), growth corresponds to $|\lambda| > 1$. Since $(F/k_o c^2 \gamma)$ is purely real, the condition for stability becomes $\psi \le 1$. The dependence of ψ on the phase advance per gap $(k_o L/2)$ is presented in Fig. 21 for several values of the parameter F. It is apparent that $F/k_o c^2 \gamma < 1/2$ is required to assure minimal growth over several periodically spaced gaps with γ constant.

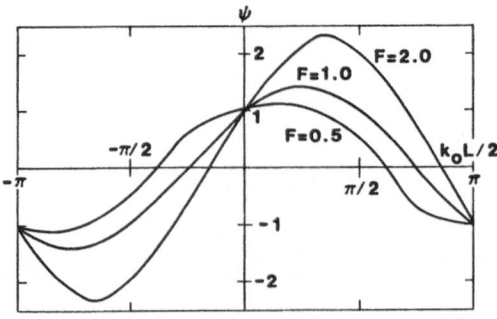

Fig. 21. Image displacement instability growth parameter ψ as a function of the phase advance per gap.

To take into account the variation in γ, the amplification after several gaps is determined by the eigenvalue of the matrix formed by multiplying the matrices for each gap. In general, the eigenvalue of the product is much smaller than the product of the eigenvalues; i.e., beam acceleration decreases image-displacement instability growth.

For high current linacs, the thin-gap approximation is not always well satisfied. To the extent that $F/k_o c^2 \gamma$ is less than $k_o \ell /2$, increasing gap width weakens the instability. Accurate treatments of this important case are now available in the literature (Adler, et al., 1983b).

BEAM BREAK-UP (BBU) INSTABILITY (Panofsky and Bander, 1968)

In addition to the zero-frequency image displacement instability, high current induction linacs with periodic accelerating cavities can be susceptible to a microwave instability termed transverse beam break-up. Physically, the instability arises from the coupling between a beam oscillating off-axis, and accelerating cavity modes with a transverse magnetic field on axis (T_{1mo} modes, for example). If the frequency of the beam oscillation occurs at a natural resonance of the cavity, then cavity mode energy will increase and be in temporal phase with the oscillations, thereby causing the oscillations to increase in amplitude during the pulse duration. Moreover, if every cavity is identical, the peak amplitude of the oscillations will also grow from cavity to cavity.

The normal modes of the accelerating cavities can be conveniently described by damped harmonic oscillator equations, driven by the off-axis beam current. For example,

$$\frac{d^2}{dt^2} A + \frac{\omega_n}{Q} \frac{d}{dt} A + \omega_n^2 = \alpha_1 \xi \tag{137}$$

where A is the axial vector potential of the mode, ω_n and Q are the mode frequency and quality factor, ξ is the transverse displacement of the beam centroid, and α is a coupling coefficient.

In turn, the beam transverse oscillation is driven by the transverse magnetic field of the accelerating cavity normal mode according to

$$\frac{d}{dt} \gamma \frac{d}{dt} \xi = \alpha_2 A \tag{138}$$

where α_2 is a second coupling coefficient. Hence, the physically significant quantity is the produce of the coupling coefficients $\alpha_1 \alpha_2$. This product is conveniently expressed in terms of the beam current I and and the scaled transverse impedance $\frac{Z\perp}{Q}$ of the particular cavity mode as (Godfrey, 1982)

349

$$\alpha_1 \, \alpha_2 = \omega_n^3 \, \frac{Z_\perp}{Q} \, \frac{Ie}{mc^2} \qquad\qquad (139)$$

Hence, the strength of the coupling between beam oscillations, cavity fields, and later beam oscillations depends on the beam current, cavity constants, and the oscillation frequency.

For a single cavity, coherent transverse oscillations can be selectively excited from transverse beam noise. What makes the BBU instability so dangerous is that subsequent identical cavities are then resonantly driven by coherent transverse oscillations. Unlike the case of image displacement, BBU always grows, apart from phase mixing.

In fact, the image displacement instability can be viewed as the low frequency limit of the beam breakup instability. Setting the time derivatives to zero in Eq. (137) and inserting the resulting expression for the vector potential into Eq. (138) yields

$$\frac{d}{dt} \gamma \frac{d}{dt} \xi = \frac{\omega_n}{Q} \frac{Z_\perp}{mc^2} \frac{eI}{mc^2} \xi \qquad\qquad (140)$$

This equation simply states that all gap modes tend to increase an initial transverse displacement. The combined effect is found by summing all the $\left(\dfrac{\omega_n \, Z_\perp}{Q}\right)$ for the various modes, which is of order ω/R^2.

Analogous to the treatment used in the case of image displacement, the thin gap approximation leads to a matrix equation for the evolution of the mode amplitudes with the matrix eigenvalues describing the growth of the gap-induced instabilities. The impulsive force F due to beam breakup can be formally obtained from the Laplace transform of Eq. (137). However, if it is assumed that the beam oscillates uniformly at $\omega_n \tau \gg Q$, then the Laplace transform variable can be replaced by $i\omega_n$. This procedure yields

$$\frac{F}{c^2 k \gamma} \approx \frac{iQ}{c} \frac{Z_\perp}{Q} \frac{\omega_n}{\Omega_c} \frac{eI}{mc^3} \, . \qquad\qquad (141)$$

Substituting Eq. (141) into Eq. (143) gives the amplification of beam breakup oscillations at the gap, in a manner similar to the case of image displacement. However, F is now purely imaginary, and $|\gamma| > 1$ for all $|F| > 0$ and $kL/2 \neq n\,\pi$, i.e., the oscillations always grow. Hence, for high current linacs, controlling beam breakup requires large focusing fields (Ω_c), and lossy cavities (low Q) with low transverse shunt impedance $(Z_\perp/Q)$. Also, the beam rise time τ_r should be long compared to the cavity field ringup time, Q/ω_n, to avoid shock excitation of the cavities.

RADLAC-II BEAM TRANSPORT ANALYSIS

To illustrate the use of the analyses presented in the preceding sections, we will now describe some important features of the design of the beam transport line of the RADLAC-II accelerator. The general goal of this device was to provide a high current (50-100 kA) electron beam at a nominal output kinetic energy of 20 MeV. Moreover, in order to satisfy a broad range of applications, including requirements for bremsstrahlung generation and atmospheric beam propagation, the equilibrium radius of the beam should be of the order of one centimeter or less after extraction from the solenoidal guide field of the accelerator.

Injector Design

The inherent simplicity and past successes (Miller, et al., 1980 and 1981) of foilless diode designs for injecting beams into vacuum regions led us to adopt this approach for the RADLAC-II injector. However, because this is an immersed source configuration, conservation of canonical angular momentum will cause the beam to rotate after extraction from the axial magnetic field, and the resulting emittance will limit the equilibrium beam radius according to (Mazarakis, 1986)

$$r_{eq} \approx \frac{\Omega_c r_c^2}{2} \left(\frac{mc}{\gamma \beta e I_n} \right)^{1/2} \tag{142}$$

where r_c is the cathode shank radius, and I_n denotes the net current after extraction from the accelerator. Depending on the density of the plasma or gas background, I_n can be substantially lower than the beam current. Based on numerical simulations of beam extraction into air at atmospheric density (Mazarakis, 1986), $I_n \approx 20$ kA was assumed for our design calculations. Hence, Eq. (142) indicates that achieving an equilibrium beam radius of approximately one centimeter would require a nominal cathode radius of ~ one centimeter and an axial magnetic field strength of ~ 20 kG. With these general constraints, our previous analysis can be used to establish the injector voltage and cathode-anode geometry which would avoid virtual cathode formation and large radial oscillations in the injector region. A nominal foilless diode design in shown in Fig. 22.

To achieve beam currents in excess of ~ 50 kA with these relatively small dimensions requires relatively high diode voltages (> 4 MeV). A systematic study of the diode performance and beam vacuum propagation was undertaken using the 4 MeV IBEX electron beam generator (Mazarakis, et

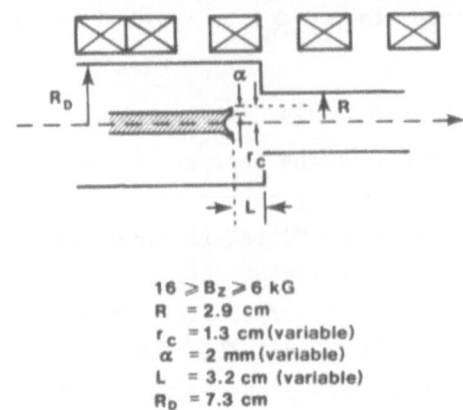

16 $\geqslant B_z \geqslant$ 6 kG
R   = 2.9 cm
r_C = 1.3 cm (variable)
α = 2 mm (variable)
L   = 3.2 cm (variable)
R_D = 7.3 cm

Fig. 22.   Schematic diagram of IBEX foilless diode injector.

al., 1984). The free variables of the experiments were anode-cathode voltage, axial magnetic field, cathode shank radius and anode-cathode spacing. Extreme care was taken to precisely align the diode axis with the magnetic field axis, as well as the vacuum pipe axis.

In all cases, the foilless diode produced a low emittance, thin annular beam with practically no radial oscillations, as indicated by beam damage patterns on a set of brass witness plates which scanned an entire cyclotron wavelength. Figure 23 shows the measured beam outer diameter ($2r_b$) at various distances from the cathode. The points lie on a straight line indicating no radial oscillations. The vacuum beam transmission to the end of the drift pipe was 100%. No diocotron instabilities were observed in accord with the growth length estimates.

At higher magnetic field strengths (~ 14 kG), a very fine structure composed of 30 to 40 filaments symmetrically located inside the beam annulus was visible on the witness targets. This is believed to result from nonuniform cathode emission and not from diocotron instability.

Figure 24 summarizes the results of our diode parametric study and shows that the most important variable for the range of B_z used in the anode-cathode spacing. During the investigation, the injector was fired over 200 times without refurbishment; the reproducibility of the beam parameters and the beam quality was very good.

Accelerating Gap Design

For the very high current ($\geq$ 50 kA), small radius ($\leq$ 1 cm) electron beams extreme care must be used in the design of the accelerating gap geometry, in order to avoid exciting large radial oscillations, while satisfying the constraints for avoiding explosive electron emission and virtual cathode formation. In general the gap geometries must employ the

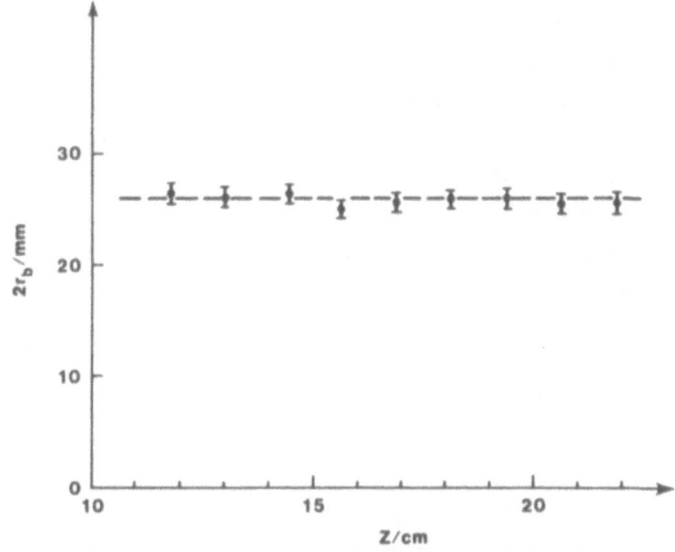

Fig. 23. Axial variation of the beam envelope. The distance
z of each measurement point is from the cathode plane.

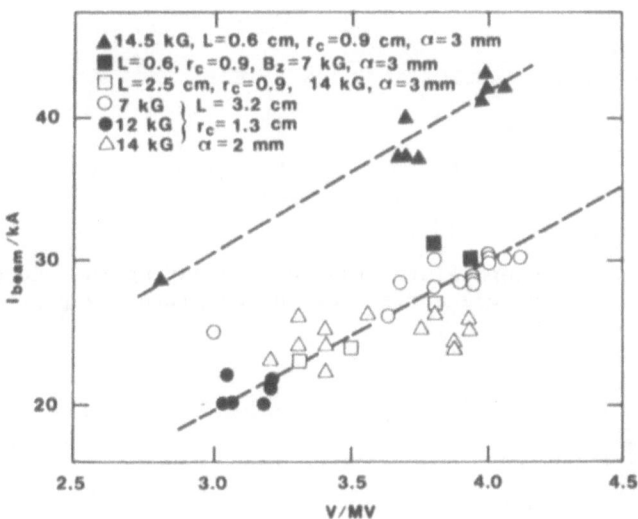

Fig. 24. Measured beam current versus diode voltage.

best features of the axial and radial gap geometries, especially at the
low voltage end of the accelerator.

To illustrate the gap design procedure, we begin with a square gap
geometry (Fig. 11, for example) with the gap spacing set by the explosive
emission criterion, Eq. (69). The behavior of the beam envelope is also
indicated. Note that for the 50 kA beam current, a virtual cathode could
form near the entrance of the gap, depending on the beam kinetic energy
[Eq. (78)]. The virtual cathode can be eliminated by contouring the

cathode and the anode of the gap as shown in Fig. 25. The Poisson solver code JASON (Sachett, 1983) is then rerun for the new gap design and the results are compared with the emission and virtual cathode criteria. If these comparisons are satisfactory, another beam envelope calculation is performed to determine the oscillation amplitudes. The procedure is then repeated until the electron emission, limiting current, and oscillation amplitude criteria are satisfied. The first two accelerating gaps of RADLAC II, designed using this procedure, are shown in Fig. 26, together

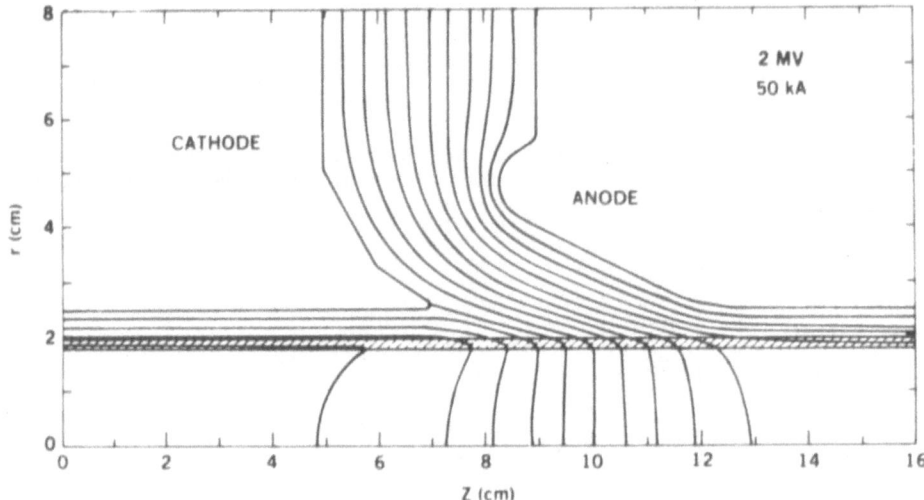

Fig. 25. Intermediate stage of contouring the anode and cathode surface of the accelerating gap.

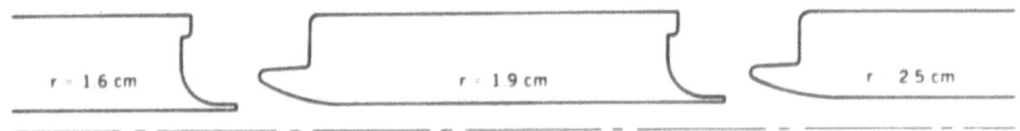

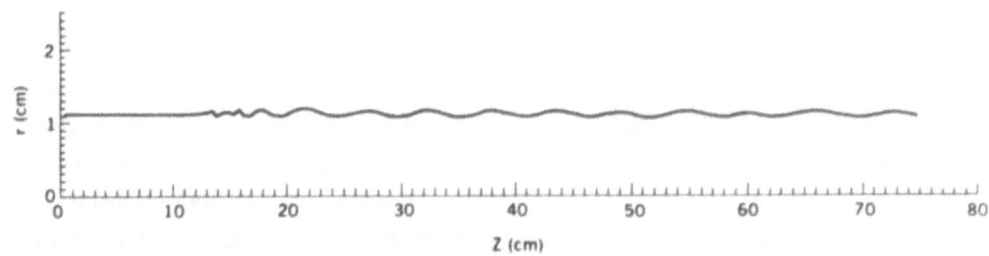

Fig. 26. Final design of the first two accelerating gaps of RADLAC II. The final oscillation amplitude is < 1 mm.

with the beam envelope solution. Note that this design also employs the radial gap concept. The predicted oscillation amplitudes are less than one millimeter (Miller, et al., 1983).

In order to check the validity of this gap design procedure, post-acceleration experiments were performed using the contoured gap concept. The accelerating assembly consisted of a radial isolated Blumlein injector and one accelerating cavity of the Pavlovskii configuration (Mazarakis, et al., 1984). An outline of the experimental setup is given in Fig. 27.

The actual gap design employed is shown in Fig. 28, illustrating the equipotential contours in the absence of beam space charge. This gap design was chosen in order to model the key features of the RADLAC-II gap design, i.e., contoured cathode and anode surfaces, and a larger drift tube radius at the gap exit; however, the assumed beam parameters were a 2 MeV, 30 kA injected beam with an outer radius of 1 cm and thickness of 2 mm. It was also assumed that the applied gap voltage would not exceed 1 MV. The radial oscillation amplitudes were not expected to exceed 1 mm.

The beam current diagnostics consisted of four Rogowski coils positioned on the cathode and anode sides of both the injector and the post-accelerating gap. The injector voltage was measured simultaneously by two capacitive probes (V), and a resistive voltage divider, while the accelerating gap voltage was monitored by a resistive divider only. The

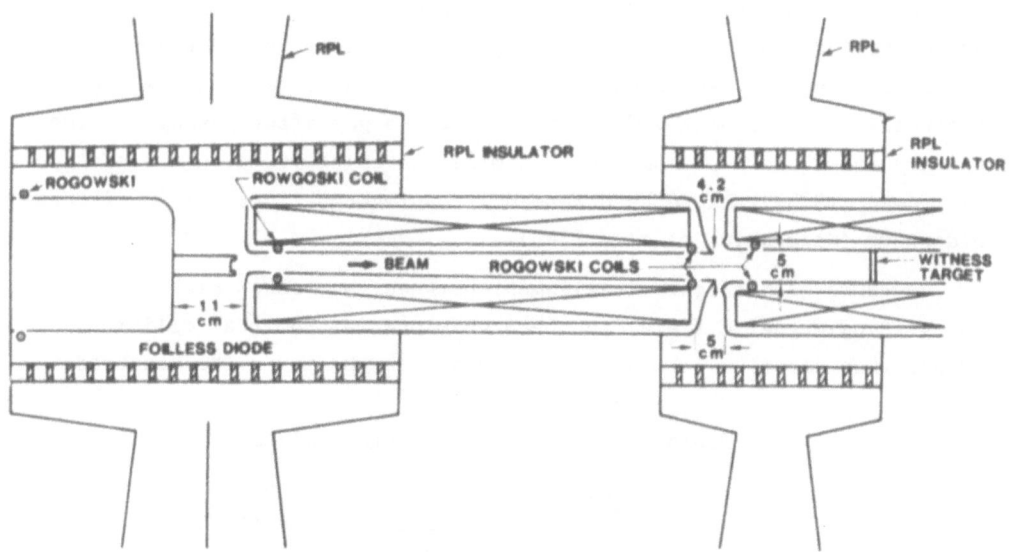

Fig. 27. Schematic diagram of the RADLAC-I contoured accelerating gap equipment.

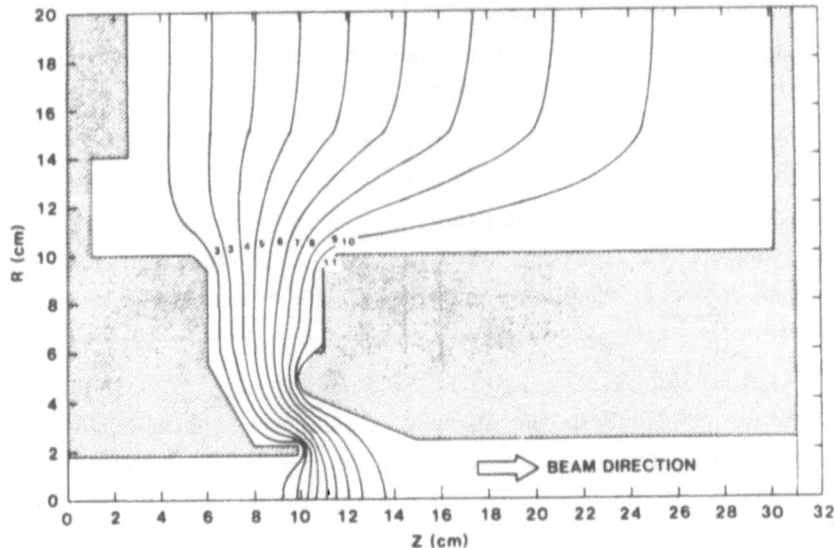

Fig. 28.   Detailed gap design used in the contoured gap beam
acceleration experiments (Equip. Plot).

beam cross section was monitored at several locations upstream and down
stream of the accelerating gap using metallic witness plates.

The experimentation was divided into two phases. In the first
phase, no accelerating voltage was applied on the post accelerating gap,
while in the second set of experiments both injector and accelerating gap
were activated. Table 1 summarizes typical values of voltage and
currents for these two phases.

Field emission from the cathode tip of the gap was not observed
during the beam passage, and the Rogowski coils upstream and downstream
of the gap gave practically identical current readings. However, during
the first stage of experimentation (zero applied voltage on the gap), a
secondary beam was observed downstream of the gap after passage of the
primary beam. This phenomenon is similar in nature to autoacceleration
(Friedman, 1973), and we have termed it "autogeneration;" it is discussed
elsewhere (Mazarakis, et al., 1983).

Because of a slight misalignment between the vacuum pipe axis and
the guiding magnetic field, the beam envelope was somewhat elliptical
rather than circular. Figure 29 (traces a and b) gives the measured

Table 1. Voltage and Current Values

	Injector Voltage	Gap Voltage	Beam Current
Stage I	2 MV	0	23 kA
Stage II	2 MV	.6 MV	40 kA

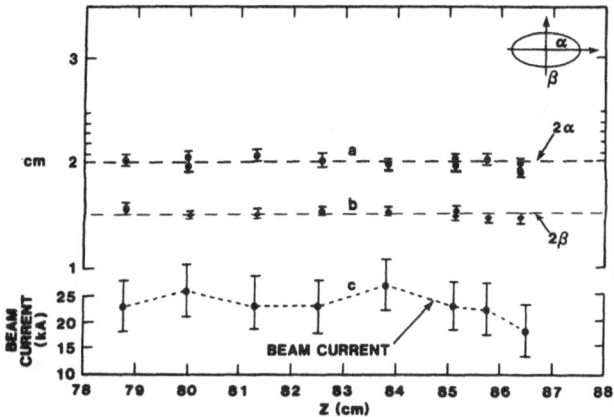

Fig. 29. Variation of the beam radial profile with distance
from the injector cathode shank (first stage of
experiment--no voltage on the post-accelerating
gap. Trace c gives the beam current for each
measurement.

major and minor axes of beam ellipses (2α and 2β) as a function of the
distance from the injector cathode tip. Trace c gives the beam current
for each particular measurement point. Figure 30 compares the variation
of the area enclosed by the beam envelope to that of the magnetic field
strength, both as functions of the distance z from the cathode tip.
Figures 31 and 32 give the same beam parameters for the second stage of
experimentation with the post accelerating gap activated. The error bars
of the major and minor axis and area of the beam ellipses include
statistical fluctuations in the geometrical measurement of the beam
imprint on the witness plate.

As evident from the figures, no radial oscillations were observed in
either phase of experimentation. The small variation of the beam cross
section area appears to follow a similar variation of the applied magnetic field, rather than indicating the onset of radial oscillations. On
the basis of these experimental results, we conclude that contoured gap
geometries appear to be the best approach for minimizing the problem of
gap-induced radial oscillations in high-current linear induction
accelerators such as RADLAC II.

Beam Stability Analysis

After the basic features of the RADLAC-II transport line design had
been established, it was analyzed with respect to growth of the various
instability mechanisms previously described (diocotron, resistive wall,
image displacement, and transverse beam breakup). The foilless diode
injector generates a hollow beam; while this beam configuration is

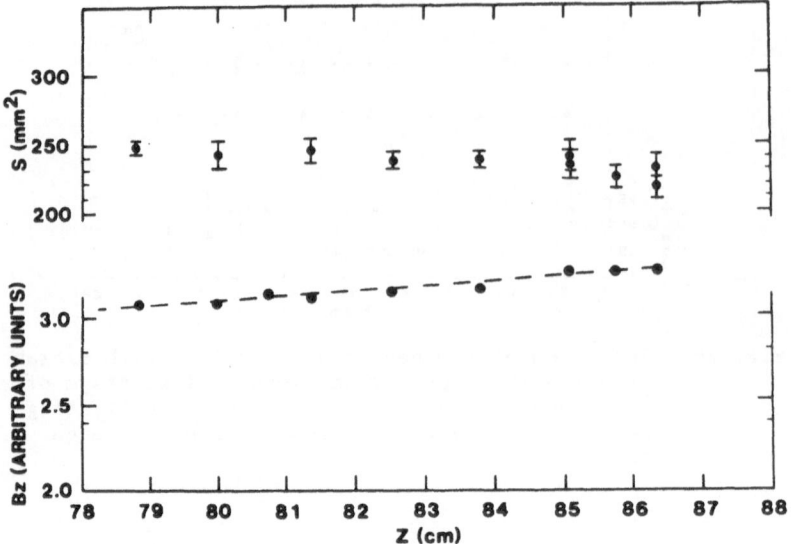

Fig. 30. Comparison of the variation of the beam envelope cross section (area) to that of the magnetic field strength along the beam pipe.

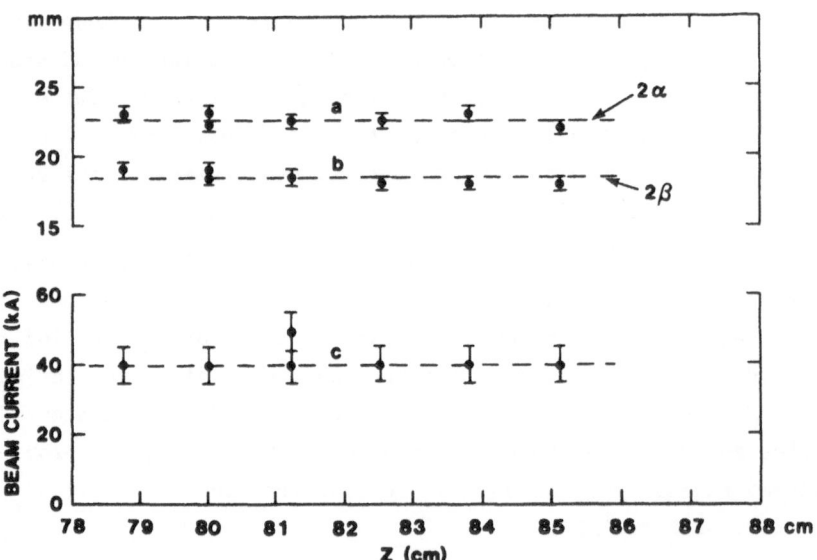

Fig. 31. Variation of the beam radial profile with distance from the injector cathode shank. Second stage of the experimentation with activated post-accelerating gap. Trace c gives the beam current for each measurement.

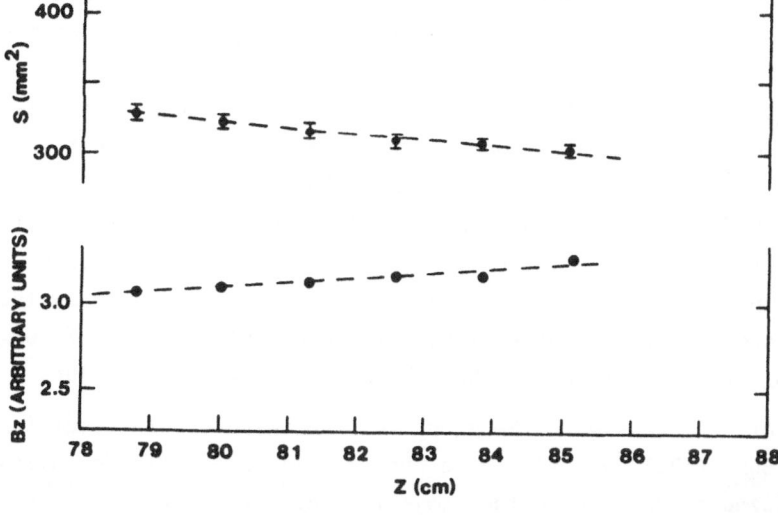

Fig. 32.   Comparison of the variation of the beam envelope
cross sectional area to that of the magnetic field
strength along the beam line. (Post-accelerating
gap activated.)

attractive from space charge limiting current considerations, very thin
beams are always diocotron unstable. Since the growth rate scales as
$(\gamma^2 B_z)^{-1}$, however, the instability should be controlled by using a high
voltage injector (4 MV), and a strong longitudinal magnetic field (20
kG). To test this possibility, the diocotron instability was investigated using the linear theory E-M code, GRADR. For diode voltages of 2-4
MV, use of a power law least squares fit for the peak diocotron growth
rate in the beam frame yields the scaling law (Miller, et al., 1981b)

$$\Gamma_D \approx I \, B_z^{-1.05} \, V_d^{-0.84} \, \delta^{-1.14} \quad f(\varepsilon) \tag{143}$$

where I is the beam current in kiloamperes, B_z is the axial field
strength in kilogauss, V_d is the diode voltage in megavolts, δ and ε are
the beam thickness and separation from the drift tube wall in centi-
meters, and $f(\varepsilon) = 1$, $\varepsilon \geq 4mm$; $f(\varepsilon) = \varepsilon^{0.42}$, $\varepsilon < 4$ mm. In the lab frame
$\Gamma_{D,lab} = \Gamma_D/\gamma$, where γ is the usual relativistic factor.

On the basis of these results, the diocotron instability does not
appear to be particularly troublesome. However, numerical simulations
clearly indicate that azimuthal asymmetries can trigger the growth of
filamentation instabilities (Poukey, 1984).

GRADR has also been used to investigate growth of the resistive wall
instability (Godfrey, 1981). Based on these results, the fastest growing
mode for annular beams in the RADLAC II parameters regime appears to be a

hybrid diocotron-resistive wall instability. Since the diocotron growth rate drops as γ^{-2} with beam acceleration, with reasonable acceleration gradients (~ 2 MeV/m) this hybrid instability can only grow at most by a few e-foldings. Thereafter, the low frequency space charge and cyclotron wave resistive wall instabilities dominate. The peak growth rates of these modes are given with good accuracy by Eq. (131), with f in the range 1-4 (for ℓ < 12). Since the corresponding group velocity can be shown to be about 2/3 of the beam velocity, the total growth of these instabilities can be estimated as $\Gamma L_a/v_g$, where L_a is the total accelerator length, assuming that the pulse length τ > L/3c. Using the largest growth rate estimate gives a maximum amplification of roughly 2.5 e-foldings for a 10 m drift tube, which is an acceptable value.

In order to estimate the growth of the image displacement instability, Eqs. (132) and (133) were numerically integrated for a few sets of idealized conditions in the projected RADLAC-II parameter regime (Godfrey, 1981). The results for a 100 kA beam are summarized in Table 2. Cases 1-5 are without acceleration, and cases 6-10 are for acceleration through gaps of nominally 2 MV each. Γ is the total growth for ten such gaps.

The phase sensitivity of the instability is clearly evident by the resonance behavior with respect to variations in magnetic field strength. While varying the intergap spacing as a function of beam kinetic energy might also help control the phase angle of the oscillation, a more straightforward approach is to reduce the separation of the image charges and currents by carefully designing the accelerating gap structures. For example, the radial accelerating gap design of Fig. 15 should substantially decrease image displacement growth. This fact is reflected in the contoured gap geometries of RADLAC II; modeling these gaps gives the reduced growths reflected in cases 8-10.

Table 2. Calculated Image Displacement Instability Growth

Case	B_z (kg)	ω/R^2 (cm^{-1})	Γ
1	20.0	0.8	7.5
2	19.88	0.8	1600.0
3	19.74	0.8	340.0
4	19.01	0.8	2.4
5	19.47	0.8	2.2
6	20.0	0.8	470.0
7	20.46	0.8	48.0
8	20.0	0.29	4.5
9	19.83	0.29	7.3
10	20.46	0.29	1.6

From the previous analysis, beam breakup growth rates are most strongly affected by the transverse impedance and the quality factor of the accelerating cavity normal modes. A typical cavity is shown in Fig. 33. High voltage standoff is provided by a vacuum interface consisting of alternating rings of metal and plastic dielectric. Inside the vacuum interface, metallic fieldshapers serve to localize and define the accelerating fields; a high dielectric constant fluid (water or ethylene glycol) is usually located outside the vacuum interface. The results of testbench measurements of the transverse impedance and cavity quality factor give very favorable values ($Z_\perp/Q$ < 10 ohms and Q < 10) for the lowest frequency, most dangerous ℓ = 1 mode (Adler, et al., 1981). Essentially, the insulator stack geometry is a very leaky cavity; since the fields extend outside the rings, they are greatly damped by the liquid dielectric medium because of the loss tangent at the frequencies of most concern (> 1 GHz).

Using the results of these measurements, the rigid disk code BALTIC has been used to evaluate growth of the beam breakup instability in RADLAC II (Godfrey, 1982). The results of these studies are summarized in Table 3 for a 100 kA beam and 5 cm gaps separated by 100 cm. Case 1 indicates acceptably small BBU growth (through ten gaps) even without acceleration; however, the sensitivity to increases in $Z_\perp/Q$ or Q are illustrated in cases 2 and 3. The damping effect of beam acceleration is evident in cases 4 and 5. Also, (as expected) there are no strong

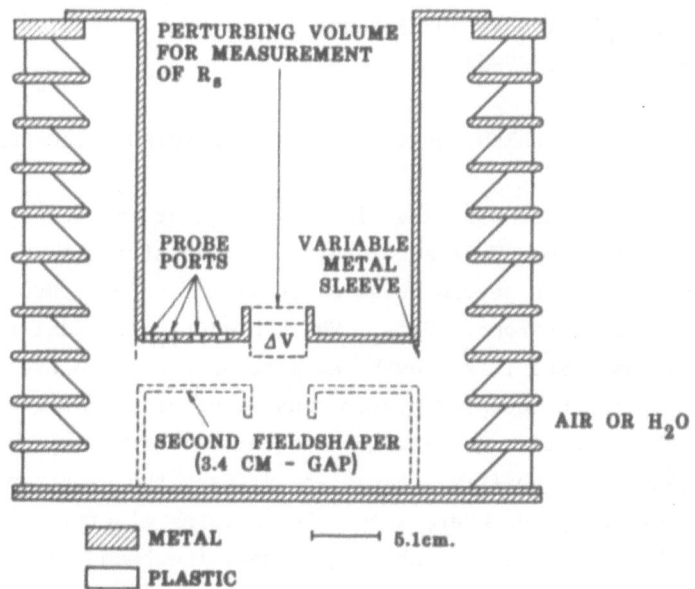

Fig. 33. Grading ring cavity with electrodes (fieldshapers) for the cavity quality factor and transverse shunt impedance measurements.

Table 3. Calculated Beam Breakup Instability Growth

Case	B_z (kg)	$Z_\perp/Q$ (ohm)	Q	$\Delta\gamma$	τ_R (cm)	Γ
1	20.	10	10	0	0	7.0
2	20.	10	20	0	0	350.
3	20.	20	10	0	0	700.
4	20.	10	10	4	0	3.3
5	20.	20	10	4	0	35.
6	19.83	10	10	4	0	4.2
7	19.66	10	10	4	0	3.1
8	20.	10	10	4	10	1.7
9	20.	20	10	4	40	3.6

magnetic field resonant effects (cases 5, 6, and 7). Finally, the effect
of nonzero risetime clearly reduces instability growth, since the
cavities are not shock excited (cases 8 and 9). In the latter cases, the
growth is sufficiently small that additional suppression mechanisms are
not required.

In summary, regarding the beam transport line design for RADLAC II,
detailed designs for the injector and accelerating gaps have been
developed which satisfy the stressing dual requirements of higher beam
currents and small beam radii. Several beam stability issues have been
analyzed, including diocotron, resistive wall, image displacement, and
transverse beam breakup. None of these appear to be particularly
serious, provided that beam misalignments are kept reasonably small (< a
few millimeters). In fact, RADLAC II and its predecessor, RIIM, have
successfully accelerated 40-50 kA beams to 8-16 MeV without evidence of
instabilities (Miller, 1985).

NEW CONCEPTS IN HIGH CURRENT BEAM TRANSPORT: ELECTROSTATIC CHANNEL
GUIDING

Although great care was taken in the design of the accelerating
cavities of the Advanced Test Accelerator, growth of the beam breakup
instability resulted in severe pulse disruption for beam currents
exceeding about 5 kA (Prono, 1985). While higher solenoidal fields
should help to decrease this growth, an alternative is to create a continuous electrostatic focusing system consisting of a positive line of
charge through the entire accelerator length. The symmetric radial
electric field of the positive charge would focus the electron beam onto
the accelerator axis; moreover, since the focusing electric field varies
inversely with radius outside the ion channel, the beam electrons are
confined with an anharmonic potential well which can lead to orbital
phase mix damping of any transverse beam oscillation (at the expense of
an emittance increase).

The first test of this concept used the beam from the experimental test accelerator (ETA) to induce positive charge on a resistive graphite thread that was supported along the length of the accelerator (Prono, et al., 1983). Although successful, this technique was not feasible for creating a positive line charge along the full 80 m length of the ATA. Instead, a small amount of benzene was introduced into the accelerator and a low energy KrF laser was used to partially photoionize a small diameter channel in the gas. The subsequent injection of the ATA electron beam rapidly expelled the plasma electrons, and the residual ion core then focused and guided the electron beam pulse. Using this technique, there was virtually no apparent growth of the BBU instability and successful operation at the full ATA design current (10 kA) was achieved (Prono, 1985). Following these initial experiments, ion channel guiding has been successfully used in beam steering and bending experiments (Frost, et al., 1985, and Shope, et al., 1985), and has been suggested for use in a recirculating linac concept (Miller, 1985). In the following the basic physics of electron beam guiding by ion focusing is briefly summarized.

If the electron beam is injected into a ionized plasma channel, the resulting motion of the plasma charges will tend to neutralize the beam self-fields, as indicated in Fig. 34. The beam space charge will become neutralized by the expulsion of plasma electrons in approximately one plasma oscillation period $\tau_p \sim (4\pi e^2 n_b/m_e)^{-1/2} = (\omega_p)^{-1}$, assuming that the beam density and the plasma electron density are approximately equal.

To avoid possible electron-electron streaming instabilities, the plasma electron line density should not exceed the beam electron line density; however, there is a minimum plasma line density necessary for a radial force balance equilibrium. To calculate this density, assume that all plasma electrons have been expelled so that the beam is propagating along a stationary ion background of linear density, $n_i r_c^2$; the fractional beam electron space charge neutralization is denoted by $f_n = n_i r_c^2/n_e r_b^2$. Solving for the radial space charge electric field and the azimuthal magnetic field and substituting these quantities into the equation of motion, yields an expression for the equilibrium radius

$$r_b^2 = \frac{\varepsilon^2}{c^4(\gamma^2-1)} \frac{I_A}{2I_b(f_n-1/\gamma^2)} \tag{144}$$

where $I_A = \beta\gamma \, mc^3/e$ is the Alfven current and $I_b = \pi r_b^2 \, en_b \, \beta_c$ is the beam current. The beam emittance is denoted by ε. It is apparent that there is no equilibrium unless $f_n \geq 1/\gamma^2$. In the limit of large beam kinetic energy, the beam self-fields essentially cancel, and the beam is

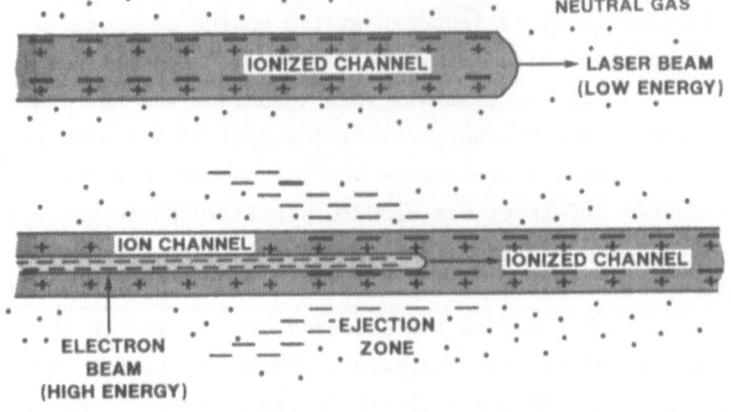

Fig. 34. A laser preionizes a plasma column; the space charge
fields of the beam expel the plasma electrons leaving
a strongly focusing positive ion channel. (For
accelerator applications, the ion channel is usually
smaller in radius than the electron beam.)

radially confined by the electrostatic force of the ion channel. In this
case, Eq. (114) becomes

$$r_b \simeq \frac{\varepsilon}{\gamma c^2} \left[\frac{I_A}{2f_n I_b} \right]^{1/2} \tag{145}$$

Having calculated the zero-order beam equilibrium requirements, we
now return to the details of how the positive ion channel becomes established and we examine the behavior of the electrons at the beam front
during this process. At least two significant effects are expected:*

(1) emittance-driven beam expansion until $f_n \sim 1/\gamma^2$,

(2) a loss of energy from beam electrons to the plasma electrons
which are expelled.

Each of these processes will cause a loss of beam particles/energy
at the beam front (erosion), decreasing the effective pulselength as the
beam propagates. We now estimate the magnitude of each of these effects,
and derive appropriate "erosion" formulae.

In the case of emittance-driven erosion, the natural scale lengths
are the length of beam pulse necessary to establish the force balance
criterion and the axial propagation distance necessary for the beam to
expand outside the channel capture radius. Roughly, the rate at which

---

* Additional erosion processes can occur if there is ion channel curvature or if transverse magnetic fields are present.

364

the beam is lost will be given by dividing the first of these scale lengths by the second. Simple estimates give

$$\frac{dx}{dz}\bigg|_{emittance} \simeq \frac{2}{\gamma R_c} \frac{f_n^{1/4}}{r_b^{3/2}} \varepsilon \tag{146}$$

The erosion resulting from beam energy loss caused by expelling electrons from the channel can also be simply estimated by calculating the difference in electrostatic potential for an unneutralized beam compared with a beam that is neutralized by the channel ion space charge. Since this difference comes only from the ions, the average electric field acting to slow the beam electrons is just $\Delta V / \ell = E \simeq (I_b f_n / \beta_o c) (1 + 2 \ln R/r_c)$, where R is the drift tube radius. Consequently, the rate at which beam electrons will lose energy is just $eE/(\gamma-1)mc^2$, or

$$\frac{dx}{dz}\bigg|_{energy\ loss} \simeq \frac{\nu f_N}{(\gamma-1)} (1 + 2 \ln R/r_c) \tag{147}$$

where $\nu = I_b/(\beta mc^3/e)$ is the dimensionless beam current.

The slowness of the beam erosion phenomena suggest that efficient propagation can be achieved by simply using long pulses. However, possible growth of various streaming instabilities between the beam electrons and the plasma species may limit the practical pulselength. If the beam itself does not generate significant ionization, then the instability most likely to occur is a transverse streaming instability between the beam electrons and the channel ions. Conventional instability analyses indicate that the characteristic growth time must scale as the ion bounce period, i.e., the time required for an ion to oscillate radially through the beam. In this case, the number of e-foldings of (linear) instability growth during the pulselength t_p should scale approximately as (Prono, 1985)

$$n_{ef} \sim t_p/r_c (I_b/m_i)^{1/2} \tag{148}$$

where m_i is the ion mass. It is also expected, however, that this instability should saturate non-linearly if the amplitude of the oscillation should become of the order of the beam radius. This instability has not yet been clearly observed (Prono, 1985). As for the cases of the various erosion models, experimental data are essential for the verification of these analytical estimates.

Because of the newness of this concept, it is reasonable to expect that its potential, as well as possible limitations, are not yet fully known. For example, the betatron oscillations and emittance growth resulting from phase mix damping may not allow this transport technique to be used for high current free electron laser applications. Yet at the same time, it was once thought that transporting currents in excess of 17 kA would be difficult because of plasma electron trapping in the magnetic self-field of the beam. By arranging the beam risetime such that $f_n \sim 1$ early in the pulse, but $f_n \sim 1/2$ at the current peak, 23 kA beams have been successfully transported via ion channels in recent autoacceleration experiments (Coleman, et al., 1986). Further work on these and several other issues is clearly necessary.

SUMMARY

In this paper we have analyzed several important issues pertaining to the transport of high current electron beams through linear induction accelerator structures. In particular several laminar flow equilibrium configurations were considered, including the effects of space charge limits and source geometries. Various aspects of the design of high voltage, high current injectors and accelerating gaps were discussed in some detail, including the problems associated with space charge limits, field emission criteria, and excitation of radial oscillations (zero frequency cyclotron waves). Several potentially disruptive beam instabilities were discussed, including the diocotron instability for annular beams, resistive wall instabilities, the image displacement instability, and the transverse beam breakup instability. To illustrate the use of these analyses, several aspects of the RADLAC II beam transport line were analyzed in detail. On the basis of this work, it appears that linear induction accelerators are capable of accelerating beam currents of tens of kiloamperes to a few tens of MeV. (RADLAC II, in fact, has successfully operated at about 50 kA and 16 MeV.) Achieving higher voltages (hundreds of MeV) at these current levels is probably possible, although the beam breakup would have to be carefully analyzed. In this regard, the new technique of ion channel guiding, which was only briefly discussed here, appears to offer considerable promise.

ACKNOWLEDGMENTS

It is a pleasure to acknowledge the many valuable contributions of several individuals. At the risk of possible serious omissions, a partial list of people I would like to thank include R. Adler, B. Baker, D. Coleman, B. DeVolder, C. Ekdahl, C. Frost, J. Freeman, T. Genoni, B. Godfrey, D. Hasti, G. Leifeste, T. Lockner, M. Mazarakis, J. Poukey, K.

Prestwich, A. Sharpe, S. Shope, L. Stevenson, D. Straw, B. Tucker, and G. Yonas.

* This work has been supported by the US Department of Energy, the US Air Force, and the Defense Advanced Research Projects Agency.

REFERENCES

Adler, R. J., Genoni, T. C., and Miller, R. B., 1981, IEEE Trans. Nucl. Sci. NS-28.

Adler, R. J., 1983, Phys. Fluids 26:1678.

Adler, R. J., Godfrey, B. B., Campbell, M. M., Sullivan, D. J., and Genoni, T. C., 1983, Part. Accel. 13:5.

Bogdankevich, L. S., and Rukhadze, A. A., 1971, Sov. Phys. Usp. 14:163.

Bogema, B. L., 1971, Univ. of Maryland Tech. Rept. 72-037.

Breizman, B., and Ryutov, D. D., 1974, Nucl. Fusion 14:873.

Buneman, O., Levy, R. H., and Linson, L. M., 1966, J. Appl. Phys. 37:3203.

Chen, J., and Lovelace, R. V., 1978, Phys. Fluids 21:1623.

Coleman, P. D., Lockner, T. R., and Poukey, J. W., March 1986, Private.

Davidson, R. C., 1974, "Theory of Nonneutral Plasmas," Benjamin, New York.

Drummond, W. E., Bourianoff, G. I., Cornet, E. P., Hasti, D. E., Rienstra, W. W., Sloan, M. L., Wong, H. V., Thompson, J. R., and Uglum, J. R., 1976, AFWL-TR-75-296 (Air Force Weapons Laboratory): App. A.

Friedman, M., and Ury, M., 1970, Rev. Sci. Instrum. 41:1334.

Friedman, M., 1973, Phys. Rev. Lett. 31:1107.

Frost, C. A., Shope, S. L., Miller, R. B., Leifeste, G. T., and Crist, C. E., 1985, IEEE Trans. Nucl. Sci. NS-32:2754.

Genoni, T. C., Franz, M. R., Epstein, B. G., Miller, R. B., and Poukey, J. W., 1981, J. Appl. Phys. 52:2646.

Godfrey, B. B., 1979a, IEEE Trans. Plasma Sci. PS-7:53.

Godfrey, B. B., and Newberger, B. S., 1979, J. Appl. Phys. 50:45.

Godfrey, B. B., Adler, R. J., Campbell, M. M., Sullivan, D. J., and Genoni, T. C., 1981, Proc. 1981 Linear Accel. Conf., Santa Fe, New Mexico.

Godfrey, B. B., 1982, AMRC-R-345 (Mission Research Corporation).

Jory, H. R., and Trivelpiece, A. W., 1969, J. Appl. Phys. 40:3924.

Langmuir, I., 1931, Phys. Rev. 3:238.

Laslett, L. J., Neil, V. K., and Sessler, A. M., 1965, Rev. Sci. Instrum. 36:436.

Lawson, J. D., 1978, "The Physics of Charged Particle Beams," Oxford University Press, Oxford.

Mazarakis, M. G., Shope, S. L., Smith, D. L., Miller, R. B., and Adler, R. J., 1983, Comments Plasma Phys. Controlled Fusion 8:23.

Mazarakis, M. G., Miller, R. B., Shope, S. L., Stevenson, L. E., Coleman, D. P., Poukey, J. W., Adler, R. J., and Genoni, T. C., 1984, SAND84-0095 (Sandia National Laboratories).

Mazarakis, M. G., Miller, R. B., Poukey, J. W., and Adler, R. J., submitted to J. Appl. Phys.

Miller, R. B., Prestwich, K. R., Poukey, J. W., and Shope, S. L., 1980, J. Appl. Phys. 51:3506.

Miller, R. B., Prestwich, K. R., Poukey, J. W., Epstein, B. G., Freeman, J. R., Sharpe, A. W., Tucker, W. K., and Shope, S. L., 1981a, J. Appl. Phys. 52:1184.

Miller, R. B., Poukey, J. W., Epstein, B. G., Shope, S. L., Genoni, T. C., Franz, M., Godfrey, B. B., Adler, R. J., and Mondelli, A., 1981b, IEEE Trans. Nucl. Sci. NS-28:3343.

Miller, R. B., 1982, "An Introduction to the Physics of Intense Charged Particle Beams," Plenum, New York.

Miller, R. B., Shope, S. L., Mazarakis, M. G., and Poukey, J. W., 1983, SAND83-0674 (Sandia National Laboratories).

Miller, R. B., 1985, IEEE Trans. Nucl. Sci. NS-32:3149.

Neil, V. K., 1978, UCID-17976 (Lawrence Livermore National Laboratory).

Ott, E., Antonsen, T. M., Jr., and Lovelace, R. V., 1977, Phys. Fluids 20:1180.

Panofsky, W. K. H., and Bander, M., 1968, Rev. Sci. Instrum. 39:2076.

Paul, A. C., 1981, UCID-19197 (Lawrence Livermore National Laboratory).

Pavlovskii, A. I., Bosamykin, V. S., Kuleshov, G. I., Gerasimov, A. I., Tananakin, V. A., and Klementev, A. P., 1975, Sov. Phys. Dokl. 70:441.

Poukey, J. W., 1984, SAND83-2511 (Sandia National Laboratories).

Poukey, J. W., and Coleman, P. D., 1985, SAND84-2652 (Sandia National Laboratories).

Prono, D. S., Caporaso, G. J., Cole, A. G., Briggs, R. J., Chang, Y. P., Clark, J. C., Hester, R. W., Lauer, E. J., Spoerlein, R. L., and Struve, K. W., 1983, Phys. Rev. Lett. 51:723.

Prono, D. S., 1985, IEEE Trans. Nucl. Sci. NS-32:3144.

Reiser, M., 1977, Phys. Fluids 20:477.

Sachett, S. J., 1983, UCID-17814 (Lawrence Livermore National Laboratory).

Shope, S. L., Frost, C. A., Leifeste, G. T., Crist, C. E., Kiekel, P. D., and Poukey, J. W., 1985, IEEE Trans. Nucl. Sci. NS-32:3092.

Wood, C. H., 1970, Rev. Sci. Instrum. 41:959.

REQUIREMENTS ON THE BEAM FOR MM-AND SUB-MM-WAVE GENERATION

J. M. Buzzi

Ecole Polytechnique
91128 Palaiseau (France)

INTRODUCTION

Since the pioneering work of J. Nation (1970) and others in the early seventies, the Intense Relativistic Electron Beam (IREB) community has given a new impulse to the development of new electronic tubes, allowing the coverage of new wavelengths with a radical change in the state of the art as regards peak power and average power.

If we look at the history of the gyrotron or synchrotron maser, one may observe that the first experimental verification of the physical mechanism was given by J. L. Hirshfield et al. (1964) using conventional hot cathode and low voltage technology. The microwave output power was then of the order of few mW or less. The subject has been asleep in western countries for about 10 years until renewed by IREB technology through the work of V. Granatstein et al. (1974).

Suddenly, in less than five years, the synchrotron maser was back in the panoply of the hot cathode business under the gyrotron form and the spectacular development that we know today.

A similar process in underway for the emergence of new ways of producing electromagnetic radiation.

The Classical Intense Electron Beam device (1 CIREB = Marx generator + pulse forming network + cold cathode, typical voltage $\approx$ 1 MV, typical drift tube current $\approx$ 1 kA, pulse duration between 10 and 100 ns) is a relatively cheap way for making experiments with beam power at gigawatt level. With this amount of power, even a small efficiency, let us say 1%, produces MW of microwave power. This is easy to produce and this rustic equipment is flexible and handy to use, compared to the hot-cathode devices, or small current high energy accelerator. Until now, the experiments with CIREB devices were ahead of practical devices by the beam power involved. But today, applications appear to be pos‑ sible at the same beam power with induction linacs.

However, "full scale" induction linacs are still expensive. This enhances the interest for CIREB because a larger class of problems may be studied with applications at hand.

At this moment, Free Electron Laser (FEL) is the leader for using CIREB or induction linacs in electromagnetic wave generation. One should not forget however, that many other schemes are still alive and may a great interest in a near future:

** – Cusptrons, peniotrons.

** – Circular FEL.

 * – Relativistic magnetrons.

 ? – Orbirons.

*** – Cerenkov Masers.

 * – Vircators.

 – etc...

The number of stars in the above list gives an idea of the required beam quality for a good efficiency. It is interesting to note that some of these schemes have prove to be working without a very high quality. This raises the question that the self-fields of intense beams are not necessarily killing microwave emission, but, on the contrary, are useful.

It should be pointed out also that some schemes allow us to reach high frequencies with relatively low accelerating voltage and this point is very important for an estimate of what the future could be.

For the conclusion of this introduction, one may summarize that the actual needs for electromagnetic waves production are:

a) high frequency.

b) high efficiency (beam quality).

c) high current (kA).

d) reliability and repetition rate.

CIREB devices are well adapted for these studies except for point d) which is not needed for a research tool working at current and voltage close to the values in induction linacs.

The following sections of this report are organized under the following headings:

How far in frequency can we go with CIREB?

What are the problems of beam "quality" with CIREB and how this beam "quality" can be measured?

Can we use the self-fields of CIREB in a positive way for microwave generation?

HOW FAR IN FREQUENCY?

For reference and if otherwise not specified, we admit that one CIREB produces a 1 kA, 1 MV electron beam.

The two candidates for high frequencies are the FEL and the Cerenkov Maser. The other devices have their frequency scaling with the plasma density or the synchrotron frequency and this appears as a limitation.

High Frequency FEL

As a first approximation, the emitted wavelength of a FEL is given by

$$\lambda_1 \approx \frac{\lambda_o}{2\gamma^2} \tag{1}$$

where λ_o is the wiggler period, and λ_1 the wavelength of the emitted frequency. Since $\gamma \approx 3$ for a CIREB, one has:

$$(\lambda_1)_{min} \approx \lambda_o/20 \ . \tag{2}$$

If one uses a helical wiggler or a permanent magnets wiggler, λ_o cannot be arbitrary small because a) $\lambda_o \gg a$ (a is the beam radium) in order to have a true sinusoidal or helical field (see for example section III), and b) for the beam "quality" requirement (see section III). Typically $\lambda_o \approx 4$ cm and therefore:

$$(\lambda_1)_{min} \approx 2 \ mm \ . \tag{3}$$

To overcome this difficulty, there are at least two possibilities: (i) a new wiggler technology and (ii) a two-stage FEL.

New Wiggler Technology

Two new wiggler technologies have been proposed: the "Mille feuilles" wiggler by V. Granatstein (Destler et al., 1986) (Fig. 1), and the "Eddy Currents" wiggler by H. J. Doucet (1982) (Fig. 2). Both wigglers have the disadvantage that the beam radius a is limited by the wiggler period λ_o and this yields in a limitation in the beam current. A practical way to overcome this difficulty is to use planar or circular flat beams (Fig. 3).

371

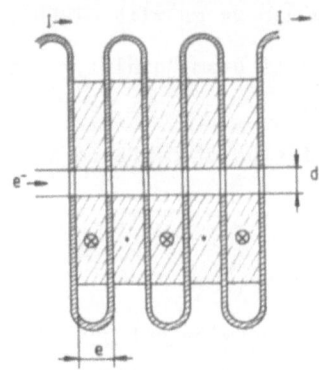

Fig. 1.   The "Mille feuilles" wiggler. The wiggler period is
λ_o = 2e, where e is the thickness of the conducting
sheet + the insulating or magnetic sheet. λ_o can be
a function of z (tapered wiggler).

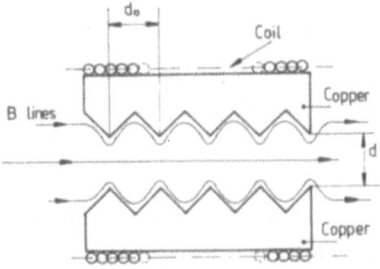

Fig. 2.   The "Eddy currents wiggler". The two gratings are good
conductors. Immersed in a fast rising B field, the field
lines are excluded from the conducting materials. Therefore, the B field inside the drift region is modulated
by the gratings.

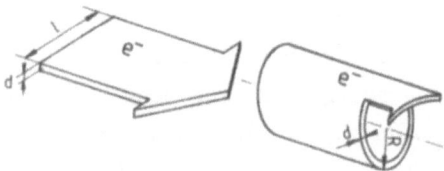

Fig. 3.   Planar and circular flat beams. Although the thickness
d is small, the width 1 or 2π R can be large to allow the
use of an intense beam

With this type of wiggler, one may expect to reach a wiggler period λ_0 of 1 mm. In this case λ_1, the emitted wavelength for a CIREB is:

$$(\lambda_1)_{min} \approx 50 \ \mu m \ . \tag{4}$$

Experiments have not yet been performed with these types of wiggler but nevertheless, one may expect to reach emission in the far-infrared domain with these wigglers.

Two-Stage FEL

This scheme is illustrated on Fig. 4. The downstream part of the beam produces radiation at λ_1. This radiation is reflected back against the beam head and in the second stage is reflected by Raman or Compton scattering at $\lambda_2 \approx \lambda_1/4\gamma^4$. The final wavelength is therefore:

$$\lambda_2 \approx \lambda_0/8\gamma^4 \tag{5}$$

or for a CIREB with $\lambda_0 = 4$ cm:

$$(\lambda_2)_{min} \approx 62 \ \mu m \ . \tag{6}$$

The validity of this scheme has not yet been demonstrated by an experiment.

Cerenkov Masers

The principle of a collective Cerenkov maser is very simple. An electron beam carries in free space two plasma waves. A fast wave whose phase velocity is greater than the beam velocity v_0 and a slow wave with a phase velocity smaller than v_0. In an empty wave guide, these waves cannot interact with the wave guide modes whose phase velocities are greater than the speed of light. However, if the wave guide is loaded with a dielectric (see Fig. 5a), then the interaction is possible as shown in Fig. 5b. This interaction yields to an instability because the slow beam wave is a negative energy wave and the waveguide mode a positive energy wave, i.e., both may grow keeping the total energy constant.

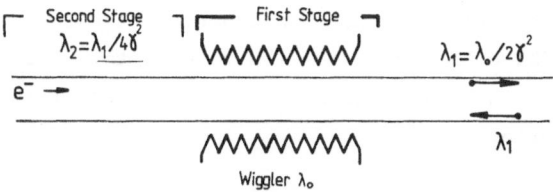

Fig. 4. Schematic of the two-stage FEL.

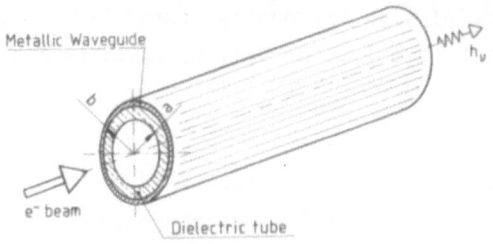

Fig. 5a.   Geometry of the circular Cerenkov Maser.

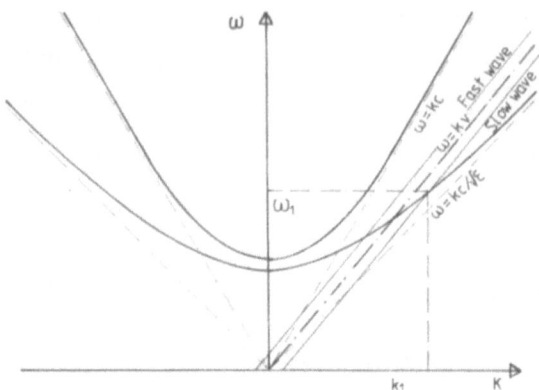

Fig. 5b.   Dispersion relation for an electron beam in a loaded
wave guide.   The wave ω_1, k_1 is unstable.

One property makes the Cerenkov maser interaction very attractive is
that the frequency goes up when the beam energy goes down.   If ω_1 is the
emitted frequency as shown on Fig. 5b, then we have:

$$\partial \omega_1 / \partial v_o < 0 \ . \qquad\qquad (7)$$

This opens the possibility of reaching high frequencies with
moderate energies.   J. Walsh and co-worker are studying, both
theoretically and experimentally, Cerenkov Masers.

Last year, an experiment was performed at the Ecole Polytechnique
(Garate et al., 1986) in collaboration with the group of J. Walsh, and
frequencies up to 300 GHz were measured.   Power up to 500 kW at 150 GHz
has been recorded.   These results were obtained with a CIREB device
(Figs. 6a and 6b) with cylindrical waveguides.

However, if higher frequencies are needed, the transverse dimensions
of the waveguide have to scale with the wavelength.   Using planar waveguide, frequencies up to 700 GHz have been measured at the Ecole Polytechnique still using a CIREB.

374

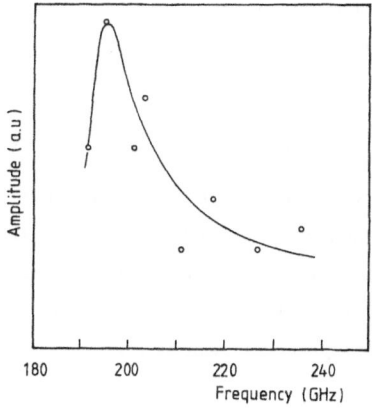

Fig. 6a. Measured wave amplitude as a function of the frequency.

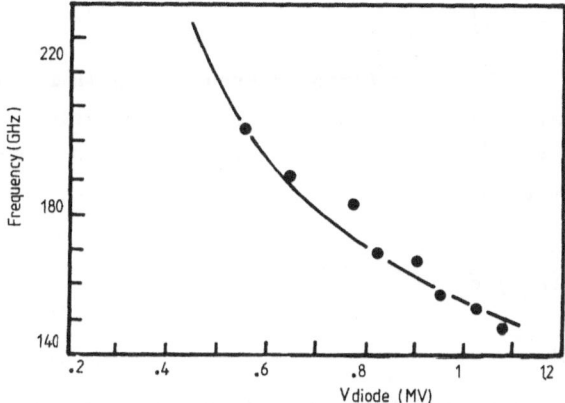

Fig. 6b. Frequency as a function of the accelerating voltage.

BEAM "QUALITY"

In this section, we will examine the requirement on beam quality of a CIREB in an FEL experiment. Because intense electron beams are supposed to work in the collective or Raman regime, we will first establish the conditions for collective process.

FEL in the Raman Regime

A plasma, i.e., a neutralized electron cloud, supports a large variety of waves. Let us consider one of the simplest equilibria for a plasma, a neutralized and cold cloud of electrons of infinite extend are field free (no static magnetic or electric field). In any direction of the wave number $\vec{k}$, the plasma can support two waves. The electromagnetic ordinary wave with the following dispersion relation

$$\omega'^2 = \omega_p'^2 + k'^2 c^2 \; , \tag{8}$$

and the plasma wave:

$$\omega'^2 = \omega_p'^2 \ . \tag{9}$$

These dispersion relations are represented in Fig. 7.

Nonlinear three-wave coupling may occur in a plasma. In particular, if a forward electromagnetic wave (ω', k') has a large amplitude, it may decay in two waves, a plasma wave and an other electromagnetic wave propagating backward. This is called Raman backscattering and the relationships between the frequencies and wavenumbers of these coupled waves are given by:

$$\omega_1' = \omega_p' + \omega_2' \tag{10}$$

$$|k_1'| = |k_p'| - |k_2'| \ . \tag{11}$$

Let us show now how this Raman backscattering is related to the FEL. Let us assume that we have a cold electron beam of velocity v in the z direction and that our beam is of infinite extend. Moreover, we consider in the laboratory frame, a modulated magnetic field of period λ_o given by:

$$\vec{B}_1 = B^\perp \sin(k_o z) \ \vec{e}_y \tag{12}$$

where $k_o = 2\pi/\lambda_o$.

In the beam frame, the frequency of this static perturbation is given by a relativistic Doppler shift:

$$\omega' = \gamma \ k_o v \ , \tag{13}$$

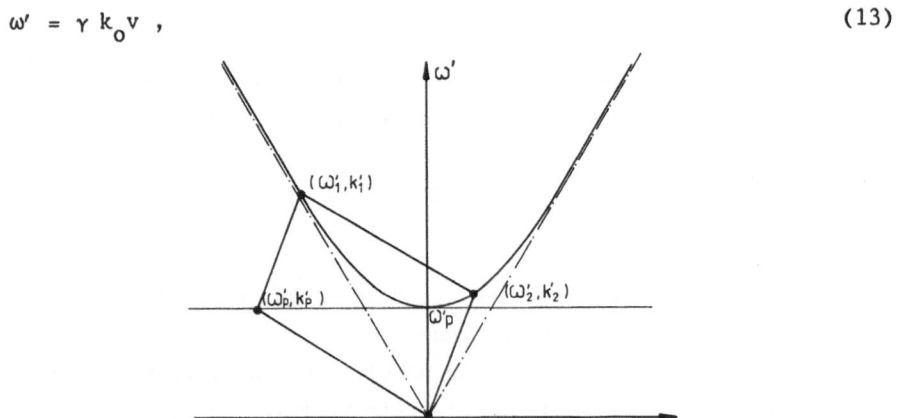

Fig. 7. Dispersion of linear waves in an infinite plasma without magnetic field. Raman backscattering is represented by the three waves (ω'_1, k'_1), (ω'_2, k'_2) and (ω'_p, k'_p).

and similarly, its wave number is given by:

$$k' = -\gamma k_o . \tag{14}$$

Therefore, in the beam frame, this perturbation is a slow wave with a phase velocity given by the beam velocity:

$$\omega'/k' = v . \tag{15}$$

By Lorentz transformation, the transverse magnetic modulation $\vec{B}_1$ becomes in the beam frame:

$$\vec{B}'_1 = \gamma B_1 \vec{e}_y \quad \text{and} \quad \vec{E}'_1 = \gamma v B_1 \vec{e}_x . \tag{16}$$

Therefore, in the beam frame of reference, the cold plasma "sees" a linearly polarized electromagnetic slow wave. Let us assume that v is close to the speed of light c, then this slow wave is very close to the ordinary electromagnetic plasma wave (ω'_1, k'_1), particularly if $\omega' \gg \omega'_p$. Like the ordinary electromagnetic wave in a plasma, this perturbation will drive a perpendicular velocity:

$$\vec{v}'_1 = \frac{ie}{m\omega'} \vec{E}_1 = \frac{ie}{m\gamma k_o} B^\perp \vec{e}_x . \tag{17}$$

If there is a second electromagnetic wave, (ω'_2, k'_2), $\vec{v}'_1$ will produce a nonlinear force in the longitudinal direction:

$$F_z = -e(\vec{v}'_1 \times \vec{B}^x_2) = -i\frac{e^2}{m\gamma k_o} B^\perp B_2 \, e^{i\phi(t,z)} \tag{18}$$

where $\phi = (\omega' t + |k'|z) - (\omega'_2 t - |k'_2|z)$. This nonlinear force F_z will strongly excite the plasma wave if:

$$\phi = \omega'_p t + |k'_p|z , \tag{19}$$

or:

$$\omega'_p = \omega' - \omega'_2, \Rightarrow \omega' = \omega'_p + \omega'_2 ,$$
$$|k'_p| = |k'| + |k'_2| \Rightarrow |k'| = |k'_p| - |k'_2| , \tag{20}$$

i.e., if we satisfy the selections rule for Raman backscattering (Eqs. 10 and 11).

In the laboratory frame, the frequency ω'_2 is upshifted by the Doppler effect:

$$\omega_2 = \gamma(\omega'_2 + k'_2 v) , \tag{21}$$

and because we have assumed $\omega' \gg \omega'_p$, then $k'_2 \approx \omega'_2/c$ and we obtain:

$$\omega_2 = (1 + \beta) \, \gamma^2 \, (k_o v - \omega'_p) \, . \tag{22}$$

This rather simplified model describes the basis of FEL in the collective regime, by opposition to the Compton regime where the electron are not bunched at the plasma frequency. Despite the rudimentary aspect of this model, one may derive two conditions for the existence of a collective FEL:

a) In order to have the electrons bunching at the plasma frequency, they must travel through the wiggler in a time longer than a plasma period:

$$\lambda_o / \gamma \, v_{//} > 1/\omega_p \tag{23}$$

b) Raman scattering is possible if the plasma wave exists, i.e., if the thermal velocity spread of the beam, $\langle \delta v \rangle$ is small compared to the relative speed Δv of the beam with the phase velocity of the slow beam wave ($v_{\phi-}$):

$$\langle \delta v \rangle < \left| v_{\phi-} - v \right| = \Delta v \, . \tag{24}$$

If this condition is not satisfied, Landau damping of the slow beam wave will occur and the three-wave nonlinear coupling mechanism is no longer possible.

We will express now these two conditions in terms of beam current and voltage.

Conditions for Collective FEL

The condition (23) can be expressed in terms of the beam current I and beam size x (see Fig. 8). We have:

$$\omega_p^2 = \bar{\omega}_p^2 / \gamma \quad \text{with} \quad \bar{\omega}_p^2 = n_o e^2 / \varepsilon_o m$$

and since

$$I = e n_o \beta_{//} c x^2 \, ,$$

where $\beta_{//} = v_{//} / c$ and $v_{//}$ is the parallel beam velocity, we obtain:

$$\omega_p^2 = \frac{4 \pi c^2}{x^2} \frac{I}{I_o} \frac{1}{\beta_{//} \gamma} \, , \tag{25}$$

where $I_o = 4 \pi \varepsilon_o m c^3 / e = 17 \, kA$.

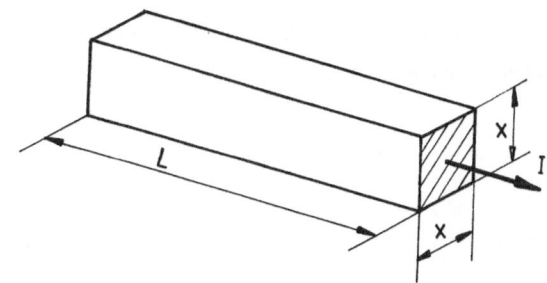

Fig. 8.   Geometry of the 'square' beam used in the beam quality
         calculations.

We may now explicate the condition (23):

$$I > I_{col} \, ,$$

$$\text{with } I_{col} = \frac{I_o}{4\pi} \left[\frac{x^2}{\lambda_o^2}\right] \gamma^3 \beta_{//}^3 \, . \tag{26}$$

Condition for the Raman Regime

A more elaborate description of the Raman regime has been given by
P. Sprangle et al. (1980). In contrast to out field free equilibrium,
where $B^\perp$ is assumed to be arbitrary small, in this model $B^\perp$ is finite.
In this case, the dispersion relation may be written:

$$D_1(\omega,k) \, D_2(\omega,k_p) = \frac{\bar{\omega}_p^2}{2\gamma} \left[\frac{eB^\perp}{m\gamma \, k_o c}\right]^2 D_1(\omega,k_p) \tag{27}$$

where $D_1(\omega,k)$ is the dispersion relation for an ordinary electromagnetic
wave and $D_2(\omega,k)$ the dispersion relation for the plasma wave. For a cold
beam we have:

$$D_1(\omega,k) = \omega^2 - k^2 c^2 - \bar{\omega}_p^2/\gamma \, ,$$

$$D_2(\omega,k) = (\omega - kv_{//})^2 - \bar{\omega}_p^2/\gamma\gamma_{//}^2 \, . \tag{28}$$

Moreover, in Eq. (27), we have $k_p = k+\gamma \, k_o$.

It can be shown that the dispersion relation (Eq. 27) describes two
regimes: the collective regime and the strong pump regime. The condition for the strong pump regime is:

$$\frac{eB^\perp}{m\gamma \, k_o v_{//}} > 4 \left[\frac{\omega_p}{k_o c \gamma_{//}^3}\right]^{1/2} \tag{29}$$

where $\gamma_{//} = (1 - \beta_{//}^2)^{-1/2}$.

This condition can be expressed with the beam current. In order to be in the collective regime the condition is:

$$I > I_{Pump} \text{ with}$$

$$I_{Pump} = I_o \, (B^\perp)^4 \, \frac{\lambda_o^2 x^2}{\Phi_o^4} \, \frac{\gamma_{//}^6}{(\gamma\beta_{//})^3} \tag{30}$$

where $\Phi_o = 8\pi^{3/4} mc/e$.

Moreover, in the Raman regime, one has to prevent the Landau damping of the plasma wave (see Eq. 24). From Eq. 27, one may derive:

$$\Delta v = |v_\phi - v| = \omega_p/2\gamma_{//}^3 k_o \, , \tag{31}$$

and the beam spread in energy, $\Delta\gamma_{//}/\gamma_{//}$ can be related to the spread in velocity $\langle\delta v\rangle$:

$$\langle\delta v\rangle = c \, \Delta\gamma_{//}/(\gamma_{//}^3 \beta_{//}) \, . \tag{32}$$

Combining Eqs. 31 and 32 in the condition 24 yields:

$$\frac{\Delta\gamma_{//}}{\gamma_{//}} < \frac{1}{2} \, \frac{\omega_p \beta_{//}}{k_o c \gamma_{//}} \, . \tag{33}$$

It is interesting now to establish a relation between the current I and the spread in energy. The simplest model one may use is to calculate the voltage potential ΔV across the beam due to its space charge. One obtains:

$$\Delta V \approx \frac{1}{2} \, \frac{I}{\varepsilon_o \beta_{//} c} \tag{34}$$

and the corresponding spread in energy is given by:

$$\frac{\Delta\gamma_{//}}{\gamma_{//}} \approx 2\pi \, \frac{I}{I_o} \, \frac{1}{\beta_{//}\gamma_{//}} \tag{35}$$

Using Eq. 35, the inequality 33, yields to the following condition for the Raman regime:

$$I < I_{Raman} , \tag{36}$$

$$\text{with } I_{Raman} = I_L \, \frac{\lambda_o^2}{x^2} \, \frac{\beta_{//}^3}{\gamma}$$

where $I_L = I_o/16\pi^3$.

In order to explicate the conditions for the collective or Raman FEL, we have to choose a relation between $\beta_{//}$ and $\beta_\perp$. This can be obtained by assuming an helical wiggler. In this case, it is well known that the electrons have helical trajectories with a constant parallel velocity. Moreover the modulus of the perpendicular velocity is given by:

$$v_\perp/c = \beta_\perp \qquad (37)$$

where $\beta_\perp = \alpha/\gamma$ with $\alpha = eB^\perp/mk_o c$, and $\gamma > \gamma_{min}$ defined by:

$$\gamma_{min} = \sqrt{1 + \alpha^2} \qquad (38)$$

It is also interesting to note that for an helical wiggler, the emitted wavelength is given by:

$$\lambda_1 = \lambda_o/2\gamma_{//} = \frac{\lambda_o}{2\gamma}(1 + \alpha^2) \qquad (39)$$

Figure 9 illustrates the conditions for the collective regime for $x = 0.5$ cm, $\lambda_o = 10$ cm and $B^\perp = 2$ kG. For this set of parameters, the two limiting currents I_{Raman} and I_{Col} intersect. This intersection point C defines also the maximum energy value where it is possible to operate in the collective regime. This value is given by:

$$I_{Col} = I_{Raman} \Rightarrow \gamma_C = \frac{1}{(2\pi)^{1/2}}\frac{\lambda_o}{x}. \qquad (40)$$

The emitted wavelength is then:

$$\lambda_{1C} = \frac{\pi x^2}{\lambda_o}(1 + \alpha^2) \qquad (41)$$

For the conditions of Fig. 9, Eqs. 42 and 43 yield $\gamma_C = 7.979$. ($V_C = 3.57$ MV) and a wavelength of 3.5 mm (frequency = 85 GHz). The current at point C is $I_C = 1550$ A.

At this point some interesting remarks can be made with regard to the collective regime:

a) The collective regime is limited in energy to low energies:

$$\gamma_{min} < \gamma < \gamma_C$$

b) In order to reach small wavelengths, one needs a small beam diameter (see Eqs. 42 and 43).

Conditions for the Strong Pump Regime

The first condition for the strong pump regime has been given by Eq. 30. To this first condition we must add a condition on the beam quality.

381

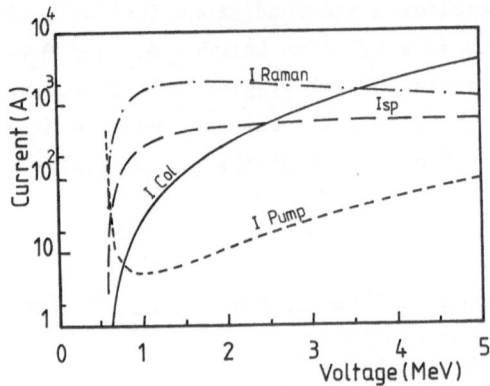

Fig. 9. Diagram showing the various regime of a low energy FEL. The dashed area corresponds to the strong pump regime. The parameters of the figure are x = 0.5 cm (Beam size), λ_0 = 10 cm (wiggler period), and $B^\perp$ = 2 kG (helical wiggler field)

The calculation is similar to the evaluation of I_{Raman}, except that, in the strong pump regime, Eq. 31 is replaced by:

$$\Delta v = \left| v_\phi - v \right| = \frac{c}{4\gamma^2_{//}2^{1/3}} \left[\frac{\omega_p^2}{k_0 c} \beta_\perp \right]^{2/3} . \qquad (42)$$

This yields a new condition:

$$I < I_{SP} ,$$

with

$$I_{SP} = \frac{I_0}{32\pi^2} \left[\frac{\beta^5_{//}\gamma^3_{//}}{\gamma} \right]^{1/2} \beta_\perp \left[\frac{\lambda_0}{x} \right] . \qquad (43)$$

I_{SP} has a finite value I_{SPmax} for large energies:

$$I_{SPmax} = \frac{I_0}{32\pi^2} (1 + \alpha^2)^{-1/2} \alpha (\lambda_0/x) . \qquad (44)$$

Figure 10 illustrates the conditions for the strong pump regime for x = 0.5 cm, λ_0 = 10 cm and $B^\perp$ = 2 kG. For this set of parameters, the two limiting currents I_{Pump} and I_{SP} intersect. This intersection point P defines also the maximum current value where it is possible to operate in the strong pump regime. This value is given by:

$$I_P = I_{SP}(\gamma = \gamma_P) \text{ where} \qquad (45)$$

$$\gamma^3_P = \frac{1}{32\pi^2} \left[\beta^{5/2}_{//} (1 + \alpha^2)^{9/4} \right] \frac{\alpha\Phi_0^4}{B_\perp^4} \beta^3_{//} \frac{1}{\lambda_0 x^3} .$$

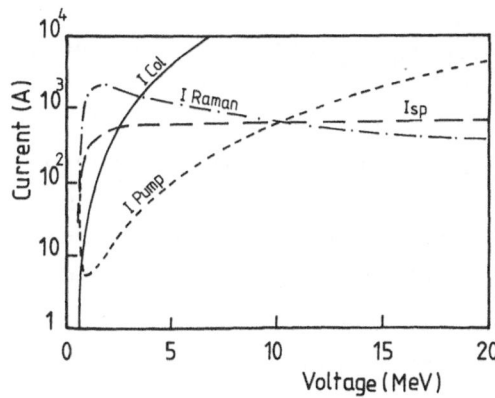

Fig. 10. Diagram showing the various regime of a high energy FEL. The dashed area corresponds to the strong pump regime. The parameters of the figure are x = 0.5 cm (Beam size), λ = 10 cm (wiggler period), and $B^{\perp}$ = 2 kG (helical wiggler field)

In Eq. 44, γ_p is easily evaluated by iteration, assuming at the first step that $\beta_{//} \approx 1$.

For the conditions of Fig. 10, γ_p = 20.8 (10.14 MV) and I_p = 646 A.

At this point one may conclude that as regards the strong pump regime:

a) There is no limitation in energy:

$$\gamma_{min} < \gamma .$$

Moreover, as long as one satisfies the condition for the strong pump regime (I > I_{Pump}), the beam quality criteria is also satisfied if $\gamma > \gamma_p$.

b) However, there is a beam current limitation given in first approximation by I_{SPmax} (Eq. 44). Like for the collective case, small beam diameters allow higher frequencies and beam currents.

Wiggler Problems

Wigglers are still a critical component of FEL, by their cost and by their effect on beam quality. We will limit the discussion here to the case of helical magnetic wiggler where the active parts are helical windings, but a similar analysis can be carried out for permanent magnet wigglers.

Inefficiency

The field produced by a double helical winding is given by (Paritsky, 1959):

$$B_r = \frac{\mu_o}{4\pi} 8k_o aI \sum_{n \text{ odd}} k_o n \ K'_n(k_o na) I'_n(k_o nr) \ \sin\{n(\theta - k_o z)\}$$

383

$$B_\theta = \frac{\mu_o}{4\pi} 8k_o a I \sum_{n \text{ odd}} k_o n \, K'_n(k_o na) \frac{I_n(k_o nr)}{k_o r} \cos\{n(\theta - k_o z)\} \tag{46}$$

$$B_z = \frac{\mu_o}{4\pi} 8k_o a I \sum_{n \text{ odd}} k_o n \, K'_n(k_o na) I_n(k_o nr) \cos\{n(\theta - k_o z)\} $$

where a is the radius of the winding, I the total current in the winding, I_n and K_n the modified Bessel function of first and second kind.

For the component of interest, i.e., n = 1, we have around the axis a pure helical field with a modulus given by:

$$B^\perp = (16\pi^2 \, Ia/\lambda_o^2) \, K'_1(2\pi a/\lambda_o) \tag{47}$$

where $B^\perp$ is in Gauss, I in kA, the radius of the winding a, and the pitch of the wiggler in meters. Figure 11 illustrates the poor efficiency of helical wigglers.

Anharmonicity

Moreover, because of the exponential growth of the harmonics (n ≠ 1) away from axis, the wiggler field is helical only close to the axis. As a matter of fact, if one computes the ratio of the modulus of the n = 1 component (RHSP) to the n = -1 component (LHSP) at a distance R from axis, one find using Graf's theorem (Abramowitz and Stegun, 1965):

$$\tau(R) = \frac{I_0(k_o R) - I_2(k_o R)}{I_0(k_o R) + I_2(k_o R)} \tag{48}$$

Figure 12 gives the variation of τ with R/λ_o and shows that bifilar wigglers produce a helical magnetic field only for small values of R/λ_o, typically $R/\lambda_o \leq 0.1$. However, this limitation is consistent with beam quality requirements studies in the previous section which all implies $R/\lambda_o \ll 1$.

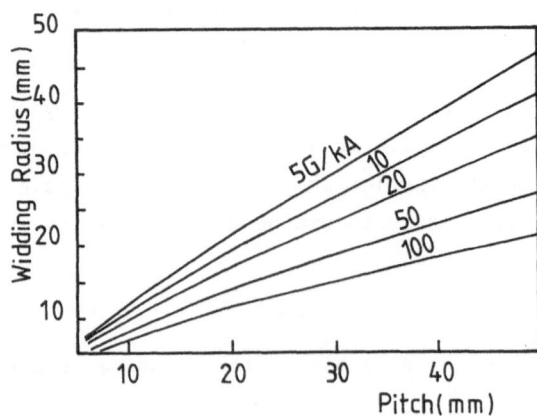

Fig. 11. Magnetic field on axis produced by an helical wiggler as a function of the winding radius and of the helix pitch.

Wiggler Instability

In most CIREB, the transport of an unneutralized beam is done using a guiding magnetic field. This circumstance makes the electron trajectories particularly unstable around a resonant value of the B_z field, B_z^r. This resonance occurs when the electrons move from a wiggler period in a synchrotron period (Drossart, 1978):

$$v_{//} \frac{2\pi}{eB_z^r/m\gamma} = \lambda_o . \tag{49}$$

As pointed out by P. Drossart (1978), working around B_z^r compensates for the inefficiency of the helical wiggler, but the wiggler entrance becomes very critical (it has been suggested to use this magnetic configuration for isotopic separation).

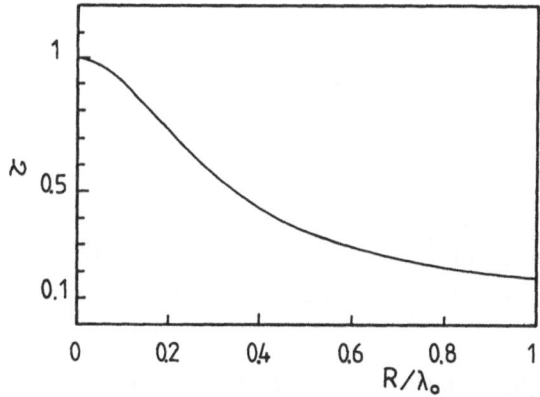

Fig. 12.   Ratio of the RHS helical field component of the LHS helical field component, τ, as a function of the radius r normalized to the wiggler period λ_o.

As shown by L. Friedland and J. L. Hirsfield (1980), in an ideal helical magnetic field, in presence of a guiding B_z field, helical trajectories are in fact an exception. For a given electron energy and helical field $B^\perp$, the ratio of $v^\perp/v_{//}$ is well defined and is a function of B_z. Figure 13 illustrates this dependence. There are two types of trajectories, Type I and Type II. Type II trajectories are always stable, but Type I are unstable for large $v^\perp$.

The electron injection and the stability can be analytically studied by using a convenient choice of variables, as shown by L. Vallier (1983). The proper variables are:

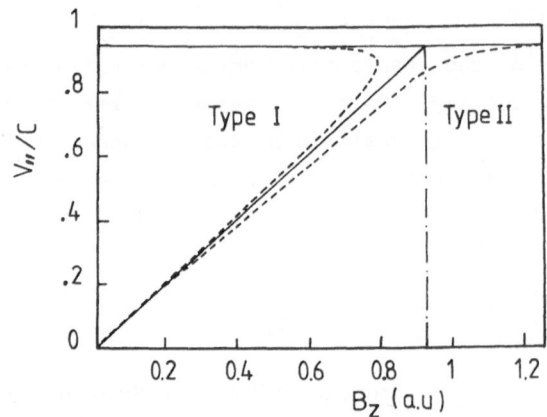

Fig. 13. Electron parallel velocity as a function of the guiding guiding field B_z for a given electron energy γ, a given wiggler period λ_o and a given helical field $B^\perp$.

$$u = \theta - k_o z$$

$$v = k_o z$$

$$P_u = \Omega R^2/2 \hspace{3cm} (50)$$

$$P_v = p_z/k_o + \Omega R^2/2$$

where $\Omega = eB_z$, $p_z = m\gamma v_z$ and the angles u, v and θ are defined on Fig. 14. R is the radius of an electron around its guiding center. The vector potential associated to the helical field is given by:

$$\vec{A} = -\frac{\vec{B}}{k_o} = A_o [\cos(k_o z)\ \underline{e}_x + \sin(k_o z)\ \underline{e}_y] \hspace{2cm} (51)$$

and the hamiltonian takes the form:

$$H/c = [m^2 c^2 + e^2 A_o^2 + 2\Omega P_u + k_o^2(P_v - P_u)^2 - 2eA_o(2\Omega P_u)^{1/2}\sin(u)]^{1/2} . \hspace{1cm} (52)$$

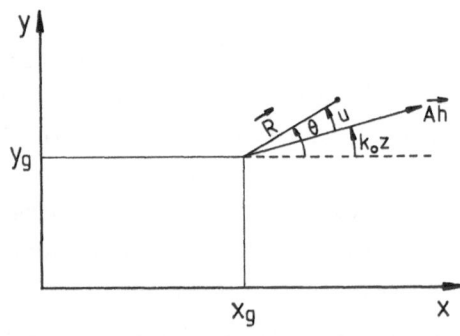

Fig. 14. Geometry of the transverse electron motion in a helical wiggler.

Because $\partial H/\partial v = 0$, $P_v = $ Cte is an integral of the motion and the interesting phase space is (u, P_u). The electron trajectories in this phase space are given by:

$$m^2 \gamma^2 c^2 = m^2 c^2 + e^2 A_0^2 + 2 \Omega P_u + k_0^2 (P_v - P_u)^2 - 2 e A_0 (2 \Omega P_u)^{1/2} \sin(u) \tag{53}$$

and are illustrated on Fig. 15. For the conditions of Fig. 15, we have two fixed points of Type I and a fixed point of Type II. Figure 15 helps to understand the difficulty of the electron injection in the wiggler: if the wiggler entrance is too sharp, then the electrons are initially distributed in u with some perpendicular given by $k_0 R$. After some time, the electrons follow the trajectories indicated in Fig. 15 and very few will be trapped around a stable helical trajectory. The result will be a strong dispersion of the electron beam in parallel and perpendicular velocities.

Tapering of the wiggler is therefore necessary. The tapering may be defined by studying the trajectories around the fixed points. By a perturbation analysis, it can be shown that the electron motion around these fixed points is an ellipse. This elliptic motion is characterized by a frequency ν given by:

$$[\nu/(eB_z/m\gamma)]^2 = \frac{B^\perp \beta^\perp}{B_z \beta_z} \left[\frac{B^\perp}{B_z} \left[\frac{\beta_z}{\beta^\perp} \right]^3 - 1 \right] \tag{54}$$

Figure 16 gives an example of the variation of ν for the two stable fixed points. The frequency ν is very important to define an adiabatic entrance in the wiggler, because if T is the traveling time along the wiggler for a small change in the field, one should have $T \gg 1/\nu$. Equation (55) shows that this condition of adiabaticity cannot always be satisfied since ν may be zero.

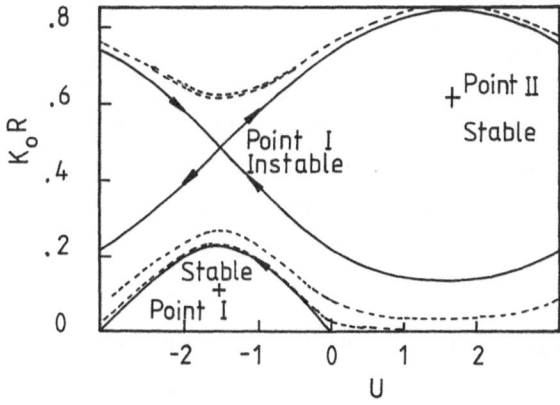

Fig. 15. The phase space $k_0 R, u$ for a value of the guiding field below B_c.

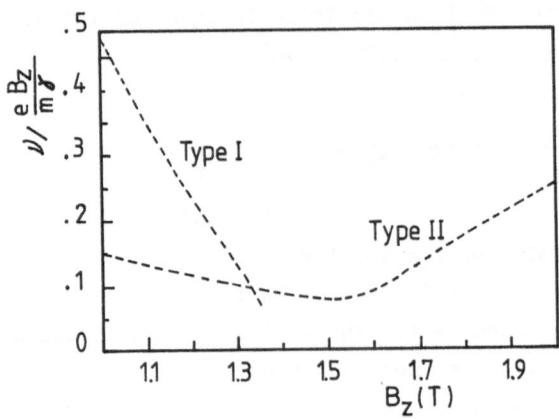

Fig. 16. Frequency of the elliptical motion around fixed
points for 1 MV electrons, $B^\perp = 200$ G, $\lambda_o = 2$ cm.

Beam Quality Measurements

Beam quality can be improved by diode design and careful electron
beam focusing and transport. But experimental diagnostics are also
desirable. Many diagnostics have been used in the past on IREB like wake
measurements and capacitive probes. However, these diagnostics are not
very accurate and also may intercept the beam and raise the question of
change in the beam properties due to the diagnostic itself.

Wake Measurements

The basic idea of the wake detector is to measure the betatron
oscillation of the beam. If the main electron beam is filtered through a
small aperture, then the betatron oscillations are given a free-streaming
calculation taking into account the beam distribution function in
velocity and the external B_z field (Felch et al., 1981). A scheme of the
apparatus is given on Fig. 17.

Assuming a given distribution function, it is possible as shown on
Fig. 18 to obtain information on the electron beam distribution itself.
In the experiment (Felch et al., 1981), the spread was assumed as
produced by the beam extraction through a Titanium foil. In this case,
the electron distribution may take the form:

$$F(v_\perp, v_{//}) = A \exp(-\theta^2/\theta_o^2)\ \delta(\sqrt{v_\perp^2 - v_{//}^2} - v_o) \ . \tag{55}$$

Even if the hypothesis of Eq. (55) is not relevant, the experimental
measurements may give an estimation of the parallel energy and of the
transverse energy spread. Moreover, because the measurement is time
dependent, it is possible to have a time resolution in the beam distribution function.

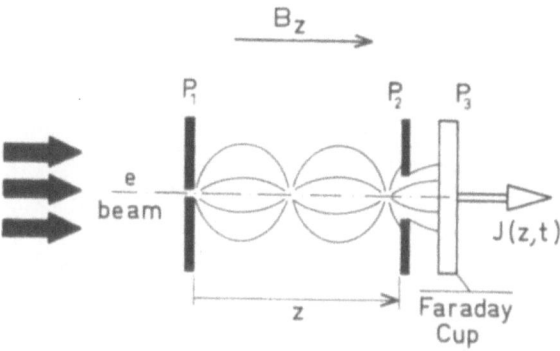

Fig. 17. Scheme of the wake detector. The plate P_1 has a hole
of radius R. Therefore the current collected by the
plate P_2 is modulated in R is the order of the beam
mean radius ρ_0.

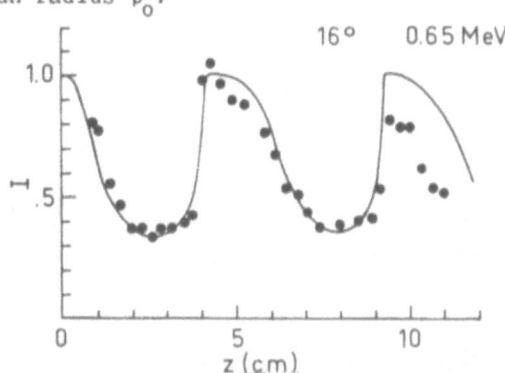

Fig. 18. Experimental measurements of the current on collector
P_2 as a function of the distance between P_0 and P_1.
The external B_z field is 3.2 kG. The theoretical
curve is a best fit drawn for θ_0 = 14° and using
Eq. 55 for the distribution function with an electron
energy of 0.6 MV.

However, in this kind of measurement, one has to stop the beam
through the entrance hole and the relation between the beam before the
stopping plate and the beamlet in the detector is not obvious.

Capacitive Probes

An other type of beam measurement has been proposed, using a capacitive probe (Shefer et al., 1983). The scheme of the experiment is given
in Fig. 19. The voltage on the probe V(t) is related to the beam
density:

$$V(t) = \frac{-eL}{C} \int_0^R n(r,t) 2\pi r dr \qquad (56)$$

where R is the beam outer radius, C is the probe capacitance and n(r,t)
is the local electron density at radius r in the beam. Since the total

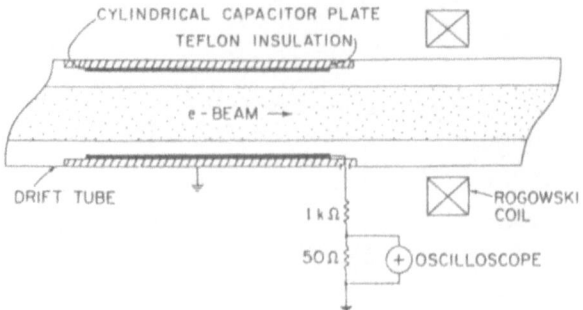

Fig. 19.   Schematic of the capacitive velocity probe.

current $I(t)$ may be measured by a Faraday cup, the beam velocity may be determined by:

$$v_{//} = \frac{L}{C} \frac{I(t)}{V(t)}$$ (57)

An advantage of this technique is that the perturbation induced on the beam is small compared to the wake diagnostic described earlier. However, the information is limited to the average beam velocity and the accuracy goes down at higher energy (> 1 MV).

Thomson Scattering

Thomson scattering appears a much more powerful technique because the perturbation induced on the beam is much smaller and because one may obtain information on the electron beam distribution itself.

Thomson scattering experiments have already been performed on IREB using a CO_2 laser as probe beam. Very interesting results have been obtained as regards the parallel energy dispersion with and without wiggle (Chen; Marshall, 1984), (Chen; Marshall, 1985). But because the relative long wavelength of this laser and of the properties of the Thomson scattering with relativistic electron beams, the beam was scanned in all its volume.

An effort is in progress now at the Ecole Polytechnique (Vallier; Buzzi, 1983), where probe beam is a neodymium laser. The choice of this wavelength allows spatial resolution of the electron beam along its direction of propagation.

A Nd-glass laser (1.06 μm, 3 J, 3 ns) is used to probe an intense relativistic electron beam (~1 kA, ~1 MV, ~20 ns) in a collective Free Electron Laser Experiment. The angle of the laser with respect to the electron beam is chosen to be 150°, because in this geometry, most of the scattered photons are pushed ahead by the snow plug relativistic effect and the Doppler shift reduces the probe laser wave length from 1.06 μm to 0.32 μm which is a convenient part of the visible spectrum. A schematic of the experiment is given on Fig. 20.

Because of the short duration of both the laser pulse and the electron beam, the synchronization of the two beams is one of the critical points of this experiment. This synchronization is done by triggering the Blumlein master switch of the electron beam with a fraction of the laser light converted in blue light with two frequency doublers. This technique has been tested with a one megavolt SF_6 switch, leading to a jitter between the electron beam and the laser beam of the order of one ns.

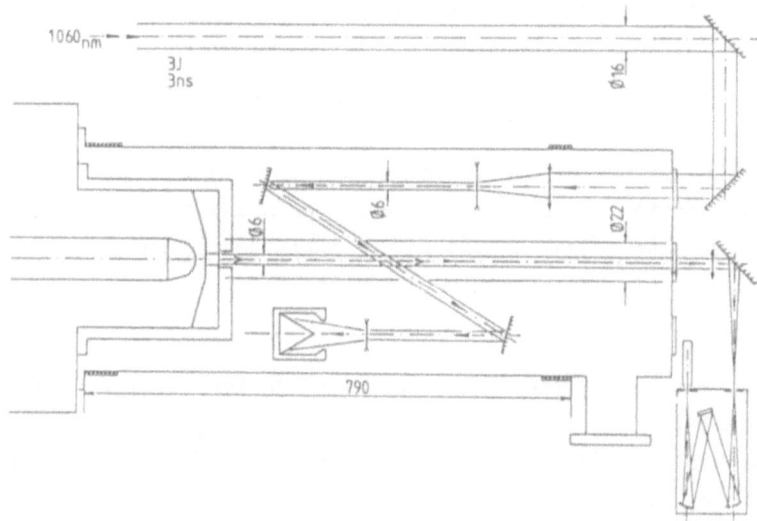

Fig. 20.   Geometry of the Thomson scattering experiment. The electron beam is guided by an external pulsed B_z field produced by the winding outside the stainless steel vacuum chamber. The two mirrors A and B as well as the telescope are movable along the electron beam.

Although no results are available today as regards the Neodynium laser, we think that this technique is very promising. In conjunction with the results of the CO_2 lasers, the Thomson scattering diagnostic is probably the best tool for checking beam quality in FEL experiment.

Because of the Doppler shift, it is limited to the MV energy range, but this range is the most critical for the beam quality. When the beam reaches higher energies, beam transport problems are reduced, essentially because the plasma frequency goes as $1/\gamma$.

SELF-FIELDS DEVICES

The needs for higher power beams and for higher electromagnetic power generation appear to be limited mainly by the beam current. High beam currents induce large self-fields and these fields are not compatible with the classical electromagnetic waves generators. However, one should not forget that in some devices, the self-fields are the necessary component to radiation. As an example, one may quote the magnetron which is a highly non linear and inhomogeneous device. In fact very impressive results have been obtained with CIREB magnetrons but these devices are limited for large currents because the B_θ of the self-magnetic field perturbs the insulation magnetic field B_z.

However, the people of radio astronomy are everyday recording the emission of gigantic electromagnetic sources, and it is not clear today if these sources are related to high quality beams, or particularly efficient emission process with natural beams and their self-fields. Moreover, we have today, in laboratory experiments, an example of a natural source of radiation, the Vircator.

Vircator

To my knowledge, the first experiment on the Vircator was done on 2 Ω device (Buzzi et al., 1978). Very quickly a confirmation was given (Mahafay et al., 1977) and the subject is still under investigation (Kwan; Thode, 1984). The basic facts are that if one injects a high current electron beam through an anode, then an emission occurs with an efficiency of about 1%. Gigawatt emissions have been observed and are interpreted only by the self-field of the device. Two emission processes have been proposed to explain the phenomena: a) the back and forward motion of the electrons around the anode and b) the vibrational motion of the virtual cathode behind the cathode.

Figure 21 illustrates the geometry of the various experiments. The most advanced theory (Kwan; Thode, 1984) clearly related the emission frequency to the beam density. This gives a limitation on the output frequency of these device types but since they are not limited in current, one may hope that gigantic emissions may be emitted with this kind of diode.

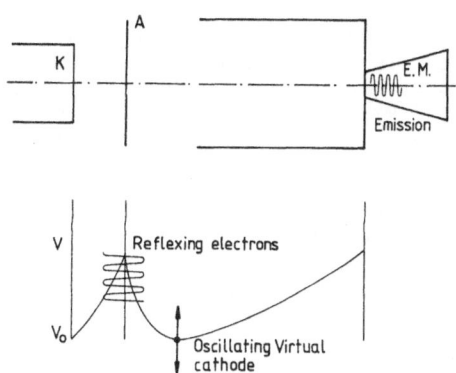

Fig. 21. Geometry of the Vircator.

CONCLUSIONS

Classical Intense Relativistic Electron Beam devices are still very interesting experimental tools for exploring new possibilities in electromagnetic waves generation.

Despite their limited electron energy in the MV range, it is still possible by new wiggler technologies or by using the Cerenkov effect to reach the far-infrared domain.

Intense beams in the kA range can meet the requirements for beam quality for FEL, if as shown in this paper, small beam sections can be realized. This emphasizes the need for careful diode design and beam diagnostics.

If processes like the FEL are limited by the self-fields of intense beams, some other devices like the Vircator are using these fields for the microwave generation. Experiments and theories in this field are promising because these generators are able to use the full power of the IREB available today.

REFERENCES

Abramowitz, M. and Stegun, I., 1965, "Handbook of Mathematical Function", p. 363, Dover Publication.
Buzzi, J. M., Doucet, H. J., Lamain, H., Rouillé, C., Delvaux, J., Jouys, J. C., 1978, J. de Phys. Lettres, L15-L18. (English translation by J. Benford from P.I. available.)
Chen, S. C., Marshall, T. C., 1984, Phys. Rev. Lett., 52(6), 425.
Chen, S. C., Marshall, T. C., 1985, IEEE J. of Quatum Electro., QE-21:924.
Destler, W. W., Granatstein, V. L., Mayergoyz, I.D., and Segalov, A., 1986, J. Appl. Phys., 60:521.
Doucet, H. J., French Patent # 7836123 (1978) and US Patent # 4.345.329 (1982).
Drossart, P., 1978, Thèse de 3.ième Cycle, University Paris XI, Orsay.

Felch, K., Vallier, L., Buzzi, J. M., Boehmer, H., Doucet, H. J.,
    Etlicher, B., Lamain, H. and Rouillé, C., 1981, Proceedings of the
    4th Topical Conference on High-Power Electron and Ion-Beam Research
    and Technology, 2:971, H. J. Doucet and J. M. Buzzi Ed., Laboratoire
    PMI, Ecole Polytechnique, Palaiseau Cedex 91128.
Friedland, L., Hirshfield, J. L., 1980, Phys. Rev. Letters, 22:1456.
Garate, E. P., Moustaiszis, S., Buzzi, J. M., Rouillé, C., Lamain, H.,
    Walsh, J., and Johnson, B., 1986, Appl. Phys. Letters, 48:1326.
Granatstein, V. L., Herndon, M., Parker, R. K., Sprangle, P., 1974, IEEE
    J. Quantum Electron, QE-10:651.
Hirshfield, J. L., Wachtel, J. M., 1964, Phys. Rev. Lett., 12:533.
Kwan, Thomas J. T. and Thode, Lester E., 1984, Phys. Fluids, 27:1570.
Mahafay, R. A., Sprangle, P., Golden, J., and Kapetanakos, C. A., 1977,
    Phys. Rev. Letters, 39:843.
Nation, J. A., 1970, Appl. Phys. Lett., 17:491.
Paritsky, H., 1959, J. Appl. Phys., 30:1828.
Shefer, R. E., Yin, Y. Z. and Bekefi, G., 1983, J. Appl. Phys., 54:6154.
Sprangle, P., Tang, Cha-Mei, Manheimer, W. M., 1980, Phys. Rev. A21:302.
Vallier, L., 1983, Thèse de 3.ième Cycle, University Paris XI, Orsay.
Vallier, L., Buzzi, J. M., 12-14 Sept. 1983, Proc. of the Fifth Int.
    Conf. on High-Power Particle Beams, U. of Calif. (USA), p. 525.

BRIGHTNESS LIMITS FOR ION SOURCES

Roderich Keller

GSI, Gesellschaft für Schwerionenforschung
Postfach 110541, 6100 Darmstadt, West Germany

INTRODUCTION

The status of high-current sources for ions from gaseous elements has by now reached quite satisfactory levels of reliability, and several basic source types are available that offer good results, see Keller (1985). Processing of materials with low vapor pressure or of corrosive gases still exhibits its inherent difficulties, but also in this area considerable progresses have been achieved, using two-gas techniques as did Shubaly et al. (1985) or hot running sources with internal oven or external supply as described by Keller et al. (1986). In consequence, it seems that the main brightness-limiting factor for ion beams as they are created by the sources is not so much determined by the plasma generator itself but rather by the extraction system used to actually form the beam.

An exception from this still applies to H- ion sources. While brightness values obtained by cesium-fed Penning sources of the Dudnikov type, see Smith et al. (1985), match those of good positive-ion sources this source type is affected by rather unstable beam creation and scarcely reproducible current values from run to run. Volume-production or cesium converter sources, on the other hand, do not yet reach high enough current density values to be competitive (Stevens et al., 1984; Prelec, 1986).

Also in the case of positive ions it must be kept in mind that the one useful ion species desired in the beam has to share the total beam current with other species whenever volatile vapors are fed into the source, the feeding gas is preferably ionized in a molecular state, sputtering or two-gas techniques are applied, or multiply charged ions are sought. Thus very often the actually achieved current and brightness values for many ion species are harder to produce than noble gas ions which are far below the maximum values obtained in the ideal case treated in the following, where the full ion current is carried by one ion species only. Throughout this paper it is assumed that a plasma generator is available yielding one ion species at whatever current density may be necessary to match the requirements imposed by the extraction conditions. As a practical example of an ion source with wide stable range of operation conditions, CHORDIS of GSI Darmstadt is shown in Fig. 1.

Another limitation of the discussion to come lies in the premise that only a single circular outlet aperture is used. The effects of employing multiple circular apertures to increase the total beam current will be explained at the end of this paper.

BEAM FORMATION

The formation of ion beams from plasma sources is governed by one basic principle: the plasma density at the outlet plane must match the extraction conditions: the higher the extraction field is, the higher the density must be. In detail, the trajectories of the extracted ions are determined by the form of the equipotential surfaces, even though the ions, due to their inertia, do not follow the field lines exactly. The detailed shape of the equipotential surfaces, however, is in turn influenced by the space-charge of the created beam, additionally to the effect caused by the imposed boundary conditions, that is, the shapes and potentials given to the various electrodes. A largely exaggerated

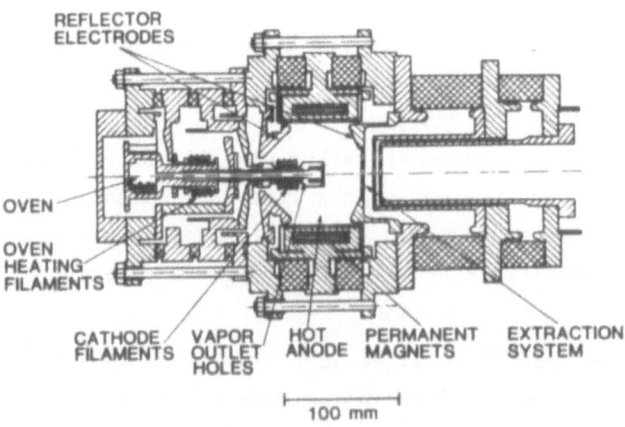

Fig. 1. Hot running multi-cusp source of the CHORDIS type (Keller et al., 1986).

illustration of the beam formation process is shown on Fig. 2. It is apparent that only a certain fraction of the total outlet aperture area contributes in supplying current to the actually transportable beam, within not too large an acceptance angle. The aim of any attempt to optimize extraction systems consists in maximizing this useful area.

Computer codes are quite helpful in understanding the problems related to ion beam formation, and there are some different codes available that seem to simulate the extraction and transport phenomena well enough (see, for example, Boers, 1979; Whitson et al., 1978; and Spädtke, 1983). A high-brightness extraction system developed by Piosczyk (1982) is shown in Fig. 3.

For an efficient transport of high-current ion beams at "low" (typically 50 keV) energy, the space-charge of the ions must be compensated by particles of the opposite charge. In the case of d.c. beams or with long enough pulse lengths of the order of some milliseconds, these compensating particles are produced by the beam itself through ionization of the residual gas in the beam line. They must, however, be kept from being accelerated backwards into the source because the normally encountered ionization rates are not high enough to allow for such extremely high losses. The induction of a screening electrode, on the other hand,

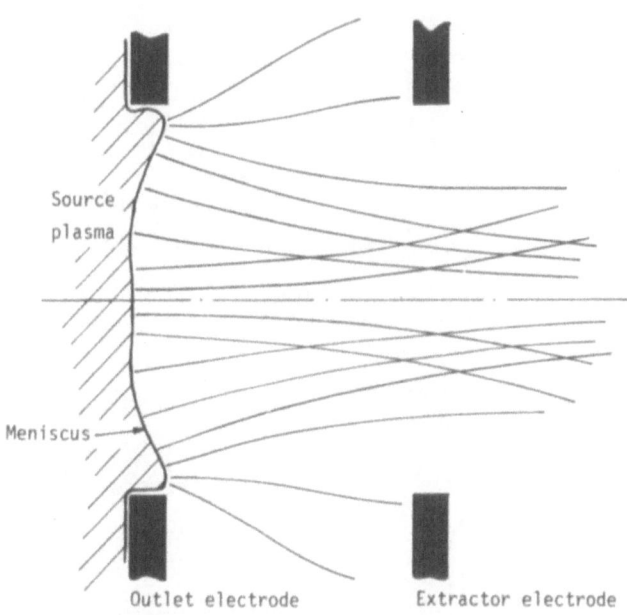

Source
plasma

Meniscus

Outlet electrode   Extractor electrode

Fig. 2. Beam formation with plasma sources (Keller, 1982).
The deformations of meniscus and ion trajectories
are by far exaggerated.

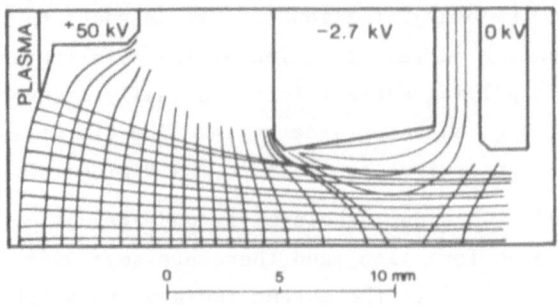

Fig. 3. Accel/decel (triode) extraction system with outlet,
screening, and ground electrodes, from left to right.
The beam current (simulated and actually measured)
amounts to 250 mA. From Piosczyk (1982).

causes a minor inconvenience due to the fact that it creates a diverging
ion optical lens. Therefore the screening electrode is commonly given the
lowest possible potential, where complete screening is just accomplished.

A quite promising technique consists of dividing the extraction gap
into two gaps as shown in Fig. 4. Thus, the total gap width can be
shortened, allowing higher fields to be applied, because of the nonlinear
relation between necessary gap width and voltage, see below. The major
benefit, however, is the possibility of using the voltage applied to the
inserted electrode as an additional free parameter to influence the shape
of the equipotentials. The technique seems to be convenient at

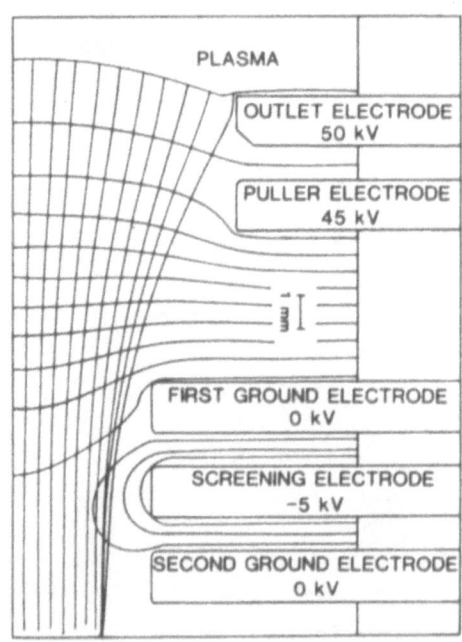

Fig. 4. Two-gap extraction system from Keller et al. (1983).
Here the screening, and ground electrodes, from left
to right. The beam current (simulated and actually
measured) amounts to 250 mA. From Piosczyk (1982).

extraction voltages above 50 kV, but in general for single-aperture systems only.

DEFINITIONS RELATING TO BEAM QUALITY

The quality of an ion beam is commonly judged by its brightness value which relates the beam current to the product of both transversal emittances. Concerning the beam brightness, frequently some misunderstanding occurs: In the accelerator community, people are familiar with the Liouville theorem which, when applicable, states that for a given beam the normalized emittance remains constant. This rule sometimes induces one to expect the maximum normalized brightness of an ion source to be constant, but that is completely wrong as will be shown below.

To avoid any misunderstandings of this kind, the exact definitions of beam quality parameters as used throughout this paper are listed in the following, and it must be stressed that every publication concerning beam quality values should contain a similar list to facilitate comparisons with results from other authors.

Here, a beam current emitted by a source is designated by I, but usually the transported part of such a beam, I_{tr}, has much more significance. For beam current comparisons with different ion species, the normalized current, $I_n = I /(A/\zeta)^{1/2}$ or proton equivalent is a convenient quantity. For a given extraction system, I_n should be constant, according to the Langmuir/Child law, see Langmuir (1929), when only the ion species is changed. The effective extraction voltage U defines the beam energy; the perveance is given by $P = I_n/U^{3/2}$. The mechanical extraction gap width is designated by d, the radius of the one circular outlet aperture is r, and the aspect ratio of the extraction gap is S = r/d. A beam waist size is given by r_{max}, the maximum radial position of any trajectory in the cross sectional plane where the waist lies. The maximum angle against the beam axis of any trajectory crossing the axis in the waist plane is α_o.

The two-dimensional, transverse emittance is $\varepsilon = A_E/\pi$ with A_E, effective area of the emittance pattern, including any portions of the phase plane that are not occupied by particles but lie within some simple convex border line encompassing the actual emittance pattern, possibly an ellipse. This quantity ε can be measured on a real beam or be derived from a beam simulation. To emphasize the distinction from ε_n, the normalized emittance, one can call ε the absolute emittance. The definition connecting both quantities is: $\varepsilon_n = \varepsilon \times \beta \times \gamma$, with β and γ, the usual relativistic parameters. As mentioned above, ε_n should be constant for a given beam even if its energy is changed. For ion sources and low-energy ion beams, $\gamma = 1$ is always fulfilled and the velocity can be evaluated from $\beta = 1.46 \times 10^{-3} \, (\zeta \times U/A)^{1/2}$.

The definition of the absolute emittance as chosen here is not the only one used in the accelerator community. Especially theoreticians treating transport problems in RF accelerators prefer RMS (root mean square) (Lapostolle, 1972) definitions which can be used in analytical calculations. The danger associated with this definition is that even the 4RMS-emittance contour, defined by

$$\varepsilon_{4rms} = 4 \, \varepsilon_{rms} = 4 \, [\overline{x^2} \, \overline{x'^2} - \overline{(x \, x')^2}]^{1/2},$$

with x, transverse position and x', angle against the beam axis for every trajectory of the beam, does not usually encompass an entire emittance pattern but rather 80% of commonly found distributions only. For arbitrary density distributions in the considered phase plane, the exact share of trajectories contained within the 4RMS ellipse is not even known.

In a different approach, the size of an absolute emittance is determined as the size of an ellipse with minimum area, just encompassing the given distribution (Keller et al., 1985). The distribution is then gradually reduced, and one can choose the desired working point on a beam current versus emittance size curve, always exactly knowing what part of the beam can still be transported in a given acceptance area. The 4RMS and Mini-ellipses for a simulated beam near its waist in the extraction region are compared in Fig. 5.

A common definition for the brightness of an axially symmetric beam reads: $B = I_{tr}/\varepsilon_n^2$. It seems indicated, however, to distinguish this quantity (better termed $B_{\varepsilon n}$, "emittance-normalized" brightness) and a newly introduced "current-normalized" brightness $B_{cn} = I_{tr}/\varepsilon^2$ that connects the normalized current with the absolute emittance.

SCALING RULES FOR HIGH-BRIGHTNESS BEAMS

As mentioned before, the following discussion is intended for conventional plasma ion sources yielding only one ion species out of one round aperture.

Two basic scaling rules for ion beam currents are throughout well established: $I_{tr} \propto (A/\zeta)^{-1/2}$ with a given system when only the ion species is varied, and $I_{tr} \propto U^{3/2}$ when nothing but the extraction voltage is changed; see Fig. 6 for an illustration.

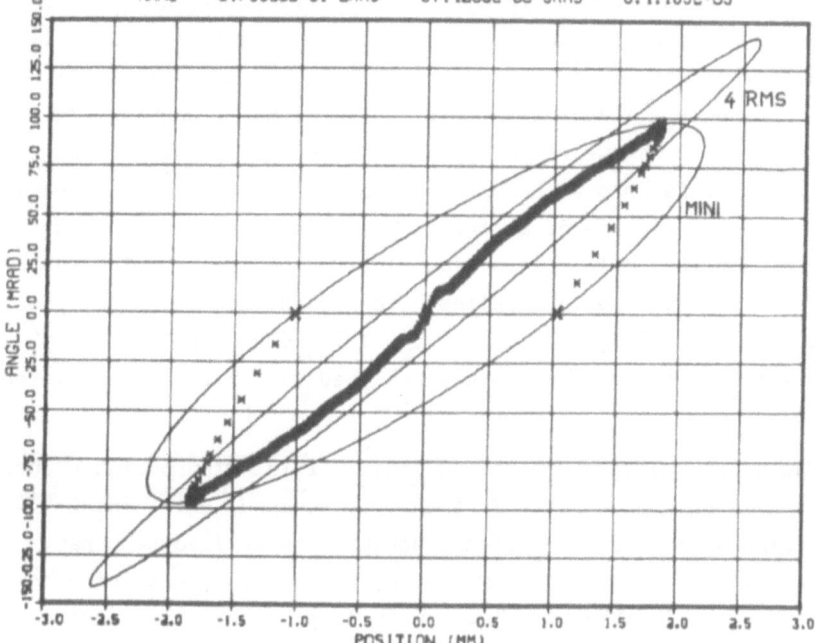

ERMS = 0.49001E+02 EMIN = 0.10009E+03 BF = 0.10000E+01
AMIN =-0.18884E+01 BMIN = 0.47885E-01 GMIN = 0.95357E+02
ARMS = -0.75980E+01 BRMS = 0.14268E+00 GRMS = 0.41163E+03

Fig. 5.   Emittance pattern of a simulated beam near the
extraction plane, with 4RMS and minimum-ellipse
contours.   After Keller et al. 1985).

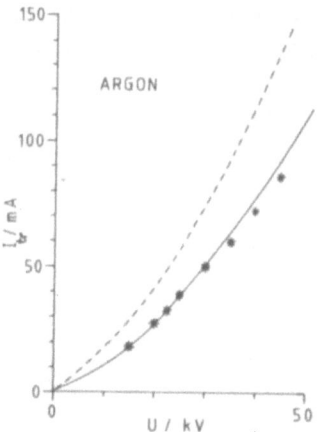

Fig. 6.   Dependence of transported ion beam currents I_{tr} from

the extraction voltage U, for a 7-aperture, 7-mm each,
triode extraction system.   The continuous curve is
fitted to current values up to 30 kV extraction voltage,

assuming a $U^{3/2}$-dependence.   Broken: empirical current
limit according to the formula derived in the text.   The
limit is not reached here because the system was optimized
for high brightness and not for maximum current.   From
Keller et al. (1986).

One might now think that by increasing the open area of the outlet aperture the beam current would rise proportionally. However, in a thorough experimental study of a duopigatron (Coupland et al., 1973) it was found that the actually transported ion currents with small divergence half-angle scale as $I_{tr,n} = P^*S^2U^{3/2}/(1+aS^2)$, see Fig. 7a, rather than $I \propto S^2 U^{3/2}$ as suggested by the Langmuir/Child law for the total currents. The factors a ("aberration factor") and P^* (low-S perveance) depend upon the divergence half-angle α accepted by the transport system, and also upon the chosen extraction electrode geometry. The quoted study states $a \simeq 3$. For the more refined two-gap extraction system presented by Keller et al. (1983) the values $a = 1.7$ and $P^* = 6 \times 10^{-8}$ (A/V$^{3/2}$) give the best fit to the current values measured within $\alpha = 20$ mrad, see Fig. 7b.

The existence of a saturation limit for the transportable current with increasing aspect ratio S strongly recommends maintaining $S \simeq 0.5$ when designing extraction systems, because the spilled beam part extracted out of larger openings causes quite unfavorable effects and the gain in current is then very low. Only sophisticated geometries, see Fig. 3, or two-gap systems permit the application of values $S \simeq 1$ without losing too much in reliability of the extraction system.

Taking the values P^* and a of the above quoted two-gap system for maximum performance reference data and allowing for $S = 1$, one can at once conclude that the maximum normalized current to be obtained from one round aperture and transported within 20-mrad divergence half-angle is a function of the voltage only and (in convenient units) amounts to

$$I_{tr,n} = 7.03 \times 10^{-4} U^{3/2} \quad (A/kV^{3/2}) \; ,$$

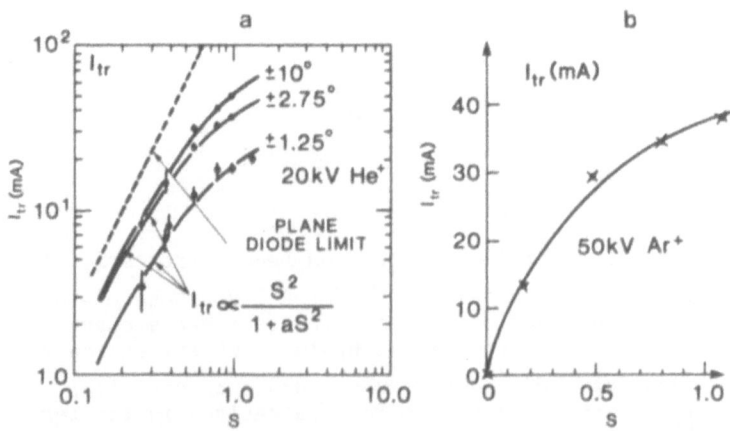

Fig. 7. Transported ion currents, I_{tr} versus aspect ratio S of extraction systems: (a) triode after Coupland et al. (1973), log-log plot, and (b) pentode after Keller et al. (1983), linear plot.

see Fig. 8. Many published beam data, reduced to one emitting outlet aperture, support this current limit formula in the sense that no current value significantly exceeds the line in Fig. 8. The references from which these data were taken are quoted elsewhere (Keller, 1982) or taken from papers published in the 1983 Kyoto ion source conference, see Takagi (1983).

The current limit formula given above does not consider the mechanical dimensions because these cancel due to the condition S = 1. In terms of beam brightness, however, these dimensions are quite important because the effective absolute emittances as defined above, taken in the plane of the beam waist, are given by the product $\varepsilon = \alpha_o x r_{max}$. It appears to be reasonably cautious to take the acceptance angle $\alpha = 20$ mrad used to derive the numerical values for the given current-limit formula as emittance-defining angle α_o in the plane of the beam waist. Further it has frequently been found (and been confirmed by computer simulations and

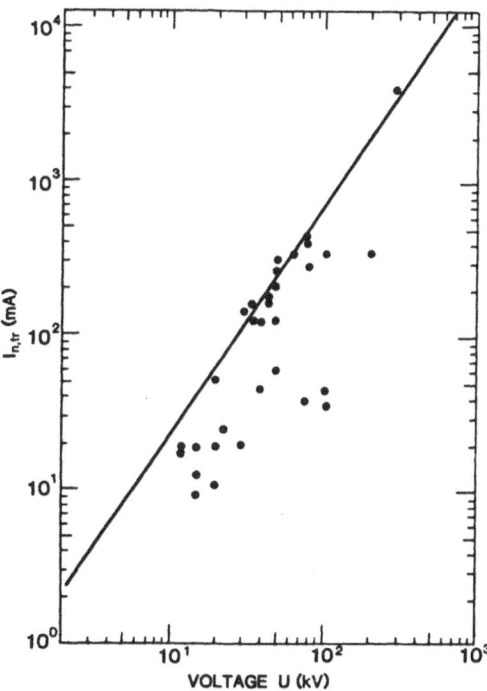

Fig. 8. Voltage dependence of normalized transported current $I_{n,tr}$. The points mark the $I_{n,tr}$ values of different actual sources, calculated for one emitting aperture only. Current values that lie significantly below the marked limit line mostly belong to neutral injection experiments, where conditions other than maximum transported current are dominant.

emittance measurements for the two-gap system treated here) that for axially symmetric, high-brightness beams the waist has one-half the extraction aperture width: $r_{max} = r/2$. Putting together the expressions for radius and angle, it now results that the absolute effective emittance of the transported beam is a function of r (which is equal to d) only, and amounts to

$$\varepsilon = 0.5 \times 20 \ r = 20 \times d \ (mrad) .$$

The question of minimizing the emittance, then is reduced to minimizing the gap width d, and the limits depend upon the high-voltage break-down law for the gap. Collected data of many existing high-current sources (Keller, 1982) show that the implicit formula given by Kilpatrick (1957) for the d.c. case indeed marks the actual break-down limits through two decades of voltages, see Fig. 9, and that a once more reasonably cautious empirical limit fitted to the displayed data is given by the explicit formula

$$d = 0.01414 \times U^{3/2} \ (mm/kV^{3/2}) ,$$

permitting 50 kV for a 5-mm gap.

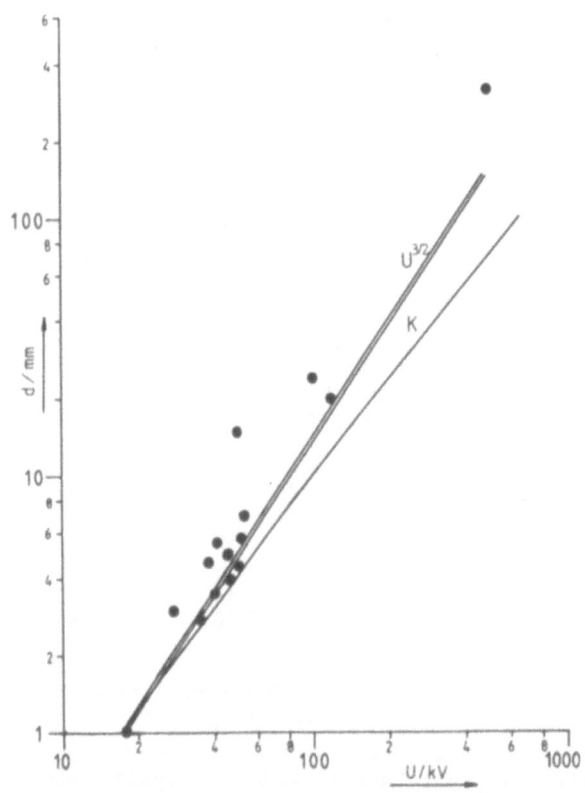

Fig. 9. Break-down limits for extraction gaps. K. Kilpatrick law (Kilpatrick, 1957). $U^{3/2}$, empirical limit.

Using this empirical scaling rule, for the absolute emittances immediately follows

$$\varepsilon = 0.1414 \times U^{3/2} \quad (\text{mm mrad/kV}^{3/2}) \; .$$

The minimum normalized emittances then scale according to

$$\varepsilon_n = 2.06 \times 10^{-4} \, (A/\zeta)^{-1/2} \times U^2 \quad (\text{mm mrad/kV}^2) \; .$$

The conclusion to be drawn for the dependence of the brightness values from the extraction voltages looks quite unfamiliar; the emittance-normalized (commonly used) as well as the newly introduced current-normalized brightness quantities strongly decrease with increasing extraction voltage, see Fig. 10:

$$B_{\varepsilon n} = 1.65 \times 10^4 \times (A/\zeta)^{1/2} \times U^{-5/2} \quad [\text{A kV}^{5/2}/(\text{mm mrad})^2]$$

and

$$B_{cn} = 3.53 \times 10^{-2} \times U^{-3/2} \quad [\text{A kV}^{3/2}/(\text{mm mrad})^2] \; .$$

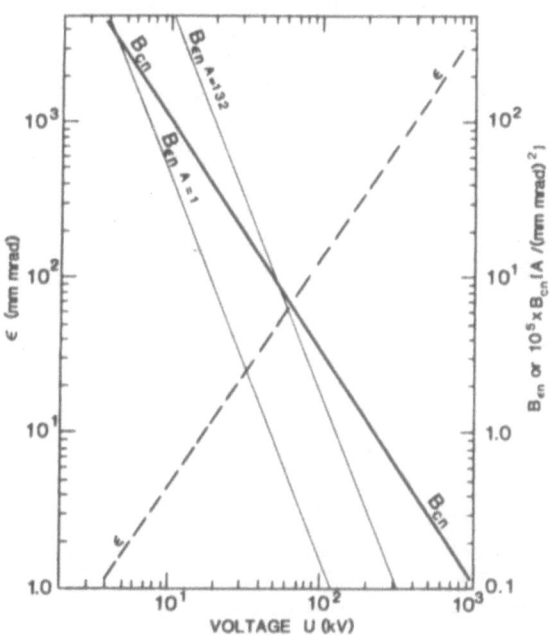

Fig. 10.   Voltage dependence of absolute emittance ε, emittance-normalized brightness $B_{\varepsilon n}$, and current-normalized brightness B_{cn}, calculated according to the formulae derived in the text.

405

By no means is this brightness scheme meant to display absolute physical constants in its numerical values. The real optimum brightness values may be quite different in various experiments, depending on different emittance definitions, the allowed acceptance angle (the 20 mrad chosen here is a completely arbitrary value), but also on the threshold values applied during the emittance measurement; further it can be advantageous to cut out only a very narrow central section of the beam, to achieve extreme brightness values as is demonstrated in Fig. 11. The numerical constants in the derived brightness scaling rules as written in this paper would best apply to cases where the maximum beam currents are sought at a given voltage, not the maximum brightness values themselves without caring what the absolute currents would be.

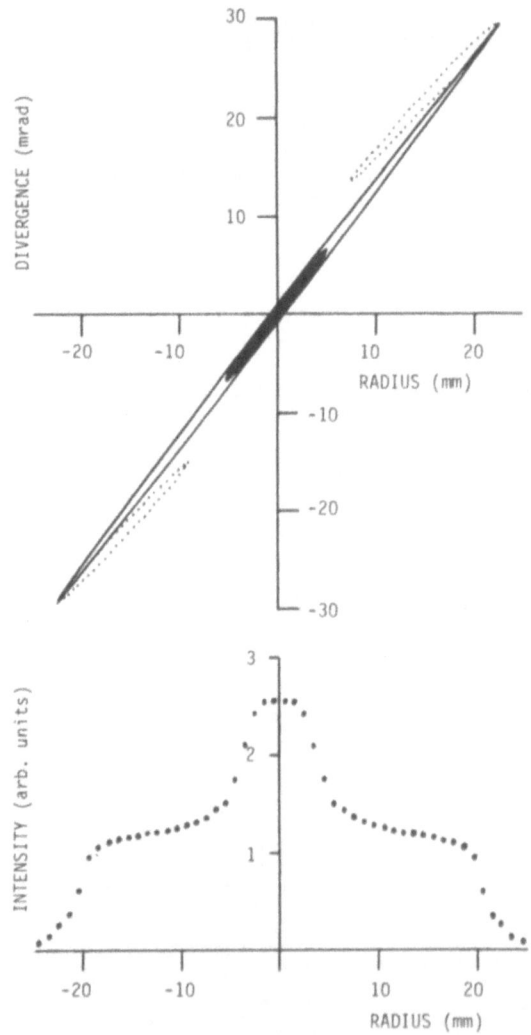

Fig. 11. Emittance pattern and profile of a 15-mA, 159 keV uranium beam extracted from the metal ion source MEVVA (Brown et al., 1985) by a compound extraction system (Keller, 1986). The intensity in the dotted wings of the emittance pattern is less than about 1% of the total intensity.

406

The emittance pattern of Fig. 11 leads to three different brightness values, depending on whether the whole pattern, the entire straight part of it, or the narrow black core only are considered. The numerical emittance-normalized brightness values for these cases are 0.22, 21, and 78 A/(mm mrad)^2, in the given order. This surprisingly large discrepancy illustrates how careful comparisons of different brightness values must be undertaken. In this sense, the brightness scaling rules just given are rather meant to indicate how the established brightness of one source would scale for the same share of beam taken if, for example, the extraction voltage had to be changed. In any case, the general tendency that the brightness decreases with increasing extraction voltage appears to be well confirmed by the available experimental data.

To eliminate a possible misunderstanding, it should be underlined that in the discussion given here the emittances are supposed to be dominated by optical effects, mostly aberrations of the aperture lenses in the extraction system. Many simulations as well as experiments clearly justify this choice as opposed to an emittance deviation from the transverse ion motions.

The current and absolute-emittance formulae given here can further be combined in another way than a brightness expression: the quotient of both (possibly to be named "emittance-specific current") does not depend on the voltage:

$$I_{tr,n}/\varepsilon = 4.97 \times 10^{-3} \quad [A/(mm\ mrad)] \ .$$

This ultimate expression appears to be a constant limit for high-brightness ion beams generated from a high-current source. It does not directly depend on other quantities, but is still subject to the premises outlined in this paper.

MULTI-APERTURE EXTRACTION SYSTEMS

In the preceding section the dependence of transported beam currents and emittances from the extraction voltage was analyzed for the case of one round emitting aperture. Very often, however, it is required to generate high ion currents at relatively low energy, exceeding by far the current limit as derived here. This can easily be achieved by employing multi-aperture extraction systems with many holes on the outlet electrode; and as long as the plasma generator yields a plasma with large enough cross section to offer constant current density to all outlet apertures the delivered ion current will simply scale proportionally to the number of these apertures.

407

The emittance of the entire beam, however, will suffer over-proportionally from the introduction of more emitting apertures because now the distances between the holes are contributing to the effective beam waist size, too. This effect leads to a brightness loss in the order of a factor of 10 when changing from a single aperture to seven apertures of equal total size, as illustrated in Figs. 12 and 13.

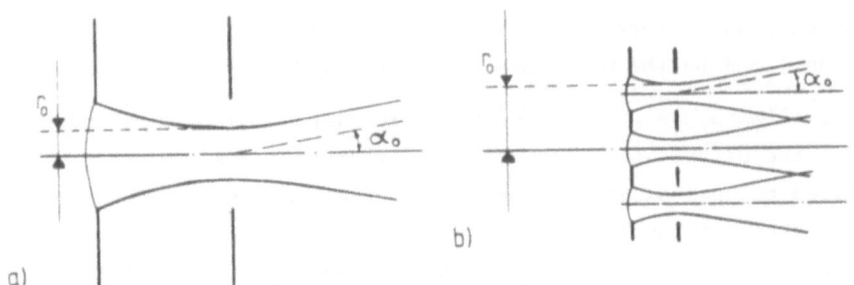

Fig. 12.   Comparison between single-aperture (a) and multi-aperture (b) extraction. Both systems have the same emitting area, assuming seven apertures in case (b). The increase of the effective beam waist radius r_0 leads to a brightness loss of a factor of 6.6 for case (b). From Keller et al. (1983).

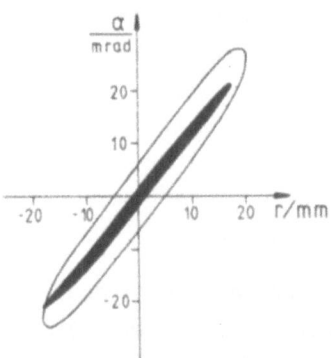

Fig. 13.   Comparison of the measured emittance patterns of two drifting argon beams, taken in 1 m distance from the source. Full black, 26-mA beam obtained from a single-aperture system with $B_{\varepsilon n}$ = 12.5 A/(mm mrad)^2.

Solid line, effective emittance contour for a 33-mA beam from a seven-aperture system, with $B_{\varepsilon n}$ = 1.1 A/(mm mrad)^2. From Keller et al. (1983).

For current requirements not too much above the stated limit, it is then worthwhile to examine if two, three, or four apertures were sufficient instead of seven, because the emittance increase would be lower than in the case of seven apertures. For aperture numbers above seven, no further brightness loss is to be expected, and in fact multi-aperture extraction systems with as many as several thousand holes have successfully been used for high-current experiments needing more than 10 A of protons.

REFERENCES

Boers, J. E., 1979, SNOW, a Digital Computer Program for the Simulation of Ion Beam Devices, Sandia Nat. Lab. report SAND 79-1027.

Brown, I. G., Galvin, J. E., and MacGill, R. A., 1985, High Current Ion Source, Appl. Phys. Letters, 47:358.

Coupland, J. R., Green, T. S., Hammond, D. P., and Riviere, A. C., 1973, A Study of the Ion Beam Intensity and Divergence Obtained from a Single Aperture Three Electrode Extraction System, Rev. Sci. Instr., 44:1258.

Keller, R., 1982, Ion Sources and Low Energy Beam Transport, Proc. Symp. on Accelerator Aspects of Heavy Ion Fusion, GSI Darmstadt report GSI-82-8:87.

Keller, R. Spädtke, P., and Hofmann, K., 1983, Optimization of a Single-Aperture Extraction System for High-Current Ion Sources, Springer Series in Electrophys., 11:69.

Keller, R., 1985, High-brightness, high-current ion sources, Proc. Workshop on High Current, High Brightness, and High Duty Factor Injectors, San Diego. Published as American Inst. Phys. Conf. Proc. (1986).

Keller, R., Sherman, J. D., and Allison, P., 1985, Use of a Minimum Ellipse Criterion in the Study of Ion Beam Extraction Systems, IEEE Trans. Nucl. Sci., 32:2579.

Keller, R., 1986, Innovations in Ion Sources and Injectors, Proc. 1986 Linac Conf., SLAC Stanford.

Keller, R., Spädtke, P., and Emig, H., 1986, Recent Results with a High-Current, Heavy-Ion Source System, Proc. 4th Conf. on Low Energy Ion Beams, Brighton. To be publ. in Vacuum TAIP.

Kilpatrick, W. D., 1957, Criterion for Vacuum Sparking Designed to Include Both RF and DC, Rev. Sci. Instr., 28:824.

Langmuir, I., 1929, The Interaction of Electron and Positive Ion Space Charges in Cathode Sheaths, Phys. Rev., 33:954.

Piosczyk, B., 1982, Preaccelerator Design and Component Development for the SNQ Linear Accelerator, Proc. 1981 Linac Conf., Santa Fe, Los Alamos Nat. Lab. report LA-9234-C.

Prelec, K., 1986, Progress in the Development of H- Ion Sources, Proc. 1986 Linac Conf., SLAC Stanford.

Schneider, J. D., Rutkowski, H. L., Meyer, E. A., Armstrong, D. D., Sherwood, B. A., and Catlin, L. L., 1979, Development of a High-Current Deuteron Injector for the FMIT Facility, Proc. 1979 Linac Conf. Brookhaven Nat. Lab. report BNL-51134:457.

Shubaly, M. R., Maggs, R. G., and Weeden, A. E., 1985, A High-Current Oxygen Ion Source, IEEE Trans Nucl. Sci., 32:1751.

Smith, H. V. Jr., Allison, P., and Sherman, J. D., 1985, The 4x Source, IEEE Trans. Nucl. Sci., 32:1797.

Spädtke, P., 1983, AXCEL-GSI, GSI Darmstadt report GSI-83-9.

Stevens, R. R., York, R. L., McConnell, J. R., and Kandarian, R., 1984, Status of the New High-Intensity H- Injector at LAMPF, Proc. 1984 Linac Conf., GSI Darmstadt report GSI-84-11:226.

Takagi, T., ed. 1983, Proc. Int. Ion Engineering Congress, Kyoto
Whitson, J. C., Smith, J., and Whealton, J. H., 1978, Calculations
    Involving an Ion Beam Source, J. Comput. Phys., 28:408.

BRIGHTNESS LIMITS IN LINEAR ION ACCELERATORS*

Thomas P. Wangler

Accelerator Technology Division, MS-H817
Los Alamos National Laboratory
Los Alamos, New Mexico 87545 USA

INTRODUCTION

Radio-frequency linear ion accelerators are attractive for high beam-current applications over a large energy range (Jameson, 1986). High peak currents are possible because relatively strong focusing can be provided, and the half-integer and integer resonances that severely limit the beam intensity in circular machines are not present in linacs. Furthermore, because there are no fundamental restrictions that limit the duty factor in linear accelerators, high average beam currents also can be delivered. Nevertheless, limits to the beam current do exist and undesired effects of the space-charge forces, which we are only beginning to understand, can cause serious degradation of the brightness. These effects usually occur at the low-velocity end of the linac, where beam density is highest, focusing is weakest, and bunches are formed from the injected dc beam.

For most rf-linac applications, a single figure or merit such as brightness is insufficient to characterize the beam. Usually, at least two variables are important; they can be chosen as beam current, I, and rms emittance, ε. For this reason, we will separately discuss limits on current and emittance, from which limits on brightness or other figures of merit can be deduced. Although brightness, defined as current density per unit of phase-space volume (usually either a 4-D of 6-D volume is specified), may be a useful figure of merit for some output-beam applications, the tune-depression ratio is probably a more relevant quantity for describing the importance of space-charge effects in linacs.

---

* Work performed under the auspices of the US Dept. of Energy.

In this context, the tune refers to the frequency of single particle oscillations about the trajectory of a reference particle, expressed in appropriate units. For periodic focusing configurations, such as those used in linacs, two tunes can be defined: (1) a phase advance σ of either transverse or longitudinal oscillations per focusing period L in the presence of space charge, and (2) σ_o, the same quantity at zero beam current. The quantity σ_o is a measure of the effective strength of external focusing. The transverse tune σ_o can be expressed in a smooth approximation as shown in the Appendix.

The tune σ is always less than or equal to σ_o because it includes the defocusing space-charge force. Although a nonzero tune spread exists in any real beam because the space-charge force is generally nonlinear, we define an effective tune σ by referring to an equivalent uniform beam. The tune depression ratio σ/σ_o is unity for zero beam current and decreases to zero as beam intensity and space charge increase. The quantity σ/σ_o determines the ratio of the space charge to emittance term in the envelope equation. It can be shown that the ratio of the space charge to emittance term in the envelope equation is equal to $\sigma_o^2/\sigma^2 - 1$. The ratio is unity when $\sigma/\sigma_o = 0.707$ and the beam is considered to be space-charge dominated for lower values of σ/σ_o. The quantity σ/σ_o is not a direct function of brightness. For example, it is a function of I/ε for a round continuous beam and of $I/\varepsilon^{3/2}$ for a spherical bunch (Wangler, Crandall, and Mills, 1986).

The problem of obtaining high-current beams with low output emittance is a challenging one for accelerator design. To solve this problem, we need to understand the limits on maximum beam current and minimum emittance. At the present time, we have a fairly good understanding of beam-current limits, at least in general terms. The peak current is limited by the focusing available to confine a space-charge defocused beam with finite emittance to within a given radial aperture a. Current limit formulas can be derived using a uniform ellipsoid model to calculate the space-charge force (Wangler, 1980, and Reiser, 1981). The transverse and longitudinal current limits, I_t and I_ℓ, are given by

$$I_t = \frac{4\varepsilon_o mc^3}{3[1 - f(p)]e} (\beta\gamma)^3 |\phi_s| \sigma_o^2 \frac{a^2}{\gamma L^2} [1 - \varepsilon_t^2/\varepsilon_{to}^2] \quad , \tag{1a}$$

and

$$I_\ell \sim (2\varepsilon_o c/\gamma^{1/2})\beta E_o Ta\phi_s^2 |\sin \phi_s| [1 - \varepsilon_\ell^2/\varepsilon_{\ell o}^2] \quad , \tag{1b}$$

for a final transverse emittance ε_t, final longitudinal emittance ε_ℓ, and zero-current acceptances ε_{to} and $\varepsilon_{\ell o}$ for the transverse and longitudinal degrees of freedom. Other accelerator parameters are defined in the Appendix. The decrease of current limits with increasing emittance reflects a competition between the space-charge and emittance terms within the available aperture and for a given focusing strength. At the current limit the final transverse tune-depression ratio can be shown to satisfy $\sigma/\sigma_0 = \varepsilon_t/\varepsilon_{to}$, and a similar expression holds in the longitudinal direction. In principle, if a lower limit on the tune-depression ratios exists, there will also be an upper limit on the emittance-dependent factors in the current limit formulas. Numerical simulation results using the computer code PARMTEQ (Wangler, 1980) have been shown to be in good agreement with Eqs. (1a) and (1b) for RFQ linacs.

For linacs that operate in a pulsed mode at low duty factor, the average current may be much smaller than the peak value. However, when the duty factor is large, limitations in the average current may become important. These limitations can arise from the beam dynamics, as a direct result of emittance growth and the consequent formation of a halo, surrounding the central core of the beam. Such a halo would lead to unwanted particle losses, which may represent a small fraction of the beam, but which, nevertheless, can produce an increased heat load, vacuum degradation, susceptibility to rf electric-field breakdown, and radio-activation of the accelerator components that makes routine maintenance very difficult.

A deterioration of the beam quality as a result of rms emittance growth has been observed both in numerical simulation studies and in experimental measurements (Jameson, Ed., 1978). This growth can occur even when Liouville's theorem is satisfied and the true phase-space volume of the beam remains invariant (Lawson, 1977). As we have discussed, a small emittance is desired not only to avoid a reduction in beam-current limit, but also for operational reasons because of the desirability of reducing halo and excessive particle loss. Furthermore, some applications place severe requirements on focusing the output beam, which can only be achieved by providing a very low emittance beam. Until recently, space-charge-induced emittance growth in rf linacs could be calculated by computer simulation, but no quantitative predictions were available to serve as guidance for high-current/low-emittance linac design, even for the ideal case of perfectly aligned beams with no nonlinear external fields and no image forces.

RMS EMITTANCE AND NONLINEAR FIELD ENERGY

A new understanding of the relationship between rms emittance and space-charge field energy, stimulated by the initial work of Struckmeier, Klabunde, and Reiser (1984), has led to some useful approximate equations

413

for emittance growth. For a round continuous beam in a linear focusing channel, a differential equation relating rms emittance and field energy can be written for an arbitrary distribution as (Wangler, Crandall, Mills, and Reiser, 1985; and Wangler, Crandall, Mills, and Reiser, 1986)

$$\frac{d\varepsilon^2}{dt} = -\frac{X^2 K}{2}\frac{dU_n}{dt} \quad , \tag{2}$$

where the rms emittance ε is defined in terms of the second moments of displacement x and divergence x' as $\varepsilon = 4(\overline{x^2}\,\overline{x'^2} - \overline{xx'}^2)^{1/2}$. The quantity K is the generalized perveance defined as $K = eI/2\pi\varepsilon_0 mc^3\beta^3\gamma^3$, and $X = 2\sqrt{\overline{x^2}}$ is the beam radius of an equivalent uniform beam (a uniform beam with the same current and same second moments $\overline{x^2}$, $\overline{x'^2}$, and $\overline{xx'}$ as the real beam). The dimensionless quantity U_n is called the nonlinear field energy, proportional to the difference between the space-charge field energies of the real beam and of the equivalent uniform beam. The nonlinear field energy is found to be independent of beam current and rms beam size and depends only on the shape of the charge density in real space. The U_n minimum is zero (for uniform charge density), and it increases as the charge density becomes more nonuniform. Thus, U_n is a measure of the nonuniformity of the charge density. Numerical values of U_n for several examples of spherical charge distributions are given in Table 1.

A generalized form of Eq. (2) for a bunched beam was derived by Hofmann (1986) and can be written for three degrees of freedom x, y, and z (with linear focusing in each plane) as

$$\frac{1}{\overline{x^2}}\frac{d\varepsilon_x^2}{dt} + \frac{1}{\overline{y^2}}\frac{d\varepsilon_y^2}{dt} + \frac{1}{\overline{z^2}}\frac{d\varepsilon_z^2}{dt} = \frac{-32}{mc^3\beta^3\gamma^3 N}\frac{d(W - W_u)}{dt} \quad , \tag{3}$$

where N is the number of particles in the bunch, and W and W_u are the space-charge field energies of the real beam and of the equivalent uniform beam. Equation (3) [and Eq. (2)] can be integrated for the case of a space-charge-dominated beam with linear continuous focusing (smooth approximation for periodic focusing), where the rms beam sizes are

Table 1. Nonlinear field energy and minimum final tune depression for a spherical bunch.

	U_n	$\sigma/\sigma_0\big|_{min}$
Gaussian	0.3083	0.32
Parabolic	0.0368	0.11
Uniform	0.0	0.00

approximately constant, independent of emittance. Equations for
emittance growth can be derived for both bunched and continuous
nonstationary beams. The result for an axially symmetric bunched beam in
the longitudinal plane can be written (Wangler, Guy, and Hofmann, 1986)

$$\frac{\varepsilon_{zf}}{\varepsilon_{zi}} = \left[1 - \frac{2}{P_i} \frac{(P_i - P_f)}{(2 + P_f)} - \frac{P_f}{P_i} \frac{G(b/a)}{(2 + P_f)} \left(\frac{\sigma_{ox}^2}{\sigma_x^2} - 1 \right) \left(U_{nf} - U_{ni} \right) \right]^{1/2}, \quad (4a)$$

and for the transverse plane we obtain

$$\frac{\varepsilon_{xf}}{\varepsilon_{xi}} = \left[1 + \frac{(P_i - P_f)}{(2 + P_f)} - \frac{G(b/a)}{(2 + P_f)} \left(\frac{\sigma_{ox}^2}{\sigma_x^2} - 1 \right) \left(U_{nf} - U_{ni} \right) \right]^{1/2}, \quad (4b)$$

where the subscripts i and f refer to the initial and final states of the
beam, a and b are the rms beam sizes in x and z, and G(b/a) is a geometry
factor equal to unity for a spherical bunch. The quantity P, called the
partition parameter, is defined as $P = \overline{z'^2} / \overline{x'^2}$ and is a nonrelativistic
measure of the kinetic-energy asymmetry in the rest frame of the bunch.
We have chosen to express Eqs. (4a) and (4b) as a function of the x-plane
tunes, although we could also have used the z-plane. Two mechanisms
contribute to the emittance growth in Eqs. (4a) and (4b). The first term
describes kinetic-energy exchange between the two planes, is zero only
when P does not change, and enters with opposite signs in the two planes
so that an emittance increase in one plane results in an emittance
decrease in the other (however, emittance increase in one plane is not
exactly balanced by the emittance decrease in the other). The second
term describes the charge-redistribution mechanism and is zero only when
U_n does not change. This term contributes either emittance growth or
decay in each plane and corresponds to an exchange between field energy
and particle kinetic energy as the charge density in real space evolves.

Although the initial values of the parameters P and U_n are known in
principle for a given initial state of the beam, we have no theory
available to allow a determination of the time dependence for P and U_n.
However, numerical simulation studies have shown that for highly
space-charge-dominated beams in linear focusing channels, the charge
density approaches a uniform distribution ($U_{nf} \simeq 0$) and the beam tends to
an equipartitioned state ($P_f = 1$). The final uniform-charge density may
be explained as a tendency for charge redistribution to shield the
interior of the beam from the linear external force, in analogy with a
cold plasma. This generally results in a matched charge-density profile
consisting of a uniform central core with a Debye sheath at the beam
edge, whose thickness is given by the Debye length, and which becomes
zero in the extreme space-charge (cold-beam) limit.

Why a beam, whose interactions predominately occur through collective fields rather than collisions, should equipartition is not yet clear. Nevertheless, these assumptions about the final state of the beam provide us with an adequate model for predicting final emittance growth. Thus, if $U_{nf} = 0$ and $P_f = 1$, Eqs. (4a) and (4b) can be written

$$\frac{\varepsilon_{zf}}{\varepsilon_{zi}} = \left[1 - \frac{2(P_i - 1)}{3P_i} - \frac{G(b/a)}{3P_i} \left(\frac{\sigma_{ox}^2}{\sigma_x^2} - 1 \right) U_{ni} \right]^{1/2} , \qquad (5a)$$

and

$$\frac{\varepsilon_{xf}}{\varepsilon_{xi}} = \left[1 + \frac{(P_i - 1)}{3} - \frac{G(b/a)}{3} \left(\frac{\sigma_{ox}^2}{\sigma_x^2} - 1 \right) U_{ni} \right]^{1/2} , \qquad (5b)$$

Numerical simulation studies have been used to test the formulas for the case of a 3-D spherical bunch (Wangler, Crandall, and Mills, 1986), using a code where the radial self-forces are calculated from Gauss' law. We use 4000 particles with rms-matched initial distributions, where, at each time step, the particles are given an impulse based on both the external and self-forces. First, we present results for a space-charge-dominated bunch with an initial Gaussian distribution, both in position and velocity space, truncated at four standard deviations, and with different values of the initial tune ratio, k_i/k_o (k and k_o are phase advances per unit length, related to the phase advances σ and σ_o in a periodic channel by $k = \sigma/L$ and $k_o = \sigma_o/L$). Figure 1 shows numerical simulation results of emittance growth $\varepsilon_f/\varepsilon_i$ at 10 plasma periods (after which the beam is relatively stable), plotted versus k_i/k_o. The curve, obtained from Eq. (5) with $P_i = 1$, $b = a$, and $G(b/a) = 1$, using the value $U_{ni} = 0.308$, is in good agreement with the simulations and shows the steep rise in emittance growth at low tune depressions (high space charge). The good agreement for low k_i/k_o is not unexpected because for this case, the assumption of final charge-density uniformity is a good approximation. An unexpected result is the agreement of Eq. (5) with the simulations at high k_i/k_o, which is due to the large error in nonlinear field energy change (by assuming a final uniform beam) that is offset by the small value of the tune ratio factor in Eq. (5). Figure 2 shows the same plot for an initial thermal or semi-Gaussian spherical bunch, which has a uniform charge density ($U_{ni} = 0$ and a Gaussian velocity distribution. A small rms emittance decrease is observed for some cases as the initially uniform charge density acquires a nonzero Debye sheath, and the sharp edge of the beam becomes a smooth rounded edge. Again the curve from Eq. (5) is in close agreement with the numerical simulation results. The numerical simulations show that this growth of emittance occurs very rapidly, in about one-quarter of a plasma period, followed by damped oscillations to a final nearly stationary beam configuration.

416

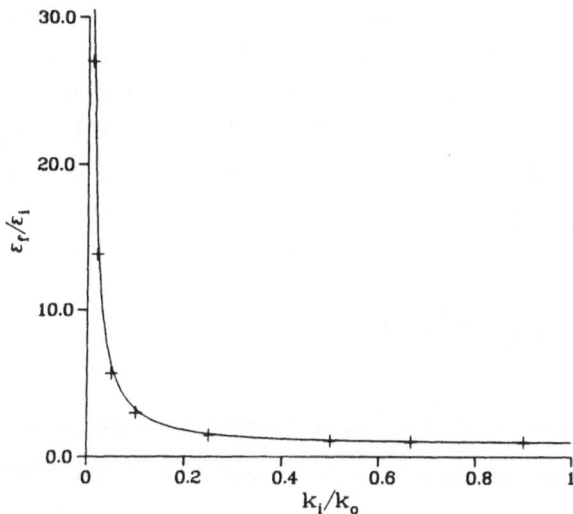

Fig. 1. Final emittance-growth ratio versus initial tune ratio
k_i/k_o for an initial Gaussian spherical bunch, truncated at
four standard deviations. The curve is generated from Eq.
(5) and the plus symbols show the results of the particle
simulations after 10 plasma periods.

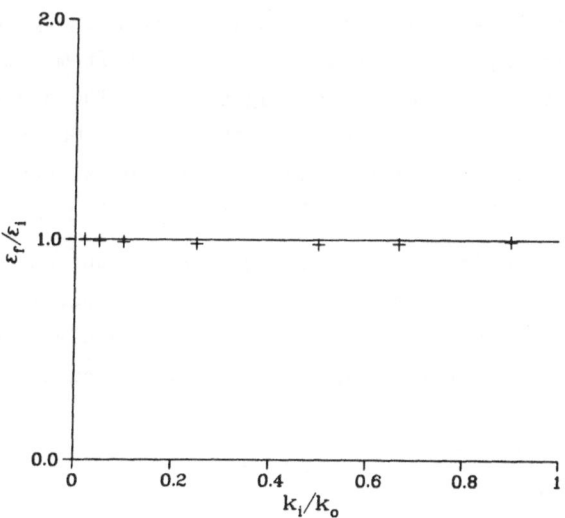

Fig. 2. Final emittance-growth ratio versus initial tune ratio
k_i/k_o for an initial thermal (semi-Gaussian) spherical
bunch, truncated at four standard deviations. The curve is
generated from Eq. (5) and the plus symbols show the
results of the particle simulations after 10 plasma
periods.

LIMITS ON EMITTANCE, TUNE DEPRESSION, AND BRIGHTNESS

The charge-redistribution mechanism of emittance growth leads to a minimum final emittance in the extreme space-charge limit (Wangler, Guy, and Hofmann, 1986; and Wangler, Crandall, and Mills, 1986). This effect was observed in early numerical studies of space-charge phenomena in high-current linacs, but never fully explained. This result can be seen by solving Eqs. (5a) and (5b) for final emittances as a function of the initial emittances assuming a space-charge-dominated beam. For the special case of a spherical bunch, the result can be written as

$$\varepsilon_f^2 = \varepsilon_i^2 + \frac{16}{3} U_{ni} \left(\frac{K_3^2 L}{\sigma_o} \right)^{2/3} , \tag{6}$$

where $K_3 = eI\lambda/20\sqrt{5}\pi\varepsilon_o mc^3 \beta^2 \gamma^3$ is a bunched-beam perveance.

As ε_i approaches zero, the final emittance approaches the value

$$\varepsilon_{f,min} = \frac{4}{\sqrt{3}} U_{ni}^{1/2} \left(\frac{K_3^2 L}{\sigma_o} \right)^{1/3} , \tag{7}$$

Equation (7) predicts that the minimum final emittance depends on the square root of the nonlinear field energy and on $I^{2/3}$ through the parameter K_3. A comparison of Eq. (6) with numerical simulations for a spherical bunch is made in Figs. 3 and 4, where $k_o \varepsilon_f$ versus $k_o \varepsilon_i$ is plotted for initial Gaussian and thermal (uniform charge density in real space) distributions, respectively. The results from the simulations and the curves from Eq. (6) are in close agreement. The nonzero value of the minimum final emittance is evident for the initial Gaussian bunch in Fig. 3, in contrast to the result for the initial thermal bunch in Fig. 4, where the minimum final emittance is zero.

The growth of emittance from charge redistribution to a minimum final value at the extreme space-charge limit reduces the brightness and increases the tune-depression ratio. For the example of a spherical bunch, one can show that the minimum tune depression is given approximately by

$$\left. \frac{\sigma}{\sigma_o} \right|_{min} = \left(\frac{U_{ni}}{3} \right)^{1/2} , \tag{8}$$

is independent of beam current, and depends only on the shape of the initial charge density through U_{ni}. Numerical values for initial Gaussian, parabolic, and uniform charge densities are given in Table 1. Other effects not yet studied, such as nonlinear focusing fields, could modify these numbers. The maximum 4-D brightness, defined as $B = K_3/\varepsilon_f^2$,min, can be written

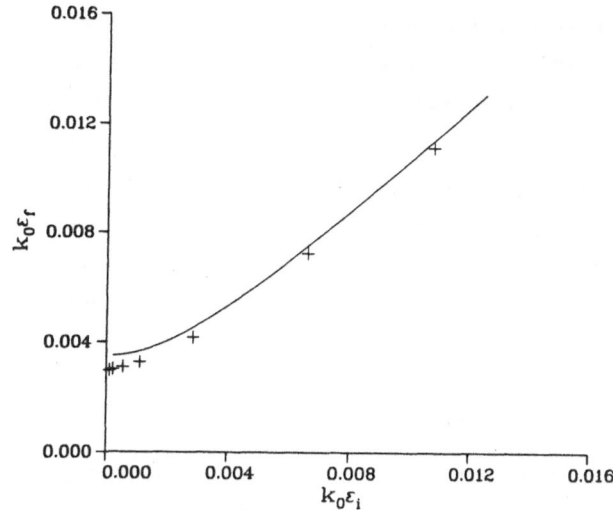

Fig. 3. Final versus initial $k_o \varepsilon$ for an initial Gaussian spherical bunch, truncated at four standard deviations. The curve is generated from Eq. (6) and the plus symbols show the results of particle simulations after 10 plasma periods.

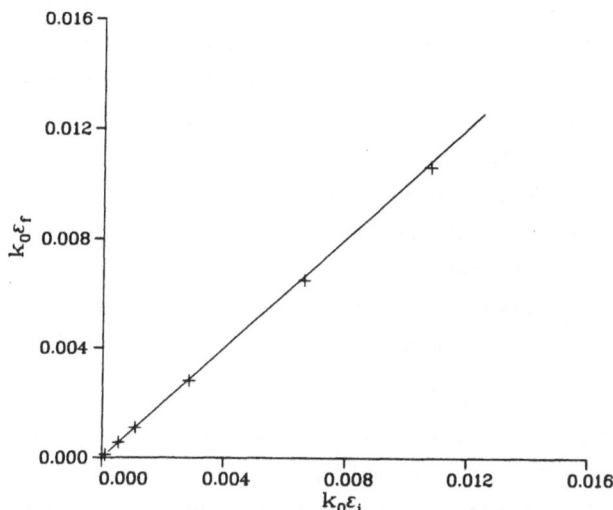

Fig. 4. Final versus initial $k_o \varepsilon$ for an initial thermal (semi-Gaussian) spherical bunch, truncated at four standard deviations. The curve is generated from Eq. (6) and the plus symbols show the results of particle simulations after 10 plasma periods.

$$B_{max} = \frac{3}{16U_{ni}} \left[\left(\frac{\sigma_o}{L} \right)^2 \frac{1}{K_3} \right]^{1/3} , \tag{9}$$

Equation (10) predicts a brightness limit that becomes infinite either as the beam intensity approaches zero or as the initial beam approaches a uniform density. Conversely, as beam intensity increases for a fixed initial distribution, the maximum attainable brightness decreases.

Numerical estimates of emittance growth from the kinetic-energy exchange and charge-redistribution mechanisms, as given by Eqs. (5a) and (5b), lead to values of the same order as have been reported experimentally (factors of 2 to 3) (Wangler, 1986). But detailed quantitative comparisons must be made with measurements from existing linacs before we can claim to have a complete understanding.

EMITTANCE GROWTH FOR QUADRUPOLE PERIODIC FOCUSING

The emittance growth formulas we have presented have been derived for continuous linear focusing. It is of great interest to determine whether the equations do represent a good approximation for periodic focusing, such as is used in real linacs. In addition, we have pointed out the advantages of an initially uniform charge density in real space to eliminate emittance growth from charge redistribution. It is important to determine whether this conclusion is also valid for periodic systems. This may be an important practical point because it would mean that an initially self-consistent distribution is not necessary to minimize rms emittance growth, and it is probably much easier to produce an initially uniform beam than an initially self-consistent one. Of course, the presence of nonlinear external forces in real linacs may modify the conclusion that the initial charge density should be exactly uniform. Finally, the designer of a linac for high-current/low-emittance applications would like to determine whether there is really no lower limit on emittance and tune depression.

In fact, the published numerical and experimental studies do appear to support the conclusion that initially uniform beams in quadrupole periodic channels give approximately no emittance growth, at least for $\sigma_0 \lesssim 60°$ to $80°$ (Hofmann, Laslett, Smith, and Haber, 1983; and Haber, 1984). In addition, for $\sigma_0 \lesssim 60°$, the emittance growth formulas for charge redistribution also seem to represent a very good approximation (Struckmeier, Klabunde, and Reiser, 1984; Struckmeier and Klabunde, 1984). For $\sigma_0 \gtrsim 90°$, significant deviations are observed from the formulas, and additional emittance growth is observed for both initially uniform and nonuniform beams (Struckmeier, Klabunde, and Reiser, 1984; and Hofmann, Laslett, Smith, and Haber, 1983). These results appear consistent with the interpretation that for $\sigma_0 \gtrsim 90°$ in periodic channels, a new mechanism becomes important, namely coherent instabilities driven by the periodic structure, which have been studied in detail for the K-V distribution (Hofmann, Laslett, Smith, and Haber, initially uniform beams in quadrupole channels appears to be approximately correct, although further studies are needed, especially for bunched beams.

Experimental support for the validity of the charge-redistribution equation for emittance growth in periodic quadrupole beam-transport systems can be seen in the published results of the GSI experiment with an initial quasi-Gaussian beam (Klabunde, Spadtke, and Schonlein, 1985), and the LBL experiment with an initial quasi-uniform charge density (Keefe, 1985). The experimentally measured emittance-growth data at low tune (high space-charge) values are significantly different, in at least qualitative agreement with the prediction of the emittance-growth equation.

CONCLUSIONS

We have identified three important emittance-growth mechanisms for high-current/low-emittance beams in linacs with linear external focusing: (1) charge redistribution towards a quasi-uniform density, which occurs very rapidly (in about one-quarter of a plasma period) and results in a transfer of space-charge field energy to particle kinetic energy; (2) kinetic-energy exchange towards equipartitioning; and (3) coherent instabilities driven by the periodic focusing system. We have obtained approximate equations for the charge-redistribution and kinetic-energy exchange mechanisms, which are expressed in terms of the change in two quantities: the nonlinear field energy, a measure of the shape of the charge density in real space; and the partition parameter, a measure of kinetic-energy asymmetry. We have learned that initially uniform and equipartitioned beams are free of these sources of emittance growth in linear continuous focusing channels. The charge-redistribution mechanism leads to a minimum final emittance in the extreme space-charge limit and to tune depression and brightness limits for nonuniform initial beams.

The results in published literature are consistent with the conclusions that coherent instabilities in periodic focusing systems can be avoided by restricting the zero current tune σ_o to values $\sigma_o \lesssim 60°$ to 80°, and that within this range, the equations for emittance growth from charge redistribution and kinetic-energy exchange represent a good approximation.

These studies must be extended to include other effects, including nonlinear external focusing and image charges (Kim, 1986). At present, the conclusion is that initially self-consistent distributions may not be necessary in practice to avoid rms emittance growth. It appears sufficient to provide initially quasi-uniform and equipartitioned beams, something that will represent a challenge for dc injector and low-energy linac design, but is probably not as difficult a requirement as an initially self-consistent distribution.

ACKNOWLEDGMENTS

The author acknowledges the work of colleagues F. W. Guy, K. R. Crandall, and R. S. Mills for numerical simulations described here and published in more detail elsewhere. Valuable discussions are acknowledged with M. Reiser, J. Struckmeier, P. M. Lapostolle, J. D. Lawson, and I. Haber. I wish to thank R. A. Jameson and J. E. Stovall for their encouragement.

APPENDIX: SPACE-CHARGE LIMITS IN LINEAR ACCELERATORS

A uniform 3-D ellipsoid model and the smooth approximation provide the information needed for current-limit formulas for an rms-matched beam in both the longitudinal and transverse degrees of freedom (Wangler, 1980).

TRANSVERSE CURRENT LIMIT

In the smooth approximation, the zero-current transverse phase advance σ_o per focusing period L (a measure of the external focusing strength) is given by

$$\sigma_o^2 = \frac{Q^2}{8\pi^2} + \Delta \quad ,$$

where Q is a dimensionless quadrupole strength given by

$$Q = \begin{cases} \dfrac{eBL^2\chi}{mc\beta\gamma a_Q} & ; \text{ magnetic quadrupoles} \\[2ex] \dfrac{eVL^2\chi}{mc^2\beta^2\gamma a_Q^2} & ; \text{ electric quadrupoles} \end{cases}$$

and Δ is the rf defocus strength given by

$$\Delta = \frac{\pi\, eE_o T \sin\phi_s\, L^2}{mc^2\beta^3\gamma^3\lambda} \quad .$$

The quantities Q and Δ are given in terms of charge e, mass m, speed of light c, velocity βc, relativistic mass factor γ, maximum pole-tip field B, interelectrode voltage V, quadrupole radial aperture a_Q, synchronous phase ϕ_s ($\phi_s = 0$ at the peak field), and rf wavelength λ. The quantity E_o is the spatially averaged accelerating field, and T is the usual

transit-time factor. For an RFQ, $E_0 = 2AV/\beta\lambda$, where A is the acceleration efficiency and $T = \pi/4$ (Crandall, Stokes, and Wangler, 1980). The
focusing period is $L = \beta\lambda$ for an RFQ and $L = 2\beta\lambda$ for a drift-tube linac
in a FODO arrangement with quadrupole lenses in each drift tube. The
quantity χ is the focusing efficiency approximately given by
$\chi = (4/\pi) \sin \pi \Lambda/2$ for FODO focusing in a drift tube linac, where Λ is
the fraction of a period that contains lenses. An envelope modulation
factor Ψ, the ratio of maximum to minimum envelope size for a matched
beam, is given approximately by $\Psi = (1 + Q/4\pi^2)/(1 - Q/4\pi^2)$. As the beam
current increases, the space-charge forces cause the matched beam
envelope to grow. Eventually the beam size is limited either by the
available radial aperture a or by nonlinear external fields that lead to
unacceptable emittance growth. The transverse current limit is

$$I_t = \frac{4\varepsilon_0 mc^3}{3[1 - f(p)]e} (\beta\gamma)^3 |\phi_s| \sigma_o^2 \frac{a^2}{\Psi L^2} \mu_t \quad ,$$

where f(p) is the ellipsoid form factor, shown in Fig. A-1, and
$p = \gamma\Psi^{1/2}(\phi_s/2\pi)(\beta\lambda/a)$. For a nearly spherical bunch, a good
approximation is $(f(p) \quad 1/3p$. The quantity $\mu_t = 1 - \sigma^2/\sigma_o^2$, where σ is
the final phase advance per focusing period including space charge (for
the equivalent uniform beam). An equivalent expression for μ_t at the
current limit is $\mu_t = 1 - \varepsilon^2/\varepsilon_{to}^2$, where ϕ is the final emittance and
$\varepsilon_{to} = a^2\sigma_o/\Psi L$ is the zero-current acceptance. For very small emittance
beams, the maximum value of μ_t may be determined by a lower limit on tune
depression σ/σ_o. Whether such a limit exists must be determined by
further study. For large emittance beams, the expression fo μ_t as a
function of ε should be used, and current limit decreases as emittance
increases, reflecting the diminished capacity for space-charge-induced
expansion of the beam envelope as the emittance term in the envelope
equation increases.

LONGITUDINAL CURRENT LIMIT

The zero-current longitudinal phase advance K_o per unit length in
the smooth approximation is given by

$$k_{\ell o}^2 = -\frac{2\pi eE_o T \sin \phi_s}{mc^2\lambda(\beta\gamma)^3} \quad ,$$

where $-90° \leq \phi_s \leq 0°$ for simultaneous acceleration, and longitudinal
focusing. The longitudinal current limit is

$$I_\ell = \frac{2\varepsilon_0 mc^3}{3f(p)e} (\beta\gamma) \quad |\phi_s| \frac{k_{\ell o}^2 a^2}{\Psi} \mu_\ell \quad ,$$

where μ_ℓ is the longitudinal equivalent of μ_t.

For a nearly spherical bunch, where $f \approx 1/3p$, we obtain

$$I \approx 2\varepsilon_o c \beta E_o Ta \phi_s^2 |\sin \phi_s| \mu_\ell / \Psi^{1/2} \quad .$$

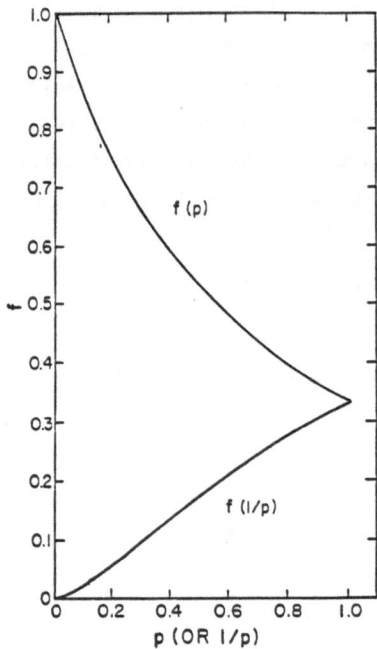

Fig. A-1.   Ellipsoid form factor f(p) vs p for p < 1 and
f(1/p) for p > 1.

REFERENCES

Crandall, K. R., Stokes, R. H., and Wangler, T. P., "RF Quadrupole Beam
    Dynamics Design Studies," Proc. 10th Linear Accelerator Conference,
    Montauk, New York, September 10-14, 1979, Brookhaven National
    Laboratory report BNL-51134, 205 (1980).
Haber, I., "Simulation of Low Emittance Transport," Proc. 1984 INS
    International Symposium on Heavy Ion Accelerators and Their
    Applications to Inertial Fusion, Institute for Nuclear Study, Tokyo,
    Japan, January 23-27, 1984, 451.
Hofmann, I., Laslett, L. J., Smith, L., and Haber, I., 1983, "Stability
    of the Kapchinskij-Vladimirskij (K-V) Distributions in Long Periodic
    Transport Systems," Particle Accelerators 13:145.
Hofmann, I., "Emittance Growth," presented at 1986 Linear Accelerator
    Conference.
Jameson, R. A., Ed., "Space-Charge In Linear Accelerators Workshop," Los
    Alamos Scientific Laboratory report LA-7265-C, May 1978.

Jameson, R. A., "Rf Linacs," these proceedings.

Keefe, Denis, "Summary for Working Group on High Current Beam Transport," Proc. Workshop on High Brightness, High Current, High Duty Factor Ion Injectors, San Diego, California, May 21-23, 1985, AIP Conference Proceedings No. 139 (1986).

Kim, Charles H., 1986, "High-Current Beam Transport With Multiple Beam Arrays," Proc. Workshop on High Brightness, High Current, High Duty Factor Ion Injectors, San Diego, California, May 21-23, 1985, AIP Conference Proceedings No. 139:133.

Klabunde, J., Spadtke, P., and Schonlein, A., 1985, "High Current Beam Transport Experiments at GSI," IEEE Trans. Nucl. Sci. 32:2462.

Lawson, J. D., "The Physics of Charged Particle Beams," Clarendon Press, Oxford (1977) 197.

Reiser, M., "Current Limits In Linear Accelerators," J. Appl. Phys., 52:555.

Struckmeier, J., Klabunde, J., and Reiser, M., 1984, "On the Stability and Emittance Growth of Different Particle Phase-Space Distributions in a Long Magnetic Quadrupole Channel," Particle Accelerators 15:47.

Struckmeier, J. and Klabunde, J., 1984, "Stability and emittance Growth of Different particle Phase Space Distributions in Periodic Quadrupole Channels," Proc. 1984 Linac Conf. Gesellschaft Fur Schwerionenforschung, Darmstadt report GSI-84-11, 359.

Wangler, T. P., "Space-Charge Limits In Linear Accelerators," Los Alamos Scientific Laboratory report LA-8388, July 1980.

Wangler, T. P., Crandall, K. R., Mills, R. S., and Reiser, M., 1985, "Relationship Between Field Energy and RMS Emittance in Intense Particle Beams," IEEE Trans. Nucl. Sci., 32(5):2196.

Wangler, T. P., Crandall, K. R., Mills, R. S., and Reiser, M., 1986, "Relationship Between Field Energy and RMS Emittance in Intense Particle Beams," Proc. Workshop on High Brightness, High Current, High Duty Factor Ion Injectors, San Diego, California, May 21-23, 1985, AIP Conference Proceedings No. 139:133.

Wangler, T. P., Guy, F. W., and Hofmann, I., "The Influence of Equipartitioning on the Emittance of Intense Charged Particle Beams," presented at the 1986 Linear Accelerator Conference, SLAC, Stanford, California.

Wangler, T. P., "Developments In The Physics of High-Current Linear Ion Accelerators," Seminar on New Techniques for Future Accelerators, Erice, Sicily, May 11-17, 1986.

Wangler, T. P., Crandall, K. R., and Mills, R. S., "Emittance Growth from Charge Density Changes in High-Current Beams," presented at the International Symposium on Heavy Ion Fusion, Washington, D.C., May 27029, 1986.

BEAM-CURRENT LIMITS IN CIRCULAR ACCELERATORS AND STORAGE RING LONGITUDINAL COASTING BEAM INSTABILITIES

Jean-Louis Laclare

James Clerk Maxwell Institute
European Synchrotron Radiation Laboratory
BP 220 - 38043 Grenoble - France

INTRODUCTION

Nowadays, performances of most particle accelerators are limited by coherent instabilities. It is a very important collective phenomenon that prevents from pushing up the beam current above the instability threshold without damaging the beam quality.

In a bright beam, particles behave differently from a single particle. The origin of the mechanism leading to this deviation is the electromagnetic field created by the beam itself. This self field in which the beam moves depends strongly on the electromagnetic properties and the geometry of the environment, namely the vacuum chamber, and all the unavoidable equipment in direct vicinity of the beam. When the intensity gets large enough, the self field starts getting sizable and cannot be neglected when compared to the main guide field.

An intense cool beam is always potentially unstable. As a matter of fact, a small initial density perturbation arising from statistical noise or previous beam manipulations can grow exponentially, involving more and more particles, and leading finally to an unstable coherent motion in which all the beam is contributing. On the other hand, a hot beam with enough spread in incoherent frequency of particles, is stable.

With a hot beam, one has to consider two conflicting effects, the driving force of the coherent motion that tends to increase the density perturbation and the spread in incoherent frequency that reacts against it by dispersing the particles. Below the instability threshold, the driving force is not strong enough to prevail over the effect of the frequency spread.

In this paper, limited to coasting beams in circular machines, we bring in step by step all the necessary ingredients to investigate coherent instabilities. The longitudinal case only will be considered.

Because of synchrotron motion, for bunched beams, coherent instabilities appear differently. However, the theoretical approach to the problem is very similar and easy to follow once the coasting beam case has been understood.

COASTING BEAM LONGITUDINAL INSTABILITIES

Single Particle Equation of Motion

We assume a circular machine with a fixed magnetic field in which a single particle would rotate at constant frequency. There is no accelerating cavity in the ring and synchrotron radiation is neglected.

Then, the period of revolution:

$$T = \frac{L}{\beta c} = \frac{2\pi}{\omega} \tag{1}$$

is a function of the particle velocity βc and of the orbit length L. Both quantities can be expanded in terms of momentum deviation with respect to a reference momentum:

$$P_{//o} = m_o \, \gamma_o \, \beta_o \, c \tag{2}$$

$$\beta = \beta_o \left(1 + \frac{1}{\gamma_o^2} \frac{P_{//} - P_{//o}}{P_{//o}} \right) \tag{3}$$

$$L = L_o \left(1 + \alpha \frac{P_{//} - P_{//o}}{P_{//o}} \right) \tag{4}$$

α is the momentum compaction. In smooth machines, an order of magnitude is given by $\alpha \sim 1/Q_x^2$, where Qx is the horizontal wave number. Higher order contributions to orbit lengthening such as betatron amplitude are neglected.

We end up with:

$$\frac{T - T_o}{T_o} = \left(\alpha - \frac{1}{\gamma_o^2} \right) \frac{P - P_{//o}}{P_{//o}} = \eta \frac{P - P_{//o}}{P_{//o}} = - \frac{\omega - \omega_o}{\omega_o} \tag{5}$$

α is often written $1/\gamma_t^2$. The transition energy is defined by:

$$E_t = m_o \, c^2 \, \gamma_t \tag{6}$$

η is negative. The angular frequency of revolution ω is larger for the particles with a positive momentum or energy deviation. The reverse applies above the transition energy. When dealing with instabilities, the sign of η is very important. It tells us whether the slow particles that arrive later at a given point of the machine have a higher or a lower energy than the fast particles that create the electromagnetic field in front.

We define a reference particle of fixed momentum $p_{//0}$ rotating at constant frequency of revolution ω_0 (rad/s). A test particle of momentum $p_{//}$ rotates at frequency ω. Its motion, with respect to the reference, is described by a pair of coordinates τ and $\dot{\tau} = d\tau/dt$.

The first coordinate τ, expressed in seconds, measures the time delay between the passing of the reference and the passing of the test particle at the same point along the ring.

The second one, $\dot{\tau}$, is the derivative of τ with respect to time t. For the unperturbed motion (no longitudinal force applied), $p_{//}$ and ω are constant. The test particle moves away from the reference with a constant speed $\dot{\tau}$. τ is a linear function of time.

$$\tau = \tau_0 + \dot{\tau}t \cdot \tag{7}$$

From equation 5, the value of $\dot{\tau}$ is given by:

$$\dot{\tau} = \frac{d\tau}{dt} = \frac{dT}{T} = -\frac{d\omega}{\omega} = \eta \frac{p_{//} - p_{//0}}{p_{//0}} \cdot \tag{8}$$

In order to take into account the effect of a longitudinal electromagnetic field at azimuthal position θ in the machine, one can use the fundamental equation of charged particle dynamics.

$$\ddot{\tau} = \frac{\eta}{p_{//0}} \frac{dp_{//}}{dt} = -\frac{\eta}{p_{//0}} e[E + v\wedge B]_{//} (t,\theta) \cdot \tag{9}$$

This last equation will be used later to describe the perturbed motion when the beam electromagnetic field is introduced.

Longitudinal Signal of a Single Particle

With the object of writing down Maxwell's equations, the solution of which leads to the beam electromagnetic field, one needs to clearly express the right hand side of these equations, namely the current density and the charge density at position θ in the machine as a function of time.

Machine physicists are used to visualizing the beam intensity on a scope by looking at the signal drawn from a longitudinal Pick-Up electrode. The P.U. electrodes are diagnostic equipment which measure the local electromagnetic field induced by particle passage.

Let us assume a perfect P.U. electrode (infinite bandwidth) located at position θ around the ring circumference and let us analyze the signal when a single particle is rotating in the machine. (Fig. 1)

When considering an unperturbed motion, the elementary signal is a series of regularly spaced impulses delivered at each revolution $T = 2\pi/\omega$.

In the time domain, the mathematical expression of the signal $s_{//}(t,\theta)$ can be written:

$$s_{//}(t,\theta) = \varepsilon \sum_{p=-\infty}^{p=+\infty} \delta\left(t - \tau - \frac{\theta}{\omega_0} - \frac{2p\pi}{\omega_0}\right)$$

$$= \frac{e\omega_0}{2\pi} \sum_{p=-\infty}^{p=+\infty} e^{jp(\omega_0(t-\tau)-\theta)} \qquad (10)$$

where τ is given by equation (7)
δ is the Dirac's function
e is the elementary charge

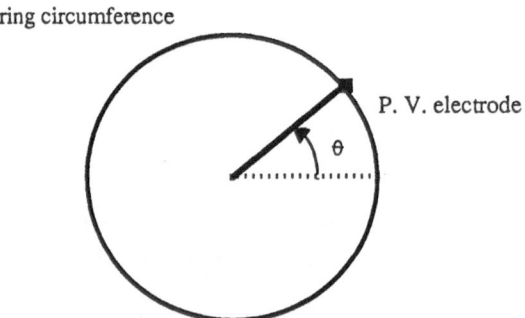

Fig. 1. Observation of test particle.

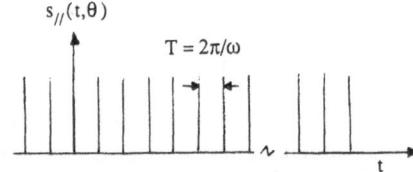

Fig. 2a. Single particle signal in time domain.

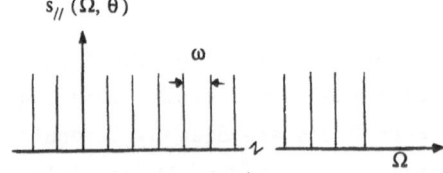

Fig. 2b. Single particle signal in frequency domain.

The frequency spectrum of the signal is obtained by Fourier analysis:

$$s_{//}(\Omega,\theta) = \frac{1}{2\pi} \int\limits_{t=-\infty}^{t=+\infty} s_{//}(t,\theta)\, e^{-j\Omega t}\, dt$$

$$= \frac{e\omega_o}{2\pi} \sum_{p=-\infty}^{p=+\infty} e^{-jp(\omega_o\tau_o+\theta)}\, \delta(\Omega-p\omega) \tag{11}$$

On a spectrum analyzer, it gives a series of infinitely sharp lines at harmonies $\Omega = p\omega$ of the revolution frequency of the particle.

Distribution Function

In order to obtain the actual signal of a beam, one needs to sum up the elementary signal of individual particles. To this end, we introduce a distribution function $\psi(\tau,\dot{\tau},t)$ suitably normalized to unity:

$$\int\limits_{\tau=0}^{\tau=2\pi/\omega} \int\limits_{\dot{\tau}=-\infty}^{\dot{\tau}=+\infty} \psi(\tau,\dot{\tau},t)\, d\tau\, d\dot{\tau} = 1 \tag{12}$$

and that describes the density in phase space at time t.

Then, the signal of the entire beam of N particles is given by:

$$S_{//}(t,\theta) = N \int\limits_{0}^{\tau=2\pi/\omega} \int\limits_{\dot{\tau}=-\infty}^{\dot{\tau}=+\infty} s_{//}(t,\theta)\, \psi(\tau,\dot{\tau},t)\, d\tau\, d\dot{\tau} \tag{13}$$

For the purpose of studying the present problem, let us suggest taking:

$$\psi(\tau,\dot{\tau},t) = g_0(\dot{\tau}) + g_o(\dot{\tau})\, e^{j(p\omega_o\tau + \omega_{//pc}t)} \quad \text{with } p \neq 0 \tag{14}$$

The first part:

$$\psi_o = g_o(\dot{\tau}) \tag{15}$$

represents a stationary coasting beam. There is no τ dependence, which means that the particule density is the same all around the ring. The $\dot{\tau}$ dependence is there to specify the momentum distribution.

The second part:

$$\psi_p = g_p(\dot{\tau})\, e^{j(p\omega_o\tau + \omega_{//pc}t)} \tag{16}$$

is a perturbation added on top of the stationary distribution. It consists in a prebunching of the beam with p wavelengths around the machines (e $^{jp\omega_o\tau}$). The perturbation does not move around the ring at the average particle frequency. A slight coherent frequency shift $\omega_{//pc}$ is assumed. $\omega_{//pc}$ is a complex number. With our notation, if the imaginary part is negative, then the perturbation will increase exponentially with time. The growth time for self bunching with p wavelengths is $-I_m(\omega_{//pc})$.

On the other hand, if $I_m(\omega_{//pc})$ is positive, the perturbation will be damped and the bunching will disappear.

Our normalization implies:

$$\int g_o(\dot\tau)d\dot\tau = \frac{\omega_o}{2\pi}$$

and

$$\int_{\tau=0}^{\tau=\frac{2\pi}{\omega_o}} \int_{\dot\tau=-\infty}^{\dot\tau=+\infty} \psi_p(\tau,\dot\tau,t)\ d\tau\ d\dot\tau = 0 \tag{17}$$

which means that the perturbation does not change the total number of particles but simply rearranges them.

Signal Induced by the Beam

When integrating $s_{//}(t,\Theta)$, one gets the signal of the entire beam:

$$S_{//}(t,\Theta) = I + S_{//p}(t,\Theta) =$$
$$I + \frac{2\pi}{\omega_o}Ie^{-j[(P\omega_o + \omega_{//pc})\ t-p\Theta]} \int_{\dot\tau} g_p(\dot\tau)\ d\dot\tau \tag{18}$$

I = average intensity, DC current
$S_{//p}(t,\Theta)$ = A.C part of the current

The signal can be Fourier analyzed.

$$S_{//}(\Omega,\Theta) = I\left[\delta(\Omega) + \frac{2\pi}{\omega_o}e^{-jp\Theta}\delta(\Omega-p\omega_o - \omega_{//pc})\int_{\dot\tau}g_p(\dot\tau)d\dot\tau\right]. \tag{19}$$

Two infinitely sharp lines would appear on a spectrum analyzer. The first one at $\Omega = 0$ is the DC component of the current induced by the stationary distribution. the second one at $\Omega = p\omega_o + \omega_{//pc}$ is the coherent line induced by the perturbation ψp.

Let us make a few remarks to justify our perturbation model.

A perfect coasting beam is stable. Mathematically, it is represented by a stationary distribution which gives a DC current and a uniform electromagnetic field. No bunching can appear.

As a matter of fact, in terms of physics, a perfect coasting beam does not exist. A beam is composed of individual particles. Every particle contributes to the current. So, on average, we get the DC current. However, riding on top of it, we have statistical noise plus some density modulation which always remains as a trace of previous beam manipulations such as injection or debunching. Therefore, there are physical reasons for considering a perfect coasting beam perturbed by a ψ_p type distribution.

Is it valid to restrict the perturbation to a single p value and probe the environment with a single frequency? The answer is yes, provided the coherent frequency shift $\omega_{//pc}$ is much smaller than ω_o. In other words, if the perturbation growth time is much longer than the revolution period, then, two adjacent harmonics $p\omega_o$ and $(p + 1)\,\omega_o$ cannot couple and oscillate independently.

Starting from the very rich spectrum induced by individual particles, with the assumed form of ψ_p, we have arranged them to concentrate the signal on a single frequency line. Let us remark that the signal is complex now. We will deal with complex current associated with complex impedances in the way we are used to in classical electricity.

Longitudinal Electric Field Created in a Perfectly Conduction Round Pipe

In order to illustrate the result of self field calculation, let us consider a very simple example of beam environment. We assume a round beam of constant radius traveling in a straight line along the axis of a circular pipe of radius b (Fig. 3).

The wall at r = b is perfectly conducting and the beam is prebunched according to equation (18). In this case, there is no AC field in the

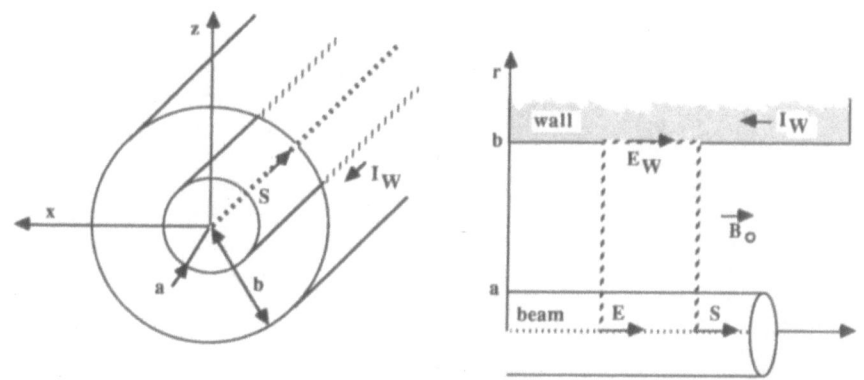

Fig. 3. Beam traveling along axis of a circular pipe.

pipe thickness and outside the wall. The AC electromagnetic field induced by the beam is completely stopped on the inner surface of the chamber wall where an image current is flowing downstream. The wall current I_w has the same magnitude (but opposite sign) as the AC beam current $S_{//p}(t, \Theta)$.

Although we are dealing with a simple type of beam environment, the exact solution of Maxwell's equation is already very complicated. It can be found in the literature if necessary.

Let us restrict ourselves to frequencies $\omega = p\omega_o + \omega_{//pc}$ well below the pipe cut off frequency.

$$\omega_c = \frac{\gamma_o R}{b} \omega_o = \frac{\gamma_o \beta c}{b} \tag{20}$$

and let us recall the expression of the space charge electric field along the beam axis with which one would end up.

$$E_{//}(t, \Theta) = - \frac{1}{2\pi R} \left[\frac{Z_o g}{j 2 \beta_o \gamma_o^2} \frac{\omega}{\omega_o} \right] S_{//p}(t, \Theta) \tag{21}$$

with $Z_o = \mu_o c = 377$ ohm

$\quad g = 1 + 2 \log(b/a)$.

Qualitative Treatment of Negative Mass Instability

In Fig. 4, we have drawn space charge electric field lines in a longitudinal cross section of the pipe and over two periods of the perturbation. The longitudinal electric field is 90° out of phase with respect to the signal. It is null on the crests and in the valleys of the signal. On the front slope of the wave crest, the force is in the forward direction and increases the energy of the particles, while at the back slope, particle energy is decreased.

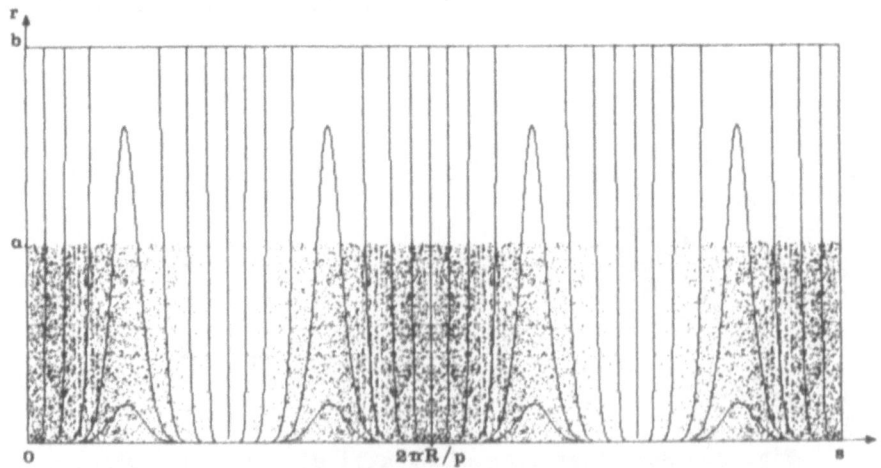

Fig. 4. Space charge field over two periods of perturbation.

Below transition energy ($\eta < 0$), acceleration means higher angular revolution frequency ω. At the front slope, particles will move ahead. At the back slope, their revolution frequency is lower, particles will move backwards. The valleys will be filled up and the crest amplitude will decrease. The initial perturbation is reduced. Space charge forces have a stabilizing affect below the transition energy.

Obviously, the reverse applies above the transition energy. The initial perturbation is reinforced by the space charge forces and the beam is unstable. This is a qualitative introduction to the negative mass instability. The appellation is related to the fact that above transition energy, particles behave as if their mass was negative. With respect to the synchronous particle, their angular velocity increases when they are decelerated.

Definition of Longitudinal Coupling Impedance

At this point, we could have started a quantitative treatment of beam stability which would not be very relevant to an actual machine.

On the other hand, solving Maxwell's equations is an impossible task when taking into account the numerous details of a vacuum chamber with the large number of cross section changes, PU electrodes, kickers or septum magnets, bellows, etc.

For our discussion about beam stability, to remain as general as possible, let us introduce a new machine parameter, the coupling impedance, which gathers all the properties of the beam electromagnetic field.

Let us go back to equation (21) and rewrite the result in the following form:

$$E_{//}(t,\theta) = -\frac{1}{2\pi R} Z_{//S.C.} \; S_{//p}(t,\theta) \tag{22}$$

$$\text{with } Z_{//S.C.} = \frac{Z_o g}{j2\beta_o \gamma_o^2} \frac{\omega}{\omega_o} \; .$$

Since $S_{//p}(t,\theta)$ is a current and $2\pi RE(t,\theta)$ is a voltage, then $Z_{//S.C.}$ has the dimension of an impedance, it can be expressed in ohms.

More precisely, in the present case, the space charge impedance is a pure inductance.

In the rest of this paper, we will assume that the form of equation (22) that expresses the longitudinal electric field in terms of beam current can be generalized to any type of environment:

$$E_{//}(t,\theta) = -\frac{1}{2\pi R} Z_{//}(\omega) S_{//p}(t,\theta) \tag{23}$$

435

where the parameter $Z_{//}(\omega)$ is the longitudinal machine impedance.

In the space charge case, we were considering that a single frequency $\omega = p\omega_o + \omega_{//pc}$ was present in the signal. The electric field has to be calculated at the same frequency anew.

One could imagine a perturbation rich in frequencies (bunched beam case). Then, in order to write the electric field on axis, one needs to Fourier analyze the signal first, then combine signal component $S_{//}(\omega, \theta)$ and impedance $Z_{//}(\omega)$ at the same frequency:

$$E(t, \theta) = -\frac{1}{2\pi R} \int_{-\infty}^{+\infty} Z_{//}(\omega) \, S_{//}(\omega, \theta) \, e^{j\omega t} d\omega \; . \tag{24}$$

The next step takes into account the fact that the impedance is not necessarily uniformly distributed around the ring. Main contributions to $Z_{//}(\omega)$ could arise from specific points in the machine where resonant objects are located. A more precise description would be obtained by using the local gradient of the impedance:

$$\frac{1}{2\pi R} Z_{//}(\omega) \rightarrow \frac{\partial Z_{//}(\omega)}{\partial s} = \frac{1}{R} \frac{\partial Z(\omega, \theta)}{\partial \theta} \; .$$

Let us remark that this last sophistication is useful in the transverse plane for taking into account beam modulation in the lattice. It is unnecessary in the longitudinal case.

Finally, since the force acting on particles involves both electric field and magnetic field effects, we include the magnetic component of the force in the left hand side.

We end up with the following definition:

$$[E + v \wedge B]_{//}(t, \theta) = -\frac{1}{R} \int_{-\infty}^{+\infty} \frac{\partial Z_{//}(\omega, \theta)}{\partial \theta} S_{//}(\omega, \theta) \, e^{j\omega t} d\omega \; . \tag{25}$$

Longitudinal Coupling Impedance of a Machine

On one hand, it is obvious that the actual electromagnetic field depends on beam environment. On the other hand, the environment cannot be made as clean and smooth as a continuous round pipe. One has to install all sorts of manipulating equipment (for injection, RF, extraction, etc.) diagnostic equipment and sometimes experimental set-ups (detectors, internal targets, etc.).

As a net result, it is essentially impossible to calculate or predict accurately enough the impedance of a machine. This is one of the major problems a machine designer has to deal with.

However, we have learned a lot from existing machines. Most of them are performance limited by coherent instabilities and many attempts for measuring and lowering the impedance have been made.

The overall measurements agree with the following qualitative description (see Fig. 5).

There are four major components of the impedance.

a) A resistive wall component which dominates at very low frequencies.

$$Z_{//R.W} = (1 + j) \frac{Z_o \beta_o}{2b} \delta_o^* \left(\frac{\omega}{\omega_o}\right)^{1/2}$$

$$\frac{Z_{//R.W}}{p} = (1 + j) \frac{Z_o \beta_o}{2b} \frac{\delta_o^*}{\sqrt{p}}$$

(26)

with $\omega = p\omega_o$ and $\delta_o^* = \sqrt{2/\mu_o \sigma \omega_o}$

b) Parasitic resonators corresponding to unwanted resonances such as higher order longitudinal modes in cavities. These narrow band, high Q resonances must be detected and damped.

c) A broad band impedance which takes into account the effect of numerous incidental cross section variations. It can be approximated by a low Q ~ 1 resonator centered around the pipe cut off frequency:

$$f_c = \frac{\omega_c}{2\pi} = \frac{\gamma_o \beta_o c}{2\pi b} \sim \text{a few GHz} .$$

This model is in agreement with experimental measurements. Its resistive part drops at high frequencies as observed in electron machines. At low frequencies, it gives an inductive contribution as observed in most machines.

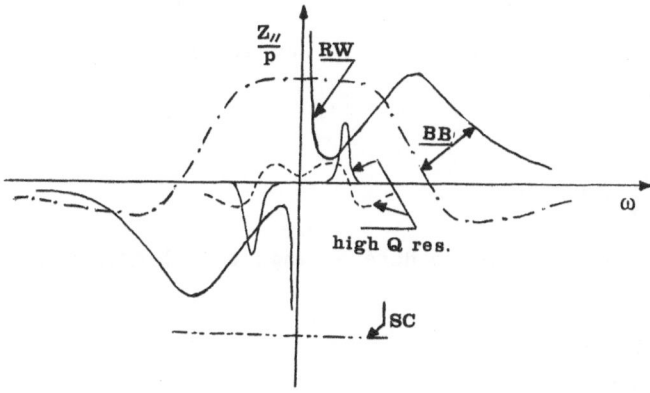

Fig. 5. Qualitative description of longitudinal impedance.

$$Z_{//B.B} = \frac{R_s}{1 + jQ\left(\frac{\omega}{\omega_c} - \frac{\omega_c}{\omega}\right)}$$ (27)

Rs = shunt impedance
Q ≃ 1 = quality factor

- Inductance at low frequencies

$$\omega \ll \omega_c \qquad Z_{//BB} = jR_s \frac{\omega}{\omega_c} \qquad \frac{Z_{//BB}}{p} = jR_s \frac{\omega_0}{\omega_c}$$

- Pure resistance at resonance

$$\omega = \omega_c \qquad Z_{//BB} = R_s \qquad \frac{Z_{//BB}}{p} = R_s \frac{\omega_0}{\omega_c}$$

- Capacitance at high frequencies

$$\omega \gg \omega_c \qquad Z_{//BB} = -jR_s \frac{\omega_c}{\omega} \qquad \frac{Z_{//BB}}{p} = -jR_s \frac{\omega_c \omega_0}{\omega^2}$$

For lowering the BB impedance, one should carefully design a smooth chamber pipe to avoid or shield abrupt changes in the cross section.

The lowest measured BB impedances are of the order of 1 ohm at peak

$$1 \ \Omega \leq \frac{Z_{//BB}}{p} = R_s \frac{\omega_0}{\omega_c} \leq 50 \ \Omega \ .$$

When no special care is taken, one can reach say 50 Ω.

d) Finally, one should not forget the standard space charge component which is very large for low β particles:

$$Z_{//S.C.} = -j \frac{Z_0 g}{2\beta_0 \gamma_0^2} \frac{\omega}{\omega_0} \qquad \frac{Z_{//SC}}{p} = -j \frac{Z_0 g}{2\beta_0 \gamma_0^2} \ .$$ (28)

As an example, let us take a 50 Mev/amu beam

$$\beta_0 = .3 \qquad \gamma_0 = 1.05 \qquad g \simeq 2.4$$

$$\left|\frac{Z_{//S.C}}{p}\right| \cong 1.37 \ K\Omega$$

This value is several orders of magnitude larger than the broad band component.

With a 7 Gev/amu beam

$$\left|\frac{Z_{//S.C}}{p}\right| = 7 \ \Omega$$

the space charge impedance is comparable with the broad band impedance.

The actual impedance seen by the beam is the sum of all the components listed above. A sketch is shown in Fig. 5 where $Re(Z_{//}/p)$ and $I_m(Z_{//}/p)$ are drawn.

We have anticipated the following and pointed out the quantity of interest $Z_{//}/p$.

Vlasov's Equation and Dispersion Relation

At this point, we have gathered all the necessary elements to start a very general discussion about stability.

- A distribution function which includes a perturbation oscillating at frequency $\omega_{//pc}$

$$\psi(\tau, \dot{\tau}, t) = g(\dot{\tau}) + g(\dot{\tau})e^{j(p\omega_o\tau + \omega_{//pc}t)}$$

- The A.C. part of the signal induced by the perturbation

$$S_{//p}(t, \theta) = \frac{2\pi}{\omega_o} I\, e^{j[(p\omega_o\tau + \omega_{//pc}t) - p\theta]} \int_{\dot{\tau}} g_p(\dot{\tau})d\dot{\tau}$$

- The impedance $Z_{//}(\omega)$ which creates the electromagnetic field on beam axis.

$$[E + v\wedge B]_{//}(t, \theta) = -\frac{1}{2\pi R} Z_{//}(p\omega_o + \omega_{//pc})S_{//p}(t, \theta)$$

$$= -\frac{I}{\beta_o c} Z_{//}(p\omega_o + \omega_{//pc})\, e^{j[(p\omega_o + \omega_{//pc})\,t - p\theta]} \int_{\dot{\tau}} g_p(\dot{\tau})d\dot{\tau} \qquad (29)$$

- The differential equation which governs single particle motion

$$\ddot{\tau} = \frac{\eta}{\left(\frac{P_{//o}}{e}\right)} [E + v\wedge B]_{//}(t, \theta) = \omega_o(t - \tau))$$

$$\ddot{\tau} = \frac{\eta I}{\left(\frac{P_{//o}}{e}\right)\beta_o c}\, e^{j(p\omega_o\tau + \omega_{//pc}t)} Z_{//}(p\omega_o + \omega_{//pc}) \int_{\dot{\tau}} g_p(\dot{\tau})d\dot{\tau} \qquad (30)$$

The electromagnetic field must be expressed at $\theta = \omega_0(t - \tau)$ when following the particle along its trajectory.

Due to the self force, individual particles will move differently and the distribution will change with time. Our task consists in finding a self consistent value of $\omega_{//pc}$ which will tell us whether the perturbation will be increased ($I_m(\omega_{//pc}) < 0$ instability) or decreased ($I_m(\omega_{//pc}) > 0$ stability).

439

The basic equation which governs the evolution of the local density distribution $\psi(\tau, \dot{\tau}, t)$ is the collision-free Boltzmann equation:

$$\frac{\partial \psi}{\partial t} + \text{div}(\psi \vec{v}) = 0 \tag{31}$$

$$\vec{v} = \begin{Bmatrix} \dfrac{d\tau}{dt} \\ \dfrac{d\dot{\tau}}{dt} \end{Bmatrix}$$

Let us remark on the similarity between this equation and the one which relates current and charge density in electromagnetism:

$$\frac{\partial \rho^*}{\partial t} + \text{div } \vec{j}^* = 0$$

When developing equation (31), we get:

$$\frac{\partial \psi}{\partial t} + \vec{v} \overrightarrow{\text{grad}} \ \psi + \psi \text{ div } \vec{v} = 0 \tag{32}$$

If one is using a set of canonical conjugate variables like τ and $\dot{\tau}$, then, div $\vec{v} = 0$ and an equivalent form of equation (31) is:

$$\frac{d\psi}{dt} = \frac{\partial \psi}{\partial t} + \frac{\partial \psi}{\partial \tau} \dot{\tau} + \frac{\partial \psi}{\partial \dot{\tau}} \ddot{\tau} = 0 \tag{33}$$

In this form, it is the Vlasov's equation. It says that the phase space density ψ does not change with time when following the motion described with canonical variables.

For the present case, it leads to:

$$0 = j(p\omega_0 \dot{\tau} + \omega_{//pc}) \ g_p(\dot{\tau}) \ e^{j*} + \ddot{\tau} \left[\frac{\partial g_0(\dot{\tau})}{\partial \dot{\tau}} + \frac{\partial g_p(\dot{\tau})}{\partial \dot{\tau}} e^{j*} \right] \tag{34}$$

where $*$ replaces $p\omega_0 \tau + \omega_{//pc} t$

The very last term is neglected as a second order effect:

$$\frac{\partial g_0}{\partial \dot{\tau}} \gg \frac{\partial g_p}{\partial \dot{\tau}}$$

and we are left with:

$$g_p(\dot{\tau}) = j\ddot{\tau} e^{-j*} \ \frac{\dfrac{\partial g_0}{\partial \dot{\tau}}}{\omega_{//pc} + p\omega_0 \dot{\tau}} \tag{35}$$

After inserting the expression of $\ddot{\tau}$ (30) and integrating over $\dot{\tau}$ values, one gets the final dispersion relation which leads to $\omega_{//pc}$ value:

$$1 = \frac{-\eta I}{\left(\frac{m_o c^2}{e}\right) \gamma_o \beta_o^2} \; j \; \frac{Z_{//}(p\omega_o + \omega_{//}pc)}{p\omega_o} \int_{\dot\tau} \frac{\frac{\partial g_o}{\partial \tau}}{\frac{\omega_{//}pc}{p\omega_o} + \dot\tau} \; d\dot\tau \tag{36}$$

Up to now, single charge particles have been assumed. The dispersion relation can be rewritten for any type of particle:

$$1 = \frac{-\eta(q/A)I}{\left(\frac{m_o c^2}{e}\right) \gamma_o \beta_o^2} \; j \; \frac{Z_{//}(p\omega_o + \omega_{//}pc)}{p\omega_o} \int_{\dot\tau} \frac{\frac{\partial g_o}{\partial \tau}}{\frac{\omega_{//}pc}{p\omega_o} + \dot\tau} \; d\dot\tau \tag{37}$$

I = electrical current

$$\frac{m_o c^2}{e} = \begin{cases} 0.511 \; 10^6 \text{ V for electrons} \\ 0.938 \; 10^9 \text{ V for protons} \\ 0.932 \; 10^9 \text{ V for heavy ions} \end{cases}$$

q/A = number of charges/number of masses

Coasting Beam Without Momentum Spread

To start discussing solutions of the dispersion relation, let us neglect the beam momentum spread. Then, the stationary distribution can be described by a Dirac function suitably normalized.

$$g_o(\dot\tau) = \frac{\omega_o}{2\pi} \delta(\dot\tau) \qquad \dot\tau = \eta \frac{p_{//} - p_{//0}}{p_{//0}} \tag{38}$$

The integration is easy to perform by parts:

$$\int_{\dot\tau} \frac{\frac{\partial g_o}{\partial \dot\tau}}{\frac{\omega_{//}pc}{p\omega_o} + \dot\tau} \; d\dot\tau = \frac{\omega_o}{2\pi} \int_{\dot\tau} \frac{\delta(\dot\tau)}{\left(\frac{\omega_{//}pc}{p\omega_o} + \dot\tau\right)^2} \; d\dot\tau = \frac{\omega_o}{2\pi} \left(\frac{\omega_{//}pc}{p\omega_o}\right)^{-2} \tag{39}$$

and the solution is given by:

$$\left(\frac{\omega_{//}pc}{p\omega_o}\right)^2 = \frac{-\eta(q/A)I}{2\pi \; \frac{m_o c^2}{e} \; \gamma_o \beta_o^2} \; j \; \frac{Z_{//}(p)}{p} = - \Lambda_{//} j \; \frac{Z_{//}(p)}{p} \tag{40}$$

Equation (40) always has two roots. It one of them corresponds to an exponential growth, then the beam is unstable.

Let us draw a few obvious consequences.

- If $Z_{//}(p)$ has a small resistive component positive or negative, the beam is unstable.

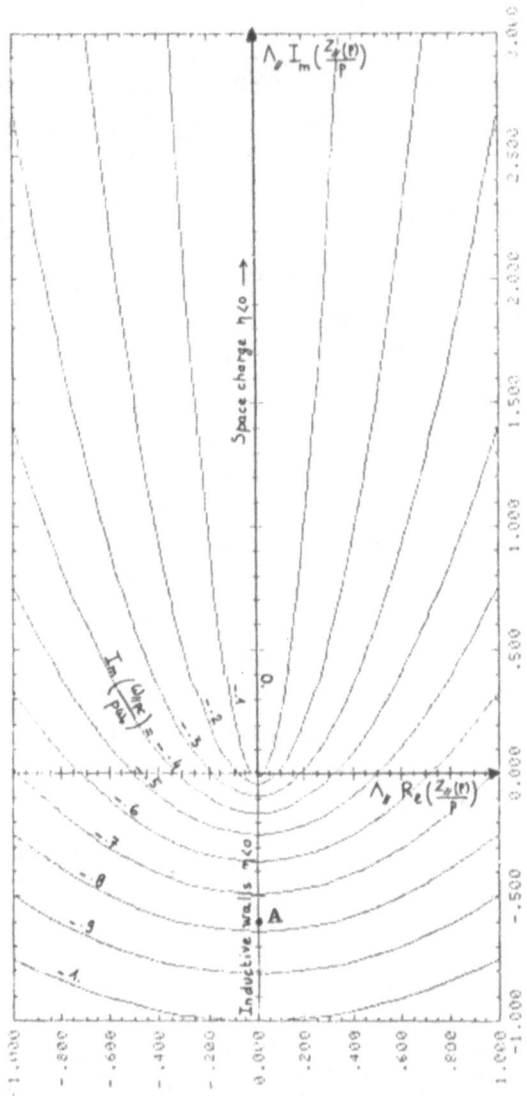

Fig. 6.   Longitudinal stability diagram for coasting beam without
momentum spread.

- For a pure inductance (broad band impedance at low frequencies for instance), $jZ_{//}(p)$ is real and negative. The beam is stable above transition energy ($\eta > 0$), unstable below transition energy ($\eta < 0$).

- When the space charge impedance, $jZ_{//}(p)$ is real and positive, it acts as a negative inductance and leads to the negative mass instability above transition energy.

The solution of Eq. (40) can be represented in a two dimension stability diagram with $\Lambda_{//} \, Re \, (Z_{//}(p)/p)$ and $\Lambda_{//} \, I_m(Z_{//}(p)/p)$ along the axes. $\Lambda_{//}$ is negative below transition energy. In Fig. 6, the curves with constant $I_m(\omega_{//pc}/p\omega_0)$ are plotted. This allows us to read the growth rate $(\tau^*)^{-1} = -I_m(\omega_{//pc})$ for a given impedance or to determine the maximum tolerable impedances for an accepted growth rate.

Let us consider a working point along the vertical axis ($Re \, (Z_{//}(p)/p) = 0$) in the unstable region (point A). The real part of the frequency shift $\omega_{//pc}$ is null, which means that the RF self field is exactly tuned to $p\omega_0$. Therefore, the self field is just trapping the beam with increasing voltage. The beam spreads in momentum.

For a different working point, the frequency shift has both real and imaginary parts. The RF self field has a frequency offset with respect to $p\omega_0$. The RF buckets are in a region free of particles. The monochromatic beam starts wiggling outside the stable area and a momentum spread appears.

Another way to look at equation (40) consists of drawing the curve $Re \, (Z_{//}(p)/p)$ corresponding to a given growth rate, in the frequency domain. In this type of diagram sketched in Fig. 7, one can realize that the beam is very sensitive to resistance at high frequency around ωc where the broad band resistance is peaked. Most of the time, it is a microwave instability that develops with hundreds or thousands of short bunches around the circumference.

Coasting Beam with Momentum Spread (Keil-Schnell Stability Criterion)

If one injects a very cool beam in a machine, the microwave insta-

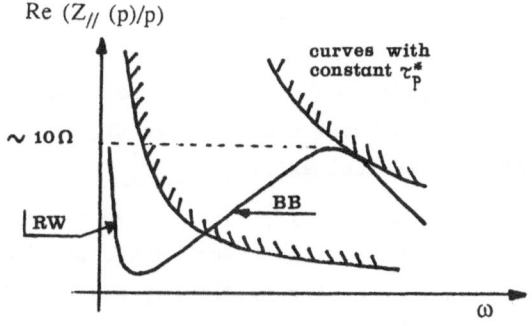

Fig. 7. Impedance diagram.

bility develops. The longitudinal emittance blows up in momentum up to a time when Landau damping gets strong enough to stop the unstable motion.

In order to illustrate the solution when momentum spread is taken into account, let us assume a parabolic stationary distribution (Fig. 8)

$$
g_0(\dot\tau) = \begin{cases} \dfrac{3\omega_0}{8\pi\dot\tau_L}\left(1 - \dfrac{\dot\tau^2}{\dot\tau_L^2}\right) & |\dot\tau| < \dot\tau_L \\[2mm] 0 & |\dot\tau| > \dot\tau_L \end{cases} \tag{41}
$$

with $\displaystyle\int_{\dot\tau} g_0(\dot\tau)d\dot\tau = \dfrac{\omega_0}{2\pi}$ as required.

The dispersion relation can be split into real and imaginary parts again.

$$
\Lambda_{//c} R_e\left(\frac{Z_{//}(p)}{p}\right) = I_m(J_{//}^{-1})
$$

$$
\Lambda_{//c} I_m\left(\frac{Z_{//}(p)}{p}\right) = -R_e(J_{//}^{-1}) \tag{42}
$$

with:

$$
\Lambda_{//c} = \frac{3(q/A)\,I}{2\pi\eta\,\dfrac{m_0 c^2}{e}\,\gamma_0\beta_0^2\left(\dfrac{\delta p}{p}\right)^2_{FWHH}}
$$

and:

$$
\tau_L = |\eta|\left(\frac{\delta p}{p}\right)_L = \frac{\eta}{\sqrt{2}}\left(\frac{\delta p}{p}\right)_{FWHH}
$$

The stability diagram is represented in Fig. 9 for small values of $\Lambda_{//c}|(Z_{//}(p)/p)|$. Curves with constant:

$$
I_m\left(\frac{\sqrt{2}\,\omega_{//pc}}{p\omega_0\,|\eta|\left(\dfrac{\delta p}{p}\right)_{FWHH}}\right)
$$

are drawn.

The stability limit corresponds to the curve with $I_m(\omega_{//pc}) = 0$. It is divided in two parts, the heart shaped curve around the origin plus the positive part of the vertical axis.

The actual shape of the stability limit curve depends strongly on the distribution edges. For coasting beams, sharp edge distributions are unstable. However, a small rounding of the edges makes the stability contour less dependent on the orientation is the impedance complex plane.

444

This is the reason why the stability criterion is so often written as a function of $|(Z_{//}(p)/p)|$

$$\left|\frac{Z_{//}(p)}{p}\right| \leq F \, \frac{\frac{m_o c^2}{e} \, \beta_o^2 \, \gamma_o \, |\eta| \, \left(\frac{\delta p}{p}\right)^2_{FWHH}}{\left(\frac{q}{A}\right) I} \tag{43}$$

where F is a form factor of the order of the unity.

With our notations, in the stability diagram:

$$\left|\Lambda_{//c} \, \frac{Z_{//}(p)}{p}\right| \leq \frac{3}{2\pi} \, F \, \tilde{=} \, 0.5 \tag{44}$$

represents a circle of radius 0.5 .

When inverted, Eq. (43), which is often called the Keil-Schnell criterion, gives the minimum spread required for stability

$$\left(\frac{\delta p}{p}\right)^2_{FWHH} \geq \frac{\left(\frac{q}{A}\right) I}{\left(\frac{m_o c^2}{2}\right) |\eta| \gamma_o \beta_o^2} \, \left|\frac{Z_{//}(p)}{p}\right| \, . \tag{45}$$

Some Remarks About Landau Damping

We have seen that a given amount of spread in momentum is necessary to stay below the instability threshold.

The stabilizing effect connected with the spread is called Landau damping. The way it acts is hidden in the mathematics of the dispersion relation. Since it is a very important mechanism without which it would be impossible to get stable beams, let us try to describe the physics involved and draw an approximate stability criterion.

We are considering a collection of particles which have a spectrum of resonant frequencies. It is intuitive that two particles with different ω do not have the same sensitivity with respect to an excita-

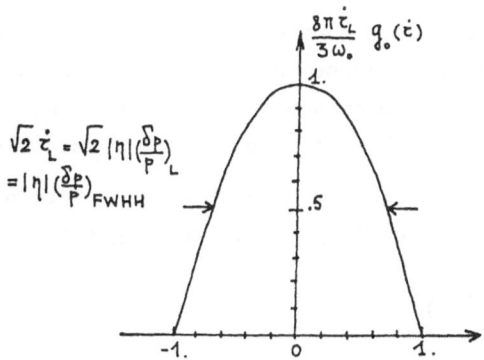

Fig. 8. Stationary parabolic distribution.

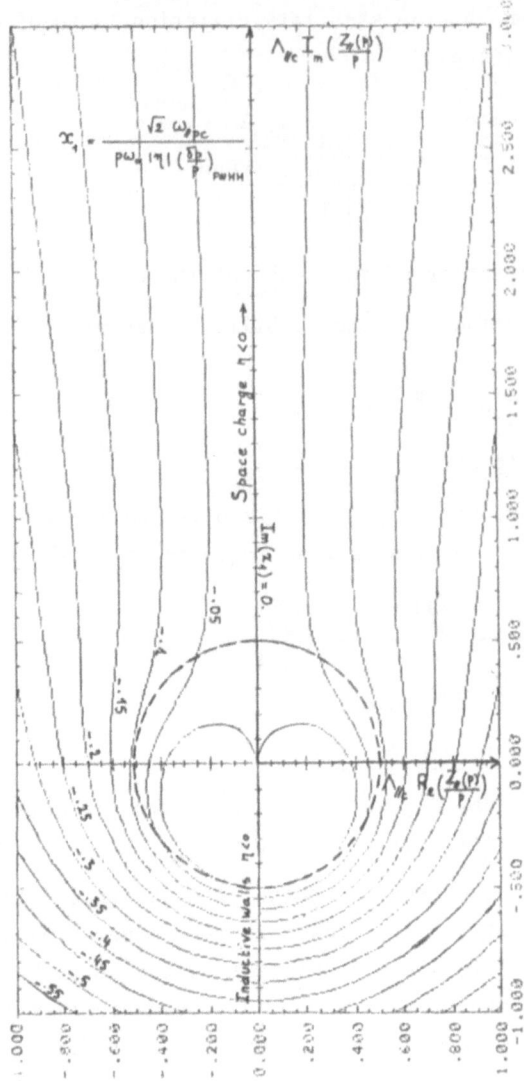

Fig. 9. Longitudinal stability diagram for coasting beam Landau damped by its momentum spread.

tion at frequency ω_{RF}. We are used to considering resonant schemes in which the excitation frequency has to be the same as the particle frequency in order to drive an amplitude growth. The important parameter is the time scale during which one observes the particle response.

For instance, we consider the equation of a pure harmonic oscillator with a perturbation at frequency ω_{RF} in the right hand side.

$$\ddot{y} + \omega^2 y = F_0 \cos\omega_{RF} t \qquad (46)$$

If one assumes $y = \dot{y} = 0$ at time $t = 0$ and $|\omega - \omega_{RF}| \ll \omega_{RF}$, the solution can be written:

446

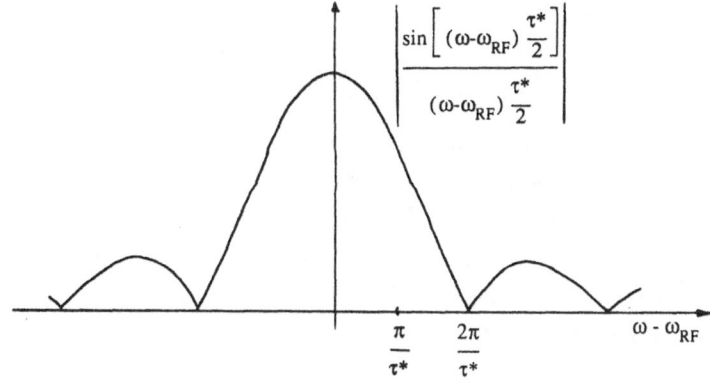

Fig. 10. Oscillator amplitude response.

$$y(\omega, t) = \frac{F_o t}{2\omega_{RF}} \frac{\sin\left[(\omega - \omega_{RF})\frac{t}{2}\right]}{\left[(\omega - \omega_{RF})\frac{t}{2}\right]} \sin\omega_{RF} t \qquad (47)$$

At a given time τ^* we compare the response $y(\omega, \tau^*)$ with the response $y(\omega_{RF}, \tau^*)$ of the resonant oscillator:

$$\frac{y(\omega, \tau^*)}{y(\omega_{RF}, \tau^*)} = \frac{\sin\left[(\omega - \omega_{RF})\frac{\tau^*}{2}\right]}{(\omega - \omega_{RF})\frac{\tau^*}{2}} \qquad (48)$$

The above quantity shows that all the oscillators with:

$$|\omega - \omega_{RF}| < \frac{\pi}{\tau^*} \qquad (49)$$

and still responding positively to the driving force (Fig. 10).

In other words, it takes a time

$$\tau^* \geq \frac{\pi}{|\omega - \omega_{RF}|}$$

before the oscillator w realizes it is being excited with a wrong frequency.

Another way to get the same result would be to analyze the excitation in the frequency domain. For $-\infty < t < +\infty$, the Fourier transform of the right hand side is a Dirac function

$$\frac{F(\omega)}{F_o} = \delta(\omega - \omega_{RF}) \qquad (50)$$

For $0 < t < \tau^*$ we get:

$$\frac{F(\omega)}{F_o} = \frac{\tau^*}{2\pi} \frac{\sin\left[(\omega - \omega_{RF})\frac{\tau}{2}\right]}{(\omega - \omega_{RF})\frac{\tau^*}{2}} \qquad (51)$$

This result is in perfect agreement with the previous one. It tells us that a pure excitation at frequency ω_{RF} can drive a range of ω

447

$$|\omega - \omega_{RF}| \geq \frac{2\pi}{\tau^*}$$

when gated during a time τ^* (see Fig. 9).

Now, let us go back to our instability problem. We start with a beam without spread. All the individual particles have the same frequency of revolution ω. Accordingly they have the same response with respect to the RF self field produced by the initial bunching. Let us consider an unstable working point (point A for instance) in the stability diagram of Fig. 6. Let $(\tau^*)^{-1}$ be the corresponding growth rate.

From now on, we increase the incoherent beam spread $\Delta\omega_{incoh}$.

- For $\Delta\omega_{incoh} \ll \pi/\tau^*$, all the particles have a positive response to the driving force, they behave coherently as a beam without spread.

- For $\Delta\omega_{incoh} \gg \pi/\tau^*$, particles at the edge of the distribution cannot follow the coherent motion very quickly. Being driven out of phase, they reduce the RF field and contribute to damping of the perturbation.

Between these two extremes, we have to guess the necessary $\Delta\omega_{incoh}$ to stay at the instability threshold. As an example, let us assume a parabolic distribution of revolution frequency ω and declare that threshold is reached when the edge particle amplitude stops growing at time τ^*.

$$|\omega - \omega_{RF}| = |\eta| p \omega_o \left(\frac{\delta p}{p}\right)_L = \frac{|\eta|}{\sqrt{2}} p \omega_o \left(\frac{\delta p}{p}\right)_{FWHH} \geq \frac{\pi}{\tau^*} \tag{52}$$

By using equation (40) to express the growth rate,

$$\tau^{*-1} = - I_m(\omega_{//pc})$$

we get a rough stability threshold

$$\left(\frac{\delta p}{p}\right)^2_{FWHH} \geq \frac{2\pi^2}{|\eta|} \left(\frac{\omega_{//pc}}{p\omega_o}\right)^2 = \frac{\pi\left(\frac{q}{A}\right) I}{\left(\frac{m_o c^2}{e}\right) \gamma_o \beta_o^2} \left|\frac{Z_{//}(p)}{p}\right| \tag{53}$$

which is π times larger than the Keil-Schnell limit.

LIMITS OF THE THEORY, TIME DOMAIN APPROACH

The theory developed here above for dealing with longitudinal instabilities is easy to apply once you know the actual impedance of the machine. It gives you the instability growth rate the sign of which tells you whether the beam is stable or unstable. The theory is a perturbation theory which uses Vlasov's equation to first order. It does

448

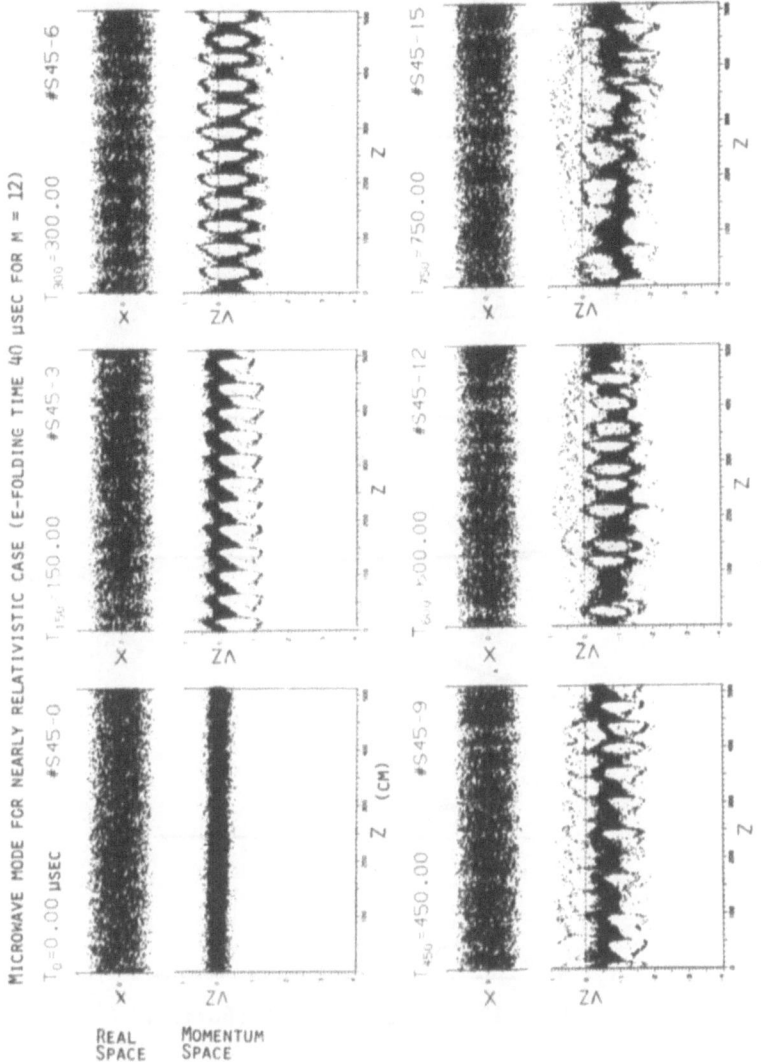

Fig. 11. Development of a longitudinal microwave instability.

not allow us to follow the unstable motion of a beam step by step. It does not tell us how much beam emittance will be spoiled by the instability. In order to answer such questions, one has to use simulation codes working in the time domain. Starting with a collection of particles which represents a perturbed coasting beam, one solves Maxwell's equations, finds the electromagnetic field to be applied to every particle and progresses step by step. In Fig. 11, the development of a longitudinal microwave instability is shown. In the case presented here, a broad band impedance resonator Q ~ 1 at pipe cut off frequency (12th harmonic of the revolution frequency) is assumed. The initial perturbation consists in a 10% density modulation at resonator frequency. One can see off-centered empty buckets appearing in phase space. Particles wind round the separatrices and spread out. On average the beam loses

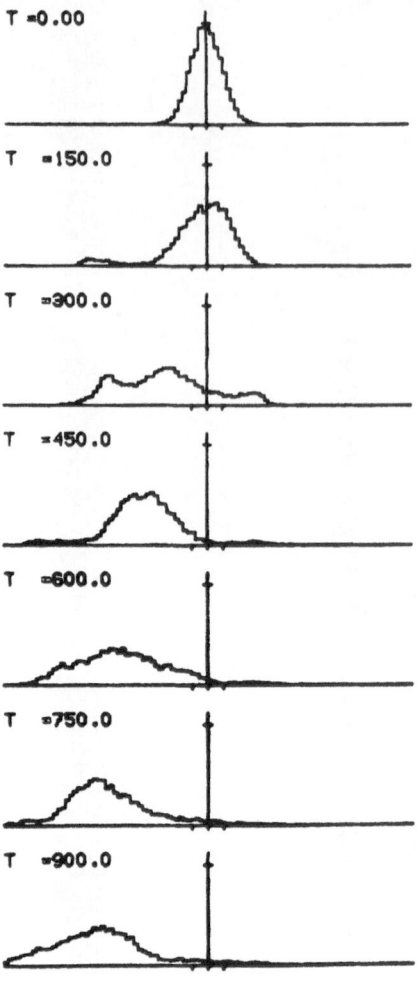

Fig. 12.   Momentum distribution.

energy. In Fig. 12 and 13, the momentum distribution and the Fourier
spectrum are shown.

CONCLUSION

The goal every machine physicist aims at, consists in delivering a
beam as bright as possible (high density, low emittance) to the users.
This requires sophisticated beam manipulations to avoid dilution or a
powerful cooling system such as radiation for electrons and stochastic
cooling or electron cooling for heavier particles. Cooling techniques
are developing very fast, and one could dream of nearly zero emittance
beams. Alas! The ultimate beam lightness one can reach is in fact
determined by the impedance seen by the beam (machine impedance plus feed
back system impedance). This is the reason why, nowadays, the impedance
is one of the main design parameters of a machine.

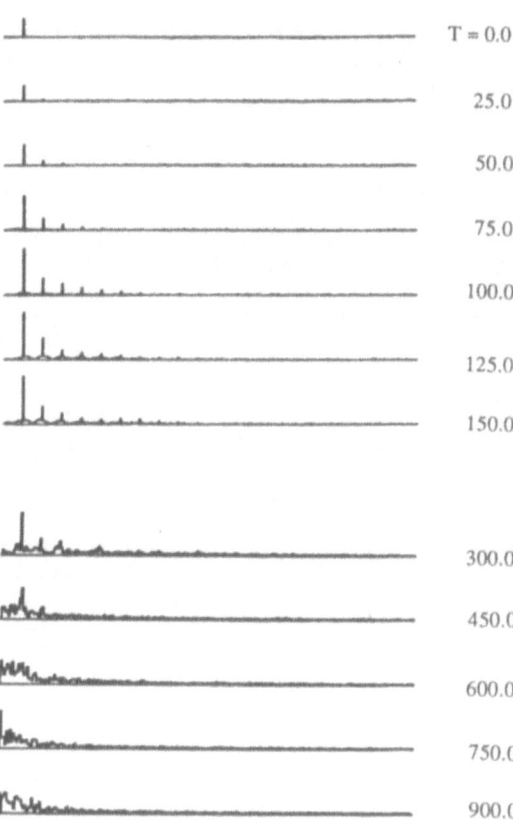

Fig. 13. Fourier spectra.

ACKNOWLEDGMENTS AND REFERENCES

In the preparation of the manuscript, I have enjoyed and benefited from numerous discussions with T. Aniel, G. Leleux, A. Ropert and A. Tkatchenko and L. N. Saturne.

I. Hoffmann from G. S. I. Darmstadt provided Figures 11, 12 and 13 of section 3.

I am very grateful to each of them for their generous help.

Concerning references, one could probably list more than one hundred valuable contributions. I feel that the list hereunder is barely enough and I apologize for having omitted many of the pioneers.

REFERENCES

Besnier, G., Nucl. Instrum. and Methods, 164, (1979).
Boussard, D., CERN Report LAB II/RF/75-2, (1975).
Channel, P. and Sessler, A., Nucl. Instrum. Methods, 136, (1976).
Chao, A. and Gareyte, J., PEP Note, 224, (1976).
Gygi-Hasnney, M., Hofmann, A., Hübner, K. and Zotter, B., CERN LEP-TH/83-2, (1983).
Hansen, S., Hereward, H. G., Hofmann, A., Hübner, K. and Myers, S., IEEE Trans. Nucl. Sci. NS-22, (1975).
Hereward, H., CERN Report, 65-20, (1965).

Hofmann, A. and Maidment, J., LEP Note, 168, (1979).

Kohaupt, R., Proc. Int. Conf. on High-Energy Accelerators, Geneva, (CERN, Geneva, 1980), (1980).

Laclare, J. L., Proc. Int. Conf. on High-Energy Accelerators, Geneva, (CERN, Geneva, 1980), (1980).

Laclare, J. L., Introducition to Coherent Instabilities Casting Beam Case, CERN Accelerator School, Gif-sur-Yvette, Paris France, (1985).

Month, M. and Messerschmid, E., Nucl. Instrum. Methods, 133, (1975).

Papiernick, A., Chatard-Moulin, M. and Jecko, B., Proc. Int. Conf. on High-Energy Accelerators, Stanford, 1974, CONF. 740522 (SLAC, Stanford, 1974), p. 375.

Pellegrini, C. and Wang, J., Proc Int. Conf. on High-Energy Accelerators, Accelerators, Accelerators, Geneva, 1980, (CERN, Geneva, 1980), p. 554.

Pellegrini, C. Frascati Report LMF-69/49, (1969).

Ruth, R. and Wang, J., IEEE Trans. Nucl. Sci. NS-26, (1981), p. 2405.

Sacherer, F., IEEE Trans. Nucl. Sci. NS-24, (1977) 1393.

Sacherer, F., IEEE Trans. Nucl. Sci. NS-20, (1973) 825.

Sacherer, F., CERN Report 77-13, (1977), p. 198.

Sands M., Slac Report TN/69-8.

Satoh, K., PEP- Note 361, (1981).

INJECTORS AND ION SOURCES

T. S. Green

UKAEA Culham Laboratory
Abingdon, Oxon, England

INTRODUCTION

Ion injectors, i.e., sources of low-to-medium energy ions, have developed rapidly in recent years. For many years designs were dominated by the duoplasmatron coupled to expansion cups which formed an effective "cathode" for an ion accelerator. Thus, the sources, accelerators, and eventually beam transport were very closely coupled. More recently the breakdown of the elements of an injector into separate components (see Fig. 1):

- plasma generator
- accelerator
- low energy beam transport

has lead progressively to optimization within each component and a fuller appreciation of the interactions between them.

The author's experience in systems for high current accelerators for use in fusion experiments has shown how new plasma generators can evolve based on better understanding of the physics within them, as discussed in the next section. Optimized high perveance accelerators have also been developed and this process is described in the third section.

The subject of space charge neutralization is low energy transport is more complex and, as stated in the final section, is still under development.

PLASMA GENERATORS

As commented in the Introduction, the need to develop high current sources for neutral beam injectors in the fusion community has been an incentive not only to achieve high current densities with high electrical and gas efficiencies but also to develop a better understanding of the

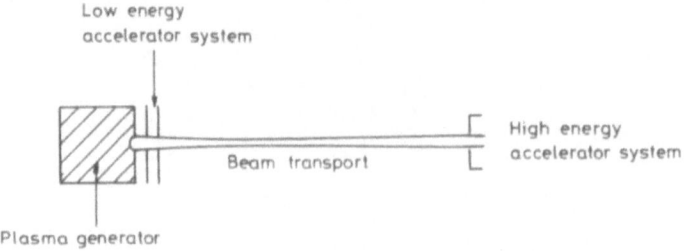

Fig. 1.   Schematic diagram of ion injector.

physics of the plasma generators. The magnetic multipole source, developed in a number of laboratories, has been extensively studied by the Culham group who first developed it as a high current injector. I will use it as an example to clarify some of the issues.

Ionization Efficiency

Figure 2 shows a schematic of a magnetic multipole source. Electrons are emitted from hot cathodes and then accelerated across a plasma sheath to gain sufficient energy to ionize the gas. The anode is covered by permanent magnets, which essentially form a magnetic wall, with leakage for the electrons to flow to the anode at the cusp lines along the magnets. The central volume is magnetic-field-free.

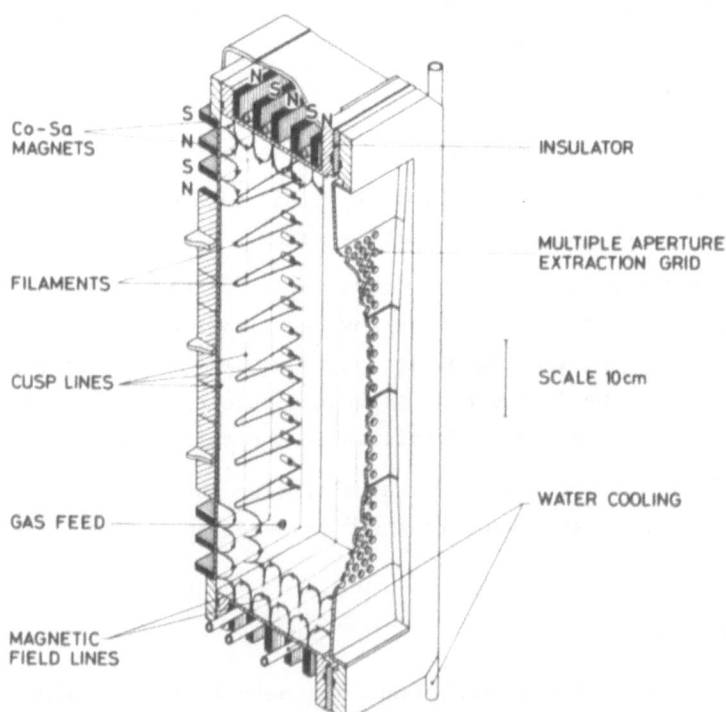

Fig. 2.   Schematic diagram of magnetic multipole source.

These primary electrons are assumed to have an energy spectrum, which experiments show to be quasi Maxwellian with a high temperature, independent of gas pressure but dependent on the arc voltage. They ionize the gas-forming ions and "cool" thermal electrons and they suffer inelastic collisions which degrade them in energy below the ionization threshold, or they flow to the anode.

Figure 3 shows a Langmuir probe characteristic, in which one can identify the primary and the thermal electron distributions.

We may write:

$$\frac{I_{ARC}}{eV} = n_p \, n_o \, S_{IN} + \frac{n_p}{\tau_p} \tag{1}$$

and

$$\frac{I_+}{eV} = \frac{n_+}{\tau_+} = n_p \, n_o \, S_{ION} \tag{2}$$

where I_{ARC} is the total arc current

e is the electron charge

V is the plasma volume

n_p is the density of primary electrons

n_+ is the density of positive ions

n_o is the density of neutral atoms

S_{IN} is the reaction rate coefficient for inelastic scattering to reduce the primary electron energy below the threshold

S_{ION} is the reaction rate coefficient for ionization

τ_p is the lifetime for loss of primary electrons to the anode and other surfaces

τ_+ is the lifetime for loss of positive ions.

We may derive the ionization efficiency, η, from these equations.

$$\frac{1}{\eta} = \frac{I_{ARC}}{I_+} = \frac{S_{IN}}{S_{ION}} + \frac{1}{n_o \, S_{ION} \, \tau_p} \tag{3}$$

In general one can only measure a fraction α of the ions which are collected on the so-called extraction electrode, and one defines as partial efficiency, η_f,

$$\frac{1}{\eta_f} = \frac{1}{\alpha} \left[\frac{S_{IN}}{S_{ION}} + \frac{1}{n_o \, S_{ION} \, \tau_p} \right] \tag{4}$$

Typical data are shown in Fig. 4 for variation of $(\eta_f)^{-1}$ with the inverse of the gas pressure in the source. Agreement between theory and experiment is good.

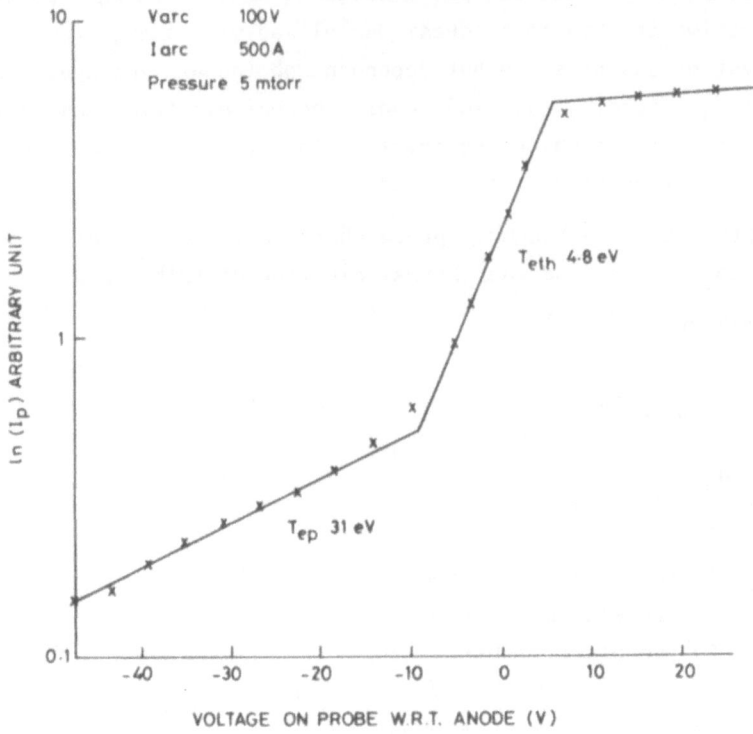

Fig. 3.   Langmuir probe characteristics in ion source showing two electron components.

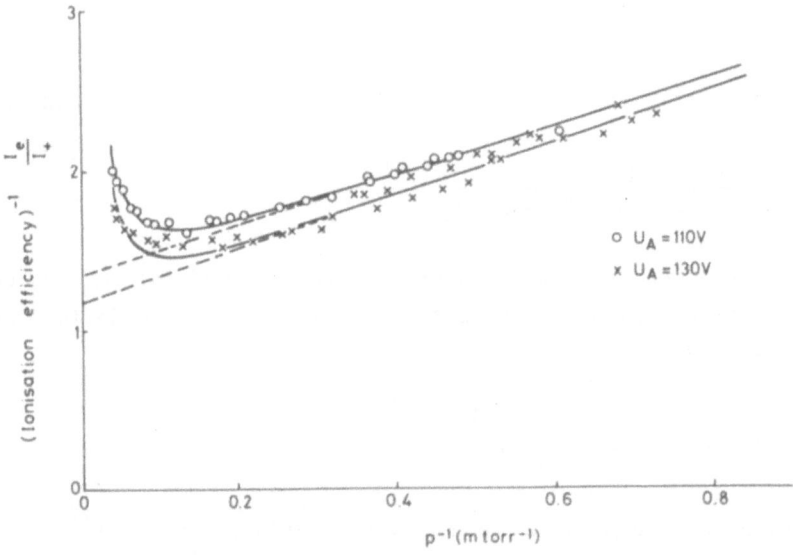

Fig. 4.   Experimental data for ionization efficiency.

Alternative tests are to measure n_+ or n_p. Data for $(n_p v_p)^{-1}$ versus pressure at constant arc current are shown in Fig. 5. This type of data leads to estimates of the value of the loss area for primary electrons, which is 170 cm^2.

Limiting Electron Containment

We may expect that we can increase the electron containment time τ_p indefinitely by increasing the magnetic field in the cusp region and reducing leakage. However, this may not be possible because the flow of thermal electrons would also be inhibited, and this flow is necessary to complete the current paths in the electrons. This condition may be derived as follows:

The thermal electron current to the anode is given by:

$$I_{ANODE} = + \frac{n_e v_e}{4} A_e \exp [-e\phi_A/kT_e] \tag{5}$$

We also have the charge neutrality condition

$$n_+ = n_e \gg n_p \tag{6}$$

where n_e is the density of thermal electrons

$\quad$ v_e is the average velocity of the thermal electrons

$\quad$ T_e is their temperature

$\quad$ k is Boltzmanns Constant

$\quad$ A_e is the loss area for thermal electrons to the anode

$\quad$ ϕ_A is the potential difference from the plasma to the anode

This current equals $I_+ \times (1 + S_{IN}/S_{ION})$, since one thermal electron is made in each ionization and also in each inelastic collision.

$$\left(1 + \frac{S_{IN}}{S_{ION}}\right) \frac{1}{\tau_+} = \frac{v_e}{4V} \cdot A_e \exp \left(-e\phi_A/kT_e\right)$$

or $\quad \dfrac{e\phi_A}{kT_e} = \log \left[\dfrac{v_e A_e \tau_+}{4V} \middle/ \left(\dfrac{S_{IN}}{S_{ION}} + 1\right)\right] \tag{7}$

which may be written

$$\frac{e\phi_A}{kT_e} = \log \frac{\tau_+}{\tau_e} - \log \left[1 + \frac{S_{IN}}{S_{ION}}\right]$$

The variation of ϕ_A with operating parameters, e.g., the arc voltage which determines the ratio, S_{ION}/S_{IN}, of the reaction rate coefficients, has been measured and found to agree with this formula.

Experimentally it is also found that as the magnetic field is increased, thus decreasing A_e, or the arc voltage is decreased, thus

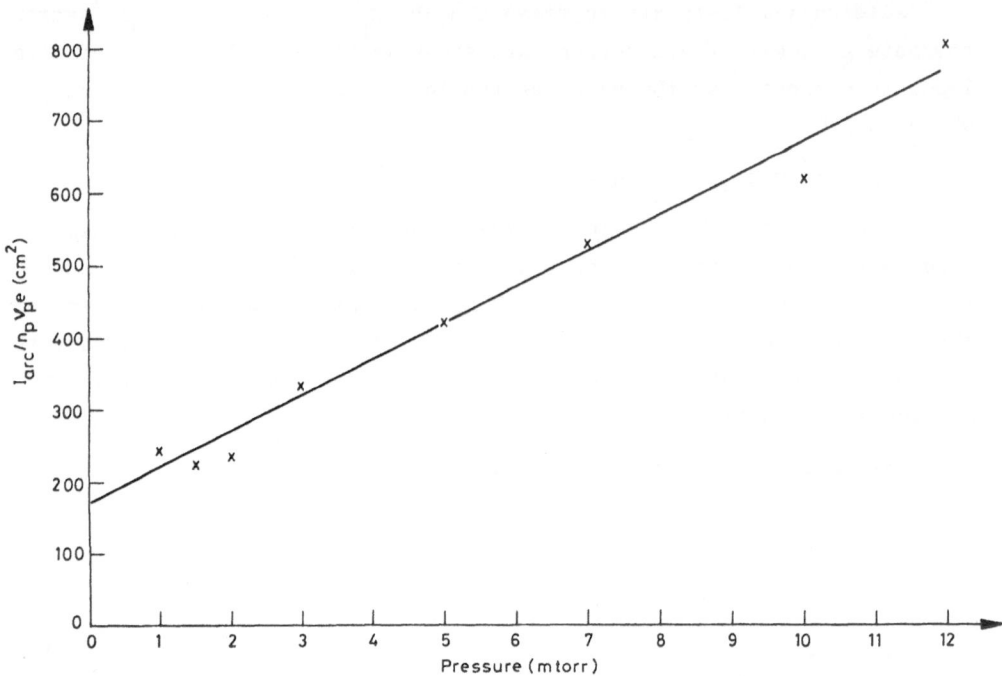

Fig. 5. Data for dependence of primary electron density, n_p, on pressure, P, to test theory. (Theory predicts $n_p^{-1} = a + bP$).

decreasing $\dfrac{S_{ION}}{S_{IN}}$, the plasma potential goes negative. The change in potential can then increase A_e since the electrons are driven into the cusp region. However, this is usually found to be a bi-stable operating condition which leads to so-called mode flipping, the alternative mode being a low efficiency one which low cathode potential drop and high anode potential difference.

This problem is generic; in all sources with arc discharges, electrons must flow to the anode which is restricted in area or shielded in some way or another by magnetic field. Violation of the condition for current continuity with positive plasma potential generally leads to instabilities or noise on the extracted beam. This further shows itself as degradation of beam transport.

Particle Heating

In this simple model we are treating, we have four particle groups: primary electrons, thermal electrons, positive ions and neutral atoms. There is energy exchange between these which is important in effecting the issues discussed above, and also - of direct relevance to this

Conference – the ion temperature which determines the emittance of any beam formed in the source.

The source of energy is the arc power which, as we say above, determines the energy of the primary electrons. These electrons exchange energy with the thermal electrons heating them and also being cooled.

The thermal electrons may then transfer energy to the positive ions. The latter then achieve a balance between this energy input and the cooling effect of ion-atom collisions. (For simplicity, I consider monatomic positive ions.)

Primary electron balance per unit volume

$$\text{Input energy} = I_{ARC} \ (U_A - e\phi_A)/eV \tag{8a}$$

$$\text{Energy loss} = n_p \ n_o \ S_{IN} \ \varepsilon_1 + \frac{n_p \ T_p}{\tau_p} + n_p \ n_e \ (T_p - T_e) \ S_{pe} \tag{8b}$$

Thermal electron balance per unit volume

$$\text{Input energy} = n_p \ n_e \ S_{pe} \ (T_p - T_e) + n_p \ n_o \ S_{IN} \ \varepsilon_2 \tag{9a}$$

$$\text{Energy loss} = \frac{n_e \ v_e \ A_e}{4V} \cdot \exp\left[- \frac{e\phi_A}{T_e}\right] \times T_e$$

$$+ \ n_e \ n_+ \ S_{e+} \ (T_e - T_+)$$

$$+ \ \frac{n_+}{\tau_+} \cdot e\phi_A \tag{9b}$$

Positive ion energy balance per unit volume

$$\text{Input energy} = n_e \ n_+ \ S_{e+} \ (T_e - T_+) \tag{10a}$$

$$\text{Energy loss} = \frac{n_+}{\tau_+} T_+ + n_o \ S_{+o} \ (T_+ - T_o) \tag{10b}$$

Atomic energy balance per unit volume

$$\text{Input energy} = n_+ \ n_o \ S_{+o} \ (T_+ - T_o) \tag{11a}$$

$$\text{Energy loss} = \frac{n_o}{\tau_o} \ (T_o - T_w) \tag{11b}$$

In principle one can solve all of these equations – and this is done in some computer simulations of discharges. Alternatively one may break up the discussion into steps and correlate observed values of T_p and T_e (from probe data) and then of T_+ (from emittance data) or of T_o (inferred from estimated changed in n_o).

As one example, we can look at variations of T_+ with T_o supposing that T_o is small compared with T_+. From Eqs. (10a) and (10b) we have have:

$$n_e \, n_+ \, S_{e+} \, (T_e - T_+) = \frac{n_+ \, T_+}{\tau_+} + n_+ \, n_o \, S_{+o} \, T_+$$

or

$$T_+ = \frac{n_+ \, S_{e+} \, T_e}{n_+ \, s_e + n_o \, S_{+o} + \frac{1}{\tau_+}} \tag{12}$$

(since $n_e = n_+$)

At high current densities $T_+ \to T_e$: at low current densities or high gas pressures $T_+ \to T_o$.

U_A = Arc voltage.

ε_1 = Average energy loss per inelastic collision as defined above.

ε_2 = Average energy transferred to thermal electron population when primary is degraded into this population.

T_p = Temperature of primary electron (eV).

T_+ = Temperature of ions (eV).

T_o = Temperature of atoms (eV).

T_w = Wall temperature (eV).

S_{pe} = Rate coefficient for energy exchange between primary and thermal electrons.

S_{e+} = Rate coefficient for energy exchange between thermal electrons and ions.

S_{+o} = Rate coefficient for energy exchange between ions and atoms.

τ_o = Lifetime for neutrals to loose energy to the walls.

Experimental Tests

In principle one may test these relationships experimentally, although in practice relatively little work has been described in the literature. One may measure T_p and T_e directly using probes.

One may also derive T_+ from measurements of the beam emittance (Fig. 6). Again one must say that very little correlation work has been done on dependence of T_+ and source operation.

It is also possible to estimate T_o. The floating potential of a probe in the plasma depends on the balance of fluxes of the primary electrons and the ions. The ratio of these fluxes depends on the gas density (this follows from Eq. 2). Consequently one can measure variations in gas density, which for a constant flow rate of gas arises from temperature variations.

These issues will be examined further in tutorial sessions.

Possible Control of Particle Temperature

Recently sources have been developed for producing high H^+, D^+ fractions on the one hand, and H^-, D^- ions on the other. Both are of the so-called filter, or tandem discharge, type. A sheet magnetic field (Fig. 7) divides a normal magnetic multipole sources into two regions. In one regions near the cathode there is a dense region of ionization. In the second region there are no primary electrons to ionize or heat the gas. Further, the magnetic field acts as a thermal barrier. Consequently electrons in this region are cooler. We see from Eq. 12 that the ions also can be cooled.

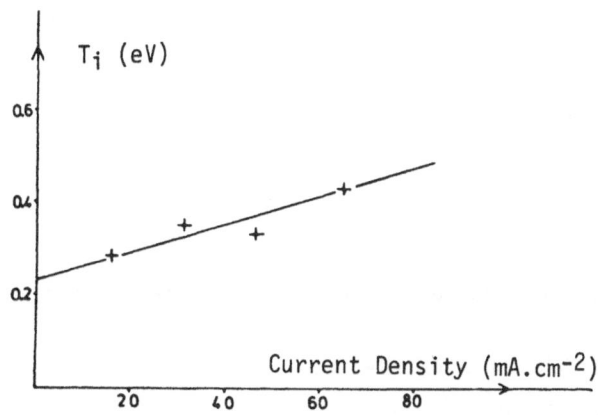

Fig. 6. Data on variation of T_+ with arc current in helium ion source (T_+ derived from emittance).

We may differentiate between various types of ion source with respect to the energy spectrum of the electrons and should recognize that the hotter the main group of electrons, the more probable it will be that the ions have high temperature and hence high emittance - although of course the detailed calculations should be done to determine the rate of energy transfer. Different types are:

i) Very high energy electrons for multiply-charged ions.

ii) Medium energy for ionization by thermal electrons.

iii) Low energy with weak high energy tail for ionization (e.g., multipole, as discussed above).

iv) Spatially separated region of high energy for ionization and very low energy near the extraction region.

461

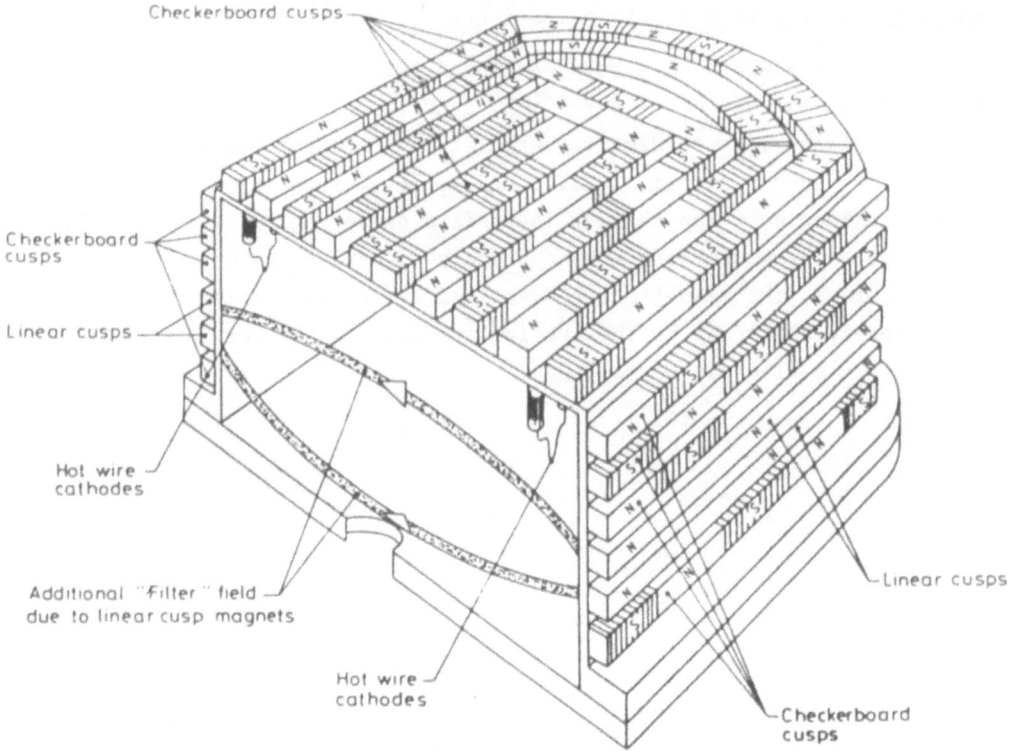

Fig. 7.   Schematic diagram of "filter" ion source.

ION BEAM EXTRACTION AND ACCELERATION

General Features

Ions are extracted from a plasma source by the application of an electric potential between electrodes and the source plasma. A single-gap accelerator with three electrodes for positive ions is shown in Fig. 8. The first electrode, at positive potential V^+, defines the equipotential plane close to the source plasma; the third is at earth potential. An intermediate electrode held $\approx -0.1V^+$ prevents electrons formed in the low energy transport section from returning to the source; it is called a suppressor electrode. Its influence on the beam extraction is small but it does play a significant role in the motion of secondary ions and electrons that may produce power loading in the electrode system.

Limitation in performance of three-electrode (triode) systems due to electrical breakdowns have been discussed extensively in the literature. These discussions indicated that high voltage, high current sources could best be operated with four-electrode (tetrode) designs. More recently, extended operation experience at high voltage has indicated that the situation is less clearcut than originally proposed. In this discussion

462

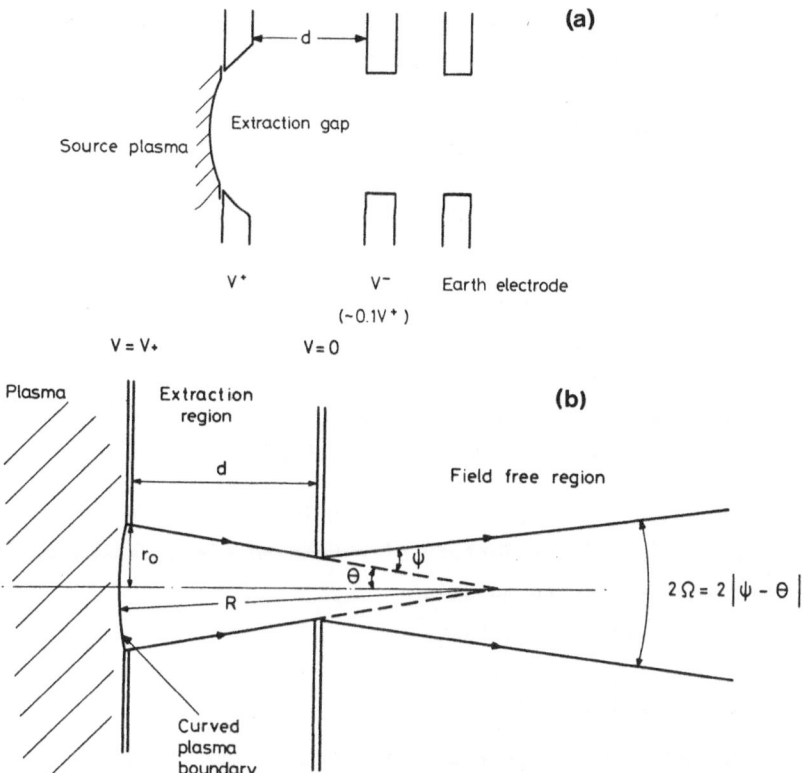

Fig. 8. "Single Gap" (triode) accelerator.
        (a) Schematic of electrodes
            (Single extraction gap).
        (b) Trajectories.

the design of both three- and four-electrode structures will be considered without reference to their limitations.

Analytical Treatment of Single-Gap Extraction

In the present discussion we are concerned with the problem of the
acceleration of ions extracted from a plasma surface, which is analogous
to that of electron acceleration in a space-charge limited diode.

The theoretical evaluations (i.e., analytical and computational) of
this problem differ in detail in their treatment of the plasma surface.
In the analytical approach it is normally assumed that the plasma
boundary is curved (either as a section of a cylinder or a section of a
sphere) with constant current density over its surface. The subsequent
analysis follows closely that presented for an electron gun with curved
emitter surfaces. It is based on the treatment by Langmuir and Blodgett,
of a spherically converging, space-charge limited flow. Figure 8b shows
schematically the trajectories of the ions.

The current of ions I, from an aperture, radius a, accelerated across a gap d is given approximately by

$$\frac{I}{V^{3/2}} = 1.72 \times 10^{-7} \frac{a^2}{d^2} \left(1 - \frac{1.6d}{R}\right) \left(A \ V^{-3/2}\right) \tag{13}$$

where V is the acceleration voltage and R the radius of curvature of the plasma surface.

The angle of convergence of the beam from the plasma surface is θ and equals a/R. The ions pass through the aperture in the negative suppressor electrode, which behaves like an electrostatic lens. The focal length is calculated from the Davisson-Calbrick formula allowing for the influence of space charge on the electric field, i.e.,

$$f = 3d . \tag{14}$$

The equation for the trajectory of a beam particle can be written using the matrix transfer formulation:

$$\begin{vmatrix} r' \\ r \end{vmatrix} = \begin{vmatrix} 1 & -1/f \\ 0 & 1 \end{vmatrix} \begin{vmatrix} 1 & 0 \\ d & 1 \end{vmatrix} \begin{vmatrix} r'(0) \\ r(0) \end{vmatrix} \tag{15}$$

where r is the radius of the particle trajectory and r' is the angle made by the particle trajectory to the beam axis, r' = dr/dz.

The equation for the beam envelope is obtained by equating r'(0) to θ and r(0) to a. The condition that the emergent beam be parallel is that R equals 4d, to a first approximation, but R = 3d is more accurate.

Inserting this relation into Eq. (13) for the perveance leads to the result that for a parallel or zero divergence beam

$$P_0 = \frac{I}{V^{3/2}} \approx 0.76 \times 10^{-7} \frac{a^2}{d^2} \ (A \ V^{-3/2})$$

$$= 0.44 P_c \tag{16}$$

where P_c is the perveance for a plane diode. The exact value of the constant depends on the approximation introduced and can vary from 0.44 to 0.60. This analysis also leads to an expression for the variation of the angular divergence Ω with perveance

$$\Omega = 0.5 \ a/d \ [(P/P_0) - 1)] . \tag{17}$$

Corresponding relations can be derived for slit apertures.

These predictions agree reasonably well with the experimental observations (Fig. 9), except that they indicate zero divergence at optimum perveance rather than the non-zero values observed experimentally, due to neglect of the finite transverse energies of the ions.

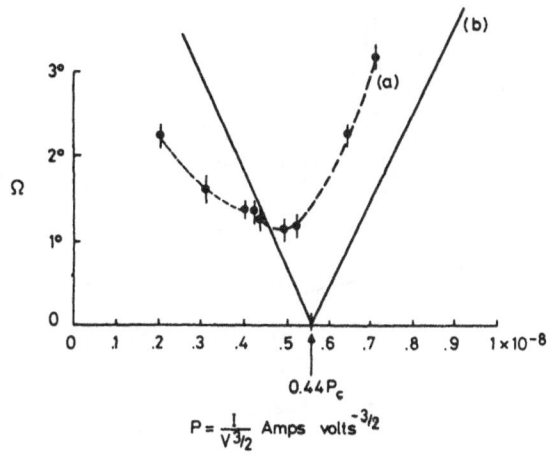

Fig. 9. Variation of angular divergence with beam perveance.

Multigap Systems

In principle the addition of more electrodes can be analyzed straightforwardly by an extension of the transfer matrix formulation:

$$\begin{vmatrix} r' \\ r \end{vmatrix} = M_{11} \cdot M_{12} \cdot M_{22} \cdot M_{23} \cdot \begin{vmatrix} r'(0) \\ r(0) \end{vmatrix}. \qquad (18)$$

where M_{11}, M_{22}, etc. represent the transfer matrices for the drift spaces between electrodes and M_{12}, M_{23} the transfer matrices of the lenses in the electrodes.

As a specific example we can consider a two-gap (or tetrode) system (Fig. 10). M_{11} is calculated assuming straight line trajectories (as above) and equals $\begin{vmatrix} 1 & 0 \\ d_1 & 1 \end{vmatrix}$. M_{22} is calculated assuming that space-charge effects are small and ions move in the constant electric field in this gap.

$$M_{22} = \begin{vmatrix} A^{-1} & 0 \\ \dfrac{2(A-1)V_1}{E_2} & 1 \end{vmatrix}. \qquad (19)$$

where $A = [1 + (V_2/V_1)]^{1/2}$, V_1 is the voltage across the first gap, and V_2 is that across the second. E_2 is the electric field in the second gap V_2/d_2. The lens transfer matrices are:

$$\begin{vmatrix} 1 & -1/f_1 \\ 0 & 1 \end{vmatrix} \quad \text{and} \quad \begin{vmatrix} 1 & -1/f_2 \\ 0 & 1 \end{vmatrix},$$

465

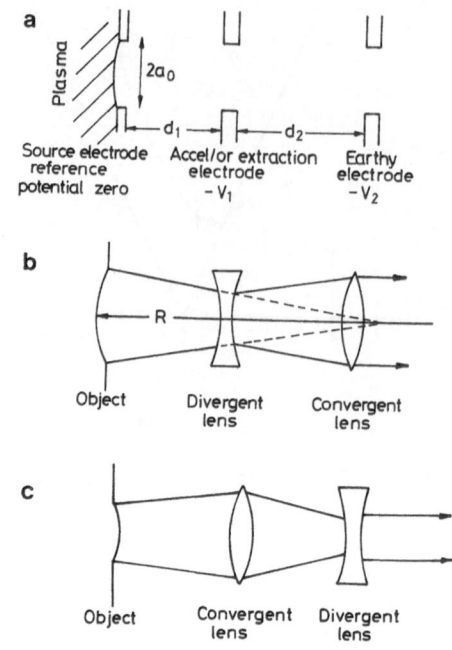

Fig. 10.   Accelerator with two gaps.
a) Disposition of electrodes
b) Equivalent optical system $V_1 > V_2$
c) Equivalent optical system $E_2^1 > E_1^2$

where the focal lengths are calculated using the Davisson–Calbrick formula. Various degrees of sophistication may be used in calculating the
electric fields used in these formulas to allow for space change and
geometry. The final trajectory equation may be written as

$$
\begin{vmatrix} r' \\ r \end{vmatrix} = \begin{vmatrix} 1 & -1/f_2 \\ 0 & 1 \end{vmatrix} \begin{vmatrix} A^{-1} & 0 \\ \dfrac{2(A-1)V_1}{E_2} & 1 \end{vmatrix}
$$

$$
\times \begin{vmatrix} 1 & -1/f_1 \\ 0 & 1 \end{vmatrix} \begin{vmatrix} 1 & 0 \\ d_1 & 1 \end{vmatrix} \begin{vmatrix} r'(0) \\ r(0) \end{vmatrix} \tag{20}
$$

The emergent beam has zero divergence ($r' = 0$) when

$$
\frac{r(0) + d_1 r'(0)}{d_1 r'(0)} \left[\frac{2V_1(A^{-1}-1)}{E_2\,f_1 f_2} - \frac{1}{f_2} - \frac{1}{f_1 A} \right] = \frac{2V_1}{E2} \frac{(A^{-1}-1)}{d_1 f_2} \frac{1}{A d_1} \tag{21}
$$

This rather complicated expression relates the possible values of
the voltage ratio (V_2/V_1 designated as Γ) and the ratio of the gap distances (d_2/d_1 designated by γ) to allowable values of the radius of
curvature of the plasma boundary [$r(0)/r'(0)$]. Particular solutions for

466

a flat boundary, giving the highest perveance, have been presented which correspond to high values of Γ/γ, i.e., of E_2/E_1.

Computational Studies of Beam Extraction

Analytical treatments suffer from a number of defects: idealization of plasma profile, neglect of electrostatic field perturbations, and neglect of finite temperature of the ions. Numerical simulations of extraction systems have been developed, to overcome these defects.

The computer programs have varied in sophistication and the degree to which they can take account of the effects omitted from the analytical treatments. All programs allow for the effect of geometry on the electrostatic fields, and some for the finite transverse ion energy. Treatment of the plasma density at the aperture in the beam-forming electrode is the most difficult task since it requires that the computations be carried out over a volume within the plasma in which a uniform flux of ions towards the aperture can be defined. Typical ion trajectories calculated using such a program are shown in Fig. 11. To minimize the computing required to obtain this result, an analytical model can be used to calculate flow towards the aperture matching the output to the computation of ion trajectories from the plasma boundary.

One of the most important applications of the computer program has been to the calculation of the motion of secondary particles created in the gap, either by ion-molecule collisions or by collisions of particles with the electrodes. These determine the power loading and halo effects.

There are limitations to these codes: perhaps the most serious is a lack of a self-consistent treatment of the plasma created by the beam in the transport region, including the earth electrode aperture. One should note, also, that there is also very much work to be done before we can treat negative ion sources correctly.

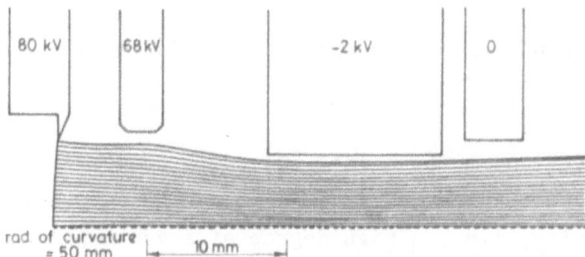

Fig. 11. Typical ion trajectories in accelerator derived from computer simulation.

LOW ENERGY BEAM TRANSPORT

General Considerations

In this section I consider transport of low energy beams, i.e., beams of energies up to 100 - 200 keV, as are commonly found in the injectors of high energy accelerators, and in particular questions arising from space charge. The use of focusing elements for zero space charge beam is extensively covered in text books, and recent developments in codes to trace finite space charge, finite emittance beams have also been covered in conferences.

My main consideration here is that of partially space charge compensated beams. One may indicate the need for neutralization by a simple analysis.

The Kaphchinsky-Valdimirsky equation gives a useful vehicle for discussing motion of an ion beam:

$$\frac{d^2a}{dz^2} = \frac{\varepsilon^2}{a^3} + \frac{K}{a} \tag{22}$$

where a is the beam radius, z the distance along the beam axis, ε the beam emittance and K the term which allows for space charge effects.

$$K = \text{Constant} \times \frac{I}{v^{3/2}} \left[(1 - f) - \left(\frac{v}{c}\right)^2 \right] \tag{23}$$

where f is the fractional compensation of the space charge, $\frac{v}{c}$ is the beam velocity divided by the velocity of light (for medium energy beams it may be neglected provided (1 - f) is not too small and one is dealing with modest beam currents << 10A).

As we have discussed before, the current which can be extracted from a single aperture depends on space charge (Eq. 16). Consequently the constant above relates to the parameters of the aperture and we find

$$K = \frac{a_o^2}{9d^2} (1 - f) \tag{24}$$

The beam expands under the space charge force with a divergence angle of $\sim \frac{14\ a_o}{3d} (1-f)^{1/2}$.

If one is designing for higher intensity beams, i.e., for higher values of a_o/d approaching unity, then it becomes essential to have f very close to unity, i.e., a high degree of space charge compensation.

[An alternative statement is that without space charge compensation one needs a lens of focal length ~ 2d to counter the space charge force (focal length ~ $\frac{a_o}{\theta}$ which would have to be a few centimeters).]

Space Charge Compensation of a Single Beamlet

In a recent discussion of beam transport the author has commented that there is no single aspect of the theory of space charge compensation of an ion beam for which there is a common unified treatment. This remark must be understood in the context of the complexity of the problem, which involves:

i) Solution of Poisson's equation or of approximations to it;

ii) Solution of the spatial dependence of electron and slow ion densities which derive from analysis of the energy distributions of these two types of particle;

iii) Equations for particle balance, both electrons and ions;

iv) Energy balance equation for the plasma; and

v) Boundary conditions if a plasma sheath exists.

A comprehensive review is beyond the scope of the present paper. Instead we will consider one model which seems consistent with published experimental data, and some implications of its results. Further aspects can be discussed in oral sessions.

(a) We assume charge neutrality on the beam axis $n_{eo} = n_{bo} + n_{io}$, where n_{eo} is the electron density on axis, n_{bo} is the density of beam ions, and n_{io} is the density of slow ions formed by beam-gas collisions. This is justified when $1 - f$ is small, as discussed above.

(b) We neglect initially the slow ions. One can show that n_{io}/n_{bo} is less than unity in most beamlet experiments, though it may be greater in multiple beams, depending on pressure and dimensions.

(c) We calculate n_e using the formulation in which we assume that the electrons bound in the potential well have a truncated Maxwellian distribution and only escape by diffusing in energy space (concepts of diffusion in energy space as the controlling mechanism for electron loss are used by several authors). When electron loss rate is balanced against production rate one derives the result:

$$\frac{n_{eo}}{\tau_{ee}} \cdot \exp(-\eta) = n_{bo} \, n_o \, \sigma_e \, v_b \tag{25}$$

where η is the normalized potential $\left(e\phi \times \left[kT_e \right]^{-1} \right)$, n_o the neutral gas density, σ_e the cross-section for electron production, τ_{ee} the electron – electron collision time, and v_b the velocity of a beam ion.

(d) Energy balance is between the energy carried out by the accelerated slow ions and the input due to beam-electron collisions. Terms in T_e are not included in the loss because of the truncated electron energy distribution.

The energy balance equation is written as

$$\text{Energy input} = \frac{n_b \, eU_b}{\tau_{be}}$$

$$\text{Energy loss} = \text{ion production rate} \times e\phi$$

$$= n_b \, n_o \, \sigma_i \, v_b \, e\phi \ .$$

Equating these terms, one has

$$\phi = U_b \times \frac{1}{n_o \, \sigma_i \, v_b \, \tau_{be}}$$

which reduces to

$$\phi = U_b \cdot \frac{m_e}{M_b} \exp \eta \qquad (26)$$

since $\sigma_i \approx \sigma_e$ and $\tau_{ee}/\tau_{be} = m_e/M_o$.

Here σ_i is the cross-section for slow ion production, and τ_{be} the time constant for energy loss by the beam ions to the electrons.

Combining Equations (25) and (26), one derives the result that

$$\phi^2 = \text{Constant} \cdot \frac{n_{bo}}{n_o} \cdot \log[\phi/\phi_o] \qquad (27)$$

where ϕ_o is approximately $\frac{m_e}{M_b} \cdot U_b$

Figure 12 shows typical experimental data for the variation of ϕ with pressure (hence n_o); the solid line shows values predicted by Eq. 27 (normalized at one point). Exact comparison is complicated by the fact that slow ions become important at the higher pressure.

An important prediction of this model is that the potential depends on the beam ion density. As a consequence the potential measured at different axial positions along a diverging beam will vary, as seen in Fig. 13. This result is not obtained from considering the neutralization fraction f to be constant, or from fluid type models of space charge

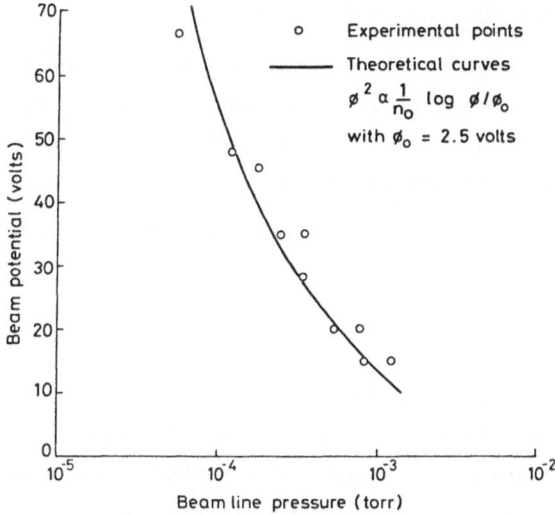

Fig. 12.   Variation of beam potential with pressure
in the beam line.

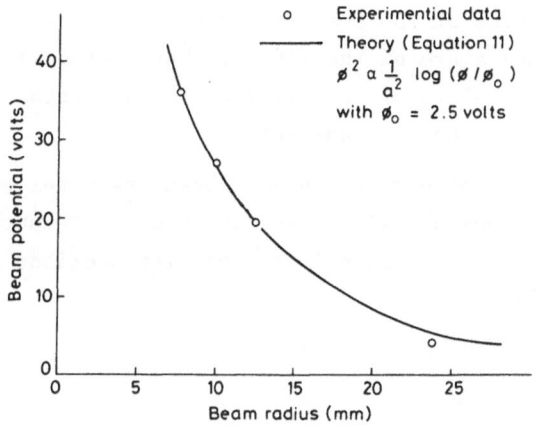

Fig. 13.   Variation of beam potential as beam propagates
along axis. ϕ is function of radius.

neutralization: it derives from the assumption that the electron loss
rate is determined by collisional processes.

We note that to a first approximation

$$\phi \ \propto \ n_b^{1/2} \ \propto \ J^{1/2} \ \propto \ \frac{I^{1/2}}{a}$$

Thus if we want to obtain high current with low space charge potentials we need to operate with large apertures.

Space Charge Neutralization of H⁻ Beams

The calculation given above is for a positive ion beam in that one assumes that quasi charge neutrality require that n_{bo} equals n_{eo}. For a negative ion beam this equality is not valid and the slow positive ions formed by ionization become important.

This theory of space charge neutralization of H⁻ beams is developing and will be treated in tutorial sessions.

BEAM EMITTANCE

The factors which determine emittance and the role of emittance in influencing beam propagation are discussed elsewhere. I wish, here, to comment only on the factors which are important in the injector.

The minimum emittance which can be achieved in an ion beam is due to the transverse energy of the ions as they leave the ion source:

$$\varepsilon_{RMS} \text{ (normalized) } = 0.016\pi \, a \, T^{1/2}$$

where a is the radius of the beam at the plasma surface in cm and T is the effective temperature of the ions at that surface in eV. As the ions pass through the accelerator, a will vary, with corresponding variation in T, since the emittance is constant.

The achievement of high brightness beams requires one to maximize I/ε^2, i.e., J/T. Thus one wishes to keep the ion temperature low at high current densities, as discussed in the earlier sections on particle heating.

RADIAL TRANSMISSION--LINE LINEAR ACCELERATORS

K. R. Prestwich

Sandia National Laboratories
Albuquerque, New Mexico 87185 USA

INTRODUCTION

Many applications of high current electron or ion beams require voltages in the 30 MeV to 100 MeV range with currents in the 10 kA to 100 kA range. The most straightforward way to achieve the high beam energies is to use a linear accelerator. RF accelerators operate at the high end of the voltage range with currents less than 10 A (Fultz and Whitten, 1971). At the low end of the voltage range, the Phermex facility at Los Alamos Scientific Laboratories has been operated at 30 MeV and 500 A (Boyd et al., 1965; Moir et al., 1985).

Ferrite-core linear induction accelerators that depend on induced voltages from rate of change of magnetic flux have operated with voltages up to 50 MeV and 10 kA currents (Christofilos et al., 1964; Avery et al., 1971; Gnalskii et al., 1970; Hester et al., 1979; Prono et al., 1985). The limitation imposed by magnetic field saturation of the ferrite while maintaining a reasonable ferrite volume has restricted the accelerating voltage to 200 kV to 300 kV per stage. The average accelerating gradient (total electron energy/total accelerator length) is about 1 MV/m.

Brief descriptions of induction accelerators without ferrite that were designed for operation at 10 MeV, 100 kA have appeared in the Soviet literature (Pavlovskii et al., 1970; Pavlovskii et al., 1974; Pavlovskii et al., 1975; Pavlovskii et at., 1980). The LIU-10 that was reported by Pavlovskii's group in 1980 produces a 13 MeV, 50 kA beam. These accelerators utilize water-dielectric, radial, pulse-forming transmission lines (PFL) for the accelerating cavities. Similar concepts utilizing vacuum insulation had been considered by Faltens, Hartwig, and Hernandez (1968) and Hartwig (1968) at the Lawrence Berkeley Laboratory for electron ring experiments, but were abandoned in favor of the iron core LIA.

473

The modular, high-current accelerators under development at Sandia Laboratories for inertial confinement fusion (ICF) experiments have lead to techniques of operating many high-voltage, oil or water insulated, low impedance, pulse-forming transmission lines (PFL) in parallel (Prestwich, 1975; Prestwich et al., 1975; Martin et al., 1975; Martin et al., 1977; Martin et al., 1981). Multichannel, liquid-dielectric, spark gaps with a jitter of 1 ns to 3 ns were developed for these accelerators (Prestwich, 1974; Johnson, 1974; VanDevender and Martin, 1975; Johnson, VanDevender, and Martin, 1980). The PFLs are operated with charge voltages in the 2 MV to 3 MV rage. The information obtained from the ICF program was applied to the development of the 9.5 MV, 28 kA RADLAC-I accelerator (Miller et at., 1981) and the 16 MV, 50 kA RADLAC-II accelerator (Mazarakis et al., 1985; Miller et al., 1985; Mazarakis et al., 1986).

Cavity Design Considerations

Radial transmission line LIAs are closed cavities containing two or more transmission lines. Accelerating voltage pulses can be generated externally and injected into the cavity as described by Smith (1979a) or induced from charged transmission lines (PFLs) which make up the cavity upon closure of a switch shorting one of the PFLs (Smith, 1979b; Eccleshall et at., 1979; Eccleshall and Temperley, 1978). To calculate the accelerating voltage waveform it is convenient to treat the cavities as PFLs. The relationship between the induced voltage of the cavity and the PFL parameter is shown with Eqs. 1-5 and Fig. 1.

$$V = \frac{\partial}{\partial t} \int B \cdot dA \ . \tag{1}$$

For a strip transmission line with a constant spacing a and width W, Eq. 1 becomes Eq. 2.

$$V = \frac{\partial}{\partial t} \left(\frac{\mu I}{W} a \ell \right) = \frac{\partial}{\partial t} (LI) \ . \tag{2}$$

If I is a constant Eq. 2 becomes Eq. 3 or Eq. 5.

$$V = \mu I \frac{a}{W} \frac{d\ell}{dt} \tag{3}$$

$$\frac{d\ell}{dt} = c = \frac{1}{\sqrt{\mu\varepsilon}} \tag{4}$$

$$V = \sqrt{\frac{\mu}{\varepsilon}} \frac{a}{W} I = Z_o I \ . \tag{5}$$

Radial Cavities

The dielectric cavity of the pulse line linear induction accelerator can be constructed with either radial or cylindrical transmission lines. The energy transfer efficiency and accelerating voltage waveforms for

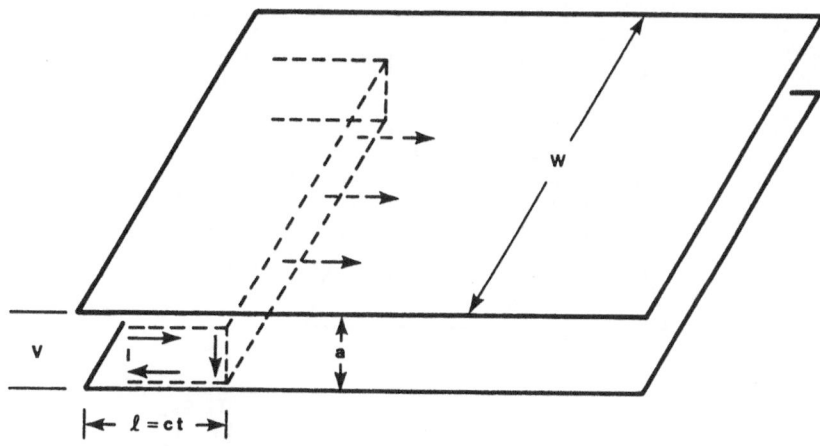

Fig. 1.   Pulse forming transmission line.

several cavity configurations are presented in this section.   To achieve
a flat topped accelerating pulse for a constant current electron beam
(i.e., a constant impedance load), the PFLs must have constant impedance
throughout their length.   Symmetric bicone transmission lines are used in
the radial cavities.   Equation 6 gives the impedance for a symmetric
bicone transmission line

$$Z = \sqrt{\frac{\mu}{\epsilon}} \frac{1}{2\pi} \ln \frac{1}{(\tan \theta_1)^{1/2}} \tag{6}$$

where θ_1 is the half angle of the conical electrode (Lewis and Wells,
1959).   An excellent approximation for this formula for θ_1 nearly equal
to $\pi/2$ radians is given by Eq. 7 where k is the slope of the cone.

$$Z = \sqrt{\frac{\mu}{\epsilon}} \frac{k}{2\pi} \cdot \tag{7}$$

Figure 2 shows a cross section of six possible cavity configurations.
The arrangements in Fig. 2(a) and 2(b) are taken from Pavlovskii (1975).

Figure 2(a) is the bicone transmission line version of this cavity
and is usually referred to as a radial transmission line cavity.   Figure
2(b) is the coaxial transmission line version.   The choice of radial or
coaxial configuration depends on the PFL impedance and pulse duration.
The configuration is usually chosen that allows the smallest volume
accelerator and produces the waveform that comes closest to the ideal
case.

In this cavity [Fig. 2(a)], if the switch is closed at t = 0, the
waveform across terminals a-b for an open circuit case is shown in Fig.
3(a).   If the beam is injected during the first half-cycle, only half of

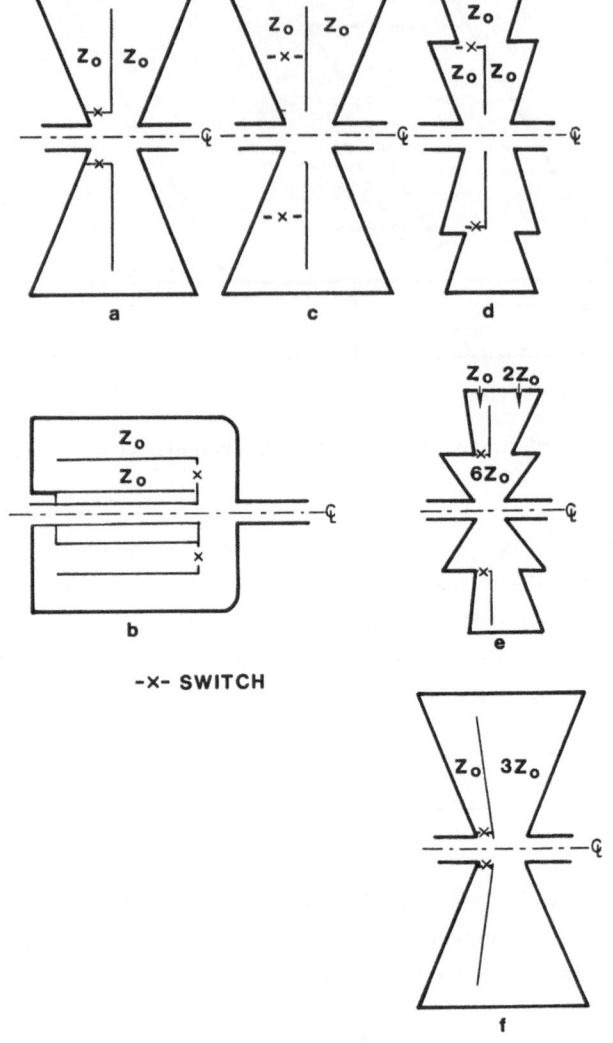

Fig. 2.   Cavity configurations:
        (a) Radial transmission line Pavlovskii (P1)
        (b) Coaxial transmission line Pavlovskii
        (c) Isolated Blumlein
        (d) S3
        (e) ET-2
        (f) ET-1

the energy can be removed. Therefore, the beam is normally injected at
the beginning of the second half-cycle and has a pulse duration of 2T,
where T is the time for the electromagnetic wave to travel from the
switch to the outer corner and back to terminal a-b. Figure 3(b) shows
the idealized waveform if the effective impedance of the injected beam
matches the impedance, Z_o, of the cavity, (i.e., $V_{a-b}/I_b = Z_o$).

The voltage versus distance along the transmission line (cavity) at
various times during the pulse is shown in Fig. 4. If we consider an

476

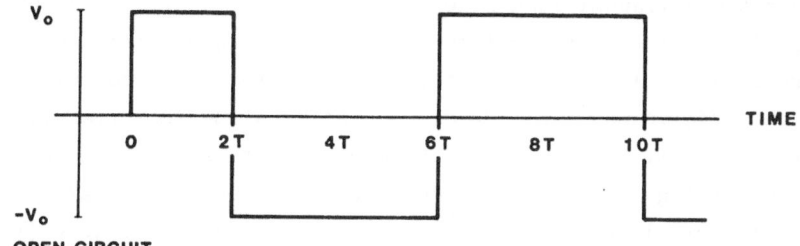

a) OPEN CIRCUIT

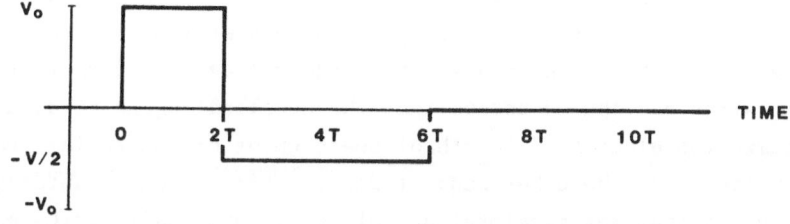

b) MATCHED LOAD

Fig. 3. Accelerating voltage waveform for the Pavlovskii
Cavity with the PFLs charged to V_o
(a) Open circuit
(b) Matched load connected at time 2T

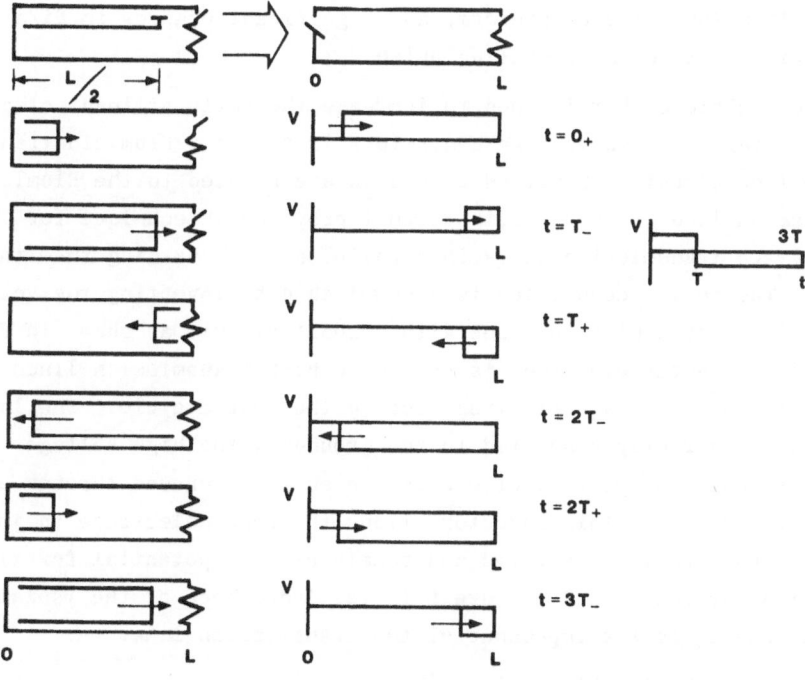

WAVES IN RADIAL PULSE LINES

Fig. 4. Ideal pulse line voltage as a function of time and line length
for an unfolded Pavlovskii cavity.

ideal corner and switch, the Pavlovskii cavity can be represented by a single-charged transmission line of length L, as shown in the upper line in Fig. 4. At t = 0, the switch is closed and an electromagnetic wave that reduces the line voltage to zero propagates from the switch. The wavefront location and voltage at each point along this transmission line at t = 0 and just before and after each reflection is shown in the remainder of Fig. 4. If a matched load is connected (resistor in Fig. 4 or beam in actual accelerator) when the wave reaches the load (t = T+), then the voltage drops to V/2 for all times shown greater than T+ and goes to zero at 3T. In this sketch, an idealized waveform is shown, and the effects of pulse rise time and reflections at the corners and output have been neglected. One can see from the voltage waveform that this configuration has the advantage that the accelerating pulse duration is four times the electrical length of one segment of the radial trans- mission line. On the other hand, three significant disadvantages exist. The first is that the accelerating voltage is only half of the charge voltage if the beam is a matched load for the cavity. The second is that any pulse distortion due to the wave going around the corner will appear in the center of the accelerating pulse and contribute to an energy spread in the accelerated electron beam. The third disadvantage is that any change in the inductive voltage drop across the switch also distorts the accelerating voltage waveform, and significant changes in di/dt occur in the center of the accelerating pulse.

One technique that is used to increase the ratio of load voltage to charge voltage in high voltage generators is to use a Blumlein trans- mission line circuit. Cavities 2c and 2d are related to the Blumlein transmission line. The Blumlein circuit provides a technique for charging two transmission lines in parallel and discharging them in series. The series connection is accomplished by inverting the voltage on one of the transmission lines with a short circuit as shown in Fig. 5. In Fig 5 the center electrode is common to both transmission lines and becomes the high voltage electrode during the charge cycle. The lower electrode is directly connected to the ground of the high voltage generator, and the upper electrode is connected to ground through an isolating inductor. This inductor allows the upper electrode to jump to high voltage when it is shorted and remain at this potential for times short with respect to L/Z_o, where L is the inductance of the isolating inductor and Z_o is the impedance of the transmission line.

As in the Pavlovskii cavity, shorting one transmission line has the effect of inverting its polarity. The switch is closed and one electrical transient time later, the open circuit output voltage becomes

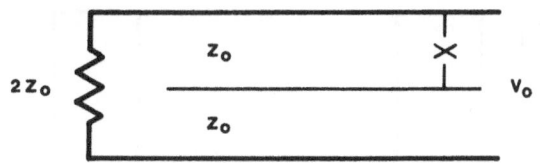

Fig. 5. Blumlein transmission line schematic--center
line is charged to V_o.

$2V_o$ and remains there for $2T$. The voltage applied to a matched load is
V_o. The output impedance is $2Z_o$.

In Figs. 2(c) and 2(d) the Blumlein isolating inductors are
transmission lines because it is not possible to place the appropriate
inductor inside a closed cavity. In the isolated Blumlein circuit (Adler
et al., 1983), Fig. 2(c), the $2V_o$ open circuit voltage is produced
without droop at the expense of decreasing the energy transfer efficiency
to 50%. The transmission line is charged and switched in the center of
one of the lines as shown. Waves that reduce the voltage across this
line to zero travel both ways from the switch. The lower half of the
bicone functions as a Blumlein with the voltage on the switched line
inverting and creating the series connection of the two PFLs and
producing $2V_o$ open circuit voltage that lasts from $T \le t \le 3T$. In this
case T is the length of the lines from the switch to the load or from the
switch to the corner. The wave that traveled radially outward from the
switch arrives at the load at $3T$ and causes the voltage to invert as
shown in Fig. 6. The isolated Blumlein waveforms for open circuit,
over-matched, matched, and under-matched loads are also included in this
figure.

Another feature of the isolated Blumlein is that it has nearly zero
prepulse. This feature has made it useful in single-stage electron beam
generators. In an ordinary Blumlein transmission line the voltage that
appears across the isolation inductor during charging of the transmission
line also appears across the load (see Fig. 7). This voltage is usually
referred to as the prepulse voltage. Although this prepulse voltage is
much smaller (~5%) than the main voltage pulse, it appears across the
anode-cathode gap for a much longer time, and it can cause deleterious
effects to the operation of cold cathode devices. As one can see from
Fig. 7, if a second inductor is added in series with the lower
transmission line, the prepulse should become zero. In practice with
high voltage generators housed in a large metal tank and using a Marx
generator to charge the transmission lines, the prepulse can not be
turned to zero because of stray capacity from the erected Marx generator

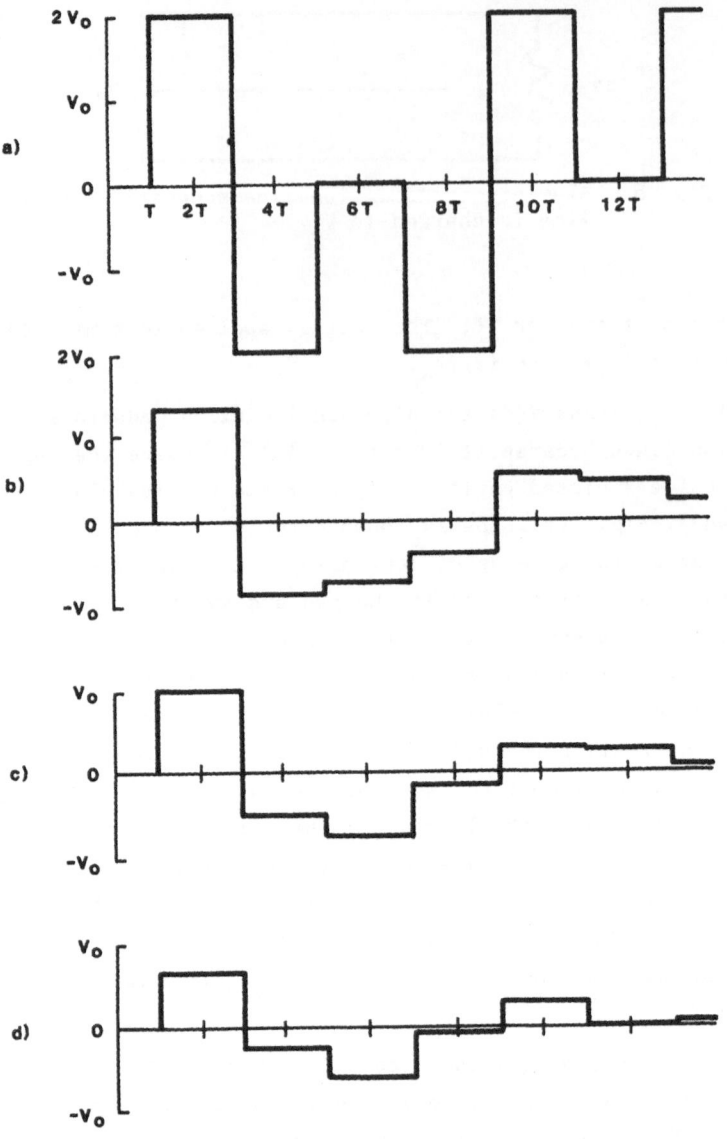

Fig. 6. Isolated Blumlein waveforms with $2Z_o$ cavity output impedance, charge voltage V_o, and the following resistive loads:

(a) $R = \infty$
(b) $R = 4Z_o$
(c) $R = 2Z_o$
(d) $R = R_o$

to the tank wall (Martin et al., 1969). Since the isolated Blumlein has two identically charged transmission lines for isolating inductors, the voltage across these are equal during charging and the prepulse is zero. This feature is the 4 MV, 100 kA IBEX generator (Ramirez et al., 1983) has allowed very reproducible electron beams to be generated.

480

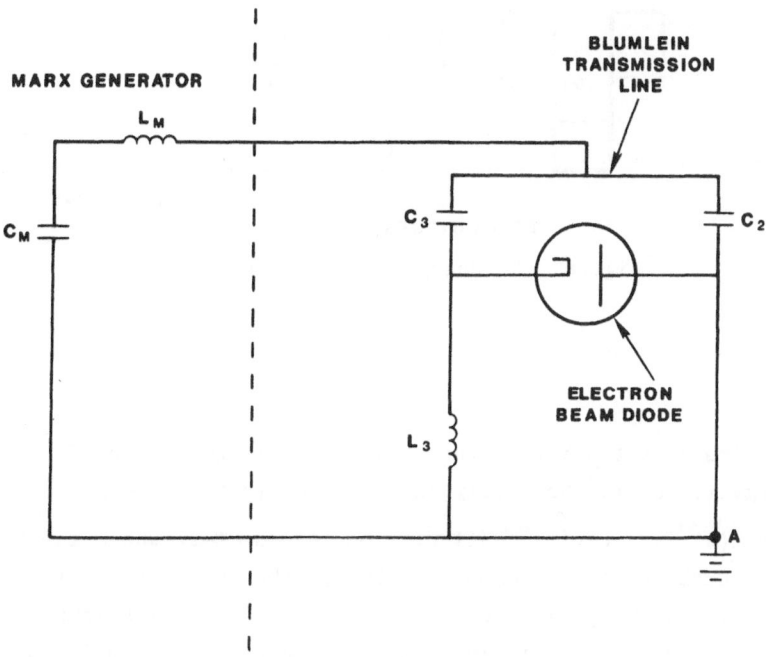

Fig. 7. Blumlein charging circuit.

The cavity configurations shown in Figs. 2(d) and 2(f) can be used
to produce accelerating voltage to charge voltage ratios equal to or
greater than one while still maintaining a high energy transfer
efficiency. In all of these configurations as well as Fig. 2(a), the
acceleration must take place on the second half-cycle of a bipolar
waveform in order to achieve high efficiency. As described by Smith
(1979b), the first half-cycle is used to set up the conditions for full
energy extraction. As in the standard Blumlein circuit, all of these
configurations obtain an output voltage gain but switch currents are
higher than load currents.

The configuration shown in Fig. 2(d) is usually referred to as S3
with the S denoting Smith (1979a), the inventor of the configuration. S3
has a strong relationship to the ordinary Blumlein transmission line
circuit. All three of the transmission lines in S3 have the same
impedance and pulse duration. Figure 8 is a sketch of the open circuit
waveform for this cavity. When the switch is closed, a wave propagates
in the PFL line that is shorted to reduce its voltage to zero as in an
ordinary Blumlein. The switch also connects the other PFL across the
uncharged transmission line and since the impedances are matched, a wave
of amplitude $V_o/2$ reduces the voltage on the PFL and increases it on the
uncharged line. The sum of the two waveforms propagating to the output
terminals gives the open-circuit accelerating voltage waveform shown in
Fig. 8. If a matched load is placed across the accelerating terminals

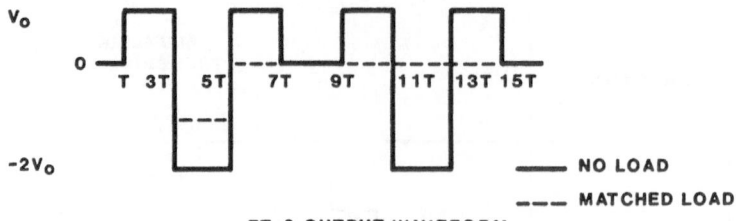

ET-2 OUTPUT WAVEFORM

Fig. 8. S3 Open-circuit waveform.

during the time $3T \leq t \leq 5T$, the accelerating voltage will be V_o and all of the energy stored in the cavity is delivered to the load.

Figure 2(e) is a cavity configuration termed ET-2, where the ET acknowledges Eccleshall and Temperley (1978), the inventors of this configuration. In this case the impedances of the transmissions lines are Z_o, $2Z_o$, and $6Z_o$ respectively. The charged PFLs are the Z_o and $2Z_o$ lines. With these ratios of impedance, the accelerating waveforms have the same shape as the waveform shown in Fig. 8, but the amplitude of each cycle is 1.5 times higher, i.e., when $T_o \leq t \leq 3T_o$, V is 1.5 V_o and from $3T_o \leq t \leq 5T_o$, V is 3.0 V_o for the open circuit case. The matched accelerating voltage for complete energy extraction is 1.5 V_o. The detailed derivation of this waveform and the waveforms for several other configurations is given in Eccleshall, Temperley, and Hollandsworth (1979); Eccleshall and Temperley, (1978); and Eccleshall and Hollandsworth (1981). With the accelerating voltage 1.5 times higher in ET-2 for a fixed charge voltage than within S3, the switch current is also 1.5 times higher in ET-2 than S3. The S3 and ET-2 switch currents are 3.0 I_B and 4.5 I_B respectively, where I_B is the matched beam current. These higher switch currents make it more difficult to produce fast-rising accelerating pulses. This difficulty arises because the switches are usually spark gaps and the inductive and resistive contributions to the rise time are related to the impedance driving the spark gaps. Higher currents imply lower impedance driving circuits.

Other considerations in selecting between S3 and ET-2 are the following. The high voltage electrode in the ET-2 PFL is not located near the accelerating gap or the insulator that separates the PFL dielectric and the vacuum. S3 has only one impedance discontinuity for the wave to transit while ET-2 has two, and one is at a corner that the waves must go around. Obtaining the desired impedance around the corner with an abrupt impedance change where the Z_o PFL begins is difficult to achieve.

The final configuration is ET-1, shown in Fig. 2(f) (Eccleshall et al., 1979). This is another two transmission line cavity, but with Z_o the impedance of the PFL that is shorted and $3Z_o$ the impedance of the PFL that is connected to the load. This configuration will produce a $2V_o$ open circuit voltage and V_o acceleration to a matched beam for a time $2T_o$. This configuration can also have 100% energy transfer efficiency. Distortions caused by the switch to the accelerating waveform occur on the leading and trailing edges of the pulse rather than in the center of the waveform as was the case with the Pavlovskii cavity (2a).

The ET-1 circuit is one of a class of circuits identified by I. Smith that give voltage amplification with 100% energy transfer efficiency. These circuits are called cumulative wave transmission lines (Smith, 1982).

The cumulative wave line has N sections of transmission line with equal wave transit time and increasing impedances Z_1, Z_2, . . . Z_N, where $Z_2 = 3Z_1$, $Z_3 = 6Z_1$, $Z_4 = 10Z_1$, etc. The general expression for Z_n, the impedance of the nth transmission line, is given in Eq. 8,

$$Z_n = \frac{n}{2} (n + 1)Z_1 . \tag{8}$$

All lines are connected in series and charged to the same voltage, V_o. When line Z_1 is shorted on the end opposite line Z_2, a wave of opposite polarity to the charge voltage is propagated towards the load. The reflection coefficients at the transition between the lines are such that all of the energy is transmitted to a load that is matched to Z_N. If a switch is closed to connect a matched load to line Z_N when this wave arrives, the output voltage across this matched load is $NV_o/2$.

Externally Fed Cavities

All of the functions of final energy storage, pulse generation, and beam acceleration are performed within the cavity discussed in the preceding section. Smith (1979a) has pointed out that it is possible to separate the energy storage and pulse generation functions from the beam acceleration and still maintain the high energy transfer efficiency. Beneficial effects from using external pulse generation could be 1) the ability to supply the accelerating pulse to several cavities from a single-pulse generator; 2) better access to the high-voltage switches; 3) the cavity dielectric is stressed to near breakdown electric fields only during the short accelerating pulse rather than the longer charge cycle and; 4) vacuum could be used for the cavity dielectric with the vacuum interface located away from the beam line.

If any of the cavities in the preceding section are operated into a matched load, they produce a bipolar pulse with $\int V_g dt = 0$, where V_g is the voltage across the accelerating gap. Similarly, cavity configurations have been identified that will transfer 100% of the energy from a bipolar, zero-integral, injected pulse to the beam. Two of these cavity configurations are shown in Fig. 9. Figure 9(a) shows a single transmission line with the feed near the accelerating gap, and with the other end of the transmission line shorted. Difficulties which could arise with this simple cavity configuration are: 1) delivering the high voltage to the feed point near the axis of the radial cavity, and 2) deflections of the beam with the magnetic field of this feed. A coaxial geometry relieves the feed problem but not the beam deflection concern. Multiple feeds will minimize the beam deflection problem. The configuration shown in Fig. 9(b) is a cavity configuration with the feed point on the outer circumference of the cavity. Although the feed points are as far away from the beam line as possible and more readily accessible, multiple feeds will still be necessary to produce symmetric energy flow to the load and a constant impedance for the drive circuit.

As in the closed cavity case there are many circuits that can be used to generate the bipolar accelerating pulse for injection into these cavities (Smith, 1979). Figure 10 is a schematic of one of these circuits. All three transmission lines have equal lengths with a one way transit time T. Lines 1 and 2 are charged to V_o. When the switch is closed, a pulse of amplitude V_o is transmitted towards the output transmission line. Simultaneously, energy flows from line 2 to line 3 causing a wave of amplitude $V_o/4$ to travel towards the load in line 2. With reflection coefficients one can readily show that a $3V_o/4$ pulse is transmitted in the output cable when these two waves reach the load. Two

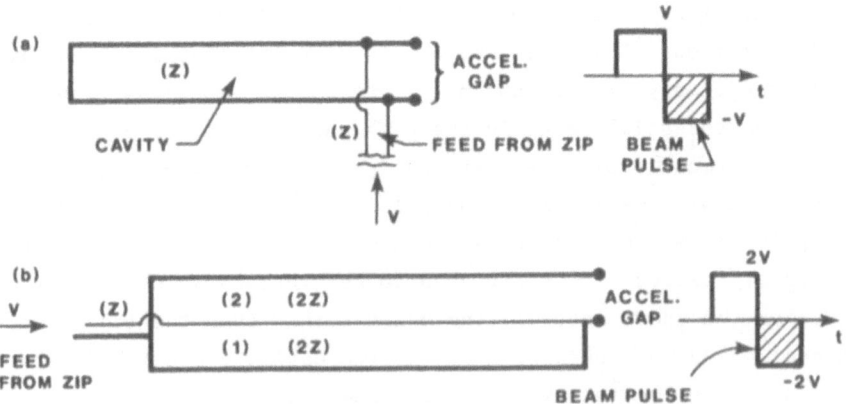

Fig. 9. External injection cavities:
(a) Single transmission line cavity
(b) Two transmission line cavity

484

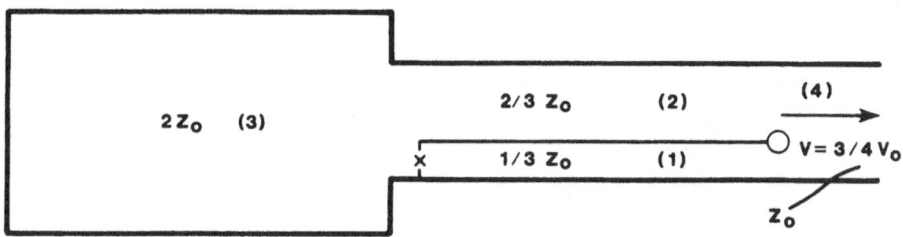

Fig. 10. Zero integral voltage generator.

transit times later the output pulse will reverse and produce another $3V_o/4$ pulse. At the end of this pulse the output voltage returns to zero. The initial energy in line 1 is $3V_o^2T/4Z_o$, and in line 2 it is $3V_o^2T/2Z_o$ for a total energy storage of $9V_o^2T/4Z_o$. The extracted energy in each half-cycle is $9V_o^2T/8Z_o$, for a total extracted energy of $9V_o^2T/4Z_o$. Thus all of the stored energy can be extracted in this bipolar pulse.

ACCELERATOR DESCRIPTIONS

The two radial pulse line accelerators that have been described in detail in the literature are the 13 MV, 50 kA, 40 ns, LIU-10 developed by A. I. Pavlovskii and his coworkers (1980) in the Soviet Union and the 9.5 MV, 28 kA, 12 ns RADLAC-I accelerator described by Miller et al. (1981). Both of these accelerators use the cavity configuration shown in Fig. 2(a).

The LIU-10 has water-dielectric cavities that are charged to 500 kV, producing 350 kV output pulses. The output of three of these cavities are connected in series to produce 1 MV across a single accelerating gap. The accelerator has 13 of these three-block modules to produce the 13 MeV electron beam. Each cavity has six gas-insulated, triggered spark gaps that must close within a few nanoseconds of each other. Each block of three cavities is charged with a Marx generator.

RADLAC I has a 2.0 MV injector and four, oil-insulated, radial transmission line cavities that produce 1.75 MV of acceleration to the 28 kA beam. Figure 11 is a schematic drawing of this accelerator, and Figs. 12 and 13 are an artist's drawing and a photograph of this accelerator. The four cavities are three meters in diameter and each of the cavity transmission lines has a 12 ns one-way wave transit time. The output parameters quoted above were achieved on the first half-cycle of the output pulse rather than the second half-cycle that produced a 24 ns pulse with similar accelerating voltages. Vacuum insulator flashover limited the pulse duration and reproducibility of the second half-cycle.

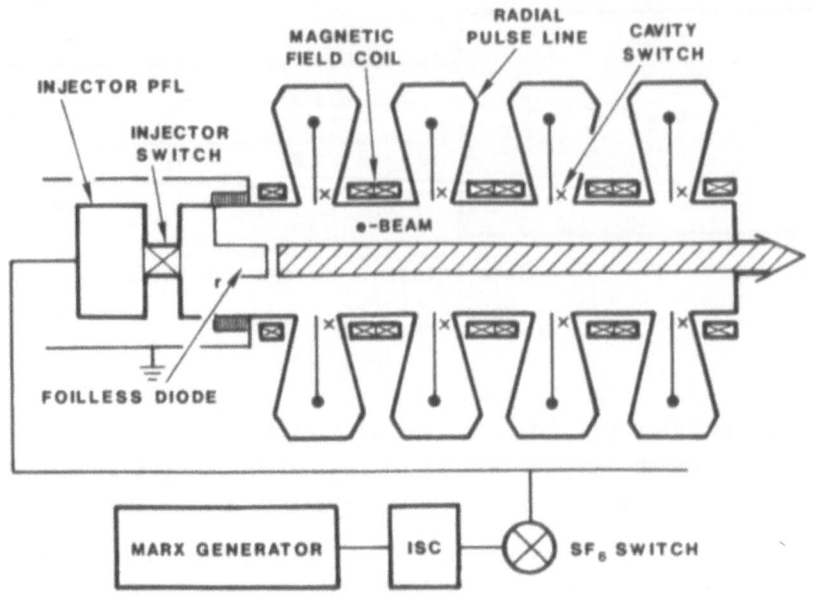

Fig. 11. RADLAC schematic design of cross-section of RADLAC I.

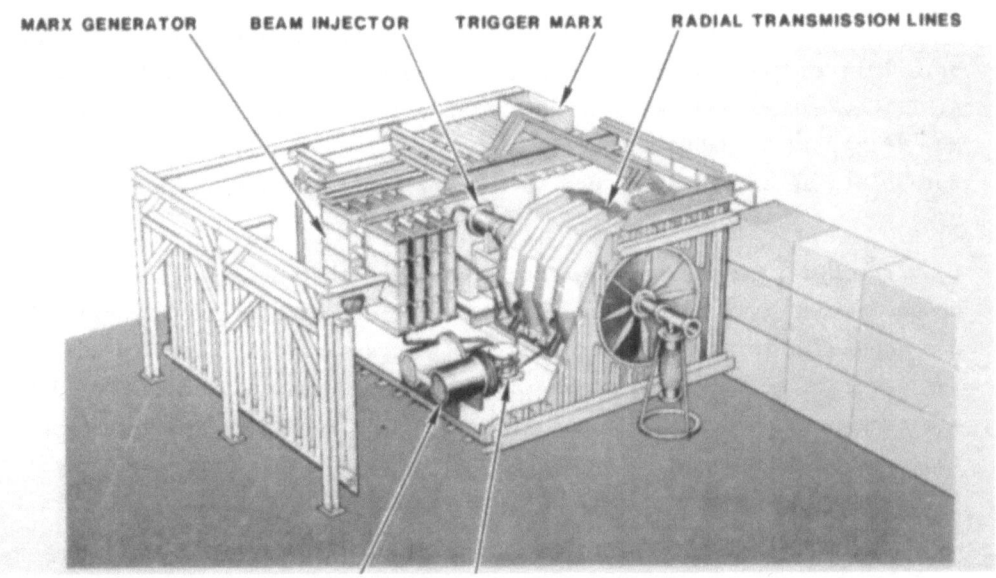

Fig. 12. Artist's sketch of RADLAC I

Therefore RADLAC I was usually operated on the first half-cycle of the accelerating pulse.

Very low jitter cavity switching was required to synchronize the arrival of the electron beam with the beginning of the 12 ns acceleration pulse. Self-closing oil-dielectric spark gaps were used to achieve less

Fig. 13.   Photograph of RADLAC I.

than 3 ns RMS jitter. The RMS jitter of these switches is usually 1% to
2% of the charge time. In RADLAC I the cavities were charged in 160 ns.
Since the inductance of multimegavolt Marx generators is too large to
achieve the rapid charging, the energy is transferred from a Marx generator to an intermediate storage capacitor and then from the intermediate
storage capacitor to the cavities. Two water-dielectric intermediate
storage capacitors were operated in parallel and simultaneously charged
the injector and the four cavities. Synchronization of the output pulse
with the beam was achieved by adjusting the oil-dielectric spark gap
electrode spacing.

These switches strongly affect the accelerating voltage pulse shape.
The accelerating voltage in this cavity is the sum of the voltage at the
output of the unswitched line and the voltage across the switch. For the
major fraction of the pulse, the inductive voltage drop across the switch
is a small contribution to the accelerating voltage. Midway through the
second half-cycle the current reverses in these switches causing a
substantial distortion of the RADLAC-I accelerating pulse as shown in
Fig. 14. This waveform should be compared to the ideal waveform shown in
Fig. 2. Although this perturbation will always be present in this
cavity, its magnitude and duration can be reduced by minimizing the
switch inductance. The rapid charging of the self-closing switches
allows each of them to operate at higher electric fields and several of
them to close in parallel. Both of these effects help to minimize the
switch inductance. RADLAC I was designed with eight switches in each

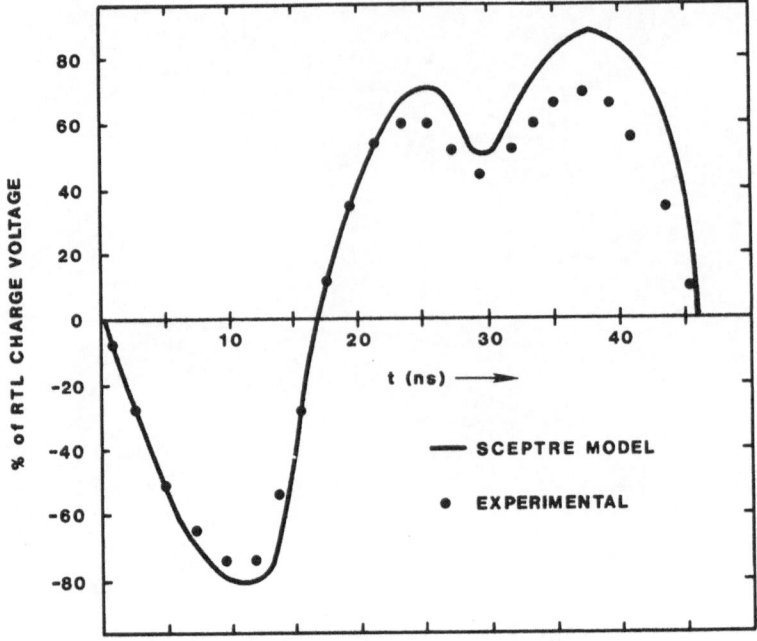

Fig. 14. RADLAC-I output waveform and SCEPTRE simulation showing distortion of accelerating pulse by switch < di/dt.

cavity and usually six of them closed. The number of channels that closed on RADLAC I (Miller et al., 1981; Adler et al. 1983) and other experiments (Prestwich, 1974; Johnson, 1974; VanDevender and Martin, 1975; and Johnson, VanDevender, and Martin, 1980) could be estimated fairly accurately with the following relationship developed by Martin (1970),

$$2\sigma_v \frac{dv}{dt} \le 0.1\ (\tau_R + \tau_L) + 0.8\ T_T, \tag{9}$$

where σ_v is the standard deviation of the switch breakdown voltage usually 0.5% to 2% of the charging voltage depending on the electrode geometry; τ_R is a measure of the time it takes for the arc to develop in the spark gap and is called the resistive phase of the rise time; τ_L is the standard inductive time constant (L/Z_o). Equation 10 is an empirical expression for τ_R that gives a fairly accurate estimate of this contribution to the rise time (Martin, 1970).

$$\tau_R = \frac{5}{E^{4/3}Z^{1/3}} . \tag{10}$$

τ_R is in ns when E is the average electric field across the switch in MV/cm and Z is the impedance (ohms) of the circuit that is supplying

488

energy to the arc. If there are n switch channels in a transmission line with Z_o impedance, Z in Eq. 10 is nZ_o.

As mentioned earlier the second half-cycle pulse duration for RADLAC I was shortened by flashover of the insulator that separates the oil and vacuum region (Adler, 1983). It was found on RADLAC I and in subsequent experiments (Tucker et al., 1985a) that the flashover strength for bipolar pulses varied as $t^{1/2}$ rather than $t^{1/6}$ as it does for unipolar pulses (Martin et al., 1970).

These insulator interfaces consist of a stack of 2-cm to 5-cm thick insulators interleaved with thin metal grading electrodes. The vacuum side of the insulator is cut at a 45° angle to the grading rings. Experiments have shown with this configuration the flashover voltage for unipolar pulses is 4-5 times higher than with straight-walled cylinders (Smith, 1964). If the polarity is such that electrons leaving the cathode at the triple junction will not impact the insulator, the breakdown strength is given by Eq. 11 (Martin, 1970),

$$Ft^{1/6}A^{1/10} = 175 , \qquad (11)$$

where F is the flashover electric field in kV/cm, t is the time the voltage is above 89% of its peak value, and A is the area of the insulator in cm^2. If the polarity is such that when electrons are emitted from the cathode and follow the electric field lines they strike the insulator, the constant in Eq. 11 becomes 105. With a bipolar waveform both polarities exist, and Adler (1983) and Tucker (1985a) have shown that the flashover electric field is given by Eq. 12

$$FT^{0.56}A^{1/10} = 46 . \qquad (12)$$

The generally accepted breakdown model (Anderson and Brainard, 1980) for these insulators consists of the following steps:

1) Electrons are field emitted from the triple junction.

2) Electrons impact the insulator surface. This surface has a secondary electron coefficient greater than one leaving a posi- tive charge on the insulator.

3) The positive charge attracts more electrons, and their impact causes gases to desorb from the insulator.

4) This gas is ionized and becomes a conducting channel.

The motion of the ions in the desorbed gas can strongly effect the time for the conducting channel to form. Anderson and Tucker (1985) surmised that if the insulator width or voltage pulse shape prevented these ions from reaching the electrodes, the plasma formation would occur much more rapidly. Experiments with an open circuit ET-2 cavity have shown that flashover strength can be increased by 45% above the values given in Eq. 12, by decreasing the thickness of each insulator form 20.6 mm to 4.67 mm (Anderson and Tucker, 1985; Tucker et al., 1985). The overall insulator grading ring stack height for these experiments was 12.6 cm.

FUTURE DIRECTIONS FOR RADIAL LINE ACCELERATORS

Experiments with RADLAC I lead us to believe that radial cavity accelerators could be designed to produce high current beams with arbitrarily high energies. ET-2 and S-3 cavities remove many of the undesirable features of RADLAC I. Using these cavities with water or ethylene glycol dielectric allows the advantage of both small volume that the LIU-10 cavities have and the high accelerating voltage that RADLAC I had. Figure 15 is a sketch comparing the size of a coaxial 1.8 MV, 25 ns, ET-2 cavity with a RADLAC I cavity. One can see the dramatic effect of the higher dielectric constant liquid. The ET-2 cavity shown in Fig. 15 has a 6 Ω output impedance. if it was operated with a 100 kA beam and an accelerating voltage to give a matched load, it would have 600 kV of acceleration. If one operates at a 3 to 1 impedance mismatch (effective load impedance 18 Ω) and a 2.4 MV open circuit voltage, the accelerating potential is 1.8 MV allowing a significant decrease in the number of accelerator cavities at the expense of a 30% reduction in energy transfer efficiency. A further reduction in the number of accelerating cavities required to achieve a particular voltage may be achievable by recirculating the beam through these cavities (Eccleshall et al., 1979; Miller, 1985).

The accelerating voltage waveform for ET-2 or S3 repeats at intervals of four times the pulse duration as shown in Fig. 8. Eccleshall and Temperley pointed out that one could use each cavity to accelerate the beam several times if the beam were recirculated through the cavity such that it arrived coincidentally with subsequent output pulses. In this case the output impedance of the cavity is made much lower than the effective beam impedance so that a substantial fraction of the energy remains to accelerate the beam on subsequent passes. Since the voltage drops and the beam current remains constant, the impedance mismatch decreases with each accelerating pulse. One should be able to design systems with the average accelerating voltage greater than 0.8 times the open circuit voltage for four or five passes and supply at least 50% of the stored energy to the beam. The switch current to beam current ratio will be even higher for these systems. Concern about pulse distortion from switch inductance or cavity corners are even greater than one is trying to produce a train of pulses for recirculation.

490

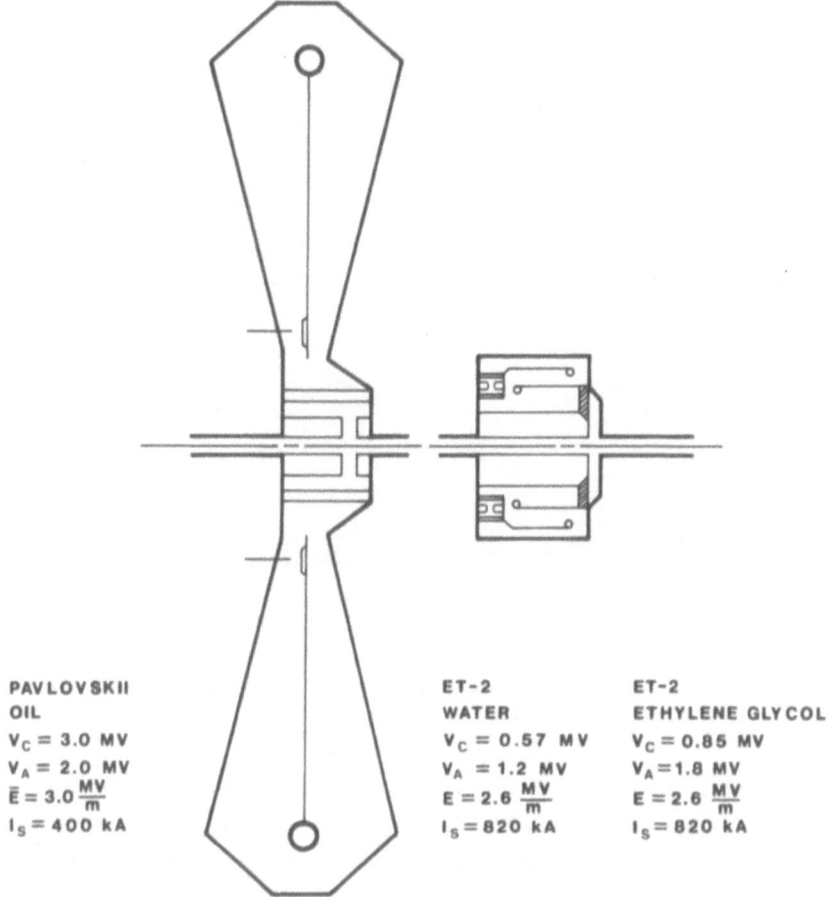

PAVLOVSKII
OIL
$V_C = 3.0$ MV
$V_A = 2.0$ MV
$\bar{E} = 3.0 \frac{MV}{m}$
$I_S = 400$ kA

ET-2
WATER
$V_C = 0.57$ MV
$V_A = 1.2$ MV
$E = 2.6 \frac{MV}{m}$
$I_S = 820$ kA

ET-2
ETHYLENE GLYCOL
$V_C = 0.85$ MV
$V_A = 1.8$ MV
$E = 2.6 \frac{MV}{m}$
$I_S = 820$ kA

Fig. 15. Comparison of oil-insulated radial cavity and coaxial ET-2
cavity with water or ethylene glycol dielectric.

Reliable, low jitter (~1 ns), low inductance, gas-dielectric spark
gaps have been developed for use in these cavities. Up to 24 of these
V/n spark gaps have been operated in parallel to achieve a very low
effective switch inductance (Tucker et al., 1985b). High voltage step-up
transformers that can be used with capacitor banks to replace the Marx
generators have been developed by Rohwein (1979 and 1980) and appear to
be ideal for charging single cavities. Completing these developments
lead to accelerator modules as shown in Fig. 16, where the cavity and its
energy storage system form a compact unit.

STRIP-TRANSMISSION LINE ACCELERATOR

Linear accelerators which have the energy supplied to the
accelerating gaps from water-dielectric strip transmission lines have
been developed. In this type of accelerator, the accelerating gap is not

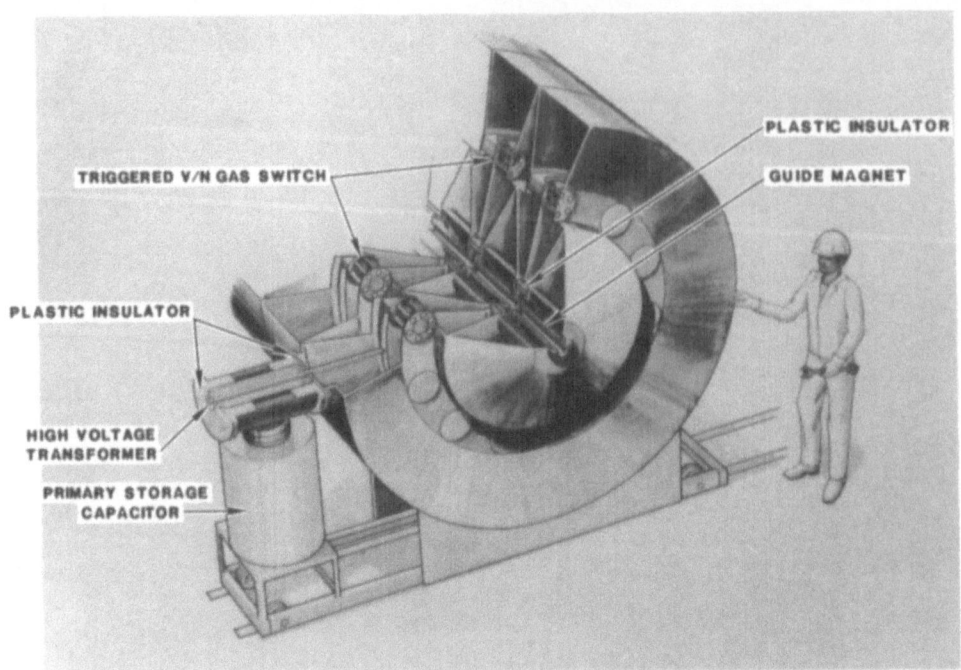

Fig. 16. S3 cavity with capacitor bank and transformer high voltage charging system (S3 compact accelerator).

completely enclosed by an induction cavity. These gaps are transient-time isolated from each other (i.e., the time it takes for a wave to travel from one accelerating gap to the next one is greater than the accelerating pulse duration). This configuration simplifies mechanical construction and the beam line is more accessible than in closed cavity accelerators. RADLAC II is a 16 MV, 50 kA, 40 ns linear electron-beam accelerator built with the strip-transmission line technology (Mazarakis et al., 1985; Mazarakis et al., 1986; Miller, 1985). Figure 17 is a schematic diagram showing the layout of this accelerator. Figure 18 is an artist's drawing of the accelerator.

The accelerator has eight pairs of tri-plate PFLs that supply energy to the beam line. The first two lines supply energy to the 4 MV injector and the remainder produce 2 MV accelerating pulses. The accelerator is composed of two modules with a Marx generator and an intermediate storage capacitor (ISC) supplying energy to four sets of PFLs. The 3.4 MV Marx generators charge the ISCs in 1.1 μs. Each intermediate storage capacitor has two output ports. Energy is fed from each of these ports through a 2.7 MV spark gap to the two sets of PFLs. The spark gaps are laser triggered and have 3 ns RMS jitter (Hamil and Smith, 1983). These gaps provide the synchronization for the accelerator. The PFLs have self-triggered, water-dielectric switches that close with about 2 ns jitter. Immediately after the PFL the tri-plate transmission lines are separated

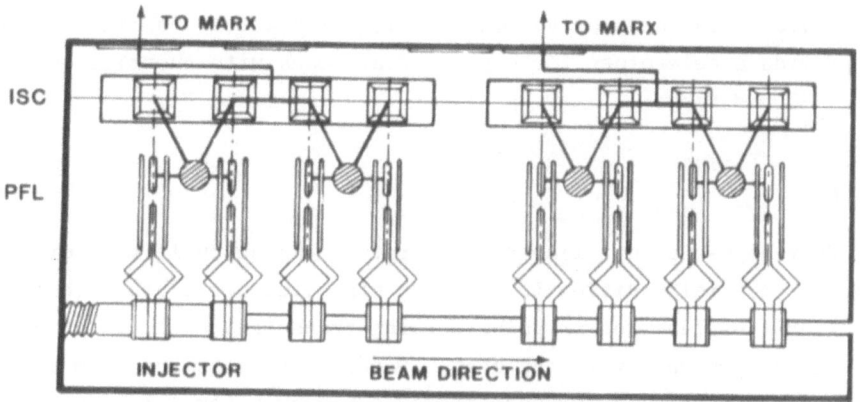

Fig. 17.   RADLAC II, PFL, ISC, and accelerating gap layout.

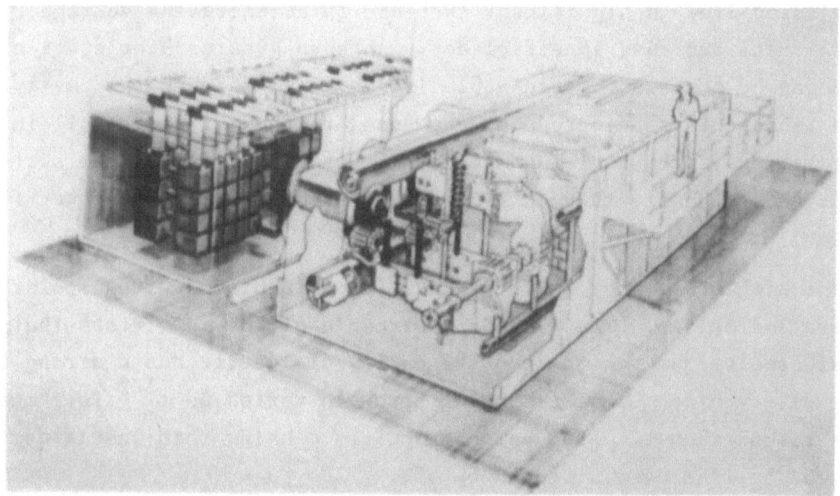

Fig. 18.   Artist's sketch of RADLAC II.

and the pulse on one of the lines inverted to effectively produce a
voltage adder. This type of inverter (Martin et al., 1981) has losses
and the output voltage is about 1.85 times the input voltage. After the
inverter the transmission line makes a 90° bend while they are being
brought back together at the insulator stack. About 10% to 15%
additional voltage amplitude is lost due to impedance mismatches as the
wave goes around the corner and while stray capacitance around the
insulator stack is being charged. This stray capacity also has the
effect of increasing the rise time and the pulse duration 10% to 20%.
Although these losses have a negative impact on the energy transfer
efficiency, our experience with RADLAC II and similar modules now allow
us to make fairly accurate predictions of the accelerating voltage
amplitude and wave shapes in the accelerator design phase. Thus this

open, relatively inexpensive construction with access to all components can provide accelerators for high current beams with energies in the 10 MeV to 30 MeV range.

SUMMARY

Very high-current, linear accelerators can be built with either radial or coaxial cavities filled with high dielectric strength liquids. These cavities function as pulse-forming transmission lines and as a means to supply energy to the beam. Several transmission line configurations are available to the accelerator designer. Some of these configurations give voltage amplification (equal to or greater accelerating voltage than charge voltage) with the possibility of 100% energy transfer efficiency. If the beam appears as an overmatched load to the output transmission line, a significant increase in accelerating voltage results with a modest decrease in efficiency. An even greater acceleration per cavity can be achieved by recirculating the beam through the cavity. All of the techniques for increasing the accelerating voltage result in higher PFL switch current than the beam current. Multichannel switching is required to handle this current and to produce symmetric power feed to the beam.

The high efficiency in these cavities is achieved with bipolar accelerating pulses. The flashover strength of the insulators that separate the vacuum region from the cavity dielectric has a strong impact of the cavity size. This flashover strength varies as $t^{1/2}$ for bipolar pulses rather than $t^{1/6}$ for unipolar pulses. Using thin insulators has improved the flashover strength by 45%.

RADLAC I and LIU-10, two high-current electron beam accelerators operating with radial transmission lines, where described and their successful operation lead to possibilities of very high energy accelerators using this technology. RADLAC II, A 16 MV, 50 kA accelerator, has been successfully implemented using water dielectric strip transmission lines to supply energy to the beam. This accelerator provides the basis for building accelerators in the 10 MV to 30 MV range with all of the components accessible for adjustment or maintenance.

REFERENCES

Adler, R. J., Miller, R. B., Prestwich, K. R., and Smith, D. L., 1983, "Radial Isolated Blumlein Electron Beam Generator," Rev. Sci. Instru., 54 (8):940.

Anderson, R. A., and Brainard, J. P., 1980, J. Appl. Phys., 51 (3):1414.

Anderson, R. A., and Tucker, W. K., 1985, "Vacuum Surface Flashover from Bipolar Stress," J. Appl. Phys., 58 (9):3346.

Avery, R., Behrsing, G., Chupp, W. W., Faltens, A., Hartwig, E. C., Hernandez, H. P. Macdonald, C., Meneghetti, J. R., Nemetz, R. G., Popenuck, W., Salsig, W., and Vanecek, D., 1971, IEEE Trans. Nucl. Sci., Vol. NS-18 (3):479.

Boyd, T. J., Jr., Rogers, B. T., Tesche, F. R., and Venable, D., 1965, "PHERMEX--A High Current Electron Accelerator for Use in Dynamic Radiography," Rev. Sci. Instru., 36 (10):1401.

Christofilos, N. C., Hester, R. E., Lamb, W. A. S., Reagan, D. D., Sherwood, W. A., and Wright, R. E., 1964, Rev. Sci. Instru., 35 (7):886.

Eccleshall, D., and Temperley, J. H., 1978, J. Appl. Phys., 49 (7):3649.

Eccleshall, D., Temperley, J. H., and Hollandsworth, C. E., 1979, IEEE Trans. Nucl. Sci., NS-26, (3):4245.

Eccleshall, D., and Hollandsworth, C. E., 1981, IEEE Trans. Nucl. Sci., 28:3386.

Faltens, A., Hartwig, E., and Hernandez, P., 1968, Proc. of ERA Symposium, UCRL 18103, p. 332.

Fultz, S. C., and Whitten, C. L., 1971, IEEE Trans. Nucl. Sci., NS-18, (3):533.

Gnalskii, A. I., Bagclamp, O. S.., Bakaev, P. V., Vakruskin, Yu. P., Malyskev, I. F., Nalivaikee, G. A., Pavlov, A. I., Suslov, V. A., and Khalchitskie, B. P., 1970, Sov. At. Energy, 28:549.

Hamil, R. A., and Smith, D. L., 1983, "Laser Triggered Switch Results from a Frequency Quadrupled ND:YAG Laser," Digest of Technical Papers 4th IEEE Pulsed Power Conf., Albuquerque, New Mexico, 1983, p. 447.

Hartwig, E. C., 1968, Proc. ERA Symposium, UCRL 18103, p. 44.

Hester, R. E., Bupb, D. G., Clark, J. C., Chesterman, A. W., Cook, E. G., Dexter, W. L., Fessenden, T. J., Reginato, L. L., Yakota, T. T., and Faltens, A. A., 1979, IEEE Trans. Nucl. Sci., NS-26 (3):4180.

Johnson, D. L., 1974, Proc. of Int'l. Conf. on Energy Storage, Compression, and Switching, Italy, p. 515.

Johnson, D. L., VanDevender, J. P., and Martin, T. H., 1980, "High Power Density Water Dielectric Switching," IEEE Trans. Plasma Sci., Vol PS-8 (3):204.

Lewis, I. A. D., and Wells, F. H., 1959, "Millimicrosecond Pulse Techniques," Pergamon Press, New York, p. 367.

Martin, J. C., 1970, "Multichannel Gaps," Atomic Weapons Research Establishment Internal Report SSWA/JCM/703/27, Aldermaston, England.

Martin, J. C. 1971, "Fast Pulse Vacuum Flashover," Atomic Weapons Research Establishment Internal Report SSWA/JCM/713/157, Aldermaston, England.

Martin, T. H., Prestwich, K. R., and Johnson, D. L., 1969, "Summary of Hermes Flash X-Ray Program," Sandia National Laboratories Report No. SC-RR-69-421.

Martin, T. H., VanDevender, J. P., Johnson, D. L., McDaniel, D. H., and Aker, M., 1975, Proc. Int'l. Topical Conf. on Electron Beam Research and Technology, Albuquerque, New Mexico, Sandia National Laboratories Report No. SAND76-5122, p. 423.

Martin, T. H., Johnson, D. L., and McDaniel, D. H., 1977, Proc. of 2nd Int'l. Topical Conf. on High Power Electron and Ion Beam Research and Tech., Cornell Univ., p. 807.

Martin, T. H., VanDevender, J. P., Barr, G. W., Goldstein, S. A., White, R. A., and Seamen, J. F., 1981, "Particle Beam Fusion Accelerator I (PBFA I)," IEEE Trans. Nucl. Sci., NS-28:3365.

Mazarakis, M. G., Smith, D. L., Miller, R. B., Clark, R. S., Hasti, D. E., Johnson, D. L. Poukey, J. W., Prestwich, K. R., and Shope, S. L., 1985, "RIIM Accelerator Beam Experiments," IEEE Trans. Nucl. Sci., NS-32 (5):3237.

Mazarakis, M. G., Leifeste, G. T., Shope, S. L., Frost, C. A., Freeman, J. R., Poukey, J. W., Stygar, W. A., Ekdahl, C. A., Smith, D. L., Johnson, D. L., Hasti, D. E., Miller, R. B., and Prestwich, K. R., 1986, to be published in the Proc. 1986 Linear Accelerator Conf., Stanford, California.

Miller, R. B., Prestwich, K. R., Poukey, J. W., Epstein, B. G., Freeman, J. R., Sharpe, A. W., Tucker, W. K., and Shope, S. L., 1981, J. Appl. Phys., 52 (3):1184.

Miller, R. B., 1985, "RADLAC Technology Review," IEEE Trans. Nucl. Sci., NS-32 (5):3149.

Moir, D. C., Builta, L. A., and Starke, T. P., 1985, "Time-Resolved Current, Current-Density, and Emittance Measurement of the PHERMEX Electron Beam," IEEE Trans. Nucl. Sci., NS-32 (5):3018.

Pavlovskii, A. I., Gerasimov, A. I., Zenkov, D. I., Bosamykin, V. S., Klement'ev, A. P., and Tananakin, V. A., 1970, Sov. At. Energy, 28 (5):432.

Pavlovskii, A. I., and Bosamykin, V. S., 1974, Sov. At. Energy, 37:228.

Pavlovskii, A. I., Bosamykin, V. S., Kuleshov, G. I., Gerasinov, A. I., Tananakin, V. A., and Klement'ev, A. P., 1975, Sov. Phys. Dokl., 70 (6):441.

Pavlovskii, A. I., Bosamykin, V. S., Savchenko, V. A., Klementev, A. P., Morunov, K. A., Nkkolskiy, V. S., Gerasinov, A. I., Tananakin, V. A., Basmonov, V. F., Zenkov, D. I., Selemir, V. D., and Fedotkin, A. S., 1980, Dokl. Akad. Nauk SSSR 250, 1118.

Prestwich, K. R., 1974, "Proc. Int'l. Conf. Energy Storage, Compression and Switching," Plenum Press, New York, p. 451.

Prestwich, K. R., 1975, IEEE Trans. Nucl. Sci., NS-22, (3):975.

Prestwich, K. R., Miller, P. A., McDaniel, D. H., Poukey, J. W., Widner, M. M., and Goldstein, S. A., 1975, Proc. Int'l. Topical Conf. on Electron Beam Research and Tech., New Mexico, Sandia National Laboratories Report SAND76-5122, p. 450.

Prono, D. S., and the Beam Research Group, 1985, IEEE Trans. Nucl. Sci., NS-32 (5):3144.

Ramirez, J. J., 1983, Proc. Fifth Int'l. Conf. High Power Particle Beams, California, p. 256.

Rohwein, G. J., 1979, "Design of the Pulse Transformer for PFL Charging," Digest of Tech. Papers of 2nd IEEE Int'l. Pulsed Power Conf., Texas, p. 87.

Rohwein, G. J., 1980, IEEE Conf. Record of 1980 Fourteenth Pulse Power Modulator Symposium, (80 Ch 1575-5).

Smith, I. D., 1964, "Impulse Flashover of Insulators in a Poor Vacuum," First Int'l. Conf. on Breakdown in Vacuum," MIT Press, Cambridge, Massachusetts, p. 261.

Smith, I. D., 1979a, "Linear Induction Accelerators Made from Pulse-Line Cavities with External Pulse Injection," Rev. Sci. Instrum. 50 (6):714.

Smith, I. D., 1979b, "Preliminary Design of 100 kA, 20 ns Staged Pulse Line Induction Linac, Sandia National Laboratories, Report No. Sand79-7043.

Smith, I. D., 1982, "A Novel Voltage Multiplication Scheme Using Transmission Lines," IEEE Conf. Record of 1982 Fifteenth Power Modulator Symposium (82) Ch 1785-5).

Tucker, W. K., Anderson, R. A., Hasti, D. E., Jones, E. E., and Bennett, L. F., 1985a, "Vacuum Interface Flashover with Bipolar Electric Fields," Digest of Technical Papers 8th IEEE Pulsed Power Conf., Arlington, Virginia, p. 323.

Tucker, W. K., Jones, E. E., Franklin, T. L., Bennett, L. F., Weber, G., 1985b, "Compact Low Filter Triggered Spark Gaps," Digest of Technical Papers of 5th IEEE Pulsed Power Conf., Arlington, Virginia, p. 254.

Uglum, J., 1977, "Use of Radial Lines and Biconic Lines for Particle Accelerators," Austin Research Associates Report I-ARAC-77-4-10 (ARAC-38).

VanDevender, J. P., and Martin, T. H., 1975, IEEE Trans. Nucl. Sci., NS-22 (3):979.

RF BREAKDOWN LIMITS

Robert A. Jameson

Accelerator Technology Division, MS H811
Los Alamos National Laboratory
Los Alamos, New Mexico 87545 USA

INTRODUCTION

The scientific explanation of high rf voltage breakdown in evacuated structures is incomplete at best. Although interest in the subject is well over 50 years old, the number of variables and the range of parameters involved are so large that no coherent picture has emerged. Effect modeling is relatively crude and often empirically weighted. Reproducability of experimental results has often been poor. The present discussion will outline some of the hypothesized models and tools and then review some of the more recent experiments and findings concerning some of the variables and parameters.

BREAKDOWN LIMITS

The classic paper in the field, and still one of the best, is Kilpatrick's 1957 RSI paper, "Criterion for Vacuum Sparking Designed to Include Both rf and dc." He defines sparking as an abrupt major dissipation of stored electrical energy across a gap between two metal electrodes. Vacuum is in terms of a gaseous mean free path greater than the electrode-gap spacing; for our purposes, almost all accelerator structures operate below 10^{-5} torr, and there seems to be little dependence of breakdown on vacuum below this level. Kilpatrick worked with metal electrodes having no special preparation, and there were no external magnetic fields present. He found that sparking occurred above a threshold but at much lower voltages than would be explained by field emission. He derived a sparking threshold formula by considering the probability of field-emitted electrons together with a linear dependence of secondary-electron emission upon the energy of a bombarding ion:

Kilpatrick Limit

$$W[E^2 \exp(-K_1/E)] = K_2, \tag{1}$$

where W = maximum possible ionic energy (dc or rf) in eV,

    E = electric cathode gradient,

    $K_1 = 1.7 \times 10^5$ volts/cm,

    $K_2 = 1.8 \times 10^{14}$;

K_1 and K_2 were determined from a fit to data at dc, 14 MHz, 200 MHz, and 2856 MHz. Curves are given for estimating W, which is a function of the gap voltage and gap dimension, the frequency, and the charge of the ion. A more useful form was derived by T. J. Boyd by fitting against frequency:

Kilpatrick Limit (frequency version):

$$f = 1.643 \ E^2 \exp(-8.5/E), \tag{2}$$

where f is in MHz and E is in MV/m.

A plot is shown in Fig. 1. In modern practice, it has been found that this limit can be exceeded by substantial factors, as high as 2-3 for cw and 5-6 for short pulses (Schriber, 1986), but that the frequency dependence is about the same. Thus it has become customary to speak of field thresholds "x times the Kilpatrick limit." It is believed that the modern threshold levels may indeed be determined by the same physical phenomena, that of ion-bombardment-aided electron emission but higher because of better vacuum techniques now that produce high vacuum without the introduction of small traces of diffusion pump oil or other contaminants.

In an accelerator structure, we are most interested in the effective accelerating gradient on the axis, which is lower by some factor than the peak surface field in the structure (Jameson, "High Brightness RF Linear Accelerators," this ASI). In Fig. 2, the Kilpatrick limit is plotted against accelerating gradient, with the assumption that the peak surface field is at twice the Kilpatrick limit and that the ratio of peak surface to accelerating field is also 2.

Electric breakdown (Palmer, 1982) based on direct secondary emission is believed to be linear with wavelength because it depends on the maximum energy attained by an electron before the field reverses. In conventional cavities at $\lambda = 10$ cm, such breakdown occurs at about 80 MeV/m. In Fig. 2, this limit is again plotted for a peak-surface to accelerating-field ratio of 2.

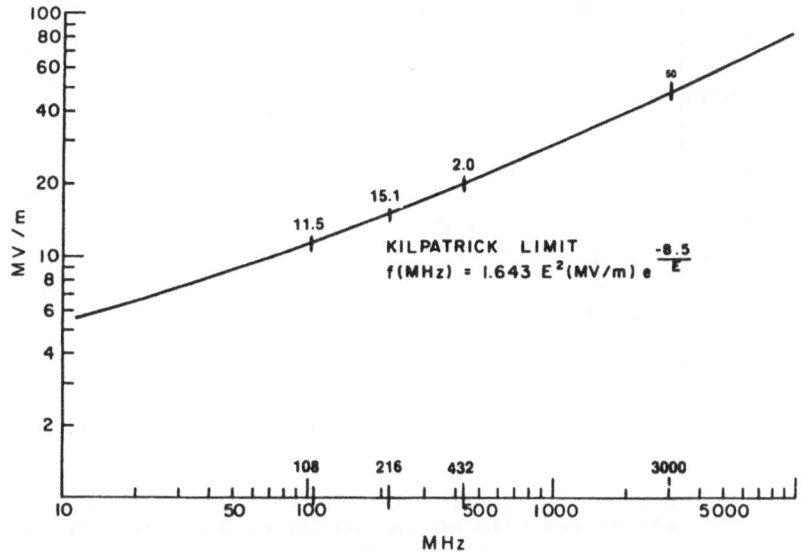

Fig. 1. Plot of the Kilpatrick Limit (frequency version).

Surface destruction by heating is predicted (Palmer, 1982) to scale with the negative square root of pulse length for long enough pulses to have effective conduction cooling. For the short pulses of interest to very high energy electron accelerators, the pulse length τ might be made equal to the structure fill time, in which case τ scales as $\lambda^{3/2}$ and the energy absorbed as $E^2\lambda^{-1/2}$; for conduction cooling $\tau^{-1/2} \propto \lambda^{-3/4}$ leading to a $\lambda^{-1/8}$ scaling of the maximum possible accelerating gradient. The points given for the curve (Figure 2) are

λ	E_o (max)	τ
10 cm	0.4 GeV/m	1 μs
1 cm	0.5 GeV/m	30 ns
10 μm	1.3 GeV/m	1 ps

A plasma limit is also noted (Palmer, 1982) for some ideas where the plasma itself might form part of the medium.

X-RAY PRODUCTION

X-ray production is usually observed during experiments on breakdown limits; measurements of the x-ray levels and energies are used to characterize the breakdown process and the highest x-ray energies seen (the "end-point-energy" of the spectrum) are used (Bolme, 1986) to calibrate the peak surface field in the structure.

Bremsstrahlung is assumed to be the source of the x-rays, and two possible electron emission sources were examined (Boyd, 1984):

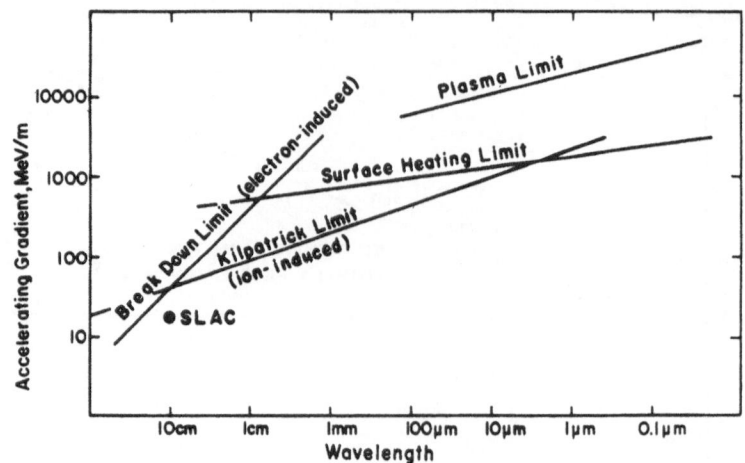

Fig. 2. Approximate limits on accelerating gradient, for
structures with assumed ratio of peak surface field
to accelerating gradient equal to 2, vs wavelength.
Kilpatrick-Limit line also assumes peak surface
field of twice KL.

Field Emission (Fowler-Nordheim) (radiation output)

$$R(mR/hr) = C_1 E^3 \exp(-C_2/\sqrt{E}) \text{ (temperature independent)}, \qquad (3)$$

where for 5 MeV, $C_1 = 3.17$, $C_2 = 213$;

for 50 MeV, $C_1 = 5.40$, $C_2 = 107.3$.

Thermionic Emission (Richardson-Schottky)

$$R(mR/hr) = C_3 E \exp(C_4\sqrt{E}) \text{ (fixed temperature)}, \qquad (4)$$

where for 5 MeV, $C_3 = 3.835 \times 10^{-8}$, $C_4 = 3.255$;

for 50 MeV, $C_3 = 28.94 \times 10^{-7}$, $C_4 = 3.354$

with both at room temperature.

Both of these expressions fit the measured data; Boyd surmised that
the best guess would probably be thermally enhanced field emission. For
a fixed field of 18.8 MV/m in a 200-MHz, 50-MeV drift-tube-linac cell
geometry, the temperature dependency would be roughly

$$R(mR/hr) = 50 + 1.25 T \text{ (deg. centigrade)}.$$

Others give the Fowler-Nordheim relations in different forms, for
example Wang et al. (1986):

Field Emission (Fowler-Nordheim) (field-emission current)

$$I_f = \frac{C}{f^2} E^{2.5} \exp\left[(-1.34 \times 10^{10})(\phi^{1.5})/\beta E\right], \qquad (5)$$

where C is a constant and ϕ is the work function in electron volts

VISIBLE-LIGHT PRODUCTION

Experiments at Chalk River (Schriber, 1986) made provisions for visual observation of the high-field surfaces during operation and used various types of instrumentation to observe and localize the many visible-light pinpoints, or glowpoints, that showed up in the conditioning and operating phases. Light levels, spectra, and geometrical location were measured. Experimental results showed that the light intensity correlated very well with the x-ray production. When the rf drive was frequency modulated, the light intensity was correspondingly modulated. The onset of glowpoints occurred at 0.3-0.5 the Kilpatrick limit. The number of the glowpoints became smaller with conditioning, but many remained during routine operation. This technique will prove to be quite useful in future experiments and may also provide important information about the physical and chemical processes that are happening at the surfaces.

OBSERVATIONS

Materials

There is fairly general agreement that the basic choice of metal does not make a large difference in the breakdown threshold (Bohne, et al., 1971; Kilpatrick, 1957; Reid and Lohsen, 1982; Tanabe et al., 1986), although a factor of 2 or 3 may be seen, and there may be a general trend toward harder materials having somewhat better voltage holdoff characteristics than softer ones.

Electrode Geometry

As noted above in the Kilpatrick Limit discussion, a variety of data points are available for different cavity geometries at different frequencies, and the breakdown thresholds are in general agreement with the frequency scaling of the Kilpatrick limit at gradients 2-3 times higher for long-pulse (greater than about 3 ms) to cw and 5-6 times higher for short pulse (≤ 1 μs). Some systematic experiments have been attempted, with other parameters fixed (Bohne, 1971; Gerhard, et al, 1985; Rabinowitz and Donaldson, 1965). It was noted that as long as the corners had radii ≥ 1 mm, they had little effect but the larger (planar) electrode area was associated with lower breakdown voltages; therefore higher breakdown voltages, as much as a factor of 2, can be obtained by using smaller electrodes of higher curvature even though this enhances the electric field. Gerhard, et al, (1985) also report achieving limits of 4X down to 2X Kilpatrick for planar electrodes as the gap increased from 0.1 to 1.0 cm, with a corresponding reduction from 6X down to 3X for round electrodes of 20 mm diam.

Thermal Properties

Most of the extant data points show breakdown below the point where the surface heating limit of Figure 2 is reached, for example, Tanabe (1986). However, it is clear that in cw operation, all surfaces must be very well cooled (Schriber, 1986); probably the surface temperatures must be kept below a few hundred degrees centigrade (Reid, Lohsen, 1982) to avoid problems. Failure to do this may be a reason why cw performance is less than that seen for pulsed service, where such thermal characteristics are not reached.

If the rf cavity is cooled to cryogenic temperatures, the Q is enhanced and less rf drive power is needed. At liquid nitrogen temperature (-197°C), the microwave surface resistivity decreases by a factor of 2.6. Recent tests (McEuen, 1985) on two geometries at 2856 MHz showed the predicted Q enhancement at low peak power was increased. At a given peak power, the Q was not a function of average power. It was found that both geometries exhibited the same behavior of Q as a function of the peak magnetic field; thus, the degradation is attributed to a peak magnetic field limit that may imply an electron multipactoring phenomenon. If this is so, proper shaping of the cavity contour and proper surface treatment might push the limit higher.

A great deal of progress has been made in the past few years in the performance of superconducting rf structures (Leemann, 1986; Tanaka, 1986). Accelerating gradients of around 5 MeV/m are being routinely achieved in multicell cavities, under actual beam accelerating conditions. Many years of research were required to reach this performance; major factors in achieving the higher gradients included curving all surfaces to avoid multipactoring paths, achieving the proper surface conditions over the large areas needed to make cavities, and developing techniques for finding and correcting problem areas in the surface. Laboratory results indicate that accelerating gradients up to around 19 MeV/m are not far off.

External Magnetic Fields

If magnetic fields are applied sufficient to turn emitted electrons and prevent them from hitting another surface, the breakdown limit can be enhanced; however, the ultimate destination of the electrons must be controlled. Evidence is mixed (Reid, Lohsen, 1982), but the literature on magnetic-shielded, high-power pulsed diodes probably should be surveyed. For the radio-frequency-quadrupole (RFQ) accelerator geometry, Swenson (1982) calculated that an external field of around 1 tesla might provide effective magnetic shielding. The performance of the RFQ improves rapidly with voltage.

Surface Properties

As already inferred, the properties of the electrode surfaces hold the secrets of the limits, or how to operate reliably near the limits. There are many variables.

The best data points at present are those of Tanabe (1986) and Wang, (1986); at about 240-Mv/m peak surface field for 4.4-μs pulses, Tanabe used an 8-microinch finish on OFHC copper electrodes, starting the tests in a clean environment (starting pressure 2 X 10^{-7} torr). This gradient reaches the electron induced breakdown limit of Fig. 2 at 3 GHz. He reports that neither diamond polishing nor electroplating enhanced the breakdown threshold level. He found no significant difference between stainless steel, aluminum, titanium, OFHC, or a titanium coating, and achieved the same threshold with plated copper. On this last point, others have had both good performance (Tanaka, 1986) and poor performance,* especially with thin plating and higher duty factor or cw operation. Tanabe found that nickel plating made the breakdown worse. The introduction of cesium has a very deleterious effect, and it has been noted that the concentration of hydrogen in the vacuum increases markedly as breakdown is approached and when it occurs (Gerhard, et al., 1985).

Many of the limit models outlined above assume that a local field enhancement occurs somewhere on the surface where breakdown begins and may be exacerbated rapidly. Earlier it was often thought that metallic whiskers were to blame, but now there are many ideas, including the effects of inclusions with dielectric or semiconductor properties, adsorbates, oxides, clumps of microparticles that might be mobile, outgassing characteristics including surface changes, grain structure, or dusts of various materials (Latham, 1981; Proc. XIth Intl. Symp., 1984).

It is possible (Peter, 1984) that carefully applied surface coatings may be useful for inhibiting breakdown at voltages far above the electron multipactor limit where the purpose of the coating would not be to reduce secondary emission as in the multipactoring case, but rather to isolate electric whiskers from the chamber and to serve as a trap for slow electrons.

The field diffusion time in the dielectric layer must be much less than half the cycle time of the rf field so that there is no field or Q perturbation; thus, the field lines terminate on the layer whiskers for only a fraction of the cavity cycle time. This is a practical thickness; perhaps 1000 A° suffices. It is also hypothesized that one reason why

---

* Los Alamos, 1986 – on the long-pulse accelerator test stand and the cw FMIT RFQ, operating at E_o ~ 1.75 – 2.0 Kilpatrick, erosion of plated surfaces occurred.

electron emission is reduced after initial operation is because adsorbed hydrocarbons on the surface are polymerized, making a layer with strong inelastic scattering for slow electrons, thereby reducing secondary emission and field emission out of excited states. Only a very thin layer ($<10A°$) would be required, and one might use multiple layers (analogous to high-power optical mirrors) to address several of these effects. Tests of dielectric coatings are scheduled at Los Alamos.

Although a general observation to date is that breakdown limits do not depend much on base vacuum level below about 10^{-5} torr, it would appear that reaching the ultimate limits will refocus on both better base pressures and very careful control of the details of local vacuum and the residual gas constituents. Experience from storage rings where 10^{-11} torr base pressures and exposure of the components to strong synchrotron radiation conditioning is yielding important information (Tanaka, 1986).

Regardless of exactly what is happening on a particular surface, experience with superconducting cavities has shown that the identification of trouble spots and specific repair of those areas is a convergent process that can reliably raise the operating threshold. Such a procedure will eventually fail, of course, but it yields valuable lessons on the relative importance of some of the variables, for example, roughness.

It seems clear enough that very pure surfaces with totally homogeneous properties would allow approach to the electric field or surface heating limits that, after all, implicitly assume such surfaces in their simple physics model. It is a great challenge to achieve such surfaces in practice.

Conditioning

It is standard to observe that the breakdown threshold improves after initial operation of a cavity. The surfaces are cleaned and perhaps changed during the outgassing processes, and protrusions or voids are modified. Operating under glow discharge conditions or purging with argon can help considerably, but careful procedure and control of the gas pressure are required (Wang, 1986).

Allowing hard sparks to occur, where significant energy is dissipated, is not recommended. Usually this results in surface damage that lowers the threshold. The preferred procedure is to operate just under the level where breakdown occurs, gradually processing upward over periods of several, often many, hours. In this regime, microdischarges occur without completely collapsing the cavity field. In this process, the problem areas get healed; perhaps whiskers are blunted by ion back bombardment or are burned off, perhaps voids or inclusions are modified.

Starting with good vacuum base pressure, aided by bakeout, helps, but the action of the rf fields or synchrotron radiation clearly produces stronger cleaning action (or the ultimate breakdown, of course).

CONCLUSION

The reason it is desirable to approach the ultimate rf breakdown limits, in practice, is simply that higher performance accelerators would result; both the accelerating cavity and the power source (or their ultimate combination) would benefit as discussed in the author's other articles at this Institute. This is a difficult and expensive research area; progress will require a great deal of commitment, but the challenge is there.

REFERENCES

Bohne, D., Karger, W., Miersch, E., Roske, W., and Stadler, B., 1971, Sparking Measurements in a Single-Gap Cavity, IEEE Trans. Nucl. Sci., 18:568.

Bolme, G. O., 1986, RF Structure Studies at High Power, Los Alamos National Laboratory, Accelerator Technology Division, AT-2 technical document.

Boyd, Jr., T. J., 22 May 1984, Los Alamos National Laboratory internal memorandum based on analysis of MURA Tech Note 535 dated 17 March 1965.

Boyd, T. J., Los Alamos National Laboratory, internal note.

Gerhard, A., Deitinghoff, H., Gruber, A., Klein, H., and Schempp, A., 1985, RF Sparking Experiments, IEEE Trans. Elec. Insulation, 20:709.

Jameson, R. A., High Brightness RF Linear Accelerators, this ASI.

Kilpatrick, W. D., October 1957, Criterion for Vacuum Sparking Designed to Include Both rf and dc, Review of Scientific Instruments, 824.

Latham, R. V., 1981, "High Voltage Vacuum Insulation: The Physical Basis," Academic Press, New York.

Leeman, C. W., to be published, The CEBAF Superconducting Accelerator-An Overview, Proc. 1986 Linear Accelerator Conf., 2-6 June 1986, Stanford Linear Accelerator Center report, Palo Alto, California.

McEuen, A. H., Lui, P., Tanabe, E., and Vaguine, V., October 1985, High Power Operation of Accelerator Structures at Liquid Nitrogen Temperatures, IEEE Trans. Nucl. Sci. (5):2972.

Palmer, R. B., 1982, Near Field Accelerators, Laser Acceleration of Particles, AIP Conf. Proc. No. 91, Los Alamos, p. 179.

Peter, W., 1 September 1984, Vacuum Breakdown and Surface Coating of rf Cavities, J. Appl. Phys. 56(5):1546.

Proc. XIth Intl. Symp. on Discharges and Electrical Insulation in Vacuum," Vols. 1 and 2, September 24-28, 1984, Berlin, GDR.

Rabinowitz, M. and Donaldson, E. E., 1965, Electrical Breakdown in Vacuum: New Experimental Observations, J. Appl. Phys. 36, 1314.

Reid, D. W. and Lohsen, R. A., October 25-28, 1982, Breakdown Criteria Due to Radio-Frequency Fields in Vacuum, 10th Intl. Symp. on Discharges and Electrical Insulation, University of South Carolina, Columbia, South Carolina.

Schriber, S. O., June 2-6, 1986, Factors Limiting the Operation of Structures Under High Gradient, Proc. 1986 Linear Accelerator Conf., op. cit.

Swenson, D. A., Los Alamos National Laboratory Memorandum, Magnetically Tuned rf Cavity Model for LAMPF II, December 4, 1982.

Tanabe, E., Wang, J., and Loew, G., June 2-6, 1986, Voltage Breakdown at X-Band and C-Band Frequencies, Proc. 1986 Linear Accelerator Conference, op cit.

Tanaka, J., June 2-6, 1986, Advanced Technology Recently Developed at KEK for Future Linear Colliders, Proc. 1986 Linear Accelerator Conference, op. cit.

Wang, J. M., Nguyen-Tuong, V. G., and Loew, A., May 1986 and June 2-6, 1986, RF Breakdown Studies in a SLAC Disk-Loaded Structure, SLAC-Pub3940, also Proc. 1986 Linear Accelerator Conf., op. cit.

RF POWER SOURCES FOR HIGH-BRIGHTNESS RF LINACS*

Robert A. Jameson and Don W. Reid

Accelerator Technology Division Office, MS H811
Los Alamos National Laboratory
Los Alamos, New Mexico 87545 USA

SUMMARY

This paper will discuss the types of devices and system architectures that show promise in providing rf power sources for future rf linear accelerators. Different types of sources will probably be required at different frequencies, but in all cases, efficiency and unit cost will be prime considerations. In many applications, high power-to-weight ratios will also be advantageous.

INTRODUCTION

For the past 25 years, most accelerators have been driven by rf sources previously developed for radar or high-power broadcast systems. The workhorse of accelerator rf power sources has been the klystron for frequencies above 400 MHz with the gridded tube being used below that frequency. Magnetrons have been used for single-structure electron accelerators.

Typical ion linac frequencies are $\leq$~1 GHz, largely because of the requirements for handling nonrelativistic particles at the desired beam current per channel and for engineering reasons at large duty factors or cw. However, the beam dynamics scaling laws prescribe a higher frequency for higher brightness; thus, if that were the primary requirement, frequencies above 1 GHz might well be used and total beam current might be increased by using multiple channels.

Typical electron linac frequencies are 500 MHz to 3 GHz, (lower for longer duty factor or cw applications, where standing-wave structures are used); 3 GHz is used for many low-duty-factor, traveling-wave accelerators, with the 2-mile SLAC linac as the premier example.

507

Future acceleration schemes might use ultra-high gradients (Jameson, "High-Brightness RF Linear Accelerators, and RF Field Limits," this ASI), resulting in frequencies up to around 30 GHz. We will take a brief look at frequency regimes, below and above about 3 GHz.

FREQUENCY $\lesssim$ 3 GHz

Typical requirements for high-brightness ion or electron accelerators at duty factors from 1% to cw are indicated in Table 1.

In the past, a single section of rf accelerator structure was usually of the appropriate length to be driven by a single rf power source, and the sources were sized as large as possible to keep the number of parts low. The Spallation Neutron Source (SNQ) project proposal (KFA, 1984) suggested the use of many, relatively small rf sources driving individual cavities. Multiple drives into a single cavity have been used on the FMIT system at Los Alamos (Fazio et al., 1979), and the HILAC at Berkeley (Smith, 1966). For applications having stringent reliability considerations and the desirability to package the rf in manageable sizes, the prospect of using multiple sources for each accelerator section is attractive, and a convenient peak power output per amplifier package appears to be between 500 and 1000 kW. This scheme has the added advantage of being able to add or substract power modules as the accelerator requirements change.

Solid-State Amplifier Systems

One of the most promising technologies for this frequency range is the solid-state amplifier. Two recent systems have been developed for radar applications (Lee et al., 1983; Brookner, 1985) with power levels that are applicable for the accelerator system designer. One of these systems was tested into a resonant cavity with excellent results (Vaughn et al., 1985).

The typical solid-state amplifier is made up of modules, or books, as some companies call them. Figure 1 shows a typical 2.5-kW modules with the new version, which uses 1986 technology. Each module contains 8 to 10 transistors and delivers about 2.5- to 3.0-kW peak power. The modules are also self-contained in that they have their own power conditioning, preamplifier, and output combiner.

Many modules are combined to develop the final output power. Some manufacturers use a Wilkinson combining scheme and others use a radial combining scheme. Figure 2 shows a 32-way radial combiner with a cw power output at 200 MHz of 2 kW. In principal, a solid-state amplifier

508

Table 1. RF power requirements.

Frequency	UHF (425 MHz typical)
Peak output power	20 to 100 MW
Pulse length	milliseconds to cw
Duty	1% to cw
Weight ratio (amplifier)	0.3g/W (1.5 kW/lb)
Volume ratio (amplifier)	15 w/cu in.
Efficiency (amplifier)	>50%
Reliability	0.9999

Fig. 1. The 2.5-kW solid-state amplifier modules showing
reduction in size using new technology.

could be built at any power level if the designer were willing to use
enough combiner stages.

The heart of the solid-state amplifier is the high-power, microwave
transistor. There are three generic types of transistor that are used
in high-power microwave amplifiers. The most commonly used type is the
silicon bipolar transistor. The silicon FET is also used, and the new-
est device is the static induction transistor or SIT (Bencuya et al.,
1985). Table 2 is a comparison of the three transistor types. A 425MHz
frequency, a 1-ms pulse length, and a 10% duty are assumed.

Fig. 2.   A 32-way radial combiner.

Table 2.   Comparison of transistor parameters.

Parameter	Silicon Bipolar	Silicon FET	SIT
Supply voltage	40 V	40 V	100 V
Efficiency	70%	70%	67%
Power out	350 W	300	100 W
Power gain	10 dB	10 dB	6 dB

Figure 3 shows the power-handling capabilities of each type of device as a function of calendar time. A pulsed condition is used for the bipolar transistor. The FET and SIT are cw devices. As can be seen, the curves are still on a rapid slope upward. If the improvement continues as indicated by the progress to date, 0.2 to 0.3 g/W for solid-state amplifiers appears achievable in the next 2 to 3 years. This means that a 1-MW power system, operating from a 40-V source would weigh 660 lbs and have a volume of 38 cu ft, a package of 3 by 3 by 4.5 ft. The anticipated efficiency, defined as source power to rf output power, is 60%. Unfortunately, the present cost of such a system is

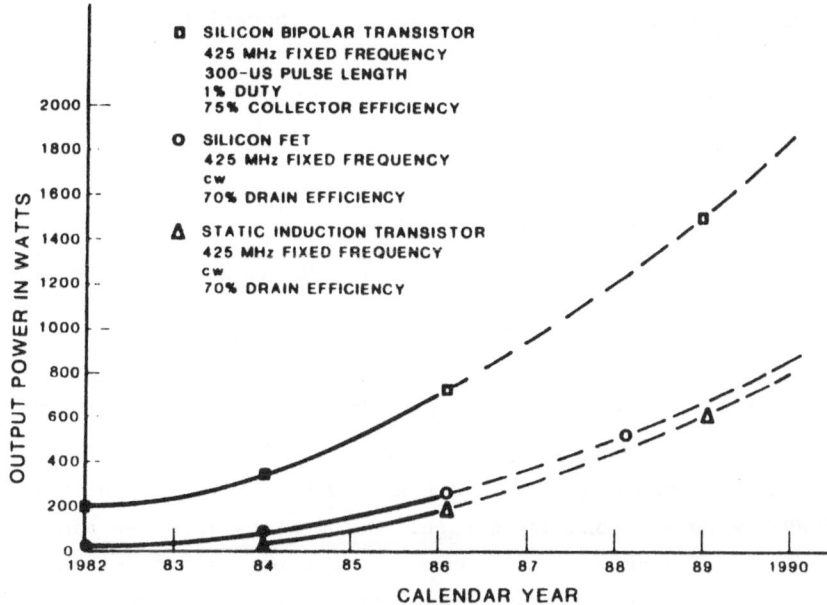

Fig. 3. Output power versus calendar year for microwave power transistors.

$5.00 to $8.00 per peak watt but should improve substantially during the next 3 to 5 years.

Another issue that must be addressed in the use of solid-state amplifiers is potential radiation damage. Tests (Martin, 1982) done on typical bipolar transistor indicate that large neutron doses cause degradation in gain and efficiency.

Tube Amplifier Systems

Historically, tubes have been the device chosen when megawatts of rf power have been required. Above 400 MHz, the workhorse has been the klystron operating at several MW peak power with crossed-field devices entering the pictures at powers up to 2 MW. Frequencies below 400 MHz have generally been the exclusive domain of the gridded tube. More recently, emission-gated devices such as the Klystrode (Shrader and Priest, 1985) and the lasertron (Wilson and Tallerico, 1982) appear promising as power sources for the 1990s at UHF and low L-band frequencies.

Klystrons are mature devices with well-understood operating parameters. They are readily available at beam efficiencies of 50 to 60%. Klystrons can be operated at any pulse length from a few microseconds to cw.

Units are available for cw operation at (power):(frequency) of

- (500 kW cw):(500 MHz),
- (500 kW cw):(353 MHz),
- (1 MW cw):(353 MHz),
- (375 kW cw):(224 MHz), and
- (600 kW cw):(2400 MHz).

Typical pulsed klystrons include:

- 1.3 GHz, 6 MW peak, 50 kW average, 300-μs pulse length, 130 kV

- 3 GHz, 37 MW peak, 20 kW average, 5-μs pulse length, 275 kV, 53-dB gain, 48% efficiency, SF_6 pressurized.

- Under development at SLAC; 3 GHz, 150 MW peak, 1-μs pulse length, 180 pps, 450 kV, 600-A beam current, 55% efficiency.

Limiting phenomena in klystrons (Faillon, 1985) include peak surface field sustainable without sparking; 30 MV/m at 10-μs pulse length and 15 MV/m a cw are considered high. Cathode loadings can range from 10-15 A/cm^2 peak or 2-3 A/cm^2 cw. Larger perveances, K, and area convergence, C, ratios make gun design more difficult; typical combinations are C = 40 and K = 2μperv, or C = 80 and K = 0.5μperv. High duty factor operation can stand heat loads of 500-1000 W/cm^2 with careful avoidance of localized heating and maintenance of surface temperature below 250-300°C except in very special circumstances. Figure 4, reproduced from the Faillon paper, indicates limiting conditions for two typical cases. Output cavity coupling is an important problem at the higher frequencies and duty factors, because of the heat removal from small components and the fairly common requirement to be able to handle all-phase mismatched loads to VSWRs as high as 2-3:1. A klystron role in accelerator applications requiring good power-to-weight or volume ratios is doubtful because the klystron size at UHF frequencies is prohibitive.

Gridded tubes operate well at UHF frequencies and show promise as compact-sized, low-duty-factor accelerator drivers, especially when use of distributed amplifiers is anticipated. These tubes operate at 50 to 60% beam efficiency and have 20-dB gain. Gridded tubes do have a pulse-length limitation. Typically, as the pulse length increases, the peak-power capability decreases, and extrapolation to cw service appears unlikely. For example, one tetrode has an advertised output power of 2 MW at a duty of 0.4% and a pulse length of 13 μs. The output power falls to 275 kW when the duty is raised to 6% and the pulse length increased to 2 ms. A typical gridded power tube is shown in Fig. 5. The tube must be housed in an external, resonant cavity.

Los Alamos has recently developed a cavity amplifier at 425 MHz that uses several planar triodes operating in parallel in the same output cavity (Hoffert, 1986). The tube arrangement is shown in Fig. 6.

512

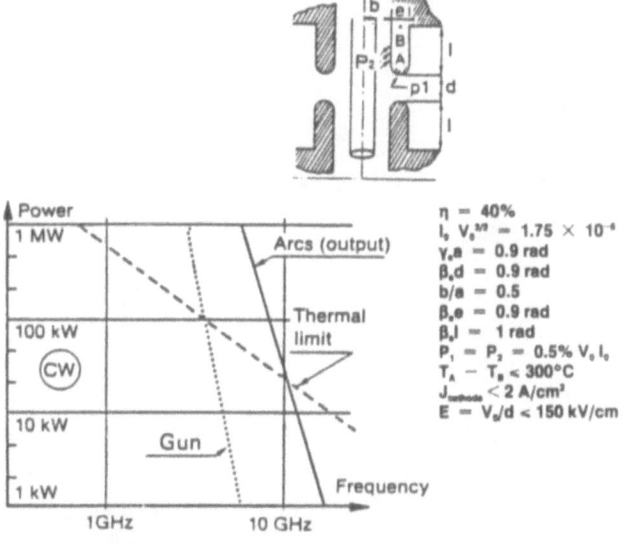

Fig. 4a. Power, frequency, and computed limit diagram for
klystron with a 40% efficiency and 1.0μperv.

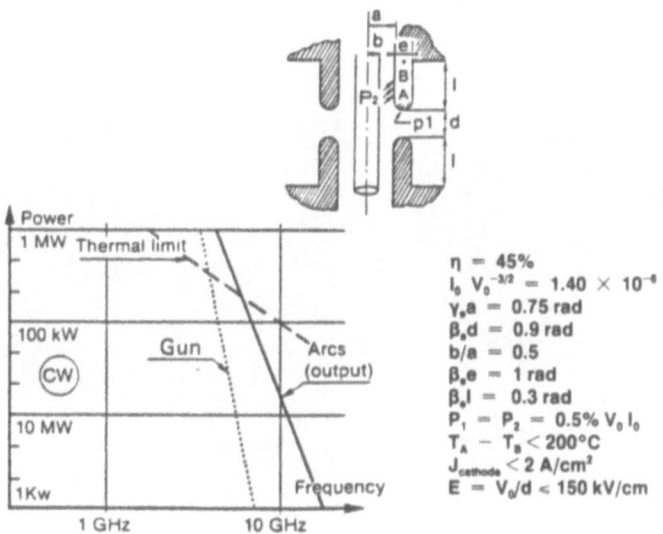

Fig. 4b. Power, frequency, and computed limit diagram for
klystron with a 45% efficiency and 1.40μperv.

The output stage operates at 60% efficiency and 13-dB gain. The output
power using nine planar triodes is 150 kW at 60-μs pulse length. Again,
this system is pulse-length and duty-cycle limited. The system is very
cost effective, however, with the entire system, including power supplies and preamplifiers, costing less than $2.00 per watt.

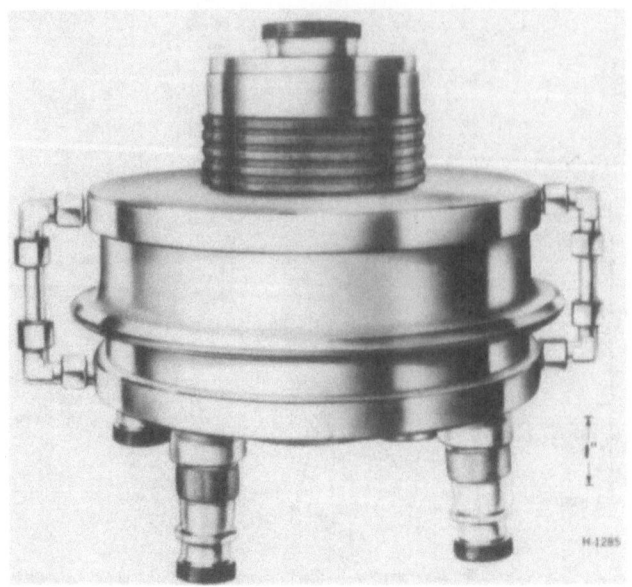

Fig. 5.   UHF tetrode power tube.

Fig. 6.   Multiple planar triode cavity amplifier.

Magnetrons have been used as accelerator drivers where the accelerator consisted of a single section.   A magnetron operating at 2.5 MW has been used successfully by several companies to combine two or more magnetrons with sufficient phase stability to drive the high-Q accelerator structure has not been successful.   Recent work at Varian-Beverly (Jenkins et al., 1985) shows promise of a solution to this problem by injection locking two magnetrons.   Magnetrons are cost effective, can be operated at a few megawatts, and do not require a preamplifier chain.

They do, however, have average power and pulse length limitations. For example, the magnetron cited above is limited to a duty cycle of 0.1%.

The gain structure of the conventional crossed-field amplifier does not provide isolation between the input and the output of the tube. This deficiency makes the device very susceptible to variations in rf load. Because an accelerator cavity has a variety of modes besides the desired accelerating mode, presents such a high VSWR during the full time, and reflects most of the stored energy back to the source after the pulse, the crossed-field amplifier could not be stably run, except possibly by using expensive circulators, which are large and have an adverse effect on system efficiency. For this reason, accelerator designers have avoided using the crossed-field amplifier. During the past few months, Raytheon has been testing a new crossed-field amplifier that is cathode driven. This technique may solve the isolation problem between the input and output circuit in that the new device has greater than 30-dB isolation between these two ports. Tests to characterize this device into a resonant load will be done soon. The present tube operates at S-band at 1.25-MW output power, 60% efficiency, and weighs 70 lbs. It has 24-dB gain. This is an exciting possibility because of the low weight and volume factors.

A relatively new device, the emission-gated amplifier tube, has been developed during the past few years. Two tube types have resulted: the lasertron and the Klystrode. In both of these devices, the beam is bunched at the required rf frequency as it is emitted from the cathode. The beam is then passed through a klystron-type output cavity where the power is extracted in the conventional way. The big advantage to this type of device is the elimination of the interaction space required in the conventional klystron, which enables the tube to be much shorter for a given frequency than that for a conventional klystron.

Eimac has developed the Klystrode, which has a triode input section and a conventional klystron output section. A cross section of the tube is shown in Fig. 7. The grid is coupled to a resonant cavity that is driven at the appropriate frequency. This scheme causes a bunched beam to be emitted from the cathode.

Eimac has proposed a pulsed Klystrode at 425 MHz with an output power of 500 kW. The pulse length is 350 μs at 1% duty, and the tube will have a gain of about 20 dB. This tube will have at least 70% beam efficiency at 85 kV and will weigh slightly over 100 lbs, with the magnet.

The lasertron is being worked on at several places including Varian, SLAC, Orsay, and Los Alamos. The basic idea is the same as the

515

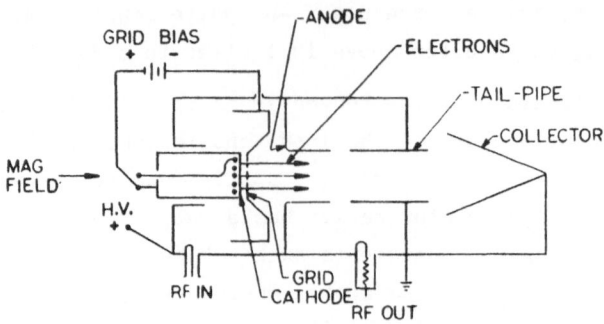

Fig. 7. Klystrode cross-section schematic.

Klystrode except that the cathode is pulsed by a laser beam. The lasertron requires a photoemissive cathode as well as the added complexity of
a laser with enough power to activate the cathode, but does have the
potential for operating at higher frequencies and faster pulse rise
times than the Klystrode because of the elimination of the capacities
associated with the grid.

FREQUENCY $\geq$ 3 GHz

Particle accelerators requiring rf above around 3 GHz are usually
aimed at accelerating electrons at low duty factor to very high energy,
so very high peak power sources are needed, such as the 150-MW SLAC
klystron mentioned in the last section. At the higher frequencies, the
accelerator structure size gets smaller, so the amount of stored energy
required gets small. Because the accelerator is a staged device
requiring multiple rf drives, rf amplifiers that can be phase-locked
together are used rather than oscillators. There are a number of development programs to make high power sources up to frequencies in the
100-GHz range, but most of these, such as gyrotrons, backward-wave
oscillators, crossed-field devices, or virtual cathode oscillators, are
oscillators and cannot yet be linked together in unison for accelerator
service.

The gyrotron oscillator can make a lot of power and can be run in
an amplifier mode (gyroklystron) that might prove suitable. Gyrotrons
are reported at 1.4-2.0 GW, 3 GHz, E_b = 900 kV, I_b = 8 kA, 60-ns pulse,
30% efficiency (Didenko et al., 1976); and at 214 kW cw, λ = 0.5 cm,
P average/λ^2 ~ 1 MW/cm^2, 50-65% efficiency (Granatstein et al., 1985).
Some initial work is being done (McAdoo, 1985) on phase-locked gyrotrons, which looks promising but is still an order of magnitude or so
away from the typical accelerator requirements of $\leq$ 1° phase difference
stage-to-stage.

Gyroklystron amplifiers are less advanced. Granatstein is now operating a three-cavity 4.5-GHz tube at 54 kW peak and 21% efficiency, and plans to build a 10-GHz unit with 30-MW peak power. The requirement for the collider application is around 300 MW peak, 25 kW average, 10 GHz, 100-ns pulse length, $\geq$50-dB gain, $\leq$50% efficiency (Wilson, 1983).

The technical boundaries for these types of tubes may be defined by characteristics of the electron gun and of the interaction circuit, with window and collector limitations also important but secondary (Huffman et al., 1984). Voltage breakdown limits restrict gun performance, thus pulsed operation below 10 μs is easier. Cathodes are not as much of a limitation now that emission of better than 10 A/cm^2 is available. In the interaction region, rf voltage breakdown is again a major limitation; performance is extended by using multiple output gaps. These extended-interaction klystrons (EIKs) have operated at 1 MW cw at both C and X-band. Around X-band frequencies, ohmic heating becomes the major problem. The mass of material heated by rf losses scales as the minus five-halves power of the frequency. Again the EIK techniques helps. Figures 8 and 9, reproduced from Huffman, 1984, show the power available today from pulsed or cw klystrons and gyroklystrons. The falloff occurs from the ohmic heating effect, but the limitations on the output windows are only a factor of 2-3 higher. Overmoded windows and output waveguides are used, which helps with voltage breakdown problems but introduces other problems with the higher-order modes.

Pulse compression schemes are another way to boost the available power in short pulses. One method (Wilson, 1983) stores energy in an auxiliary cavity, which is released by triggering a 180° phase shift in the klystron drive and sent to the accelerating cavity. Voltage enhancements of around 1.4 are achieved in practice. Another new method (Farkas, 1986) uses an elegant arrangement of couplers fed with n phase-coded rf drives in such a was that the output pulse is n-times higher and n-times narrower than the input drive pulses.

Future rf sources for high-power, high-brightness rf linacs will have to be more efficient and have low cost per the desired unit of power, whether peak or average or both. In this respect, it is not clear whether the types of tubes discussed here so far are the answer. In a companion article at this ASI (Jameson, High-Brightness RF Linear Accelerators), high-brightness economics is discussed and examples are given to show the necessary system tradeoffs, cost considerations, and possible new schemes for coupling the driving and accelerated beams more closely together. Also, an innovative scheme is outlined using a sequential arrangement of free-electron-laser and induction-linac modules to generate microwave power at 30 GHz in enormous quantities. The

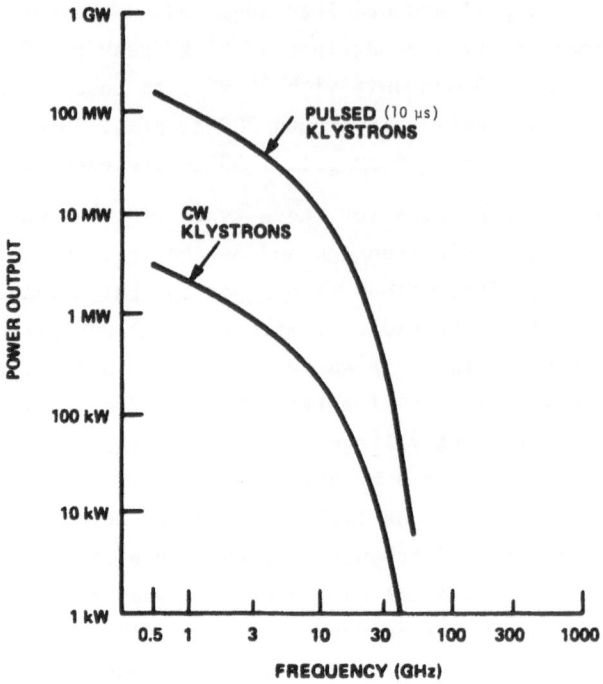

Fig. 8. Technology limits for pulsed and cw klystrons.

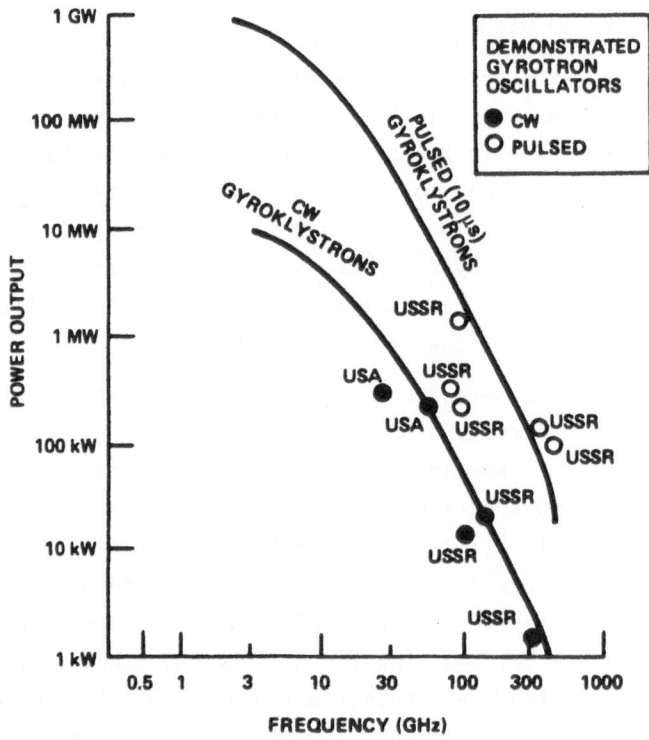

Fig. 9. Technology limits for pulsed and cw gyroklystrons.

reader of this paper should now read that section of the other paper. The ideas expressed there have not yet been realized practically, and may in fact prove impractical. The point is that we do not have in hand any really viable, economically attractive sources to power some of the envisioned particle accelerator applications of the future, and hence those accelerators may not be built. The rf power field needs some inventions.

ACKNOWLEDGMENTS

For permission to use specific figures used in this paper, the author thanks the following:

- Phil Kammerman, Westinghouse Electric Corp. (Fig. 1)
- Zaven Guiragossian, TRW (Fig. 2)
- Jere Stabley, RCA (Fig. 5)
- Merald Schrader, EIMAC (Fig. 7)
- Jeff Meyer, MACOM-PHI, for the data used to generate Fig. 3

REFERENCES

Bencuya, I., Cogan, A. I., Butler, S. J., and Regan, R. J., 1985, "Static Induction Transistors Optimized for High-Voltage Operation and High Microwave Power Output," IEEE Trans. Elec. Dev., 32(7):1321.

Brookner, E., 1985, "Phased-Array Radars," Sci., Am., 252(2):94.

Didenko, A. N., et al., 1976, Sov. J. Plasma Physics, 2:283.

Faillon, G., 1985, "New Klystron Technology," IEEE Trans. Nucl. Sci., 32(5):2945.

Farkas, Z. D., "Binary Peak-Power Multiplier and Its Application to Linear Accelerator Design," SLAC Pub 3694, May 1985 and to be published in IEEE Trans. on Microwave Theory and Techniques, MTP-16, October 1986 (Special Issue on New and Future Applications of Microwave Systems).

Fazio, M. V., Johnson, H. P., Hoffert, W. J., and Boyd, T. J., 1979, "A Proposed RF System for the Fusion Materials Irradiation Test Facility," IEEE Trans. Nucl. Sci., 26:3018.

Granatstein, V. L., Vitello, P., Chu, K. R., Ko, K., Latham, P. E., Lawson, W., Striffler, C. D., Drobot, A., 1985, "Design of Gyrotron Amplifiers for Driving 1 TeVe$^-$e$^+$ Linear Colliders," IEEE Trans. Nucl. Sci., 32(5):2957.

Hoffert, W. J., 1986, Los Alamos National Laboratory, personal communication.

Huffman, G., Boilard, D., Stone, D., 1984, "Power Limits for Accelerator Tubes from UHF to Ka Band," Proc. 1984 Linear Accelerator Conference, May 7-11, 1984, GSI, Darmstadt, 180.

FUNDAMENTAL FEATURES OF SUPERCONDUCTING CAVITIES FOR HIGH-BRIGHTNESS ACCELERATORS

Guenter Mueller

Department of Physics
University of Wuppertal
D5600 Wuppertal 1, West Germany

INTRODUCTION

It is now more than 20 years ago since the first electrons were accelerated in a superconducting lead-plated copper cavity at Stanford (Pierce et al, 1965). Between 1968 and 1970 very successful experiments were performed with X-band resonators fabricated from bulk niobium (Turneaure et al, 1968 and 1970) which is the element with the highest critical temperature. These experiments together with the predictions based on the microscopic BCS theory of superconductivity (Bardeen, Cooper and Schrieffer, 1957) lead to the early expectation that accelerating fields of at least 30 MV/m and quality factors above 10^{10} could be achieved with superconducting cavities in a continuous wave mode (cw).

Consequently in the seventies several applications of superconducting accelerating or deflecting structures were started, as for instance for the Stanford Superconducting Recyclotron (Lyneis et al., 1981), the Illinois Microtron (Axel et al., 1979), the CERN-Karlsruhe Particle Separator (Citron et al., 1979) and the Argonne Superconducting Heavy Ion Postaccelerator (Bollinger et al., 1983). As a first application of superconducting cavities, a beam was successfully accelerated with a prototype structure to 4 GeV in the Cornell Synchrotron (Sundelin et al., 1974). It became obvious, however, that in complex structures like the one shown in Fig. 1 moderate accelerating fields of less than 3 MV/m could be obtained due to anomalous field limitations such as electron multipacting and quenching. However high quality factors (up to 7×10^9) were achieved over long operating periods reliably and under routine conditions. In accordance with the theme of this Institute, it should be mentioned that HEPL operated the first free-electron laser using the high brightness beam of the Stanford Recyclotron (Deacon et al., 1977).

521

The experiences with these complex superconducting accelerating structures have shown that the key to successful large scale application lies in a better understanding of the anomalous loss mechanisms and in the development of adequate fabrication and surface preparation methods. These factors guarantee a high cavity performance for reasonable costs. Therefore, after an introduction to the fundamental features of superconductors in rf fields, the main part of this lecture will deal with the anomalous losses which are responsible for performance limitations observed today, and the progress we have achieved during the past few years. This discussion will be concentrated on velocity of light cavities, because the geometry of structures for low β proton and heavy

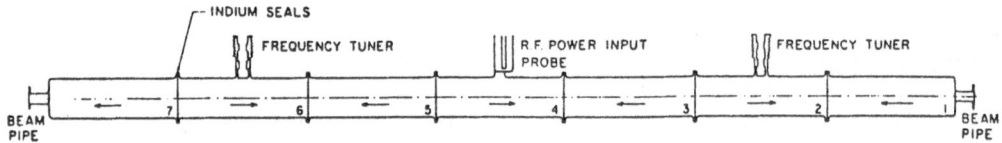

6 m ACCELERATOR STRUCTURE COMPOSED OF SEVEN 7/2λ SUBSTRUCTURES

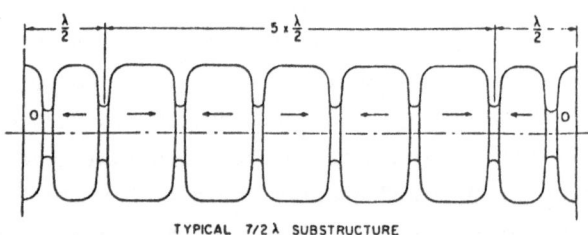

TYPICAL 7/2 λ SUBSTRUCTURE

Fig. 1. Schematic diagram of the Stanford-HEPL superconducting 6 m structure and a typical substructure (Turneaure et al., 1974).

ion acceleration differ completely and is less suitable for systematic investigations on the loss distribution of the superconducting surface. Nevertheless any result on improved methods can be transferred between both communities, and increased accelerating fields have been reported for low β structures also (for a recent review see Bollinger, 1986). In the last section the design criteria for superconducting cavities for electron accelerators are outlined, and a status report on current projects is given, including some considerations about couplers, tuners and cryostats. Finally the question will be discussed; are superconducting continuous wave accelerating systems suitable for high brightness accelerators?

522

Surface Impedance

The interaction between an rf electromagnetic wave and a metal surface is described by the surface impedance. Because of the finite current density of all conductors there is a small penetration of the tangential magnetic field into the metal surface and therefore also a small induced tangetial electrical field, resulting in a Poynting vector

$$\vec{S} = \vec{E} \times \vec{H}* \ . \tag{1}$$

In the frame of reference of Fig. 2 the surface impedance is defined as

$$Z_s = R_s + iX_s = E_x(o,t)/H_y(o,t) \ . \tag{2}$$

Its real part, the surface resistance R_s, leads to the dissipated power density per unit area

$$\frac{dP}{dA} = \frac{1}{2} \, \mathrm{Re}|E_x(o) \cdot H_y^{\ *}(o)| = \frac{1}{2} \, R_s |H_y(o)|^2 \ , \tag{3}$$

while the surface reactance X_s describes the non-dissipative energy exchange and is connected to the penetration depth λ by

$$X_s = \omega \mu \lambda \ , \tag{4}$$

where μ is the magnetic permeability of the metal.

For the calculation of Z_s, the material equation between the current density j and the electromagnetic fields in the conductor is needed. In the case of a normal conductor Ohm's law is valid as long as the mean free path ℓ of the electrons is small compared to λ, and leads to the formulas of the normal skin effect:

$$Z_s = (\frac{\omega \mu}{2 \sigma})^{1/2}(1+i) = \frac{1}{\sigma \delta}(1+i) \ , \tag{5}$$

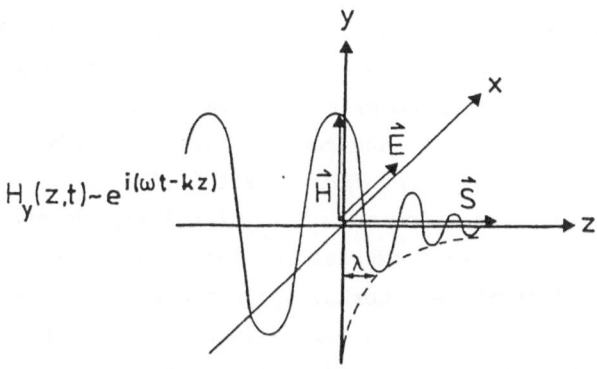

$$H_y(z,t) \sim e^{i(\omega t - kz)}$$

Fig. 2.   Penetration of an electromagnetic wave into a metal surface.

where σ is the electrical conductivity and δ is the skin depth

$$\delta = (\frac{2}{\omega\mu\sigma})^{1/2} = 2\lambda \ . \tag{6}$$

At 500 MHz the resulting values for copper at room temperature are R_s = 5.8 mΩ and δ = 3μm. In very pure metals ℓ which is proportional to σ can be increased several orders of magnitude by cooling to cryogenic temperatures. Then Ohm's law must be replaced by a non-local material equation j(E) described by the anomalous skin effect which leads in the limit $\ell \gg \lambda$ to (Reuter and Sondheimer, 1948)

$$Z_s = \frac{8}{9}(\frac{\sqrt{3}}{16\pi} \frac{\omega^2\mu^2\ell}{\sigma})^{1/3}(1+\sqrt{3}i) \ . \tag{7}$$

Therefore R_s becomes independent of ℓ, and the benefit of cooling a normal conducting cavity is limited to a factor of about five.

In the case of a superconductor, the conduction electrons form into the so called "Cooper pairs" which carry the electric current without any losses. Cooper pairs are ordered states in momentum space with the two electrons having opposite but equal momenta and opposite spins. The pairing energy per electron $\Delta(T)$ is very weak and in the BCS theory at T = 0 K correlated to the critical temperature T_c by

$$\Delta(o) = 1.76 \ kT_c \ , \tag{8}$$

where k is the Boltzmann constant. A measure for the range of the attractive force between the two electrons is given by the coherence length ξ_0 which depends on ℓ by

$$\frac{1}{\xi_0(\ell)} = \frac{1}{\xi_0(\infty)} + \frac{1}{\ell} \ . \tag{9}$$

Due to the finite density of Cooper pairs $n_c(o)$ there is a minimum penetration depth of an electromagnetic field into the superconductor – the so called London penetration depth, λ_L. The calculation of Z_s for superconductors depends on the material equation between j and the magnetic vector potential, A, and can be performed analytically only for both extreme limits, namely the local limit ($\xi_0(\ell) \ll \lambda_L$: London superconductors) and the anomalous limit ($\xi_0(\ell) \gg \lambda_L$: Pippard superconductors). The exact expressions for Z_s derived by Mattis and Bardeen (1958) and Abrikosov, Gorkov and Khalatnikov (1958) are rather complex and must be calculated numerically for most of the superconductors. Codes have been developed by Turneaure (1967), Halbritter (1970) and Blaschke (1981). The main result of these computations is that R_s can be reduced drastically into the nΩ range for high T_c superconductors at reasonable operation temperatures.

Frequency and Temperature Dependence of the Surface Resistance

Instead of a more detailed description of the calculation of the absolute value of Z_s, the scaling laws of R_s with frequency and temperature can be derived by a simple two fluid model approximation of the superconducting state introduced by H. London in 1934. This model is based on the assumption that at finite temperatures, in addition to the carriers of the supercurrent (Cooper pairs), some "normal" electrons are present in superconductors leading to rf losses. With knowledge of the BCS theory we can easily estimate the density of these "normal" electrons, $n_e(T)$, from the probability that a Cooper pair is broken up. This is given by the Boltzman factor

$$n_e(T) = 2 \, n_c(T)e^{-\Delta(T)/kT} \qquad \text{for } n_e \ll n_{co} \,. \qquad (10)$$

At temperatures below $T_c/2$, $n_c(t)$ and $\Delta(T)$ are close to their values at $T=0$

$$n_c(T) = 2 \, n_c(o)e^{-(\Delta(o)/kT_c)T_c/T} \qquad \text{for } T < T_c/2 \,. \qquad (11)$$

The corresponding electrical conductivity associated with these "normal" electrons should be proportional to $n_e(T)$. As already explained in the beginning of this chapter, there is a small induced electrical field in the superconductor due to the time dependent magnetic field

$$E(t) \propto \frac{dH(t)}{dt} \propto \omega \, H(t) \,. \qquad (12)$$

The rf power dissipated in moving the "normal" electrons will be

$$\frac{dP}{dA} \propto \sigma \, E^2(+) \propto n_e(T)\omega^2 H^2(t) \,. \qquad (13)$$

Comparing this result with Eq. 3 and inserting Eq. 11 leads to the scaling law

$$R_s(\omega,T) \propto \omega^2 \, e^{-(\Delta(o)/kT_c)T_c/T} \qquad \text{for } T < T_c/2 \,. \qquad (14)$$

In Fig. 3 experimental data on the surface resistance of reactor grade niobium by Klein (1981) and Mueller (1983) are compared to the computational results of Blaschke (1981). These calculations include the influence of impurity scattering due to the presence of the anisotropic paring energy caused by the anisotropic fermi energy of niobium. This extension of the BCS theory removed a long existing discrepancy between the experimentally measured quadratic dependence and the theoretically (BCS) predicted $\omega^{1.8}$ dependence at around 1 GHz. Therefore the two fluid model approximation leads to the correct frequency dependence for $\omega < 2\Delta/h$ but cannot account for the change of the slope at frequencies above 10 GHz.

The exponential temperature dependence of R_s described by Eq. 14 has been confirmed by many measurements between 100 MHz and 100 GHz. For real cavities the existence of a temperature independent residual resistance R_{res} has to be taken into account (see Fig. 4):

$$R_s(T) = R_{bcs}(T) + R_{res} \cdot \qquad (15)$$

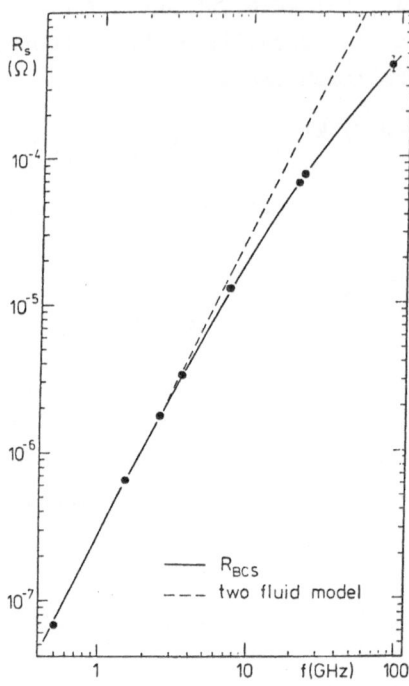

Fig. 3. Frequency dependence of the surface resistance of reactor grade Nb at 4.2 K.

For the temperature dependent part, often referred to as the BCS surface resistance, microscopic theory results in

$$R_{bcs} = A(\lambda_L, \xi_o, \ell) \; f(\omega,T) \; \frac{\omega^2}{T} \; e^{-(\Delta(o)/kT_c)T_c/T} \quad \text{for } T < T_c/2 \; , \qquad (16)$$

where $A(\lambda_L, \xi_o, \ell)$ varies up to a factor two mainly depending on ℓ being larger than A (for higher purity) and $f(\omega,T)$ is a correction function weakly dependent on ω and T. Comparing Eq. 16 to measured curves like Fig. 4 the reduced energy gap $\Delta(o)/kT_c$ can be determined – which is, for most superconductors, slightly larger than predicted from the BCS theory Eq. 8.

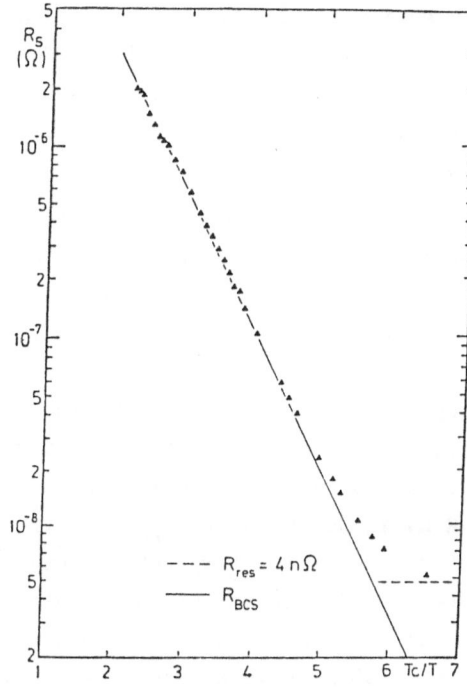

Fig. 4.   Temperature dependence of the surface resistance
of reactor grade Nb at 3 GHz.

For the discussion of the residual resistance it should be noted
that R_s is normally determined from the quality factor of a cavity which
is defined as

$$Q_o = \frac{\omega\, u}{P} = \omega\, \frac{\int \mu_o H^2 dV}{\int R_s H^2 dA} \, ,$$ (17)

where u is the stored energy and P the dissipated rf power in the cavity.
Therefore only for a homogeneously assumed R_s (Eq. 17) can be reduced to

$$Q_o = G/R_s \approx 290 \; Q/R_s \, ,$$ (18)

where G is the so called geometry factor which is typically 290 Q for the
accelerating mode (TM 010) of a cavity. Since there are various origins
for the more or less inhomogeneously distributed residual losses on the
superconducting surface, R_{res} determined this way gives only an average
measure for the quality of the superconducting surface. At present the
residual Q_o's of real superconducting accelerating cavities are limited
by rather local residual losses. These are caused by trapped magnetic
flux or by nonsuperconducting impurities deposited on, or embedded in,
the surface. This has been shown by measurements of the spatial
distribution of rf losses using a temperature mapping technique in
subcooled helium (Piel, 1980). If, for example, only ten parts per

million of the cavity surface consists of normal conducting matter ($R_s \sim 10^{-2} \Omega$), Q_o will be limited to about $3 \cdot 10^9$. Consequently improved fabrication and surface preparation methods as well as sufficient shielding of the earth's magnetic field must be applied to achieve Q_o values approaching 10^{10}.

For large scale application of superconducting cavities in accelerators, an operating temperature of 4.2 K is much more convenient than lower temperatures. Therefore high T_c superconductors (see Tab. 1) are favoured to keep the BCS surface resistance (Eq. 16) at 4.2 K as low as possible. While Pb is considered as a useful material especially in the very low frequency range around 100 MHz, Nb is the most frequently used material for velocity of light structures between 350 MHz and 10 GHz. Since typical values of R_{res} for well-prepared Nb cavities are below 100 nΩ, Q_o (4.2 K) is dominated by R_{BCS} for frequencies above 1 GHz (see Fig. 3). Therefore, operating temperatures near 2 K are necessary to get Q_o values above 10^9. Alternatively A 15 superconductors with much higher critical temperatures like Nb_3Sn can be employed. Promising R_{res} values in the range 10 to 50 nΩ have already been achieved with single cell Nb_3Sn cavities at 500 MHz (Arnolds-Mayer and Chiaveri, 1986) and 3 GHz (Peiniger and Piel, 1985). However, much work on the Nb_3Sn cavities is desirable in order to clarify the observed strong decrease of Q_o for increasing field levels.

Fundamental Field Limitations

The magnetic field is completely expelled from a superconductor only in the true Meissner state (Meissner and Ochsenfeld, 1933), i.e. below the critical magnetic field H_c or H_{c1} in a type I or a type II super-conductor, respectively. The temperature dependence of the thermo-dynamically defined H_c is given by

$$H_c(T) = H_c(0)(1-(T/T_c)^2) . \tag{19}$$

While for type I superconductors like Pb the phase transition to the normal conducting state occurs abruptly at H_c, a dc magnetic field starts to penetrate into a type II superconductor like Nb at the lower critical field H_{c1}. However, superconductivity still remains up to the upper critical field H_{c2}. From the phenomenological theory of Ginzburg and Landau (1950) correlations between H_c, H_{c1} and H_{c2} have been derived using the Ginzburg-Landau parameter $\chi = \lambda / \xi_{GL}$, where ξ_{GL} is a measure of the spatial variation of the Cooper pair density, n_c".

$$H_{c2} = \sqrt{2} \; \chi \; H_c \tag{20}$$

$$H_{c1} = 1/2 \; \chi \; (\ln\chi + 0.08) \; H_c \qquad \text{for } \chi \gg 1 \text{ (Abrikosov, 1957) .} \tag{21}$$

Therefore a superconductor will be of type II for $\chi > 0.7$. χ increases for decreasing l, i.e. for increasing impurity level of the metal.

So far all considerations have been proven to be valid only for the equilibrium condition of dc magnetic field penetration and may not apply to microwave cavities, if the nucleation centers for the phase transition to the normal conducting state cannot be created fast enough. In this case the metastable persistence of the Meissner state up to a, so called, superheated critical field H_{sh} is proposed. The result of the calculations of Matricon and James (1967) based on the Ginzburg Landau equations have the limiting form (Tigner and Padamsee, 1982)

$$
\begin{aligned}
H_{sh} &\sim 0.75 \ H_c && \text{for } \chi \gg 1 \\
H_{sh} &\sim 1.2 \ H_c && \text{for } \chi \sim 1 \\
H_{sh} &\sim 1/\sqrt{\chi} \ H_c && \text{for } \chi \ll 1
\end{aligned}
\tag{22}
$$

with smooth transitions of the function $H_{sh}(\chi)$. Although there is some experimental evidence for the relevance of H_{sh} as the rf critical field H_c^{rf} in the case of some type I superconductors like Pb (Hogi et al., 1977) and for the surpassing of H_{c1} in the case of some type II superconductors (Pfister, 1976), even the best experimental results achieved with Nb and Nb_3Sn cavities do not yet attain the theoretical expectation given by H_c or H_{sh} (see Table 1). Since the fundamental limits for H_c^{rf} for type II superconductors is unknown, the most optimistic values are assumed in Table 1 in order to estimate the maximum achievable accelerating field E_a^{max}. A peak magnetic surface field enhancement factor of $H_p/E_a = 45$ Oe/MV/m is assumed. This is a good "rule of thumb" for a "typical" accelerating cavity excited in the TM 010-π-mode. Further investigations are needed, especially for Nb_3Sn, to reduce the gap between experimental results and theoretical promises.

Table 1: Characteristic parameters of the most favoured materials for superconducting cavities.

Material	T_c [K]	Δ_0/kT_c	$H_c(0)$ [Oe]	typical χ	$H_{sh}(0)$ [Oe]	$H_{max}^{exp}(2K)$ [Oe]	$E_a^{max}(2K)$ [MV/m]
Pb	7.2	2.15	800	0.5	1050	900	22
Nb	9.2	1.85	2000	1	2400	1590	50
Nb_3Sn	18.2	2.2	5400	20	4000	1060	120

ANOMALOUS LOSSES

The origin of field limitations well below the critical magnetic surface field and the causes of residual losses are the main areas of interest for research on superconducting cavities. Several diagnostic techniques have been developed to study these questions (for a review see Piel, 1980). The main result of all these investigations is that the performance of superconducting cavities is limited by anomalous losses which are, in general, of a very local nature. We distinquish mainly among three kinds of anomalous losses due to their effect on the loading of the cavity Q_o which finally leads to a field limitation. The first is electron multipacting, which is a resonant multiplication of free electron currents. This effect was a very annoying field limitation, especially in low frequency superconducting cavities, before 1979. Computer simulations by Lyneis et al., (1977) and Klein and Proch (1979), has resulted in the proposal that multipacting should be suppressed drastically by the use of spherically or elliptically (Kneisel et al., 1981) shaped cavities instead of cylindrically shaped (Fig. 1) cavities. This technology has solved the multipacting problem up to the highest fields reached to date, so that today all superconducting accelerating structures for electrons are of spherical or elliptical design. More detailed discussions about multipacting will be found in the review papers of Lyneis (1980), Weingarten (1984) and Piel (1986).

In this chapter the progress achieved on both actual anomalous loss mechanisms, namely quenching induced by defects and non-resonant electron loading by field emission, will be reviewed. These obstacles can be partially overcome by localization and subsequent removal of corresponding defects or by thermal stabilization and improved preparation of the cavity surface.

Temperature Mapping

As each energy loss mechanism will finally lead to an increase of the temperature of the cavity wall, temperature measurements are a powerful means to get a more detailed insight into the causes for field and Q_o limitations. Obviously the temperature increase of the inner cavity surface, ΔT_1, depend on the heat production as well as on the cooling conditions determined by the thermal conductivity, the thickness of the cavity wall and the heat transfer to the helium bath. As a result that will always be a temperature difference ΔT_o between the outer cavity surface and the helium bath due to the Kapitza resistance. Quenching occurs, if a locally enhanced power dissipation (due to a normal conducting defect for example) drives the superconducting environment above its critical temperature. The result is that a macroscopic region of the cavity becomes normal conducting. Consequently, a sharp breakdown of the cavity field and Q_o is observed at a certain threshold field level. In such cases a substantial heat flux develops which leads to film boiling and a marked increase of the temperature of the helium film close to the quench area. Therefore quench areas can be easily detected even in superfluid helium with temperature sensors not in contact with the cavity wall. This technique was first demonstrated at Stanford (Lyneis et al., 1972). The measurement of the spatial distribution of rf losses well below the quench field by temperature mapping of the outer cavity surface, however, reveals much more information about the nature of high loss areas.

A very important breakthrough was achieved by the development of the temperature mapping technique in subcooled helium (Piel and Romijn, 1980), i.e. at a bath temperature slightly above the λ-temperature (2.2K) and a bath pressure of about 1000 mbar. In such a subcooled bath, bubbles are absent and therefore microconvection produced by bubbles rising from the heated surface is avoided. This reduces the cooling capability of liquid helium substantially and enhances the temperature increase at the outer cavity surface. Furthermore a convection stream from the warmed-up helium layer builds up which is not disturbed by bubbles as in the case of nucleate boiling helium.

Since its first application on a spherically shaped 500 MHz single cell cavity at CERN (Bernard et al., 1980), temperature mapping in subcooled helium has become a standard tool for the investigation of anomalous losses in superconducting cavities. Moreover, this technique is of great practical importance for the quality control of multicell cavities, the performance of which is limited by the worst cell. As an example a recently built thermometer scanning system for the superconducting accelerating structure proposed for the CEBAF project (Leemann, 1986) is shown in Fig. 5. Eighty nine carbon thermometers (100Ω, 1/8 W Allen Bradley) can be rotated around the cavity sliding on the surface while under spring tension. The resistance values and the angular position are read by a computer controlled data acquisition system which takes a full scan of the cavity surface (5° step width) in about 15 minutes. In Fig. 6 a typical temperature map measured with this system is shown. Beside a background of more or less homogenous losses with respect to the magnetic surface field distribution, clearly several regions of enhanced losses are present. These are normally associated with microscopically small defects and can be localized from the temperature map to within a few cm^2.

For a quantitative investigation of the energy loss mechanisms from

Fig. 5.   General configuration and close up view of the
carbon thermometer for a 5 cell 1.5 GHz cavity.

temperature maps at different field levels the measured temperature
increase of the resistor thermometers has to be calibrated against the
heat flux density at the outer cavity surface. Obviously the reduced
efficiency of unshielded thermometers, which measure only a fraction of
the real surface temperature increase, must be considered as well as the
local heat convection flow. These influence the cooling conditions in
subcooled helium. Therefore, careful calibration measurements have been
performed using a special heater set up (Romijn et el., 1983) or the
scanning system has been calibrated in situ at higher temperatures, where
the BCS losses dominate (Huppelsberg, 1985). From the measured spatial
distribution of the heat flux density (a so called Q-map) conclusions
about the rf losses inside the cavity can be made, if the thermal
conductivity and the thickness of the cavity wall are taken into account.
First results of these investigations have been reviewed at the last
workshop on rf superconductivity (Mueller, 1984).

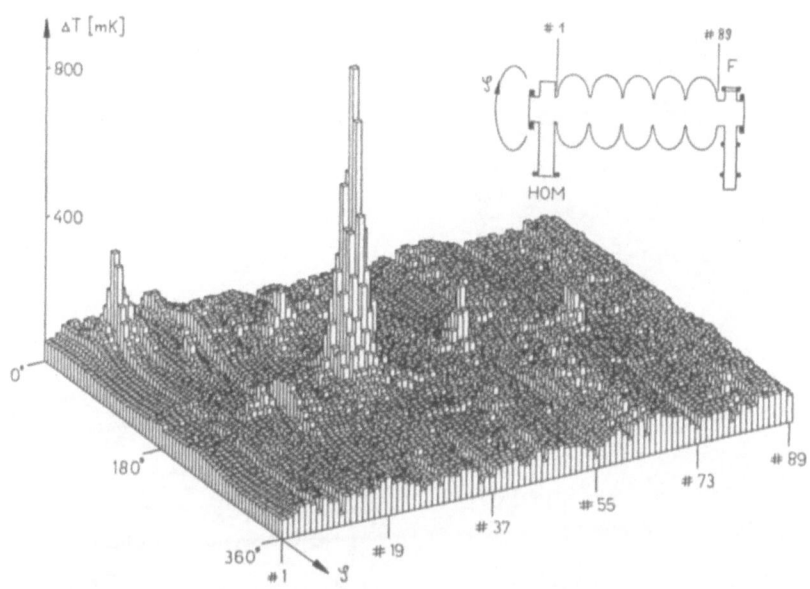

Fig. 6.   Temperature map of a 5 cell 1.5 GHz cavity measured at E_a= 5.6
MV/m in subcooled helium (2.5 K).   The maximum temperature
increase of ΔT = 900 mK occurs in the second cell (No. 35 at ϕ =
150°) (Bensieck et al., 1986).

The usefulness of the temperature mapping technique in subcooled
helium is limited by being restricted to bath temperatures above 2.2 K.
For Nb cavities the resulting BCS losses (Eq. 16) cause in subcooled
helium a maximum achievable surface field of about 250 Oe at 3 GHz and
scales inversely with the rf frequency.   Therefore the development of
improved thermometers for the extension of temperature mapping into the
superfluid helium regime is under way (Kneisel and Mueller, 1985).

Classification of Defects

In a wide sense defects are all those irregularities of a real
cavity surface, which lead to additional losses when compared to an ideal
superconductor.   Since the systematic application of temperature mapping
we have learned a lot about the physical nature of defects by the
subsequent inspection of high loss regions.   Some impressive photographs
of defects in quench areas (Fig. 7 a-c) have been obtained in a series of
experiments on 3 GHz single-cell cavities at CERN (Padamsee et al.,
1983).   The photographs were taken with a scanning electron microscope on
specimens cut from the corresponding wall pieces.

Such defects are typically between 50µm and 1 mm and cause quenching
of reactor grade Nb cavities at accelerating field levels from 2 to 10
MV/m.   Generally higher field levels are achieved for smaller defects,
but there is also a dependence on the location and the physical nature of

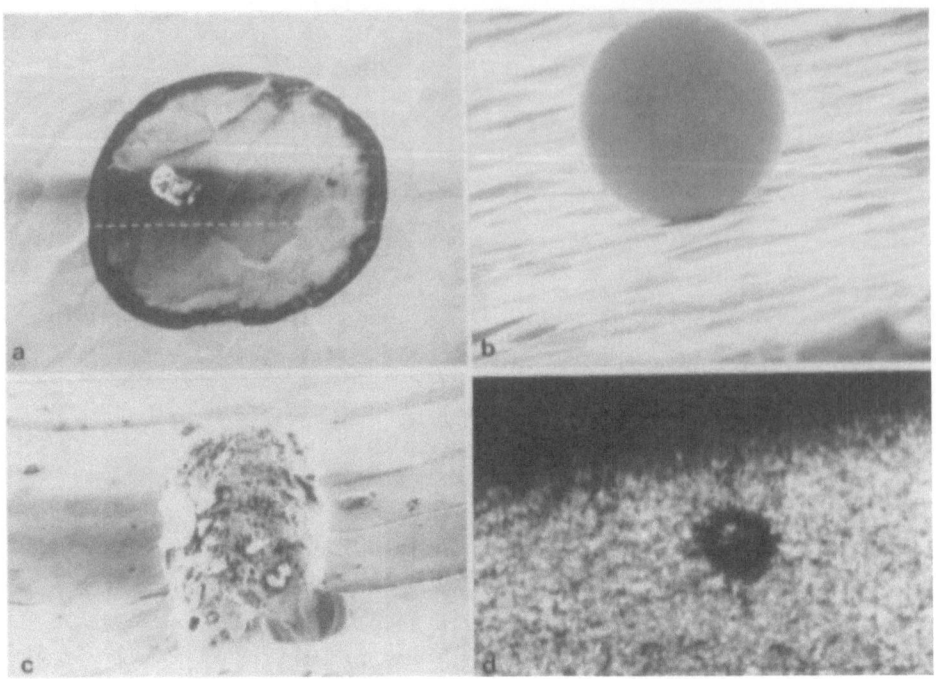

Fig. 7 a) 440 μm diameter drying from chemical residues;
b) 80 μm Nb sphere stain presumably originating from
a welding bead; c) 360 μm Tungsten inclusion produced
by TIG welding; d) Central spot with 1.5 mm diameter
halo, which was produced during quenching of a 500 MHz
cavity (Bernard et al., 1983).

the defects. Various kinds of defects have been identified, most of
which are of a rather trivial nature. Two main classes of defects can be
distinguished, namely accumulations of foreign material and geometrical
irregularities of the cavity surface. Examples of the first type are
chemical residues, dust particles and particles inclusions pressed into
the surface or segregated from the bulk (Fig. 8). The other class of
defects is assumed to result from bad cooling conditions or field
enhancement. Welding beads, holes with sharp edges, whiskers, scratches,
microfissures and delaminations have been found.

The identification of defects in high loss regions has given
valuable hints for the improvement of the fabrication and surface
preparation techniques for superconducting cavities. Examples are the
careful inspection of the raw material, defocused electron beam welding,
tumbling or grinding of the welds, final rinsing procedures with

534

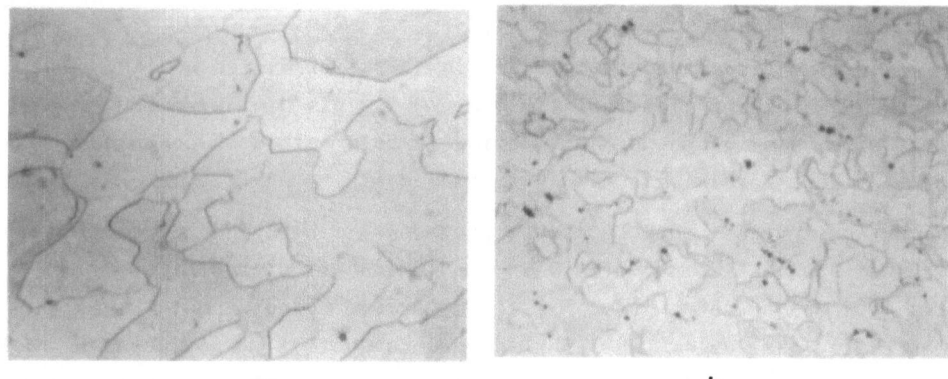

a	b

Fig. 8. Microscopic inclusions of foreign material of a few μm diameter for material of a) high and b) low purity. Inclusions were made visible by etching the mechanically polished Nb samples.

demineralized and dustfiltered water, and assembly of the cavities in clean rooms. All of these steps help to achieve high residual Q_o values and high accelerating fields more reliably-especially in multicell cavities. In case of a large defect the field level of a cavity can be gradually increased by guided repair using local grinding or local chemical treatments.

Thermal Stabilization

Increased quench field levels can be achieved not only by the reduction of the number or the size of defects but also by their thermal stabilization. Model calculations (Padamsee, 1982) have shown that the threshold field H_q for a thermal instability caused by a large normal conducting defect of radius r_D and surface resistance R_D scales approximately as

$$H_q \propto \left(\frac{\lambda(T_c - T_B)}{r_D R_D}\right)^{1/2} , \tag{23}$$

where T_B is the bath temperature and λ is the thermal conductivity of the cavity wall. In the superconducting state Cooper pairs do not contribute to heat transport. Therefore the electronic thermal conductivity increases strongly for increasing temperature and dominates λ above Tc/2. At lower temperatures phonon conductivity can become important. The electronic thermal conductivity is proportional to the residual resistivity ratio RRR = $\rho(300k)/\rho(4.2K)$ of the normal conducting metal. For niobium the convenient relationship

$$RRR \approx 4 \ \lambda(4.2) \ mK/W \qquad\qquad\qquad (24)$$

can be derived (Padamsee, 1984). In Fig. 9 the predicted scaling of H_q towards an assumed fundamental limit of 2000 Oe is displayed for reduced defect sizes and increased RRR. Obviously both methods become less efficient for increasing breakdown field levels.

During the last few years much progress has been made concerning the thermal stabilization of defects in niobium cavities (see Fig. 10). Until 1983 only reactor grade Nb with a maximum RRR of about 40 was commercially available. The poor thermal conductivity was caused by high impurity levels from interstitially dissolved 0, C and N atoms (Schulze, 1981). These impurities can be controlled to a great extent during the electron beam melting of the raw niobium and during consecutive manufacturing steps of the sheet material. Due to a refinement in production techniques, especially multiple electron beam melting of ingots under improved vacuum conditions, industry can now deliver Nb sheets up to RRR $\approx$ 200. A further effective procedure developed at Cornell to clean niobium cavities, especially of the most critical oxygen impurity, is solid state gettering with yttrium (Padamsee, 1983) or titanium (Kneisel, 1984). During this process a yttrium or titanium film is vapor deposited on the niobium surface from a surrounding foil at a temperature of 1250°C and 10^{-5} mbar pressure. This surface layer acts as a sink for the interstitial diffusion of impurities due to its higher affinity, and can afterwards be chemically dissolved. Depending on the

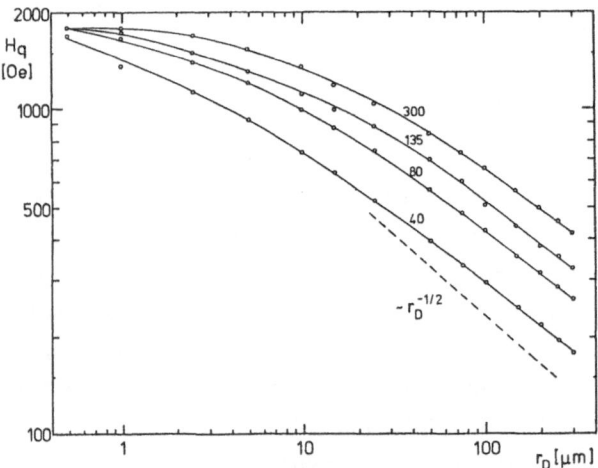

Fig. 9. Dependence of the quench field on defect size for 3 GHz niobium cavities (typical $R_d = 8 \cdot 10^{-3} \Omega$) of different RRR at T_B = 1.4 K. Results are from advanced model calculations (Elias, 1985).

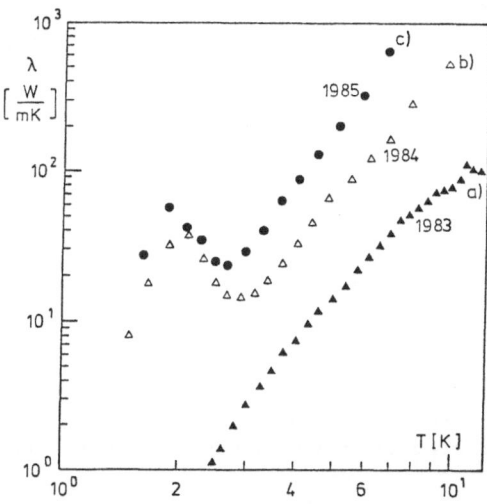

Fig. 10. Temperature dependence of the thermal conductivity of typical
samples from a) reactor grade, b) multiple electron beam melted
and c) yttrified, niobium.

initial impurity contents, RRR values approaching 600 have been obtained
using this purification technique.

The influence of the thermal conductivity of niobium on the
performance of superconducting cavities has been proven in several ways.
The easiest way is to compare statistical results achieved with a certain
type of cavity when fabricated from niobium with different purity levels.
In Fig. 11 this is done for all measurements on single-cell 3 GHz
cavities of spherical shape excited in the TM_{010} mode. This work was
performed at CERN and Wuppertal (Lengeler et al., 1985). Each data point
corresponds to a different cavity or to a fresh cavity surface obtained
after chemical polishing of about 20 μm. The expected dependence of the
achieved accelerating fields on the RRR of the niobium can clearly be
seen. Similar results have been obtained in a series of measurements on
eight X-band elliptical cavities with RRR values between 25 and 1400
(Padamsee, 1985). The higher purity of the niobium was produced by
outgassing at high temperatures under ultra high vacuum conditions.

The virtue of high thermal conductivity has been investigated more
directly by the post-purification of a 3 GHz single-cell cavity, which
originally was fabricated from medium purity niobium (RRR = 80). Quench
field levels in four tests of at most E_a = 12 MV/m were observed. After
its yttrification at Cornell the same cavity (RRR=350) is no longer
limited by defect induced thermal instabilities up to about 23 MV/m.
Heavy electron loading by field emission begins at this level (see Fig.
12).

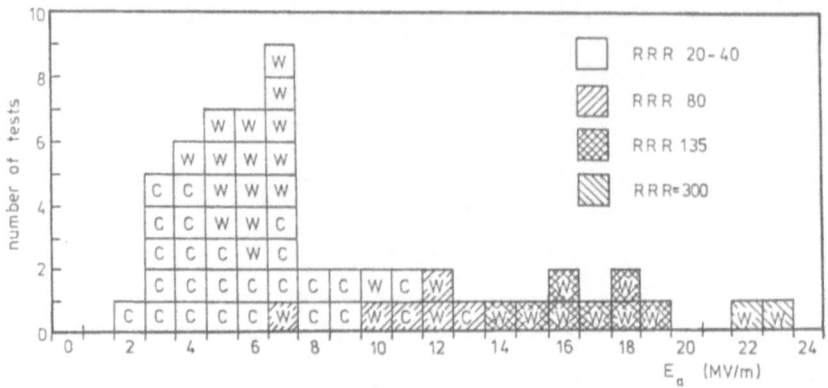

Fig. 11.   Performance of superconducting 3 GHz single-cell cavities
fabricated from niobium of different purity levels.

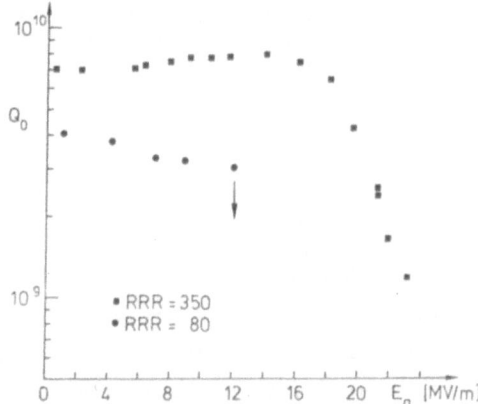

Fig. 12. Q_o versus E_a dependence of a 3 GHz single cell cavity before
and after yttrification

The most important issue associated with commercially available high
purity niobium can be seen from the improvement of multicell cavity
performance. Not only have the obtainable fields increased, but also the
reliability with which the present design fields of Ea = 5 MV/m can be
reached at all frequencies. This is due to the thermal stabilization of
the remaining defects.

Field Emission

Since quenching due to large defects has been shifted reliably
towards higher accelerating field levels (often above 10 MV/m which
corresponds to electric surface fields typically more than 25 MV/m),
field emission induced electron loading constitutes another important
field limitation of superconducting cavities. One of the very first
temperature maps obtained at CERN (Bernard et al., 1980) proved the

existence of point like electron sources which emit at anomalously low electric surface fields (below 10 MV/m) (see Fig. 13). Each significant electron emitter can be clearly identified by the line like loss regions resulting from the impacting electrons. These electrons stay in the same azimutal plane of electric field lines but follow different trajectories due to varying rf phase. Most of the energy of the impacting electrons is converted in heat. However, a few percent cause X-rays which can also be detected.

More detailed information about the individual electron emitters can be extracted from the quantitative analysis of such temperature maps together with electron trajectory calculations (Weingarten, 1984). Not only the location of an electron emitter but also its local field enhancement factor β and emitting area A can be determined in this way. The measured field emission currents I_e can be described by the Fowler-Nordheim relation

$$I_e \propto A\, E^{2.5}\, e^{-1.5\phi/\beta E} \,, \tag{25}$$

where ϕ is the work function and β reflects enhanced emission compared to an ideal surface. Because of the exponential increase of the electron current for increasing fields, the loading of the cavity Q_o is normally dominated by only one emitter. In this case β can be determined alternatively from integral quantities like the Q_o degradation, the X-ray intensity measured inside or outside the cryostat, or the electron current picked up by a probe inside the cavity. For superconducting cavities typical values of β between 100 and 500 have been measured.

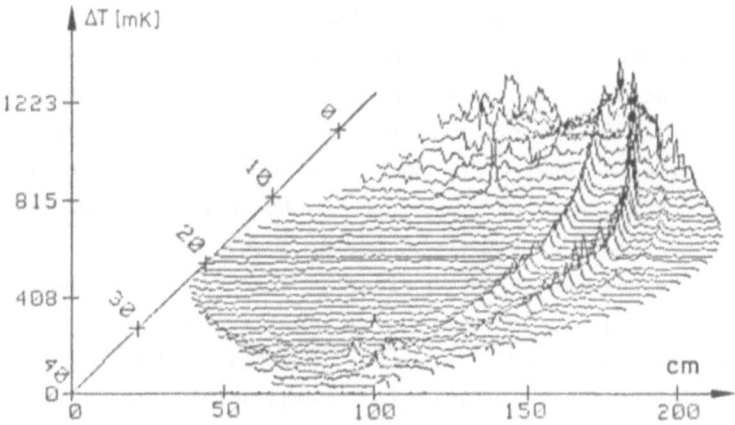

Fig. 13. Temperature map of a 500 MHz single cell cavity at $E_a = 3.2$ MV/m showing line like loss regions typical of electron emission from pointlike sources.

Normally lower values correspond to higher field levels. Experiments with multimode reentrant cavities have shown that there is no intrinsic frequency dependence of β (Klein and Turneaure, 1983). Nevertheless stronger field emission loading is usually observed in low frequency cavities which is to be expected from statistical models because of the larger cavity surface area.

The origin of the anomalous enhanced field emission in superconducting rf cavities is still unknown. A correlation with the dc field emission from broad area cathodes (Lathan, 1981) can be assumed. At the University of Geneva the dc field emission properties of niobium samples prepared in a similar way as a cavity surface has been investigated using a UHV field emission scanning microscope. Parallel electron microscopy and microfocus Auger analysis further clarifies the physical nature of the emitting sites (Niedermann et al., 1984). The main result of these and similar (Lathan 1984) investigations is the fact that localized emission sites frequently consist of micron-size particles of foreign elements some of them sitting rather loosely on the surface. A pure electrostatic field enhancement by metallic surface protrusions is generally not found. On the other hand only a few of the millions of particles present are normally active emitters, so that very exceptional conditions must be fulfilled at emission sites. Various elemental compositions of emitting particles have been observed, but relatively often C is involved. These observations confirm the assumption that dust particles are one of the most probable sources of enhanced field emission in superconducting cavities.

Systematic investigations of the effect of high temperature treatments on the field emission from niobium samples have recently uncovered a special type of field emitters, the density of which can be varied by different annealing temperatures (Niedermann et al., 1986). In Fig. 14 the number of field emitting sites on a 1 cm^2 Nb sample which exceed a threshold current of 40 nA at a field of 9 MV/m is given for three consecutive cycles of heat treatments. The samples were not removed from the UHV system for 30 minutes at all temperatures. Surprisingly the density of sites for the as introduced sample, increases up to heat treatments at 800°C whereafter a sharp decrease is observed. Emission free surfaces can be obtained by annealing at temperatures above 1200°C. The activation or creation of field emitters by heat treatments at temperatures around 800°C is clearly demonstrated by the reappearance of a large number of sites during the repetition cycles. The sites consist mostly of C, S or Mo. This spectacular result can be attributed to the surface segregation of impurities which diffuse into the bulk at very high temperatures.

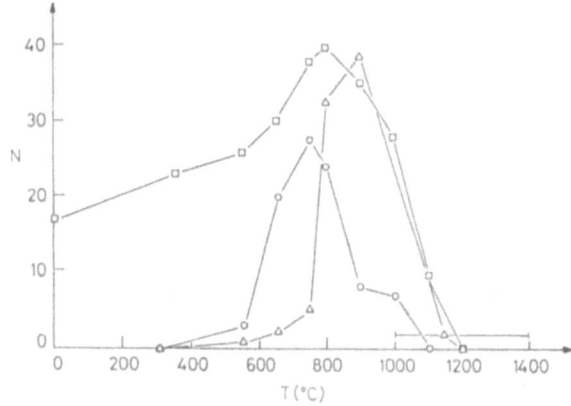

Fig. 14. Number of field emitting sites as a function of the heat treatment temperature for a Nb sample during the first (□), second
(Δ) and third (o) thermal cycle. (Sankarramann et al., 1986).

All these observations have given valuable hints for the possible
reduction of the field limitation of superconducting cavities due to
field emission. Obviously dust-filtered chemicals and demineralized
water as well as clean room environments during assembly of the cavities
must be applied to reduce the number of micron-size surface contaminants.
Moreover the use of high purity material may be beneficial by reducing
field emission due to impurity segregations. The production of field
emission free surfaces by high temperature firing of cavities is quite
challenging but not always feasible. Last but not least, in situ helium
processing of superconducting cavities is a well-known powerful technique
(Schwettman et al., 1974) for reducing the current from field emission
sites. It becomes much more difficult to apply for increasing field
levels.

Progress in Single-Cell Cavity Performances

Summarizing the knowledge gained on the anomalous loss mechanisms of
superconducting cavities and the discussed cures, the present state of
the art can be well demonstrated by the best results of single cell
niobium cavities achieved at different laboratories. The results given
in Fig. 15 are restricted to the time period after the introduction of
spherically or elliptically shaped cavities for three reasons. First,
this cavity shape is exclusively used today to avoid multipacting.
Moreover, modern cavities are fabricated from sheet metal to reduce costs
contrary to former practice, when cavities had been machined from bulk
material. Last but not least, it should be mentioned that the best
results published during the seventies (for a review see Citron, 1980)
were either not reproduced or only for a very small percentage of large

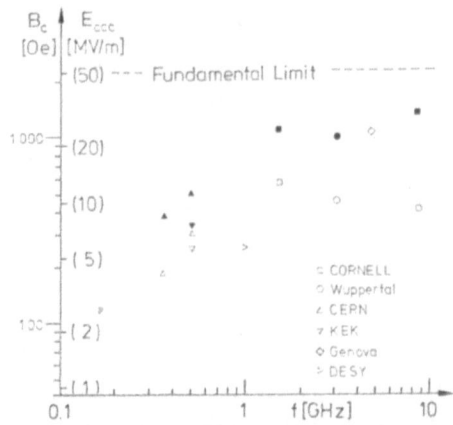

Fig. 15. Maximum achieved magnetic surface fields in single-cell
          cavities at different frequencies built from low purity
          (RRR≤40, open marks) and high purity (RRR≥100, full marks)
          niobium sheet. The scale of the corresponding accelerating
          field of multicell cavities is only approximatively valid.

numbers of cavities. Today, good results are obtained much more
reliably.

The obvious dependence of the best results at all frequencies on the
purity of the niobium illustrates the benefit of improved thermal
conductivity for progress in cavity performance. Furthermore, the field
limitation for low purity niobium cavities is generally governed by
quenching. Some of the values for high purity cavities are already
limited by field emission loading. This observation can be explained by
improved thermal stabilization of defects as well as by the assumption of
a reduced number or size of defects. This observation is supported by
increased reliability for high cavity performance. At frequencies above
1 GHz, more than half of the expected fundamental field limit has been
achieved with single-cell cavities. Excellent progress has been made at
lower frequencies so that accelerating field levels of 10 MV/m are within
reach at all frequencies.

It should be mentioned that at high field levels, high Q_o values
become essential to keep cryogenic losses low. Residual Q values above
10^{10} have been achieved with single-cell cavities at all frequencies.
However, the niobium cavities must operate at temperatures below 2 K.
Coated, spherically shaped Nb_3Sn single-cell cavities have achieved
sufficient Q_o values at 4.2 K. Maximum operating fields have not
exceeded 7 MV/m. Further improvements in the performance of coated Nb_3Sn
are possible by thermally stabilizing defects and improving cavity
cleanliness. Further work is necessary.

SUPERCONDUCTING ACCELERATING SYSTEMS

Design Criteria for Accelerating Structures

Superconducting accelerator structures for velocity of light particles usually consist of a chain of weakly coupled cavity resonators operated in the highest TM_{010} passband mode. In this so called π-mode, the accelerating field is equal in magnitude and apposite in direction in the individual cells (see Fig. 1). The length of each cell for maximum acceleration of $\beta = 1$ particles is just half of the wave length. The field flatness of such a π-mode structure depends sensitively on the tuning errors of the individual cells and scales inversely with the square of the number of cells. Therefore, short total lengths less than 2 m are preferred for actual designs of superconducting structures.

The rf power per unit length, P, needed to establish an accelerating field in an unloaded cavity is characterized by the shunt impedance per unit length, r, of the accelerating structure

$$P = E_a^2/r \; . \tag{26}$$

The power per unit length is proportional to Q_o. The specific shunt impedance r/Q_o depends on the cavity shape and scales linearly with frequency due to the increasing number of cells per unit length. For spherically or elliptically shaped cavities with large beam apertures, typical values of

$$r/Q_o \approx 700 \; \Omega/m \cdot f/GHz \tag{27}$$

are obtained. The dissipated power per unit length can therefore be estimated therefore by

$$P = \frac{E_a^2}{(r/Q_o)Q_o} \approx \frac{E_a^2 R_s}{700\Omega/m \cdot f/GHz \cdot 290\Omega} \; , \tag{28}$$

where Q_o has been replaced by (Eq. 18). High frequencies are favoured for the operation of normal conducting linear accelerators, since the surface resistance at room temperature scales with the square root of the frequency (Eq. 5). By contrast, for superconducting structures (Eq. 28) results in $P \propto f$, and low frequencies are preferred as long as R_s is dominated by the BCS surface resistance (Eq. 15,16), as it is the case for niobium at 4.2 K. At temperatures below 2 K and for Nb_3Sn, scaling of the dissipated power will depend more on the frequency dependence of the residual resistance, if the presently achieved R_{res} values of 10–100 nΩ cannot be reduced drastically.

This important difference between normal conducting and super-conducting accelerating structures becomes interesting for those large scale applications where the dissipated rf power contributes significantly to the operating costs of a machine. For storage ring cavities at a typical frequency of 500 MHz, copper structures with $Q_o \approx 50,000$ can be operated in a continuous wave mode up to accelerating field levels near 2 MV/m. The dissipated power resulting from (Eq. 28) (1.4 MW/m for E_a = 5 MV/m) becomes too high for practical use. A 500 MHz superconducting structure with $Q_o = 3 \cdot 10^9$ consumes (at E_a = 5 MV/m) a cw power of only 24 W/m. This level must be corrected for the Carnot and technical efficiency of a refrigerator by a factor of about 500 for 4.2 K operation. Nevertheless, much less main power is needed for super-conducting accelerating systems and the available rf power can be concentrated to the beam.

The main advantage of superconducting accelerating structures is the fact that the dissipated power (Eq. 28) is dominated by the value of Q_o and no longer by the specific shunt impedance of the cavity. There-fore r/Q_o can be treated as an almost free parameter for the optimization of the cavity shape with respect to performance limitations and the interaction with high current beams. As a consequence different design criteria have been developed for superconducting accelerating structures when compared to normal conducting ones. Nonsuperconducting cavities are optimized with respect to high r/Q_o values.

For superconducting accelerating structures, the design of choice is a tapered spherical or elliptical shape for the individual cells without nosecones. In addition to the avoidance of multipacting, such cavity shapes obviously reduce the risk of quenching or enhanced field emission at low field levels due to residues from surface treatments. Further-more, the cells are usually kept free from any ports for couplers, tuners or probes to avoid field distortions, which can lead to multipacting. Input and output coupling is therefore provided from the beam tubes.

Additional design criteria for the acceleration of high currents are somewhat more complex. The main problem is to keep the interaction between the structure and the bunched particle beam under control. The suppression of multi-pass instabilities is best controlled by external loading of all dangerous higher order modes with special HOM couplers. High frequency eigenmodes of the structure are dangerous. The Fourier spectrum of the beam can excite resonantly high fields, which may lead to longitudinal or transverse beam instabilities. As a minimum it leads to an unwanted heating of the cavity wall or to field limitations. Unfortunately, sufficient damping of all higher order modes is complicated by the constraint that the HOM couplers are located at the beam tubes. Obviously the best solution to this problem is to use a small number of cells per structure which results in less passband modes. Moreover large iris apertures and beam tube diameters are quite helpful in most cases due to high intercell coupling. Both result in less interaction with the beam as expressed by reduced shunt impedances. Last but not least, higher threshold currents can be achieved in single bunches for larger apertures due to lower transverse impedances for the production of wake fields. This can result in single-pass instabilities. Summarizing it is quite obvious that the beam stability requirements can be fulfilled better with rf cavities at lower frequencies.

The optimization of the design for superconducting accelerating structures depends on the operating conditions for the specific application and is usually performed with computer programs like SUPERFISH (Halbach and Holsinger, 1976), URMEL (Weiland, 1983) and TBCI (Weiland, 1983). As an example, the optimized superconducting cavity design for the LEP storage ring at CERN is shown in the photograph in Fig. 16. The main improvement of this four-cell cavity compared to earlier designs consists in the multimode end-cell compensation provided by the wide portion of the beam tubes (Haebel et al., 1984).

A very similar cavity design (Fig. 17) has been chosen at DESY for a possible energy upgrade of the electron ring of HERA (Proch, 1986). The superconducting accelerating structures will be built either from thick niobium sheet material or from thin niobium covered with an electroplated silver layer. By means of brazed copper tubing the latter alternative should allow pipe cooling due to the high thermal conductivity of silver (Susta, 1984).

One of the most advanced large scale applications of superconducting cavities is presently under development for the CEBAF project (Leemann, 1986). The cw electron linac is based on elliptically shaped five-cell cavities with waveguide couplers (Fig. 18) which have been originally developed at Cornell for the CESR storage ring (Ineisel et al., 1983). This accelerating structure has been tested successfully in CESR (Sundelin, 1985) up to a beam current of 22 mA. This current level is much higher than the 0.8 mA required for CEBAF.

Fig. 16.   352 MHz niobium prototype structure used for the energy upgrade of LEP (Bernard et al., 1985).

Fig. 17.   Copper model of the 500 MHz four-cell structure proposed for
HERA.   TM011 and TM012 modes cause sufficiently high field
levels at the location of the couplers.

Fig. 18.   1500 MHz niobium prototype structure for CEBAF (developed at
Cornell).

In Fig. 19, the 3 GHz accelerating structure for the superconducting
recyclotron at Darmstadt (Heinrichs et al., 1979) is shown. It is a
spherically shaped twenty-cell structure with wide beam tubes and was
developed at Wuppertal. Its large beam tube diameter allows the
propagation of all dangerous higher order modes. Sufficient HOM damping
is expected for the low design current of 20 µA.

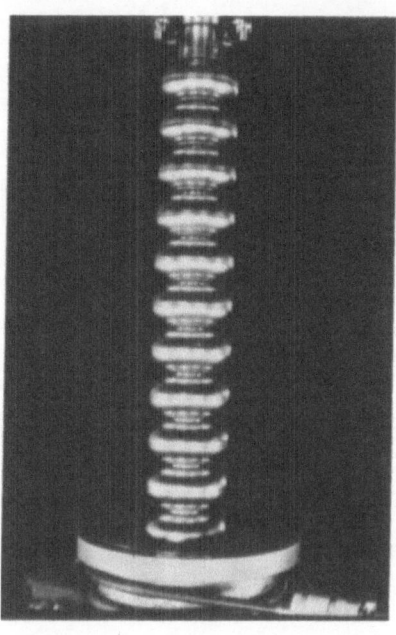

Fig. 19. Sectional view of a twenty-cell niobium structure (3 GHz).

Status of Current Projects

Since the late seventies several new activities have been started to
apply the benefits of superconducting rf cavities to large electron
storage rings such as LEP, TRISTAN, and HERA as well as to cw electron
linear accelerators with multiple recirculation like the Darmstadt
recyclotron and CEBAF.   All of these current projects are based on
multicell accelerating structures of rounded shape (see Fig. 16-19).   In
Table 2 the best results achieved with prototype structures in both
laboratory and beam tests are summarized.

It is not surprising that these best results have been obtained for
cavities built from high purity niobium with typical RRR values between
100 and 140.   One can understand the enthusiasm of the community much
better from the fact that these results have been obtained with only one
exemption (Cornell, Kneisel et al., 1985), on the first attempt.
Moreover, a further increase in the thermal conductivity of multicell
niobium cavities can be achieved.   Sheet material with RRR = 200 is now
commercially available and post-purification methods have been applied
successfully to single cell cavities.   As a consequence design values of
E_a up to 10MV/m and Q_o in the low 10^9 can be explored across the whole
frequency range from 350 MHz to 3GHz.

The beam test results given in Table 2 have been achieved as first
efforts under rather aggravating circumstances in existing machines at
KEK (Furuya et al., 1984), Cornell and Darmstadt (Grundey et al., 1984),

Table 2. Best performances of superconducting accelerating structures for current projects.

Laboratory Accelerator	CERN LEP	KEK TRISTAN	CORNELL CEBAF	WUPPERTAL/ DARMSTADT RECYCLOTRON	
f[MHZ] T[K]	350 4.2	500 4.2	1500 2.0	3000 2.0	
number of cells	4	3	5	4	20
E_a[MHz] Lab. Beam	7.5 –	5.2 4.3	15.3 6.5	12.3 5.6	7.4 –
Q_o(5 MV/m)[10^9]	3.2	0.5	4.0	4.5	1.5

respectively. Less performance degradation can be expected therefore under improved routine conditions for large scale applications. Nevertheless the long term operation of some of these and other superconducting resonators (Shepard, 1984) have shown so far that accelerating fields and quality factors remain stable under the environmental conditions of real accelerators.

In view of these large scale applications, industrial production of superconducting niobium cavities has started at several companies. For the 130 MeV recyclotron presently under construction at Darmstadt (Alrutz-Ziemssen et al., 1986), all accelerating structures are ready for installation. Most of them have been fabricated from stock niobium of low purity (RRR = 40), but on average the design gradient of 5 MV/m can be reached. All of the Cornell type cavities recently built by industry for the CEBAF project have clearly exceeded the specified values of E_a = 5 MV/m and Q_o = $2.4 \cdot 10^9$ (Bensiek et al., 1986). Up to 400 of these structures will be needed for CEBAF, the approval and funding of which is expected soon. For large scale application of superconducting cavities in storage rings LEP, TRISTAN and HERA some parameters have not yet been fixed. In any case, it is planned to install a few superconducting structures in the first stage of these accelerators. At CERN in parallel with the construction of a fully equipped accelerating module based on a high purity niobium structure, a development program of niobium coated copper cavities (Benvenuti et al., 1985) is under way. This technology is considered as an alternative because of cavity costs (Lengeler, 1986).

At DESY two different prototype accelerating modules using bath cooling or pipe cooling of the superconducting cavities are presently under construction and will be tested next year in PETRA (Proch, 1986). Last but not least, a new project to employ superconducting 500 MHz single-cell cavities for a free electron laser has been started in a collaboration between HEPL and TRW.

Since the design accelerating field of 5 MV/m can be achieved reliably with cavities built from high purity niobium, more efforts are now spent on the optimization of all components which contribute significantly to the costs of superconducting accelerating systems -- especially couplers, tuners and cryostats.

The main requirements on couplers for superconducting cavities are low rf losses in parallel with low cryogenic losses even at high power levels. Simple mechanical layouts should be preferred in order to minimize vacuum failures and contamination of the superconducting cavity. Generally wave guide couplers can be considered as well as coaxial couplers. The latter have to be chosen for frequencies below 1 GHz to stay within a reasonable size for cryostats. At higher frequencies a coaxial coupler design would simplify the construction of the cryostats and the cavities. Therefore the development of coaxial beam tube couplers is highly desirable for superconducting accelerating structures.

Fundamental mode couplers have to supply up to 100 kW rf power to the high current beams of storage rings. They supply about 1 kW rf power to low current cw linear accelerators. This corresponds to external Q values in the range 10^5 to 10^8 depending on the special application, illustrating that the band width of the superconducting accelerating structures will remain somewhat smaller than for normal conducting cavities. At CERN, coaxial fundamental mode couplers capable of the power needs for LEP have been developed (Haebel, 1984), which make use of cylindrical rf windows (see Fig. 21). It is hoped that such windows located outside of the cryostat will show much increased lifetimes when compared to standard coaxial or rectangular windows. Window failures will be more troublesome for superconducting cavities.

The maximum power which can be deposited by the beam in the higher order modes of the accelerating structures has to be reduced by a strong damping of these modes with HOM couplers. If the bandwidth $\Delta f \propto 1/Q_{ext}$ of HOM resonance is small compared to the repetition rate of bunches, the dissipated power for each HOM will be maximally (Haebel and Sekutowicz, 1986)

$$P_{max} = I_B^{\,2} \, (r/Q_o) \, Q_{ext} \, , \tag{29}$$

where I_B is the beam current and r/Q_o is the specific shunt impedance of the HOM. Therefore low external Q values have to be provided for all dangerous (high r/Q_o) modes by additional HOM couplers, which must not load the fundamental mode. Waveguide couplers with their low frequency cut-off offer an easy microwave, but complex cryogenic solution. Therefore compact coaxial HOM couplers have been developed recently by Haebel and Sekutowicz (1986) for the superconducting LEP and HERA prototype structures. In Fig. 20 the most advanced DESY version is displayed, which is based on a tunable parallel filter at low current level instead of the usual series filter for the fundamental mode rejection. This coupler offers the most convenient cryogenic solution, but the trade-off is the rather tight tolerance condition for the filter. Measurements on copper models have shown that similar low Q_{ext} for waveguide couplers can be achieved, which are already smaller than those of naturally damped copper cavities. Therefore superconducting accelerating systems with such reduced HOM shunt impedances will fulfill the beam stability requirements of LEP and HERA better than normal conducting ones.

The frequency tuning of all presented accelerating structures is based on the change of the cavity length L and provides typically

$$\Delta f/f \sim (0.3 \pm 0.2) \; \Delta L/L . \tag{30}$$

Usually two-stage tuning systems are considered in which the totally needed tuning range is covered by a slow coarse tune, while the required precision is achieved by a fast fine tune. The methods usually discussed are motordrives or thermal expansion systems for the coarse tuning and piezoelectric or magnetostrictive rods for the fine tuning.

The cryostats for large scale applications are always based on a modular design to achieve high flexibility. At CERN a straightforward approach for a cryostat with a very high accessibility to all critical parts has been constructed for LEP (Stierlin, 1986). The heart of this cryostat (Fig. 21) is a "skin-shaped" helium vessel welded around the cavity, the length of which is stabilized by three magnetostrictively and thermally tuned rods. The thermal radiation shield consists of gas-cooled copper sheets surrounded by superinsulation. An aluminum alloy supporting frame wrapped in a thin stainless steel sealing envelope serves as an easy demountable vacuum tank. This cryostat can be opened without removal from the beam line, and no axial movement is necessary to exchange cavities. Reasonable cryogenic losses have been measured with the first prototype of this cryostat (Fig. 22).

Summarizing, all major components for superconducting accelerating systems have been developed, but further improvements and simplifications

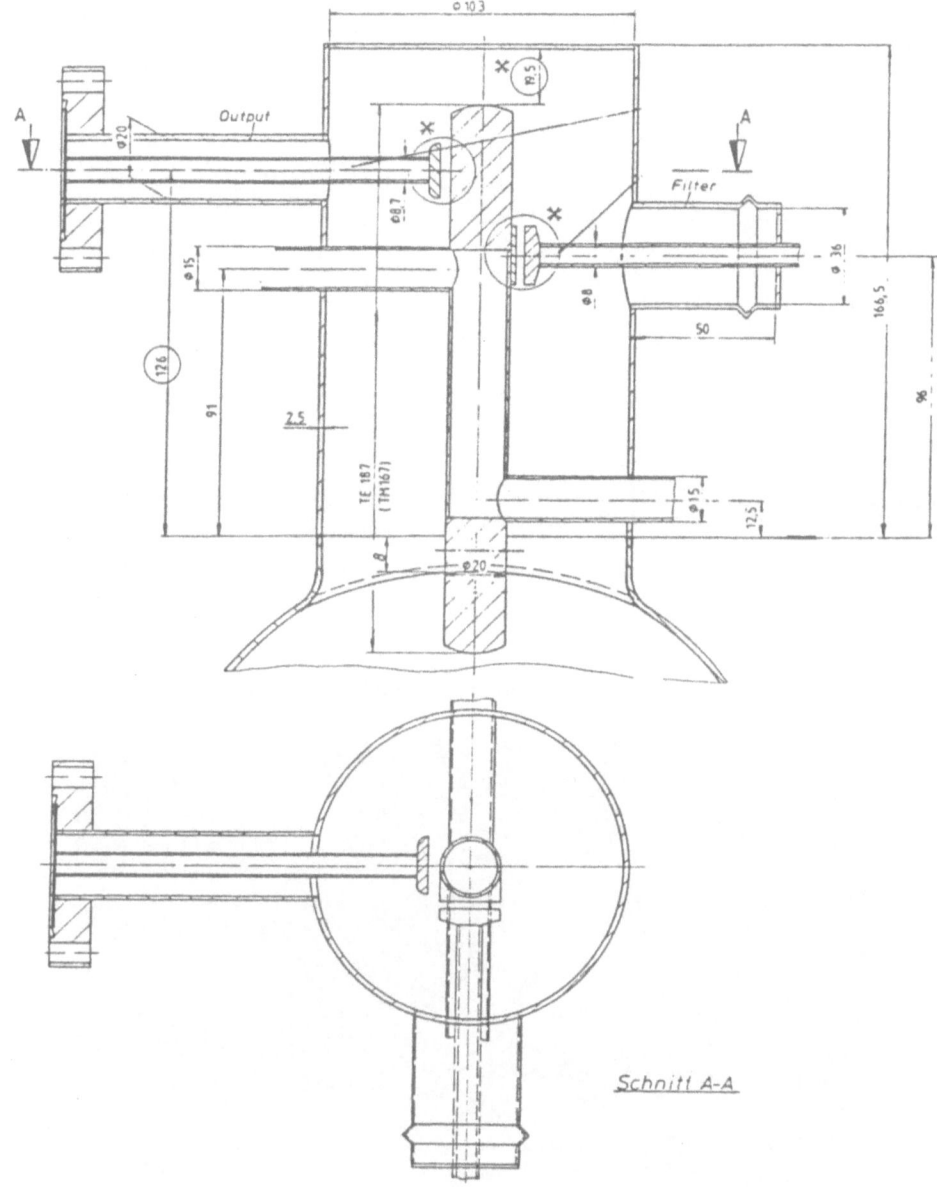

Fig. 20. Coaxial HOM coupler developed at DESY. Three couplers with different orientations of the stubs provide sufficient damping of all modes.

will be welcomed for all components especially items concerning the reduction of component costs.

High Brightness Accelerators

In the previous chapter it has been shown that superconducting accelerating structures offer some outstanding features such as high cw accelerating gradients and low rf power loss, especially at low

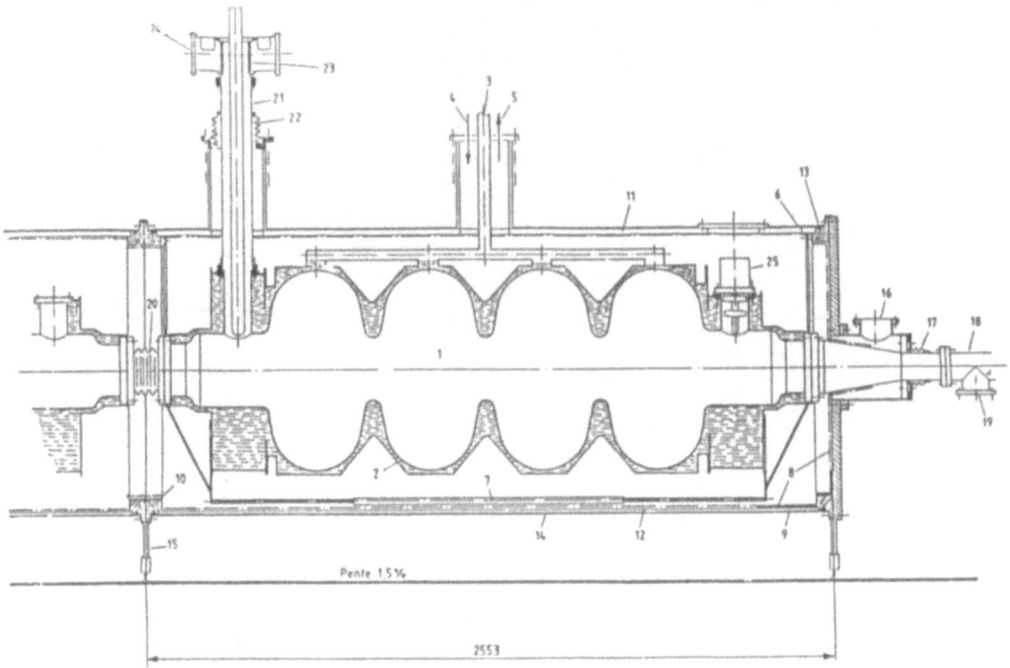

Fig. 21. Cryogenic module of the superconducting accelerating system
for LEP.

Fig. 22. Photography if the first prototye cryostst for LEP.

frequencies. Therefore the question should be discussed; are superconducting accelerating systems suitable for high brightness
accelerators? More precisely, are they suitable for the acceleration of
high currents while preserving beam quality and beam stability? Beam
quality must be considered for most applications in six-dimensional phase
space, i.e. low transverse emittances and low energy spread are both

important. Since the background of these parameters will be described in much more detail by other lecturers at this Institute, only a few arguments will be given here concerning the influence of superconducting cavities on the brightness of electron beams.

The first set of advantages arises from the actually available high accelerating fields. Obviously this allows fewer cavities for a given beam energy resulting in less beam cavity interactions. Moreover, these high field levels can be achieved at low frequencies which are favourable in the case of superconducting niobium cavities because of the reduction of rf losses at low frequencies (28). Less cavities per unit length, as well as large beamtube diameters, will drastically reduce beam instabilities. Lower frequencies lead to the attractive solution of having less cells per structure. This reduces the number of passband modes and makes their damping from the beam tubes of the structures much easier. Because of these reasons the use of high gradient single cell resonators has been suggested for high brightness accelerators (Schwettman et al., 1985).

An important advantage of superconducting cavities in the interesting frequency range of 300 MHz to 3 GHz is the fact that the cavity shape can be optimized for minimum interaction with the beam. The minor importance of the specific shunt impedance r/Q_o compared to its decisive importance in normal conducting cavities allows effective optimization. Therefore superconducting cavities of rounded shape and with large beamtubes and iris apertures should result in much higher threshold currents for beam instabilities.

Continuous wave mode operation results in further advantages for beam quality. Alternatively for applications where high peak currents are important (as for instance free-electron lasers) each rf bucket is filled with a reduced number of particles. This should allow higher average currents combined with low emittances. Furthermore cw feedback systems could be employed to reduce the energy spread of the beam. An example of these possibilities is given by the unique beam properties of the Stanford Superconducting Recyclotron (Lyneis et al., 1981), where an energy resolution of $\Delta E/E \sim 10^{-4}$ has been reached. For further improvements of beam quality, harmonically resonant cavities have been cossidered (Hess et al., 1985).

Obviously superconducting cavities have great potential for application in high brightness accelerators. Low frequency structures should be preferred for high power beams. For these applications the overall efficiency becomes an important parameter. Superconducting cavities are very competitive in that respect, especially if energy recovery from the beam within the same structures can be achieved.

CONCLUSION

During the past few years superconducting accelerator technology has advanced towards large scale applications by considerable improvements in cavity performances. Since the solution of the multipacting problem by the use of cavities of rounded shape, high accelerating fields can be achieved at low frequencies. Much progress has been made concerning the understanding of field limitations due to anomalous loss mechanisms using improved diagnostic techniques. Quenching caused by defects has been successfully reduced by improved fabrication and surface preparation methods for the cavities. Based on commercially available high purity niobium, accelerating fields up to 10 MV/m and quality factors of $3 \cdot 10^9$ are at present reasonable design values for superconducting accelerating structures.

Further improvement of accelerating field levels will require more research on thermal stabilization, the avoidance of defects and on field emission sources in the cavities. Higher Q_o values are quite desirable especially at high field levels and operating temperatures of 4.2 K. Further investigations on high T_c superconductors like Nb_3Sn is desirable.

Superconducting accelerating systems are expected to be superior to room temperature systems for high brightness beams. Large scale applications in electron storage rings like LEP, TRISTAN and HERA as well as in linear accelerators like CEBAF are foreseen. All necessary components for the accelerating structures like couplers, tuners and cryostats have to be optimized to reduce cost. As a result of this progress, the production of superconducting accelerator modules based on high purity niobium cavities by industry has started.

REFERENCES

Abrikosov, A.A., 1957, Zh. Eksp. Teor. Fiz., 32:1442.
Abrikosov, A.A., Gorkov, L.P., and Khalatnikov, I.M., 1958, Sov. Phys. JETP 8:182.
Alrutz-Ziemssen, K., Gräf, H.D., Huck, V., Hummel, K.O., Kaster, G., Knirsch, M., Richter, A., Schanz, M., Simrock, S., Spamer, E., Titze, O., Grundey, Th., Heinrichs, H., and Piel, H., 1986, Proc. of the Lin. Acc. Conf., Stanford.
Arnolds-Mayer, G., and Chiaveri, E., 1986, CERN/ER/RF 86-2.
Axel, P., Cardman, L.S., Gräf, H.D., Hanson, A.O., Hoffswell, R.A., Jamnik, D., Sutton, D.C., Taylor, R.H., and Young, L.H., 1979, IEEE Trans. NS-26:3143.
Bardeen, J., Cooper, L.N., and Schrieffer, J.R., 1957, Phys. Rev., 108:1175.

Bensiek, W., Garner, G., Hager, J., Biallas, G., Chargin, A., Moss, B., Parkinson, J., Amato, J., Brawley, J., Coulombe, D., Morse, D., Heidt, A., Herb, S., Kirchgessner, J., Nakajima, K., Padamsee, H., Palmer, F., Phillips, H.L., Reece, C., Sundelin, R., Tigner, M., Schulz, K., Klein, U., Pallussek, A., Vogel, H.P., Conner, E., Loer, S., Weller, P., Fleck, R., Grundey, Th., Heinrichs, H., Huppelsberg, D., Matheisen, A., Peiniger, M., Piel, H., and Roeth, R., 1986, Proc. of the Lin. Acc. Conf., Stanford.

Benvenuti, C., Circelli, N., Hauer, M., and Weingarten, W., 1985, IEEE Trans. MAG-21:153.

Bernard, Ph., Cavallari, G., Chiaveri, E., Haebel, E., Heinrichs, H., Lengeler, H., Picasso, E., Picciarelli, V., and Piel, H., 1980, Proc. of the 11th Int. Conf. on High Energy Acc., Geneva ,p.878.

Bernard, Ph., Cavallari, G., Chiaveri, E., Haebel, E., Lengeler, H., Padamsee, H., Picciarelli, V., Proch, D., Schwettman, A., Tückmantel, J., Weingarten W., and Piel, H., 1983, Nucl. Instr. Meth. 206:47.

Bernard, Ph., Lengeler, H., and Picasso, E., 1985, CERN LEP note 524 and CERN/EF/RF 85-1.

Blaschke, R., 1981, Recent Developments in Condensed Matter Physics Vol. 4:425, ed. by J.T. Devreese et al., (Plenum Press, New York).

Bollinger, L.M., 1983, IEE Trans. NS-30:2065.

Bollinger, L.M., 1986, Nucl. Instr. Meth., A244:246.

Citron, A., Dammertz, G., Grundner, M., Husson, L., Lehm, R., and Lengeler, H., 1979, Nucl. Instr. Meth., 164:31.

Citron, A., 1980, Proc. of the 1st Workshop on RF Superconductivity, Karlsruhe, KfK 3019.

Deacon, D.A.G., Elias, L.R., Madey, J.M.J., Ramian, G.J., Schwettman, H.A., and Smith, T.I., 1977, Phys. Rev. Lett., 38:892.

Elias H., 1985, University of Wuppertal, WUD 85-12.

Furuya, T., Hara, K., Hosoyama, K., Kojima, Y., Mitsonubo, S., Noguchi, S., Nakazato, T., and Saito, K., 1984, Proc. of the 5th Symp. on Acc. Sc, and Techn., KEK.

Ginzburg, V.L., and Landau, L.D., 1950, Zh. Eks. Teor. Fiz., 20:1064.

Grundey, Th., Heinrichs, H., Klein, U., Mueller, G., Nissen, G., Piel, H., Genz, H., Gräf, H.D., Janke, M., Richter, A., Schanz, M., Spamer, E., and Titze, O., 1984, Nucl. Instr. Meth., 224:5.

Haebel, E., 1984, Proc. of the 2nd Workshop on RF Superconductivity, CERN, ed. by H. Lengeler.

Haebel, E., Marchard, P., and Tückmantel, J., 1984, CERN/EF/RF 84-2.

Haebel, E., Stierlin, R., and Tückmantel, J., 1985, CERN/EF/RF 85-7.

Haebel, E., and Sekutowicz, J., 1986, DESY M-86-06.

Halbach, K., and Holsinger, R.F., 1976, Part. Acc. 7:213.

Halbritter, J., 1970, Kernforschungszentrum Karlsruhe KfK 3/70-6.

Heinrichs, H., Klein, U., Mueller, G., Piel, H., Proch, D., Weigarten, W., Genz, H., Gräf, H.D., Grundey, Th., Richter, A., and Spamer, E., 1979, Proc. of the Conf. on Nucl. Phys. with Electromagn.Interact., Mainz 1979, Lecture Notes in Phys. 108:176.

Hess, C.E., Schwettman, H.A., and Smith, T.I., 1985, IEE Trans. NS-32:2924.

Huppelsberg, D., 1985, University of Wuppertal, Wu D 85-3.

Klein, U., 1981, University of Wuppertal, WU DI 81-2.

Klein, U., and Proch, D., 1979, Proc. of the Conf. on Future Poss. of Electron Acc., Charlottesville/USA.

Klein, U., and Turneaure, J.P., 1983, IEEE Trans. MAG-19:1330.

Kneisel, P., Vincon, R., and Halbritter, J., 1981, Nucl. Instr. Meth., 188:669.

Kneisel, P., Nakajima, K., Kirchgessner, J., Mioduszewski, J., Pickup, M., Sundelin, R., and Tignor, M., 1983, IEEE Trans. NS-30:3348.

Kneisel, P., 1984, Cornell University SRF 840702.

Kneisel, P., and Mueller, G., 1985, Cornell University SRF 851201EX.

Kneisel, P., Amato, J., Kirchgessner, J., Nakajima, K., Padamsee, H., Philipps, H.L., Reece, C., Sundelin, R., and Tigner, M., 1985, IEEE Trans. Mag-21:1000.

Latham, R.V., "High Voltage Vacuum Insulation, The Physical Basis" (Acad.Press, London, New York, 1981).

Latham, R.V., 1984, Proc. of the 2nd Workshop on RF Superconductivity, CERN ed. by H. Lengeler.

Leeman, C.W., 1986, Proc. of the Lin. Acc. Conf., Stanford.

Lengeler, H., Weingarten, W., Mueller, G. and Piel, H., 1985, IEEE Trans. MAG-21:1014.

Lengeler, H., 1986, Proc. of the Lin. Acc. Conf., Stanford.

London, H., 1934, Nature 133:497.

Lyneis, C.M., McAshan, M.S., and Nguyen Tuong Viet, 1972, Proc. of the Proton Lin.Acc.Conf., Los Alamos, p.98.

Lyneis, C.M., Schwettman, H.A. and Turneaure, J.P., 1977, Appl. Phys. Lett, 31,541).

Lyneis, C.M., 1980, Proc. of 1st Workshop on RF Superconductivity, Karlsruhe, KfK 3019.

Matricon, J., and James, D.S., 1967, Phys. Lett., 24A:241.

Mattis, D.C., and Bardeen, J., 1958, Phys. Rev., 111:412.

Meissner, W., and Ochsenfeld, R., 1933, Naturwiss., 21:787.

Mueller, G., 1983, University of Wuppertal, WU DI 83-1.

Mueller, G., 1984, Proc. of the 2nd Workshop on RF Superconductivity, CERN, ed. by H. Lengeler.

Niedermann, Ph., Sankarraman, N., and Fischer, Ø., 1984, Proc. of the 2nd Workshop on RF Superconductivity, CERN, ed. by H. Lengeler.

Niedermann, Ph., Sankarraman, N., Noer, R.J., and Fischer, Ø., 1986, J. Appl. Phys., 59:892.

Padamsee, H., 1983, CERN/EF/RF 82-5 91982 and IEE Trans. MAG-19:1322.

Padamsee, H., 1985, Cornell University SRF 830902 (1983) and IEEE Trans. Mag-21:1007.

Padamsee, H., Tückmantel, J., and Weingarten, W., 1983, IEEE Trans. Mag-19:1308.

Padamsee, H., 1984, Proc. of the 2nd Workshop on RF Superconductivity, CERN, ed. by H. Lengeler.

Padamsee, H ., 1985, IEEE Trans. MAG-21:149.

Peiniger, M., and Piel, H., 1985, IEEE Trans. NS-32:3612.

Pfister, H., 1976, Cryogenics 16:17.

Piel, H., 1980, Proc. of the 1st Workshop on RF Superconductivity, Karlsruhe, KfK 3019.

Piel, H., and Romijn, R., 1980, CERN/EF/RF 80-3.

Piel, H., 1985, IEEE Trans. NS-32:3565.

Piel, H., Proc. of the CERN Acc. School "Advanced Accelerator Physics", Oxford (1985) and University of Wuppertal Wu B 96-14(1986).

Pierce, J.M., Schwettman, H.A., Fairbank, W.M., and Wilson, P.B., 1965, Proc. of the 9th Int.Conf. on Low Temp.Phys. Part A, p.396.

Proch, D., 1986, priv. communication.

Reuter, G.E.H., and Sondheimer, E.H., 1948, Proc.Roy.Soc., A195:336.

Romijn, R., Weingarten, W., and Piel, H., 1983, IEEE Trans. MAG-19:1318.

Sankarraman, N., Niedermann, Ph., Noer, R.J., and Fischer, Ø., 1986, Proc.of the 33rd Int. Field Emission Symp., Berlin.

Schulze, K.K., 1981, J. of Metals 33:33.

Schwettman, H.A., Turneaure, J.P., and Waites, R.F., 1974, J. Appl., Phys. 45:914.

Schwettman, H.A., Smith, T.I., and Hess, C.E., 1985, IEEE Trans. NS-32:2927.

Shepard, K.W., 1984, Proc. of the 2nd Workshop on RF Superconductivity, CERN, ed. by H. Lengeler.

Stierlin, R.G., 1986, CERN/EF/3233H/RGS/ed.

Sundelin, R., Kirchgessner, J., Padamsee, H., Philipps, H.L., Rice, D., Tigner, M., and von Borstel, E., 1974, Proc. of the Int.Conf. on High Energy Acc., Stanford, p.128.

Sundelin, R.M., 1985, IEEE Trans. NS-32:3570.

Susta, J., 1984, Proc. of the 2nd Workshop on RF Superconductivity, CERN, ed. by H. Lengeler.

Tigner, M., and Padamsee, H., 1982, AIP Conf.Proc. of the SLAC Summer School, 1982, ed. by M. Month, p.801 (AIP, New York,1983) and Cornell University CLNS 82/553.

Turneaure, J.P., 1967, Stanford University HEPL 507.

Turneaure, J.P., and Weissman, I., 1968, J. Appl. Phys. 39:4417.

Turneaure, J.P., and Nguyen Tuong Viet, 1970, Appl. Phys. Lett. 16:333.

Weiland, T., 1983, Nucl. Instr. Meth. 216:329.

Weiland, T., 1983, DESY M-83-17.

Weingarten, W., 1984, Proc. of the 2nd Workshop on Rf Superconductivity, CERN, ed. by H. Lengeler.

Yogi, T., Dick, G.J., and Mercereau, J.E., 1977, Phys. Rev. Lett. 39:826.

FREE-ELECTRON LASER AMPLIFIER DRIVEN BY AN INDUCTION LINAC*

V. Kelvin Neil

Lawrence Livermore National Laboratory
P. O. Box 808, L-626
Livermore, California 94550 USA

INTRODUCTION

A free-electron laser (FEL) directly converts the kinetic energy of
a high-brightness, relativistic electron beam into coherent radiation.
It is a classical device, much like a traveling wave tube. In an FEL
amplifier the conversion takes place in a single pass through a wiggler
magnet, therefore, the fraction of kinetic energy converted must be high
if the device is to be efficient. Since the conversion is a rapidly
increasing function of the electron beam current, a current of 1 kA or
greater is desired. The particle energy required depends on the
wavelength of coherent radiation. For a given wiggler wavelength, λ_w,
wiggler magnetic field B, and radiation wavelength, λ_s, there is a
resonance condition that determines the proper electron energy. This
condition will be derived later in this paper and is

$$\lambda_s = \frac{\lambda_w}{2\gamma^2} \left[1 + \frac{1}{2} \left(\frac{eB\lambda_w}{2\pi\, mc^2} \right)^2 \right] \, , \tag{1}$$

in which γ is the electron energy in units of the rest energy, mc^2. If B
and λ_w do not vary with position as the particle's energy is converted
and γ decreases, the condition is no longer satisfied at some position
and saturation results. However, if B or λ_w or some combination varies
with axial position, the resonance condition can be maintained. The
conversion efficiency (also called extraction efficiency) is greatly

* Work performed jointly under the auspices of the U. S. Department of
Energy by Lawrence Livermore National Laboratory under contract W-7405-
ENG-48, for the Strategic Defense Initiative Organization and the U. S.
Army Strategic Defense Command in support of SDIO/SDC-ATC MIPR No. W31-
RPD-63-A072.

enhanced and the output radiation energy increased. A device in which B/λ_w varies is called a variable parameter wiggler or a tapered wiggler. An experiment called ELF (electron laser facility) is a device with $\lambda_w = 9.8$ cm, $\lambda_s = 8.6$ mm and $\gamma = 8$. The experiment is described in Sec. 4. Experimental results for both untapered and tapered configurations are presented.

The linear induction accelerator is ideally suited for the production of an electron beam with several kA current and energy of a few MeV to several 10's of MeV. An induction accelerator is basically a one-to-one transformer in which the electron beam acts as the secondary. Its invention was stimulated by the desire to accelerate very high currents at high efficiency, which requires a low impedance device. To accomplish this the impedance of the transmission line cables is fed directly to the beam using magnetic material as the isolation cores. The cores can then be stacked as shown in Fig. 1. This figure also indicates schematically the components of an FEL amplifier driven by an induction linac. The only challenge in producing electron beams suitable for driving an FEL amplified is the condition of high-brightness. High current and high average power have been accomplished, as discussed later. In this paper we define the brightness in terms of the volume, δV^4, in 4D transverse phase space occupied by the beam to be

$$[amp/(cm - rad)^2] = \frac{\pi^2 I}{\gamma^2 \delta V^4} \quad . \tag{2}$$

If the volume is a 4D ellipsoid, $\delta V^4 = \pi^2 \varepsilon_x \varepsilon_y / 2$, with ε_x and ε_y the emittance in the two transverse planes. This definition is not universally employed.

THE INDUCTION ACCELERATOR

History

The first linear induction accelerator was constructed at Lawrence Livermore National Laboratory (then known as Lawrence Radiation Laboratory, Livermore) in the early 1960's. This machine was replaced by a new Astron accelerator in 1969 (Beal, et al., 1969). The purpose of these machines was to supply electrons for plasma confinement through magnetic field reversal from a cylindrical layer of circulating electrons. The second Astron accelerator and all other linear induction accelerators ever built in the USA are listed in Table 1, along with the output beam parameters. The Astron accelerator has long since been retired, as has the ERA injector (Avery, et al., 1971). In fact, the ETA (Experimental Test Accelerator, Hester, et al., 1979) was constructed in the building

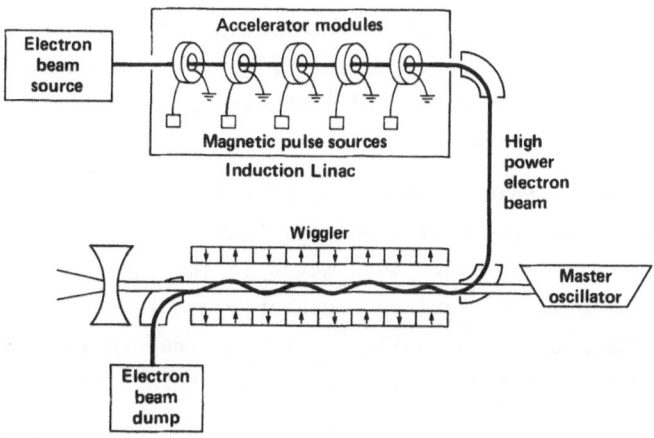

Fig. 1. The components of an FEL amplifier driven by an induction linac.

Table 1. Electron induction linacs in USA.

	Kinetic energy	Beam Current	Pulse Length	Avg. rep. rate (max)	Burst rep. rate
Astron injector, LLNL Original (1963)	3.7 MeV	350 A	300 ns	60 Hz	360 Hz for 7 pulses
Upgrade	6 MeV	800 A	300 ns	60 Hz	800 Hz for 100 pulses
NBS prototype (1971)	0.8 MeV	1,000 A	2,000 ns	1 Hz	--
ERA injector LBL (1971)	4 Mev	1,000 A	45 ns	5 Hz	--
ETA, LLNL (1979)	4.5 MeV	10,000 A	30 ns	2 Hz	900 Hz for 5 pulses
FXR, LLNL (1982)	18 MeV	3,000 A	70 ns	0.3 Hz	--
ATA, LLNL (1983)	45 MeV	10,000 A	60 ns	5 Hz	(1,000 Hz for 10 pulses)
HBTS, LLNL (1984)	1.5 MeV	2,000 A	60 ns	100 Hz	(1,000 Hz)

previously containing the Astron accelerator. In retrospect, the Astron accelerator would have been a fine machine for FEL experiments with radiation wavelength of a few mm. At 500 A, 5 MeV, the energy variation at the head and tail of the 300 ns pulse was 1/2%. During the pulse the energy spread and variation with time were not measurable. The emittance at 5 MeV was about 10 mrad-cm, so the brightness was of the order of 10^5 A/(rad-cm)^2. Even with antiquated technology, the repetition rate was impressive for short periods of time.

The National Bureau of Standards prototype consisted of one soft iron induction core providing a pulse length of 2 µS (Leiss, et al., 1980). This device has since been moved to the Naval Research Laboratory where it is employed for various research projects including FEL experiments. The FXR machine was built for flash x-ray radiography and is presently operated at Livermore (Kulke and Kihara, 1983).

The ETA operating at an energy of about 3.5 MeV and with output current reduced to ~ 3 kA is presently being used as the driver for the ELF experiment described later. The machine is to be replaced by an accelerator employing more modern technology, in particular a high continuous pulse rate, reduced current, and high brightness.

The ATA (Advanced Test Accelerator, Reginato, 1983) is presently being prepared to drive an FEL amplifier experiment called PALADIN. This experiment will use a wiggler with λ_w = 8 cm and a radiation wavelength of 10.6µm. It should be noted that high brightness was not a design goal in either ETA or ATA. The principle design goal was high current: an order-of-magnitude higher than previously achieved in induction linacs. The High Brightness Test Stand (HBTS, Caporaso and Birx, 1985) is an experimental test bed for pulsed power technology as well as cathode-anode geometry necessary to maximize beam brightness. In order to enhance the brightness in ATA, the cathode-anode geometry has been painstakingly redesigned (Boyd, et al., 1985). Only a small portion of the electron gun hardware need to be replaced in order to implement this design.

Beam Breakup

The high current in these machines push against a very fundamental limit in all linacs, namely the beam breakup instability. The phenomenon exhibits unstable coherent transverse motion of the beam resulting from the excitation of transverse modes in the accelerating cells (R. K. Cooper, this conference). The induction machine has considerable advantage over rf linacs with regard to suppressing this instability because the cells are inherently very low Q structures. There is no necessity for highly resonant structures as in an rf linacs; the cells merely couple the drive cables to the beam. The induction core is ferrite, a very good microwave absorber. The cells are carefully shaped to suppress resonances and additional ferrite inserted in strategic locations. But even if all resonances are destroyed, the very presence of the accelerating gaps lead to a resistive, or radiative, instability.

This problem limited the current in Astron to less than 1 kA, and was a well-recognized obstacle when ATA was constructed. It turned out that the focusing provided by the solenoidal transport coils are not sufficient to suppress the instability at full current (Chong, et al., 1985) but could probably do so at 2 or 3 kA required for the PALADIN experiment. We were seeking innovative solutions, and developed a technique called laser guiding.

Laser Guiding

In this technique (Martin, et al., 1985), a small diameter KrF laser beam passes through the center of the cathode and down the entire length of the machine. A background of benzene gas is fed into the machine from about the 5 MeV point at a pressure of approximately 10^{-4} torr. A small percentage of the gas is ionized by the KrF laser, the background electrons are expelled by the beam, and the remaining low density column of ions provides focusing forces that are far stronger than those provided by the solenoids. In addition, the focusing forces are nonlinear and introduce Landau damping, or phase mixing, which is sufficient to completely suppress the instability. The instability cannot be completely suppressed without phase mixing; it can only be made tolerable.

But laser guiding presents some undesirable side effects. The very phase mixing causes an inevitable emittance growth. The laser ionizes only a small fraction of the benzene molecules, and the beam electrons can ionize more. So the focusing forces are time-dependent during the beam pulse, and a time-dependent emittance can result. But with a lower benzene pressure, a more powerful laser could ionize a much larger fraction of the molecules and leave few if any to be ionized by the beam electrons. Techniques for matching the beam onto and off of the ion channel have not been fully perfected. So although laser guiding reduced the instability in ATA to the extent that transverse oscillations cannot be detected, and the technique is undoubtedly a real breakthrough, it is not clear how useful it is when extremely high-brightness is essential.

ONE DIMENSIONAL FEL THEORY

A comprehensive ID theory of the tapered-wiggler FEL amplifier was presented by Kroll, Morton and Rosenbluth (1981). the notation used here follows Prosnitz, et al., (1981). Here we will briefly review the theory with emphasis on the importance of small energy spread and high brightness. We employ Gaussian units in this section.

Particle Motion

The electric and magnetic fields we consider are the wiggler magnetic field B and the radiation electric field E of the form

$$\bar{B} = - B_w \sin k_w z \cosh k_w y \; \hat{y} \tag{3}$$

$$\bar{E} = E \sin (kz - \omega t + \phi) \; \hat{x} \; , \tag{4}$$

in which $k_w = 2\pi/\lambda_w$, $k = 2\pi/\lambda$, $\omega = kc$ and ϕ is the phase of the radiation field. There are, of course, other field components, but these expressions will suffice for this simple treatment. We also set $\cosh k_w y = 1$. For electrons the equation of motion is (with m the rest mass)

$$\frac{dv_x}{dz} = \frac{eB_y}{\gamma \, mc} \; , \tag{5}$$

and therefore

$$v_x = \frac{eB_w}{\gamma \, mck_w} \cos k_w z \; . \tag{6}$$

The change in energy of the electrons is determined by the equation

$$\frac{d\gamma}{dz} = - \frac{ev_x E_x}{mc^2 v_z}$$

$$= - \left(\frac{e}{mc^2}\right)^2 \frac{E_s B_w}{2\gamma \, k} \left\{ \sin[(k + k_w)z - \omega t + \phi] \right.$$

$$\left. + \sin [(k - k_w)z - \omega t + \phi] \right\} \; . \tag{7}$$

We have set $v_z = c$ in this expression. The second term in { } in Eq. (6) represents a fast wave that can contribute little to the average energy change and therefore will be neglected.

We define the relative phase ψ to be

$$\psi = (k + k_w)z - \omega t + \phi \; , \tag{8}$$

so that

$$\frac{d\psi}{dz} = k(1 - \frac{c}{v_z}) + k_w + \frac{d\phi}{dz} \; , \tag{9}$$

and now v_z must be accurately determined. The radiation phase ϕ does change slowly with z as we shall see below, but we ignore $d\phi/dz$ in Eq. (8) to obtain the resonance condition

$$k[1 - (c/v_z)] + k_w = 0 \quad . \tag{10}$$

Physically, Eq. (9) states that as a particle travels one wiggler wavelength in z, the radiation travels a distance $\lambda_w + \lambda_s$. In terms of $\beta_z \equiv v_z/c$, the equation may be rewritten in the form

$$k_w = \frac{k(1 - \beta_z^2)}{\beta_z(1 + \beta_z)}$$

$$\frac{k}{2}(1 - \beta_z^2) \quad . \tag{11}$$

A particle with v_z satisfying Eq. (11) maintains constant phase relative to the radiation wave, continually gaining or losing energy according to Eq. (7). We further have $1 - \beta_z^2 = 1 - \beta^2 + (v_x/c)^2$, where $\beta = v/c$ and v the total electron speed. Inserting Eq. (6) and averaging over one wiggler period, we have

$$1 - \beta_z^2 = \frac{(1 + a_w^2)}{\gamma^2} \quad , \tag{12}$$

in which we revert to conventional notation by introducing the quantity a_w by the definition

$$a_w = eB_w/\sqrt{2}mc^2 k_w \quad . \tag{13}$$

Equations (11) and (12) provide the resonance condition given in the introduction.

Energy Spread

We may write Eq. (9) in the form

$$\frac{d\psi}{dz} = k_w - \frac{k}{2\gamma^2}(1 + a_w^2) \quad . \tag{14}$$

Introducing the quantity e_s by the definition

$$e_s = eE_s/\sqrt{2}mc^2 \quad , \tag{15}$$

we may write Eq. (7) as

$$\frac{d\gamma}{dz} = -\frac{e_s a_w}{\gamma} \sin \psi \quad . \tag{16}$$

We define a resonant particle and attach a subscript r to quantities pertaining to this particle. We set $\psi_r = 0$ and $d\psi_r/dz = 0$, so that $2k_w\gamma_r^2 = k(1 + a_w^2)$. For particles with $\gamma \neq \gamma_r$, we set $\gamma = \gamma_r + \delta$ and linearize Eq. (14) to obtain

$$\frac{d\psi}{dz} = 2k_w \frac{\delta}{\gamma_r} \quad . \tag{17}$$

Taking the derivative of this equation and employing Eq. (16) with $\gamma = \gamma_r$ we obtain the so-called pendulum equation, namely

$$\frac{d^2\psi}{dz^2} + \kappa^2 \sin \psi = 0 \quad , \tag{18}$$

in which

$$\kappa^2 = 2e_s a_w k_w / \gamma_r^2 \quad . \tag{19}$$

Equations (16), (17) and (18) describe the motion in $\gamma - \psi$ phase of particles in an untapered wiggler. The description is completely analogous to rf accelerator theory. Particles lying within a closed curve $P(\psi, \delta)$, called the separatrix, execute periodic oscillations (called synchrotron oscillations) about the synchronous particle. They undergo no average energy change. For an untapered wiggler as treated here, the separatrix is given by

$$P(\psi, \delta) = \pm \left(\frac{e_s a_w}{k_w}\right)^{1/2} (1 + \cos \psi)^{1/2} \quad . \tag{20}$$

Particles within the separatrix are said to be trapped. The separatrix is completely analogous to a stationary "bucket" in rf accelerator theory. From Eq. (20) we see that the maximum energy spread for trapped particles is $\Delta\delta = 4(e_s a_w/k_w)^{1/2}$, which varies as the square root of the radiation field amplitude E_s. The input radiation power density ($\propto E_s^2$) necessary to trap particles in the bucket varies as the 4th power of the energy spread, thus small energy spread is definitely desirable.

If particles are uniformly distributed in the bucket, no net kinetic energy is converted to radiation. But if particles initially are all in the upper half of the bucket, they will oscillate around to the bottom, losing energy which is converted to radiation. Saturation in an untapered wiggler occurs when oscillation continues until some particles are gaining energy ($\psi < 0$) while others are still losing energy ($\psi > 0$) so that the net energy conversion rate goes to zero. The saturation phenomena is considerably more complex if, as in the ELF experiment, the conversion is large and E_s increases rapidly. The bucket grows and more particles are trapped.

Without going through the mathematical details we simply state that in a tapered wiggler, we can still define a resonant particle with $\psi_r \neq 0$ and $d\psi_r/dz < 0$. The separatrix takes the familiar fish shape of a decelerating bucket in an rf accelerator. Particles still execute oscillations about the synchronous particle, but the entire bucket is moving down in energy, so that there is average energy conversion regardless of the distribution of trapped particles.

The Radiation Field

The energy converted to radiation can be calculated from the wave equation for E_y, namely

$$\frac{\partial^2 E_y}{\partial z^2} - \frac{1}{c^2} \frac{\partial^2 E_y}{dt^2} = \frac{4\pi}{c^2} \frac{\partial J_y}{\partial t} \tag{21}$$

This equation is solved in the slowly varying envelope approximation. That is, we assume that the amplitude E_s and the phase ϕ vary slowly with z. For E_y given by Eq. (4) and $\psi_s = kz - \omega t + \phi$, Eq. (21) becomes

$$2k \left[\frac{dE_s}{dz} \cos \psi_s - E_s \frac{d\phi}{dz} \sin \psi_s \right] = \frac{4\pi}{c^2} \frac{\partial J_y}{\partial t} \; . \tag{22}$$

We note that in these units the impedance of free space is $4\pi/c$. To convert to practical units, replace this factor by $Z_o = 120\pi\Omega$. We do that in the following.

Only the Fourier components of J_y varying as $\sin \psi_s$ and $\cos \psi_s$ contribute to Eq. (22). In simulation codes such as FRED (which is briefly described in the following) these components are found by Fourier decomposing a particle distribution. In any case, the solutions to Eq. (22) are

$$\frac{dE_s}{dz} = \frac{Z_o}{\sqrt{2}} a_w \, J \left\langle \frac{\sin \psi}{\gamma} \right\rangle \; , \tag{23}$$

$$\frac{d\phi}{dz} = \frac{Z_o}{\sqrt{2}} a_w \frac{J}{E_s} \left\langle \frac{\cos \psi}{\gamma} \right\rangle \; . \tag{24}$$

In these expressions J is the total beam current density and $\langle \; \rangle$ indicates an average over all particles in the interval $-\pi < \psi < \pi$. Equations (23) and (24) describe the change in amplitude and phase of the radiation electric field. The change in phase is such as to produce a focusing of the radiation. There has been experimental verification of the phase change in the ELF experiment, but the focusing can be definitively observed only in an experiment that is longer than the Rayleigh range.

Emittance

In all the above discussions, the transverse motion of particles arises only from the wiggler magnetic field. In practice particles have additional transverse velocity components from finite temperature, or emittance. A spread in transverse velocity has the same effect as a spread in energy in that both produce a spread in axial speed v_z, making it more difficult for particles to remain in phase with the radiation. The ramifications of finite emittance can be illustrated from equation $1 - \beta_z^2 = 1 - \beta^2 + (v_t/c)^2$. In the above v_t was merely v_x from the motion in the wiggler, and led to Eq. (14). But with finite emittance we have $\bar{v}_t = v_x \hat{x} + \bar{v}_e$, with $\bar{v}_e$ the random transverse velocity. If the betatron wavelength is much longer than λ_w, $\bar{v}_e$ does not change much over one wiggler period, the average of $v_x v_{ex}$ over a period is negligible, and Eq. (14) becomes

$$\frac{d\psi}{dz} = k_w - \frac{k}{2\gamma^2} \left[1 + a_w^2 + (\gamma \beta_e)^2 \right] \quad . \tag{25}$$

There are now some obvious complications. The resonance condition is different for particles with different β_e. Furthermore, depending on the focusing of transverse motions, β_e may change during a betatron oscillation. The particle may be inside the bucket during part of a betatron period and outside during the remainder. This observation brings us to the question of focusing within the wiggler.

Along with the wiggler magnetic field given by Eq. (3) there is a B_z given by

$$B_z = - B_w \cos k_w z \sinh k_w y \quad . \tag{26}$$

The force in the y direction resulting from $v_x B_z$, with v_x given by Eq. (6), gives a net focusing force in the y direction. This focusing, known as edge focusing in beam transport theory, was recognized by Phillips (1960). There is no focusing in the x direction. However, a_w varies with y through the factor $\cosh k_w y$. For betatron motion in the y direction only, the variation of a_w and β_e are such that $a_w^2 + \gamma^2 \beta_e^2$ is constant over a betatron oscillation. A particle with velocity $v_{ye} \neq 0$, $v_{xe} = 0$ may have a different resonance condition than one with $v_{ye} = v_{xe} = 0$, but at least the condition remains constant over a betatron period.

Focusing in the x direction can be accomplished with quadrupole magnets, as is done in the ELF wiggler. Over the entire 3m length of the wiggler there is an air-core quadrupole focusing in the x direction. The defocusing in the y direction is small compared to the very strong edge focusing in that device. But the resulting betatron oscillations in the x direction are sinusoidal, and $\gamma \beta_{ex}$ is not constant over a betatron period, which causes little problem in ELF, but is generally undesirable.

There is a method for providing focusing in both x and y such that $a_w^2 + (\gamma\beta_e)^2$ remains constant over a betatron period. Again the method was recognized by Phillips, but invented independently by Scharlemann (1985). The details are a bit complicated, but the feat is accomplished by parabolic shaping of the wiggler pole faces to introduce a sextupole component into the wiggler magnetic field. This pole-face shaping is implemented in the PALADIN wiggler. Even though a particle will not move into and out of the bucket during a betatron period, finite emittance smears out the resonance, increases the input power required to trap particles, and decreases the efficiency of energy conversion.

THE ELF EXPERIMENT

The parameters of the ELF experiment are given in Table 2. These parameters were chosen in part because ETA was available as a driver. Originally plans were to do experiments at λ_s = 4 mm and 2 mm as well as 8.6 mm. The application of the device to heat plasma in a magnetic confinement fusion experiment gave the device practical justification.

Theoretical support for FEL experiments is primarily provided by the computer code FRED (Fawley, et al., 1984 and Scharlemann, 1985). This code follows particles in one slice of the beam $-\pi < \psi < \pi$, or length $\sim \lambda_s$, through the wiggler. Particle trajectories are calculated

Table 2. ELF operating parameters.

Wiggler
Period (cm)	9.8
Length (m)	2.94
Number of periods	30
Peak magnetic field (kG)	0-5 (adjustable)
Horizontal focusing quadrupole strength (G cm^{-1})	30-60

Electron beam
Kinetic energy (MeV)	3.0 - 3.5
Current (A)	850
Unnormalized edge emittance (mrad-cm) ($\varepsilon \equiv x_{max}x'_{max}$)	70
Equilibrium beam cross-section (cm) (in wiggler)	0.6 x 1.2
Current density profile	$\propto \left[1 - \dfrac{x^2}{x_{max}^2} - \dfrac{y^2}{y_{max}^2}\right]$

Microwaves
Frequency (GHz)	34.6
Waveguide size (cm)	3 x 10 cm
Input power (kW)	50
Design mode	TE_{01}

in 3 dimensions, and the radiation field equations are solved in two
dimensions, r and z in cylindrical coordinates for free space propagation, or x and z for propagation in a waveguide. In addition to the
radiation fields, electrostatic forces from the spatially modulated
charge density are included. These electrostatic forces are significant
with the high current and low electron energy in the ELF experiment, but
not in the forthcoming PALADIN experiment using ATA. In the following
theoretical predictions are compared to experimental results, and
generally found to agree very well.

Hardware

The experimental setup is described by Orzechowski, et al. (1983).
A schematic drawing of the wiggler magnet surrounded by the air-core
quadrupoles is shown in Fig. 2. Experiments use a 3 kA current out of
ETA at about 3.5 MeV. An emittance filter in the beamline selects a
known value of δV^4 and transmits up to 1.2 kA with a brightness of
$2 \times 10^4 A/(cm-rad)^2$. The electron beam pulse through the wiggler is 12 to
15 ns duration. The wiggler wavelength is uniform in z, but each two
periods of the wiggler are individually controlled, allowing tapering and
variation of the interaction length in the untapered experiments. The
input signal is generated by a 35 GHz magnetron, delivering up to 50 kW
for a period of 500 ns.

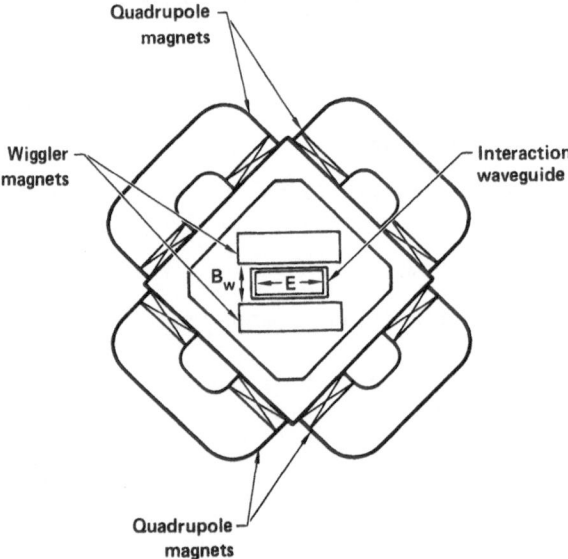

Fig. 2. Cross section of the interaction region showing
orientation of wiggler field and electron
oscillation. The TE_{10} mode is excited in the
waveguide. Quadrupole magnets stabilize the
electron orbits in the horizontal plane.

Untapered Wiggler

Experimental techniques and results of the experiments with the untapered wiggler are given by Orzechowski, et al. (1986). The small signal gain (growth from noise) is shown in Fig. 3, which indicates a gain of 26.6 dB/m. The extrapolated input noise level is 1.5 mW.

Amplifier performance was studied as a function of wiggler length. The length is varied by tuning the downstream portion of the wiggler field so that it is significantly off resonance and does not contribute to the FEL interaction. (A low magnetic field is necessary to focus the beam.) Experimental results are indicated by the circles in Fig. 4. The initial exponential growth is about 34dB/m, somewhat higher than the small signal gain. Saturation occurs at about 1.4 m at an output power level of 150 MW. Up to 180 MW has been achieved. The saturated output power shows very little dependence on input power from 3 to 30 kW. By varying the electron beam current, the saturated power is found to vary as I^2, and this is one of the aspects of saturation that is not fully understood.

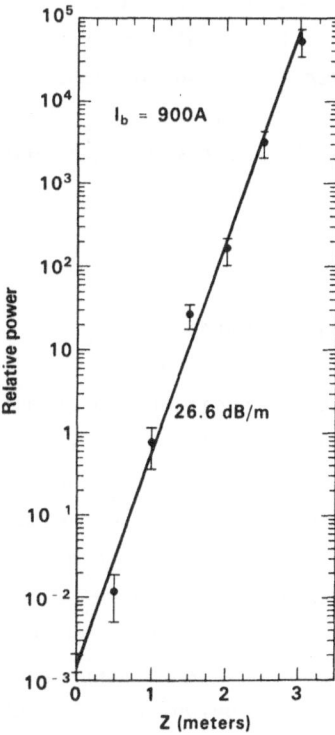

Fig. 3.   Small signal gain in the super-radiant mode as a function of wiggler length. Extrapolating the signal back to the origin gives an effective input signal of 1.3m W.

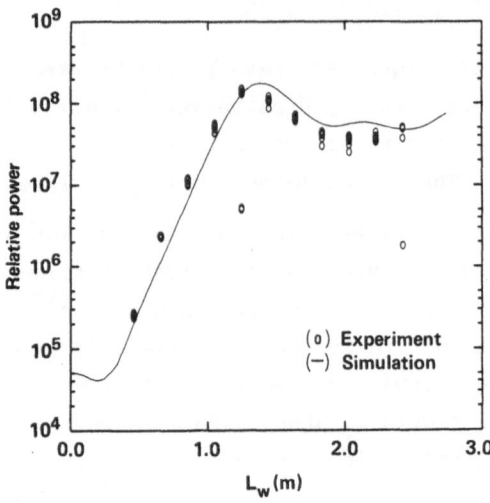

Fig. 4. TE$_{01}$ power vs. wiggler length. The circles are experimental data; the solid line is the result of simulation.

Comparison of experimental results with code calculations for the untapered wiggler has been presented by Scharlemann, et al., (1986). Results of simulation of the amplifier are shown as the solid curve in Fig. 4. There is one caveat in that the experiment used B$_w$ = 3.8 kG while the code results were obtained for 3.65 kG. There is a discrepancy in the post-saturation power oscillations. These oscillations are probably a manifestation of coherent synchrotron oscillations, and the discrepancy probably arises from inadequate treatment of the electro-static field by the simulations.

Tapered Wiggler

Saturation occurs in the untapered wiggler because particles lose energy and fall out of resonance. By tapering (reducing) the wiggler magnetic field the resonance condition can be maintained. Experimentally the optimum taper was found by maintaining the untapered configuration to the saturation point (taking advantage of the exponential gain) and then turning up B in each subsequent two-period segment of the wiggler to maximize the output power. This procedure determines the taper while increasing the wiggler length. The tapered wiggler experiment is described by Orzechowski, et al., (1986) and the results are shown by the + in Fig. 5. The peak electron beam current was 1.1 kA in this experiment. Experimentally the peak output power reached 1.5 GW, an order of magnitude more than achieved in the untapered wiggler, and corresponding to a conversion (extraction) efficiency of 40%.

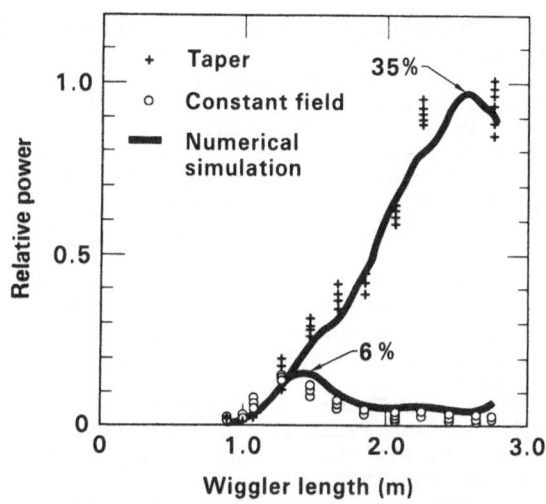

Fig. 5. Results of experiment and simulation for the tapered and untapered wiggler, the latter taken from Fig. 4.

Using the experimentally determined taper, the simulation produced results shown by the upper solid curve in Fig. 5. The lower circles and solid curve in Fig. 5 are the same as shown in Fig. 4 for the untapered wiggler.

REFERENCES

Avery, R., Behrsing, G., Chupp, W., Faltens, A., Hartwig, E. C.,
    Hernandez H. P., Macdonald, C., Meneghetti, J. R., Nemetz, R. G.,
    Popenuck, W., Salsig, W. and Vanecek, D., 1971, Proc. of the 1971
    Part. Acc. Conf. IEEE Proc. Nuc. Sci. NS-18 #3, 479.
Beal, J. W., Christofilos, N. C. and Hester, R. E., 1969, Proc. of the
    1969 Part. Acc. Conf. IEEE Trans. Nuc. Sci. NS-16 #3, 294.
Boyd, J. K., Caporaso, G. J. and Cole, A. G., 1985, Proc. of the 1985
    Part. Acc. Conf. IEEE Trans. Nuc. Sci. NS-12 #5, 2602.
Caporaso, G. J., and Birx, D. L., 1985, Proc. of the 1985 Part. Acc.
    Conf. IEEE Trans. Nuc. Sci. NS-32 #5, 2608.
Chong, Y. P., Caporaso, G. J. and Struve, K. W., 1985, Proc. of the 1985
    Part. Acc. Conf. IEEE Trans. Nuc. Sci. NS-12 #5, 3210.
Fawley, W. M., Prosnitz, D. and Scharlemann, 1984, E. T., Phys. Rev. A30,
    2472.
Hester, R. E., Bubp, D. G., Clark, C., Chesterman, A. W., Dexter, W. L.,
    Fessenden, T. J., Reginato, L. L., Yakota, T. T. and Faltens, A. A.,
    1979, Proc. of 1979 Part. Acc. Conf. IEEE Trans. Nuc. Sci. NS-26 #3,
    4180.
Kroll, N. M., Morton, P. L. and Rosenbluth, M. N., 1981, IEEE Journal
    Quantum Electronics 17, 1436.
Kulke, B. and Kihara, R., 1983, Proc. of the 1983 Part. Acc. Conf. IEEE
    Trans. Nuc. Sci. NS-30 #4, 3030.
Leiss, J. E., Norris, N. J. and Wilson, M. A., 1980, Part. Acc. 10 223).
Martin, W. E., Caporaso, G. J., Fawley, W. M., Prosnitz, D., and Cole, A.
    G., 1985, Phys. Rev. Letters 54, 685).

Orzechowski, T. J., Prosnitz, D., Halbach, K. Kuenning, R. W., Paul, A. C., Hopkins, D. B., Sessler, A. M., Stover, G., Tanabe, J., and Wurtele, J. S., 1983, "A High Gain Free Electron Laser at ETA" Lawrence Livermore National Laboratory Report UCRL-88705).

Orzechowski, T. J., Anderson, B. R., Fawley, W. M., Prosnitz, D., Scharlemann, E. T., Yarema, S. M., Sessler, A. M., Hopkins, D. B., Paul, A. C., and Wurtele, J. S., Proceedings of the Seventh International Free Electron Laser Conference, E. T. Scharlemann and D. Prosnitz, eds., 1986, Nuclear Instruments & Methods A249, in press.

Orzechowski, T. J., Anderson, B. R., Clark, J. C., Fawley, W. M., Paul, A. C., Prosnitz, D., Scharlemann, E. T., Yarema, S. M., Hopkins, D. B., Sessler, A. M., and Wurtele, J. S., 1986, Lawrence Livermore National Laboratory Report UCRL-94841. Submitted for publication in Phys. Rev. Letters.

Phillips, R. M., 1960, IRE Trans. on Electron Devices $\underline{7}$, 231.

Prosnitz, D., Szöke, A. and Neil, V. K., 1981, Phys. Rev. $\underline{A24}$, 1436.

Reginato, L. L., for the ATA Staff, 1983, Proc. of the 1983 Part. Acc. Conf. IEEE Trans. Nuc. Sci. $\underline{NS-30}$ #4, 2970.

Scharlemann, E. T., 1985, Jour. App. Phys. $\underline{58}$, 2154.

Scharlemann, E. T., Fawley, W. M., Anderson, B. R., and Orzechowski, T. J., Proc. of the Seventh International Free Electron Laser Conference, E. T. Scharlemann and D. Prosnitz, eds., 1986, Nuclear Instruments & Methods A249, in press.

FEL OSCILLATORS (MICROTRONS)

Elio Sabia

ENEA, Dip. TIB., Divisione Fisica Applicata, C.R.E.
Frascati, C.P. 65 - 00044 Frascati, Rome, Italy

INTRODUCTION

The free electron laser (FEL) is certainly one of the most interesting sources of coherent radiation to have been developed over the
last few years. The working principle of this device is completely
different from that of "conventional" lasers. The active medium,
indeed, does not consist of atoms of molecules but is a beam of relativistic electrons. In addition, the gain mechanism is essentially
classical and arises from the interaction between electrons, laser
field, and a spatially periodic magnetic field provided by a special
device usually called "Undulator Magnet" (UM).

A SHORT REVIEW OF FEL THEORY

In a FEL device an ultrarelativistic electron beam (e-beam) interacts with an undulator magnet (Fig. 1) in which it undergoes transverse
oscillations and emits radiation at a fixed wavelength. In this way the
e-beam can amplify a copropagating laser beam, or, once the radiation is
stored in an optical resonator and reinteracts with the e-beam, it is
reinforced and the system works as a self-sustained oscillator.

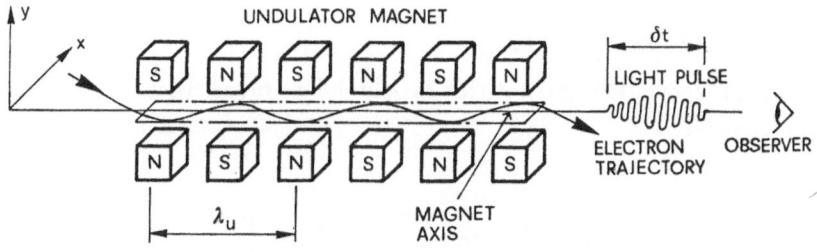

Fig. 1. Undulator magnet geometry.

According to the Weizsacker-Williams approximation of the FEL, the undulator field can be treated as a radiation field with wavelength

$$\lambda^* = 2\lambda_u \tag{1}$$

and photon density

$$\bar{n} = \frac{\alpha}{4} \frac{K^2}{\lambda_u r_0^2} , \tag{2}$$

where

$$K = \frac{e\bar{B}\lambda_u}{2\pi \, m_0 c^2} .$$

α = fine structure constant

r_0 = classical electron radius

λ = undulator wavelength

$\bar{B}^u = B_0$ for helical undulator

$B_0 / \sqrt{2}$ for linear undulator

B_0 = on-axis magnetic field

Within this framework, the interaction of the electrons with the magnetic undulator can be understood as the interaction with a very intense electromagnetic wave. It is now very easy to evaluate the wavelength of the scattered light. According to the Compton scattering formula we get

$$\lambda = 2 \, \lambda_u \, \frac{1-\beta\cos\Theta}{1+\beta} \tag{3}$$

which for small angles and ultrarelativistic energies reads

$$\lambda = \frac{\lambda_u}{2\gamma^2} (1+K^2+\gamma^2\Theta^2) , \tag{4}$$

where

$$\gamma = E/mc^2$$

E = electron beam energy

The well-known FEL gain formula can be deduced in a classical sense and reads:

$$g(\omega) = - \, \frac{4\pi^2}{\gamma} \, \frac{\lambda L}{\Sigma_E} \, \frac{\hat{I}}{I_0} \, \mathcal{J} \, \frac{K^2}{1+K^2} \left(\frac{\Delta\omega}{\omega}\right)_0^{-2} \frac{d}{d\nu} \left(\frac{\sin\nu/2}{\nu/2}\right)^2 \tag{5}$$

where Σ_E is the e-beam cross section, $\mathcal{J}$ is the filling factor

$$\mathcal{J} = \begin{cases} 1 & \text{if } \Sigma_E > \Sigma_L \\ \Sigma_E/\Sigma_L & \text{if } \Sigma_E < \Sigma_L \end{cases}$$

and Σ_L is the laser-beam cross section, L is the undulator length, $\hat{I}$ the e-beam peak current, $I_0 = ec/r_0$ ($\equiv 1.7 \times 10^4 A$) is the Alfvén current, λ the laser wavelength, $\Delta\omega/\omega$ the relative homogeneous bandwidth and

$$\nu = 2 \pi N \frac{\omega_0 - \omega}{\omega_0} \qquad (\omega_0 = \frac{2 \pi c}{\lambda})$$

N = number of periods of the undulator.

A rigorous analysis of the gain has been made by Dattoli and Renieri (1985a). Relation (5) is relevant to a small signal, single mode, homogeneously broadened FEL operation. By homogeneous broadening we mean a FEL operating with an e-beam whose energy spread and emittances produce negligibly small effects. The beam qualities produce both a broadening of the emission line and a reduction of the gain. The value of the inhomogeneous linewidth in terms of the beam emittances and energy spread reads (Dattoli and Renieri, 1985a):

$$\left(\frac{\Delta\omega}{\omega}\right) = \left(\frac{\Delta\omega}{\omega}\right)_0 \sqrt{1 + \mu_x^2 + \mu_y^2 + \mu_\varepsilon^2} \ . \tag{6}$$

The μ –coefficients are the ratio between the inhomogeneous and homogeneous widths.

In particular,

$$\mu_\varepsilon = 4N\sigma_\varepsilon \ , \quad \sigma_\varepsilon \equiv \text{r.m.s. energy spread}$$

$$\mu_{x,y} = \frac{4N\gamma^2}{1+K^2}\left\{\frac{1}{2\sigma_{x,y}^4}\left(\frac{\varepsilon_{x,y}}{2\pi}\right)^4 + 2\left(\frac{K\pi}{\gamma\lambda_u}\right)^4 h_{x,y}^2 \sigma_{x,y}^4\right\}^{1/2} \ , \tag{7}$$

where $\varepsilon_{x,y}$ are the radial and vertical emittances, $\sigma_{x,y}$ the transverse e-beam dimensions, $h_{x,y}$ are coefficients depending on the undulator geometry, namely, $h_x = h_y = 1$ for helical undulators and $h_x = -\delta$, $h_y = 2+\delta$ ($\delta \ll 1$) for the linear case, with polarization along the y-axis. Physically δ is the magnitude of the sextupolar term along the x-direction (Dattoli and Renieri (1985a)). (For reference frame definition see Fig. 1).

It is worth noting that the inhomogeneous broadening due to the emittances consists of two distinct contributions: the first due to the angular divergence; the second due to the transverse size of the beam which explores regions of different magnetic field. The effects of the inhomogenities on the gain are shown in Fig. 2.

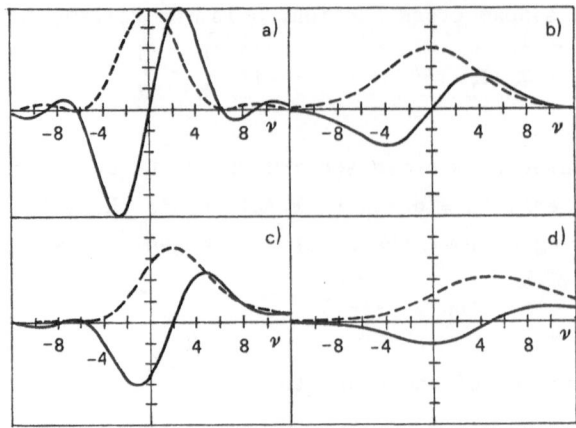

Fig. 2. Inhomogeneous broadened gain:
a)$\mu_\varepsilon=\mu_x=\mu_y=0$; b)$\mu_\varepsilon=1,\mu_x=\mu_y=0$;
c)$\mu_\varepsilon=0,\mu_x=1,\mu_y=0$; d)$\mu_\varepsilon=\mu_x=\mu_y=1$.

Another important parameter plays a crucial role in FEL operation
with a bunched e-beam: the longitudinal bunch length produces a kind of
mode-locking in the laser operation (Dattoli and Renieri, 1981a; 1980a;
1983). It can indeed be shown that the strength of the coupling between
longitudinal modes is given by the parameter

$$\mu_c = \frac{N\lambda}{\sigma_z} \tag{8}$$

where σ_z is the r.m.s. bunch longitudinal length. The larger μ_c the
greater the number of the coupled longitudinal modes. The bunched
e-beam structure (Fig. 3) is also responsible for the so-called
lethargic FEL behaviour, i.e., the slow-down of the light pulse due to
the interaction and the necessity of shortening the cavity length with
respect to the nominal round-trip period to have tuning between light
and e-bunches (Dattoli and Renieri, 1985a).

It is possible to combine all the above effects in a single gain
formula developed in the framework of the FEL supermode theory (Dattoli
and Renieri, 1985a; Dattoli et al., 1981b, 1980b; CNEN Rep. 80.35/cc.,
1980) which includes the strong signal and multimode behaviour. If we
limit ourselves to a single pass FEL, the gain may be defined as

$$G_h = g_h^\circ \; [Reg_\gamma[\Theta;\mu_c;\mu_\varepsilon,\mu_x,\mu_y] \tag{9}$$

$$\text{for helical undulator}$$

$$G_{\ell,n} = g_{\ell,n}^\circ \; |Req_\gamma[\Theta_n;\mu_c;n\mu_x,n\mu_y,n\mu_\varepsilon] \tag{10}$$

$$\text{for linear undulator}$$

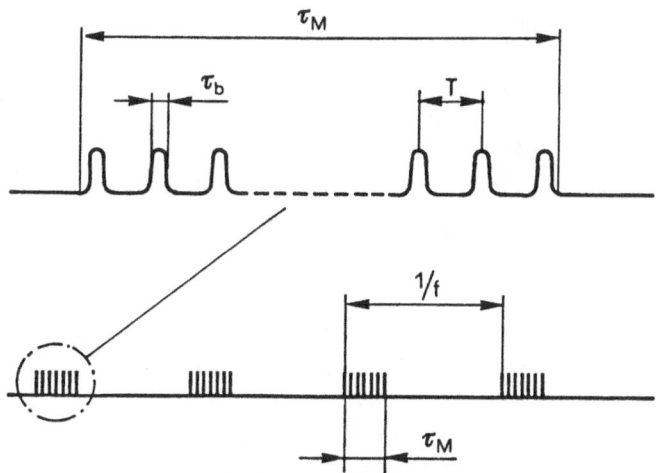

Fig. 3. e-beam structure from a RF machine: τ_b ≡ microbunch time duration; τ_M ≡ macropulse time duration; T ≡ microbunches time separation; f ≡ repetition frequency.

where

$$g_h = 22 \times 10^{-4} N^2 \frac{K^2}{1+K^2} \frac{\hat{I}[A]}{\gamma}$$

$$g_{\ell,n} = 22 \times 10^{-4} n^2 N^2 \frac{K^2}{1+K} \frac{\hat{I}[A]}{\gamma} [J_{(n-1)/2}(\frac{n}{2} \frac{K^2}{1+K^2}) - J_{(n+1)/2}(\frac{n}{2} \frac{K^2}{1+K^2})]^2$$

(11)

n is the harmonic number, J(·) the n-th cylindrical Bessel function.

The function $|Req_\gamma$ contains the most important features relevant to the gain, the most remarkable being the dependence on the θ variable which is named "delay parameter",

$$\Theta = - \frac{\omega_b \delta T}{\pi g_h^0 N}$$

where $\delta T = T_c - T_e$ T_c ≡ cavity round trip period

$$\Theta_n = - \frac{\omega_b \delta T}{\pi g_{\ell,n}^0 N}$$

(12)

In Fig. 4 we have plotted the function $|Req_\gamma$ versus θ for difference values of μ_ϵ, μ_x and μ_y (at fixed μ_c).

In Fig. 5 report typical behaviour of Req_γ vs θ together with the dimensionless peak laser power χ (Dattoli and Renieri, 1985a). It should be noted that the maximum gain and the maximum output laser power do not correspond to the same value of θ; therefore the maximization of the gain does not mean the maximum output laser power. The average laser power can be evaluated according to the following formula (Dattoli et al., 1985b):

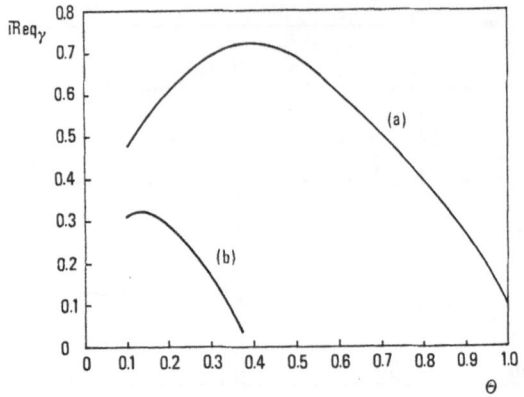

Fig. 4. Req_γ vs Θ. a)$\mu_c=0.5,\mu_\varepsilon=\mu_x=\mu_y=0$;
b)$\mu_c=0.5,\mu_\varepsilon=\mu_x=\mu_y=0.5$.

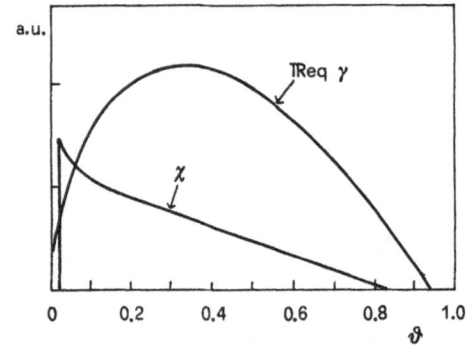

Fig. 5. Gain function and dimensionless laser power vs Θ.

$$\bar{P}_L[KW] = \eta_A \frac{\chi(\Theta)}{2N} \left[1 - \frac{\tau_R[\mu s]}{\tau_M[\mu s]}\right] \frac{\gamma_M}{\gamma_T} \bar{P}_K[KW] \ , \tag{13}$$

where η_A is the acceleration efficiency [*] and $\bar{P}_K$ is the average power of the RF supplier, τ_M is the macropulse time duration, N is the number of undulator periods, γ_M and γ_T are the mirror transmissivity and total cavity losses respectively, and τ_R is the laser signal rise time linked to the gain by the following expression

$$\tau_R[\mu s] \simeq 0.14 \frac{L_c[m]}{g-\gamma_T} \tag{14}$$

where g is the gain and L_c the cavity length.

---

[*] The ratio of the e-beam power to that of the supplier.

Let us discuss a simple criterion to optimize the output optical power.

The gain (and thus the rise time) is, among other things, a rather complicated function of the number of undulator periods which enter the gain function through the inhomogeneous and coupling parameters [see Eqs. (9) and (10)]. All the effects of gain reduction due to energy spread, emittances, and slippage, and optimum number of undulator periods can be included. A typical dependence of the gain on N, for fixed machine parameters, is shown in Fig. 6. It is remarkable that in this case too the maximum gain does not coincide with the maximum average laser power as shown in Fig. 7 where we have reported the behaviour of $\eta_T = \overline{P}_L/\eta_A P_K$ (transfer efficiency from the RF supply to the laser) vs N. It should be pointed out that a properly chosen number of periods may lead to an improvement on the power of even a factor about 2. From Fig. 8 it is evident that the effect is more pronounced with decreasing cavity losses and also for decreasing values of energy spread and emittances (see Fig. 9).

The width ΔN of the "transfer efficiency curve" η_T shown in Figs. 8, 9 is linked to the "threshold" periods N_S^{min}, N_S^{max}, i.e., the values of periods for which $\tau_R = \tau_M$. In order to get the maximum average output laser power one must choose a number of periods near N_S^{min} so that $\tau_R < \tau_M$. (Ciocci et al., to be published)

Another physical variable which plays an important role in the gain mechanism (see Eqs. (9, 10, 11)) is the e-beam peak current. In a single pass FEL, which utilizes a microtron as e-beam source, the peak current value attainable may be less than the one needed to reach the saturation threshold. In order to overcome this drawback, it is possible to realize a mechanism of bunch compression. This mechanism has three effects:

1) an increase of the energy spread;
2) a reduction of the micro-bunch length;
3) an increase of the peak current.

The first two effects increase the coefficients μ_ε and μ_c [see Eqs. (7, 8)] and thus reduce the gain function [see Eqs. (9, 10, 11)]; the third effect increases the gain; the total effect may result in an effective enhancement of the gain.

A significant parameter which describes the effect of the micro-bunch compression on the gain is the following quantity (the ratio of the gain to the gain with bunch compression) (Ciocci et al., 1986).

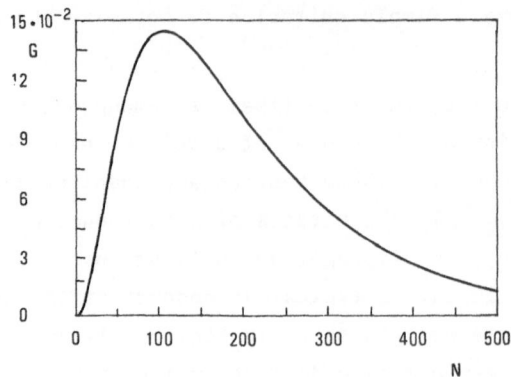

Fig. 6. Gain vs N at fixed machine parameters:
E=88 MeV, σ_ε=0.1%; I=30 A,
$\varepsilon_{x,y}$=0.6 mm mrad; K=0.5.

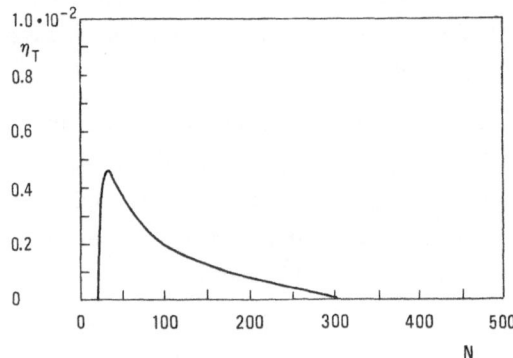

Fig. 7. $\eta_T = \overline{P}_L / \eta_A \overline{P}_A$ vs N; γ_M=1%, γ_T=2%;
τ_M=15µs.

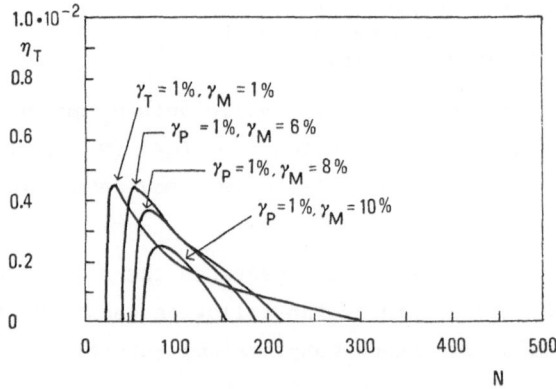

$\gamma_T = 1\%, \gamma_M = 1\%$

$\gamma_P = 1\%, \gamma_M = 6\%$

$\gamma_P = 1\%, \gamma_M = 8\%$

$\gamma_P = 1\%, \gamma_M = 10\%$

Fig. 8. η_T vs N at different values of γ_M and $\gamma_p = \gamma_T - \gamma_M$.

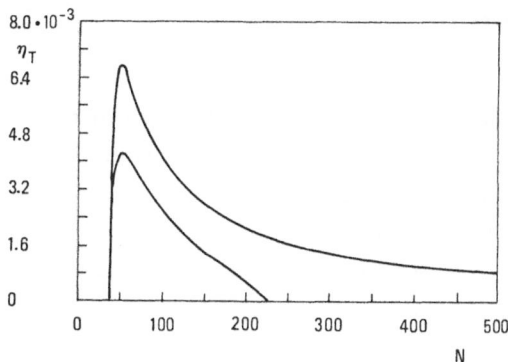

Fig. 9. η_T vs N at different values of energy spread
and emittances ($\gamma_p=1\%$, $\gamma_M=8\%$) a)$\varepsilon_{x,y}=0$, $\sigma_\varepsilon=0$;
b)$\varepsilon_{x,y}=0.7$ mm mrad, $\sigma_\varepsilon=0.6\times10^{-3}$.

$$R = \frac{\mu_\varepsilon}{\mu_\varepsilon^o} \frac{(1+\mu_c^o/3)(1+1.7\mu_\varepsilon^{o2})}{(1+\frac{\mu_c^o}{3}\frac{\mu_\varepsilon}{\mu_\varepsilon^o})(1+1.7\mu^2)} \qquad (15)$$

where the superscript means the the μ coefficients are evaluated before
the bunch compression.

In Fig. 10 we have plotted the quantity R vs the coefficient μ_ε
at the laser wavelength $\lambda = 19$ μm, $\mu_\varepsilon^o = 0.216$, $\sigma_\varepsilon^o = 7$ mm, N = 45. The
maximum of R (= 1.75) is obtained at a value $\mu_\varepsilon/\mu_\varepsilon^o \simeq 3$ which corresponds to a micro-bunch compression of a factor 3.

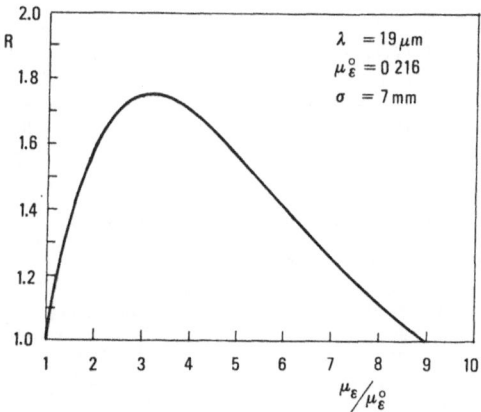

Fig. 10. R vs $\mu_\varepsilon/\mu_\varepsilon^o$ at laser wavelength $\lambda=19$ μm.

Introduction

The operating wavelength of FEL sources developed in recent years or under realization, is spread over wide regions of the spectrum, from the visible up to the far infrared. A chart of the operating and proposed experiments is shown in Figure 11 where the FEL sources are characterized by a wavelength-energy plot. The energy is relevant to the e-beam accelerator while the wavelength is the operating laser one. In Fig. 12 the accelerating devices which can be used for FEL are reported in a current-energy plot. We limit ourselves to low energy single passage accelerators, in particular to circular microtrons. Low energy machines are indeed the widely accepted experimental solution for operation in the infrared. Furthermore, the circular microtron has a number of advantages, as listed below.

a) It is a compact machine which can reach relatively high energies.

b) It can reach relatively high peak current.

c) It produces an e-beam with a good energy spread ($\sigma_\varepsilon \sim 10^{-3}$) and relatively good emittances (comparable to those of a Linac).

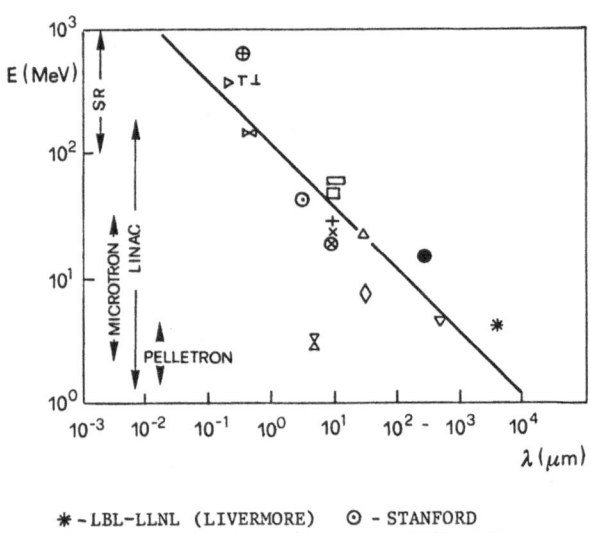

✱	– LBL–LLNL (LIVERMORE)	⊙	– STANFORD
▽	– UCSB (SINGLE STAGE)	⊥	– NOVOSIBIRSK
⋈	– UCSB (TWO STAGES)	⋈	– ORSAY
●	– AT&T–BELL	⊕	– FRASCATI (INFN)
△	– FRASCATI (ENEA)	▷	– BROOKHAVEN
▭	– UK PROJECT	▢	– MK III–STANFORD
+	– TRW STANFORD	◊	– YEREVAN
×	– LANL (LOS ALAMOS)	⊤	– SOR–RING
⊗	– MSNW–BAC		

Fig. 11.   FEL scenario.

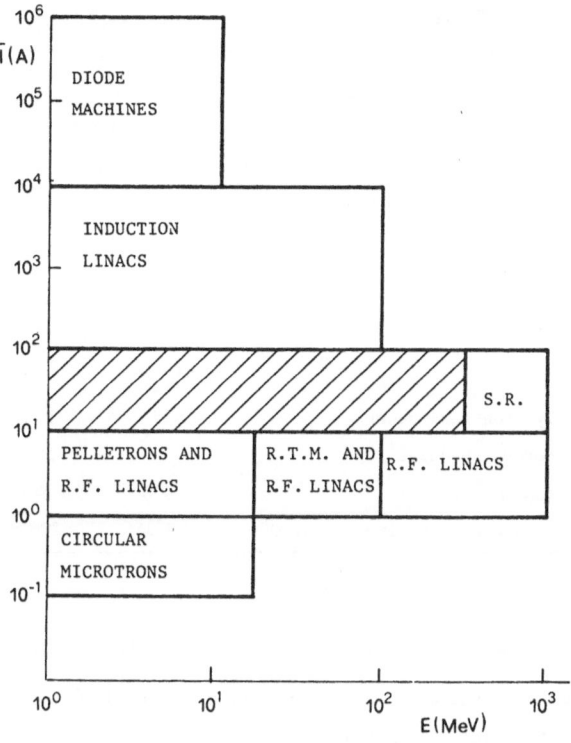

Fig. 12. Current vs Energy for existing electron accelerators.

At the ENEA Frascati centre, a single pass FEL utilizes a 20 MeV circular microtron as electron beam source. The main characteristics of this device have been described in a number of papers (Bizzari et al., 1983; 1985a). (See Table 1.)

Let us briefly discuss the basic features of the microtron. In a circular microtron the electrons move in circular orbits in a constant magnetic field with constant frequency of the accelerating RF field, provided by a powerful klystron or magnetron (operating typically in the S-band). The electrons follow circular orbits with common tangent point where the accelerating cavity needed to provide energy to the electrons is placed (see Fig. 13). After each passage through the cavity, the electrons gain a certain amount of energy and pass to the next orbit. After a certain number of orbits, they pass through a shielding magnetic channel and are extracted. The synchronism of the electron motion with the accelerating field is achieved by the fact that each succeeding orbit is longer than the preceeding one by an integer number of the period of the accelerating field. The ratio of the orbit period to the accelerating field period is called the "harmonic number" of the accelerating mode. The circular microtron is an accelerator with a changing harmonic number.

Table 1. Microtron parameters.

Electron beam energy (MeV)	20
Relative energy spread	0.12%
Electron bunch duration (ps)	20
Horizontal emittance (mm. mrad)	6π
Vertical emittance (mm. mrad)	2π

	Now in operation	Planned for 1986
Klystron peak power	5 MW	15 MW
Macropulse duration	5 µs	12 µs
Peak current	2.4 A	4 A
Average current (macropulse)	120 mA	200 mA

The experimental activity is developed in three successive steps:

1) analysis of the spontaneous emission at $\lambda = 10.6$ µm employing a pulsed helical undulator (UM1);

2) oscillation experiment at $\lambda = 10.6$ µm employing a pulsed linear undulator (UM2);

3) oscillation experiment at $\lambda = 32$ µm and at higher order harmonics employing a $SmCo_5$ permanent magnet linear undulator (UM3).

The first two steps have been carried out and the 10.6 µm oscillator was successfully operated in September 1985. The third stage of experiment is at an advanced stage of development.

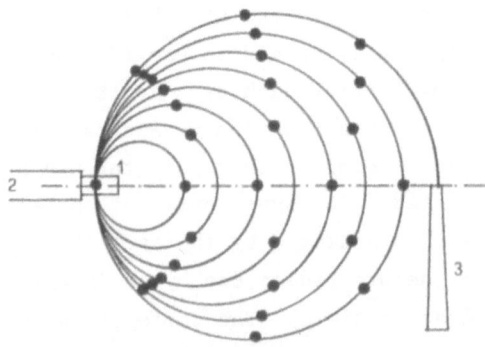

Fig 13. Circular microtron schematic layout (the dots represent the electron bunches). 1) accelerating cavity; 2) waveguide; 3) shielded magnetic channel for electron extraction.

The experimental layouts relevant to the stages of development mentioned above are sketched in Figs. 14a and 14b where two different configurations of the e-beam transport channel are shown.

With reference to Fig. 14a, the e-beam transport channel realized for the use with both helical and linear pulsed undulators is provided by a couple of quadrupole magnets $Q_1 - Q_2$, by the bending magnet M_1, by another couple of quadrupoles, $Q_3 - Q_4$, and by the "extracting" magnet M_2. It should be pointed out that, in this case, due to the dispersion of off-energy electrons generated by the magnet M_1, the e-beam horizontal dimension is not constant through the undulator. This drawback will be overcome by employing the achromatic bending triplet (M_1, M_2, M_3) in the realization of the e-beam transport channel which will be used with the undulator UM3. This will also simplify the e-beam alignment procedure when the microtron energy tunability is exploited for the tuning of the laser wavelength.

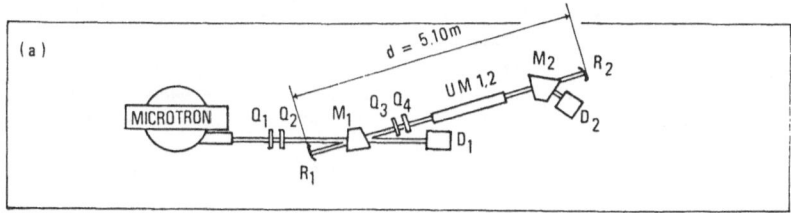

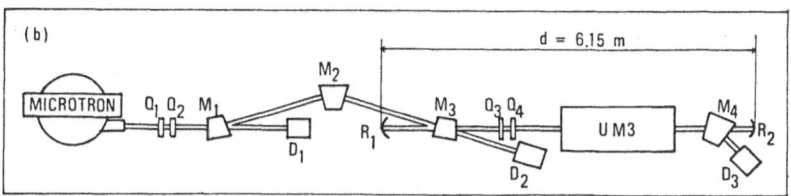

Fig. 14. Layouts of the Frascati-ENEA experiment.
a) pulsed undulator magnet (UM1,2);
b) permanent magnet linear undulator (UM3).

Experimental Results

At an early stage of the experiment (Bizzari et al., 1985a), accurate spectroscopic measurements on the spontaneous radiation emitted in a pulsed helical undulator (see Table 2) were performed which showed a strong dependence of the emission spectral width on the alignment of both e-beam and optical axis with respect to the undulator. The observed spectral width was strongly affected by the contribution of the e-beam emittances to the inhomogeneous broadening. Furthermore, a small average e-beam current (~ 30 mA) made it impossible to achieve oscillation.

Much effort was concentrated on improving the performance and reliability of the microtron (Bizzari et al., 1985b). The improvements enabled us to obtain a stable operation of the microtron up to 120 mA average current. To achieve higher values of the gain, a linear pulsed undulator was designed and realized. Indeed, for this undulator the horizontal emittance does not affect the inhomogeneous broadening (see Table 2).

Due to the good quality and stability of the e-beam at an average current of 80 mA together with the narrow emission width (Fig. 15) and an expected gain per pass of 2.1% at λ = 10.6 μm, careful measurements with the optical resonator installed were performed.

Table 2. Radiation characteristics as a function of undulator type.

	UM1 Pulsed helical undulator	UM2 Pulsed linear undulator	UM3 Permanent magnet linear undulator
Period λ_u [cm]	2.4	2.4	5
Length L[m]	1.2	1.2	2.25
Undulator parameter $K= B\lambda_u/2\pi mc^2$	0.1–0.5	0.2–1.0	1–2
Repetition frequency f[Hz]	0.2	0.2	____

The optical resonator was 4.937 m long, and was composed of two spherical mirrors positioned slightly asymmetrically around the undulator. This value of the resonator length fulfils the requirement for the cavity round trip to the multiple of the electron bunch-bunch distance. The output coupling mirror R_1 had a radius of curvature ρ_1 = 3 m and was realized on a ZnSe substrate with a multilayer dielectric coating of 99.7% reflectivity in the range 10 - 12.5 μm. A calorimetric measurement had shown an absorption of 0.11% so that about 0.2% of the intracavity power was coupled out. The end mirror R_2 was realized with a dielectric enhanced silver coating deposited on a silicon substrate with a radius of curvature ρ_2 = 2.5 m and a reflectivity of 99.7%.

The stability parameter of the resonator was $g_1 g_2 = 0.7$. The fundamental TEM_{oo} optical mode had a waist at the middle of the undulator with a diameter $2w_o = 0.32$ cm.

The high value of the mirror reflectivity permitted evaluation of the resonator losses by measurement the decay time of the light stored in the optical cavity. A total loss $\gamma_T = 1.4\%$ was evaluated. The optical pulse changed dramatically as the electron beam was carefully aligned on the undulator axis (see Fig. 16). The radiated power increased almost linearly during the e-beam pulse up to the end of the pulse, showing the "growing up" of the stimulated radiation inside the resonator. The slope of the rising optical pulse at an average current of 120 mA indicated a net gain of about 1%. The duration of the e-beam pulse $\tau_M = 5$ μs, which corresponds to about 150 passes of the optical bunches in the resonator, did not allow saturation to be reached. A plot of the intensity of the optical pulse vs the average e-beam current is shown in Fig. 17.

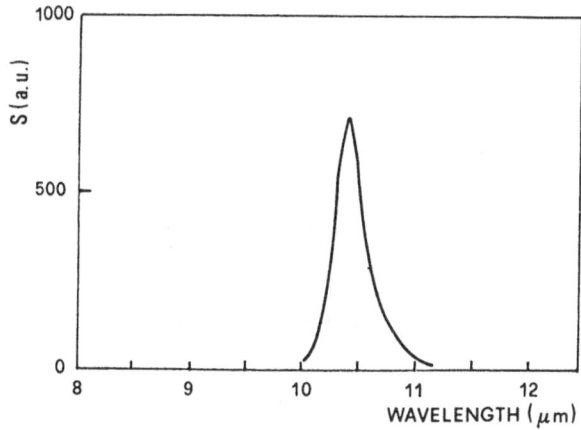

Fig 15. Spontaneous radiation spectrum emitted in a cone of aperture $\Delta\theta = 8$ mrad for the pulsed linear undulator UM2.

Tunable Oscillator

The oscillation experiment with the permanent magnet undulator, at $\lambda = 32$ μm and higher harmonics, will be achieved with the presently operative microtron whose energy may be turned in the range 15-20 MeV. The characteristics of the 20 MeV, first harmonic operation are summarized in Table 3.

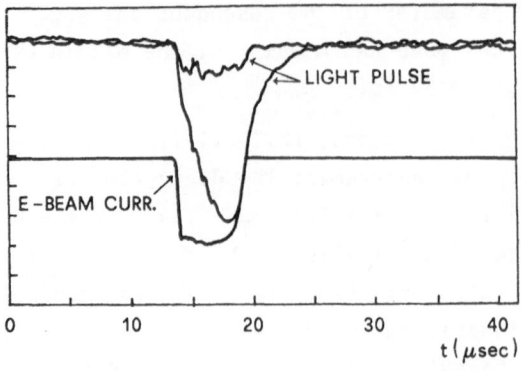

Fig 16.   Light pulse: a) below threshold (e-beam misaligned),
b) above threshold (e-beam aligned).

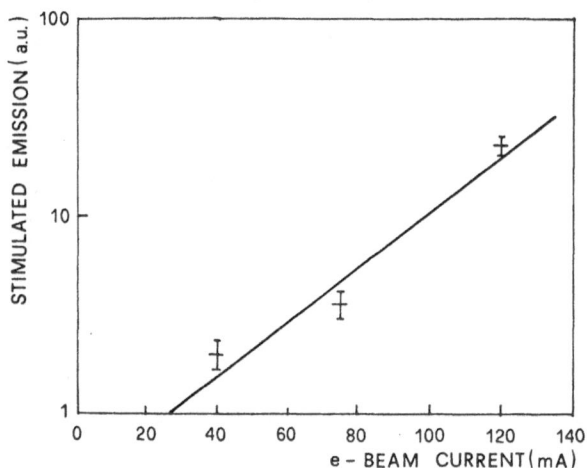

Fig. 17.   Intensity of the optical pulse vs the
average e-beam current.

Table 3.   First harmonic operation.

Wavelength λ	32.6 µm
e.b. energy E	20 MeV
Gain G	27%
Cavity losses γ_T	5%
Net gain α	22%
Pulse rise time τ_R	3.5 µs
Energy per pulse	0.7 J
Average power $\overline{P}_L$ (at 150 Hz)	100 W

Operation at higher harmonics to get laser action at λ = 16 μm is possible, in principle, in a number of different experimental configurations. The most significant ones (i.e., 16 MeV – 3rd harmonic; 20 MeV – 5th harmonic) are summarized in Table 4.

It must be stressed that the main trouble with the higher harmonics operation comes from the efficiency which is lowered by a factor of n (order of the harmonic) with respect to the operation of the fundamental harmonic.

The laser cavity for operation at λ = 32 μm consists of two spherical mirrors (R_1, R_2) asymmetrically positioned about the modulator. The mirror spacing, which must be a multiple of the electron bunch-bunch distance, is 6.35 m. A shorter spacing is not allowed due to the position and dimension of the bending magnets M_3, M_4 and the undulator. Both mirrors R_1, R_2 are 5 cm in diameter. The radii of curvature are ρ_1 = 4 m and ρ_2 = 3 m respectively. The stability parameter of the resonator is $g_1 g_2$ = 0.66. The distribution of the fundamental transverse mode in the resonator has a waist at the middle of the undulator with a diameter of $2w_0$ = 0.68 cm. The laser beam is confocal over the undulator length. The Fresnel number of the undulator is $N_u \approx 2$. Thus, the diffraction losses per pass in the undulator are

Table 4.   Third and fifth harmonic operation.

E.b. energy	(MeV)	16	20
λ	(μm)	16	16
n		3	5
γ		32	40
I_p	(A)	8	6.5
K		1	2
μ_c		0.1	0.1
$n\mu_\varepsilon$		0.8	1
$n\mu_y$		0.5	0.9
$n\mu_x$		negligible	negligible
G	(%)	11	13
γ_T	(%)	3	3
Net gain	α(%)	8	10
Rise time	τ_R(μs)	9	7
Energy/pulse	(J)	0.08	0.07
Rep. frequency	(Hz)	150	150
$\overline{P}_L$	(W)	12	10.5

less than 0.1%. The mirror R_1 is a diamond machined, copper mirror with a reflectivity better than 98%. The output coupling mirror R_2 is a multilayer, dielectric coated, silicon susbtrate with 97% reflectivity and about 0.5% transmittivity in the wavelength range 30 to 40 μm.

REFERENCES

Bizzari U., Ciocci, F., Dattoli, G., De Angelis, A., Ercolani, M., Fiorentio E., Gallerano, G. P., Letardi, T., Marino, A., Messina, G., Renieri, A., Sabia, E., Vignati, A., 1983, Bendor Free Electron Laser Conf., J. Physique Coll. Cl., 44:313.
Bizzari, U., Ciocci, F., Dattoli, G., De Angelis, A., Fiorentino, E., Gallerano, G. P., Letardi, T., Marino, A., Messina, G., Renieri, A., Sabia, E., Vignati, A., 1985a, Proc. 1984 Free Electron Laser Conf., Nucl. Instrum. Methods, A237:213.
Bizzari, U., Ciocci, F., Dattoli, G., De Angelis, A., Fiorentino, E., Gallerano, G. P., Giabbai, I., Giordano, G., Letardi, T., Marino, A., Messina, G., Mola, A., Picardi, L., Renieri, A., Sabia, E., Vignati, A., 1985b, Proc. FEL Conf. (Tahoe City, USA), Nucl. Instrum. Methods, to be published.
Ciocci, F., Dattoli, G., Gallerano, G. P., Renieri, A., Sabia, E., IEEE J. Quantum Electron., to be published.
Ciocci, F., Dattoli, G., Gallerano, G. P., Sabia, E., 1986, Bunch compression mechanism and FEL gain optimization, ENEA, TIB-FIS, Internal memorandum n.55.
Dattoli, G., Renieri, A., 1985a, Experimental and theoretical aspects of the free electron laser, in: "Laser Handbook, Vol. IV", M. L. Stich and M. S. Bass, eds., North Holland, Amsterdam.
Dattoli, G., Renieri, A., 1981a, Nuovo Cimento, 61B:153.
Dattoli, G., Marino, A., Renieri, A., 1980a, Opt. Commun., 35:407.
Dattoli, G., Marino, A., Renieri, A., 1983, in: "Physics and Technology of Free Electron Lasers", S. Martellucci, A. N. Chester, eds., Plenum, London.
Dattoli, G., Marino, A., Renieri, A., Romanelli, F., 1981b, "Progress in the Hamiltonian picture of the free electron laser", IEEE J. Quantum Electron., QE-17.
Dattoli, G., Marino, A., Renieri, A., 1980b, A multimode small signal analysis of the free electron laser, Opt. Commun., 35.
Dattoli, G., Marino, A., Renieri, A., 1980a, "Analysis of the single pass free electron laser: the multimode small signal regime", Proc. Centro di Frascati, Rome Italy, CNEN Rep. 80.35/cc.
Dattoli, G., Letardi, T., Madey, J. M. J., Renieri, A., 1985b, Nucl. Instrum. Methods, A237:326.

THE U. K. FREE-ELECTRON LASER

J. M. Reid

Kelvin Laboratory: University of Glasgow
Glasgow, G750QU Scotland

INTRODUCTION

The United Kingdom Free Electron Laser facility was organised by a collaboration, involving the Universities of Glasgow and Heriot-Watt and the Science and Engineering Research Council Daresbury Laboratory.

The facility has four essential components which are:

(a) a source of relativistic electrons, the Linear Accelerator operated by Glasgow University at East Kilbride.

(b) a "wiggler" magnet to produce a radiating perturbation system, designed and built by Daresbury Laboratory;

(c) a cavity with an infra-red detector and data acquisition system, designed and assembled by Heriot-Watt University;

(d) an electron spectrometer to measure the effect of the electron-photon interactions in the cavity on the electron energy spectrum. A novel wire detector was designed and constructed by Dundee Technical College for this purpose.

THE EQUIPMENT

The layout of these individual elements is shown in Fig. 1.

Linear Accelerator

The Linear Accelerator was constructed by Vickers Engineering Radiation Division, Swindon, England in 1966/67 and installed at the Kelvin Laboratory, East Kilbride, Scotland. It was handed over to the Physics Department of the University of Glasgow in March 1967 and from then until 1983 it was the basic facility supporting a Nuclear Structure Research Programme funded by the Nuclear Structure Committee of the

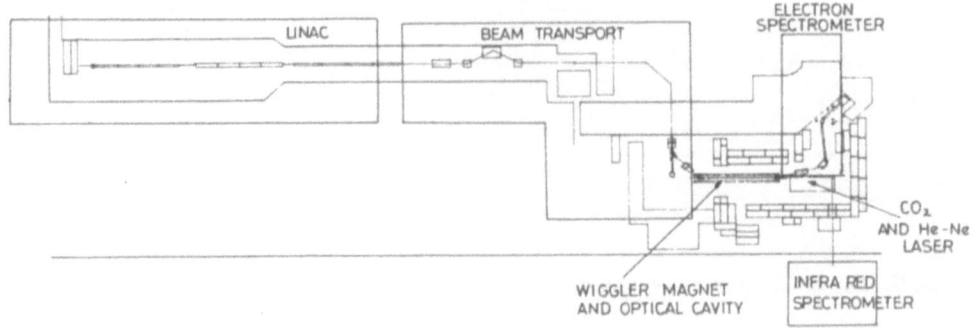

Fig. 1. Plan of U. K. FEL facility.

U. K. Science and Engineering Research Council. With the transfer of the Nuclear Structure Research to the High Duty Cycle Microtron at Mainz in West Germany, the Kelvin Laboratory accelerator became available as a suitable source for the Free Electron Laser.

The accelerator consists of three sections (Fig. 2), each supplied with radio frequency power by a Thomson-Varian 2015 Klystron. Each klystron is pulsed by an oil-filled modulator and provides 25 MW, 2856 MHz, 4 microsecond pulses of radio frequency power to its section of iris-loaded accelerating structure. The repetition rate can range from 50 to 250 Hz, with a maximum duty cycle of 0.001. Each accelerating section is divided into 4 lengths. Length 1 of Section 1 has its phase velocity graded from 0.3c to 1.0c and acts as an "electron buncher"; all the other 11 lengths have phase velocities close to $\underline{c}$.

There are seven iron-cored dipoles along the length of the accelerator to provide horizontal and vertical steering of the beam, and there are solenoidal windings on Sections 1 and 2 to provide axial focusing.

At low currents, the radio frequency in the sections can be phased to give an output electron energy of 160 MeV. As the current increases, beam loading reduces the energy, until, at 250 mA current, averaged over

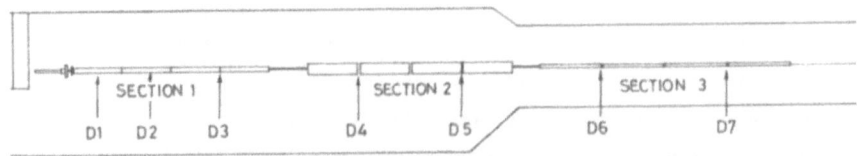

Fig. 2. 160 MeV electron linear accelerator.

the 4 microsecond macropulse, the energy is about 100 MeV. At higher currents still, defocusing R. F. modes are set up in the structure and "beam blow-up" occurs.

As a consequence of the action of the "buncher", the electrons occupy a R. F. phase angle of about 6 degrees, which at 2856 MHz means that each output electron micropulse is of 5.8 picoseconds duration and is 1.75 mm long.

The measured energy spread at the exit of the accelerator is about 1% full width half maximum at 100 MeV, i.e. +/- 0.5 MeV. This spread arises from the different trajectories of the electrons in the early part of the accelerator and is independent of the final energy, i.e. it remains +/- 0.5 MeV when the R. F. phases are adjusted to give an output energy of 50 MeV. This was too wide an energy spread for the electron scattering experiments carried out in the 1970's. It was much reduced by an Energy Compression System (ECS), illustrated in Fig. 3, which, by sending the electrons round a magnetic system in which the distance traveled is dependent on electron momentum, and then recombining them, sorts the electrons out according to their momentum, the slower being delayed with respect to the faster. The dispersion is about 45 degrees of phase angle per % energy difference. At 50 MeV with the energy spread of 2% the phase angle is about 90 degrees and the micropulse duration thus about 90 picoseconds. The pulse is then passed through an R. F. cavity, phased to accelerate the low momentum and decelerate the high momentum populations. This reduces the energy spread to +/- 0.15 MeV. As the electrons are relativistic, the operation of the cavity does not change the length of the bunch. It leaves the electrons occupying 90 degrees phase angle, i.e. 90 picoseconds in duration and 2.7 cm long.

The Macropulse

Because of the filling time of the accelerating structure, which is about 0.75 microseconds, the effective electron macropulse is shorter than the 4 microsecond duration of the R. F. pulse from the klystrons. Assuming the macropulse to be effectively 2 microseconds long, we find that it will be a train 600 metres long, containing 5700 equally spaced micropulses of electrons separated by 10.5 cm. The number is reduced to 950 and the separation is increased to 63 cm when the Sub-harmonic Buncher described below is used.

The Micropulse

Without the Energy Compression System in operation, the micropulses are 5.8 picoseconds in duration and 1.75 mm long. Corresponding to an

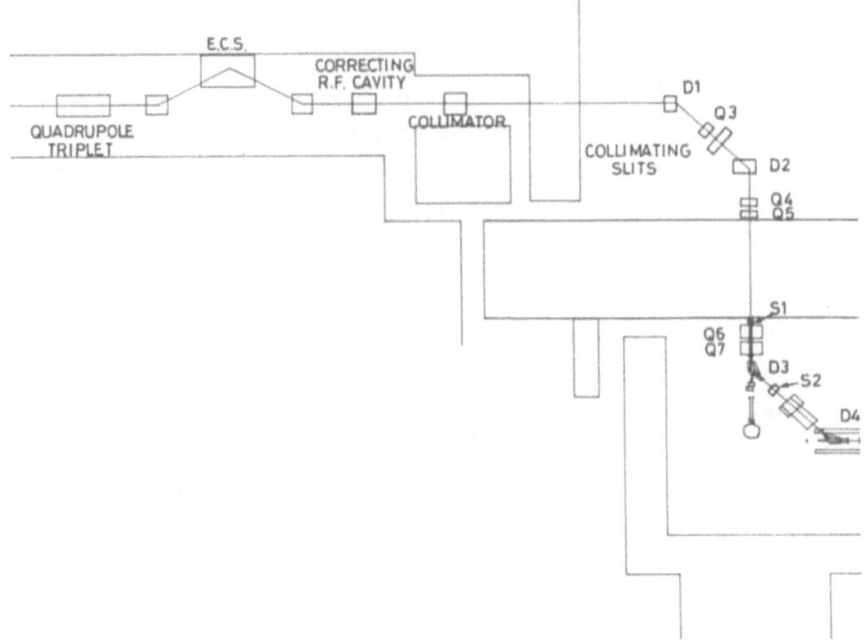

Fig. 3. Layout of beam transport components including
the energy compression system.

average current of 250 mA in the 2 microsecond macropulse, the current
averaged over the micropulse will be 15 Amps.

When the ECS is brought into operation the duration of the micropulse is increased to 90 picoseconds and its length to 27.2 mm; the
average current in the micropulse is reduced to about 1 Amp.

The Triode Gun

When operating for Nuclear Structure Research a diode gun was used
to inject electrons into the accelerator. For FEL operation a triode
gun has been installed. The cathode/grid structure is that of a TV306
valve with a cathode emitting surface of 12 mm in diameter. Its grid to
cathode cut-off voltage is less than 20 volts.

The Sub-harmonic Buncher

It is found that the beam blow-up which sets the limit to the
achievable beam current is dependent on the current averaged over several electron macropulses. If some of the electron pulses in the train
are regularly left unpopulated, it is possible to achieve higher average
micropulse currents. Every sixth pulse is populated by the following
arrangement. A pulse of 5 microseconds duration is applied to the cathode of the triode gun. A radio-frequency signal of 476 MHz is applied,
via a coaxial resonant line, to the grid. When suitably phased with

respect to the 2856 MHz filling the accelerator structure, the desired "one-in-six" operation is achieved. Despite beam blow-up, with the Sub-harmonic Buncher (SHB) in operation, currents averaged over the micropulse of 6 Amps should be available with the ECS in operation.

The Beam Transport System

As shown in Fig. 3, the electrons are transported in vacuo some 30 metres from the exit of the accelerator, through the ECS and Beam Deflection Room to the wiggler magnet in the Laser Hall. Along the way, the beam has to be steered, its size controlled and diagnostic facilities provided.

Between the accelerator and the ECS, there is a quadrupole triplet which provides general control over the divergence of the electron beam, focusing it through a water-cooled collimator following the ECS. This collimator forms the small area source for the beam handling system that follows, decoupling it from any beam disturbances occurring upstream. Two 90 degree deflecting assemblies, D1/Q3/D2 and D3/Q8/D4, provide a parallel displacement of the beam. These assemblies are achromatic, i.e. the beam emerges from the second with the same geometrical distribution and divergence as it entered the first; in particular there is no energy dispersion as between input and output. There is however, energy dispersion between dipoles D1 and D2 and between D3 and D4. Between D1 and D2, adjustable horizontal slits determine the momentum band of electrons transmitted to the Wiggler. This band can be as narrow as 0.1% FWHM. Quadrupoles Q4 and Q7 allow adjustment of the beam divergence in both the horizontal and vertical planes. Quadrupoles Q4 and Q5 are wired to steer as well as to alter the beam divergence. Additional steering is provided in the horizontal and vertical planes by S1 and S2. The second achromatic assembly is fitted with an air coil steering facility in the vertical plane and a trim coil on d4 for fine adjustment in the horizontal plane.

Six ports are provided, one at the centre of the circle after D4, the other five along the wiggler magnet, through which television cameras view fluorescent screens; these permit the operator in the Control Room to monitor the beam position and size as he adjusts the various magnetic parameters. Some eight non-intercepting current toroids give a pulse presentation of the electron current at strategic points along the beam line so that the loss of electrons can be quickly observed as the steering is adjusted.

The vacuum in the beam handling system is maintained at about 0.1 microtorr by two turbo-molecular pumps, one sited near quadrupole Q8, the other after the cavity en route to the electron spectrometer.

The Wiggler Magnet

A perturbing magnetic field of planar, as distinct from helical, geometry is used.

The wiggler magnet (Fig. 4) produces a near sinusoidal magnetic field with spatial wavelength 65 mm in the horizontal plane. It extends for 76 wavelengths, i.e. 4.94 metres, and is constructed in four equal sections. Each section consists of two linear arrays of blocks of samarium cobalt, permanently magnetised, dimensions 70x15.9x15.9 mm arranged as in Fig. 5. To prevent the displacement of the centre of the beam oscillation from the axis that would otherwise occur, the regular block arrangement is departed from by starting and ending each of the eight arrays with a half-width block (70x15.9x7.95 mm). There are thus 75 full and 2 half-blocks in each of the four arrays and 600 full blocks and 16 half blocks in the wiggler. The field of each block was measured and the blocks were then assembled in an order that minimised the accumulation of error in each section. The block dimensions allow for a gap of about one third of a millimetre between blocks.

There are alignment monuments at the ends of each of the four wiggler magnet sections, and five TV beam monitors (V2 to V6), one at each end of the wiggler and one between each section.

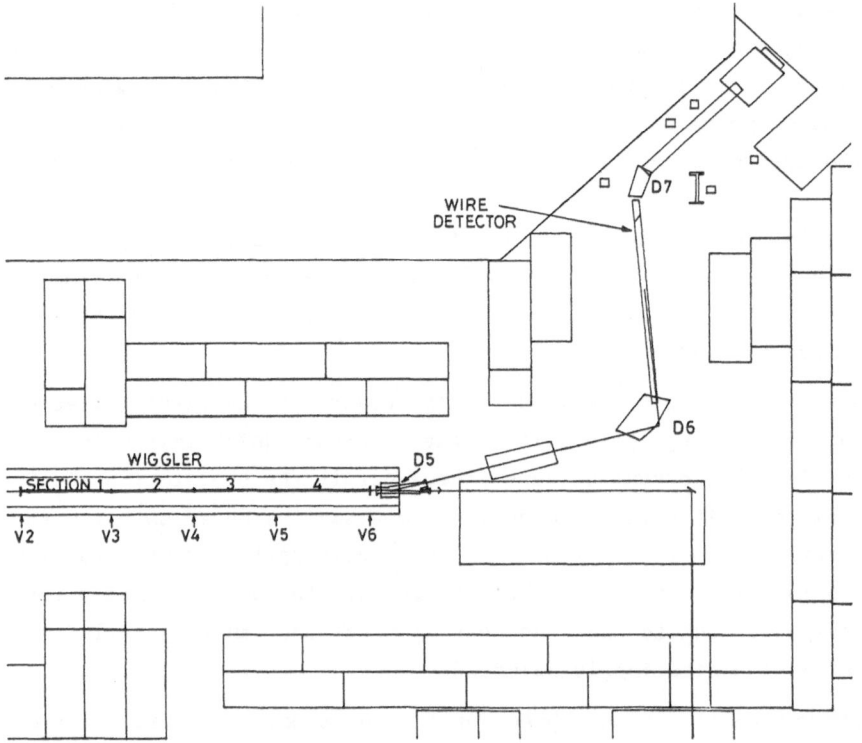

Fig. 4. Arrangement of wiggler magnet and the electron spectrometer.

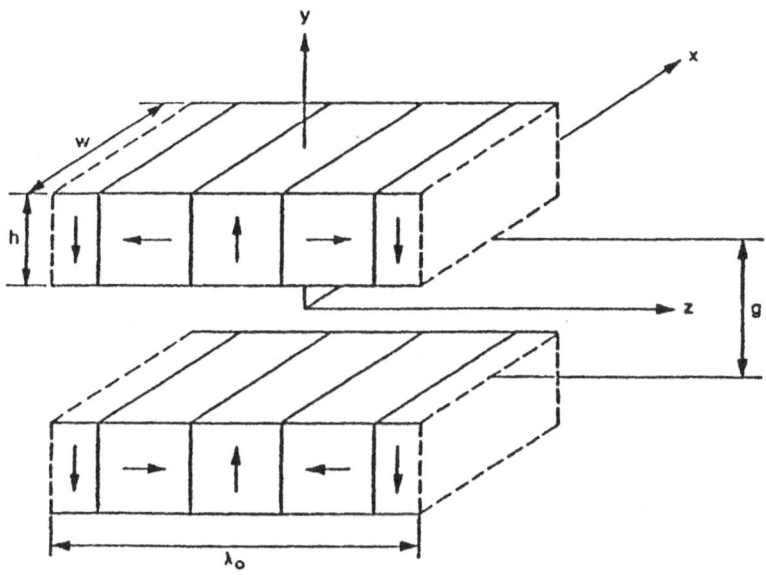

Fig. 5. Arrangement of permanent magnets in the wiggler.

The Wiggler Vacuum Vessel

The stainless steel (permeability about 1.1), rectangular section, vacuum envelope in the gap between the wiggler arrays is 20 mm outside dimension, 1 mm wall thickness and 60 mm high (Fig. 6). This limits the minimum gap between arrays to 22 mm. At this separation the field on the axis is designed to be 0.46 Tessla. The gap can be opened out to 200 mm, when the field is reduced to less than 0.2 mT. With the minimum gap setting the maximum K of 2.7 is attained.

The on axis field is adjusted by carefully altering the gap, there being a variation of 4.8% per mm gap change.

With lamda wiggler=65 mm, at electron energy of 50 MeV(gamma=100), the maximum amplitude of "wiggle" is less than half a millimeter. Adjustment of the gap alters K, the wiggler field parameter, and hence the wavelength of the spontaneous radiation. In the gain experiments discussed below, the electron energy was 58.4 MeV and K was carefully adjusted to be 2.4, giving a spontaneous emission wavelength of 9.65 microns and providing gain on the 9.7 micron 9P(36) emission line of the CO_2 laser.

Electron Spectrometer

A high resolution electron spectrometer, based on magnets D5 and D6 (Fig. 4), analyses the electron beam after its interaction with the photon beam in the cavity in the wiggler. A wire monitor is positioned at the focus of D6, inclined at 42.4 degrees to the beam direction with

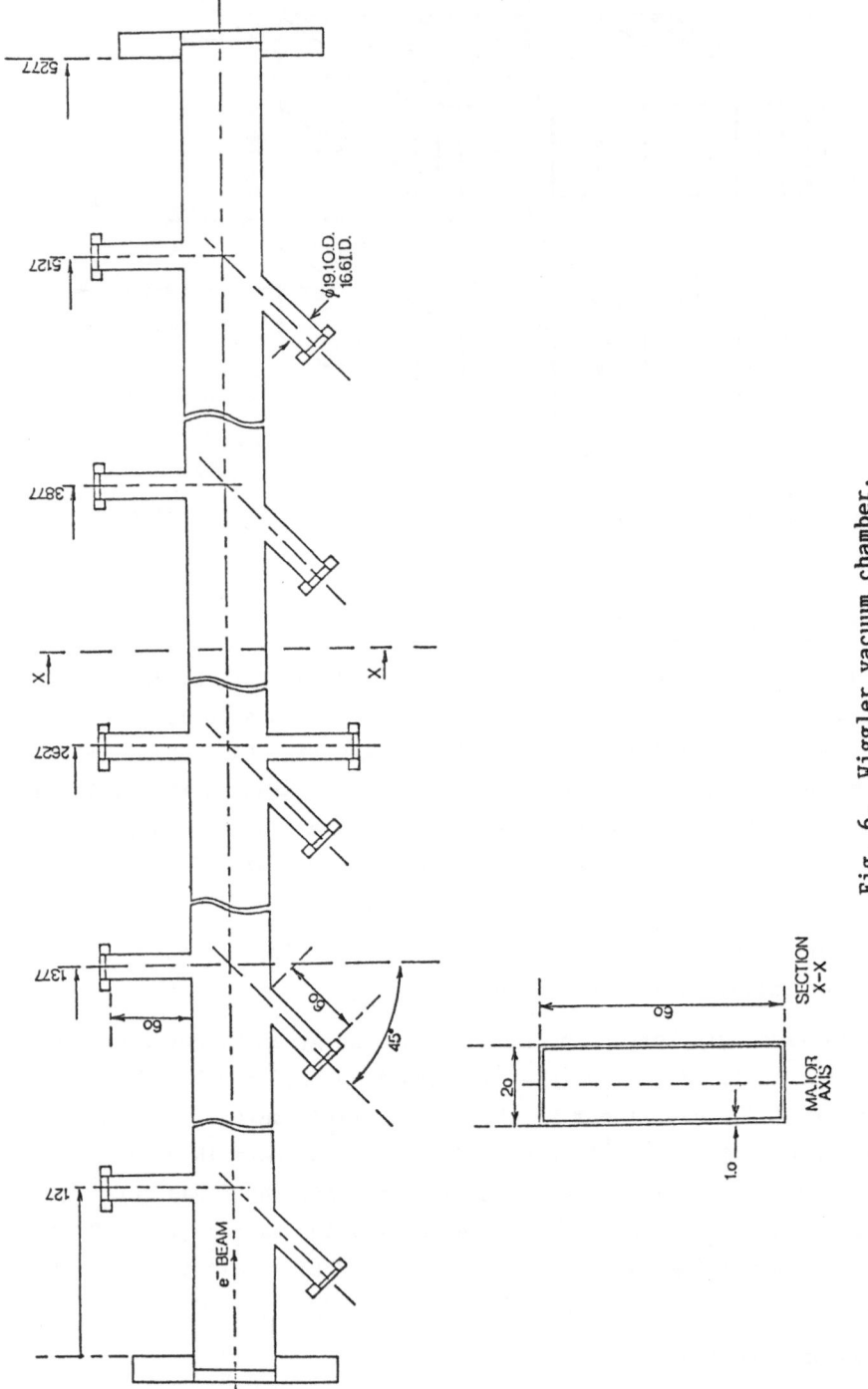

Fig. 6. Wiggler vacuum chamber.

a design dispersion of 44.3 mm per %. Focusing in the vertical plane is avoided to reduce the possibility of wire burnout. 96x40 micron thick tungsten strips tensioned on a stainless steel frame and insulated from it by a machined ceramic support are used as secondary emission detectors. The signals from the individual foils are processed by a microprocessor, situated below the detector, capable of "freezing" the spectrum during one macropulse and transmitting it to a microcomputer in the Control Room. This computer produces a spectrum on a real time display and can also produce "hard copy" on a simple printer. It is extremely useful for monitoring the energy of the electrons in the linac pulses.

A 30 foil prototype wire detector of identical design is used for standard emittance measurements in the 0-degree line with D1 switched off.

It is proposed to make retractable miniature detectors on the same principle for monitoring beam position at chosen locations.

Beam Dump

D7 bends the beam in the direction of a water-filled beam dump, which the beam enters after passing out of the vacuum and crossing an air gap of 30 cm. There is a thick concrete surround to this whole area.

Laser Cavity

The cavity is formed by two mirrors, 7872 mm apart, radiussed to about 4.5 metres to give near-concentric geometry (Fig. 7). The overlap of optical and electron beams was optimised by minimising the Gaussian beam area over the cavity length of 7.872 metres. This procedure led to a waist radius of 6.5 mm at the mirrors and 4.4 mm at the wiggler termini. With WO=2.2 mm and lamda=10.6 microns, ZR/min=1.44 metres, where ZR= the Rayleigh range. It was calculated that the 18 mm dimension of the vacuum system between the poles of the wiggler would not clip more than 0.03% of the beam over the whole tuning range (2-20 microns) of the FEL.

This mode was chosen as offering the prospect of maximum overlap of the infra-red beam with the electron beam. The mirrors, which are inside the vacuum system, sit on heavy plinths and their separation and individual orientations can be altered remotely. The mirror separation is variable over a range of 20 mm in steps of 0.05 microns.

The central 6000 mm of the cavity consists of the stainless steel vacuum system sitting between the poles of the "wiggler" magnet.

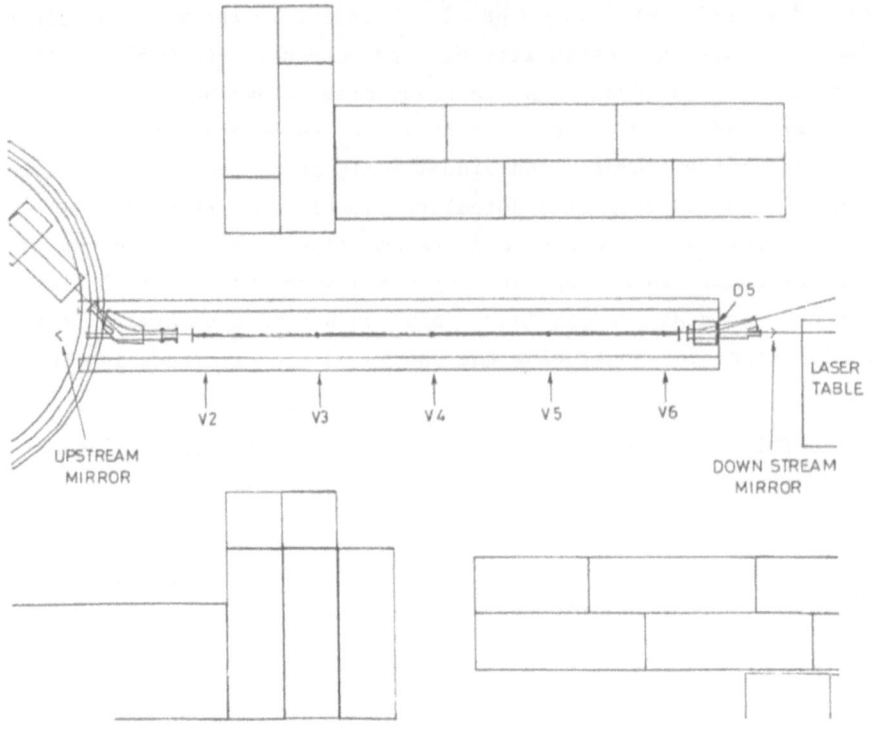

Fig. 7. Details of wiggler magnet.

The round trip time for an optical pulse of the 7.8723 metre long cavity is 52.5 nanoseconds and so within the 2 microseconds of the macropulse, an optical pulse from the first electron bunch makes 38 round trips.

To enable the electron beam to be accurately aligned and so guard against betatron oscillations, arising from the beam entering the wiggler with incorrect entry parameters, causing collisions with the walls of the vacuum tube, 5 beam position monitors were inserted at the ends of the wiggler and between each section.

The alignment of the cavity optics is checked with a Hughes 3235HPC He-Ne laser, set up to produce wavefront interference "ring" structure.

Optical Components

It was planned originally that, to minimise optical losses, the upstream mirror would be gold-coated, copper-nickel (50 mm diameter, 4.5 metre radius of curvature). However, for the recent series of experiments where there was a requirement to send a beam from the CO_2 laser traveling with the electron beam through the wiggler, partially transmitting mirrors have been installed at both ends of the cavity. These are plano-concave lenses of zinc selenide. The plane surface has an anti-reflection coating. The concave surface, radius of curvature 4.5 metres, is coated with a dielectric multilayer stack with reflectivities of 98% and 99% in the mirrors used. To permit alignment with the He-Ne laser (lamda=632.8 nm) high surface precision is required.

In the near future higher reflectivity mirrors (reflectivity 99.7%), also on ZnSe substrates, but with confocal geometry, will be installed.

The vacuum windows on the cavity are optically flat barium fluoride plates which have 70% to 90% transmission from the UV to the mid IR and have substantial immunity from radiation induced discoloration.

CO_2 Laser

The CO_2 laser used for "seeding" the cavity and for gain measurements is an Edinburgh Instruments (EI) PL3 CO_2 Laser. It is line-tunable and offers high gain at 10.6 microns, the nominal FEL operating wavelength.

Capable of more than 10 watts CW output, its frequency, operated in the sealed-off mode, was stable to 100 kHz over 1 second and its amplitude stable to better than 0.5%.

The laser was set up to produce far-field TEM_{oo} mode, vertical polarised output. There is a rotating slit in the system which chops the infra-red output into 3 millisec pulses, 25 pulses per second.

Optical Tables

At the exit from the cavity there is a very heavy steel table (12'x4'6") on which sits the CO_2 Laser, together with a He-Ne Laser and other diagnostic equipment. Measurements were made to confirm that the table was free from vibrational disturbance. From a mirror on this table, the optical beam from the cavity is reflected through a hole in the screening wall and across a second table on which the infra-red detectors and spectrometer are mounted. The mirrors used on the CO_2 beam path are all aluminium coated. A large rectangular ZnSe flat is used as a beam splitter. It is coated for 50% reflectivity at 10.6 microns and the highest possible transmission at 633 nm(He-Ne) to reduce alignment laser losses.

Alignment of the Output Optics

The He-Ne beam (Fig. 8) is first injected into the 8 metre long cavity and, using the "pop-down" targets, aligned along the axis. The wavefront interference rings are monitored in front of the laser, a portion of the retro-reflected beam being at the same time separated at

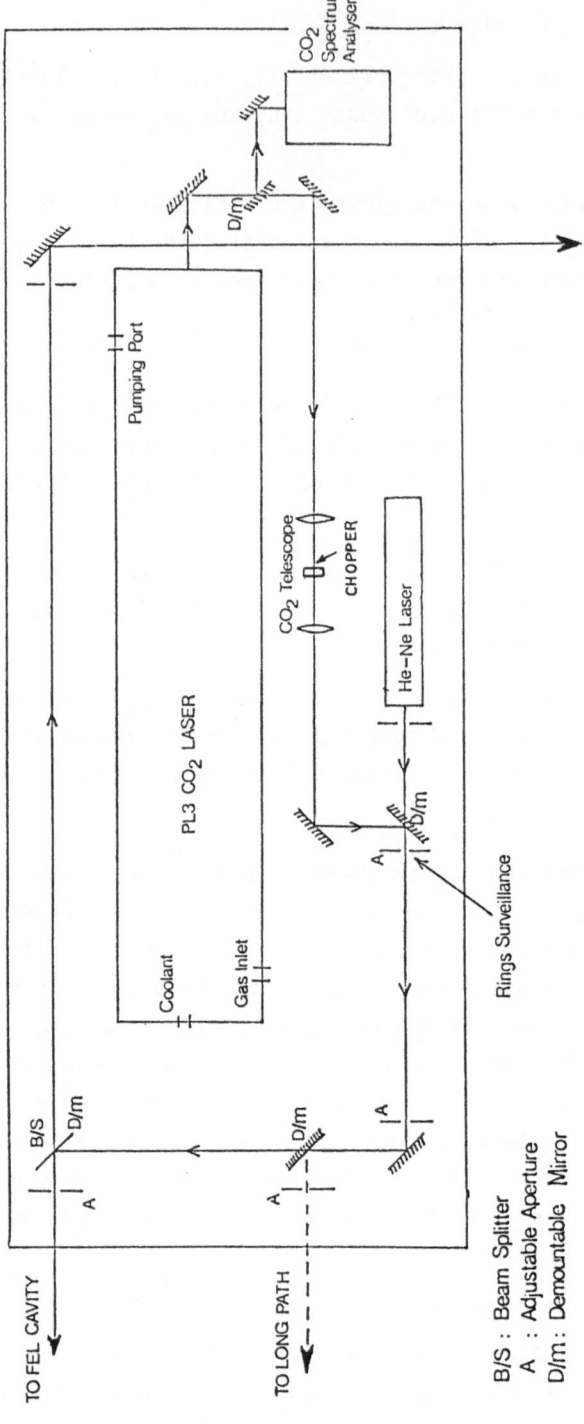

Fig. 8. Details of arrangement of diagnostic laser beams.

B/S : Beam Splitter
A : Adjustable Aperture
D/m : Demountable Mirror

the beam splitter and deflected into the diagnostic area for equipment alignment.

The CO_2 laser and the He-Ne beams were coaligned over a 10 metre long path.

Optical Diagnostics

The optical diagnostics used has three main components, an X-Y translator with pinhole, monochromator and detector. These permit the measurement of spatial and spectral properties of the emitted radiation. With the addition of the data acquisition system, the temporal properties can also be investigated.

The monochromator is a Spex dual-grating model with fast optical speed (f/4), high transmission efficiency (>70%) at blaze wavelength, and low scattered light level. The grating blaze wavelengths of 500 nm and 10.6 microns are suited to the visible-harmonics and IR outputs.

A variety of photodiode detectors were used in the visible-harmonic diagnostic work, silicon avalanche photodiode, large area photodiode and vacuum photodiode. Detectors for the IR were limited to cooled (77K) CdHgTe(MCT) photoconductive/photovoltaic models.

Visible Harmonic Spontaneous Radiation

For the analysis of the visible harmonic spontaneous radiation, the grating monochromator and photodiode were used. Spatial and spectral profiles of the spontaneous radiation were also observed with a colour TV camera.

Data Acquisition and Analysis System

The system is based on a BBC(Acorn) Computer model B, which is situated in the control room. It is fitted with a disc drive for data storage and programme archiving, and is interfaced via a remote-control system (REMDACS Intersil) sited in the diagnostic area, to a variety of optical beam steering and diagnostic components. The position of each instrument is recorded and continuously updated in the memory.

A microcomputer, a TDS900, sited in the diagnostic area deals with the chopper control, the timing, sampling, digitising, temporary storage of data and data manipulation (Fig. 9). Synchronisation to within 5 microseconds between the system and the linac macropulse enables the chopper to be controlled to allow each alternate macropulse to be overlapped with an injected chopped pulse from the laser (Fig. 10). The signal from the detector was sampled and digitised a total of six times, three in each macropulse. The first three measurements, 27 microseconds apart, established the laser intensity with and without gain and also

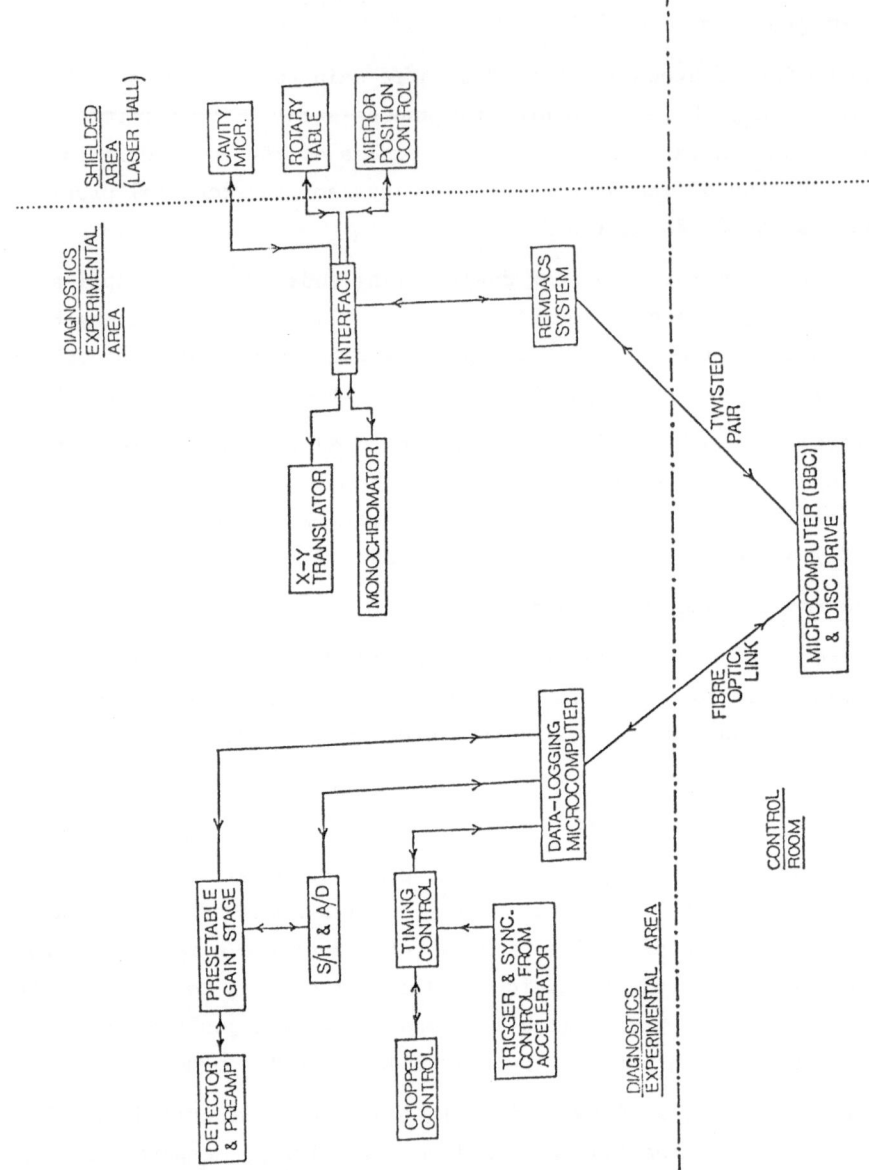

Fig. 9. Block diagram of computer control and data acquisition system.

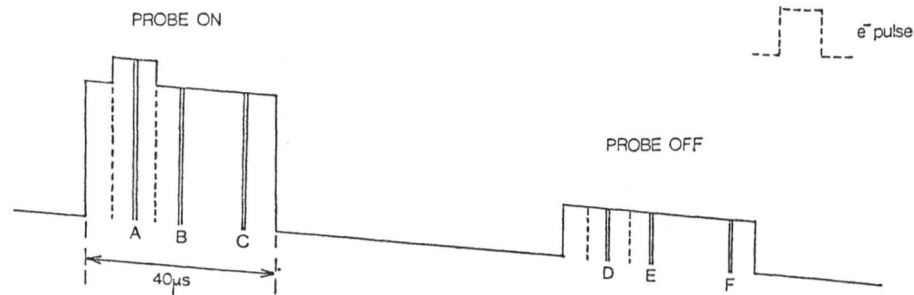

Fig. 10. Arrangement of data acquisition system for
gain measurement.

subtracted errors introduced by a sloping or fluctuating probe pulse
intensity. The second set of measurements taken when only FEL spontaneous emission and background were present, enabled the errors introduced
by spontaneous emission and DC voltage offset of the detector system to
be subtracted. The gain, G, could then be calculated from the relation

$$G = (A-2B+C-D+F)/(B-E)$$

The minimum measurable gain was established as 0.01% by analysing
the slope of an injected, chopped He-Ne laser pulse.

It was possible to study the transient properties of the gain with
the mirrors adjusted to form a cavity with 0.1 microsecond resolution by
adjusting the timing of the sample-hold and analogue-digital converter
in 100 nanosecond steps. Both raw and processed data were processed via
the fibre-optic link to the master control computer in the linac control
room. Integration of control and data logging enables complicated syn-
chronised experiments to be carried out with all relevant data and parameter settings logged together for later analysis.

MEASUREMENTS MADE WITH THE EQUIPMENT

Spontaneous Radiation

With the electron energy set to 100 MeV, the energy for optimum
linac performance, with the wiggler set to a high K-value and the mirrors removed from the cavity, there was copious spontaneous emission.
It was found, as expected, that upwards of 80% of the spontaneous radiation was emitted in the higher harmonics. These were readily observable up to about the thirtieth (Fig. 11) when they pass out of the
visible.

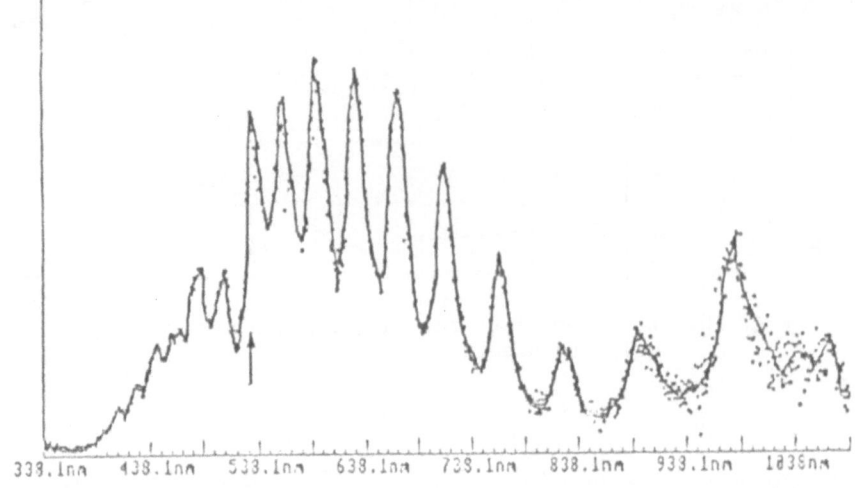

Fig. 11. Spectrum of spontaneous radiation in the visible.

Advantage was taken of the third harmonic, at a visible wavelength
of about 600 nm, in the initial setting up of the wiggler magnet. Spontaneous radiation spectra from each section in turn, the other three set
to maximum gap, were measured (Fig. 12) to confirm the alignment of the
sections and the calibration of the wiggler magnetic field.

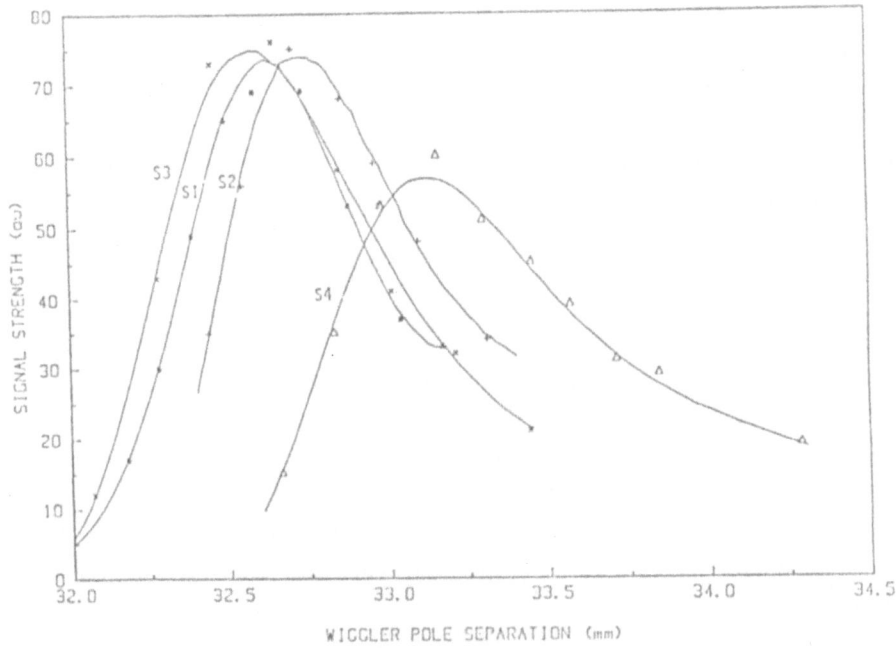

Fig. 12. Variation in intensity of third harmonic with
gap separation of each wiggler section.

At a reduced electron energy of 58 MeV and with the wiggler field set to give a K-value of 2.53, corresponding to the nominal fundamental operating wavelength of 10.6 microns, scans of the higher harmonic spectra were made and found to compare favourably with theoretical predictions.

Fundamental wavelength spontaneous emission at 10.6 microns was observed and its linewidth carefully measured. The linewidth was only about 20% greater than expected.

When the cavity mirrors were installed and aligned and their separation set to synchronise the IR pulses being reflected backwards and forwards between the mirrors, with the single transit electron pulses, a considerable enhancement of the spontaneous radiation resulted (Fig. 13).

The exponential decay of the energy from the cavity on resonance provided a check on the optical losses in the cavity. At this stage the non-observance of the expected onset of oscillatory behaviour was thought to lie with insufficient gain per pass and gain measurements in the infra-red were put in hand.

Gain Measurement

With the cavity mirrors removed, FEL gain was measured at 10.6 micron in March 1986. Better laser stability was achieved using the 9P(36) 9.7 micro emission line of the CO_2 laser and this line was used in later gain measurements.

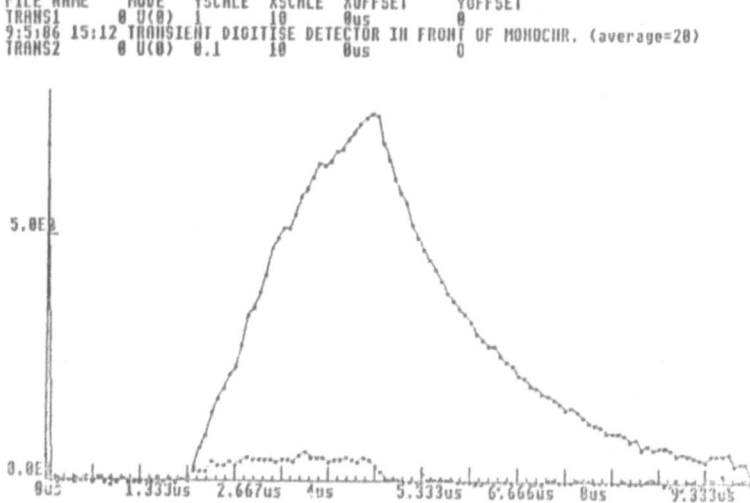

Fig. 13. Enhancement of spontaneous radiation on cavity resonance.

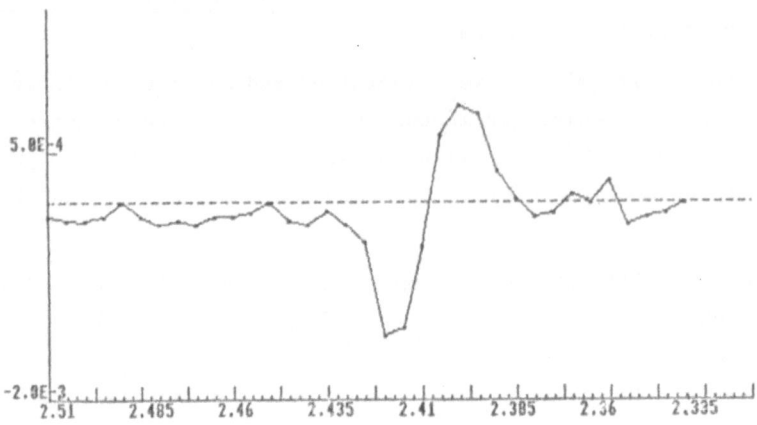

FILE NAME   MODE   YSCALE XSCALE XOFFSET   YOFFSET
GAIN5   9 R(7) 5E-4   32   2.51   2E-3
9:4:86 14:45 GAIN MEASUREMENT K=2.51 TO 2.35 STEP .005 AVERAGE OF 1 SCAN TERMINA
TED @ K=2.34 (average=2000)

Fig. 14.   Single pass gain at 9.7 microns.

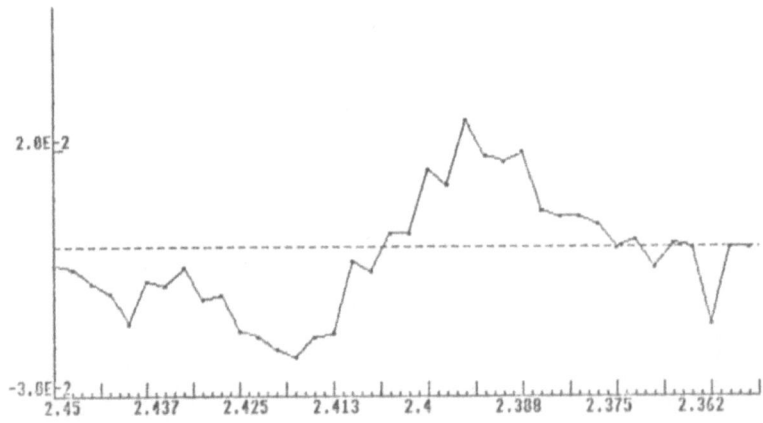

FILE NAME   MODE   YSCALE XSCALE XOFFSET   YOFFSET
GAIN6   9 R(29) 1E-2   32   2.45   3E-2
8:7:86 16:35 GAIN MEASUREMENT WITH CAVITY Q HIGH AND ON RESONANCE K=2.45 TO 2.38
STEP .005 2 SAMPLES .5MM SLITS (average=2000)

Fig. 15.   Gain with mirrors installed to form cavity.

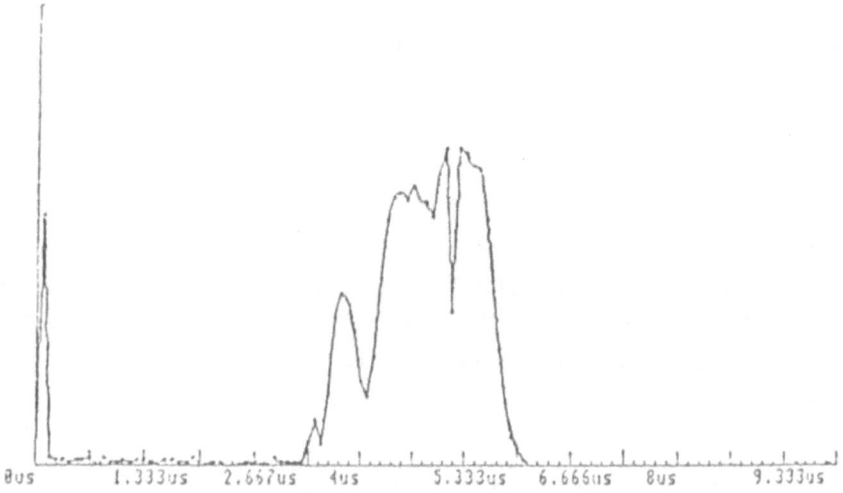

Fig. 16.   Time variation of electron pulse.

FILE NAME   MODE   YSCALE XSCALE XOFFSET   YOFFSET
GAIN4   9 U(7) 3   32   2.455   -6
26:3:86 15:41 GAIN MEASUREMENT K=2.455 TO 2.355 STEP .005 CO2 LASER 9P36 ENERGY
58.4MEV SLITS 1.25MM CURRENT 120MA 2 AVERAGES (average=2000)
GAIN4   9 U(8) 3.3333 32   2.455   -6

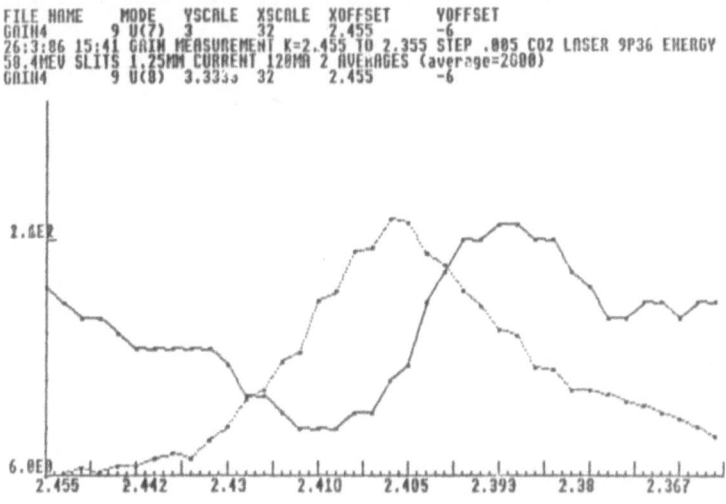

Fig. 17.   Gain and spontaneous radiation curves.

The predicted curve of gain against K was observed (Fig. 14).

The gain measured with a sampling time of about 100 nanoseconds has been steadily improved over recent weeks by careful electron beam alignment and by the achievement of more satisfactorily shaped electron current pulses.

The mirrors were installed and gain measurements made. The best gain measured with this configuration is now about 0.7% at 9.7 microns (Fig. 15).

Allowing for this being the average over the micropulse and the space between pulses and assuming a bunch length of 80 picoseconds and therefore a peak gain multiplication factor of 20, a single pass gain of 15% is arrived at. It is hoped in the near future to improve the time profile of the electron pulse (Fig. 16) by programming the grid voltage on the electron gun to remove pulse ripples from the modulator delay lines and so achieve further improvements in gain.

A measurement of gain and spontaneous radiation under the same conditions (Fig. 17) shows the expected time relationship.

Work will continue to try to improve the gain by more careful adjustment of the many parameters involved in setting up the accelerator, the wiggler, the cavity and the CO_2 Laser. The repeatability of the present measurements and the improved stability of the overall system give hope that values large enough to see the onset of oscillation growing from IR injection is achievable.

THE ACO STORAGE RING FREE ELECTRON LASER

Pascal Elleaume

Laboratoire pour l'Utilisation du Rayonnement
Electromagnetique, LP CNRS 008, Bâtiment 209D,
Universite de Paris-Sud, 91404 Orsay Cedex, France

INTRODUCTION

This paper is divided into two parts. The first part describes the elements constituting the free electron Laser: the storage ring, the undulator/optical klystron, and the optical cavity. This is followed by a summary of the main characteristics of the FEL beam. The second part shows how crucial it is to use a high brightness electron beam.

GENERAL DESCRIPTION OF THE FEL

The ACO Storage Ring

Figure 1 presents a view of the ACO storage ring. It is a small ring with a perimeter of 22 m. The experimental section is new occupied by the undulator/optical klystron. The ring has 8 dipole magnets and 4 quadrupole triplets periodically located between the bending magnets. The present operation of the ring requires three main power supplies to run the dipoles and quadrupoles. The 8 dipoles are connected in series to one supply. The 4 central quadrupoles of each triplet are connected in series to a second supply. The 8 other quadrupoles are connected in series to the third supply. The focusing is therefore determined by two values, K_L and K_C, which are proportional to the current circulating in the lateral and central quadrupole, respectively. The linear operating point of the storage ring is essentially determined by the energy of the beam and the focusing of the bending and quadrupole magnets, i.e. K_L and K_C. In addition smaller supplies are used to power the rf cavity, sextupole magnets (in order to correct the chromaticity) and horizontal and vertical dipole corrections (to tilt the closed orbit in the vacuum chamber).

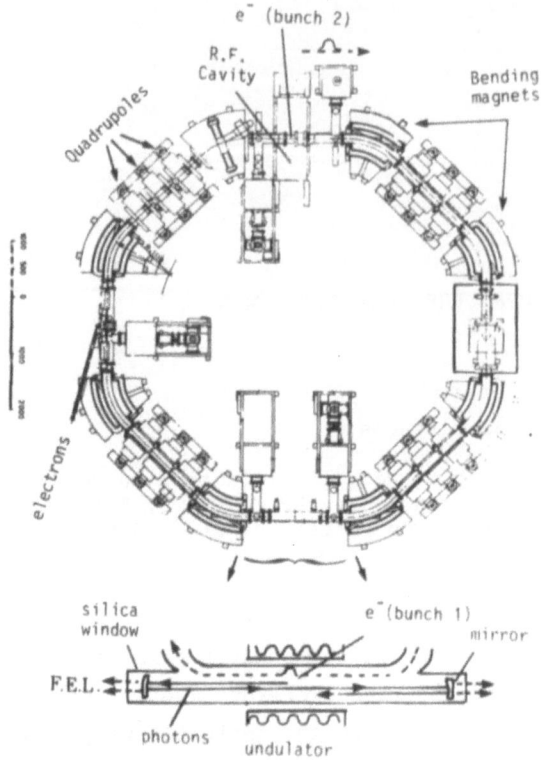

Fig. 1. Schematic of the ACO storage ring.

Very important focusing parameters are the horizontal and vertical
tunes ν_X and ν_Z, which represent the number of betatron oscillations per
revolution in the ring. Beam cross sections and angular spread are
determined by the so-called β functions related to ν_X and ν_Z by the
relations

$$\nu_X = \frac{1}{2\pi} \oint \frac{ds}{\beta_X(s)} \,. \tag{1}$$

$$\nu_Z = \frac{1}{2\pi} \oint \frac{ds}{\beta_Z(s)} \,. \tag{2}$$

where $\beta_X(s)$ and $\beta_Z(s)$ are the horizontal and vertical beta-functions at
position s on the circumference. Since there are only two sets of independent quadrupoles, there is a one to one relationship between (K_L, KC)
and (ν_X, ν_Z). In the following we shall discuss the operating point
in terms of (ν_X, ν_Z) instead of (K_L, K_C).

Within the approximation of no collective effects (sufficiently
small stored current) the rms transverse dimensions σ_X and σ_Z and rms
angular spread (σ_X' and σ_Z') are:

$$\sigma_X^2 = \varepsilon_X \beta_X + \left(\eta \frac{\sigma_\gamma^2}{\gamma} \right) \,. \tag{3}$$

$$\sigma_Z^2 = \varepsilon_Z \beta_Z \quad . \tag{4}$$

$$\sigma'^2_X = \frac{\varepsilon_X}{\beta_X} \left(1 + \frac{\beta'^2_X}{4} \right) \quad . \tag{5}$$

$$\sigma'^2_Z = \frac{\varepsilon_Z}{\beta_X} \left(1 + \frac{\beta'^2_Z}{4} \right) \quad . \tag{6}$$

where ε_X and ε_Z are the horizontal and vertical emittance independent of s, $\eta(s)$ is the dispersion function at position s, σ_γ/γ is the relative rms energy spread of the beam and β'_X (β'_Z) is the derivative of β_X (β_Z) with respect to s. A detailed derivation and explanation of Eqs. (3) through (6) are given by M. Sands (1971). And I shall only deal with their consequence to the FEL operation on ACO.

The operating point used for the FEL experiment corresponds to

$$\nu_X = 2.82 \tag{7}$$

$$\nu_Z = 1.63 \tag{8}$$

$$\left. \begin{aligned} \beta_X &= 1.8 \text{ m} \\ \beta_Z &= 1 \text{ m} \\ \beta'_X &= \beta'_Z = 0 \\ \eta &\simeq 0.1 \text{ m} \end{aligned} \right\} \quad \text{at undulator location} \tag{9}$$

$$\left. \begin{aligned} \frac{\sigma_\gamma}{\gamma} &= 2 \text{ to } 12 \times 10^{-4} \\[1em] \varepsilon_X &\simeq \varepsilon_Z \simeq 1 \text{ to } 2 \times 10^{-8} \text{ mrad} \end{aligned} \right\} \quad \text{at 220 MeV} \qquad \begin{aligned} &(10) \\[1em] &(11) \end{aligned}$$

The longitudinal profil of the electron beam and the energy spread are largely dominated by turbulent bunch lengthening. As a consequence the RMS bunch length σ_ℓ and energy spread are strong function of the stored current. The operating point is located on a second order coupling resonance $2\nu_X - \nu_Z = 4$.

The coupling resonance permits partition of the total emittance ε into:

$$\varepsilon_X \simeq \varepsilon_Z \simeq \frac{\varepsilon}{2} \quad . \tag{12}$$

The beam cross section in the undulator is therefore almost circular $\left(\sigma_X/\sigma_Z \sim \sqrt{\beta_Z/\beta_X} = 0.75 \right)$ as opposed to a noncoupled case where usually $\varepsilon_Z \ll \varepsilon_X$ leading to a flat beam cross section. For FEL operation the coupled case presents two advantages over the noncoupled case: a better beam Touschek lifetime (smaller intrabeam scattering rate) and a better overlap with the TEM_{oo} mode of the optical cavity.

Let us point out the crucial importance of sweeping the ions trapped in the electron beam. Their presence results in an important lifetime reduction and vertical dimension enhancement. Trapped ions were removed by applying 500-1000 V to the long electrodes in the vacuum chamber. The stable operation of the ring then required the use of sextupoles to shift both chromaticities to a positive value and therefore avoid the head-tail instability.

An important feature concerning the lifetime of the beam stored in the ring, was observed for the first time and could be of great importance for the future of storage ring FELs. The lifetime was greatly improved by turning on the laser (1 h 10 min to 2 h 30 min) in one case. This is easily explained by the strong bunch lengthening observed with the laser saturation. A decrease of the electron volume density produces less intrabeam scattering and a higher Touschek lifetime. Coherent intrabunch and bunch to bunch oscillations were also seen to be stabilized as the laser was turned on.

For the FEL experiment electron energies ranging from 150 to 240 MeV have been used. The optimum energy is probably around 220 MeV. In the higher energies one has less gain and mirrors reflectivity drops quickly because of high VUV-X ray flux. At the lower energies, collective effects dominate the beam decreasing the lifetime of the stored beam and its brightness, thus resulting in an optical gain reduction.

The Undulator/Optical Klystron

Figure 2 presents the vertical magnetic field and the calculated horizontal trajectory of the undulator/optical klystron that we have used for the FEL. Its characteristics are summarized in Table 1. The central part called dispersive section has a higher magnetic field. The use of such a dispersive section was originally proposed by N. A. Vinokurov and A. N. Skrinsky (1977). A detailed calculation of the gain and its dependence on energy spread, beam divergence and transverse dimensions has been done by P. Elleaume (1983).

Figure 3 presents a typical spectrum of the synchrotron radiation (also called spontaneous emission) emitted on the axis of the trajectory. The fringes are entirely due to the dispersive section. The higher the field or the longer the dispersive section, the more fringes one obtains. J. Madey (1979) has proved that the gain versus wavelength is approximately equal to the derivative of the spontaneous emission versus wavelength. An important result follows, namely the stronger the dispersive section the higher the gain. However the fringes are eventually blurred by the energy spread which reduces the gain. This results in an optimal field in the dispersive section

616

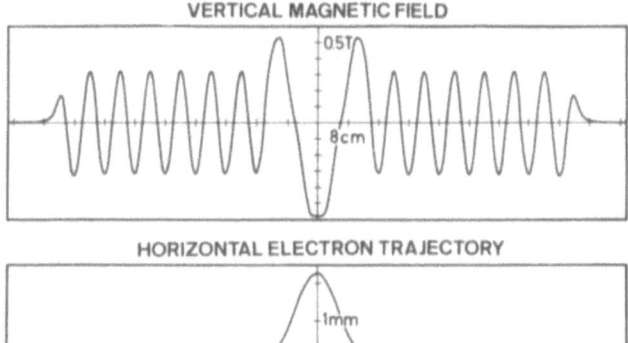

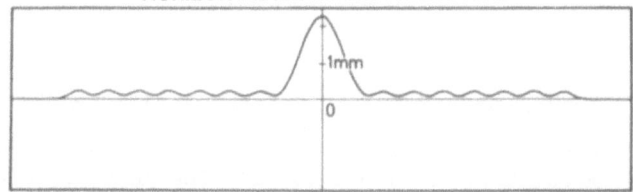

Fig. 2. Vertical magnetic field calculated for the Orsay
optical klystron (gap: 33 mm) and the corresponding
calculated horizontal electron trajectory at an
energy of 240 MeV.

Table 1. Undulator/optical klystron characteristics.

Optical Klystron Characteristics	
Overall length	1.3 m
Type	SmCo5 permanent magnet
Undulators	
Number of period	2x(N = 7)
Period	78 mm
Transverse pole width	100 mm
Minimum magnetic gap	33 mm
Maximum field	0.31 T
K	0-2.3
K in the FEL experiments	1.1-2
Dispersive section	
Length	240 mm
Maximum field	0.58 T

depending on the energy spread which gives the highest optical gain. A
detailed experimental investigation of the blurring of the fringes is
given by D. Deacon et al. (1984).

The Optical Cavity

The optical cavity is made of two identical mirrors with a 3 m
radius of curvature located on the undulator axis 5.5 m apart. The
undulator is centered between these two mirrors. Each mirror has a
physical aperture of 17 mm securing a high fresnel Number (~ 25) at

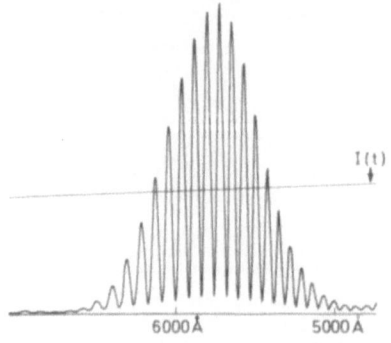

Fig. 3. Spontaneous emission spectrum dI/dλdΩ measured for
an electron energy of 238 MeV and a magnetic field
parameter of K = 2.09 at low current where the
modulation is almost total. The current decay I(t)
is superimposed.

visible optical wavelengths. This results in an easy alignment procedure allowing starting up the FEL in a very short time (1 to 10
minutes) once the ring has been injected. The maximum observed optical
gain has always been less than 0.5% par pass. Laser oscillation requires the use of very high reflectivity mirrors located within the
ultra high vacuum of the ring. So far we have only used dielectric
TiO_2/SiO_2 mirrors. They suffer from the VUV-X ray emission of the
undulator and their reflectivity usually stabilizes around 99.9%. A
detailed description of the diagnostic techniques and mirror degradation
observations is given by P. Elleaume et al. (1985).

Main Characteristics of the FEL Optical Beam

A general review of the FEL characteristics can be found in the
papers of M. Billardon et al. (1983, 1985) and P. Elleaume et al.
(1984). I shall briefly summarize them.

The total average power has been found to be in excellent agreement
with the so-called Renieri's limit:

$$P \simeq \eta_c \, \frac{\sigma_\gamma}{\gamma} \, P_s \tag{13}$$

where P_s is the total synchrotron radiation power radiated in the whole
ring, and η_c the efficiency of the optical cavity (ratio between mirror
transmission and all the other losses). An average power of 2 to 3 mW
has been obtained at 220 MeV a stored current of 200 mA and an efficiency $\eta_c \simeq 0.1$. Equation (13) follows essentially from the fact
that the saturation occurs through the laser induced energy spread.

Tunability has been obtained in the optical range of wavelength
0.47 to 0.65 μm using several sets of mirrors. The shorter wavelength

is presently limited by the availability of high reflectivity mirrors. The longest wavelength of 0.65 μm is not a physical limit. Infrared should be easily obtained with a slightly more complicated procedure for starting the oscillation because of the invisibility of the radiation. The longest laser wavelength is probably determined by the undulator period λ_o necessitating the operation of the ring at a too low energy to maintain the resonance:

$$\lambda = \frac{\lambda_o}{2\gamma^2} \left(1 + K^2/2\right) \tag{14}$$

with $K = 93.4\ B_o\lambda_o$.

B_o is the peak undulator field and γ the electron energy divided by the rest mass energy. When maintaining the resonance condition [Eq. (14)] for longer wavelength λ, the undulator field becomes a stronger perturbation to the storage ring lattice. Field errors and inhomogeneities become very important; in fact we observe a reduction of the lifetime when turning on the undulator field at low energy of the ring.

The temporal and spectral structure of the laser crucially depends on the distance between the mirrors which must be perfectly matched with the perimeter of the ring to assure long time overlap of the electron and light pulses in the undulator every turn. Laser oscillation is typically obtained within a variation of the 5.5 m distance not exceeding 10 μm. Such as accurate matching is performed through the electronic adjustment of the frequency of revolution. At perfect matching, one obtains the minimum relative spectral width of 5×10^{-4}.

Relaxation oscillation is routinely observed at perfect matching, energy spread acting as an energy reservoir as inverted population does in normal lasers. Q-switched operation is easily performed by a quick transverse displacement of the electron beam or a change of revolution frequency. One obtains pulses of 0.5 ms duration with peak power 200 times larger than the average power (insensitive to the Q-switch frequency in a wide frequency range: several Hz to a kHz).

THE CONSEQUENCE OF THE HIGH BRIGHTNESS ON THE ACO FEL

The following discussion is applied to the case of small gain (no exponential growth within a single pass through the undulator) and constant period undulator (no taper). I first deal with short undulators and apply it to the ACO FEL. I then give a short description of the long undulator case.

General Treatment for Short Undulators

First, let split the Brightness into three quantities: the electron current and the horizontal and vertical emittances. Obviously one needs the highest current since the optical gain is proportional to it. Next, the use of a small emittance is also crucial for maximizing the gain. As discussed above the working emittance in our case is very low: $\varepsilon = \varepsilon_X = \varepsilon_Z \simeq 10^{-8}$ mrad with the convention of storage ring for emittance; namely $\varepsilon = \sigma^2/\beta$ where σ is the rms size and β the usual β function.

Essentially there are two emittance thresholds. A substantial drop of optical gain is experienced as either is crossed. The first threshold is connected to the so-called inhomogeneous broadening. Electrons traveling off axis or with an angle may not be resonant and not contribute to the gain. The second threshold is connected to the so-called filling factor. The electrons propagating out of the core of the TEM_{oo} mode of the optical cavity do not contribute to the gain. As we shall see later the inhomogeneous broadening emittance threshold is higher than the filling factor threshold (in the limit of short undulator) and I shall start dealing with the former.

The FEL interaction between a light and electron beam is based on a resonant interaction. The general relation of resonance is:

$$\lambda = \frac{\lambda_o}{2\gamma^2} \left(1 + \frac{1}{2} K_X^2 + \frac{1}{2} K_Z^2 + \gamma^2 \theta_X^2 + \gamma^2 \theta_Z^2 \right) \tag{16}$$

where λ is the light wavelength, λ_o is the undulator period, $K_X = 93.4\, B_X \lambda_o$, $K_Z = 93.4\, B_Z \lambda_o$, (SI units) with B_X (B_Z) the peak undulator field in the X (Z) plane. θ_X (θ_Z) is the angle of propagation between the electron and light beam in the X (Z) plane. γ is the electron energy divided by its rest energy. Generally K_X (K_Z) depends on the transverse position X (Z) through field variation. Let $X = 0$, $\theta_X = 0$, $Z = 0$, $\theta_Z = 0$ the initial conditions for an electron propagating exactly on axis of the undulator. The shift of resonant wavelength associated to an electron propagation off axis is:

$$\delta\lambda(X, \theta_Z, Z, \theta_Z) = \frac{\lambda_o}{2\gamma^2} \left[\frac{1}{2} \left(\frac{2\pi\, K_X X}{\lambda_o} \right)^2 + \frac{1}{2} \left(\frac{2\pi\, K_Z Z}{\lambda_o} \right)^2 + \gamma^2 \theta_X^2 + \gamma^2 \theta_Z^2 \right] . \tag{17}$$

To establish Eq. (17) I have used the dependence of the undulator field

$$B_X \sim B_{oX} \cosh\left(2\pi \frac{X}{\lambda_o} \right) . \tag{18}$$

Eq. (18) follows from Maxwells equations.

A significative gain drop occurs when the resonance is not fulfilled in the whole undulator length namely:

$$\delta\lambda \sim \frac{\lambda}{\pi N} \qquad (19)$$

where N is the number of periods in the undulator.

From Eq. (17) and Eq. (19) it is straightforward to calculate the sensitivity of the gain to angular divergence or beam transverse dimensions:

$$\theta_X, \ \theta_Z \lesssim \sqrt{\frac{2\lambda}{\pi L}} \qquad (20)$$

$$X \lesssim \frac{\lambda_o}{2\pi} \frac{2\gamma}{K_X} \sqrt{\frac{\lambda}{\pi L}} \qquad (21)$$

$$Z \lesssim \frac{\lambda_o}{2\pi} \frac{2\gamma}{K_Z} \sqrt{\frac{\lambda}{\pi L}} \qquad (22)$$

where $L = N\lambda_o$ is the total undulator length. Now lets minimize $\delta\lambda$ with the emittance ε_X and ε_Z fixed. In the limit $\beta_X, \ \beta_Z \gg L$ (short undulator) one has:

$$X = \sqrt{\varepsilon_X \beta_X} \qquad (23)$$

$$Z = \sqrt{\varepsilon_Z \beta_Z} \qquad (24)$$

$$\theta_X = \sqrt{\frac{\varepsilon_X}{\beta_X}} \qquad (25)$$

$$\theta_Z = \sqrt{\frac{\varepsilon_Z}{\beta_Z}} \qquad (26)$$

Introducing Eq. (23) to Eq. (26) into Eq. (17), one obtains the β_X and β_Z values which minimize $\delta\lambda$:

$$\beta_X = \sqrt{2} \frac{\lambda_o \gamma}{2\pi K_X} = 2.4\times10^{-3} \frac{\gamma}{B_X} \qquad (27)$$

$$\beta_Z = \sqrt{2} \frac{\lambda_o \gamma}{2\pi K_Z} = 2.4\times10^{-3} \frac{\gamma}{B_Z} \qquad (28)$$

The emittance threshold is then given by:

$$\varepsilon_X = \frac{\beta_X}{L} \frac{\lambda}{\pi} \qquad (29)$$

$$\varepsilon_Z = \frac{\beta_Z}{L} \frac{\lambda}{\pi} \qquad (30)$$

This calculation has been made by J. Madey (1984). In his paper he mentioned that the β values Eq. (27) and Eq. (28) minimizing the inhomogeneous broadening are exactly equal to the natural β values of the undulator originating from the focusing properties of the magnetic field. These two nice properties of Eq. (27) and Eq. (28) make us think

that these are the right β values for a FEL design. (To wit, a warm
fuzzy feeling, Eds.) However, if this may be true in the case of a long
undulator, it is not true for short undulators because of the second
emittance threshold which I will now discuss.

Let's suppose a cylindrical cavity mode having a beam waist ω interacts with a cylindrical electron beam having a RMS width σ. Neglecting the divergence of both beams P. Elleaume and D. Deacon (1984) have
shown that the gain is proportional to a so-called filling factor F:

$$F = \frac{1}{\omega^2 + 4\sigma^2} \cdot \qquad (31)$$

Maximization of the filling factor leads to minimizing both ω and 2σ.
However, for a fixed emittance, this would be done at the expense of
angular divergence and Eq. (31) would fails. An exact calculation of
the filling factor taking into account angular divergence of both beams
has not yet been done except in the particular case of a filament
electron beam (W. Colson and P. Elleaume, 1982). Although it is
possible to make an approximate theory which is very close to the exact
theory in the filament beam case. For this, one defines the filling
factor as:

$$F = \frac{1}{V} = \frac{1}{\int \left(\omega^2 + 4\sigma^2\right) dZ} \qquad (32)$$

where V is the 3D interaction volume. The divergence is taken into
account by the dependence of ω and σ on the longitudinal coordinate s.
The questions I am now going to address are what is the β value which
maximizes F and what is the emittance threshold. The dependence of ω on
s is given by A. Yariv (1975):

$$\omega^2 = \omega_o^2 \left(1 + \frac{s^2}{z_o^2}\right) \qquad (33)$$

with

$$Z_o = \frac{\pi \omega_o^2}{\lambda} \qquad (34)$$

where the s = 0 point achieves the smallest waist ω_o. Z_o is called the
Rayleigh range. The dependence of σ on s can be calculated from M.
Sands (1971):

$$\sigma = \sqrt{\varepsilon \beta} \qquad (35)$$

$$\beta = \beta_o \left(1 + \frac{s^2}{\beta_o^2}\right) \qquad (36)$$

Relation Eq. (36) is valid for a free field section which is not our case but is a good approximation if the natural β values Eq. (27) and Eq. (28) are much larger than the undulator length (short length approximation). Obviously the maximization of F requires having the waist of both beams located at the middle of the undulator.

Inserting Eq. (33) to Eq. (36) into Eq. (32) one calculates the Z_o and β_o values maximizing F:

$$Z_o = \beta_o = \frac{1}{2\sqrt{3}} L \quad .$$

(37)

Inserting Eq. (37) into Eq. (32) gives

$$F = \frac{\sqrt{3}}{4L^2 \left(\frac{\lambda}{4\pi} + \varepsilon\right)} \quad .$$

(38)

The interpretation of Eq. (38) is the following. F is inversally proportional to the sum of two emittances, an optical one $\frac{\lambda}{4\pi}$ and an electric one ε. The optical one is the **diffraction limited** emittance of the light beam. The electron emittance threshold for filling factor drop is obviously:

$$\varepsilon = \frac{\lambda}{4\pi} \quad .$$

(39)

At this point one can precisely state the short length approximation as

$$\frac{\beta_{oX}}{L} \ll 1 \text{ and } \frac{\beta_{oZ}}{L} \ll 1$$

(40)

where β_{oX} and β_{oZ} are the natural undulator β functions given by Eq. (27) and Eq. (28). This approximation is not only valid for a short undulator but also for small undulator field or large electron energy. In this case the emittance threshold Eq. (39) due to the filling factor is smaller than the one due to inhomogeneous broadening Eq. (29) and Eq. (30). Let's call ε_F and ε_I those two emittance (respectively). Let β_F and β_I be the associated optimum β values. Taking into account both gain reduction processes, one ends up with an optimum β value between β_F and β_I depending on $\frac{\varepsilon}{\varepsilon_F}$ and $\frac{\varepsilon_I}{\varepsilon_F}$.

Application to the ACO Storage Ring FEL

Taking
$\qquad \lambda = 0.5 \ \mu$
$\qquad \gamma = 440 \ (220 \text{ MeV})$
$\qquad K = 2$
$\qquad \lambda_o = 8 \text{ cm}$
$\qquad L = 1.3 \text{ m}$

one calculates:

$$\varepsilon_F = 4 \times 10^{-8} \text{ mrad}$$

$$\beta_F = 0.46 \text{ m}$$

$$\left. \begin{array}{l} \varepsilon_I = 4.8 \times 10^{-7} \text{ mrad} \\ \\ \beta_I = 4 \text{ m} \end{array} \right\} \quad \text{in the vertical plane}$$

$$\left. \begin{array}{l} \varepsilon_I = \infty \\ \\ \beta_I = \infty \end{array} \right\} \quad \begin{array}{l} \text{in the horizontal plane} \\ \\ \text{(planar undulator)} \end{array}$$

The operating point currently used satisfies

$$\varepsilon_X \simeq \varepsilon_Z \simeq 1 \text{ to } 2 \times 10^{-8} \text{ mrad}$$

$$\beta_X = 1.8 \text{ m}$$

$$\beta_Z = 1 \text{ m}.$$

From Eq. (20) to Eq. (22) one easily verifies that almost no gain reduction occurs by inhomogeneous broadening although some reduction (15 to 25%) occurs through the filling factor. The optimum β_X value is 0.46 m and the optimum β_Z is somewhere between 0.46 and 4 m.

As a matter of fact the first operation of the FEL above threshold was done in 1983 with the usual operating point of ACO having $\beta_Z = 4$ m. A significant improvement of the gain has been obtained by decreasing β_Z to 1 m in a subsequent experiment allowing larger tunability and improve stability.

The Long Undulator Case

In the long undulator case the whole discussion concerning the inhomogeneous broadening is still valid, the one concerning the filling factor is not. This comes from the fact that Eq. (36) is only true for a field free (or almost free) straight section. If the natural β_N value of the undulator is small compared to the undulator length, the observed β function oscillates around β_N with a period $\pi\beta_N$ (Sands, 1971). In this limit the maximization of F leads to:

$$\beta = \beta_N \quad \text{all along the undulator}$$

$$Z_o = L/\sqrt{12}$$

the optimum filling factor is:

$$F = \frac{1}{4L\beta_N \left(\frac{\lambda}{4\pi} \frac{L}{\sqrt{3}\beta_N} + \varepsilon \right)} \quad .$$

The emittance threshold is

$$\varepsilon = \frac{\lambda}{4\pi} \frac{L}{\sqrt{3}\beta_N} \gg \frac{\lambda}{4\pi} \gg \frac{\beta_N}{L} \frac{\lambda}{\pi} \ .$$

This threshold becomes larger than the inhomogeneous broadening one. In the long undulator approximation ($\beta_N/L \ll 1$) the optimization clearly leads to $\beta = \beta_N$, the main gain reduction occurs through inhomogeneous effects and the filling factor problem is negligible.

For practical reasons, most of undulators that have been build are planar instead of helical. If the field is vertical, one then faces a mixed situation where $\frac{\beta_N \text{ vertical}}{L} \ll 1$ but $\frac{\beta_N \text{ horizontal}}{L} = \infty$. People have suggested adding linear focusing all along the undulator. The idea is to decrease the vertical focusing by shifting it into the horizontal plane. The final device has equal vertical and horizontal focusing. Several solutions have been proposed such as periodically adding quadrupoles, canting the pole faces (D. Quimby, J. Slater, 1984) of the undulator, or designing parabolic pole faces (E. T. Scharlemann, 1985). All of three solutions lead to an increase of the gain by increasing the horizontal emittance threshold. Since focusing in the horizontal is done at the expense of a vertical defocusing, the vertical filling factor emittance threshold decreases which is not a problem as long as it stays above the inhomogeneous one.

Application to Future X-UV Storage Ring FEL

When going to shorter wavelength ε_F and ε_I drop, requiring an extremely small emittance. For example a VUV FEL of 1250 A° would require $\varepsilon_I = 10^{-8}$ mrad, which is close to the state of the art on a 1 GeV storage ring. To cross the threshold for lasing one also requires a high peak current. Within the present technology, I believe that the ACO storage ring FEL is in a unique situation in the sense that $\varepsilon \ll \varepsilon_F$ which allows **almost no gain reduction**. Future storage ring FELs are likely to suffer much larger gain reductions when going toward very short wavelengths and achieving high brightness will be a major task.

REFERENCES

Billardon, M., Elleaume, P., Ortéga, J. M., Bazin, C., Bergher, M., Velghe, M., Petroff, Y., Deacon, D. A. G., Robinson, K. E., Madey, J. M. J., 1983. First Operation of a Storage Ring Free-Electron Laser, Physical Review Letters, 51, p. 1652.
Billardon, M., Elleaume, P., Ortéga, J. M., Bazin, C., Bergher, M., Petroff, Y., Velghe, M., 1985. Results of the Orsay storage ring free electron laser oscillation. Nuclear Instruments and Methods, A237.

Billardon, M., Elleaume, P., Ortéga, J. M., Bazin, C., Bergher, M., Velghe, M., Deacon, D. A. G., Petroff, Y., 1985. Free electron laser experiment at Orsay: A Review, IEEE J. Quant. Elect., QE-21, p. 805.

Colson, W. B., Elleaume, P., 1982. Electron dynamics in free electron laser resonator modes, Appl. Phys., vol. B.29, p. 101.

Deacon, D. A. G., Billardon, M., Elleaume, P., Ortéga, J. M., Robinson, K. E., Bazin, C., Bergher, M., Velghe, M., Madey, J. M. J., Petroff, Y., 1984. Optical klystron experiments for the ACO storage ring free electron laser, Appl. Phys., B34, p. 207.

Elleaume, P., Ortéga, J. M., Billardon, M., Bazin, C., Bergher, M., Velghe, M., Petroff, Y., 1984. Results of the free electron laser experiments on the ACO storage ring. Journal de Physique, 45, p. 989.

Elleaume, P., 1983. Optical klystrons, Proc. Bendor Free Electron Laser Conference, Journal de Physique, vol. 44, C1-353.

Elleaume, P., Deacon, D. A. G., 1984. Transverse mode dynamics in free electron lasers, Appl. Phys., B33, p. 9.

Elleaume, P., Velghe, M., Billardon, M., Ortéga, J. M., 1985. Diagnostic techniques and VUV induced degradation of the mirrors used in the Orsay storage ring free electron laser, Appl. Optics, 24, p. 2762.

Madey, J. M. J., 1979. Relationship between mean radiated energy, mean squared radiated energy and spontaneous power spectrum in a power series expansion of the equations of motion in a free electron laser. Il Nuovo Cimento, 50B, p. 64.

Madey, J. M. J., 1984. Conceptual system design of XUV FEL's, in "Free Electron Generation of Extreme Ultraviolet Coherent Radiation", AIP Number 118, p. 12.

Quimby, D., Slater, J., 1984. Emittance acceptance in tapered-wiggler free-electron lasers, Proc. SPIE 453. (Proc. of Orcas Island FEL Conference).

Sands, M., 1971. Physics of electron storage ring, Proc. of the International School of Physics "Enrico Fermi", p. 257.

Scharlemann, E. T., 1985. Wiggle plane focusing in linear wigglers, J. Appl. Phys., 58, p. 2154.

Vinokurov, N. A., Skrinskii, A. N., 1977. The optical klystron, Preprint INP 77, 59, Novosibirsk, USSR.

Vinokurov, N. A., 1977. The optical klystron, Proc. of the 10th Int. Conf. on High Energy Charged Particle Accelerators, Serpukhov, p. 454, 1977.

Yariv, A., 1975. "Physics of Quantum Electronics", (Wiley, N.Y.).

EMITTANCE, BRIGHTNESS, FREE-ELECTRON LASER BEAM QUALITY, AND THE SCALED THERMAL VELOCITY

C. W. Roberson
Office of Naval Research
Arlington, Virginia 22217 USA

Y. Y. Lau and H. P. Freund*
Naval Research Laboratory
Washington, DC 20375 USA

We discuss the role of emittance, brightness, and other factors affecting the quality of beams in the free electron laser. Since the key parameters in the dynamics of the free electron laser interaction depend on the beam density and axial energy spread of the beam, the ratio of the current density to the relative axial energy spread is taken as a figure of merit for comparing the quality of electron beam sources as potential drivers of free electron lasers. The scaled thermal velocity is introduced as a parameter in estimating the effect of thermal spread in a high gain free electron laser. We find that the effective thermal spread regime can be made small near the cyclotron resonance in field immersed free electron lasers, and the unusual possibility that the efficiency of the fastest growing mode with warm beams can be higher than cold beams. A detailed dispersion relation for the free electron laser interaction where the beam energy distribution is determined by pitch angle scattering and the total beam energy is conserved is presented. Numerical solutions for the growth rate and efficiency are calculated which show a shift to lower phase velocity as a function of momentum spread and an increase in efficiency of the fastest growing mode.

INTRODUCTION

The free electron laser (FEL) has become <u>the</u> alternative radiation source from microwave to soft x-ray wavelengths. The principal advantages of the free electron laser are its high power capability and

---

* Permanent Address: Science Applications, Inc.
McLean, VA 22102

tunability, while the disadvantages are its requirements on high voltage and high quality electron beams.

Figure 1 shows the peak power of some representative microwave devices and bound electron lasers. The conventional microwave devices cover the wavelength range down to mm continuously, but the power capability drops rapidly at shorter wavelengths. These devices depend on slow wave or cavity structures to obtain an interaction of the electro-magnetic wave with the electron beam, and the dimensions of these struc-tures are proportional to the wavelength. At high powers and short wavelengths, they are limited both by the power density of the electron beams and by electric field breakdown on the structures. These diffi-culties are also shared by gyrotrons, whose operation frequencies are by the tuning magnetic field and by the electromagnetic tubes, whose dimension is also proportional to the wavelength.

The free electron laser uses a periodic transverse magnetic field (i.e., a wiggler) to give the beam a velocity component in the transverse direction (Fig. 2). This axial periodicity eliminates the need for a slow wave structure, while retaining the tunability of coherent free electron radiation sources. As described below, the transverse component of the beam velocity interacts with the electric and magnetic fields to produce a ponderomotive force in the propagation direction. This pon-deromotive force bunches the electron beam parametrically at a frequency and wavelength that re-enforces the radiation (i.e., coherent radiation).

RADIATION AND EMITTANCE

Although the free electron laser is not constrained in its cross section by a slow wave structure one must be concerned about the matching of the radiation and electron beam profiles to ensure efficient inter-action. This places a condition on the beam emittance in terms of the wavelength.

To get a "rule of thumb" relation between the emittance and radia-tion wavelength we will compare the profile of Gaussian radiation beam and an emittance dominated electron beam with no focusing force. The profile of both beams is hyperbolic. The effective radius of the radia-tion beam is

$$R^2 = R_r^2 \left(1 + \frac{\lambda^2 z^2}{\pi^2 R_r^4} \right) \tag{1}$$

where R_r is the minimum spot size or waist, λ is the radiation wavelength and Z is the distance in the propagation direction (Yariv, 1976).

628

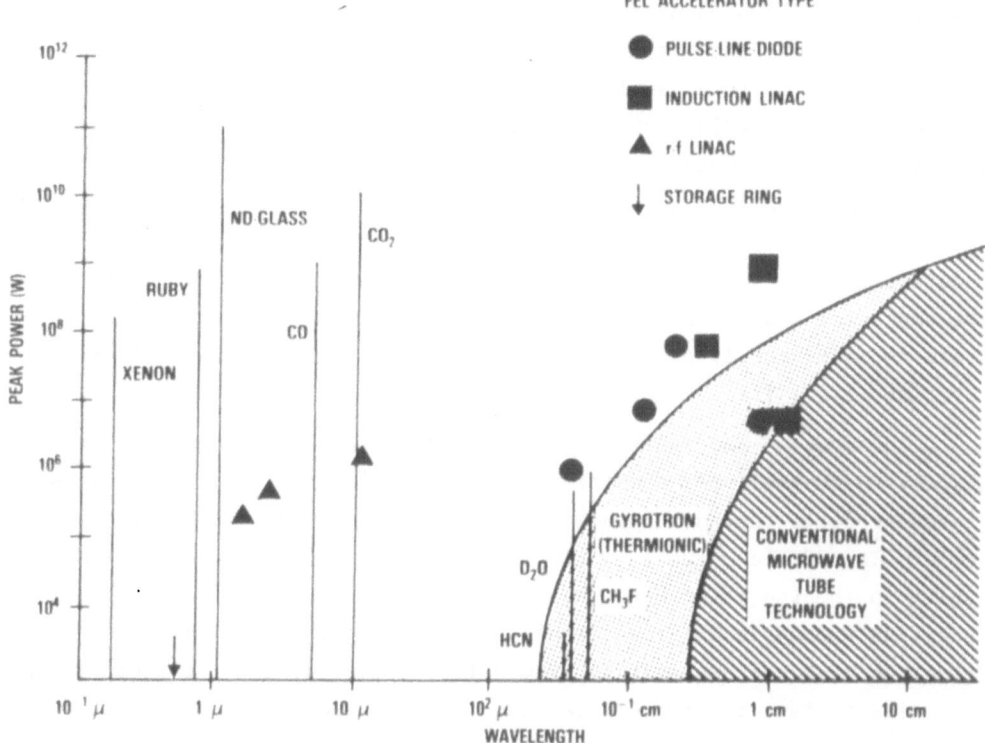

Fig. 1. Performance of various types of radiation sources.

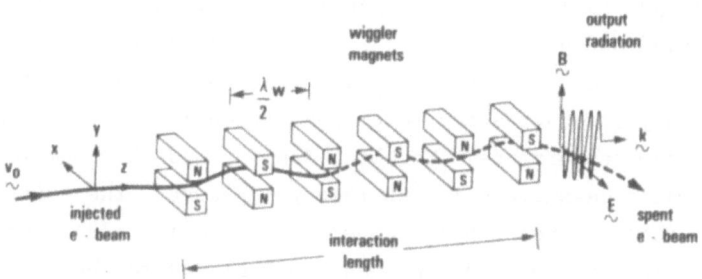

Fig. 2. FEL interaction model.

The profile of the electron beam is

$$R^2 = R_e^2 \left(1 + \frac{\varepsilon^2 z^2}{R_e^4}\right) \tag{2}$$

where ε is the emittance and R_e the radius of the waist.

If we require that $R_e = R_r$ to ensure a maximum geometrical overlap of the radiation and electron beams we find the relation

$$\lambda = \pi\varepsilon \tag{3}$$

The dimensional units of the wavelength and emittance are the same. For example, an emittance of 1 mm-mrad beam would be matched with a Gaussian radiation beam with a wavelength of 3.14 microns under ideal conditions. Without even going into the dynamic processes of energy conversion, this condition clearly suggests the demand on a high quality beam for short wavelengths. On the other hand, it should be anticipated at this point that the dynamical processes, which include axial bunching, would place additional constraints on the longitudinal properties of the beam. Here we emphasize that this simple relation on the emittance of the electron beam is based merely on good coupling due to the transverse geometrical overlap (i.e., filling factor).

There are other factors which change this simple picture significantly under realistic conditions. For example, the wiggler provides a focusing force that determines the equilibrium radius of the electron beam. In addition, the free electron laser interaction can cause a phase shift in the radiation beam that can result in its "self-focusing" behavior [Sprangle and Tang (1981)]. Hence, neither the radiation nor electron beam are necessarily hyperbolic, a more detailed calculation of the filling factor and accelerator acceptance must be carried out.

EMITTANCE, BRIGHTNESS, AND FREE ELECTRON LASER BEAM QUALITY

The output radiation wavelength is determined primarily by the wiggler period and beam voltage. The radiation power is determined by the beam current and by the intrinsic efficiency. In general, the gain and efficiency are increasing functions of density, but decrease with the effective beam temperature. The beam density and the beam temperature are independent parameters in FEL theory. However, the emittance is related to the transverse beam temperature. Neil (1979) suggested (in an unpublished report) that the emittance be related to the current by

$$\varepsilon_n \ (\text{mrad} - \text{cm}) = 320 \ \sqrt{I \ (\text{amp})} \tag{4}$$

where ε_n is the normalized emittance. This relation between the density and transverse beam temperature was based on an empirical scaling of data from certain linear accelerators. This equation is generally referred to as the Lawson-Penner relation. It is equivalent to the statement that the brightness is the same for various accelerators (Roberson, 1983). However, there are many exceptions to this relation (Roberson 1985, Lawson and Penner 1985, Roberson 1983).

The emittance is an important parameter in the envelope of a beam. The envelope of an electron beam born in an axial magnetic field may be calculated from (Lee and Cooper 1976, Lawson 1977)

$$R^{11} + \frac{\Omega^2 R}{4\beta^2 c^2 \gamma^2} - \frac{2\nu}{\beta^2 \gamma^3 R} - \frac{\varepsilon_n^2 + (P_\theta/mc)^2}{\beta^2 \gamma^2 R^2} = 0 \tag{5}$$

where Ω is the nonrelativistic cyclotron frequency, $P_\theta = \dfrac{q\psi_o}{2\pi}$, where ψ_o is the flux linking the cathode, and

$$\nu = \frac{R_o^2 \omega_p^2}{4c^2} \tag{6}$$

(If the cathode is not field-immersed, $P_\theta = \Omega = 0$.) When the third term in this equation is larger than the fourth, the waist will be determined by space charge forces, rather than emittance. When the space charge term is small, the waist is determined by the effective emittance

$$\varepsilon_{eff}^2 = \varepsilon_n^2 + \left(\frac{P_\theta}{mc}\right) \tag{7}$$

Although the emittance is a key parameter in determining the radial electron beam profile, the dynamics of the FEL interaction is determined by ponderomotive/space charge wave and the <u>axial</u> component of the velocity spread in the electron beam.

A definition of free electron laser beam quality that emphasizes the axial energy spread is (Roberson 1985)

$$B_Q = J/(\Delta\gamma_z/\gamma) \tag{8}$$

where J is the current density,

$$\gamma_z = \frac{1}{\sqrt{1 - \beta_z^2}}$$

and $\Delta\gamma_z$ is the spread in the axial energy from all the sources.

Freund and Ganguly (1986) carried out computer simulations of efficiency of different FEL configurations in the high gain regime as a function of the relative momentum spread. In case 1 (Fig. 3), the parameters were beam voltage 250 kV, beam current 35 A, wiggler magnetic field 2.0 kG, axial magnetic field 2.0 kG and wiggler wavelength 1.175 cm. The cold beam efficiency was twenty per cent and decreased an order of magnitude as a result of a 2.0% spread in axial momentum. In case 2 (Fig. 4), the beam voltage was 750 kV, the current was 200 A, the wiggler magnetic field 1 kG and the wiggler wavelength 4 cm. There was no axial magnetic field in case 2. In case 2 the cold beam efficiency was seven percent and decreased an order of magnitude with a 0.2% spread in the axial momentum. That is, the efficiency in case 2 is ten times as sensitive as case 1 to the spread in axial momentum.

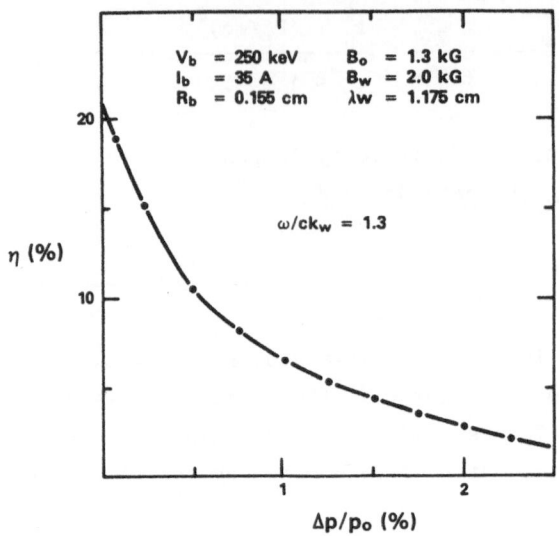

Fig. 3. Case 1. Efficiency vs axial momentum spread for a
low voltage FEL in a magnetic field
(TE$_{11}$ Mode (R$_g$ =0.36626 cm).

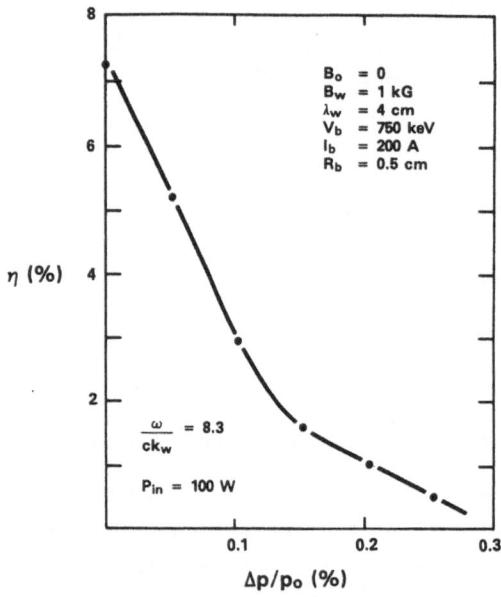

Fig. 4. Case 2. Efficiency vs axial momentum spread for a
electron laser not immersed in an axial magnetic field
(TE$_{11}$ Mode (R$_g$ = 1.5 cm).

In Fig. 5 the normalized efficiency n/n_0 (where n_0 is the efficiency
in the limit of zero thermal spread) versus the inverse of the beam
quality. The normalized efficiency in these two cases scale in a similar
way with the inverse of the beam quality, even though they differ significantly in their operating parameters. Hence the FEL beam quality is a
useful quantity for comparing electron beams without knowledge of the
details of the FEL interaction.

632

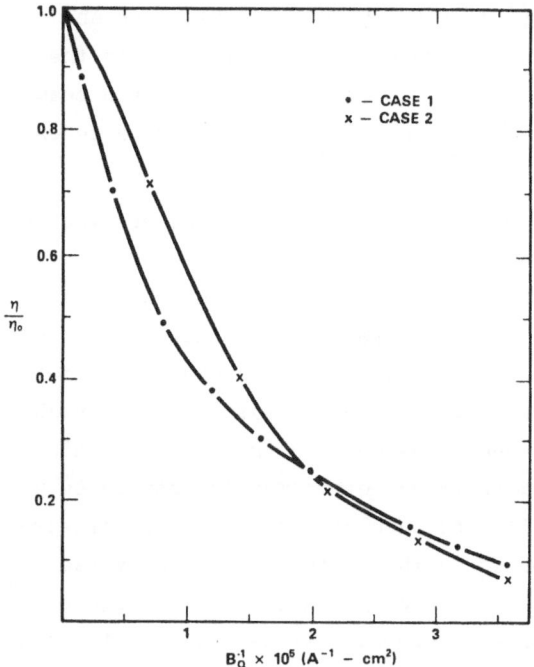

Fig. 5. Normalized efficiency vs the inverse of the beam quality.

An important point can be made here. The term $\varepsilon_n + (p_\theta/mc)^2$ is often treated as an effective transverse emittance in the envelope equation when calculating the radial profile of the electron beam. That is, both the transverse emittance and magnetic field limit the waist of the beam.

However, the dynamics of the ponderomotive space charge wave in the FEL interaction is sensitive to the axial velocity spread, hence this replacement of an effective transverse emittance is relevant in determining the geometrical overlap (filling factor) of the radiation and electron beams.

The dynamics of the FEL interaction are strongly effected by the magnetic field near cyclotron resonance and scaling of efficiency with beam quality for those cases is quite different. A more complete description is required to include such things as resonant effects.

The axial energy spread can come from a number of sources as the electrons emerge from the cathode, are accelerated, and participate in the free electron laser interaction. For example, (1) There is an initial velocity spread due to the temperature of the cathode. (2) The roughness of the emitting surface can cause an energy spread. (3) The non-uniform emission from the emitting cathode surface can cause a growth in the energy spread of the beam. (4) Electric and magnetic fields that are asymmetric, non-uniform, or non-adiabatic in the axial direction can lead to effective energy spreads in the electron beam. (5) The self space charge fields of the electron beam can cause energy spreads in the axial direction.

Among the various factors cited in the previous paragraph, the thermal velocities due to cathode temperatures are considered unimportant for FELs driven by induction accelerators. Non-uniform and non-adiabatic fields may be considerably reduced with careful gun design. Space charge effects may be corrected by the well-known Pierce method, at least for one set of parameters. It then appears that the problems which are cathode-related such as patchiness, non-uniform emission, roughness, and non-constant work function, etc., may in many cases place a fundamental limit on the quality of an electron beam. An assessment of the effects of surface roughness on the ratio of the transverse velocity spread to the axial velocity has been given recently. Figure 6 shows the model for the cathode surface roughness. It is found that the maximum values of $\gamma\beta\theta$ in the temperature and space charge limits are (Lau 1985, 1987):

$$
\gamma\beta\theta = \begin{cases} 0.16 \ \sqrt{\phi/D} \ h/(h^2 + w^2)^{1/4} & \text{(Temp. ltd.)} \\[2ex] 0.079\sqrt{\phi/D} \ \dfrac{h}{(h^2 + w^2)^{1/4}} \ \left(\dfrac{h}{D}\right)^{1/6} & \text{(Sp. ch. ltd.)} \end{cases} \tag{9}
$$

Here h and w are height and width of the "bump" on the cathode in units of 100 microns, D is the anode cathode spacing in cm and ϕ the voltage in megavolts. For example, for an anode-cathode voltage drop of 250 keV over a distance of 5 cm, say, surface roughness of 5μm size leads to an equivalent maximum transverse temperature of 10 eV if the cathode is operated in the temperature limited regime, but of 0.55 eV in the space charge limited regime. Note that these equivalent temperatures are considerably higher than the normal operating temperature of cathodes.

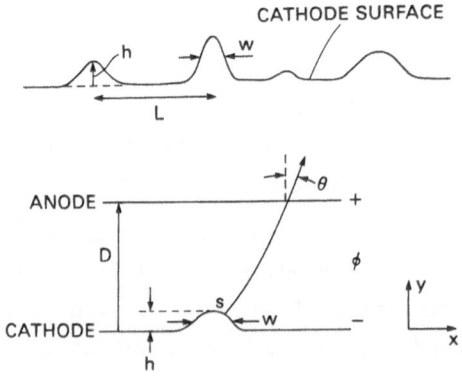

Fig. 6. Model of cathode surface roughness.

Equation (9) gives the following ratio of roughness-induced emittances for the temperature-limited and the space charge-limited regime

$$\frac{\varepsilon_n \ (\text{sp. ch. ltd.})}{\varepsilon_n \ (\text{temp. ltd.})} = 0.5 \left(\frac{h}{D}\right)^{1/6} \tag{10}$$

This ratio assumes a value between 0.2 and 0.5 over a wide range of parameters because of the weak 1/6 power dependence in Eq. (10).

The relative energy spread due to the emittance, self-potential and wiggler gradients is

emittance $\qquad \dfrac{\Delta \gamma_z}{\gamma} = \dfrac{1}{2} \dfrac{\varepsilon_n^2}{R^2}$

self potential $\qquad \dfrac{\Delta \gamma_z}{\gamma} = \dfrac{1}{\gamma} \left(\dfrac{R \omega_b}{2c}\right)^2 = I/I_A$

wiggler gradients $\qquad \dfrac{\Delta \gamma_z}{\gamma} = \left(\dfrac{R \Omega_\omega}{2c}\right)^2$

where R is the beam radius, W_b the beam plasma frequency, Ω_ω the electron cyclotron frequency due to the wiggler magnetic field, c the velocity of light, and I_A the Alfven current.

Figure 7 is a plot of the beam quality of some representative accelerators from Roberson (1985). These include an electrostatic accelerator (EA), radio frequency linear accelerators (rfl), induction linear accelerators (IL) and high current field immersed diodes (solid squares). In general, the beam quality appears as a decreasing function of current. High quality electron beams can be obtained from diodes immersed in a magnetic field, but there is limited experience in the acceleration and transport of such beams to high energies. Another method is the technique in rf linear accelerators of compressing the beam in the axial direction by subharmonic bunching. Although this technique may increase the peak current without increasing the transverse energy spread of the beam, it must increase the axial energy spread (since the phase space volume is conserved).

The Ponderomotive Force

The energy spread of the electrons affect the FEL through the ponderomotive force.

To understand the origin and central role played by the ponderomotive force in bunching the beam, let us consider a low current electron beam in which the space-charge forces can be neglected. We assure that the initial velocity of the electron beam is in the axial

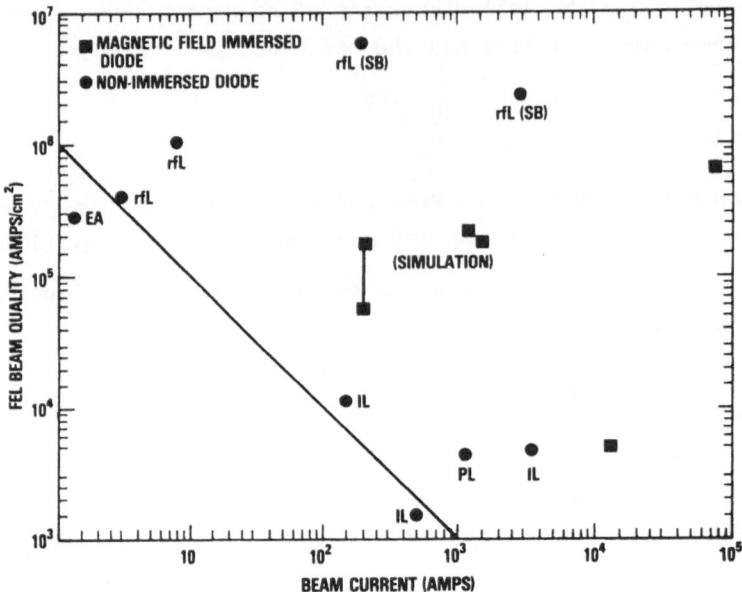

Fig. 7. FEL beam quality as a function of current or various types of accelerators.

direction and that we have a simple linear wiggler in the y direction (Fig. 2) That is $\underline{v}_z = v_{oz} \, \hat{e}_z$ and $\underline{B}_w = B_w \sin(k_w z)\hat{e}_y$ where v_{oz} is the axial beam velocity, B_w the wiggler magnetic field and k_w is the wiggler wavenumber.

As a result of the Lorentz force, the beam electrons will acquire a wiggler velocity in a direction perpendicular to $\underline{v}_z$ and $\underline{B}_w$ which is given by $\underline{v}_w = v_w \cos(k_w z)\hat{e}_x$ where $v_w = \Omega_w / \gamma_0 k_w$. We now assume the presence of a linearly polarized radiation field $\underline{E} + \underline{B} = E \cos(kz - \omega t)\hat{e}_x + B \sin(kz - \omega t)\hat{e}_y$ where $k = \omega/c$ is the radiation wave number. This radiation field can exist as part of the noise spectrum in the case of the oscillator or is supplied externally in the case of an amplifier. As a result of the wiggler field and the radiation field, a ponderomotive force in the z direction develops:

$$F_z = -(|e|/c) \, (\underline{v}_w \times \underline{B})_z \sin[(k + k_w)z - \omega t]$$

where $k_w = 2\pi/\lambda_w$ and λ_w is the wiggler period. For relativistic beams, the ponderomotive force in the z direction arises primarily from the interaction between $\underline{v}_w$ and the magnetic component of the radiation field. This ponderomotive force drives a longitudinal current density δJ_z and in turn a density modulation δn which are related by $|e| \delta \frac{\partial n}{\partial t} = \nabla \cdot \delta J \, \hat{e}$. The density modulation has a form given by $\cos[(k + k_w)z - \omega t]$. Note that the density modulation on the electron beam, driven by the ponderomotive force, is at the effective wave number $k + k_w$, and results in a

636

transverse current that can excite radiation which is in phase with the existing radiation field. The transverse current has the form, $\delta \underline{J} = - |e| \delta n \underline{v}_w \cos (kz - \omega t) e_x$. That is, the perturbed beam density coupled with the wiggler velocity generates a transverse component of current with a wave number and frequency equal to that of the original radiation field. This component of the perturbed current density is a source of coherent radiation which is in phase with the existing radiation. This process is illustrated in Fig. 2 where the electron beam is shown entering the wiggler, acquiring a component of velocity in the direction perpendicular to the wiggler field and being modulated by the ponderomotive force.

If the ponderomotive force drives the density modulation $\delta n(k + k_w)$, at a frequency and wave number that is also a beam mode, then the free electron laser is said to operate in the collective regime.

To determine the radiation wavelength we assume that the interaction is the strongest when the phase velocity of the ponderomotive wave is near the axial beam velocity; $v_{ph} = \omega/(k + k_w) \simeq v_{oz}$. Taking $\omega = ck$ for the radiation, the output wavelength becomes $\lambda = [(1 - \beta_z)/\beta_z]\lambda_w \simeq \lambda_w/2\gamma_z^2$, where $\beta_z \simeq v_{oz}/c$. This is the wavelength scaling relation that makes the FEL a tunable radiation source. This also suggests the potential importance of axial velocity spread to beam-wave interaction.

THE FREE ELECTRON LASER DISPERSION RELATION AND SCALED THERMAL VELOCITY

To quantify the FEL interaction mechanism, we write the dispersion relation for a cold beam: (Sprangle et al., 1985)

$$\left[\omega - k^2 c^2 - \frac{\omega_b^2}{\gamma_0} \right] \left[(\omega - v_{zo}(k + k_w))^2 - \frac{\omega_b^2}{\gamma_0 \gamma_z^2} \right] = \frac{\omega_b^2 \beta_w^2 ckk_w^2}{\gamma_0} \qquad (11)$$

Here β_w is the perpendicular velocity given to the beam by the wiggler, divided by the velocity of light. The right hand side of this equation is the coupling term due to the wiggler.

The individual modes are readily identified in this dispersion relation [Fig. 8]. The wiggler motion β_w couples the transverse electromagnetic modes (the first bracket on the left hand side of the dispersion relation) to the longitudinal beam modes (second bracket on the left hand side of the dispersion relation).

The $\omega_b^2/\gamma_0 \gamma_z^2$ term of the longitudinal beam modes is due to the self consistent AC space charge potential contribution to the ponderomotive force. When this term is significant the free electron laser is said to operate in the Raman regime.

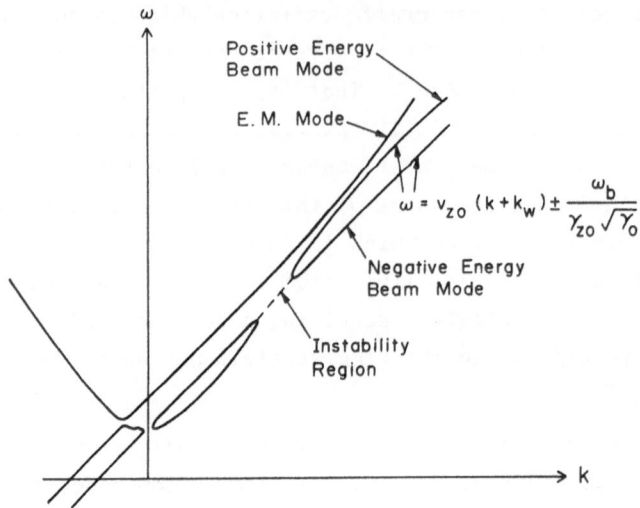

Fig. 8. Dispersion diagrams for the interacting modes.

If the amplitude of the wiggler magnetic field is sufficiently strong that

$$\beta_w > \left[\frac{2\omega_b}{c\ \gamma_0^{1/2}\ k_w\ \gamma_z^2} \right]^{1/2} \qquad (12)$$

the space charge term $\omega_b^2/\gamma_0\gamma_z^2$ can be neglected and the FEL is said to operate in the high gain, or strong pump regime in which the spatial growth rate is given by

$$\Gamma = \frac{\sqrt{3}}{2} \left[\frac{\beta_w^2\ \omega_b^2\ k_w}{\gamma_0 c^2} \right]^{1/3} \qquad (13)$$

The efficiency, determined by the trapping of the electron beam by the ponderomotive/space charge wave, is

$$\eta = \left[\frac{\omega_b \beta_w}{4\ \sqrt{\gamma_0}\ c\ k_w} \right]^{2/3} \qquad (14)$$

The dispersion relation with finite velocity spread is now considered. Unfortunately, it cannot be reduced to a simple form if the more realistic Gaussian distribution function is used for the beam electrons. To obtain a rough estimate and to establish a scaling parameters for the velocity spread, one may use a Lorentzian distribution for the axial velocity, the dispersion equation for the strong pump regime is expected to be of the form

$$D (\omega, k) \left[\omega - v_{oz} (k + k_w) + ik \, \bar{v}_z \right] = \frac{\omega_b^2 \, \beta_w^2 \, c^2 k k_w}{\omega^2 \gamma_0} \tag{15}$$

to lowest order in $\bar{v}_z$, where

$$D (\omega, k) = 1 - \frac{c^2 k^2}{\omega^2} - \frac{\omega_b^2}{\omega^2 \gamma_0} , \tag{16}$$

and $\bar{v}_z$ is the axial thermal velocity of the beam. This dispersion relation is analogous to the dispersion relation for the well-known two stream interaction of an electron beam with a plasma [O'Neil and Malmberg (1968)]. In fact, the structures of both dispersion relation are identical and we can define a scaled thermal velocity, which in the high gain regime becomes

$$S = \frac{\bar{v}_z}{v_{oz}} \left[\frac{4\omega^2 \gamma_0 \gamma_z}{\omega_b^2 \, \beta_w^2} \right]^{1/3} \tag{17}$$

Physically the scaled thermal velocity determines when the thermal velocity plays an important role in the FEL interaction. Specifically, with the introduction of the dimensionless parameter S, the dispersion relation (17) may be case in a normalized form:

$$y(y - x + is)^2 = 1 \tag{18}$$

Here, the dimensionless variables x and y are related to ω and k in such a way that $y = 0$ corresponds to the electromagnetic mode $\omega - kc = 0$ and $y - x = 0$ corresponds to the beam mode $\omega - (k + k_w) v_{zo} = 0$. From this equation, one can easily see that a beam can be considered cold if $S \ll 1$ as far as FEL interaction is concerned. Thermal effect becomes important as $S \simeq 0(1)$. Indeed, according to (18), the topology of the dispersion curves change markedly if $S \rightarrow 0(1)$ from that of the cold beam [O'Neil and Malmberg (1968)]. In passing, we express the cold beam efficiency η of Eq. (14) in terms of S as

$$S = \frac{\gamma_z^2 \, \bar{v}_z}{\eta \, v_{oz}} \tag{19}$$

In Fig. 9 we plot the ratio of the electron beam thermal velocity in the axial direction to the difference between the axial velocity of the electron beam and the phase velocity of the ponderomotive space charge wave, as a function of the scaled thermal velocity, S. As S is increased from zero (cold beam) to greater than one, the ratio becomes approximately equal to one. Then the phase velocity of the ponderomotive space charge wave is resonant with beam thermal electrons. There is a change in the topology of the dispersion relation as the interaction goes from the cold beam nonresonant to the warm beam resonant interaction. This change in the topology of the dispersion relation results in a decrease in phase velocity of the fastest growing mode. Hence to the extent that the efficiency of the FEL is determined by trapping of the beam particles, there is the usual possibility of higher efficiency with a warm beam than a cold beam. In this regime a more refined kinetic description must be used to calculate the growth rate and efficiency.

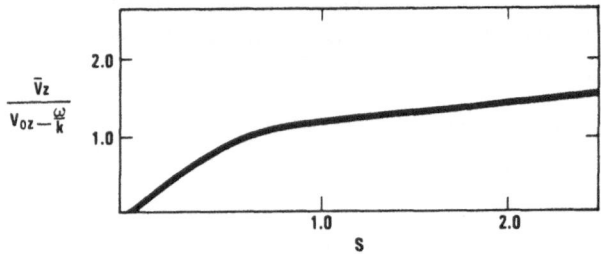

Fig. 9.   Ratio of beam velocity spread to velocity mismatch as a function of the normalized parameter S.

When an axial magnetic field is included, the dispersion relation is modified to become (Freund et al., 1982)

$$\left[\omega^2 - k^2 c^2 - \frac{\omega_b^2(\omega - kv_{oz})}{\gamma(\omega - \Omega_o - kv_{oz})} \right] \left\{ \left[\omega - (k + k_w)v_{oz} \right]^2 - \right.$$

$$\left. \Phi_o \omega_b^2 / \left(\gamma_z^2 \gamma \right) \right\} \equiv \Phi_o \beta_w^2 \frac{\omega_b^2}{\gamma} c^2 k k_w \tag{20}$$

where

$$\Phi_o = 1 - \frac{\beta_w^2 \gamma_z^2 \Omega_o}{\left(1 + \beta_w^2 \right) \Omega_o - k_w v_{oz}} . \tag{21}$$

At wavelengths far from cutoff (i.e., $\omega \sim kc$) we expect the effective scaled thermal velocity to be

$$S \simeq \frac{\bar{v}_z}{v_{oz}} \left[\frac{4 \, \omega^2 \gamma_o^2 \gamma_z^2}{\Phi_o \, \omega_b^2 \, \beta_w^2} \right]^{1/3} \tag{22}$$

640

where the beam plasma frequency is now multiplied by the function Φ_o. It should be noted that near the resonance, $\Omega_o \sim k_w v_{oz}$, both Φ_o and β_w can be significantly enhanced. As a result, a higher thermal spread $\bar{v}_z$ can be tolerated to yield the same value of S. In fact, such increased tolerance to the effects of thermal spread near the resonance has been found in simulation (Freund and Ganguly, 1986).

In the regime below resonance $(\Omega_o < k_w v_{oz})$ the ponderomotive/space change wave couples to the electromagnetic wave in a fashion similar to the FEL without an axial magnetic field. Above resonance, the ponderomotive/space charge wave couples to the electron cyclotron mode of the beam. Near resonance the Φ_o function can become quite large, resulting in a small scaled thermal velocity. That is, the phase velocity of the ponderomotive space charge wave is reduced so that it "sees" the beam as a cold beam.

Growth Rate and Efficiency – Some Numerical Calculations

Long wavelength free electron lasers (i.e., $\lambda \approx 1$ cm) are frequently driven by an electron beam accelerated through a diode, in such cases it is appropriate to assume the total energy of the beam is conserved (i.e., $P_\perp^2 + P_z^2 =$ constant). As a result, we choose a distribution function

$$
F_o\left(P_x, P_y, P\right) = n_b \frac{P_z}{\pi \Delta P_\perp^2 P_o} e^{-P_\perp^2/\Delta P_\perp^2} \delta\left(P - P_o\right) \tag{23}
$$

where n_b is the ambient density of the electron beam, p is the total momentum, P_x and P_y are the canonical momenta in the x and y directions, $P_\perp^2 = P_x^2 + P_y^2$ and $P_z = \left(p^2 - p_x^2 - p_y^2\right)^{1/2}$ is the axial momentum. This distribution describes a monoenergetic beam (i.e., $p = p_o$) but with a pitch angle spread characterized by $\Delta P_\perp$ which describes a nonzero emittance. A complete derivation of the dispersion equation will be published elsewhere (Freund and Ganguly, 1986). For the purposes of the present work, we note that the dispersion equation is of the form $\left(k = k_+ + k_w\right)$

$$
\left[\left(\omega - kv_{11}\right)^2 - \frac{\omega_b^2}{\gamma_0 \gamma_{110}} T\left(\zeta\right)\right]\left[\omega^2 - c^2 k_+^2 - \frac{\omega_b^2}{\gamma_0}\right] =
$$

$$
- \frac{\omega_b^2}{2\gamma_0^3}\left(\frac{\Omega_w}{ck_w}\right)^2 T\left(\zeta\right)\left[\omega^2 - c^2 k^2 - \frac{\omega_b^2}{\gamma_0} T\left(\zeta\right)\right], \tag{24}
$$

where

$$\gamma_{110} \equiv \left(1 - v_{110}^2/c^2 \right)^{-1/2} ,$$

$$T(\zeta) = -\zeta \left[1 + \zeta e^{-\zeta} E_1(-\zeta) \right] \tag{25}$$

describes the effect of the spread in transverse canonical momenta,

$$\zeta \equiv \frac{\gamma_0^2 m^2}{\Delta P^2} \left(v_{110}^2 - \frac{\omega^2}{k^2} \right) , \tag{26}$$

and E_1 is the exponential integral function. It should be remarked that the zero emittance limit is recovered in the limit as $\Delta P_\perp \to 0$, for which $\zeta \to \infty$ and $T(\zeta) \to 1$.

Equation (24) describes the dispersion equation in both the (Raman) and strong-pump (Compton) regimes. The effect of the thermal spread has been studied by numerical solution of the dispersion equation for the case of the propagation of a 750 keV, 255 A-cm^{-2} electron beam through a wiggler field of 1 KG amplitude and 4 cm period. For this choice of parameters $\Omega_w/\gamma_0 ck_w = 0.151$, $\omega_b/\gamma_0^{1/2} ck_w = 0.185$, $v_{110}/c = 0.902$, which corresponds to the Raman regime with a peak growth rate of $(\mathrm{Im}k/k_w)_{max} = 0.052$ in the absence of a thermal spread (i.e., $\Delta P_\perp = 0$). The numerical solution of Eq. (24) for these parameters is shown in Fig. 10 in which we plot the frequency corresponding to peak growth and the peak growth rate versus ΔP. It is suggested from the figure that the transition between the cold and and thermal beam regimes occur for $\Delta P_\perp/p_0 \simeq 0.09\text{--}0.11$. This transition is expected to occur when $(\mathrm{Re}k_+ + K_w)\Delta v_{11} = I_m(k_\perp v_{oz})$, where for our choice of distribution $\frac{\Delta v_{11}}{v_{110}} \simeq \Delta P_\perp^2/2P_{\perp 0}^2$. Using the real and imaginary values for K_+ in the $\Delta P_\perp = 0$ limit we obtain an estimate of $\Delta v_{11}/v_{110} \simeq 0.005$ which corresponds to a $\Delta P_\perp/p_0 \simeq 0.10$. This is in substantial agreement with the numerical solution of the dispersion equation. Finally, we observe that ζ is related to the S-parameter introduced earlier by

$$|\zeta| \simeq \left(\frac{v_{110}^2}{\beta_0^2 c^2} \right) \frac{1}{s} \simeq \frac{1}{s} \tag{27}$$

In order to estimate the effect of the thermal spread on the efficiency, we observe that for saturation via particle trapping each electron loses an energy given by $\Delta \varepsilon \simeq 2\gamma_0 \gamma_{110}^2 m v_{110} \left(v_{110} - v_{ph} \right)$, where $v_{ph} = \omega/\left(k_\perp + k_w \right)$ is the phase velocity of the ponderomotive wave. Averaging this over the distribution function (23), we find that the saturation efficiency is given by

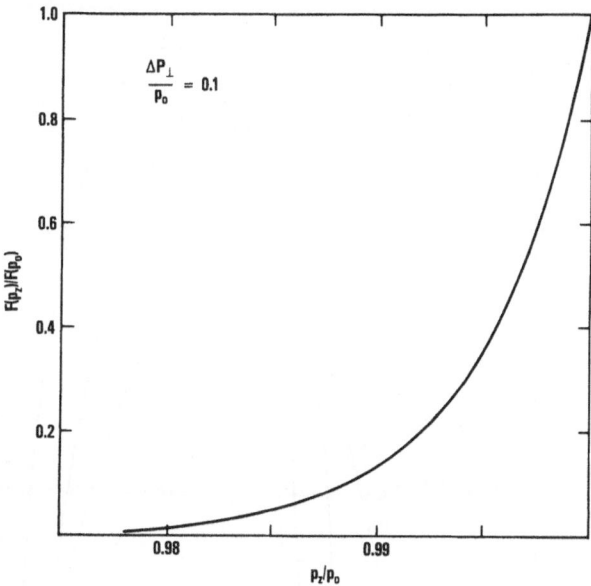

Fig. 10.   Momentum distribution function.

$$\eta \simeq \eta_0 \left[1 - \frac{\Delta P_\perp^2}{P_{110}^2} \left(\gamma_{110}^2 + \frac{v_{ph}}{v_{110} - v_{ph}} \right) \right] \qquad (28)$$

to lowest order in $\Delta P_\perp$, where

$$\eta_0 \simeq \frac{2 \gamma_0 \gamma_{110}^2}{\gamma_0 - 1} \frac{v_{110} \left(v_{110} - v_{ph} \right)}{c} \qquad (29)$$

is the saturation efficiency in the zero emittance limit. The efficiency
is plotted as a function of $\Delta P_\perp$ in Fig. 11 for two cases: (1) at a
fixed frequency corresponding to peak growth (for $\Delta P_\perp = 0$), and (2) at
the frequency corresponding to peak growth. In each case the assumption
that all particles are trapped is implicit.

It is of interest to note that the efficiency at peak growth rate
increases with energy spread when S is sufficiently small. This result
is contrary to intuition. It happens because the energy spread in the
axial component of the beam causes a change in the real part of the
dispersion relation that reduces the phase velocity of the ponderomotive
wave, thus increasing the efficiency due to trapping of the beam
particles by the fastest growing wave. Such a change in the phase
velocity is also observed in the Lorentzian model [cf. Eq. (18)] as S
increases from zero. Specifically, as S increases from zero to 0.4,
$\left| (v_{110} - v_{ph})/v_{110} \right| = \left| \text{Re } y - x \right|$ increases from 0.5 to 0.532, according
to Eq. (18) at maximum gain [i.e., when $\text{Im} y \propto \text{Im} \omega$ is maximized for real
values of x.] This change in the topology of the dispersion relation was

643

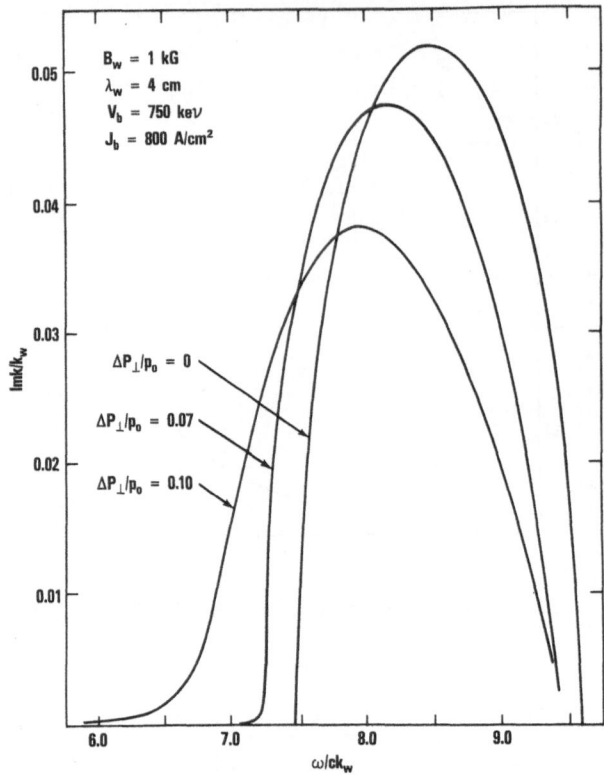

Fig. 11. Growth rate vs frequency.

predicted by O'Neil and Malmberg (1968) and observed by Roberson and Gentle (1971) experimentally on the transition from a cold to warm beam interaction with a plasma.

The increase of the efficiency η with S ($\Delta P_\perp$) is also reflected in Eq. (28). For small values of $\Delta P_\perp$, $\eta \simeq \eta_o \propto \left(v_{110} - v_{ph}\right)$ which increases with $\Delta P_\perp$, as explained above. However, for layers of $\Delta P_\perp$, the second term in Eq. (28) becomes important, and the efficiency begins to decrease, as shown in Fig. 12. In addition, when the beam becomes sufficiently warm, only those electrons that are at the resonant phase velocity of the ponderomotive wave are trapped, resulting in a further decrease in the efficiency.

ACKNOWLEDGMENTS

We have had useful discussions with P. Sprangle and C. M. Tang. One of us (CWR) would like to acknowledge a useful discussion with T. Smith on the relation between wavelength and emittance. This work was supported by the Office of Naval Research.

644

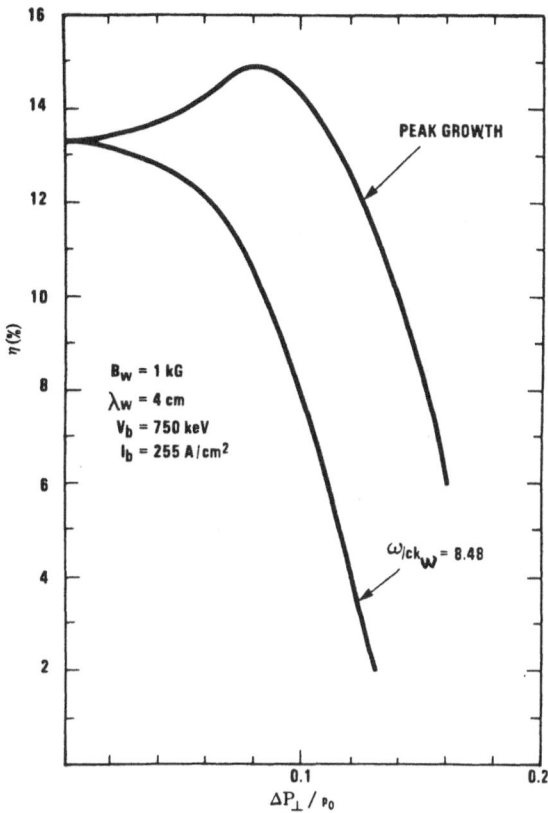

Fig. 12. Efficiency as a function of perpendicular momentum spread.

REFERENCES

Freund, H. P., and Ganguly, A. K., 1986, Phys. Rev. (to be published)
Freund, H. P., Sprangle, P., Dillenburg, D., da Jornada, E. H.,
    Schneider, R. S., and Liberman, B., 1982, Phys. Rev. A 26, 2004
Lau, Y. Y., 1985, "Effects of Cathode Surface Roughness on the Quality of
    Electron Beams," Bull. Am. Phys. Soc. 30, 1583. Also, J. Appl.
    Phys. (Jan 1987 issue)
Lawson, J. D., 1977, The Physics of charged Particle Beams. (Oxford
    University Press, Oxford, UK) Chap. 4
Lawson, J. D. and Penner, S., 1985, IEEE J. Quantum Electron., QE-21, 174
Lee, E. P. and Cooper, R. K., 1976, Particle Acc., 7:83
Neil, V. K., 1979, SRI Tech. Rep. JSR-79-10
O'Neil, T. M. and Malmberg, J. H., 1968, Phys. Fluids 11, 1754
Roberson, C. W., 1983, Proc. of Soc. of Photo-Optical Instr. Eng., 453,
    320
Roberson, C. W., 1985, J. of Quant. Electron. QE-21, 860
Roberson, C. W. and Gentle, K. W., 1971, Phys. Fluids 14 2462
Roberson, c. W., Pasour, J. A., Mako, F., Lucey, R. F. and Sprangle, P.,
    1983, Chapter 7, Infrared and Millimeter Waves 10, 361
Sprangle, P., Tang, C. M. and Roberson, C. W., Nucl. Instr. and Methods,
    1985, Phys. Research A239, 1; and Laser Handbook, Vol. 4, eds. M.
    Stitch and M. S. Bass, North Holland, Amsterdam (to be published)
Sprangle, P., Tang, C. M., 1981, Appl. Phys. Lett. 39:677
Yariv, A., 1976 Introduction to Optical Electronics (Holt, Rinehart, and
    Winston) Chap. 3

INDUCTION LINACS FOR HEAVY-ION FUSION*

Denis Keefe

Lawrence Berkeley Laboratory
University of California
Berkeley, California 94720 USA

INTRODUCTION

Inertial Confinement Fusion (ICF) is an alternative approach to magnetic Confinement Fusion for a future source of fusion electrical energy based on virtually inexhaustible fuel sources. The ICF method relies on supplying a large beam energy (3 MJ) in a short time (10 nsec) to ignite and burn a spherical capsule containing a few milligrams of deuterium and tritium; ablation of the surface--as a plasma--drives an implosion of the fuel, and leads to ignition at the center when the compression has reached an appropriate value (about 1000 times normal liquid density). The energy can be supplied directly; in that case a large number of beams must be brought in to illuminate the capsule in a spherically symmetric manner, which complicates the final focusing. Alternatively, the beam energy can be used to produce high temperature radiation which can be contained within a hohlraum to implode a separately placed capsule.

A multigap accelerator for heavy ions, relying on the physics and engineering base of research accelerators, combines several advantages as a driver for fusion energy over the other two driver candidates, laser and light ion beams, in the following regards:

    i) Efficiency

    ii) Repetition rate

---

* This work was supported by the Office of Energy Research, Office of Basic Energy Sciences, U.S. Department of Energy, under Contract No. DE-AC03-76SF0098.

iii) Reliability

    iv) Long stand-off distance for the final focus.

A discussion of target needs, reactor systems and the choice among driver systems is given by Keefe (1982).

The enormous current (20 particle-kiloamperes) that must be delivered in the final short pulse, and the preservation of a low beam emittance, are the two features that lie well beyond the experience of today's research accelerator experience.

The ion range must lie in the neighborhood of 0.1-0.2 gm cm^{-2}; thus the kinetic energy of a heavy ion is about 10 GeV (i.e., 50 MeV/amu) and the ion speed about 0.3 c. The focal spot at the target is about 2 mm radius at a stand-off distance of 10 m, and the normalized emittance should not exceed 20 mm-mrad. For details of the progress in studies of Heavy Ion Fusion (HIF) see the proceedings of the recent Heavy Ion Fusion Symposium in Washington, DC, May 1986 (Bangerter et al., 1986).

DRIVER CONFIGURATIONS

Two heavy-ion accelerator driver systems to deliver high current beams of heavy ions (A = 200) with kinetic energy about 10 GeV are shown schematically in Fig. 1.

The rf/storage ring method starts with eight low-β accelerators, the beams being sequentially combined in pairs--after some stages of acceleration--to deliver a high current beam (160 mA) to the main linac (Badger et al., 1984). In an r.f. linac, the current remains constant since the length of the bunch expands in direct proportion to the speed during acceleration. When acceleration to 10 GeV is complete, the current is amplified from 160 mA in a sequence of manipulations in storage rings, including multiturn injection and bunching, to 20 kA to be finally delivered to the target in some ten to twenty separate beams.

The induction linac system, by contrast, relies on amplifying the current simultaneously with acceleration to keep pace with the kinematic change in the space-charge limit (Keefe, 1976; Faltens et al., 1981). Sixteen multiple beams can be accelerated in the same structure with independent transport systems from source to target; this approach would represent the simplest single-pass system.

A knowledge of the space-charge limit for beam current is crucial in the design of just the low-β parts of the rf/storage ring system, but is clearly central to the design of the induction linac at every point along its length.

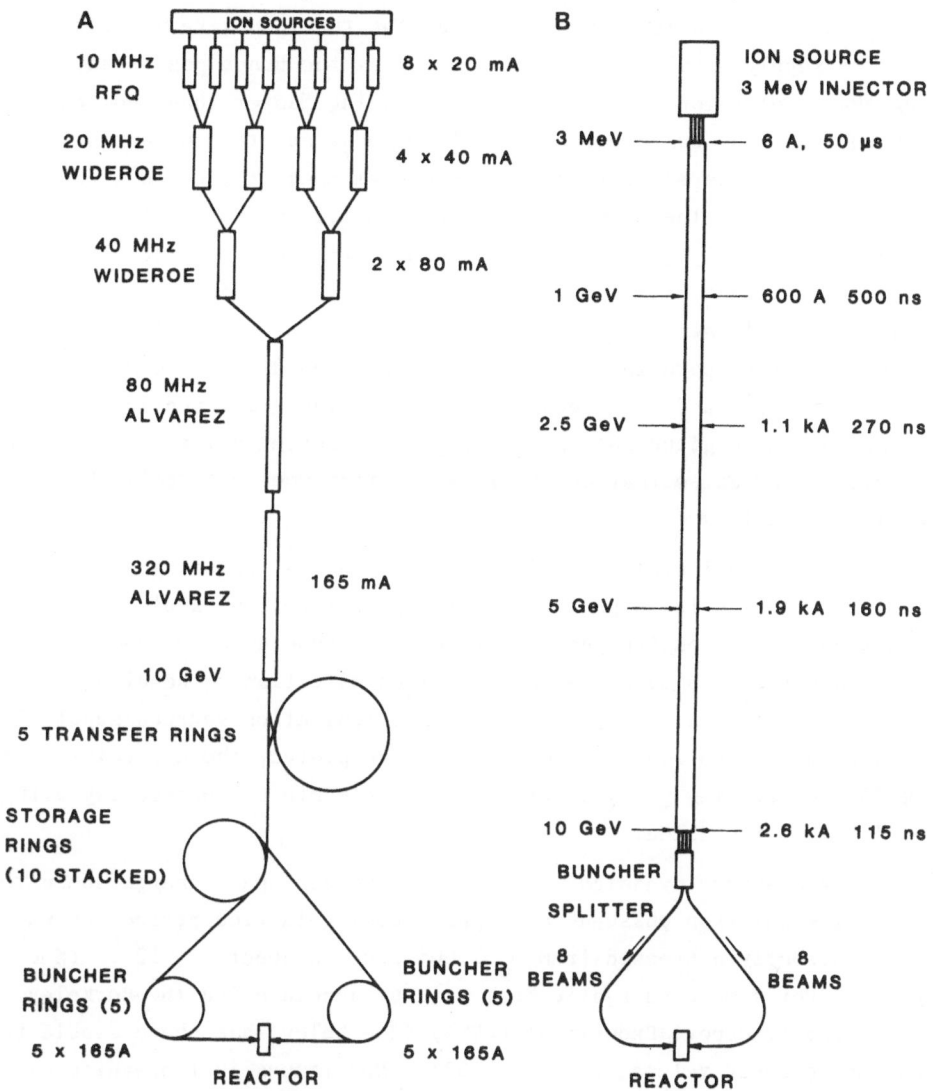

Fig. 1. Two designs for heavy-ion accelerator drivers at present under study. (a) In the rf linac scenario the current is amplified by a factor of 10^5 in a sequence of multiple storage rings (rf - linac storage rings). (b) The induction linac uses current amplifications during acceleration and final bunching in a transport-drift section to raise the current by a factor of 3,000 (induction linac).

THE CURRENT-AMPLIFYING INDUCTION LINAC

The basic idea of a heavy-ion induction linac using current amplification is to inject a long beam bunch (many meters in length, several microseconds in duration) and to arrange for the inductive accelerating fields to supply a velocity shear so that, as the bunch passes any point along the accelerator, the bunch tail is moving faster than the head. As a consequence, the bunch duration will decrease and the current will be amplified from amperes at injection to kiloamperes at the end of the driver (~10 GeV). The current is further amplified by a factor of about 10, and the pulse length further shortened correspondingly to about 10 nanoseconds, in the drift section between the accelerator exit and the final focusing lenses. Transverse space-charge forces are large enough that some sixteen parallel beams are needed to handle the beam in the drift-compression and the focus sections. In the drift section one is relying on the longitudinal space-charge self-force in the beam to remove the velocity shear so that chromatic aberration does not spoil the final focusing conditions.

A proof-of-principle experiment, called MBE-4, is being assembled at Berkeley. The aim is to prove the principle of current amplification while keeping the longitudinal and transverse beam dynamics under control and, in addition, to face the additional complication of handling multiple beams (four in MBE-4). Four surface ionization sources supply 20 mA apiece of cesium ions at 200 kV. When completed, the apparatus will have 24 accelerating gaps; at present, experiments are proceeding with the eight now installed.

The transverse dynamics in MBE-4 is strongly space-charge dominated in that the betatron phase-advance per focusing-lattice period for each beam is strongly depressed-from $\sigma_0 = 60°$ down to about $\sigma \sim 12°$. (See Fig. 2.) For a mono-energetic beam without acceleration the Berkeley Single Beam Transport Experiment (SBTE) (see below) has shown stable beam behavior to lower values of σ (7° – 8°). New issues in transverse dynamics, however, arise in MBE-4 because of (a) the difference in velocity along the bunch as it passes through a given lens, which results in values for σ_0 and σ that vary along the bunch length, and (b) the discrete accelerating kicks which can cause envelope-mismatch oscillations.

For the longitudinal dynamics, two separate features arise in MBE-4. Space charge effects throughout the body of each long bunch (about 100 cm long and 1 cm radius) are strong enough that the dynamical response to velocity kicks or acceleration errors is described in terms of space-charge (Langmuir) waves rather than in single-particle terms. Secondly, the tapered charge density that occurs at the end of the bunch will result in collective forces that are accelerating at the head and decelerating at the tail and, if not counteracted, will make the ends of the bunch spread both in length and in momentum. A major part of the

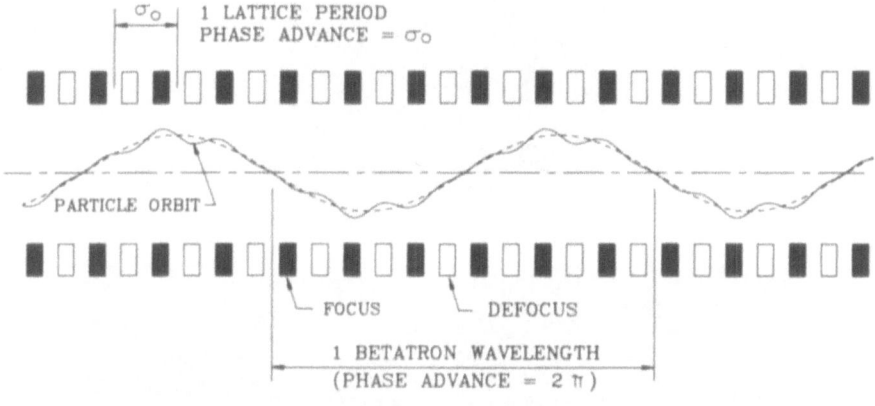

WITHOUT SPACE CHARGE

1 LATTICE PERIOD
PHASE ADVANCE = σ_O

PARTICLE ORBIT

FOCUS   DEFOCUS

1 BETATRON WAVELENGTH
(PHASE ADVANCE = 2π)

WITH SPACE CHARGE

1 LATTICE PERIOD
PHASE ADVANCE = σ BEAM ENVELOPE

PARTICLE ORBIT

FOCUS   DEFOCUS

1 BETATRON WAVELENGTH

Fig. 2.   In a strong-focusing lattice (alternating focusing and
defocusing quadrupoles) a single particle executes
quasi-sinsusoidal betatron oscillations (upper). Its
motion is characterized by the phase advance of the
sinusoid per repeat length of the structure, σ_o. With
space-charge present--a defocusing force--(lower) the
phase advance, σ, (or oscillation frequency) is
decreased.

experimental effort is centered on designing and successfully deploying
the electrical pulsers to handle the correcting fields at the bunch ends.

Figure 3 shows an example of current amplification results obtained
to date (June 1986), where it can be seen that the pulse duration has
been shortened by nearly a factor of two and the current correspondingly
increased (Fessenden et al., 1986). Because MBE-4 operates at relatively
low energy (accelerating from 200 keV to 1 MeV), we can try rather
aggressive schedules for current amplification, which correspond to
setting up a large velocity shear, $\Delta\beta/\beta$. We do not have a firm argument
for exactly how high a velocity-shear may be and still be considered

651

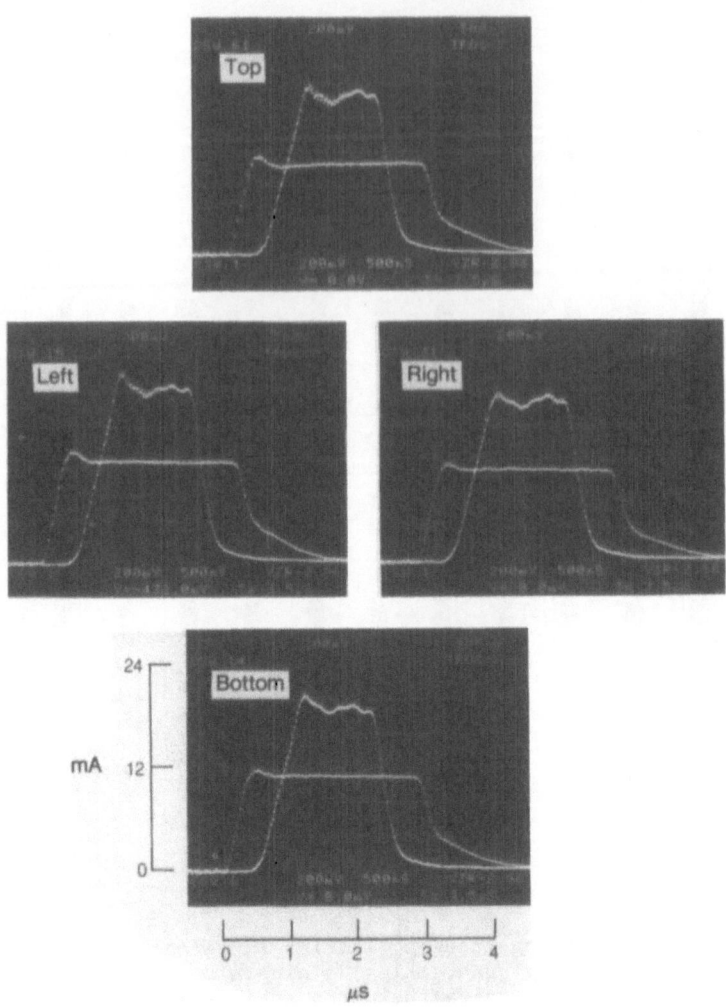

Fig. 3. Oscillograms for all four beams in MBE-4 show the
injected current trances (lower amplitude, longer
duration) and the amplified current traces after
eight accelerating units.

tolerable. An experiment with $\Delta\beta/\beta = 0.4$ has been completed; this is
more than will be needed in a driver.

HIGH CURRENT BEAM BEHAVIOR AND EMITTANCE GROWTH

The Single Beam Transport Experiment (SBTE)

Since the IEEE Particle Accelerator Conference in Vancouver in May
1985, the results on high-current beam transport limits (Tiefenback and
Keefe, 1985) in the 87-quadrupole SBTE have been refined and more
careful calibrations made. The results, shown in Fig. 4, are
substantially unaltered; at the highest currents and lowest emittance
values obtainable from the 120-200 kV cesium injector, no detectable
growth in emittance was observed in the 41-period transport section,

652

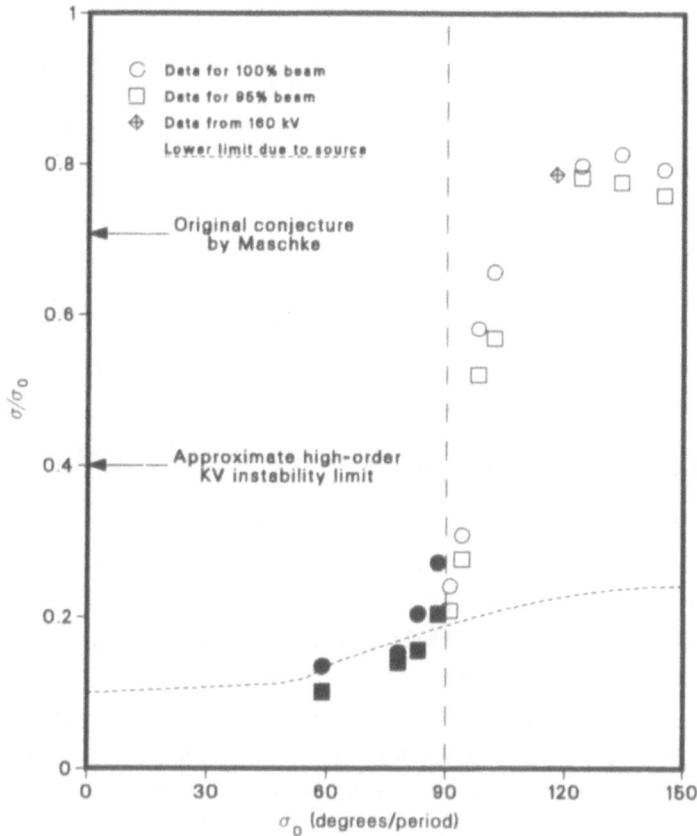

Fig. 4. Results from the Single Beam Transport Experiment. The solid
data points are for cases where no emittance growth nor
current loss could be detected. The dashed curve indicates
the lower limit on σ/σ_0 that could be reached because of
ion-source limitations. Above $\sigma_0 = 88°$, emittance growth and
current loss can be avoided only for values of σ/σ_0 lying
above the open data points.

provided σ_0 did not exceed 88°. A threshold value of current above which
emittance growth occurs could, however, be measured for values of σ_0 in
excess of 88°. Since the transportable current is greatest for $\sigma_0 < 88°$,
the design of drivers will be restricted to σ_0 values in this range.
[Tiefenback (1986) has found that beyond $\sigma_0 = 88°$ the threshold
corresponds rather well to the empirical condition that the beam plasma
period equals the beam transit time through three lattice periods.]

Earlier theoretical work on beam current limits in AG focusing
systems utilizing an idealized distribution (the Kapchinskij-Vladimirskij
or K-V) indicated that it could be dangerous to use σ_0 greater than 60°,
and that σ could probably be depressed from that value down to 24°, but
not below (Hofmann et al., 1983). The experimental limits from SBTE
shown in Table 1 can be seen to be much more encouraging.

Table 1. Experimental limits on σ_0, σ.

σ_0	60°	78°	83°
σ	< 7°	<11°	<15°

Is There a Lower Limit on σ_0?

In his original consideration of high current limits in magnetic AG systems Maschke showed that the limiting particle current could be written (nonrelativistically) as:

$$I_p = K(\eta B)^{2/3}(\varepsilon)N^{2/3}V^{5/6}\Big/q^{1/2}A^{1/2} , \qquad (1)$$

with B the limiting pole-tip field, η the fraction of length occupied by magnetic lenses, qV the ion kinetic energy, and A,q, the ion mass and charge state respectively. [Two other equations, involving lattice-period and radius, must be simultaneously obeyed for Eq. (1) to hold.] The coefficient, K, originally selected by Maschke was for an implicit assumption that σ/σ_0 could not be less than 0.7. In light of the improved knowledge from experiment and simulation it is useful to use the "smooth approximation" (Reiser, 1978) to write the explicit dependence of K on σ_0 and σ, viz:

$$K \propto \sigma_0^{2/3}\Big/(\sigma/\sigma_0)^{2/3} . \qquad (2)$$

Thus, in fact, if there were no lower bound on σ/σ_0 the transportable current could grow very large (the required aperture, however, would do likewise).

Just as the SBTE measurements were beginning, Hofmann and Haber, each using simulation codes for well-centered beams without images, reported that for σ_0 = 60°, σ could be allowed to go lower than 24° without emittance growth occurring. During the course of SBTE measurements, further simulations showed that values of σ down to 1° or 2° might be alright.

Our view of the situation changed, however, with the simulation studies by Celata et al. (1985) of an off-axis beam--which corresponds to the real-world situation. For a beam with σ/σ_0 = 6°/60°, no growth was detected. If _either_ a dodecapole component in the field _or_ the effect of images in the electrodes was introduced, however, steady growth in the rms emittance accompanied by oscillations showed up clearly. When _both_ images and the right amount of dodecapole component were included, how ever, the surprising result emerged that the emittance did not grow.

RE-DISTRIBUTION IN FIELD ENERGY IN HIGH-CURRENT LOW-EMITTANCE BEAMS

The growth in emittance due to a change in the beam distribution in configuration space alone has been a topic of much discussion in the past few years. An intense beam with a non-uniform spatial distribution will usually readjust itself in a fraction of a plasma period to an almost exactly uniform distribution. The change in electrostatic field energy is always such that energy is fed into the thermal motion of the beam particles thus causing emittance growth. For given initial and final distributions, Wangler has given a prescription for determining the amount of growth, which we can re-write in the form

$$\frac{\varepsilon_f}{\varepsilon_i} = \left(1 + \frac{1}{4} \frac{\langle x^2 \rangle}{\lambda_D^2} \frac{\Delta E}{E}\right)^{1/2} ,$$

where $\langle x^2 \rangle$ is the mean square radius of the beam, λ_D is the Debye length (proportional to $\varepsilon/\sqrt{I}$), and $\Delta E/E$ is the fractional increase in electrostatic field energy as the beam relaxes from a non-uniform to a uniform shape. $\Delta E/E$ is of the order of 0.1 or less. Equation (3) is simply a statement of conservation of energy, i.e., the excess field energy shows up as increased thermal motion of the beam particles. In contrast to the case of a beam in a synchrotron, for instance, where $\langle x^2 \rangle \ll \lambda_D^2$ and the effect is negligible, we are concerned with space-charge dominated beams with $\langle x^2 \rangle \gg \lambda_D^2$, and the effect can be quite serious.

This mechanism for emittance growth clearly can occur just after an ion source which is emitting a non-uniform beam. But it is also of importance in combining (or splitting) beams that are round or elliptical by means of a septum. Simulation results on emittance growth in the case where four beams are stacked side by side by septa to form a single larger beam were given by Celata (1986).

NEW CONSIDERATIONS FOR DRIVER DESIGN

Much of the early design work for induction linac drivers was restricted to considering that ions with charge state $q = 1$ were most suitable and, also, that $\sigma/\sigma_o = 24°/60° = 0.4$ was an optimum value. The driver design program, LIACEP (Faltens et al., 1979) did, however, indicate that capital savings could ensue if either condition could be relaxed, but at the cost of additional complications, namely:

i) Reduced current at any point (V) in the driver (see Eq. 1).

ii) Generating ions with q > 1, which was visualized to be done by stripping a beam with q = 1 at some intermediate energy.

iii) An increased number of beam lines in the drift-compression section.

The results from SBTE and simulations have altered our thinking and encouraged us to re-open the matter of using ions with charge state q > 1. As an illustration, consider the reference case given in 1981 for V = 10 GV, q = 1. (See Fig. 1.) We could build only the first 5 GV part and use charge state q = 2 to give the same final kinetic energy, 10 GeV. We could still maintain the same particle current at each voltage point provided the product $q^{1/2}(\sigma/\sigma_o)^{2/3}$ is kept constant, i.e., $\sigma/\sigma_o \propto q^{-3/4}$. [This can be seen from Eqs. (1) and (2).] Since we know that very low values are permitted for σ/σ_o, we can in principle continue this argument to higher charge states, dropping σ/σ_o in value and shortening the accelerator at each step. A limitation occurs, however, beyond q = 3 (for A = 200) because the increased perveance (i.e., space-charge) in the final drift lines rises as q^2 and the increased cost of the very large number of final beam lines that will be needed overrides the cost reduction in the accelerator. This argument is given in more detail by Lee (1986).

It now appears that the direct generation of adequately high currents of ions with q > 1 from a source is possible as a result of work by Brown and Galvin with the MEVVA source (Brown and Galvin, 1986). Using a similar source, Humphries (1986) has shown how to avoid plasma pre-fill of the extraction region, and thus has solved the problem of rapid turn-on of the source (< 1 μsec) needed for an induction linac driver (Humphries, 1986).

Since the SBTE has shown that σ_o can exceed 60° safely (but not 88°), present driver design have benefited by using σ_o = 80°, resulting in a somewhat greater beam current limit [see Eq. (2)].

With ions of q = 1, the low velocity end of the linac (< 250 MeV) represented only 10% of the cost (Faltens et al., 1981). With ions of q = 3 the bulk of the accelerator has been shorted from 10 GV down to 3.3 GV and the cost of the front-end represents a much more significant fraction of the overall cost; hence, it is now receiving more design attention. If electrostatic lenses are used in the low velocity end, the mapping argument given earlier (for magnetic transport), from equal voltage points in a q = 1 to a q > 1 case, no longer holds unless the number of beams is increased. With higher charge-state, therefore, we visualize a driver starting with as many as 64 beamlets that undergo the bulk of the acceleration. (See Fig. 5.) Before this strategy can be established as a viable one, however, the emittance growth in combining high current beams must be understood better.

THE HEAVY ION FUSION SYSTEMS STUDY (HIFSA)

The first systems assessment for a power plant based on an induction linac driver has been in progress for a year and a half under the auspices of EPRI and the DOE Office of Program Analysis and Office of Basic Energy Sciences. The major participants include McDonnell-Douglas (MDAC), LANL, LBL, and LLNL. The main emphasis as expressed in the term "Assessment" is not on developing a point design such as HIBALL (Badger et al., 1984) but on exploring a broad range of parameters to establish general conclusions. (A wide variety of point designs can, of course, be generated from the results.)

Four different reactor types and five different target designs are included in the examination. The driver parameters range from a kinetic energy of 5 GeV to 20 GeV and a beam energy from 1 MJ to 10 MJ. Results to date show that a cost of electricity of 5.5 cents/kW-hr seems quite reasonable to expect for a 1000 MWe plant that uses ions with A = 200, q = 3. The familiar "economy-of-scale" effect is also apparent, with the cost of electricity being less (4.5 cents/kW-hr) if a 1500 MWe plant is considered, or more (9.5 cents/kW-hr) for a 500 MWe plant. One of the more interesting results is that such values of electric energy cost can

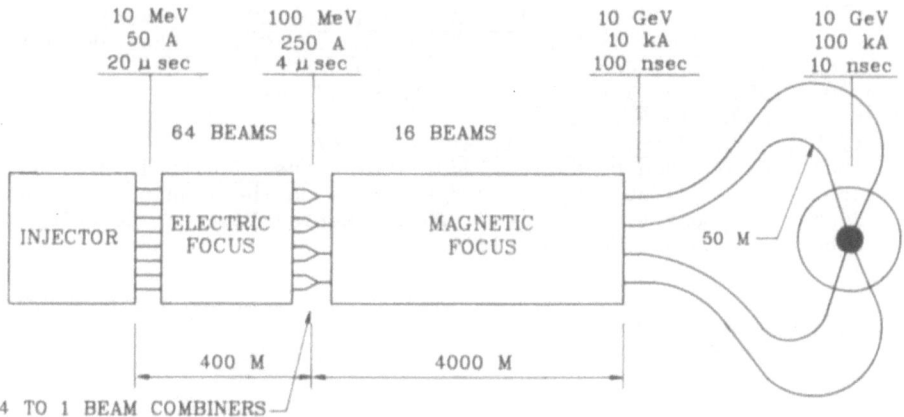

Fig. 5. Schematic of present concept for a driver using ions with charge state 3. The total beam current shown is in electrical (not particle) amperes.

be realized for a very broad range of driver parameters and for several choices of both reactor and target designs.

SUMMARY

Experimental progress to date has strengthened our belief in the soundness and attractiveness of the heavy ion method for fusion. What surprises that have shown up in the laboratory (e.g., in SBTE) have all been of the pleasant kind so far.

The systems assessment has supported the view that the heavy ion approach can lead to economically attractive electric power and that a wide variety of options exists in all parameters. The systems work has also been of great help in pointing the way for the research and development activities.

REFERENCES

Badger, B. et al., 1984, "HIBALL-II, An Improved Heavy Ion Beam Driven Fusion Reactor Study," Univ. of Wisc. Rep. No. UWFDM-625.

Bangerter, R. O., T. G. Godlove, and M. P. Reiser, 1986, Proc. Int. Symp. on Heavy Ion Fusion, AIP Conference Proceedings No. 152.

Brown, I., and J. Galvin, 1986, in Ref. 2.

Celata, C. M., I. Haber, L. J. Laslett, L. Smith, and M. G. Tiefenback, 1985, IEEE Trans. Nuc. Sci. 32:2480.

Celata, C. M., 1986, in Ref. 2.

Faltens, A., E. Hoyer, D. Keefe and L. J. Laslett, 1979, Proc. Workshop on Heavy Ion Fusion, Argonne 1978, Argonne Natl. Lab. Rep. ANL-79-41, p. 31.

Faltens, A., E. Hoyer, D. Keefe, 1981, Proc. 4th Int. Top. Conf. on High-Power Electron and Ion-Beam Res. and Techn., Palaiseau, (ed. H. J. Doucet and J. M. Buzzi) 751.

Fessenden, T. J., D. L. Judd, D. Keefe, C. Kim, L. J. Laslett, L. Smith and A. I. Warwick, 1986, in Ref. 2.

Hofmann, I., L. J. Laslett, L. Smith and I. Haber, 1983, Particle Accelerators 13:165.

Humphries, S., Jr., 1986, Particle Accelerators, 20 (in press).

Keefe, D., 1976, Proc. 1976 Proton Linear Accel. Conf. (Chalk River) Atomic Energy of Canada, Ltd., Rep. No. AECL-5677, 272.

Keefe, D., 1982, Ann. Rev. Nuc. Part. Phys. 32:391.

Lee, E. P., 1986, in Ref. 2.

Reiser, M., 1978, Particle Accelerators, 8:167.

Tiefenback, M. G., and D. Keefe, 1985, IEEE Trans. Nuc. Sci. 32:2483.

Tiefenback, M. G., 1986, "Space-charge Limits on the Transport of Ion Beams in a Long A. G. System" (Ph.D. Thesis) Lawrence Berkeley Laboratory Report No. LBL-21611.

HIGH-AVERAGE-POWER ELECTRON ACCELERATORS FOR FOOD PROCESSING

Stanley Humphries, Jr.

Institute for Accelerator and Plasma Beam Technology
University of New Mexico, Albuquerque, New Mexico 87131   USA

In the last decade, dramatic advances have been made in the physics and technology of accelerators for high power relativistic electron beams (Humphries). Research in this area has been largely directed toward defense applications. Although the technological feasibility of accelerators for defense is uncertain, the technology could lead to valuable commercial applications. This report reviews the field of food processing by accelerator generated radiation. Advances in induction linear accelerators and high power RF accelerators could significantly impact this area.

The food processing application is particularly exciting because 1) requirements for average power and efficiency are challenging yet within the bounds of present technology, and 2) successful development could lead to clearly identified economic benefits and social advances. It is noteworthy that fully one quarter of the world's food production is lost after harvesting. The use of accelerator-generated radiation to reduce spoilage and facilitate the transport of food could have a profound effect on world nutrition, particularly in Third World countries.

Radiation processing of chemicals and plastics by electrons from low energy electrostatic accelerators has long been a commercially viable field (Scharf, 1986). In contrast, food processing by accelerator-generated radiation is in its nascent stage. This reflects in part the necessary delays in approval of commercial operations to determine safe limits for the process. There are also technical and economic problems to be solved. Existing and past projects in food processing by accelerator-generated radiation are listed in Table 1. Most work has been performed at moderate beam energy (a few MeV) or power (10-20 kW). In the first section, we will show that the overall economy of

Table 1. Accelerators for food processing (February, 1985).*

COUNTRY	LOCATION	TYPE	APPLICATION	STATUS
China	Beijing Radiation Center	LINAC 5 MeV	Experimental, demonstration	Operation to terminate
Ecuador	Escuela Politecnica Nacional, Quito	LINAC 6 MeV, 2 kW	Pilot plant	Expected, 1986
France	SODETEG, Paris	LINAC 6MeV, 7 kW	Demonstration multipurpose	Operation since 1968
	Societe GUYOMARCH, Britanny	CASSITRON 10 MeV, 10 kW	Commercial, frozen poultry	July, 1985
FRG	Karlsruhe	LINAC 10 MeV, 6 kW	Experimental, demonatration, multipurpose	Program terminating
	Hamburg "Anton Dohrn"	X-ray, 200 kV, 30 kW	Demonstration multipurpose	Out of use
GDR	Leipzig	LINAC	Food, multi-purpose	Planning
Israel	Sorcy NRC, Yavme	ICT, 1.5 MeV 75 kW	Pilot plant, poultry feed	Operating
Mexico	Inst. of Physics, UNAM, Mexico City	Dynamitron, 3 MeV, 25 kW	Exper. maize disinfestation	Program terminating
Malaysia	Puspati, Selangor	ICT	Multipurpose, planning scale	Planning
Nether-lands	ITAL, Wageningen	Van de Graaf	Experimental	Program terminating
Poland	Techn. Univ. Lodz	LINAC	Multipurpose	Completed, 1983
USA	LLNL, Livermore, Cal.	Induction linac	Experimental, demonstration	Operating 1985
	Ford Laboratory Dublin, Cal.	Induction linac	Commercial, demonstration	Expected, 1986
USSR	Odessa Port Elev. RDU, Odessa	2 ELV-2, 1.4 MeV, 20 kW	Grain disinfestation, 200 t/h	Operating since 1980
	VNIIKOP, Birjulovo Moscow	LINAC	Multipurpose	Program terminating

* Based on a compilation by J. Farkas, IFFIT, Wageningen (IAEA)

accelerator processing facilities improves at high average power and high output beam energy. This fact motivates consideration of accelerators with beam power in the range 200–500 kW.

The utility of accelerators also improves at high beam kinetic energy. Present regulations permit operation at 5 MeV for γ-ray production and 10 MeV for direct electron deposition. The long range of high energy electrons in material may permit irradiation of food in its final shipping container, reducing handling costs. Larger packages or full pallets of food can be treated if electrons are converted to γ-rays in a target. The efficiency of the conversion process increases with kinetic energy, and the γ-rays have improved directionality. The development of robust electron accelerators with high average power is a challenge to the accelerator scientist. The machines must also have high reliability, long component lifetime, and high electrical efficiency.

The report is divided into two sections. The physical and chemical principles of radiation processing of food are reviewed in the first section. The relative benefits of accelerators compared to radiosotopes are emphasized. Parameters of practical accelerators are reviewed; they are bracketed by considerations of processing costs and food wholesomeness. In the second section, four concepts for high power processing accelerators are reviewed. The examples are long-pulse induction linacs (Ford Laboratories, Dublin, California), RF linacs with continuous duty cycle (Chalk River Nuclear Laboratory), high-frequency magnetically-switched induction linacs (Lawrence Livermore National Laboratory) and pulsed standing-wave RF linacs (the author). The parameters and underlying physical principles of the accelerators are reviewed. The devices differ in terms of size, efficiency, and required technological development. Taken together, the studies confirm that there are a number of attractive approaches to economical food processing accelerators.

PRINCIPLES OF FOOD PROCESSING BY ACCELERATOR-GENERATED RADIATION

Introduction of Food Irradiation

This section gives an overview of the physical, chemical, biological and economic principles that underlie the processing of food by accelerator-generated ionizing radiation. Sufficient information is provided to understand design considerations for irradiation facilities and acceptable parameters for accelerator output energy, power levels, net efficiency, and total cost.

It should be recognized that radiation-induced modifications to improve food are already common. Two examples are preservation by thermal energy (canning) and microwave cooking. The novel feature of the high energy radiation generated by accelerators is that it can ionize atoms in the irradiated medium. In contract to thermal radiation, ionizing radiation can achieve the same biological effect (such as the reduction the number of microorganisms in food) with much smaller net deposited energy.

The two types of ionizing radiation useful for food processing are high energy electrons (β-rays) and photons (γ-rays). These elementary particles transfer energy to materials mainly through collisions with atomic electrons. It is essential to emphasize that ionizing radiation (in the energy range of interest) does not produce radioactivity in food; rather, it induces the formation of short-lived reactive chemicals. These chemicals inactivate microorganisms that can spoil food or cause diseases. The long-term chemical changes in food following irradiation are extremely small compared to commonly accepted chemical preservation processes such as salting and smoking. Chemical preservation methods can degrade the nutritional content of food and can exacerbate metabolic disorders such as high blood pressure. In contrast, irradiation of food within the guidelines set by the United Nations World Health Organization (WHO), the International Atomic Energy Agency (IAEA) and the Food and Agriculture Organization (FAO) (Joint FAO/IAEA/WHO Expert Committee) effectively destroys microoganisms, causes only small chemical changes, introduces negligible radioactivity, and maintains nutritional value.

Present applications of accelerator-generated radiation to a variety of industrial processes are listed in Table 2 in order of increasing absorbed dose. The dose, defined in a later section, indicates the amount of energy deposited by ionizing radiation per unit mass of the sample. Many of the applications listed, such as curing of surface coatings, are already commercially viable. Existing accelerators for these applications generate low energy beams (<2 MeV). Electron beams from these machines have a short range and produce a high surface dose. Food processing applications demand higher energy accelerators with more penetrating radiation. Although food treatment requires lower doses than industrial processes, the radiation must extend uniformly through the larger volume. Table 3 lists food-related applications and associated dose ranges.

The food irradiation process is illustrated schematically in Fig. 1. A high-energy electron beam is generated in the vacuum environment of an accelerator. The beam can be used in two ways for radiation processing. In the first (Fig. 1a), the electron beam is expanded and passed through a thin metal vacuum foil directly into the material to be treated. Foods are composed mainly of elements with low atomic number so that the energetic electrons transfer energy through Coulomb collisions with

Table 2. Industrial applications of accelerator generated radiation.*

PROCESS	DOSE (kGy)	PROCESSING CAPACITY (kg/kw-hr) (e-)
Destruction of insects	0.25-1	3600-14000
Food preservation	1.0-25.0	14403600
Cellulose depolymerization	5-10	360-720
Graft copolymerization	10-20	180-360
Medical sterilization	20-30	120-180
Curing of coatings	20-50	72-180
Polymerization of emulsion	50-100	36-72
Vulcanization of silicones	50-150	24-72
Cross-linking of polymers	100-300	12-36
Vulcanization of rubber	100-300	12-36

* Adapted from W. Scharf, Particle Accelerators and Their Uses (Harwood, Chur, Switzerland, 1986), 839.

Table 3. Applications of radiation to food processing.

Inhibition of sprouting and delay of ripening of fruits and vegetables to extend shelf life

Insect disinfestation to facilitate the international movement of fruits and vegetables without the use of dangerous chemicals (ethylene dibromide or methyl bromide)

MEDIUM DOSE (1-10 kGy)

Reduction of microbial load to retard spoilage

Reduction in number of non-sporing pathogenic microorganisms to reduce diseases caused by bacteria (salmonellosis) and parasites (trichinosis)

Improvement in technological properties of food

HIGH DOSE (10-50 kGy)

Sterilization for commercial purposes

Elimination of viruses

electrons in the medium, as described in a later section. The penetration distance of electrons is short (~5 cm in water at 5 MeV). This implies that direct electron irradiation is best suited to fluid media (like wheat) that can be exposed to the beam in a narrow sheet. Direct electron deposition gives a high dose rate (power deposited per mass), allowing a high throughput of the fluid media.

In order to irradiate large containers, the electron beam can be converted to a stream of γ-rays, as shown in Fig. 1b. Conversion occurs

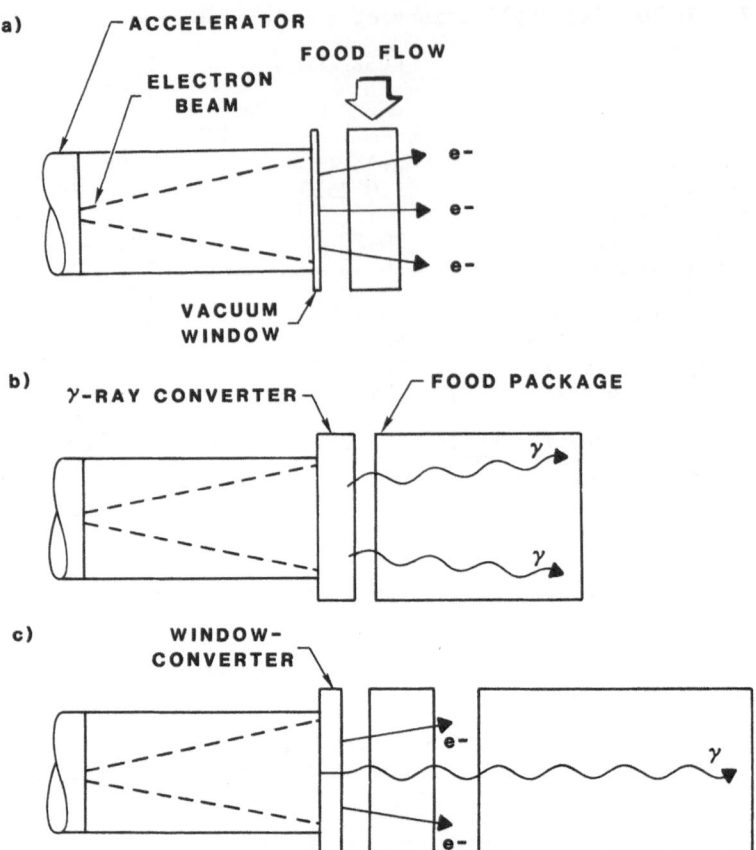

Fig. 1. Food processing by accelerator-generator radiation.
a) Direct electron deposition. b) γ-ray conversion.
c) Hybrid mode.

when the beam impinges on a block of metal with high atomic number (such
as tungsten). A portion of the beam energy is transferred to photon
energy (bremsstrahlung radiation). The γ-rays emitted from the target
have a spread in energy up to the maximum energy of the electron beam.
The γ-rays have a longer energy transfer range in Fig. 1c material;
therefore, useful doses can be obtained in packages with greater
depth. Another advantage of photon conversion is that a thin vacuum
window is not necessary. Hybrid systems [Fig. 1(c)] have also been
proposed (McKeown, 1985). In hybrid mode, a window of high atomic number
material with thickness less that the electron range is used so that both
electrons and γ-rays are generated. For some accelerator irradiators, it
may be advantageous to modify the operational mode depending on seasonal
variations of available foods. The addition of thin and thick movable
in-vacuum tungsten targets would increase the operating-mode flexibility
of an accelerator.

Radioisotopes are the main alternative to accelerators for the generation of radiation for food processing. Radioisotopes have unstable nuclei; nuclear disintegration leads to the emission of radiation over an extended periods of time. Commercial radioistopes are produced by the exposure of stable elements to neutrons in a fission reactor. The two isotopes of interest for food irradiation are Co^{60} and Cs^{137}. The properties of emitted radiation are listed in Table 4. In a single disintegration, a Co^{60} nucleus emits a β^- particle and changes to an excited Ni^{60} nucleus. This nucleus immediately emits two γ-rays of energy 1.17 and 1.33 MeV. Since the β^- usually does not escape from the containing vessel, Co^{60} acts as a γ source.

A diagram of a typical Co^{60} irradiator is shown in Fig. 2. The facility is set up with reentrant shielding and a conveyer system for continuous processing. Batch processing is prohibitively expensive and time consuming. The radioisotope emits radiation continuously; therefore, it must be stored in a water pool to gain access to the cell when the facility is not in operation. Only a fraction of the γ-rays emitted isotropically from the source are absorbed in the processed material because of solid angle limitations. One of the advantages of accelerators is that the radiation is emitted with good directionality, allowing more effective utilization. Other advantages for large-scale systems are discussed in the section on Comparison of Accelerators and Radioisotopes for Food Processing Applications. Despite accelerator advantages, most present small-scale commercial irradiators use radioisotopes because of the relative simplicity of setting up such a facility (IAEA, 1984). A partial list of existing commercial facilities is given in Table 5.

Chemical and Biological Effects of Ionizing Radiation

1. Definition of dose.

A complete description of the interaction of ionizing radiation with matter can be quite complex. For example, an energetic γ-ray can give

Table 4. Properties of radioisotopes for food irradiation.

SPECIES	RADIATION EMITTED	DECAY HALFLIFE
Co^{60}	β^- (0.3 MeV) γ (1.17 MeV) γ (1.33 MeV)	5.26 years
Cs^{137}	β^- (1.18 MeV, 6.5%) β^- (0.52 MeV, 93.5%) γ (0.662 MeV, 93.5%)	30 years

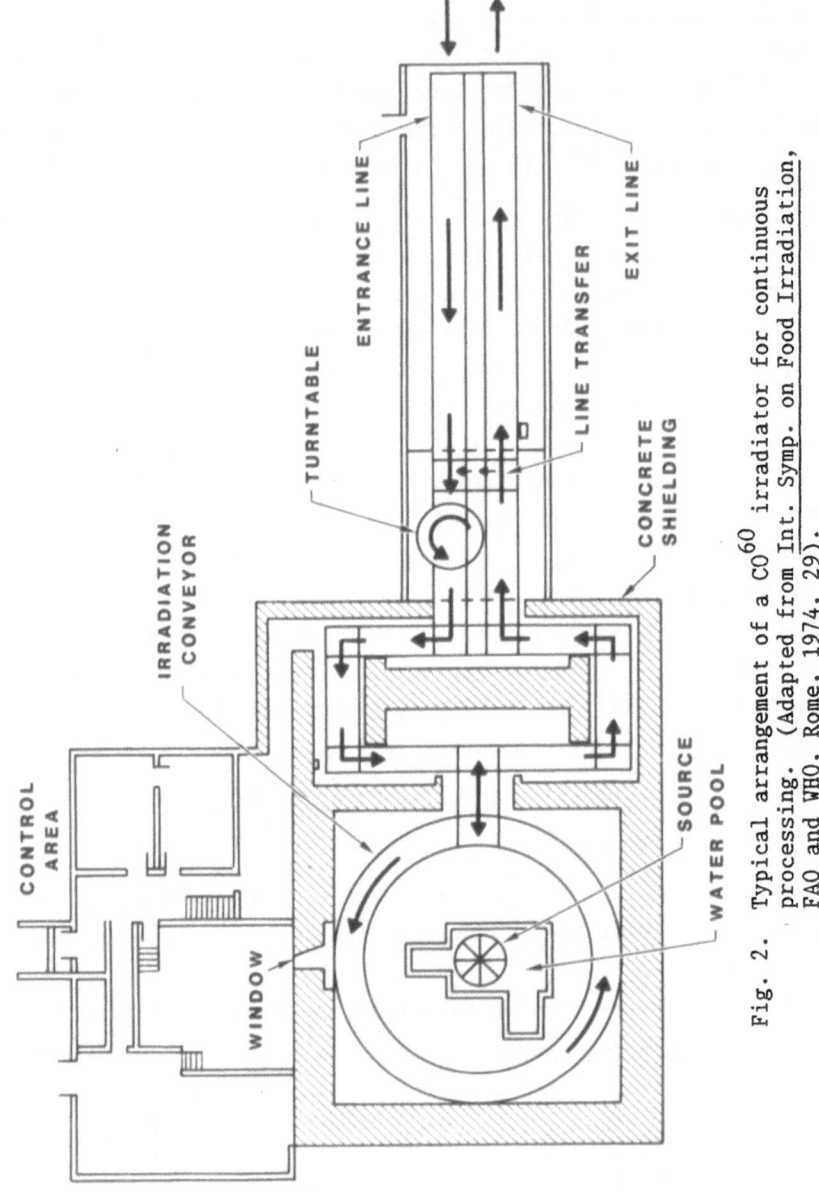

Fig. 2. Typical arrangement of a CO^{60} irradiator for continuous processing. (Adapted from Int. Symp. on Food Irradiation, FAO and WHO, Rome, 1974, 29).

CONTROL AREA

IRRADIATION CONVEYOR

TURNTABLE

ENTRANCE LINE

LINE TRANSFER

EXIT LINE

CONCRETE SHIELDING

WINDOW

SOURCE

WATER POOL

Table 5. Commercial food irradiation activities.*

LOCATION	STATUS	PRODUCTS TREATED	APPROX. CAPACITY
Dhaka, Bangladesh	Planned	Potatoes, onions, fish	–
Fleurus, Belgium	Completed (1980)	Spices, animal feed	100 m^3/month
France	Planned	Food in general	–
Budapest, Hungary	Planned	Spices, onions, potatoes	–
Tel Aviv, Israel	Completed (1983)	Animal feed	–
Fucino, Italy	Under construction	Potatoes, onions, garlic	25000 tons/season
Shihoro, Japan	Completed (1973)	Potatoes	10000 tons/month
Wageningen, Netherlands	Pilot plant (1968)	Frozen chicken, frogs legs,...	1500 tons/year
Ede, Netherlands	GAMMASTER-1 (1972) GAMMASTER-2 (1982)	Spices, frogs legs, shrimp	1000 tons/year
Tzaneen, South Africa	Completed (1982)	Mangoes, straw-berries, onions,...	7000 tons/year
Kempton Part, South Africa	Completed (1981)	Fruits, vegetables coconut powder,...	–
Pretoria, South Africa	Completed	Fruits, vegetables, chicken	–
Bangkok, Thailand	Planned	Food in general	–
Rockwaway, NJ U.S.A	Completed	Spices, seasonings	500 tons/year

* Adapted from Food Irradiation Processing (International Atomic Energy Agency, 83-06554, Vienna), 1984.

rise to an electron-position pair. These particles in turn generate knock-on electrons through collisions and an additional γ-ray when the positron annihilates. The general trend is toward the division of the available energy among lower energy electrons. Ultimately, all the energy appears as either atomic excitation, reactive chemicals, or heat in the medium. Experimental observations indicate that chemical and biological effects in medium irradiated by β^- or γ-rays are mainly deter-mined by the total energy deposited per unit mass of the medium (dose), with little dependence on the identity or kinetic energy of the primary particle.

The standard SI unit of dose is the gray (Gy), corresponding to 1 J/kg of absorbed energy:

$$1 \text{ Gy} = 1 \text{ J/kg} . \tag{1}$$

An absorbed does of 4.2 kGy would raise the temperature of water by 1 °C. Another commonly used unit of dose is the rad, defined as

$$1 \text{ rad} = 100 \text{ ergs/gm} = 10^{-2} \text{ Gy} . \tag{2}$$

For consistency, all doses in this report will be quoted in Gy.

Although absorption of β and γ-rays lead to similar chemical effects in an irradiated medium for the same dose, calculation of the dose depends on the characteristics of the primary particles. Given the same incident energy flux of electrons and γ-rays, the electrons will produce a higher dose rate (power/mass) because they have a shorter range in material.

The unit most often used to characterize the amount of radiation available from the radioisotope source (activity) is the Curie (Ci):

$$1 \text{ Curie} = 3.7 \times 10^{10} \text{ nuclear disintegrations/s} . \tag{3}$$

The SI standard unit for activity, the becquerel (1 Bq = 1 disintegration/s), is rarely used. As an applications example, consider the output from a 1 MCi (10^6 Ci) Co^{60} source. This is a typical activity level for a small processing plant. The γ-ray power released by nuclear activity is

$$P = (2.50 \text{ MeV/dist}) \times (10^6 \text{ eV/MeV}) \times (1.6 \times 10^{-19} \text{ J/eV}) \times (3.7 \times 10^{10}$$
$$\text{dist/Ci}) \times (10^6 \text{ MCi/Ci}) = 14.8 \text{ kW} . \tag{4}$$

2. Radiation chemistry of water.

Biological matter consists of about 80 per cent water. Therefore, a knowledge of chemical changes induced in water by ionizing radiation is a useful starting point to understand the chemical effects on living organisms (Ebert and Howard, 1963). High energy particles transfer their energy to a medium by ionizing or exciting atoms. On the average, one ionization occurs for each 30 eV of energy lost by the primary particle. The free electrons ejected are rapidly captured by atoms in the medium to form negative ions. The chemical reactivity of the positive and negative ions is not high. They can become reactive if they combine with other ions or atoms to form free radicals. A free radical is an electrically neutral molecule with unpaired electrons in the outer shell; it is denoted by the superscript "0", as in OH^0. Because of the upaired electron, free radicals are highly effective at inducing chemical changes in biological systems. For example, if a free radical is generated in the nucleus of a microorganism, it may break and disable the DNA chain, preventing reproduction. The production of free radicals accounts for the fact that energy deposited in a medium by ionizing radiation is much more effective at inducing chemical changes than an equivalent amount of thermal energy.

There are many possible interactions that can generate free radicals. The following is a common series of events. An electron is ejected from a water molecule by an ionizing collision:

$$H_2O \rightarrow H_2O^+ + e^- \ . \tag{5}$$

The electron can then be captured by another water molecule,

$$e^- + H_2O \rightarrow H_2O^- \ . \tag{6}$$

The positive H_2O^+ and negative H_2O^- ions can decompose into an ion and a free radical in the presence of another water molecule,

$$H_2O^+ \rightarrow H^+ + OH^\circ \ , \tag{7}$$

$$H_2O^- \rightarrow OH^- + H^\circ \ . \tag{8}$$

The H^+ and OH^- ions have little excess energy and usually recombine to form water. The free radicals, on the other hand, are highly reactive. Depending on the spatial distribution of free radicals along an ionization path, they can react with biological molecules or can combine with each other, according to

$$H^\circ + OH^\circ \rightarrow H_2O \ , \tag{9}$$

$$H^\circ + H^\circ \rightarrow H_2 \ , \tag{10}$$

$$OH^\circ + OH^\circ \rightarrow H_2O_2 \ . \tag{11}$$

The reactions of Eqs. (9) and (10) deactivate the radicals, while the reaction of Eq. (11) leads to the production of hydrogen peroxide, an active oxidizing agent.

Energetic ions (such as α particles) have a short range in material and generate an intense ionization path. In this circumstance, the hydroxyl radicals are closely grouped along the ionization path, favoring the production of hydrogen peroxide. In consequence, an amount of energy deposited in a medium by α particles or energetic ions is roughly an order of magnitude more effective for bringing about chemical and biological changes than β^- or γ^- particles. Unfortunately, the short range of energetic ions makes them impractical to use for commercial irradiation. Electrons and γ-rays produce a fairly uniform distribution of H° and OH° radicals, generating little H_2O_2 in pure water. The biological effectiveness of electrons and γ-rays is considerably enhanced if

669

molecular oxygen is present in irradiated water. In this case the oxygen can combine with the hydrogen radical to form the peroxyl radical:

$$H° + O_2 \rightarrow HO_2° .$$ (12)

This radical is not highly reactive, but it has a long lifetime and can diffuse through the medium. Ultimately, two peroxyl radicals can combine to form hydrogen peroxide,

$$HO_2° + HO_2° \rightarrow H_2O_2 + O_2 .$$ (13)

Radiation achieves its biological effect mainly through free radicals and reactive chemicals such as hydrogen peroxide. These chemicals have short lifetimes and dissipate before the food is consumed. In actual biological systems, more complex chemicals can be formed (radiological products), but these are produced in extremely small quantities. The net chemical change to the food is almost insignificant compared to changes induced by common cooking methods such as broiling. It should be recognized that the dose to achieve a desired biological affect (such as reduction of microbes in a food) through electron or gamma ray irradiation depends sensitively on the conditions of irradiation. For example, higher doses are required if the irradiation is carried out under anaerobic conditions or if the food is frozen. Optimum doses for the treatment of food products in their normal state are determined empirically.

3. Biological effects and required dose.

Effect of ionizing radiation on multicellular organisms depends on complex interactions between cells as well as inactivation of individual cells. For instance, damage to one organ can lead to the death of a multicellular animal. Generally, the higher an organism is on the evolutionary scale, the lower the required lethal dose. This is the reason that low doses are adequate for insect disinfestation and destruction of parasites. In contrast, direct hits are necessary to deactivate microorganisms, such as bacteria or protozoa. In this context, a hit implies an ionization event resulting in a local concentration of reactive chemicals. The hit could either kill the organism outright or prevent its reproduction. This is the reason why microorganisms are resistant to radiation, especially in the spore state when metabolism is greatly reduced.

The deactivation of microorganisms is well described by target theory (Bacq and Alexander, 1961; Casarett, 1968). It is assumed that microorganisms are distributed randomly in a medium. The probability of a hit can be estimated by the cross-sectional area of the sensitive part of the cell (for example, the cell nucleus). The number of organisms removed from the living population (ΔN) by an increment of dose (ΔD) is proportional to number of surviving microorganisms, N:

$$\Delta N/\Delta D = -(kN) \ . \tag{14}$$

where k is a proportionality constant that depends on the target area in the cell. Equation (14) implies that the fraction of surviving microorganisms (S) is given by:

$$S = \exp(-D/D_{37}) \ , \tag{15}$$

where the D_{37} dose corresponds to 37 per cent survival. There is a wide variation of D_{37} among microorganisms, depending on the state of the organism and the size of the biological target. For instance, D_{37} equals 30–80 Gy for Eschercheria coli, 1 kGy for bacillus mesentericus spores, and 4 kGy for tobacco mosaic virus (Casarett, 1968). Figure 3 is a semilog plot of survival versus dose of Eschercheria coli, a common laboratory bacterium. The data closely follows an exponential variation. The figure also shows the large variation in D_{37} resulting from differences in environment during irradiation.

Irradiation can a serve a number of biological purposes in the treatment of food (see Table 3). Low doses can inhibit metabolic activities in vegetables. This process has been used to inhibit sprouting of stored onions and potatoes. For disease control, the dose is chosen to reduce the population of pathogens to levels well below those

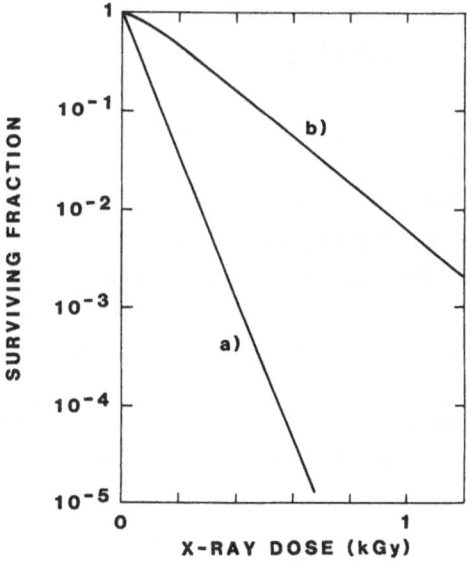

Fig. 3. Semi-log plot of survival fraction of Eschercheria coli as a function of dose and conditions of medium. [Adapted from E. L. Powers, "Considerations of Survival Curves and Target Theory," Phys. Med. Biol 7, (1962), 3.]

dangerous to people in ordinary health. An example is the treatment of poultry for salomnella at levels of 3-4 kGy. Finally, radiation can be used to reduce the microbiological load to retard food spoilage. A typical reduction factor is 10^6. Experimental results on the shelf life extension of seafood are shown in Table 6.

Interactions of Electrons and γ rays with Matter

1. Collisional energy losses of electrons.

Energetic electrons passing through matter loose energy primarily through Coulomb collisions with atomic electrons. The energy lost per unit distance of travel by collisions is given in MKS units by the well known formula (Bethe, 1933):

$$-\left(\frac{dT}{dx}\right)_c = \frac{2\pi e^4 NZ}{m_e \beta^2 c^2} \left[\ln \frac{m_e \beta^2 c^2 T}{2I^2(1-\beta^2)} \right. \tag{16}$$

$$\left. - \ln 2 \, (2\sqrt{1-\beta^2} -1 + \beta^2) + (1-\beta^2) + \frac{[1 - \sqrt{1-\beta^2}]^2}{8} \right]$$

where the quantities T, m_e, and β refer to the kinetic energy, rest mass, and (v/c) of the electron. The quantities N, Z and I refer to the atomic density, atomic number, and average ionization potential of the medium. Notice that the stopping power is approximately proportional to the density of the medium and weakly dependent on I. Equation (16) implies an energy loss rate of about 200 MeV/m in typical foods in the energy interval 4-16 MeV. The range of an electron in matter is determined from Eq. (16) by taking the integral:

$$R = \int_{T_o}^{0} \frac{TdT}{-(dT/dx)} \tag{17}$$

where T_o is the initial kinetic energy of the electron. Electrons slowing down in materials suffer significant angular scattering because their mass is comparable to that of the atomic electrons. The quantity R in

Table 6. Typical shelf life extensions.*

PRODUCT	DOSE (kGy)	NORMAL SHELF LIFE (days)	SHELF LIFE WITH IRRADIATION (days)
Shrimp	1.5	14-21	21-40
Crab 2	7	35	
Haddock	2	12-14	30
Clams	4	5	30

* Adapted from Fish and Shellfish Hygiene, (United Nations Food and Agriculture Organization, 1966), 50.

Eq. (17) must be interpreted as the integrated distance along a path that is not a straight line. For low energy electrons, there is a significant probability that electrons will be scattered enough to leave the target through the entrance face before losing all their energy (back-scattering). Backscattering is undesirable for radiation processing applications because it reduces the efficiency of energy transfer from the beam to the medium and increases the heat load on windows. The probability of backscattering is reduced at higher energies because the electrons have greater relativistic mass and penetrate the target before undergoing substantial angular deflection.

2. Bremsstrahlung radiation.

The violent transverse acceleration resulting from scattering of energetic electrons gives rise to the emission of electromagnetic radiation (bremsstrahlung) in the form of X-rays and γ-rays (Koch and Motz, 1959). The term X-ray refers to photons in the energy range below 100 keV, while γ-ray refers to higher energy photons. The radiation stopping power for bremsstrahlung is given approximately as (Knoll, 1979):

$$-\left(\frac{dT}{dx}\right)_r = \frac{NTZ(Z+1)e^4}{137m_e^2c^4}\left[4\ln\left(\frac{2T}{m_ec^2}\right) - \frac{4}{3}\right] \tag{18}$$

Note that the energy loss rate in Eq. (18) is proportional to the electron kinetic energy (neglecting in the weak variation of the log factor). The kinetic energy decreases exponentially with distance in the medium. The radiation stopping range is determined mainly by the properties of the medium with little dependence on the incident electron kinetic energy.

3. Comparison of collisional and radiative losses.

The total stopping power for energetic electrons is the sum of contributions from collisions and bremsstrahlung. The ratio of radiative to collisional stopping power is approximately (Knoll, 1979):

$$(dT/dx)_r/(dT/dx)_c = TZ/700 , \tag{19}$$

where T is expressed in MeV. Radiation production is favored by high incident electron energy and by targets with high atomic number. Equation (19) demonstrates the advantage of accelerators with high output energy for industrial γ-ray processing. The equation also explains why dense materials like tungsten are used as targets form conversion of electron energy to photon energy. On the other hand, if electrons are injected directly into food, the fraction of electron energy converted to γ-rays is small because foods consists largely of low-Z elements. This is advantageous because 1) beam energy converted to photons in the food is wasted because the γ-rays have a high probability of escaping, and 2) energetic γ-rays have a higher probability than electrons for producing nuclear activation.

Energy variations of collisional and radiative stopping powers are illustrated in Fig. 4 for tungsten and water. The water data typify electron interactions with most foods. The stopping power is higher in tungsten because of its greater density. At 10 MeV, half the energy of electrons is converted to photon energy in tungsten, while the fraction is only 10 per cent in water. A plot of total electron range (including contributions from collisions and radiation) is given in Fig. 5 for tungsten and water. Figure 5 shows that a tungsten converter must be 2 mm thick to stop all electrons at 5 MeV. The optimum converter thickness to stop electrons and allow photons to escape may be smaller if there is significant electron scattering. Clearly, beams at power levels

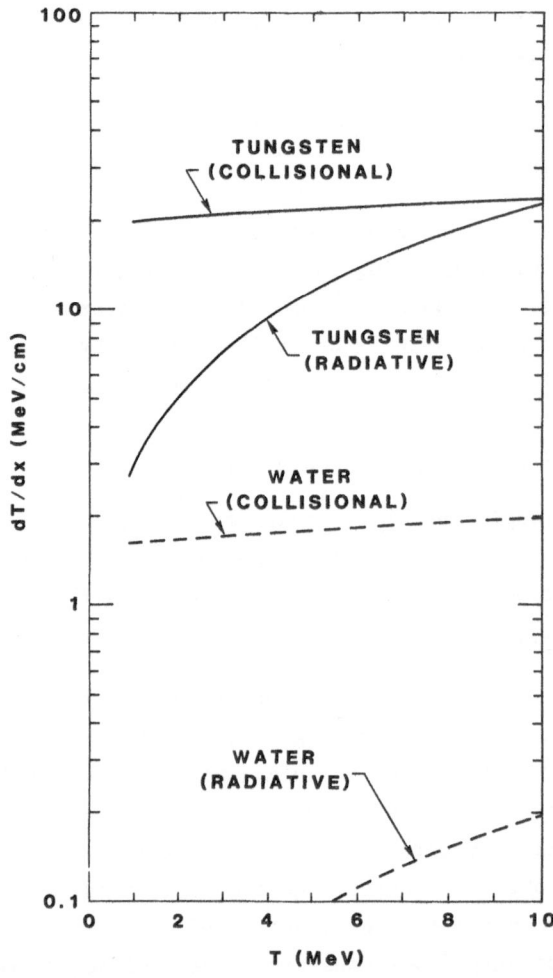

Fig. 4. Collisional and radiative stopping powers for electrons as a function of kinetic energy in tungsten and water.

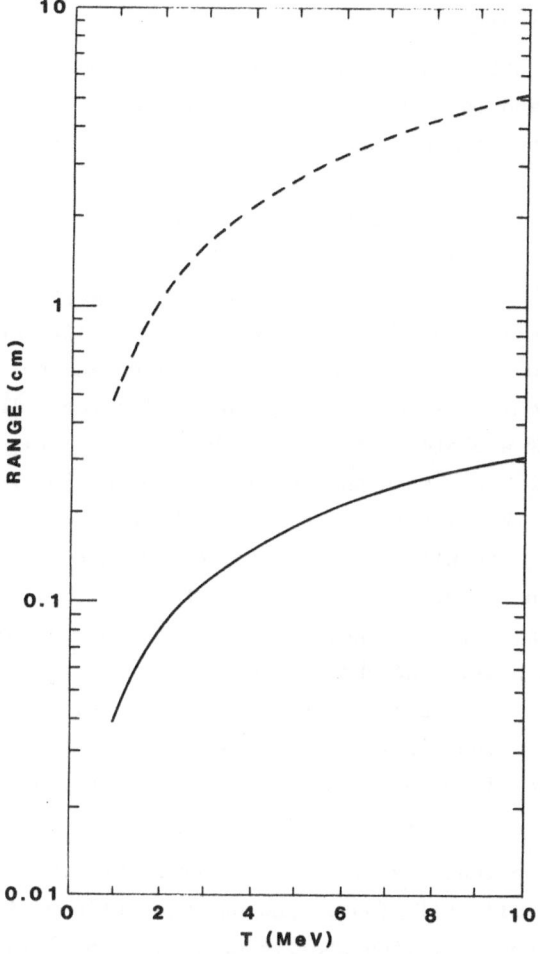

Fig. 5. Electron range in tungsten and water as a
function of energy.

exceeding 100 kW must be defocused or scanned over a large area to avoid
melting the thin converter.

The large area vacuum window for direct electron bombardment presents a major mechanical problem. Ideally, the window material should
have low atomic number to minimize collisional loss and bremsstrahlung
production. At the same time, the window material should be strong and
have a high melting point. Titanium, with $Z = 22$ and $\rho = 4.54$ gm/cm^3, is
a good compromise. In practical windows, fractional energy loss is small
so that dT/dx can be taken as constant. The stopping power for 10 MeV
electrons in titanium is about 9 MeV/cm. Each electron transfers and
energy 0.34 MeV passing through a window of thickness 3.8×10^{-2} cm (15
mil), corresponding to a 3.4 per cent energy loss. For electrons with
kinetic energy T_o, the window loss scales as $1/T_o$. The reduced fractional energy loss is another advantage of high energy beams.

4. Attenuation of γ rays.

Although bremsstrahlung from energetic electrons consists of a broad spectrum of high and low energy photons, only the γ-ray production is significant because of X-ray absorption in the converter. In the energy range 1-10 MeV, γ rays transfer energy to media by three processes (Evans, 1967): 1) photoelectric absorption, 2) Compton scattering, and 3) pair production. There is also a small probability that the γ ray will interact with a nucleus in the material. This process, which can lead to nuclear activation, is discussed in the section on Nuclear Activation. Photoelectric absorption occurs when an energetic photon is absorbed by an atom, exciting an inner shell electron. The process leads to the ejection of a photoelectron from the atom. At higher photon energy, a collision between a photon and an atomic electron can be treated as though the electron were free. The associated process, Compton scattering, results in deflection of the photon and transfer of energy directly to the electron. At multi-MeV photon energies, pair production is possible. The photon disappears and is replaced by an electron-position pair. The photon must supply the rest energy of theparticles, 1.02 MeV. Excess energy appears as kinetic energy of the charged particles. The kinetic energy is transferred to the medium through charged particle collisions. The rest energy of the positron is available following annihilation.

Photon interactions are discrete events rather than the sum of a large number of small collisions. Photon interactions with matter are parametrized by the absorption coefficient, μ, with dimension cm^{-1}. The flux of <u>full energy</u> photons moving in the z-direction, F(z) is related to the absorption coefficient by

$$F(z)/F_0 = \exp(-\mu z) \ . \tag{20}$$

where F_0 is the initial flux. The quantity $1/\mu$ gives the mean free path for an interaction. The effective depth of irradiation may be significantly larger than $1/\mu$; for example, a reduced energy photon may continue forward after a Compton scattering event. The absorption coefficient for lead showing the relative division between different collisional processes is plotted in Fig. 6(a). Absorption coefficients for water and lead in the energy range of interest for food processing are plotted in Fig. 6(b).

Thick-Target Interactions

The calculation of bremsstrahlung production by electrons in a practical target is quite involved. The target must be thick, so that electrons lose most of their energy passing through it. Scattering and energy loss strongly influence the output spectrum and angular

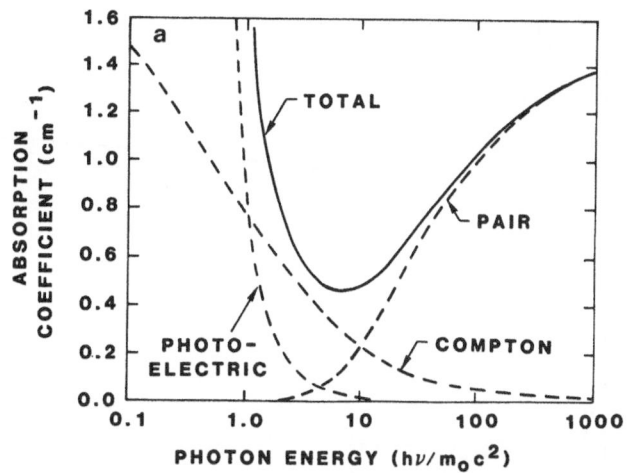

Fig. 6a. Attenuation coefficient, μ, for γ rays as a function of energy. Contribution of different photon interactions to the attenuation coefficient in lead. [Adapted from I. Kaplan, Nuclear Physics (Addision-Wesley, Reading, Mass., 1962).]

distribution of γ rays. Absorption in the target eliminates the low energy portion of the γ-ray spectrum. Despite the complexity, the problem of thick target bremsstrahlung can be solved numerically since the angular scattering process, the differential bremsstrahlung cross-sections, and the γ-ray absorption coefficients are well known. The calculations are usually performed by Monte Carlo codes (Hableib and Melhorn, 1984) that follow the diverse histories of primary electrons and secondary particles using weighted random probabilities to choose the interactions. Total γ-ray spectra are estimated by taking averages over a large number of incident electron histories. Calculated and measured γ-ray spectra in the forward direction for 11.3 MeV electrons incident on thick tungsten targets are shown in Fig. 7. There is a spread of γ-ray energies. The distribution approaches zero at an energy equal to the incident electron energy. There is a cutoff of low energy photons from absorption in the target.

Bremsstrahlung radiation from relativistic electrons is emitted primarily in the forward direction. This is a major advantage of beam-generated γ radiation. Thin target bremsstrahlung from electrons with energy $\gamma m_o c^2$ is confined to an angle (Jackson, 1975).

$$\tau < 1/\gamma , \qquad (21)$$

Equation (21) implies an angle of 3° for 10 MeV electrons. The angular distribution in thick targets is broadened because of electron scattering and the decrease of γ as the electron slows down. Theoretical

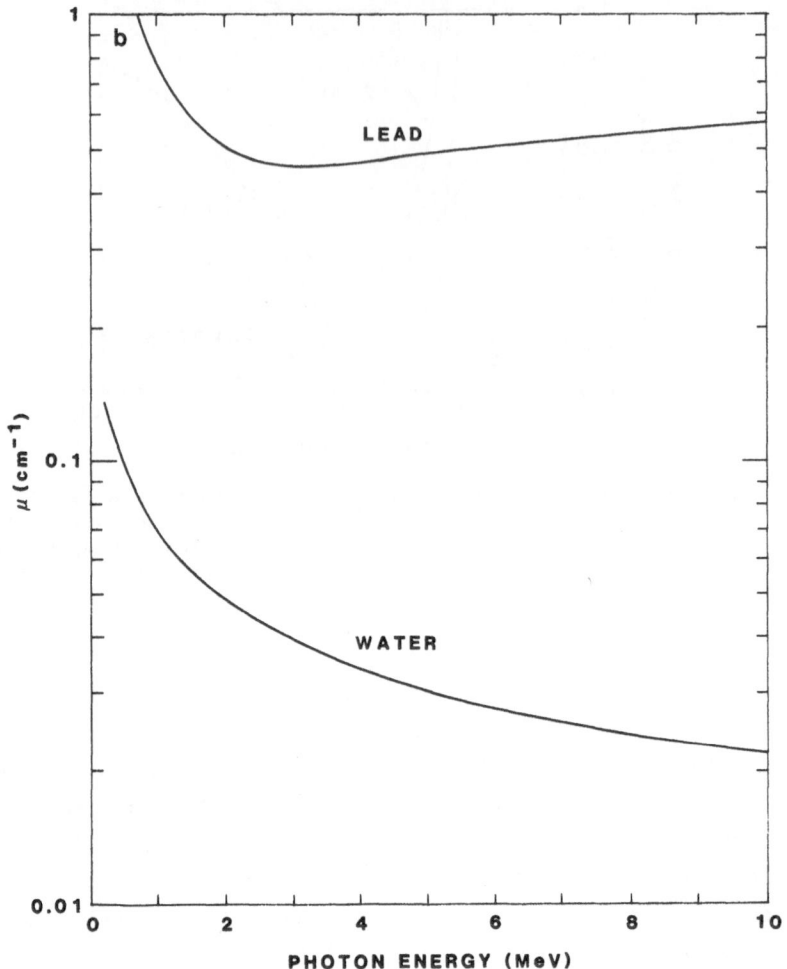

Fig. 6b. Attenuation coefficient, μ, for γ rays as a function of
energy. Values of μ in lead and water over the photon
energy range of interest for food processing. [Adapted
from I. Kaplan, <u>Nuclear Physics</u> (Addision-Wesley, Reading,
Mass., 1962).]

predictions of angular distributions (Koch and Motz, 1959) in tungsten
targets appropriate for T > 2 MeV are illustrated in Fig. 8. At 5 MeV,
the emitted radiation is spread over an angle of about 30°.

We have seen in the section on Collisional and Radiative Losses that
bremsstrahlung production becomes relatively more important as electron
energy increases. A rough scaling law for thick target bremsstrahlung is
that the total photon output energy scales as T_o^2. The photon flux in
the forward direction scales as T_o^3 because of the smaller angle of
divergence. The total efficiency for thick-target bremsstrahlung
production at relativistic energies (neglecting target absorption) is
given approximatfly by:

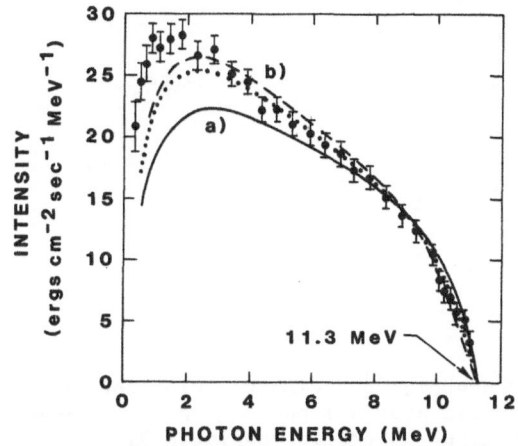

Fig. 7. Measured and predicted γ-ray energy spectrum for
11.3 MeV electrons incident on a thick tungsten
target. [Adapted from H. W. Motz, et. al,
Phys. Rev. <u>89</u>, 968 (1953).]

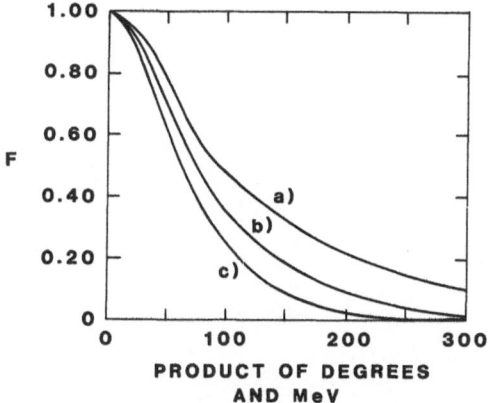

Fig. 8. Angular distribution of bremsstrahlung radiation
generated in a thick tungsten target. [Adapted
from H. W. Koch and J. W. Motz, Rev. Mod. Phys.
<u>31</u>, 920 (1959).]

$$\varepsilon = 1 - (\text{collision loss})/E_o = \tag{22}$$

$$= 6\times10^{-4} Z T_o / (1 + 6\times10^{-4} Z T_o) \ ,$$

where T_o is kinetic energy in MeV. Predictions of Eq. (22) for lead and
carbon are plotted in Fig. 9. The conversion efficiency in lead is about
20 per cent at 5 MeV and 33 per cent at 10 MeV. The fraction of energy
converted to useful photons for applications is lower than the prediction
of Eq. (22) because of target absorption and electron backscattering.
Figure 10 shows calculations of actual bremsstrahlung yield in the

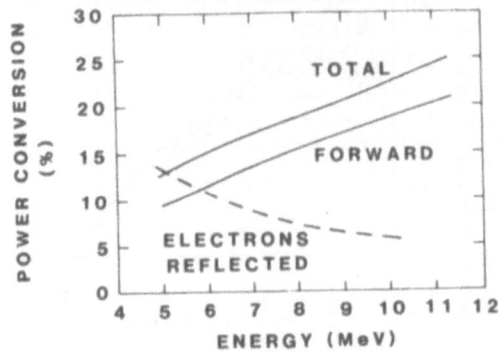

Fig. 9. Conversion efficiency: electron to γ-ray energy, summed over all photon angles and energies (Target absorption neglected).

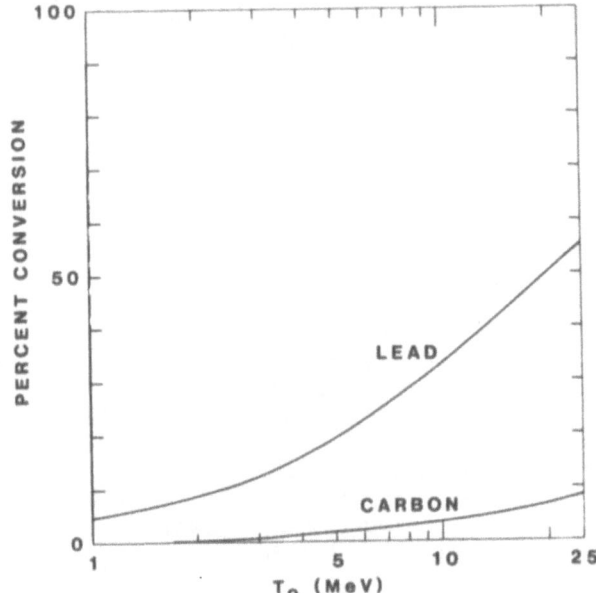

Fig. 10. Fraction of electrons backscattered, fraction of electron energy converted to bremsstrahlung radiation, and fraction of electron energy converted to forward directed γ rays as a function of electron kinetic energy. Thick tungsten target. [Adapted from J. McKeown, IEEE Trans. Nucl. Sci. NS-32, 3292 (1985).]

forward direction from a tungsten converter. Typical values are 10 per cent at 5 MeV and 17 per cent at 10 MeV. Figure 10 further demonstrates the advantages of using higher electron energy.

Monte Carlo codes can also yield information on relative energy deposition in a thick target to calculate the absorbed dose. For instance, calculations of dose distributions from direct electron deposition in food must include collisional energy transfer and scattering of both the primary and knock-on electrons. Detailed calculations may include bremsstrahlung production; in this case, the computer code must follow the history of the resulting photons. Results from such a calculation are shown in Fig. 11(a) for 10 MeV electrons incident on a slab of water from one side. Energy deposition is non-uniform, peaking near the end of the electron range. This is a disadvantage since food regulations typically call for a minimum dose and set a maximum overdose. In the case of Fig. 11(a), the product of beam power and irradiation time must be large enough so that the surface dose equals the minimum value. There is a 43 per cent overdose at the maximum point. The shaded area of Fig. 11(a) represents energy that has been deposited usefully. The light areas represent wasted energy since the overdose is not required for the application. The situation is improved by using two-sided irradiation, as shown in Fig. 11(b). There is about the same maximum excess dose, but the amount of energy wasted is cut in half. Figure 11 shows that a sample width of 8 cm is typical for media at the density of water. A wider depth can be irradiated with lower density fluid media such as wheat.

Dual-sided exposure is more critical for γ irradiation, as shown in Fig. 11(c). Results are plotted for absorption of a bremsstrahlung spectrum of γ rays generated by 5 MeV electrons incident on tungsten. Both single-sided and dual-sided irradiation results are plotted. The shaded area indicates the useful dose for single-sided irradiation with the assumption of a 70 per cent overdose at the surface. With two-sided irradiation, energy waste is reduced by a factor of 5. The useful dose region extends over 33 cm for water. Typical packages have reduced density because of air spaces; a package size of 1 m may be typical for this case.

Nuclear Activation of Irradiated Food

Gamma rays passing through matter can induce a nuclear reaction if the photon energy exceeds a threshold value. Such a reaction can leave behind nuclei that will decay at a latter time releasing energy; the medium is said to be radioactive. There are a wide variety of γ-induced reactions with thresholds below 10 MeV. Threshold energies tend to decrease with the atomic number of the target atom. The most significant reactions are photoneutron reactions in which an incident γ ray ejects a neutron from a nucleus. The transformed nucleus may be radioactive. In addition, the neutron migrates and may excite another nucleus when it is captured. Sodium reactions are typical of those that may occur in food. The radioactive isotope Na^{22} (1.24 MeV and 0.51 MeV γ-ray emissions, 2.58 year halflife) is produced when a neutron is ejected from the nucleus of

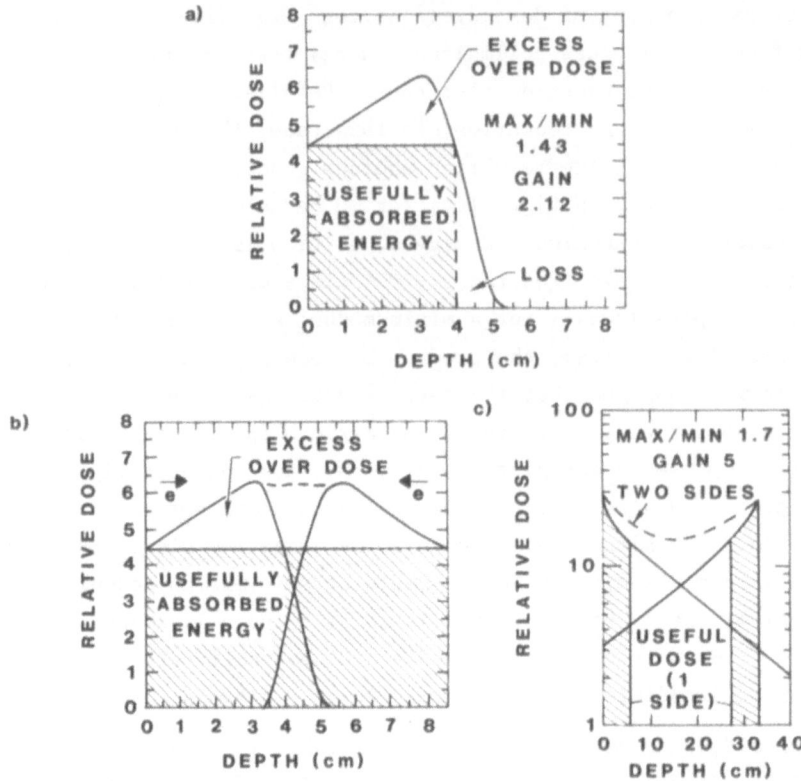

Fig. 11. Energy deposition profiles for ionizing particles incident
on water. a) Relative dose as a function of depth, 10 MeV
electrons incident from one side. b) Relative dose, 10 MeV
electrons, dual-sided irradiation. c) Relative dose, single
and dual-sided irradiation by thick target bremsstrahlung
photons generated by 5 MeV electrons on tungsten. [Adapted
from J. McKeown, IEEE Trans. Nucl. Sci. NS-32, 3292 (1985).]

Na^{23}. The γ-ray threshold energy for this reaction is 12.4 MeV. Capture
of a neutron by Na^{23} produces Na^{24} (2.75 and 1.37 Mev γ-ray emission, 15
hour halflife).

It is essential that treatment with ionizing radiation does not
leave the food radioactive. Some nuclear reactions inevitably occur, but
the probability must be small enough so that the induced radioactivity in
the food is small compared to natural background levels. Based on over
30 years of research on accelerators for food processing, the Codex
Alimentarius Commission of the United Nations has made recommendations on
dose and set limits on electron beam energy (Joint FAO/IAEA/WHO Expert
Committee). The energy limits are 10 MeV for direct electron irradiation
and 5 MeV for bremsstrahlung conversion. The higher limit on direct use

of electrons reflects the fact that there are few electron induced nuclear reactions in the light elements characteristic of food. These energy limits have been adopted by most nations involved in food irradiation.

Experimental and theoretical work performed on food activation by irradiation show that the accelerator energy limits are extremely conservative at the recommended dose (Kock and Eisenhower, 1965; Brynjolfsson, 1978). To illustrate this fact, Fig. 12 shows experimental and theoretical results on the Na^{22} activity induced in meats by direct electron deposition of a 50 kGy dose. Activity is immeasurable for electron energy below an effective threshold of 13 MeV. The activity is higher for ham because of higher salt content. The activity rises rapidly with energy, but the measured levels are quite low. For comparison, the Maximum Permissible Concentration (MPC) of Na^{22} in water for non-radiation workers in the United States is 40 pCi/cc. The activity induced in ham at an electron energy of 24 MeV is only 1.4 pCi/gm. Thus, a person consuming as much irradiated ham as water would ingest an amount of radiation resulting from the treatment equal to about 3 per cent of the amount allowed in water. Activation levels for irradiation from a bremsstrahlung spectrum are only about a factor of ten higher; therefore, irradiation by γ-rays generated by 10 MeV electrons would produce a negligible amount of Na^{22} activity.

An extensive study of activation by food irradiation (Becker, 1983) concludes that there is no expected increase in body radioactivity levels for a person consuming food irradiated by 10 MeV electrons over the course of a lifetime. Activity levels are extremely small if there is a normal storage time between irradiation and consumption. Even immediately after irradiation, food samples are safe to handle. A typical level is 0.25 µCi, about the same level as the natural radioactivity of the food. It is interesting to note that food irradiation can actually reduce the radioactivity level of food when it is consumed. This unusual result arises from the decay of naturally occurring C^{14} (halflife, 5800 years) following harvesting. If food irradiation allows the food to be stored longer, the C^{14} levels will decrease. For example, the net activity of foods irradiated by 10 MeV electrons will be lowered if the radiation treatment allows a one day storage extension. A comparison of radioactivity levels associated with food irradiation to natural background levels is given in Table 7. It is clear that food is quite safe when irradiated by accelerator-generated radiation at the recommended beam energy levels. There is a good possibility that beam energy limits may rise when more data and operational experience have accumulated.

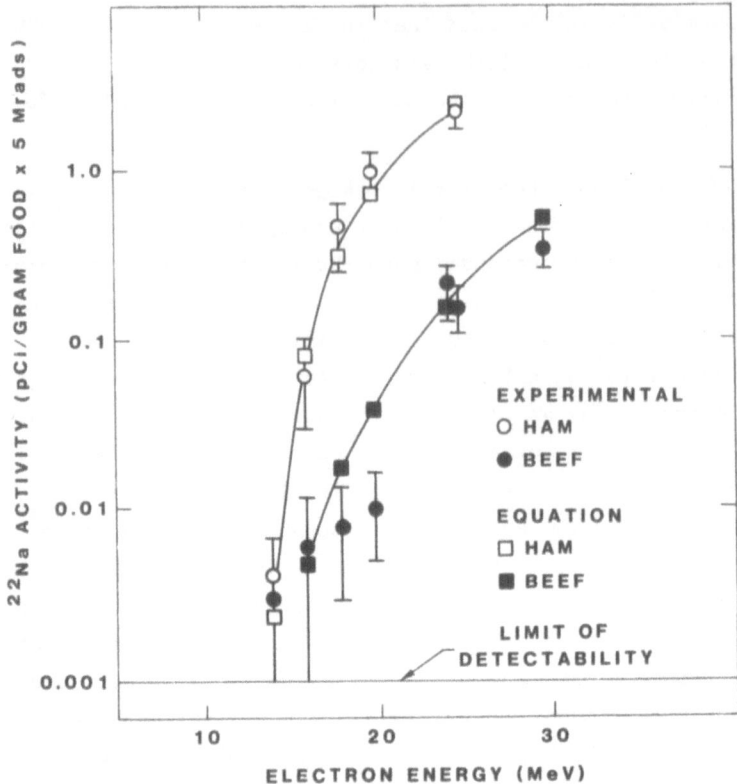

Fig. 12. Theoretical predictions and experimental results on the generation of Na^{22} in beef and ham by direct electron irradiation as a function of electron energy. [Adapted from R. A. Glass and H. D. Smith, "Radioactivity Produced in Foods by High-energy Electrons," (Stanford Research Institute, S-572, 1959).]

Comparison of Accelerators and Radioisotopes for Food Processing Applications

The two main options for the generation of radiation for commercial food treatment facilities are accelerators and radioisotopes. In this section, some of the operational and economic advantages of accelerators will be discussed. The application of accelerators to food processing has the potential to improve the economics of the process and to make it more acceptable to the general public.

The main operational advantage of an accelerator is that both the machine and radiation it produces can be turned off. There is no possibility of radiation escaping from the facility. Despite the fact that isotope irradiators will be well engineered and regulated, there is a strong negative public perception of such facilities. For example, plans for a small Cs^{137} irradiator at the National Food Laboratory in

Table 7. Comparisons of radioactivity levels.*

Activity	Comparative quantity	Induced activity per kg of meat irradiated by electrons of energy
1 µCi		
0.1 µCi	Natural activity in human body	
0.01 µCi	Natural activity/kg of meat	
1 nCi	Allowed I^{131}/liter of milk	
0.1 nCi	Increase from fallout/kg of meat	
0.01 nCi		
1 pCi	Reported increased activity in milk, Three Mile Island incident	
0.1 pCi		16 MeV
0.01 pCi		14 MeV
1 fCi	1 disintegration/hour	12 MeV
0.1 fCi		
0.01 fCi	1 disintegration/week	10 MeV
1 aCi	1 disintegration/year	

* Adapted from R. L. Becker, "Absence of Induced Radioactivity in Irradiated Foods" in Recent Advances in Food Irradiation, edited by P. S. Elias and A. J. Cohen (Elsevier Biomedical, 1983).

Dublin, California, were recently canceled because of protests from local residents (Committee on Radiation Application Information, 1986). Accelerator irradiators can be safely sited in populated areas and mobile environments. For instance, accelerator irradiators could be located on factory fishing ships.

Another operational advantage of accelerators is that high dose rates can be achieved with direct electron deposition. This allows a high throughput of liquid medium. In the γ-ray mode, accelerators can produce a bremsstrahlung spectrum with a higher average energy than Co^{60}, allowing irradiation of larger packages. For some types of food products, it may be possible to irradiate the food on its original pallet. The directionality of accelerator-generated radiation can lead to a higher dose uniformity and radiation utilization fraction (radiation usefully absorbed/radiation produced). A computer-controlled beam scanning system could emphasize the edges of packages, omit spaces between packages, and adjust to different size containers. Dose uniformity is a problem for radioisotope systems because dual-sided irradiation is impractical. The food packages must surround the source because the γ-rays are emitted isotropically. The only options for dual sided irradiation are to rotate the packages or pass them through twice. Radioisotope irradiators require a complex carousel system (Fig. 2) to fill the maximum solid angle surrounding the source with packages, while accelerator facilities can use a simple straight through conveyer. If the cost and size of accelerators can be reduced, dual-sided irradiation can be achieved with two small machines.

Accelerators in the 100 kW range can provide high dose rates for processing larger amounts of food. There is general agreement that the unit cost of irradiation facilities (cost per mass of food processed) decreases with the facility throughout (mass processed per time). A recent study of the economics of Co^{60} irradiators (Morrison, 1985) includes such factors as the purchase of the radioisotope, radiation shielding, buildings, product handling equipment, insurance, and personnel salaries. Unit cost for the irradiation of fish fillets and young chicken (2 kGy) as a function of annual plant throughput are summarized in Fig. 13. The figures may be somewhat optimistic because they are based on a 24 hour/day schedule and a radiation utilization fraction of 0.25. Large scale operations ($\geq$50 Mkg/year) have a clear advantage. Note that existing commercial irradiation facilities are well below the optimum range; the largest plant listed in Table 5 processes 5.5 Mkg/year.

At first glance, intuition might lead one to believe an irradiator based on a complex accelerator system would be more expensive than one with a passive radioisotope source. This may not be true for a number of reasons. Chief among them is the fact that the simple radioisotope source depends on the existence of a complex fission reactor. The expense of operating the reactor is reflected in the initial and replacement costs of the radioisotope. In the remainder of this section, some of the economic advantages of accelerators will be considered. There is a good probability that unit costs below those for isotope irradiators can be achieved. Comparisons will be made only with Co^{60} irradiators, since the world production rate of Cs^{137} is far too small to support major commercial operations.

For an accelerator irradiator, the amount of beam power needed to achieve a specified annual throughput of food is given by

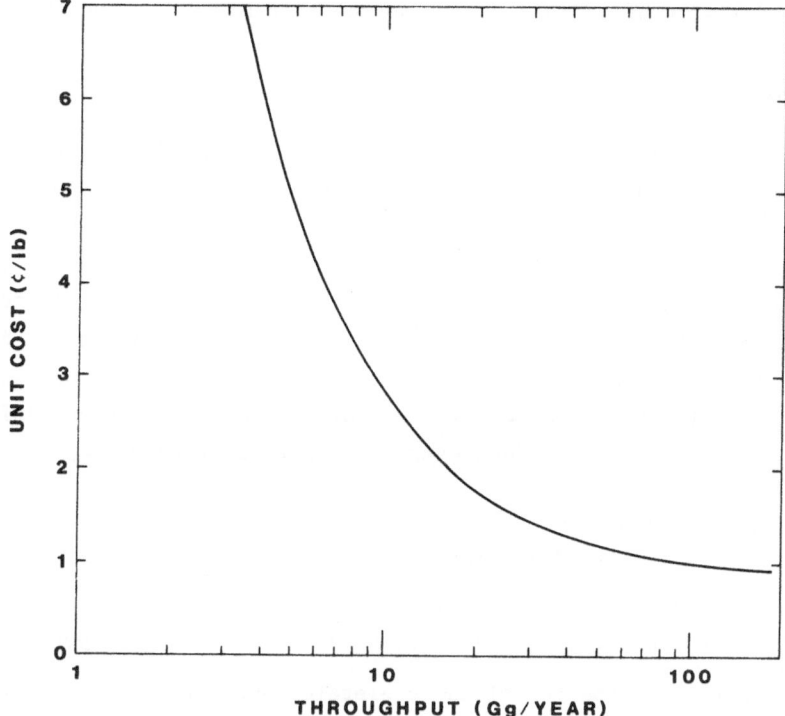

Fig. 13. Unit costs for irradiation of fish fillets and young
chicken, 2 kGy dose, as a function of annual plant
throughput. [Adapted from R. M. Morrison, "Economics
of Scale in Single Purpose Food Irradiators," International Symposium on Food Irradiation (International
Atomic Energy Agency, IAEA-SM-271, 1985).]

$$P(W) = MD/T\varepsilon_c\varepsilon_t, \qquad (23)$$

where M is the throughput in kg/year, T is the number of seconds of
operation per year, D is the dose in Gy, ε_c is the conversion efficiency
of electrons to useful γ rays, and ε_t is the target utilization fraction.
The final figure includes effects of the vacuum window, dose non-
uniformity, and γ rays that miss the package. For example, assume direct
electron deposition (ε_c = 1) to achieve a dose of 2 kGy in a throughput
of M = 100 Mkg/year. Assume further that ε_t = 0.7, reflecting the good
directionality of the electrons and the option of double sided irradiation. Sixteen hours of operation per day for 5 days per week corresponds
to T = 1.5 x 10^7 s/year. Substitution of the factors into Eq. (23) gives
a beam power of 19 kW. For the same parameters, a 5 MeV γ-ray irradiation accelerator should have a beam power of about 200 kW. This power
figure is reflected in the accelerator designs discussed later.

The activity of a radioisotope source to achieve a given throughput is given by

$$A(MCi) = MD/T\varepsilon_r(1.48 \times 10^4),\tag{24}$$

where the quantities M, D and T are defined in Eq. (23). The factor ε_r is that radiation utilization fraction, taken as 0.2. Approximately 4.5 MCi of Co^{60} are required for a throughput of 100 Mkg/year. The initial purchase price of the radioisotope (neglecting shipping costs) is about $5 million. A more relevant quantity is the cost of the radioisotope that must be added over the lifetime of the plant to maintain the initial activity level. With continuous replenishment, a quantity equal to the initial activity is available for reuse at the end of the plant life. The activity that must be added over a time ΔT is given by

$$\Delta A = 0.693\, A_o \Delta T/t_{1/2},\tag{25}$$

where A_o is the initial activity and $t_{1/2}$ is the radioisotope halflife. For Co^{60}, approximately $6.6 million worth of radioisotope must be added over a 10 year period. The figure is $8.9 million for a 15 year period.

The operating expenses of an accelerator-based food processing plant and the cost of items (other than the radiation source) will probably not be higher than those for a radioisotope plant. Shielding costs may be reduced in an accelerator facility with direct electron irradiation because there is little generation of γ-rays. Although the average γ-ray ray energy from an accelerator with a converter is higher than that from Co^{60}, the total shielding requirement may not be higher because of the directionality of the photons. The package handling system for an accelerator-based irradiator is simpler than the carousel essential for a radioisotope plant. Given that the bulk of the plant, handling costs, and salaries are similar, the accelerator irradiator has an economic advantage if the cost of the radiation source is significantly lower.

From the above discussions, the accelerator option is attractive if a 10 MeV electron machine with 19 kW beam power (for direct electron deposition) of a 5 MeV machine with 200 kW beam power (for γ conversion) can be fabricated in quantity for less than $6.5 million. The inherent assumptions of the comparison are: 1) the electron accelerator can operate for 10 years without replacement of major parts, and 2) the accelerator is reliable. The latter condition means that the accelerator can operate continuously for 16 hours each day, 5 days per week. Based on present experience with accelerator irradiators, it is likely that the capital cost of production accelerators will be lower than Co^{60}.

One factor that must be included in the comparison is the cost of electricity to run the accelerator. For a given electrical efficiency,

ε_e, and dose, D, the electricity requirement is

$$U \text{ (kW-hr/lb)} = (1.3 \times 10^{-7})D/(\varepsilon_c \varepsilon_t \varepsilon_e), \qquad (26)$$

where ε_c and ε_t have been defined previously. For direct electron irradiation with $\varepsilon_e = 0.5$, $\varepsilon_t = 0.7$, $\varepsilon_c = 1.0$, and $D = 2$ kGy, Eq. (26) implies that $U = 7.43 \times 10^{-4}$ kW-hr/lb. With an electricity rate of 7 cents/kW-hr, the unit cost of electricity is 5×10^{-3} cents/lb. For a γ-ray conversion system with $\varepsilon_c = 0.1$, the electricity cost is only about 0.05 cents/lb. The implication is that electricity represents a small fraction of the irradiation cost. With regard to the design of accelerators, there is little penalty for accelerator inefficiency as long as the additional power conditioning equipment does not add substantially to the capital cost.

Accelerator irradiators are more tolerant than radioisotopes to fluctuations of plant throughput. These fluctuations are inevitable if there are seasonal variations of food production within the shipping range of the plant. There are severe unit cost penalties for isotope irradiators operating below their peak level. This is a consequence of the fact that the amount of radioisotope purchased must be appropriate for the peak load and that the γ-ray flux is constant, whether utilized or not. With accelerators, the increase in unit costs for below peak operation may be less serious. Although a full-time operator must still be present for low power operation, the handling costs are reduced. In contrast to radioisotopes, the lifetime of the accelerator is extended by a reduced duty cycle. Another advantage of accelerator-based irradiators is that they have the flexibility to process different size packages or different density media by varying the beam energy or operating mode (Fig. 1). Product flexibility may be essential in regions with strong climate variations and limited growing seasons such as the U.S. Midwest.

Status of Food Processing by Radiation

At present, over twenty-five countries have approved for consumption one or more food items processed by ionizing radiation. The FAO/IAEA/WHO Expert Committee on Irradiated Food has reviewed extensive data and has stated unequivocally that the process can meet its microbiological aims (Joint FAO/IAEA/WHO Expert Committee, 1981). Furthermore, there are no observed problems of activation, chemical changes, or loss of nutrition at the recommended doses. On the basis of recommendations by the Expert Committees, the United Nations FAO/WHO Codex Alementarius Commission has adopted a general standard for irradiated foods as well as a code of practice relating to food irradiation facilities. Several countries, including Canada, Chile, Denmark, and Thailand are considering approval of all irradiated food up to specified dose limits (Food Irradiation

Processing, 1984). Recently, irradiation of fresh fruits and vegetables up to 1 kGy for inhibition of growth and maturation was approved in the United States (Code of Federal Regulations, 1986, Federal Register, 1984). Previous to this, approval was granted for irradiation of pork (to 1 kGy) for prevention of trichinosis and the treatment of some low volume foods (such as onion powder) up to 30 kGy. Since irradiation is presently classified as a food additive in the United States, a review must be conducted for each type of food before the process is approved.

Despite encouraging developments and evidence that treatment of food by radiation is safer than most chemical treatments, the commercial future of the process is uncertain. At present, the market for high power food irradiation accelerators is not clearly defined, and there are few evident sources of venture capital to develop them. A number of factors contribute to the slow pace of radiation food processing commercialization (Committee on Radiation Application Information, 1986):

1. There is a lack of commitment and investment from the food industry.

2. Radiation processing has not yet been approved by many national governments.

3. There is uncertainty about consumer reaction.

4. There is little operational information on the economic feasibility of the process; demonstration facilities to the present have been subcritical in terms of throughput.

5. There is insufficient experience to determine how the process may be controlled and regulated, particularly with regard to international trade.

Radiation processing must compete directly with chemical methods of preservation. While portions of the general public have a negative opinion of any process involving radiation, there is little realization of the long-term health hazards of common chemical food treatments. This problem of perception can be solved only through increased public education. It is particularly important that the public be made aware that treatment of food by accelerator-generated radiation has no technological overlap with nuclear fission reactors or nuclear weapons. Progress in food processing by accelerator-generated radiation can occur only when consumers can clearly identify advantages of general health, nutrition, convenience, and cost.

FOOD PROCESSING ACCELERATORS

Introduction

In this section, four concepts for food processing accelerators will be discussed. The intent is to illustrate a variety of possible approaches and the physical principles of high power electron accelerators rather than to review the field in detail. Three of the accelerators are in the study stage while the fourth, an induction accelerator, is under construction. At present, there is no existing accelerator or food processing facility in the optimum parameter range (5-10 MeV, >100 kW). It is premature to make comparisons between accelerator options; the best approach will be determined by reliability and overall economics as well as by technical considerations.

The options discussed include a long-pulse, low-repetition-rate induction linac, a continuous duty cycle RF linac based on conventional microwave technology, a high repetition-rate induction linac using pulsed power compression by saturable core magnetic switches, and a pulsed high-gradient RF accelerator based on advanced high power microwave technology. The presentation of the four accelerators is ordered according to the degree of extrapolation over present experience. The long pulse induction linac and the CW RF linac illustrate parameters that can be achieved with existing technology, while the other devices show the potential advantages of accelerator innovation.

The goal of this report is to emphasize that food processing accelerators are technically feasible. It should be noted that moderate energy (≤ 4 MeV) electron beam accelerators for radiation processing have been successfully operated and are commercially available in the power range approaching 200 kW. Existing machines are electrostatic. A survey of the physical principles and typical parameters of these devices is given below.

Electrostatic Accelerators for Radiation Processing

A review of the field of radiation processing at moderate energy is given in by Scharf (1986). Present applications include cross-linking of cable insulators for improved electrical properties, production of packaging materials, and the curing of lacquer surface coatings. In the United States, the total installed capacity of radiation processing accelerators exceeds 10 MW (Silvermann, 1978).

Three types of high power electrostatic generators are illustrated in Figs. 14, 15 and 16. The cascade generator (Fig. 14) is well known for its use in the Cockcroft-Walton accelerator. The diode-capacitor string boosts and rectifies the alternating voltage from a high voltage step-up transformer. Operation of circuit is described by Humphries (1986). The steady-state voltage levels in the circuit with no loading are listed in Fig. 14. The circuit shown produces a positive voltage six times that of the peak AC amplitude. A commercial unit produced by Haefely has operated at 4 MeV with a load current of 5 mA (20 kW).

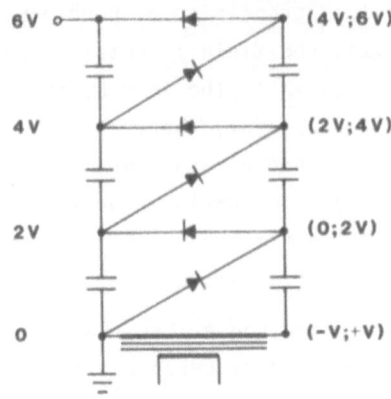

Fig. 14. Principle of operation: cascade generator.

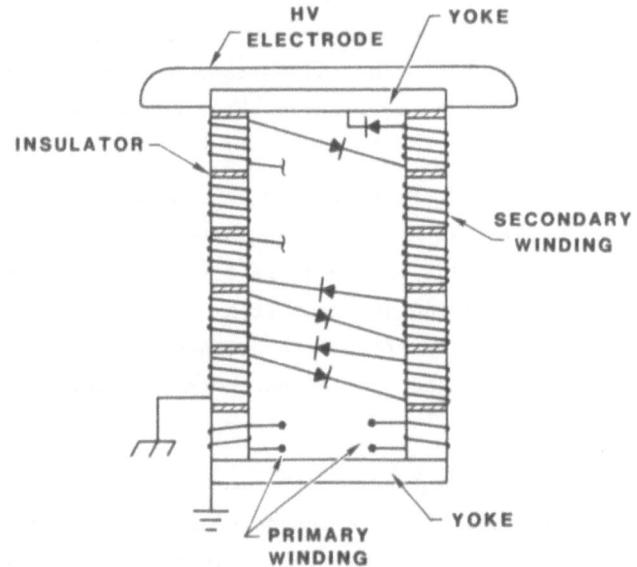

Fig. 15. High voltage transformers. a) Conventional
transformer with a secondary winding linking
a single core. b) Insulated core transformer.

Insulation sets the voltage limit in electrostatic generators. The
cascade generator achieves high voltage because the circuit is spatially
extended and can be designed for good voltage grading. An ordinary
high-voltage step-up transformer (Fig. 15a) has a secondary winding that
links the laminated steel core. Insulation of the secondary limits
attainable voltage with moderate size cores to the 100 kV range. Insulation problems are reduced in an insulating-core transformer (ICT)
(Fig. 15b). The device was invented by R. Van de Graaff. The flux core
is divided into a number of voltage graded sections separated by sheets

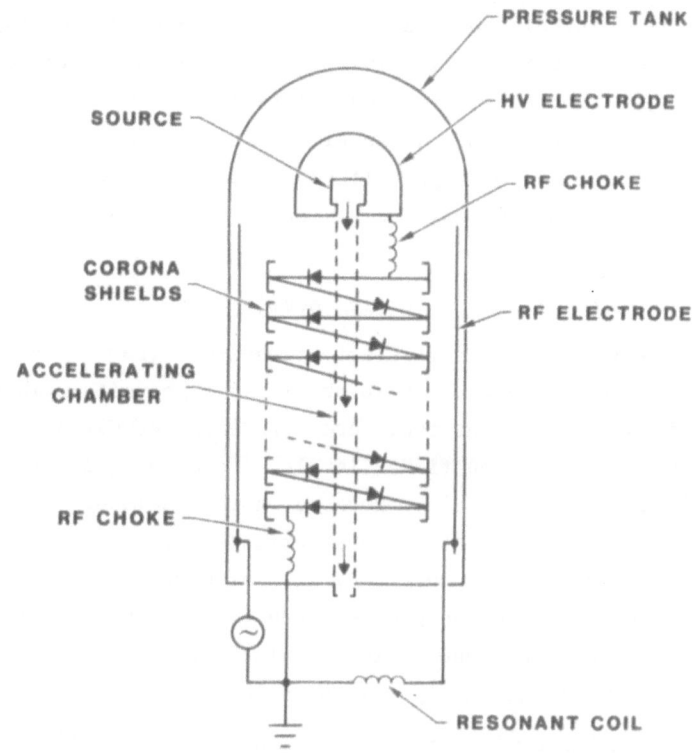

Fig. 16. Principle of operation: the Dynamitron.

of insulator. Each section has its own secondary. The secondaries are linked together by a diode string to produce rectified high voltage. ICT accelerators manufactured by High Voltage Engineering have achieved beam powers of 50 kW at voltages up to 3 MV.

The cascade generator and the insulated core transformer are series circuits. The available current drops off rapidly with increasing voltage. This problem is solved by a third type of generator, the Dynamitron (Fig. 16), manufactured by Radiation Dynamics, Inc. The device has similarities to the cascade generator; the main difference is that the stages are charged in parallel. This is accomplished by capacitive coupling to electrodes resonantly excited by a high frequency AC voltage. Parallel charging allows high current generation at high voltage.

The generator is housed in a pressurized metal chamber with the RF electrodes mounted as shown in Fig. 16. The electrodes, in combination with an external inductor, form a low frequency (~100 kHz) resonant circuit. The circuit is driven by a high power vacuum tube oscillator. A graded high voltage rectifier stack with equipotential electrodes is located on the axis of the tank. Capacitive coupling between the RF electrodes and the stack electrodes induces a high DC voltage on the HV

693

electrode. Operation of the Dynamitron can be understood by inspection of the equivalent circuit of Fig. 16. The orientation of the diodes shown gives a positive output voltage. Although the Dynamitron uses RF technology, it must be classed as an electrostatic generator since it shares the same problems of high voltage insulation. Dynamitrons with power levels of 200 kW (100 mA at 2 Mev or 50 mA at 4 MeV) are available commercially.

Long-Pulse, Iron-Core Linear Induction Accelerator

The size and cost of electrostatic generators grow rapidly with voltage above 1 MeV because of the problems of high voltage insulation. The volume of the accelerator enclosure increases roughly as the cube of the operating voltage. A multi-MeV levels, a complex geometry of equipotential electrodes must be incorporated in the enclosure for electric field grading. Vacuum insulators in an extended acceleration column present breakdown problems, especially with high current beams. For beam energy above 5 MeV, insulation problems can be avoided by multi-stage acceleration of beams in inductively isolated structures. Both induction accelerators and RF linacs apply this principle.

The physical basis of an induction linac cavity is illustrated in Fig. 17 (Humphries, 1986). Voltage from a pulse power generator is applied to an acceleration gap enclosed within a toroidal metal cavity. Current from the generator can flow through the on-axis beam load or around the outside of the cavity as leakage current. An empty cavity has a small inductance; therefore, the leakage current is large and the generator output is almost shorted. The remedy is to include a ferromagnetic core, as shown in Fig. 17, to increase substantially the leakage circuit inductance. The advantage of this arrangement is that even though a high voltage is applied across the acceleration gap, there is no voltage around the outside of the cavity. The reason for this is that the inductive contribution to the electric field on the periphery cancels the applied electrostatic voltage.

Any number of induction cavities can be added in series to achieve high beam energy. For example, if the peak electrostatic voltage is limited to 250 kV, a 10 MeV beam can be generated with 40 cavities. The main limitation imposed by the ferromagnetic cores is on the pulselength. The cavity can function only when the cores are unsaturated with high magnetic permeability. Faraday's law implies that the product of voltage (V_o) and pulselength (Δt) in an induction cavity is constrained by

$$V_o \Delta t \leq \Delta B A_c, \tag{27}$$

where ΔB is the maximum flux swing in the magnetic material and A_c is the cross sectional area of the core. The two common materials for fast

694

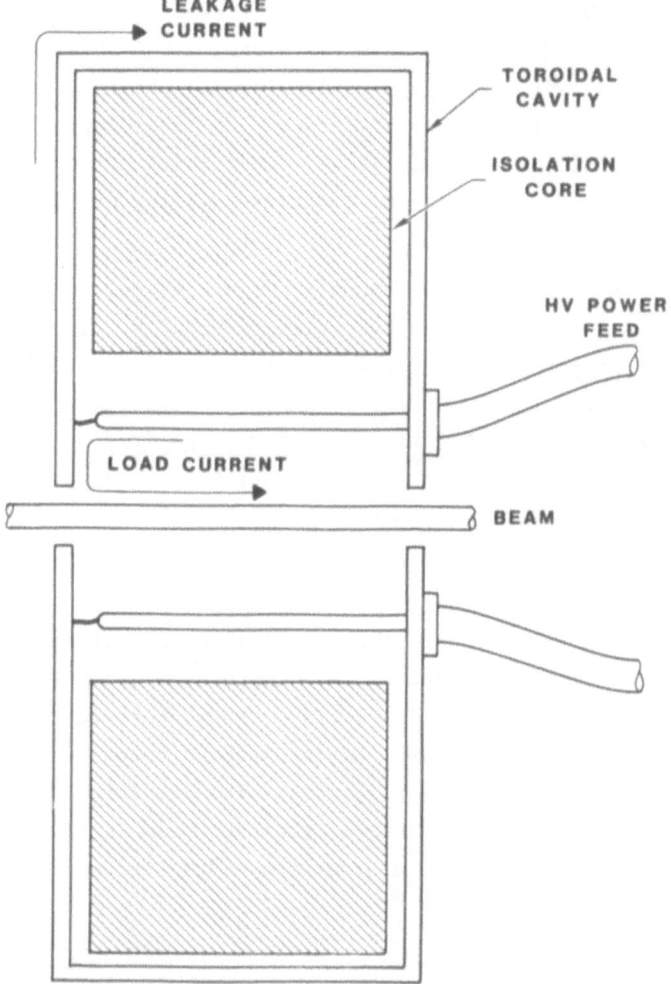

Fig. 17.   Schematic view of an induction linac cavity.

pulse isolation cores are laminated silicon steel and ferrites. Silicon steel has a high flux swing (3 tesla) but low resistivity. In consequence, the magnetic skin depth is small, usually comparable to practical lamination thicknesses. Skin depth effects limit pulselengths to the range ≥ 1 µs. In contrast, ferrites have a high resistivity and a large skin depth but a small flux swing (0.6 tesla). They are best suited to short pulses (≤ 100 ns).

The first accelerator to consider is a long pulse induction linac. The author participated in the design of the machine, ILINAC-1. It is being developed for commercial operations at Ford Laboratories of Dublin, California. The goal of the design was to achieve high voltage and high average power in a simple, conservative and robust system with a short development time. Therefore, the accelerator utilizes conventional pulsed power technology (spark gaps) at low repetition-rate (10 Hz). In order to achieve high average power with reasonable beam current (a few kA), a long pulse was necessary, motivating the choice of silicon steel isolation cores. The accelerator follows the technology developed at the National Bureau of Standards (Wilson). Modifications were made to achieve reduced cost and high net efficiency in the drive and reset circuits.

The accelerator (Fig. 18) consists of a 1 MV pulsed electrostatic injector and nine inductive post-acceleration units. The injector is driven by a Marx generator. Characteristics of the system are listed in Table 8. The cores consist of 2 mil silicon steel laminations insulated with mylar to withstand the turn-to-turn voltage drop of 15 V/turn. Special windings, flux forcing loops, penetrate the core to equalize magnetic saturation over the core radius.

The accelerator has a number of novel features to reduce cost and maximize net electrical efficiency. The beamline, cores, and pulsed power system of a module share the same insulating oil enclosure. In contrast to previous induction linacs, components are not housed in discrete cavities. As shown in Fig. 19, the support for the acceleration tube is mechanically independent of the massive cores to reduce construction costs and to aid in alignment. The acceleration tube is an open structure to reduce problems of resonant beam instabilities. Beam focusing is performed by intermittent duty-cycle solenoidal coils.

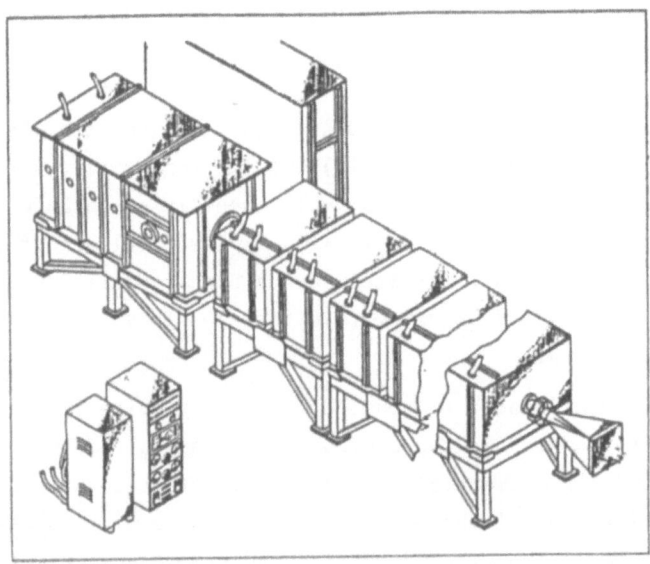

Fig. 18.   ILINAC-1, overall view (Courtesy, F. Ford).

Table 8. ILINAC-1.*

Type: Induction linac, silicon steel cores

Configuration: 1 MV electrostatic injector. Nine 1 MV induction units with four 250 kV gaps/unit and 16 isolation cores/unit

Beam energy: 10 MeV, average

Beam current: 2 kA

Pulselength: 1 μs

Power source: Parallel single switch Marx generators, 120 kV open circuit, directly coupled to beam

Repetition rate: 10 Hz

Average power: 200 kW

Average gradient: 0.42 MV/m

Total length: 24 m

Electrical efficiency: 50–60%

* Parameters courtesy of F. Ford

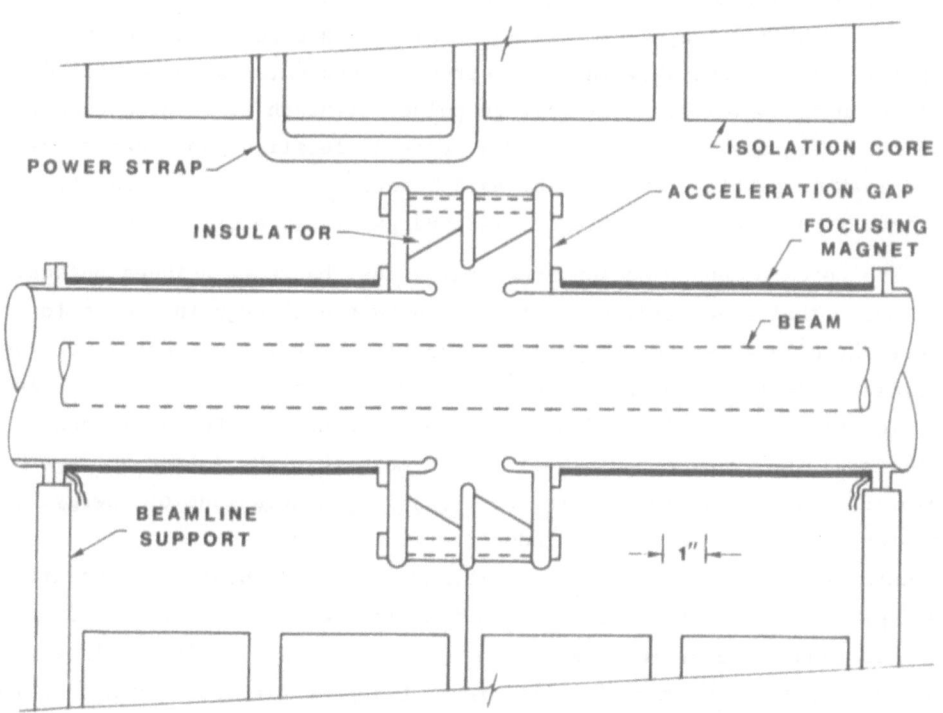

Fig. 19. Detail of beamline, ILINAC-1.

In the low repetition-rate system, the power required for the focusing coils is reduced by more than an order of magnitude compared to DC excitation.

In previous induction linacs, a constant accelerating voltage results when energy is transferred from a high voltage transmission line to a matched constant current beam load. This method involves multiple stages of power conditioning. A simpler method is used in ILINAC-1 to achieve an approximately constant voltage pulse. It is based on the fact that the beam need not have constant current over the pulse since beam quality is a minor concern. In particular, for a variation of beam current of the form

$$i(t) = i_o \sin(\pi t/t_p), \quad (0 \le t \le t_p) \tag{28}$$

a constant voltage pulse can be obtained directly from a Marx generator. The equivalent circuit is illustrated in Fig. 20. If a Marx generator with open circuit voltage V_o has series inductance (L) and capacitance (C) that satisfy the relationships

$$t_p = \pi(LC)^{1/2}, \tag{29}$$

$$V_o/i_o = 2(L/C)^{1/2}, \tag{30}$$

the voltage applied to the beam load with current given by Eq. (28) has the constant value of $V_o/2$. Furthermore, there is 100 per cent energy transfer over a time t_p. The switched current waveform from the 1 MV injector of ILINAC-1 approximates Eq. (28). With core leakage currents, computer calculations show that the circuit components can be modified to achieve an approximately flat voltage pulse, although at reduced energy transfer efficiency. In the ILINAC-1 post-acceleration gap, four cores are driven in parallel by a 140 kV Marx generator. A step-up transformer configuration is used to achieve a matched gap voltage of 250 kV.

The cores of an induction accelerator must be reset between pulses. This means that a negative current must be passed through the cores to invert the orientation of the magnetic domains. If this is not done, the saturated core has a low relative μ. The reset circuit must be isolated from the high voltage induced by the main power pulse. For a silicon steel core, the reset circuit must supply more than 500 A at a voltage and over a time determined by the Eq. (27). Considerations of energy efficiency preclude the use of inductors or resistors as isolation elements. The problem is solved in ILINAC-1 with a synchronized rotating spark gap. The gap provides complete isolation during the power pulse and a low impedance during reset. The main spark gaps in ILINAC-1 are field distortion gaps for long lifetime. Using an electrode erosion rate of 0.5×10^{-4} gm/coulomb (Bel'kov, 1972), an average current of 12 kA for

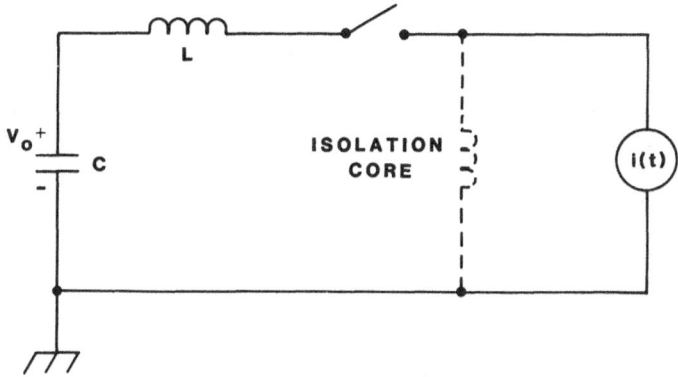

Fig. 20. Equivalent circuit for beam switching of
          pulsed power.

each Marx generator switch, and a repetition rate of 10 Hz, the erosion
lifetime of molybdenum electrodes is about one year of 260 days with
16 hours/day.

The long-pulse accelerator is the circuit equivalent of a high
voltage pulse transformer. The laminated steel cores lead to a straightforward system with the scale size and appearance of a large piece of
conventional electrical equipment. The requisite pulsed power technology
is conservative; there is little problem with synchronization and jitter.
The main drawback of a steel core induction linac is the relatively low
average gradient and large leakage currents. Net system energy efficiency is estimated to be from 50 to 60 per cent.

Continuous-Wave RF Electron Linac

The acceleration field in most existing RF electron linacs is
applied by traveling waves, as shown in Fig. 21a (Humphries, 1986). High
power microwaves are injected into a waveguide that supports an electromagnetic wave with phase velocity matched to the electron velocity. Such
a waveguide, with $v_{phase} \leq c$, is called a slow wave structure. Particles
are accelerated by the longitudinal electric component of the wave.
Traveling wave accelerators can achieve high gradient (5-10 MV/m) but
require very high input power to overcome resistive losses in the
structure; generally, they have low electrical efficiency. They must be
driven by short macropulses (a few μs) at low duty cycle. Their main
applications are in high energy physics research. For commercial
accelerator applications, standing wave linacs are more practical. These
devices, successfully applied to high power ion acceleration, can be
operated in the CW (continuous wave) mode or with a pulsed duty cycle.

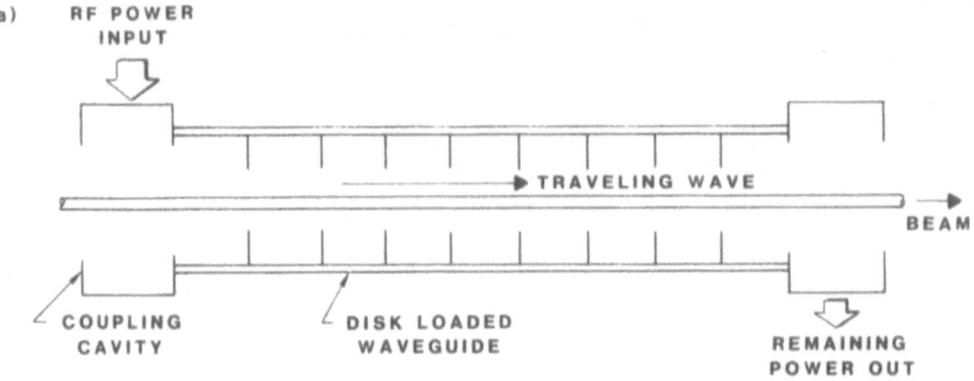

a) RF POWER INPUT

TRAVELING WAVE

BEAM

COUPLING CAVITY

DISK LOADED WAVEGUIDE

REMAINING POWER OUT

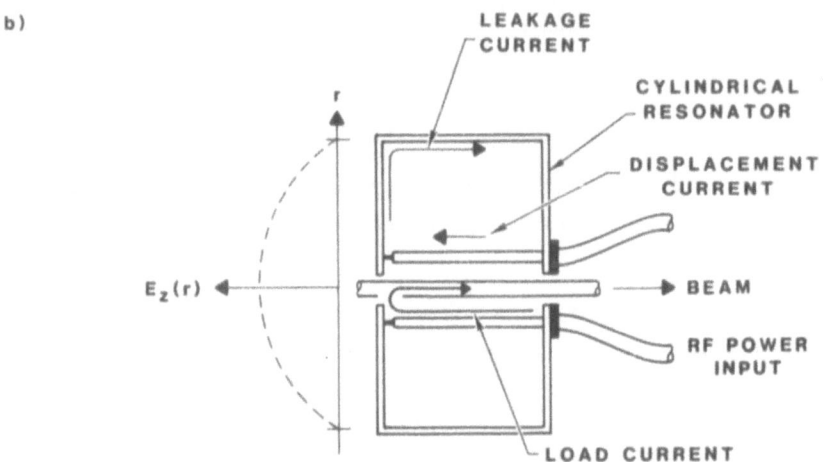

b)

LEAKAGE CURRENT

CYLINDRICAL RESONATOR

DISPLACEMENT CURRENT

$E_z(r)$

BEAM

RF POWER INPUT

LOAD CURRENT

Fig. 21. Physical bases of RF linear accelerators.
a) Traveling wave linac.
b) Inductive isolation in a resonant cavity
for acceleration.

A standing wave RF linac consists of a series of resonant cavities. Inductive isolation results from the large displacement currents that flow when the cavity is excited at the resonant frequency. This fact is illustrated by Fig. 21b. To begin, assume that power lines carrying AC voltage at the resonant frequency enter at the center of a cylindrical cavity; the goal is to transfer the power efficiently to a load on axis. As in the case of the induction linac, current from the source can flow either through the load or around the cavity wall as leakage current. If the frequency of the applied power equals the resonant frequency of the TM_{010} mode in the cavity, the leakage circuit appears to have infinite impedance. This reflects the fact that at resonance the leakage current to maintain an on-axis voltage is supplied totally by the cavity displacement current.

The resonant cavity has an additional useful characteristic; it can act as a step-up transformer. In the standard geometry (Fig. 21c), power

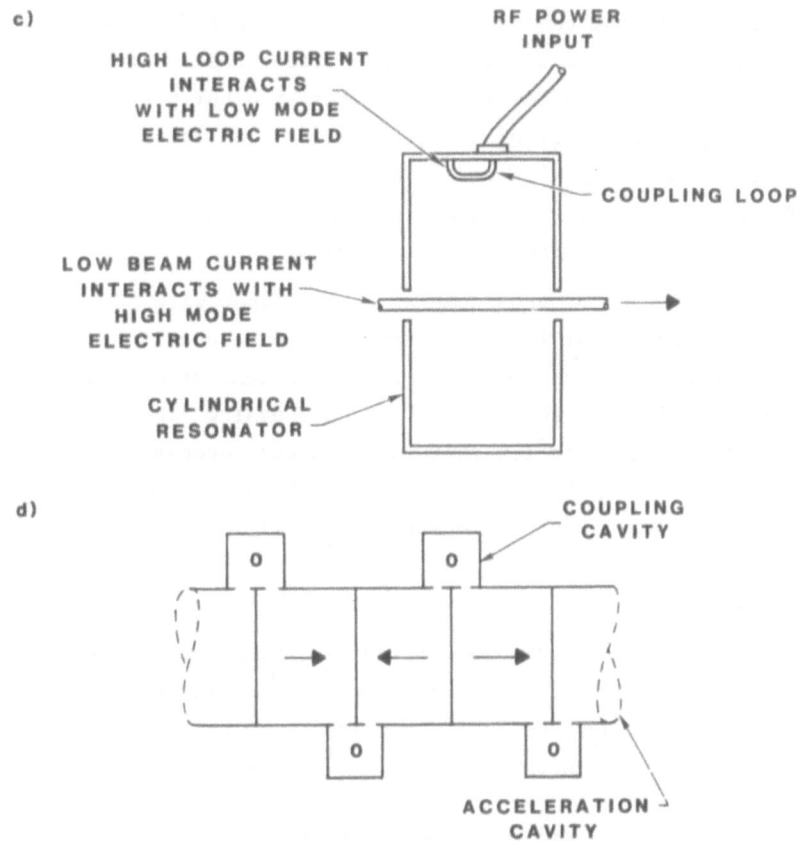

c)

HIGH LOOP CURRENT
INTERACTS
WITH LOW MODE
ELECTRIC FIELD

RF POWER
INPUT

COUPLING LOOP

LOW BEAM CURRENT
INTERACTS WITH
HIGH MODE
ELECTRIC FIELD

CYLINDRICAL
RESONATOR

d)

COUPLING
CAVITY

ACCELERATION
CAVITY

Fig. 21. Physical bases of RF linear accelerators (Cont.).
c) Resonant cavity as a RF transformer.
d) Side-coupled linac.

is coupled magnetically into the cavity by a loop at large radius and
withdrawn by the load beam on the axis. The coupling loop is driven by a
microwave source at low voltage and high current, while the relatively
low current beam is driven at high voltage. For net acceleration of
beams, it is essential that the particles cross the cavity only during
the accelerating phase of the voltage. In RF accelerators, the entering
electron beam must be bunched into clumps with arrival times synchronized
to the cavity oscillations.

High beam kinetic energy is achieved by connecting multiple acceleration cavities in series. The phase of oscillations in a series of
cavities must vary to match the changing velocity of the electron
bunches. It is possible to build a multi-cavity RF linac with individual
phased power feeds for each cavity (as discussed in Section 2.e). A more
common approach is to use a single power source and couple cavities
together by slots which allow shared electromagnetic fields. The side-
coupled linac, and advantageous geometry for high power beams, is shown

in Fig. 21d. Half the cavities are located on-axis while the other half are displaced. The on-axis cavities are used for beam acceleration. The off-axis cavities have a low level of excitation but play an important role in energy transfer along the structure. With the relative field polarities shown in Fig. 21d ($\pi/2$ mode), the side-coupled linac has effective energy coupling and good frequency stability. These properties make it well suited to high power beam loads.

Detailed studies of a CW coupled-cavity linac with 500 kW beam power have been carried out at the Chalk River Nuclear Laboratory of Atomic Energy of Canada, Limited (McKeown, 1985; McKeown, et al., 1985; McKeown and Sherman, 1985). In addition to food processing, the machine was also considered as a candidate for radiation-induced saccharification to cellulose from wood by-products. The Chalk River Laboratory has a long history of electron linac development; presently, a 4 MeV, 20 mA CW linac is operational. Parameters of the proposed 500 kW accelerator are listed in Table 9.

The accelerator is composed of four 2.5 MV structures containing 21 accelerating cavities each. The cavities are excited by two commercial klystrons. Similar devices have demonstrated lifetimes of 47,000 hours at the Stanford Linear Accelerator. Cavity losses are minimized by

Table 9. AECL 500 kW RF linac.*

Type: On-axis coupled cavity CW RF linac.

Configuration: Injector, buncher, four 2.5 MeV RF structures (first with graded β).

Cavity configuration: Coupled cavity array with on-axis coupling, 21 cells per structure.

Beam energy: 10 Mev, average

Beam current (average): 50 mA

Power source: Dual 500 kW klystrons

Frequency: 2.45 GHz

Repetition rate: CW

Average beam power: 500 kW

Electrical efficiency: 30 per cent

Structure length: 1.25 m

Total length: Approximately 9 m

* Parameters courtesy of J. McKeown

---

operating at the highest possible frequency. The 2.45 GHz was motivated by this concern and by the availability of high power klystrons. Klystrons have an electrical efficiency ranging from 60 to 70 percent. The expected power balance for RF energy in the accelerator is 500 kW to the beam, 300 kW to the cavity structure, and 200 kW to waveguides and other RF structures. The net accelerator efficiency is about 30 percent.

The accelerating structures are composed of a series of accelerating cavities separated by on-axis coupling cavities, as shown in Fig. 22. The cylindrically symmetric geometry is easier to fabricate than the side-coupled structure. The first structure is a graded β structure; it is fabricated with a longitudinal variation of cavity geometry so that the phase of the oscillating cavity electric fields are matched to the increasing velocity of the electrons. The beam is highly relativistic in the remaining cavities, so it is sufficient to use a uniform geometry corresponding to a particle velocity equal to the speed of light.

Considerable attention was devoted to cooling channels for the RF structure. At a gradient of 2 MV/m, the power dissipation in the structure is 60 kW/m. Higher gradients are unlikely in a nonsuperconducting, CW accelerator. Thermal expansion of the cavities can cause 900 kHz shifts of the resonant cavity frequency. Active tuners must by included in the structures to maintain equal resonant frequencies.

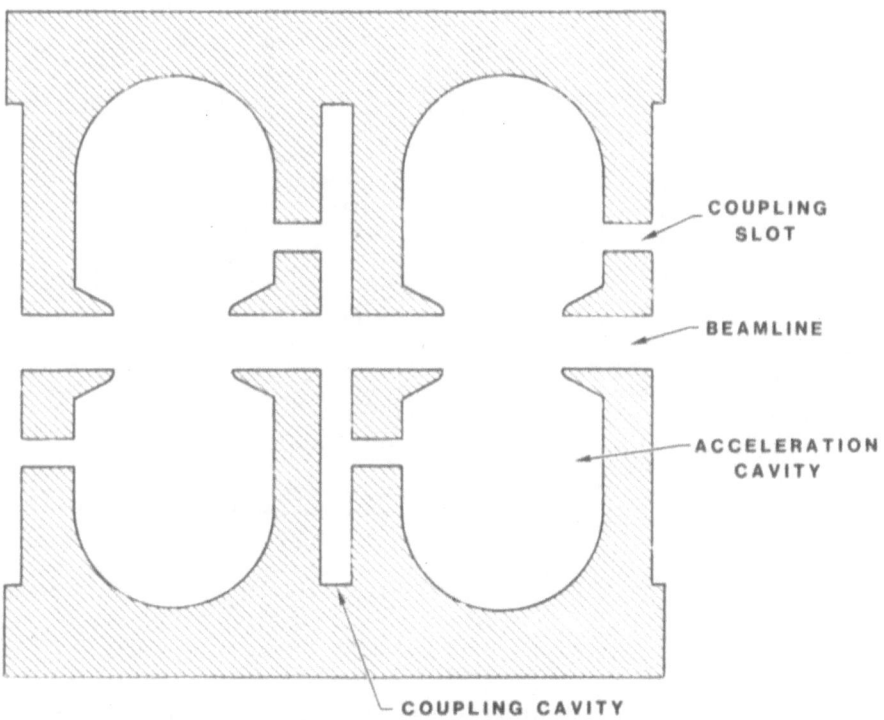

Fig. 22. On-axis coupled linac.

An area in which the AECL design extends past present accelerator experience is in beam loading. Existing electron linacs are not designed for high efficiency; most of the RF energy is wasted through resistive dissipation in the cavity walls. In the low efficiency regime, the beam constitutes a small fraction of the RF generator loading. The generator is matched to the structure independent of the presence of the beam. The AECL design assumes high beam loading; in the absence of a beam, a good proportional of the RF energy will be reflected back to the source. This situation presents a potential danger to the klystrons. The solution is to use a recirculator, which shunts reflected energy to a microwave absorber. In this case, the total system always presents a matched load to the klystron. There is also a concern that the beam loading will affect the RF modes in the coupled cavity structure. This question can only be resolved by operational experience with a high gradient, heavily loaded structure.

Figure 23 is a scale drawing of the 500 kW accelerator and a facility for single sided irradiation. Note the linear conveyer system.

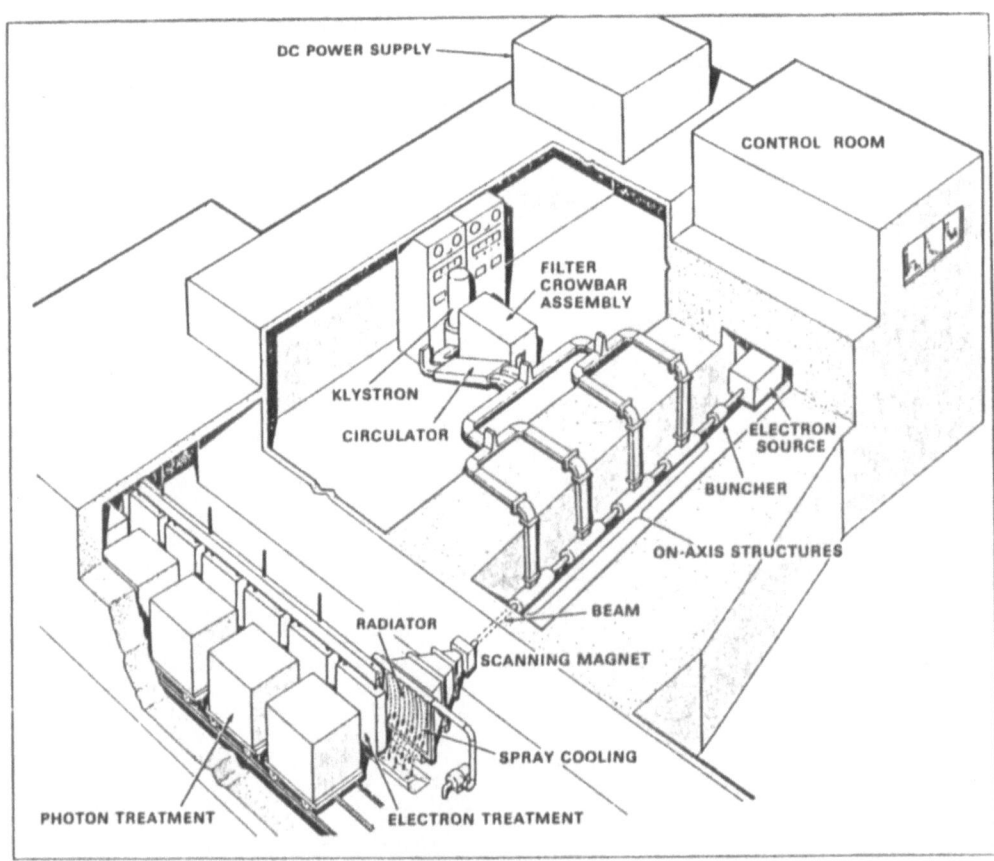

Fig. 23. Scale drawing of AECL 500 kW linac and food processing facility (Courtesy, J. McKeown).

704

Operation with simultaneous electron and γ-ray irradiation is illustrated. The cooled vacuum window also acts as a radiator. In addition to an available electron beam power of 350 kW, the system produces γ radiation equivalent to 6 MCi of Co^{60}.

High-Repetition-Rate Linear Induction Accelerator

High-current ferrite-core linear induction accelerators have been developed in a long-term project at the Lawrence Livermore National Laboratory. The project is directed towards particle beam weapons. As part of an effort to demonstrate applications feasibility of such accelerators, recent work has concentrated on raising the repetition-rate (and hence average power) of induction accelerators. Parallel work has been carried out to investigate civilian applications as part of a technology transfer program (Birx et al; Birx, 1986). Parameters for a 650 kW, 10 MeV linear induction accelerator based on exptrapolations of current technology are listed in Table 10.

The cavity geometry of a ferrite core induction linac is equivalent to that of Fig. 17. The optimum parameter range is quite different from that of accelerators with laminated steel cores. Ferrites have a volume resistivity about seven orders of magnitude higher than silicon steel. The skin depth for a 10 ns pulse is on the order of 1 μm. Therefore, laminated construction is unnecessary and eddy current losses are small. The drawback of ferrites is that the maximum flux swing is a factor of five lower than silicon steel. This means that the cross-sectional area of a ferrite core must be five times greater than that of a steel core for the same volt-second product [Eq. (27)].

In order to achieve the highest average gradient in a linear induction accelerator, the shortest possible pulse should be used. Eddy current losses limit silicon steel cores to the μs range, but ferrites can be used with small losses for pulses as short as 0.01 μs. In ferrite core accelerators, the lower limit on pulselength is determined by limitations of how power switches. The practical range for multi-kA beams is 50-100 ns. Although ferrites have lower relative permeability compared to steel, cavity leakage currents are low because of the short pulse-length.

The LLNL accelerator design is based on a 75 ns pulse and a 1 kA beam current. The relatively low beam current can be supplied by a dispenser cathode of diameter 12.7 cm; furthermore, low current minimizes problems of transverse instabilities. In order to achieve high average power output, the accelerator must operate at a high repetition rate (~1 kHz). The pulse frequency is beyond the capabilities of conventional pulsed power technology which is limited by the recovery time of spark

Table 10. LLNL high repetition-rate ferrite core induction linac.*

Type: Induction linac, ferrite cores

Configuration: 600 kV inductive injector, 4:1 step-up. 64 induction cavities operating at 150 kV.

Beam energy: 10 MeV

Beam current: 1 kA

Pulse length: 75 ns

Power source: Thyratron controlled capacitor bank driving magnetic pulse compression circuit, 150 kV output at 2 Ω, driving 68 core sets in parallel.

Repetition rate: 1 kHz

Average power: 650 kW

Average gradient: 0.39 MV/m

Total length: 26 m

Electrical efficiency: 65%

* Parameters courtesy of D. Birx

gaps. Thyratrons can achieve 1 kHz repetition-rate and can switch average power greater than 100 kW, but they have relatively low peak current and voltage limits.

The solution to the switching problem developed at LLNL is based on the use of a thyratron as the primary switch to transfer energy to a passive magnetic pulse compression circuit. The thyratron operates at moderate current over a long time scale (~5 μs). The magnetic circuit steps up the voltage and compresses the power; the output current is boosted while the pulselength is shortened. The principle of the magnetic pulse compression circuit is illustrated in Fig. 24. The circuit consists of capacitors and inductors wound on ferrite cores. Before a pulse, all cores are reverse biased so that initially they have a high relative permeability. Referring to Fig. 24, in the first stage the thyratron transfers energy through a step-up transformer to charge C_1 to 150 kV. The volt-second product of the core in L_1 is chosen so that the core saturates near the peak charge voltage on C_1. The inductance of L_1 drops rapidly, and the circuit energy is transferred to C_2. The circuit has high efficiency for energy transfer if all the capacitances are equal. The number of turns in each inductor decreases in a geometric progression along the circuit so that the energy is transferred on an

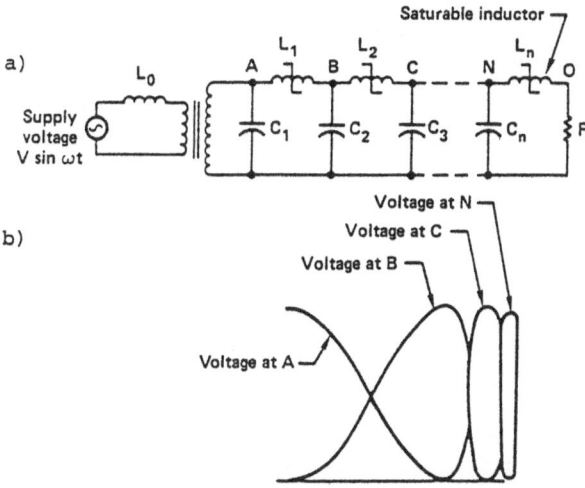

Fig. 24. Magnetic power compression. a) Equivalent circuit.
b) Voltage waveforms following switching of primary
capacitor bank (Courtesy, D. Birx).

increasingly faster time scale. Energy is conserved, but power is
multiplied. The final stage is switched out to the induction cavity on a
75 ns time scale. With careful circuit design, some of the energy in the
main voltage pulse can be diverted to reset the cores in the switching
circuit. High power magnetic pulse compression circuits have been
operated in the burst mode at LLNL at frequencies as high as 5 kHz. In
the 650 kW accelerator, a single compact magnetic compression circuit is
sufficient to drive the complete machine.

The LLNL design was optimized for peak electrical efficiency. In
order to minimize energy losses from cavity leakage currents, a low value
of average accelerating gradient was chosen. This choice leads to the
rather large system of Fig. 25. Only about 9 per cent of the system
energy is lost in the inductors leading to a 62% net electrical efficiency. With adequate cooling, higher core losses can be sustained. Since
electrical power is a minor factor in the economic balance, it is con-
ceivable that the gradient of the accelerator could be doubled with small
operating cost penalty, significantly reducing the size and capital cost
of the machine.

Figure 26 shows a cavity of the low gradient accelerator. The
focusing solenoidal provides a 300 G field, sufficient to focus the 1 kA
beam. The vacuum insulator is shielded from the beam. The acceleration
gap of the cavity is made as small and non-perturbing as possible to
minimize the growth of resonant transverse oscillations.

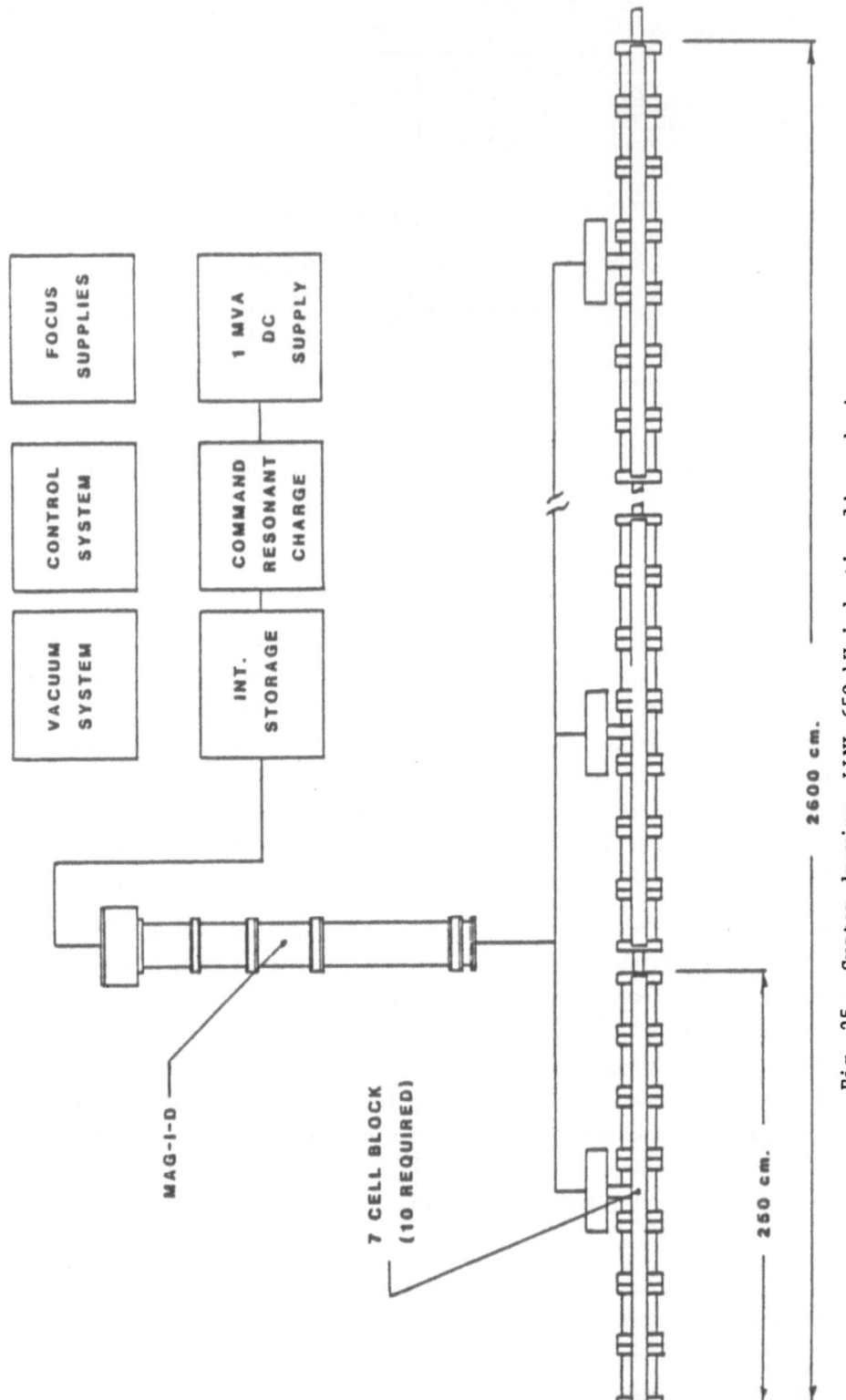

Fig. 25. System drawing, LLNL 650 kW induction linac design (Courtesy, D. Birx).

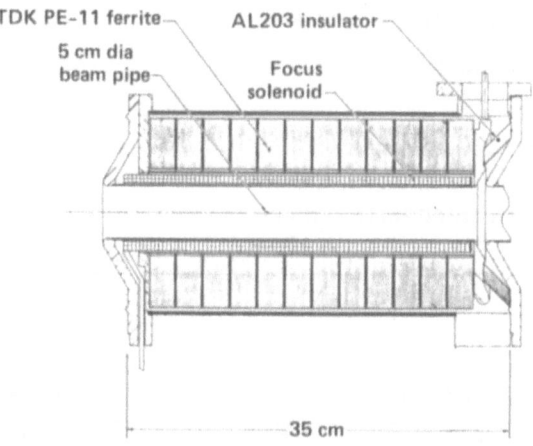

Fig. 26.   Cavity of LLNL 650 kW accelerator (Courtesy, D. Birx).

The LLNL approach has many attractive features that could lead to a
compact, reliable accelerator. Induction accelerators appeared suitable
for food processing applications; experiments carried out on the LLNL
High Brightness Test Stand have shown no demonstrable differences in the
quality of food irradiated by radioisotopes or by a low-duty-cycle pulsed
beam. With careful circuit design, high power thyratrons have quoted
lifetimes exceeding 10^9 shots, corresponding to over a month of operation. Although the technology has been successful for moderate times in
burst mode operation, demonstrations of long term reliability are essential. The main unresolved technical question is the reset of the cavity
cores at high repetition-rate. Actively switched circuits are necessary
since passive reset circuits waste too much energy.

COMPACC - Compact Pulsed RF Linac

The relative sizes of the three 10 MV accelerators discussed are
shown in Fig. 27. Generally, the systems are not compact. Large
accelerators increase the cost of buildings and shielding. Excessive
accelerator length precludes the use of opposing accelerators for dual-
sided irradiation. COMPACC, an approach to low-cost compact accelerators
developed by the author, will be discussed in this section. The accelerator is a pulsed standing wave RF linac with novel power supplies to
achieve very high gradient with good electrical efficiency. The relative
size of a 10 MeV, 250 kW machine is displayed in Fig. 27. The short
length is achieved by operating at a gradient of 7.5 MV/m in the acceleration cavities. A number of accelerator modules can be grouped in a
single facility, allowing dual sided irradiation.

Compact accelerators demand high average accelerating gradient. It
is unlikely that gradients exceeding 1 MV/m can be achieved in induction

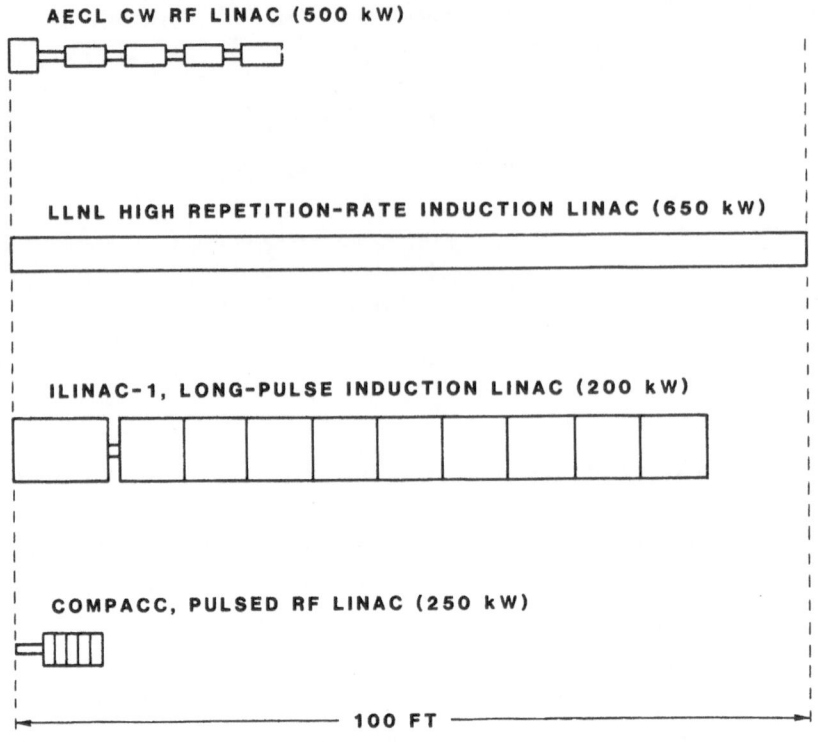

Fig. 27. Relative dimensions of 10 MeV accelerator
concepts discussed in report.

accelerators. The characteristics of the isolation cores present a
limitation expressed by the volt-second relationship of Eq. (27).
Furthermore, induction accelerators have vacuum insulators exposed to
high electric fields. Insulators can fill only a limited fraction of the
accelerator length; gradients on the insulator are limited to the range
2-5 MV/m in continuous operation.

Radio-frequency linacs do not have exposed vacuum insulators. The
metal surfaces of the RF cavities can sustain electric field in the range
10-20 MV/m without breakdown. The gradient limitation in RF electron
accelerators is usually set by resistive losses in cavity structures; for
an average gradient E_o, cavity power loss scales as E_o^2. As shown in the
previous section, CW standing-wave linacs are limited to $E_o \leq 2$ MV/m by
1) the availability of power supplies, and 2) cooling of cavity
structures. It is important to note that raising E_o does not automatically lead to a decrease in accelerator efficiency. Increased cavity
losses can be balanced by higher beam current to maintain the same
relative beam loading and efficiency. This is possible as long as the
beam current is low enough so that 1) the beam can be easily transported,
2) energy transferred to the beam per RF cycle is small compared to the

stored RF energy, and 3) the beam does not excite unwanted RF modes leading to instabilities.

COMPACC achieves high gradient by operating on an intermittent duty cycle with high beam current. This is, in fact, the operating mode of most present RF accelerators. A scale view of the accelerator is shown in Fig. 28 and parameters are summarized in Table 11. The cavities are simple TM_{010} structures. They are uncoupled; each cavity has its own power feed. This geometry is referred to as an independently phased cavity array. The small volume cavities have no internal structures and a large exposed surface area for effective cooling. The accelerator is comprised of only five cavities. The peak on-axis voltage is 2 MV, corresponding to a peak accelerating gradient of 7.5 MV/m. The cavities have a radius of 50 cm and a resonant frequency of 400 MHz. The estimated cavity Q is 3.5×10^4. At the high gradient, each cavity dissipates 500 kW during the acceleration pulse. With 67 per cent beam loading, a power of 1 MW/cavity is transferred to the beam.

The power input to COMPACC during an acceleration pulse is 7.5 MW. Average beam power during a macropulse is 5 MW. For an average beam power of 250 kW, the RF supply must supply an average power of 75 kW/cavity at 400 MHz. The duty cycle is 5 per cent; a 500 μs macropulse

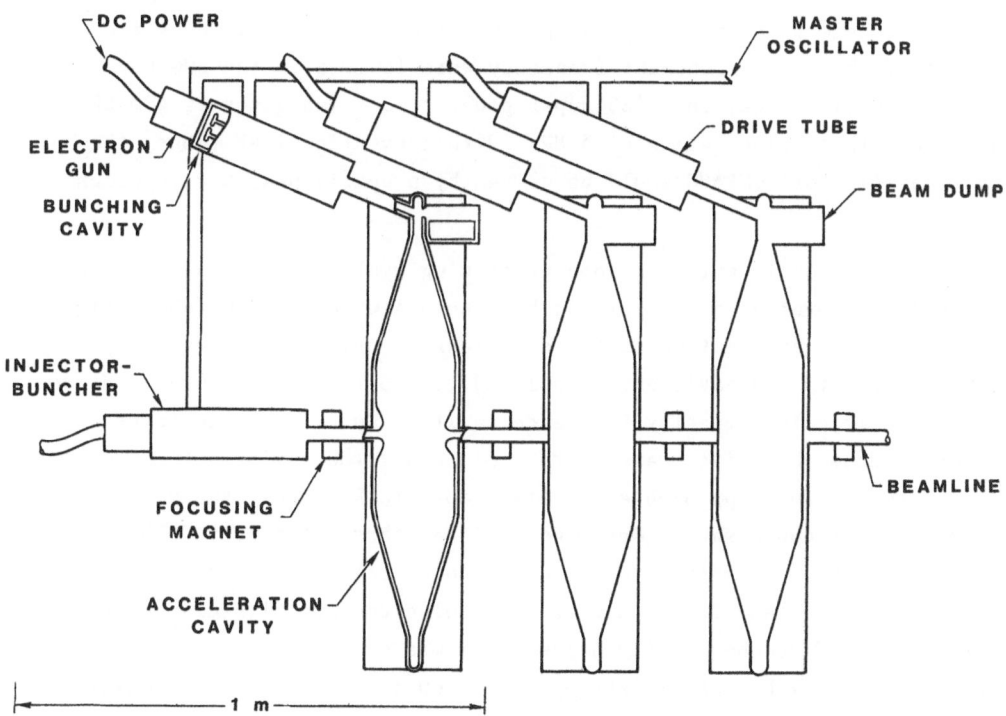

Fig. 28. Scale view of COMPACC (three cavities illustrated).

711

Table 11. COMPACC - high gradient pulsed RF linac.

Type: Pulsed RF linear accelerator, individually phased cavities.

Configuration: Injector, buncher, 5 independent cavities

Beam energy: 10 MeV

Beam average current (during pulse): 5 A

Macropulselength: 500 μs

Pulse repetition frequency: 100 Hz

Power source: Five drive tubes: 85 kV, 30 A electron guns, bunching cavity

Average beam power: 250 kW

Peak beam power: 5 MW

Average gradient: 2.9 MV/m

Total length: 3.5 m

Electrical efficiency: 40%

implies a pulse repetition-rate of 100 Hz. These requirements are within the capabilities of present conventional RF tubes (Tellerico, 1979). For example, a low frequency klystron is commercially available can generate 500 kW average power with a pulse power rating of 3 MW. Triodes are also useful if the cavity resonant frequency were lowered to the 200 MHz range. For instance, the 7835 super power triode can generate 450 kW average with a pulse rating of 5 MW. With conventional RF tubes., the five cavities of COMPACC would be driven by properly phased waveguides in 23).

In order to investigate food processing systems with minimum size and cost, innovative drive methods have been studied for COMPACC. Figure 28 illustrates application of the direct-drive principle (Humphries, 1984) to excite the cavities. Modulated low-voltage, high-current drive beams are injected directly into each cavity. Energy is transferred directly from the drive beams to the load beam; the cavity acts as an RF transformer. This approach eliminates separate RF tubes, vacuum RF windows, high power waveguides, and recirculators. The drive tubes are simple devices, equivalent to the electron source and modulation cavity of a klystron. Each tube consists of an electron gun with dispenser cathode, an RF buncher cavity driven by a master oscillator, and a bunching tube with magnetic transport. Figure 28 illustrates coaxial bunching cavities in a unitized package with the bunching tube.

The drive beams, injected near the cavity periphery, interact with the TM_{010} mode in a region of low electric field. Saturation of power transfer occurs when the voltage at the periphery is comparable to the drive beam voltage. The on-axis acceleration voltage is much higher than that acting on the drive beam. The ratio is the voltage step-up of the cavity. The cavity geometry of Fig. 28 is the result of computer studies to maximize voltage step-up; it achieves an amplification of 40:1.

In accordance with klystron experience, about 60 to 70 per cent of the driving beam energy can be converted to RF energy. The remainder is dissipated in beam dumps. There are two possible fault modes that could lead to interruption of operation. If the load beam current drops, the RF energy in the acceleration cavity will rise. This situation could lead to reflection of drive electrons and damage to the drive tube. Conversely, if the RF energy in the load cavity drops, an excessive amount of energy may be deposited in the beam dumps. These conditions can be rapidly corrected by a feedback system and grid to control the current from the drive or load guns.

The following parameters illustrate typical operating conditions. The DC electron guns of the drive tubes generate 30 A at 85 kV. The guns are within the present capability of klystron technology. Electrons are produced by a low current density dispenser cathode, 2.5 cm in diameter. Transport of the 30 A beams in the small diameter bunching tube requires a 1 kG solenoidal magnetic field. Net power requirements are quite low if the magnets are driven by high power transistors at an intermittent duty cycle. The perveance of the drive beams is low enough so that they traverse the load cavity with little expansion. The peak voltage in the bunching cavity is 12 kV. The master oscillator supplies less than 5 kW of RF energy to the five drive tubes. The load beam injector and buncher are identical to the drive beam structure. The load beam injector operates at a reduced current of 5 A.

The total length of COMPACC with injector is only 3.5 m. The device illustrates some of the potential advantages of accelerator innovation. The physics and technology of direct-drive cavities is currently under investigation at Westinghouse Research and Development Center (Nahemow). One potential problem area is the excitation of high frequency cavity modes by the bunched, asymmetric drive beams. Higher order modes waste energy and could lead to deflections of the load beam. There is a good chance this problem can be avoided by adding waveguides, leading to matched dump resistors, have a cutoff above the frequency of the fundamental mode. This method has been used with considerable success in superconducting linac cavities (Sundelin, 1985).

ACKNOWLEDGEMENTS

The opinions and conclusions expressed in this report are entirely those of the author. I would like to thank F. Ford, N. Neilson, J. McKeown, and D. Birx for supplying information on their work. I would also like to thank S. J. Hernandez for his help in assembling material.

REFERENCES

Bacq, Z. M. and Alexander, P., 1961, Fundamentals of Radiobiology, Second Edition, (Pergamon Press, New York).

Becker, R. L., 1983, "Absence of Induced Radioactivity in Irradiated Foods," in Recent Advances in Food Irradiation, edited by Elias, P. S. and Cohen, A. J. (Elsevier Biomedical, Netherlands).

Bel'kov, Jan. 1972, E. P., Pribory i Tekhnica Eksperimenta 1.

Bethe, H. A., 1933, Handbuch der Physik 24 (Springer, Berlin).

Birx, D. L., Hawkins, S. A., Poor, S. E., Reginato, L. L., Smith, M. W., "Optimization of Magnetically Driven Induction Linacs for the Purpose of Radiation Processing," submitted to Rev. Sci. Instrum.

Birx, D. L., 1986, "Induction Linacs as Radiation Processors,".

Brynjolfsson, A., 1978, "Energy and Food Irradiation," in Food Preservation by Irradiation (IAEA, Vienna), 285.

Casarett, A. P., 1968, Radiation Biology (Prentice-Hall, Englewood Cliffs, NJ).

Code of Federal Regulations, April, 1986, Part 179, Docket No. 81N-0004.

Committee on Radiation Application Info., June, 1986, (Atomic Industrial Forum, Bethesda).

Ebert, M. and Howard, A., editors, 1963, See, for instance, Radiation Effects in Physics, Chemistry and Biology, (North-Holland, Amsterdam).

Evans, R. D., 1967, See, for instance, The Atomic Nucleus (McGraw-Hill, New York).

Food Irradiation Processing 1984, (IAEA, Vienna).

Hableib, J. A. and Melhorn, T. A., 1984, See, for instance, "ITS: The Integrated TIGER Series of Coupled Electron/Photon Monte Carlo Transport Codes," (Sandia National Laboratories, SAND84-0573).

Humphries, S., 1986, For a review of accelerator physics, see Principles of Charged Particle Acceleration (Wiley, J., New York).

Humphries, Jr., S., "High Current Electron Beam Accelerators," to be published, Nucl. Instrum. and Methods.

Humphries, Jr., S., 1984, Proc. 5th Int'l. Conf. High-Power Beams, edited by Briggs, R. J. and Toepfer, A. J., (Lawrence Livermore National Laboratory, CONF-830911), 454.

"Irradiation in the Production, Processing and Handling of Food," 1984, Federal Register 49, 5714.

Jackson, J. D., 1975, See, for instance, Classical Electrodynamics, Second Edition (Wiley, J., New York), 705.

Joint FAO/IAEA/WHO Expert Committee, 1981, Wholesomeness of Irradiated Food, (World Health Organization, Tech. Report Series 659, Geneva).

Knoll, G., 1979, See for instance, Radiation Detection and Measurement (Wiley, J., New York).

Koch H. W. and Motz, J. W., 1959, See, for instance, "Bremsstrahlung CrossSection Formulas and Related Data," Rev. Mod. Physics 31, 920.

Koch, H. W. and Eisenhower, E. H., September, 1965, American Chemical Society Symposium, Atlantic City.

McKeown, J., 1985, "Radiation Processing Using Electron Linacs," IEEE Trans. Nucl. Sci. NS-32, 3292.

McKeown, J., and Sherman, N. K., 1985, "Linac Based Irradiators," Radiat. Phys. Chem. 25, 103.

McKeown, J., Labrie, J. P., and Funk, L. W., 1985, Nucl. Instrum. and Methods B10/11, 846.

Morrison, R. M., 1985, "Economics of Scale in Single Purpose Food Irradiators," Proc. Int. Symp. on Food Irradiation Processing (Int. Atomic Energy Agency, IAEA-SM-271-63).

Nahemow, M., Private communication.

Scharf, W., 1986, For a review of accelerator applications, see Particle Accelerators and Their Uses (Harwood, Chur, Switzerland).

Silvermann, J., 1978, Radiat. Phys. Chem. 14, 17.

Sundelin, R. M., 1985, IEEE Trans. Nucl. Sci. NS-32, 3570.

Tellerico, P. J., 1979, "Advances in High Power RF Amplifiers," IEEE Trans. Nucl. Sci. NS-26, 3877.

Wilson, M., (Ref. on NBS induction linac).

SUMMARY OF LINEAR-BEAM TRANSPORT

Terry F. Godlove

Department of Energy
Office of Basic Energy Sciences
Washington, D.C. 20545 USA

ABSTRACT

The paraxial beam envelope equation is briefly reviewed to point out the qualitative features of beam transport in external focusing structures and linear accelerators. Recent developments in understanding high-current beams are discussed. The importance of generalized perveance and depressed phase advance are emphasized. Comments are offered on the use of the term 'brightness' in the context of current applications.

INTRODUCTION

Progress in understanding high-current particle beams in cases where space charge is important has been rapid during the last few years. In particular, the heavy ion fusion (HIF) program, with it's reliance on the highest possible beam currents, has spawned a whole new way of looking at non-relativistic and semi-relativistic beams. Concepts common to plasma physics have become useful in describing high-current beams, and both theoretical and experimental work has converged on understanding the equilibrium dynamics of such beams. The purpose of this review is to summarize relevant portions of other material given at this Advanced Study Institute (ASI) in such a way as to provide a tutorial for those not familiar with the subject. It should be emphasized that readers interested in rigorous treatments should consult the Bibliography below as well as other pertinent chapters in this Proceedings.

It should be noted that circular accelerators and storage rings are beyond the scope of this paper. These topics have been covered by other speakers at this ASI. Of primary interest here are beam currents applicable to RF and induction linear accelerators, to electron and ion sources and injectors in general, and to linear beam transport.

Equation (1) is our starting point. It describes the motion of the "envelope" of the beam with mean edge radius a. The transverse coordinates are x and y, and primes denote derivative with respect to the axial coordinate z. Paraxial trajectories are assumed throughout, enabling separation of the longitudinal and transverse motion

$$a'' + ka - (\varepsilon^2/a^3) - (K/a) = 0. \tag{1}$$

Here k is the force constant of the external focusing system, which in general is a function of z. For alternating gradient (AG) or thinlens solenoidal focusing, k is rapidly varying. Later a "smooth approximation" is used in which case k represents the average net inward focusing force. Sinusoidal solutions around a are assumed, and the "betatron" wavelength λ_o is defined by $(2\pi/\lambda_o)^2 = k$. The high frequency ripple associated with AG focusing is superimposed on the basic envelope solutions and is ignored when the smooth approximation treatment is invoked.

The last two terms of Eq. (1) are due to the transverse temperature of the beam, symbolized by the emittance ε, and the radially outward force due to the space charge of the beam, symbolized by the generalized perveance K. The rms emittance ε_x (or ε_y with y substituted for x) is given by

$$\varepsilon_x = 4 \; (\langle x^2\rangle\langle x'^2\rangle - \langle xx'\rangle^2)^{1/2} \tag{2}$$

where the averages include each particle in the beam. The normalized emittance, separately for each plane, is given by

$$\varepsilon_N = \beta\gamma\varepsilon \tag{3}$$

where the subscripts x, or y, have been omitted for simplicity. Equation (3) exhibits the familiar reduction of the absolute emittance as the longitudinal momentum $\beta\gamma$ is increased, in systems in which ε_N is conserved.

The dimensionless perveance is given by

$$K = 2(I/I_o)(m/M)(Q^2/\beta^3\gamma^3) \tag{4}$$

where I is the particle current (hence the square of the charge state Q rather than the first power), $I_o = 10^7$ mc/e (MKS) = 17.0 kiloamps, m is the mass of the electron, M the mass of the particle under consideration, $\beta = v/c$, and $\gamma = (1-\beta^2)^{-1/2}$ is the total energy of the particle normalized to MC^2.

In the above form the perveance is useful for ions of any mass and charge state and for electrons. For convenient reference it is plotted in Fig. 1 as a function of kinetic energy. Note that the ordinate must be multiplied by the particle current in amperes and by the square of the charge state. Perveance is a very useful concept because in the limiting case of negligible emittance and aberrations it determines the trajectories. Beams having the same perveance and the same initial radius will expand due to their own space charge in the same way, or in a focusing geometry will focus to the same minimum waist. In the heavy ion fusion program, for example, a small accelerator can model much of the beam physics of a large fusion driver provided the perveance is maintained constant.

For ions in the non-relativistic limit the general perveance reduces to $K = (I/15.5 \text{ megamp})Q^2 A^{1/2}(466/T)^{3/2}$, where T is the kinetic energy of the particle in MeV and A is the atomic weight in amu. In this form it is readily identified with the perveance used for decades in the microwave tube industry. The latter definition is simply $I/(V)^{3/2}$ (with amps and volts), related to the Child-Langmuir formula for a space-charge

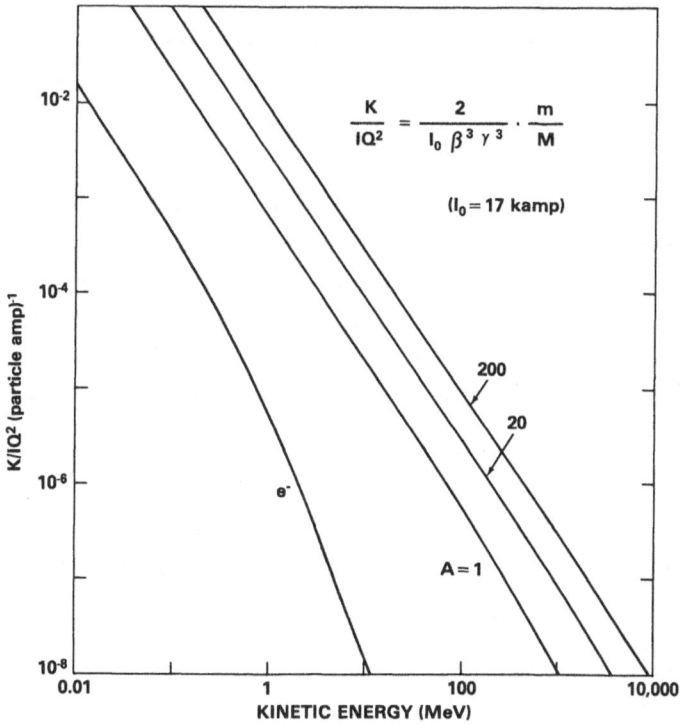

Fig. 1. Generalized perveance K, divided by IQ^2, vs. particle kinetic energy, for four values of particle mass M. Other symbols are particle current I, charge state Q and electron mass m.

limited diode. (The term 'microperveance' refers to multiplying the result by 10^6.)

Perveance is a useful concept for electron and ion sources in a wide variety of applications. Table 1 gives the perveance, Eq. (4), for a number of example sources ranging over nearly six orders of magnitude in current, from the proposed proton source for the European hadron accelerator (HERA - reported by H. Klein) to the electron injector used in the Advanced Test Accelerator (ATA) at Livermore. Here it can be seen that all of these sources have a perveance of the order of 0.01, give or take a small factor. In the microwave tube industry, this would correspond to a microperveance of 0.7.

Keller has summarized a number of ion sources and finds what amounts to a perveance limit of 0.0146 corresponding, for example, to a 250 mA proton source at 50 kV (example proton source in Table 1).

It is interesting, but not surprising, that high-current electron guns have roughly the same normalized perveance as the best ion sources. The klystron gun chosen as an example has the highest perveance in the Table, but of course has had decades of development. At lower voltage, other electron guns can be found with even higher perveance. The ATA (Advanced Test Accelerator, at Livermore) injector is included primarily for comparison of the output beam. A strict comparison is difficult because of the more complex electrode structure.

Evidently a perveance limit of about 0.02 is about all that nature will allow in a space-charge limited converging gun geometry where a waist is desired just beyond the source aperture. The beam would expand due to space charge in a few diameters if it were not captured in a focusing environment. This statement, of course, must be qualified. It applies to cylindrically symmetric geometry, with no magnetic field at the emission surface, and minimal neutralization of the beam.

Although it is beyond the scope of this summary to include a detailed discussion of the design of these sources for low emittance, a few

Table 1. Perveance of some electron and ion sources

	V(kV)	I(amp)	K($\times 10^{-3}$)	$\beta\gamma$
HERA (proton)	18	0.02	5	0.006
Example proton	50	0.2	12	0.010
Example H$^-$	100	0.1	2	0.015
LBL Cs^{+1}	160	0.02	2.4	0.0016
Klystron gun (e$^-$)	300	300	19	1.2
ATA Injector (e$^-$)	2500	10,000	6	5.8

remarks are in order. The emittance of the resulting beam in general depends on the radial uniformity of the beam (at given z), on details of the electrode shapes to reduce aberrations and minimize edge effects, and on the source transverse temperature. Keller, incidentally, finds that the source brightness as defined by I/ε^2 decreases rather dramatically as the voltage on the source is increased, for a wide variety of ion sources. The designer who needs beam current is thus caught in a bind between absolute current and beam brightness. For this reason some applications will require the use of multiple beams to obtain both high current and high brightness. A. Maschke pointed this out in connection with the HIF program some years ago.

In other geometries in which longer distances are involved a much more stringent condition is placed on the maximum perveance. For example the HIF group at LBL has found that the normalized perveance of the heavy ion beam which is to be focused on a fusion target 10 meters from the last lens must be less than about 2×10^{-5}. Calculations indicate that neutralization can relax this limit by perhaps one order of magnitude. In any case rather stringent limits can be found for each application, and use of the normalized perveance allows rapid comparison of different designs where the mass, charge state, and kinetic energy are all variable parameters.

LIMITING SOLUTIONS

We consider now various limiting solutions to Eq. (1). For no external focusing and $K = 0$, $a'' = \varepsilon/a^2$, leading to expansion of the beam envelope along a hyperbolic trajectory from a minimum waist. Similarly, for negligible emittance and no external focusing, $a'' = K/a$, leading to expansion of the beam due to its own space charge along a more complex envelope.

For equilibrium transport in which the external force equals, on the average, the outward pressure of emittance and/or space charge then $a'' = 0$ and again two limiting cases present themselves: emittance dominated, when the ratio of the last two terms, Ka^2/ε^2, is much less than one; and space charge dominated, when this ratio is >>1. The two terms are equal, for example, for a typical ion source ($K = 0.01$), when $\varepsilon = 50$ cm-mrad for a 1 cm diameter beam. Similarly, a 1 mm beam would require that $\varepsilon = 5$ cm-mrad for this condition. For comparison to practical experience, a proton source having a normalized emittance of 0.1 cm-mrad at 100 kV has an absolute emittance of 7 cm-mrad. On the other hand the ATA injector has an absolute emittance of about 50 cm-mrad at 2.5 MeV, so that the terms are equal when the beam diameter is about 1 cm.

It should be mentioned at this point that in real systems the aperture must always be made larger than the ideal calculated beam diameter. We have neglected, for example, the high-frequency ripple associated with AG focusing, the beam halo always present to some degree, and system aberrations.

The solution for the emittance dominated case is well known:

$$a^2 = \varepsilon/(k)^{1/2} = \varepsilon\lambda_0/2\pi = \varepsilon_N\lambda_0/2\pi\beta\gamma \ . \tag{5}$$

Solutions for the space-charge dominated case are not so well known and have been the subject of considerable research during the last decade, starting with the Maschke formula of 1976. The picture which emerges can be summarized as follows: Space charge tends to shield an individual particle from the external focusing and causes an increase in the betatron wavelength of particle motion compared to that which would obtain without space charge. The change in wavelength is conveniently expressed by referring to the phase advance of the sinusoidal motion per period of the external focusing. The phase advance without space charge is defined as

$$\sigma_0 = 2\pi L/\lambda_0 = L(k)^{1/2} \tag{6}$$

where L is the period of the external AG focusing lattice. Thus σ_0 depends only on the lattice period and the focusing strength.

A solution is then required of Eq. (1) which gives the phase advance with space charge included. This has been done by assuming that an individual particle moves within the boundary of a well defined beam with uniform charge density, and corresponds to the assumption of a linear outward force proportional to Kr/a^2. With this smooth approximation a simple solution has been derived for the "depressed" phase advance σ, as follows:

$$(\sigma/L)^2 = (\sigma_0/L)^2 - K/a^2 \ . \tag{7}$$

In the space-charge dominated limit $\sigma_0^2 \gg \sigma^2$ and it follows that

$$a^2 = K(L/\sigma_0)^2 \ . \tag{8}$$

Equations (7) and (8) show that what matters in these beams is the generalized current density, or "perveance density" K/a^2, and that in fact the maximum possible value in this type of linear transport is $(\sigma_0/L)^2$, independent of the emittance.

The final step in this treatment is to eliminate the radius by noting that in general $a^2 = \varepsilon L/\sigma$ on geometric grounds. Substitution in

722

Eq. (7) then yields - in the same limit - the simple relation

$$\sigma/\sigma_o = \sigma_o \varepsilon/KL .$$ (9)

Equation (7) can be written in another form, instructive from the point of view of plasma physics. Substituting $\sigma = L\omega/\beta c$, where ω is the particle angular frequency, one obtains

$$\omega^2 = \omega_o^2 - (1/2)\omega_p^2, \text{ where } \omega_p^2 = 2K(\beta c/a)^2$$ (10)

is the beam plasma frequency. Thus the depression of the transverse oscillation frequency from its its zero-current value is directly related to the beam plasma frequency, which in turn is directly related to the normalized current density.

Another useful parameter, also from Plasma physics, is the Debye length, or "shielding" length. If one defines the Debye length λ_D in terms of maximum transverse velocity v_x as $\lambda_D = v_x/\omega_p$, then substituting $v_x = \varepsilon\beta c/a$ and ω_p from above gives $\lambda_D = \varepsilon/(2K)^{1/2}$. From this it follows that the emittance and space-charge terms in Eq. (1) are equal when $\lambda_D = a/(2)^{1/2}$.

At this point a numerical example of a space-charge dominated beam may be instructive. Assume that a beam from a source is matched and captured into an AG channel and has a normalized perveance of 0.006, based on the observations discusses above, and an absolute emittance of 7 cm-mrad. Further assume that the zero-current phase advance is 90° (four quadrupole pairs per wavelength) and is depressed by a factor of 10 due to space-charge. Equation (9) yields a required L of 18 cm and Eq. (8) gives a beam radius of 0.9 cm. For this case the ratio of the space-charge term to the emittance term in Eq. (1) is 100 and the Debye length is only 0.6 mm. Of course if this beam is accelerated at constant current then the perveance rapidly decreases with energy and space-charge becomes less important. But these considerations clearly show the importance of handling the beam with great care at low energy. Moreover they show the relative magnitude of space charge forces at higher energy.

EMITTANCE GROWTH

Of equally great importance is the subject of emittance growth. For space-charge dominated beams, there was not much progress in the subject until the heavy ion fusion program participants began attacking the problem on three fronts: analytic theory, large-scale computer simulation, and experiment. The principal conclusions are that the growth of emittance is small provided that the zero-current phase shift σ_o is about 90° or less and that the beam density is uniform across the beam, with reasonably sharp edges. In particular, the so-called Single Beam Test Experiment (SBTE) at LBL has set a new standard for such experiments. Although the energy and current are low, what matters is the perveance, which is high in the experiments (LBL Cs+1 in Table 1). The experiments have demonstrated stable, 100% beam transmission through 40 periods (80 quadrupoles) for beams with a phase depression of about one order of magnitude. No emittance growth is observed.

The conclusions stated above apply to rather idealized beams. Effects of lens misalignment, quadrupole shape (higher order terms), image forces, and the bending of such beams are important and are the subject of continuing study. Some details and references may be found in the article by D. Keefe.

Recently, a new theory has been formulated which yields predictions for the emittance growth of space-charge dominated beams. It is based on the idea that the electrostatic field energy present in non-ideal beams is transferred rapidly to particle kinetic energy, heating up the beam and causing emittance growth until an equilibrium is established. The theory is summarized in an accompanying article by Wangler. Although the theory is still relatively new, it appears to give excellent agreement with the few experiments available so far. In particular, the LBL beams with uniform density are predicted to have no emittance growth, in agreement with their results; and the Gaussian-like beams employed at GSI are predicted to have a final emittance ranging up to three times their initial emittance, increasing as the phase advance is depressed. This also is in agreement with the GSI findings. When combined with computer particle simulation, the impact of the theory is impressive and gives considerable credence to the concept of uniform, Debye-edged beams as being ideal from the point of view of stable transport with minimum emittance growth.

BRIGHTNESS

Beam brightness, defined here as proportional to I/ε^2, has served as a useful figure of merit for electron and ion sources for a long time. Lawson's accompanying article gives a more precise definition and includes some discussion of its meaning, and Keller gives semi-empirical formulas for both current and brightness based on a survey of many ion sources. However, as implied by Lawson, it is not at all clear that this definition of brightness has much meaning for particle beam system applications. To the extent that the source brightness determines the ultimate system performance, then it is indeed useful; but in most interesting applications of particle beams, including HIF, FEL's, and linear

724

colliders, the ultimate system performance is a function of several major variables, and the dependence on I and ε cannot be so simply parameterized.

A good example of a different dependence on I and ε can be found directly from Eq. (9). The depressed phase advance given by this formula provides a very direct and meaningful figure of merit for a wide variety of space-charge dominated beams. Since σ_0 and L are constants of the system, or at least slowly varying, the fundamental scaling parameter is K/ε, or I/ε. This interesting result is consistent with the following argument: fixing on the depressed phase advance as a scaling parameter fixes the current density, since they are directly related. This means that K (or I) is proportional to a^2. But the radius in general is $a = (\varepsilon L/\sigma)^{1/2}$ as noted above. Thus K is proportional to ε and the appropriate figure of merit is K/ε.

For highly bunched beams in an RF linac, Eq. (9) has a different form which can be found in the paper by T. Wangler. He reports that an appropriate scaling parameter for RF linacs is approximately $K/\varepsilon^{3/2}$.

For the design of linacs for free electron lasers the situation is also not easily characterized. While a useful figure of merit for the light output of any laser, including a FEL, is the intensity per unit area per unit solid angle, which is proportional to I/ε^2, it does not necessarily follow that the best system performance will result from an optimization of the linac for maximum I/ε^2. It would appear that a more comprehensive characterization should be performed, taking into account the appropriate limitations including space charge, before concluding that a particular combination of linac parameters should be optimized. As pointed out by several speakers at this ASI, the crucial importance of the longitudinal momentum distribution as well as the beam current density must be fully taken into account.

Heavy ion fusion also involves a system which cannot be optimized by simply optimizing I/ε^2. Although this parameter applies to the ion source, the total system involves a fairly complex interplay of about half a dozen key parameters, including K, ε, number of beams (given that more than one is necessary), and momentum spread.

Finally, it should be emphasized again that this summary is meant only to be a primer in the subject. Details may be found in other articles in this ASI proceedings as well as in the Bibliography below.

ACKNOWLEDGMENTS

I am indebted to a number of colleagues who have been working in this field for many years, in particular J. D. Lawson, D. Keefe, and T.

Wangler for recent discussions, to Edward P. Lee for a critical review of the manuscript, and to many others at LBL, NRL, the University of Maryland, LANL, GSI and SLAC for continuing discussions.

REFERENCES

The American Institute of Physics Conference Proceedings No. 139 entitles "High Current, High Brightness and High Duty Factor Ion Injectors" Ed's G. Gillespie, Y. Kuo, D. Keefe and T. Wangler, Series Editor R. G. Lerner (A.I.P., New York, 1986) contains a number of chapters pertinent to this topic. Notable are those by R. Keller on ion sources, M. Reiser on beam transport, C. H. Kim on pertinent portions of the LBL program, I. Haber on computer simulation, and a chapter on the field energy theory by T. P. Wangler, K. R. Crandall, R. S. Mills and M. Reiser. These articles contain numerous references to earlier papers. For example the origin of the beam envelope equation may be found in the chapters by Reiser and by Wangler et al.

With the exception of the more recent work, J. D. Lawson's book, "The Physics of Charged-Particle Beams" (Clarendon Press, Oxford, 1977) is a unique reference in this field. Also, the Proceedings of the 1986 Heavy Ion Fusion Symposium, to be published as an A.I.P. Conference Proceedings, is expected to contain several articles pertinent to this topic.

APPENDIX A: POSTER PAPERS

VACUUM ARC ARRAY ION INJECTOR*

C. Burkhart and S. Humphries, Jr.

Institute of Accelerator and Plasma Beam Technology
University of New Mexico, Albuquerque, NM 87131

ABSTRACT

The development of a novel plasma source for ion extraction is presented. This source utilizes an array of six vacuum metal arcs to form a plasma anode for ion extraction. The vacuum metal arc offers many advantages as a plasma source. By varying the cathode material, a wide range of ions may be produced. The vacuum arc will produce these ions at high fluence for pulse durations ranging from microseconds to dc. Additionally, the use of an array offers advantages over a single arc source. These advantages include, large extraction area ($> 50 \text{ cm}^2$) and reduction of temporal fluctuations. Ion extraction from this source has been achieved using a multi-aperture, three element extractor. Currents of C^+ in excess of 1 A at an energy of 15 kV are typical. Characteristic beam divergence is approximately 100 mrad.

INTRODUCTION

In recent years, the vacuum metal arc (Lafferty, 1980) has received attention as a plasma source for extraction of intense ion beams. Several approaches have been employed to ignite and sustain these arc discharges for plasma formation (Burkhart et al., 1985; Adler and Picraux, 1985; Brown et al., 1985). Differing methods of extraction have also been used (Adler and Picraux, 1985; Brown et al., 1985; Humphries et al., 1986). In all of the work to date however, the plasma source has consisted of a single vacuum metal arc. While the vacuum arc has been clearly demonstrated to be a versatile, high fluence source capable of

---

* Work supported by U.S. Department of Energy under contract DE AC0383 ER13138

long pulse operation, some deficiencies have been encountered. The most prominent of these are the large temporal fluctuations in the plasma density. These fluctuations appear to arise from cathode spot migration (Len et al., 1986). Through the use of an array of parallel arcs these fluctuations can be significantly reduced.

This paper focuses on the development of the vacuum arc array plasma source and its advantages over a single arc. Its performance with three dissimilar cathode materials is documented. Finally, the results of initial extraction experiments are presented.

EXPERIMENTAL APPARATUS

The vacuum arc array ion injector is illustrated in Fig. 1. It consists of the vacuum arc array plasma source and a three element multi-aperture extraction gap. The injector is mounted on a 10^5 cm^3 stainless steel vacuum chamber which is maintained at 10^{-6} torr with a cryo-pump.

The plasma source consists of six vacuum metal arcs uniformly arranged on a 3.8 cm radius. The independent arc cathodes are mounted on insulated stalks which may be moved axially to adjust the arc gap. To facilitate changing cathode materials, the cathode tips are threaded onto the stalks for easy removal. Each annular cathode contains a small Teflon flashover plasma source. These small sources are used to trigger the main gap. All six cathodes share a common stainless steel anode. The anode is recessed around each arc location to provide line of sight shielding between the independent arc locations.

The extractor is multi-aperture, the aperture pattern is illustrated in Fig. 1. Each aperture is of 1.27 mm radius. The aperture spacing is approximately twice that value to assure the beamlets are extracted independently. A three element, accel-deccel configuration is used. The geometric configuration is based on the single aperture optimization of Coupland et al. (Coupland et al., 1973).

The associated electronics for the source are shown schematically in Fig. 2. Dc voltage is applied to the arcs via a six stage pfn. Separate ballast resistors for each arc assure independent operation. The vacuum arcs are fired by discharging a small capacitor across the Teflon insulator housed within each cathode tip. The small burst of plasma formed by this discharge provides a conductive channel from anode to cathode to initiate the arcs. The extraction potential is supplied by a resistively limited, switched capacitor. To facilitate beam diagnostics, the gap cathode is held at ground and the anode, plasma source, and its

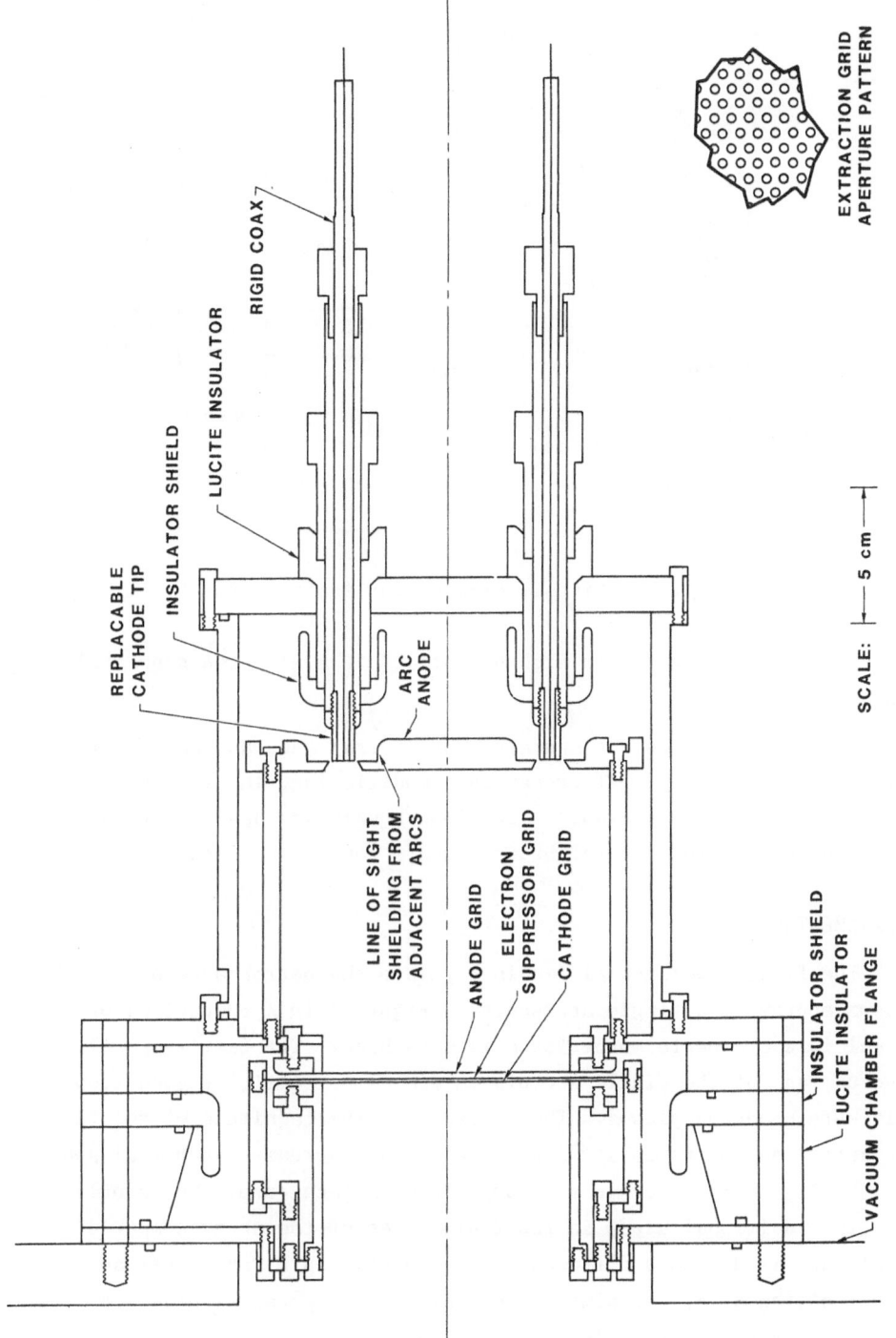

REPLACABLE
CATHODE TIP

INSULATOR SHIELD

LUCITE INSULATOR

RIGID COAX

ARC ANODE

LINE OF SIGHT
SHIELDING FROM
ADJACENT ARCS

ANODE GRID

ELECTRON
SUPPRESSOR GRID

CATHODE GRID

INSULATOR SHIELD

LUCITE INSULATOR

VACUUM CHAMBER FLANGE

EXTRACTION GRID
APERTURE PATTERN

SCALE: |—— 5 cm ——|

Fig. 1.   Vacuum arc array ion injector.

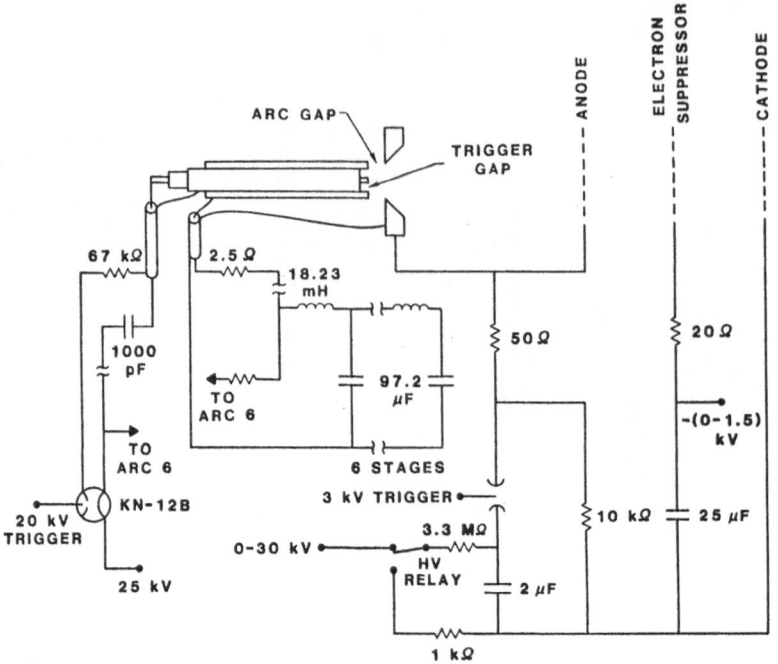

Fig. 2. Circuit diagram for ion injector.

electronics are elevated to high potential. The electron suppression bias is applied dc.

The plasma source has operated for several thousand shots with excellent reliability. After careful conditioning, the extractor has also performed quite reliably. Spurious breakdowns have occurred on only a few percent of the several hundred shots now on the machine.

EXPERIMENTAL RESULTS

The first experimental findings relate the performance of the array plasma source to a single arc source. Figure 3 is a comparison of single vs multiple arc performance for a lead cathode. The fast temporal fluctuations of the single arc plasma are almost entirely gone from the multi-arc produced plasma. The increase in the magnitude of the flux in the array source traces is simply due to the increased number of sources. The driving current levels are comparable in each case. One should note that the additional droop in the flux traces of the array compared to the single arc is due to a difference in respective pfn wave forms and not an effect of the array. A similar smoothing of temporal fluctuations was observed for carbon and aluminum cathodes.

Next, the scaling of extractible ion flux from the plasma with arc current is investigated. Figure 4 is a plot of the data for carbon,

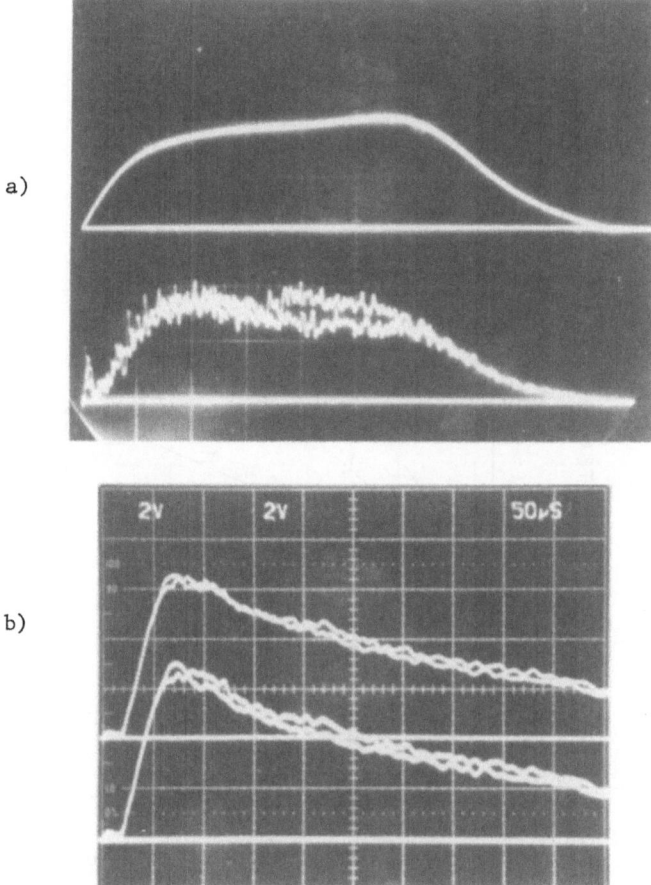

Fig. 3. Comparison of single and multiple arc fluxes: a) Single arc performance, upper trace: Arc current (100 A/div), lower trace: Extractible ion flux (120 mA/cm^2/div), b) Arc array performance, extractible ion flux (90 mA/cm^2/V), upper trace: on device centerline, lower trace: at 1.5 cm radius.

aluminum, and lead. The results indicate a linear relationship in each case. As previously noted, the flux level falls off over the length of the pulse. This is in part due to a similar fall off in driving current. The reduction in ion flux is however greater than that for the current. This results in a mismatch between plasma density and extractor potential which is discussed in the next section.

The final experimental results to be presented are of the extracted beam. Data for a typical 15 kV carbon beam is presented in Fig. 5. The beam current is measured with an ion current density probe. This probe uses an electrostatic trap to prevent the loss of secondary electrons from the collector surface. The measurement is taken 20 cm downstream

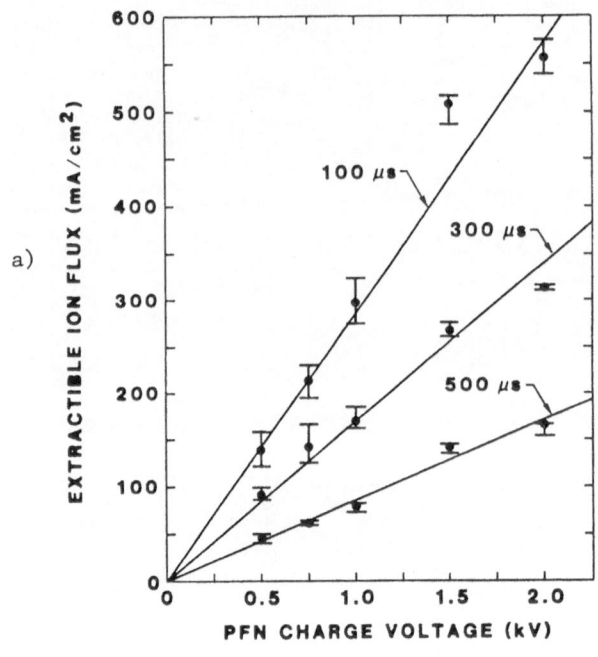

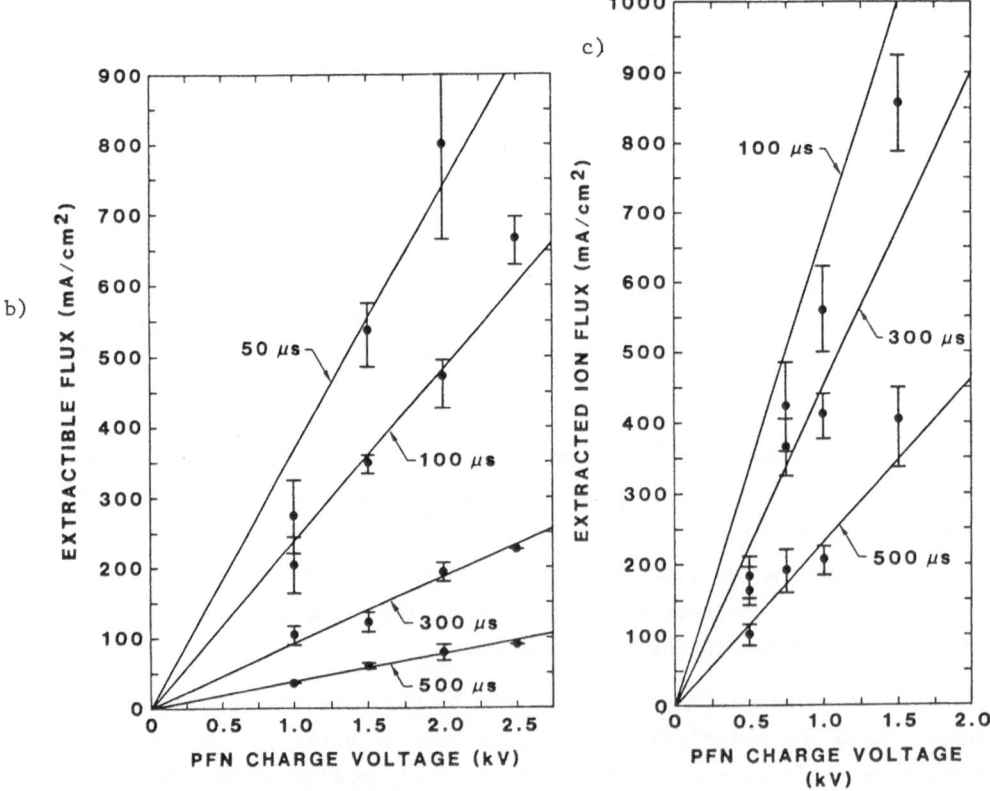

Fig. 4. Scaling of extractible ion flux with arc current: a) lead
         cathode, b) aluminum cathodes, c) carbon cathodes.

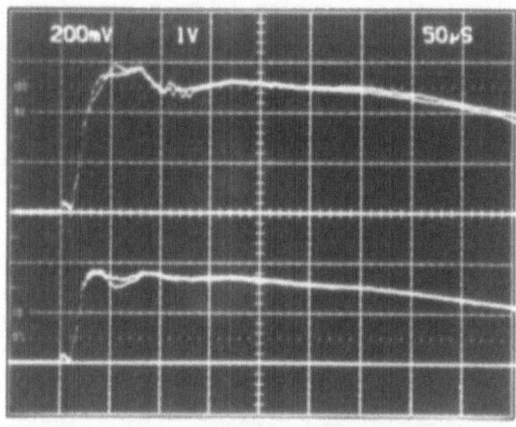

Fig. 5. Typical extraction data, 15 kV, C$^+$ beam upper trace: total gap current (10 A/V), lower trace: total beam current (0.5 A/V).

from the extractor cathode. Comparison of the beam current of nearly one ampere to the gap current indicates a gap efficiency of 15%. It should also be noted that the beam current falls off less rapidly than the available plasma density. This is due to decreasing beam divergence during the length of the shot. The beam divergence is calculated to be 135 mrad early in the pulse. This decreases to 90 mrad near the end of the extraction pulse.

CONCLUSIONS

It has been shown that an array of vacuum metal arcs may be driven in close proximity to produce a versatile plasma source for ion extraction. This array is driven by a single pfn current source, resistive isolation is sufficient to assure independent operation of the individual arcs. This source is capable of supplying an extractible ion fluence in excess of 1 A/cm^2. Further, the bulk of the fast temporal fluctuations in plasma density characteristic of vacuum arcs are eliminated. The extractible flux from this device scales linearly with the driving current. Finally, by changing cathode materials a wide variety of ions may be produced.

Coupling a multi-aperture, three element extractor to this plasma source resulted in successful beam extraction. Beams of 15 kV C$^+$ in excess of 1 A total current are typical. Additional measurements indicate that this system can be extrapolated to higher beam energies and currents.

The main problem with the system is beam matching in the extractor
to achieve a low divergence beam. Although this is in part due to a
droop in the extractor potential as the capacitor is drained, this can be
easily remedied. More important is the drop in the available plasma flux
late in the pulse. Correcting this problem will require a tunable pfn to
boost the current as the pulse proceeds.

REFERENCES

Adler, R. J. and Picraux, S. T., 1985, Nucl. Instum. and Methods, B6:123.
Brown, I. G., Galvin, J. E., and MacGill, R. A., 1985, Appl. Phys. Lett.,
     47:358.
Burkhart, C., Coffey, S., Cooper, G., Humphries, Jr. S., Len, L. K.,
     Logan, A. D., Savage, M. and Woodall, D. M., 1985, Nucl. Instum. and
     Methods, B10/11:792.
Coupland, J. R., Green, T. S., Hammond, D. P., and Riviere, A. C., 1973,
     Rev. Sci. Instrum., 44:1258.
Humphries, Jr. S., Burkhart, C., and Savage, M., 1986, Part. Accel., 17.
Lafferty, J. M., Ed., 1980, Vacuum Arcs: Theory and Application, New
     York.
Len, L. K., Burkhart, C., Cooper, G. W., Humphries, Jr. S, Savabe, M.,
     and Woodall, D. M., 1986, IEEE Trans. Plasma Sci., PS-14:256.

IMPEDANCE VARIATIONS IN THE LOAD OF A THYRATRON-SWITCHED DISCHARGE CIRCUIT

Graeme L. Clark

Department of Physics
University of St. Andrews
St. Andrews, Fife, KY 16 955 Scotland

A discharge circuit capable of running at up to 10 kHz into a time-varying load impedance has been developed. This type of circuit has applications in running discharge heated metal vapour lasers.

COPPER VAPOUR LASER

The copper vapour laser (CVL) is a high repetition rate, high efficiency laser, capable of scaling to powers of over 100W. A longitudinal electrical discharge is used both to heat the active zone to temperatures sufficiently high for the production of Cu vapour (1550°C) and to pump the Cu atoms to the required energy levels for lasing to occur.

Figure 1 shows a schematic outline of the CVL. The thermal insulation is used both to provide a temperature gradient between the discharge zone and the outer walls of the laser and to confine the discharge to the region between the electrodes. Ne is used as a buffer gas to carry the discharge between the electrodes until the discharge wall temperature is high enough for Cu vapour to be produced. Once lasing has started, the Ne aids recombination in the interpulse period as well as the deactivation of the (metastable) lower laser levels of Cu.

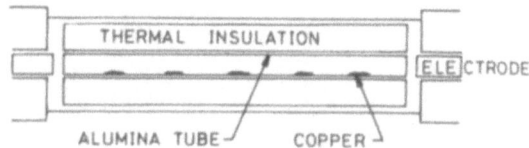

Fig. 1. Schematic outline of CVL.

Figure 2 shows the discharge circuit used to run the CVL. Two types of thyratron have been successfully used: a CX1535 and a CX1625 (both oil-cooled). The results shown in photographs 1 to 8 are from a circuit using the CX1535 thyratron.

IMPEDANCE VARIATION

Photographs 1 to 6 show the evolution of the voltage pulse across the laser as the discharge tube heats up. Photograph 1 is taken 5 minutes after the discharge is switched on. Initially, as the discharge heats up, the impedance of the plasma falls and the voltage pulse narrows. However, as the alumina discharge tube heats up, surface contaminants outgas and reverse the fall in impedance. This causes a larger reflection from the load to the thyratron anode, so the charging voltage on the storage capacitors increases. this increases the peak voltage across the laser (photograph 2). As more contaminants outgas, the discharge pulse widens (photograph 3) and the proportion of power dissipated in the circuit components (such as the thyratron) increases. By flowing the buffer gas through the laser, contaminants are quickly flushed out of the system and so the discharge impedance starts to fall again (photograph 4). At the same time, the current drawn from the power supply drops as the size of the reflection on the thyratron anode falls. In photograph 5, 30 minutes after switching on, the effect of the contaminants has disappeared. The impedance of the plasma continues to fall, however, as the gas temperature rises and Cu vapour starts to diffuse through the discharge. Photograph 6 shows the discharge pulse 50 minutes after switching on as the Cu density reaches the threshold level for lasing to occur.

The presence of contaminants in the discharge increases the impedance of the plasma, so that a significant amount of power must be dissipated by the circuit components rather than in the discharge. Photograph 7 shows the effect on the voltage across the laser. The main

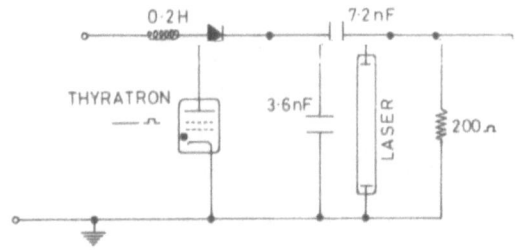

Fig. 2. CVL discharge circuit.

738

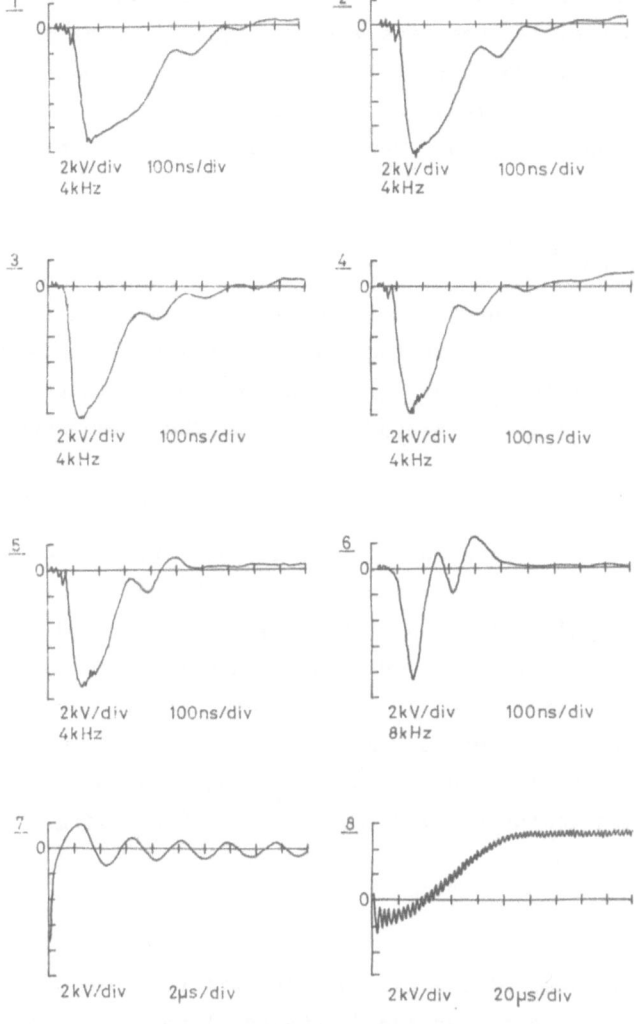

Photographs 1 to 8.

discharge pulse can just be seen in the left hand side of the photograph. the oscillations occupying the rest of the photograph are caused by the bypass inductor – peaking capacitor loop. These oscillations can also be superimposed on the charging voltage of the storage capacitor (photograph 8). Similarly, the height of the oscillations across the laser affects the size of the negative voltage on the thyratron anode.

THYRATRON DISSIPATION

If the reflection from the load becomes too large, dissipation at the thyratron anode can become a serious problem. A large negative voltage appearing at the anode before the thyratron plasma has fully recovered will accelerate ions from the plasma to the anode, causing heating effects. The peak value of the inverse current is proportional to the amplitude of the negative voltage present during this current flow. The peak inverse current is also directly proportional to the peak forward current pulse, since the forward current pulse determines the ion density in the anode region. The power dissipated by this inverse current is given by

$$P_{inv} = k\ e_{px}^2\ i_b\ f$$

where e_{px} is the peak inverse voltage, k is an empirically determined constant which increases with the volume of the grid-anode region in different types of thyratron, i_b is the peak forward current, which is determined by the external circuit, and f is the pulse repetition rate.

CONTAMINANTS

A high purity alumina tube is usually used to confine the discharge in a CVL. The main problem associated with this material is the presence of large amounts of contaminants which require prolonged outgassing procedures for their removal. When a new alumina tube is inserted into the laser, the initially red discharge characteristic of Ne soon becomes white as atmospheric gases outgas from the alumina. A mass spectrometer connected up to the laser detected H, H_2O, N_2, O_2, and CO_2. The concentration of these contaminants decays slowly, with a large variation between alumina tubes.

COMPUTER MODEL

A computer model of the CVL has been written to describe the kinetics of the discharge as the discharge plasma heats up. Figures 3 and 4 show the variation in the voltage across the discharge and the plasma resistance respectively as the gas temperature increases and then as Cu vapour diffuses through the discharge plasma.

The assumptions made about this model are:

1. The electron energy distribution is Maxwellian.

2. The discharge is divided into a hot and a cool zone, with different gas and wall temperatures in each.

3. Only Cu and Ne are included.

4. Diffusion and recombination are assumed to be negligible during the period of interest.

5. The thyratron is modeled by a reverse-biased voltage which decays exponentially. Current reversal in the thyratron is not allowed.

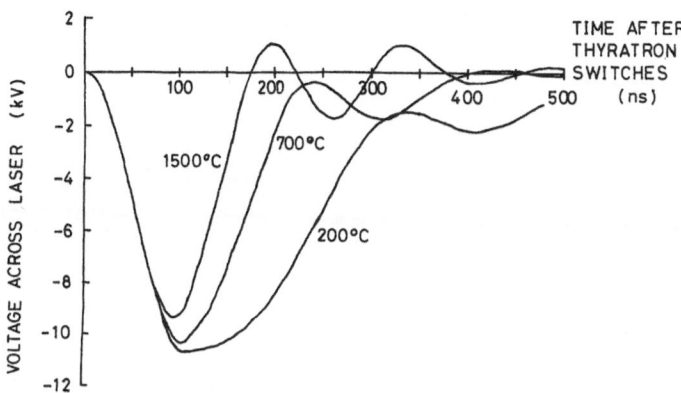

Fig. 3. Variation of discharge voltage with time.

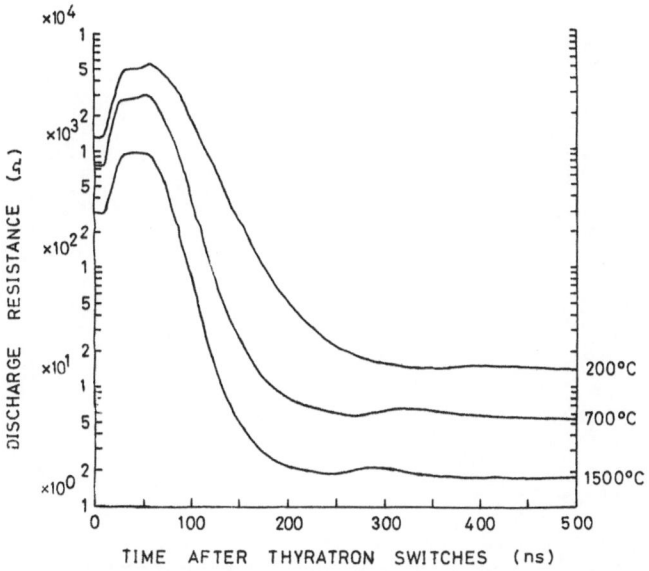

Fig. 4. Variation of plasma resistance with time.

GYRAC, A PHOTON FACTORY?

O. Gal

Laboratoire de Physique des Milieux Ionisés
Laboratoire du C. N. R. S.
Ecole Polytechnique, 91128 Palaiseau Cedex, France

ABSTRACT

The principle of the Gyromagnetic Autoresonant Accelerator (GYRAC) is presented: it is based on the autoresonance which appears if one tries to destroy Cyclotron Resonance conditions by "slowly" increasing the magnetic field. This indeed produces a corresponding variation of the electrons energy which maintains the resonance.

From a simplified model (K. GOLOVANIVSKY 1982, modified), a system of non-linear differential equations are derived, which provide us with the basic mechanisms such as a potential wall for the phase, and yields to interesting limitations such as critical profiles for the magnetic field variation.

A more realistic model is defined (cylindrical mode, paraxial development for the magnetic field, 3-D equations) which allows us to precise physical limitations and to give a numerical example. One can thus design a 12 cm radius photon factory which produces synchrotron radiation of 850 A with 80 MeV electrons (for B_{max} = 5T). Calculations have been made with this model which don't refute the results obtained with the first one. But really important flaws are mentioned which are to be overcome.

PRINCIPLE

Cyclotron Resonance

Let us consider the interaction of an electron rotating in an homogeneous magnetic field $\vec{B}_o$ at the angular frequency

$$\omega = eB_o/m_o\gamma \tag{1}$$

where $\gamma = 1/\sqrt{1 - v^2/c^2}$; e and m_0 are the charge and the rest mass of electron with a perpendicular electric field $\vec{E}_0$ rotating at the constant frequency ω_0 (see Fig. 1). The tuning is achieved at the resonance normalized energy γ_0:

$$\omega_0 = eB_0/m_0\gamma_0. \tag{2}$$

It is well known [see reference (1)] that the electrons whose energy is close to γ_0 are trapped in oscillations: those with $\gamma \gtrsim \gamma_0$ (that is $\omega \lesssim \omega_0$) are decelerated by $\vec{E}_0$ (cf. GYROTRON) and those with $\gamma \lesssim \gamma_0$ are accelerated. This corresponds to oscillations of the phase shift between the momentum and $\vec{E}_0$ around $+ \pi/2$ (non-accelerating phase).

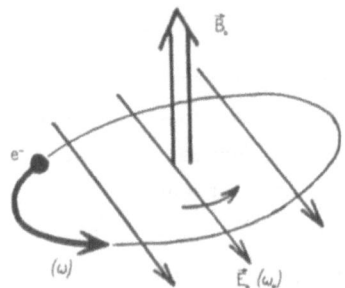

Fig. 1. Cyclotron resonance.

Autoresonance

Let us now "slowly" increase the magnetic field from the initial resonance condition:

$$\vec{B}(t) = \vec{B}_0 [1 + b(t)] \tag{3}$$

As we shall see, that only moves the phase oscillations up to an accelerating phase, so that the synchronism $\omega \approx \omega_0$ is maintained. That means that the energy oscillates now around a new, "slowly" increasing resonance energy

$$\gamma(t) \sim \gamma_0 [1 + b(t)]. \tag{4}$$

It is thus possible, provided that $\vec{B}$ increases more slowly than the critical profile given by the maximum electrical power available, to give energy to the electrons following the arbitrary profile chosen for B(t).

SIMPLIFIED MODEL (K. GOLOVANIVSKY 1982, MODIFIED O. GAL 1986)

Equation

Plane motion only, in homogeneous fields E_0 and $\vec{B}(t)$ is considered (see Fig. 2). It is then described by two variables: the phase shift ψ between the electron's momentum $\vec{p}$ and $\vec{E}_0$, and the electron's normalized energy γ.

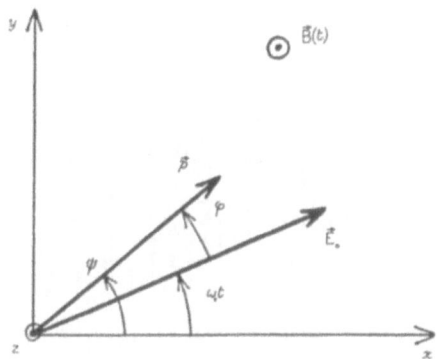

Fig. 2. Simplified model.

The equation of motion written for the components perpendicular to $\vec{p}$ gives the effectives frequency of "rotating" motion:

$$\Omega = \omega_0 \left[\frac{\gamma_0(1+b)-\gamma}{\gamma} + \frac{E_0}{cB_0} \frac{\gamma_0 \sin \psi}{(\gamma^2-1)^{1/2}} \right] = \frac{d\psi}{dt} \tag{5}$$

(ψ is the absolute phase of $\vec{p}$ see Fig. 2). The equation for ψ can be easily derived:

$$\frac{d\psi}{dt} = \omega_0 \frac{\gamma_0(1+b)-\gamma}{\gamma} + \frac{E_0}{cB_0} \omega_0 \frac{\gamma_0 \sin \psi}{(\gamma^2-1)^{1/2}} . \tag{6}$$

Taking account of the betatron acceleration due to the electric field induced by $\vec{B}(t)$, and synchrotron damping (simply expressed with Ω and γ), the equation for parallel components becomes:

$$\frac{d\gamma}{dt} = \frac{E_0}{cB_0}\, \omega_0 \gamma_0\, (1-1/\gamma^2)^{1/2} \cdot \cos\psi + \frac{1}{2}\, \gamma_0\, (1-1/\gamma^2)\, \frac{db}{dt}\, \frac{\omega_0}{\Omega}$$

$$- \tau\, \omega_0^2\, \gamma^2\, (\gamma^2 - 1)\, \left(\frac{\Omega}{\omega_0}\right)^2$$

(7)

where Ω/ω_0 is given by (5) and $\tau = e^2/6\pi\, \varepsilon_0\, m_0\, c^3 = 6.252.10^{-24}$ s

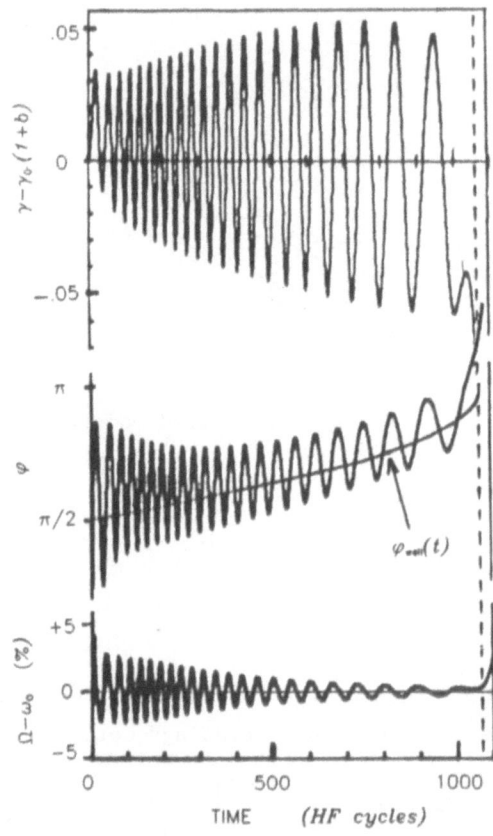

Fig. 3a. Autoresonance in simplified model. The autoresonant regime is cut-off when $db/dt \sim db_{crit2}/dt$.

These two equations can be numerically solved. The results (see Fig. 3.) show clearly the oscillations of Ω around ω_0, those of γ around $\gamma_0(1+b)$ and those of ψ around the accelerating phase $\gamma_{well}(t)$.

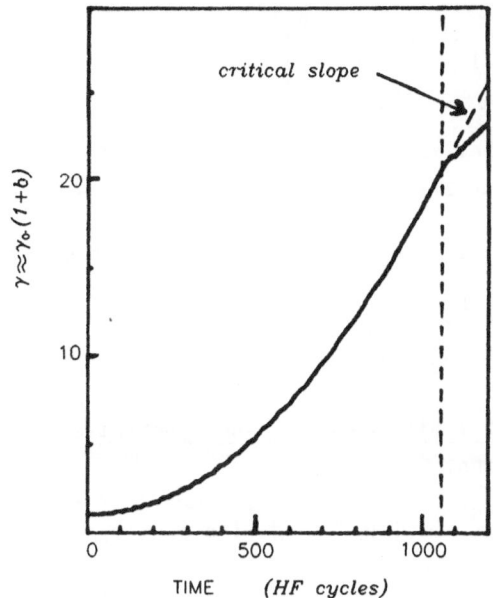

Fig. 3b. Autoresonance in simplified model. The autoresonant regime is cut-off when $db/dt \sim db_{crit2}/dt$.

$$\left[\omega_o/2\pi = 2.4 \text{ GHz}, \quad g_o = \frac{E_o}{cB_o} = 3.10^{-3}, \right.$$

$$\left. b(t) = 0.1 \times \frac{(g_o \omega_o t)^2}{2}, \quad \gamma_o = 1 \right]$$

Critical Profiles for $B_z(t)$

As said above, the finite electrical power available to accelerate the electron limits the slope of the b(t) function if one wants to keep the autoresonant regime. The critical profile $b_{crit}(t)$ is given by the equation:

$$\gamma_o \frac{d}{dt} b_{crit} = \left(\frac{d\gamma}{dt} \right)_{max} \tag{8}$$

where the maximum is obtained from (7) with $\cos \psi = -1$, and γ is replaced according to (4).

Case of $\gamma \approx 1$ (with $q\vec{E}_o$ force only): one takes $\gamma_o = 1$, and Eq. (8) gives:

$$b_{crit1}(t) = \left[1 + \left(\frac{E_o}{cB_o} \omega_o t \right)^2 \right]^{1/2} - 1. \tag{9}$$

This shows that for low energy electrons ($\gamma \approx 1$) an horizontal initial tangent for B(t) is necessary (see Fig. 4).

Case of $\gamma^2 \gg 1$ with betatron term this gives simply:

$$b_{crit2}(t) = 2 \frac{E_o}{cB_o} \omega_o t. \tag{10}$$

Case of $\gamma^2 \gg 1$ with betatron and synchrotron terms integrating Eq. (8), one obtains:

$$4\tau \omega_o^2 \gamma_o^3 b_\infty^3 t = [th^{-1} + tg^{-1}](b_{crit3}/b_\infty) \tag{11}$$

where: $b_\infty = \frac{1}{\gamma_o} \left(\frac{\gamma_o E_o/cB_o}{\omega_o \tau} \right)^{1/4}$ (see Fig. 4) $\tag{12}$

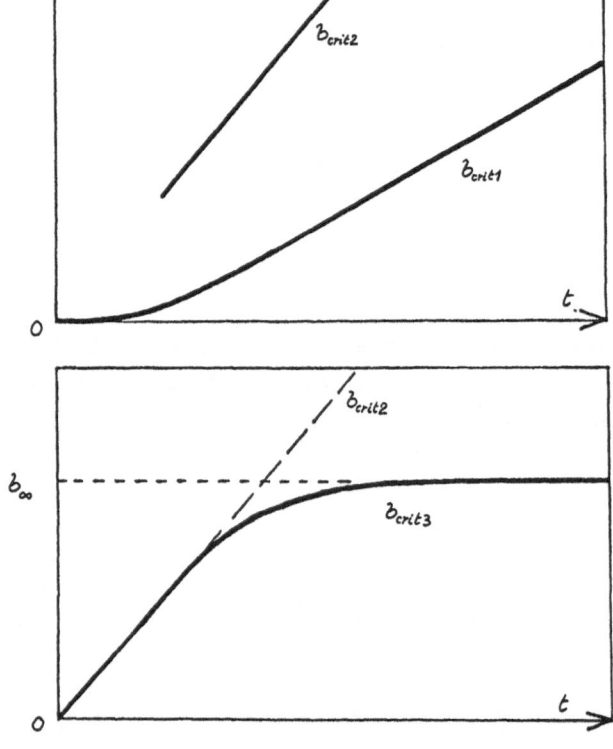

Fig. 4. Critical profiles for b(t) (in arbitrary units).

Limit for γ

The limit b_∞ of $b_{crit3}(t)$ corresponds to a limit for γ:

$$\gamma_{\ell im} = \left(\frac{\gamma_0 E_0/cB_0}{\omega_0 \tau}\right)^{1/4}.$$ (13)

When γ approaches this value, synchrotron damping tends to balance exactly the maximum electrical power.

Potential Well for the Phase

With the resonance conditions for high γ

$$\gamma_0(1 + b) - \gamma \ll \gamma \text{ and } \gamma^2 \gg 1,$$

γ can be eliminated from the differentiated Eq. (6), by using Eq. (7) without synchrotron term, which gives:

$$\frac{d^2\psi}{d(\omega_0 t)^2} = -\frac{\delta U}{\delta \psi}$$ (14)

where:

$$U(\psi, t) = -\frac{1}{2}\frac{db/d(\omega_0 t)}{b}\psi - \frac{E_0/cB_0 \cdot \sin \psi}{b}$$ (15)

(see Fig. 5), assuming $db/dt = \text{const}$).

This potential is locally minimum at:

$$\psi_{well} = \cos^{-1}\left(-\frac{db/dt}{2\omega_0 \cdot E_0/cB_0}\right)$$ (16)

provided that $db/dt < db_{crit2}/dt$. When db/dt varies with time, $\psi_{well}(t)$ is a good guiding function for $\psi(t)$ (see Fig. 3).

REALISTIC MODEL

The rotating electric field is performed by a stationary TE11 mode in a cylindrical resonator. The magnetic field B_z is a magnetic mirror (to confine the z-motion) calculated with its B_r component according to a paraxial development. The slow growth of B induces an electric field E_ϕ.

Dimensionless Parameters

The main quantities may be expressed through five dimensionless parameters with high physical significance:

$$g_o = \frac{E_o}{cB_o}$$ compares electrical force with initial magnetic force;

$$\alpha = \omega_o/\omega_{cut-off}$$ security factor with respect to the cut-off frequency ($\alpha = \omega a/j'_{11}c$ where a is the cavity's radius, $j'_{11} = 1.84118$);

$$\gamma_o = eB_o/m_o\omega_o$$ initial resonance energy; $\qquad$ (17)

$$\gamma_{\ell im} = (g_o\gamma_o/\omega_o\tau)^{1/4}$$ limit imposed by synchrotron damping;

$$\gamma_{max} = \gamma_o\frac{B_{max}}{B_o}$$ energy reached at the end of the growth, when $B_z = B_{max}$

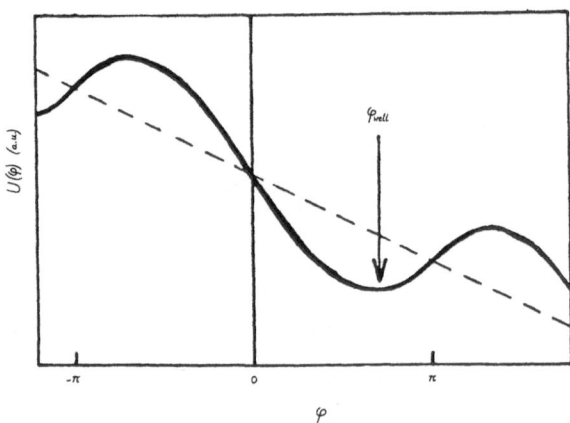

Fig. 5. Potential well for the phase (drawn for db/dt = const).

The concrete parameters which can be derived are the following:

ω_o $\qquad$: frequency of the TE11 mode;

a and 1 : radius and length of the cavity

$$\text{(with } 1/a = \frac{\pi}{j'_{11}} \frac{1}{(\alpha^2-1)^{1/2}}\text{)};$$

B_o $\qquad$: nominal initial B_z; $\qquad$ (18)

E_o $\qquad$: nominal rotating $E_\perp$;

B_{max} $\qquad$: nominal B_z at the end of the growth.

Equation (7) gives also the synchrotron power per electron produced when $B_z = B_{max}$:

$$P_S = \tau\, m_o\, c^2\, \omega_o^2\, \gamma_{max}^4. \tag{19}$$

This radiation is essentially emitted around the wavelength:

$$\lambda_S = \frac{2\pi\, c}{\omega_o\, \gamma_{max}^3} \tag{20}$$

Scaling

From Eqs. (17) to (20), a simple relation can be easily derived which clearly show where are the constraints limiting the choice of parameters. From the designer's point of view, the most significant parameters are B_{max} and a. So the fundamental relation is

$$\gamma_{max} = 3.18 \cdot \frac{1}{\alpha} \left(\frac{a}{1\,cm}\right)\left(\frac{B_{max}}{1\,T}\right), \tag{21}$$

which shows that is will be really difficult (or expensive) to exceed few hundreds for γ_{max}.

The synchrotron radiation is then characterized by

$$P_S[W/el] = 1.608 \times 10^{-13} \cdot \frac{1}{\alpha^2} \left(\frac{a}{1\,cm}\right)^2 \left(\frac{B_{max}}{1\,T}\right)^4$$

$$\lambda_S\,[\overset{o}{A}] = 1.057 \times 10^7 \cdot \alpha^2 \left(\frac{a}{1\,cm}\right)^{-2} \cdot \left(\frac{B_{max}}{1\,T}\right)^{-2}. \tag{22}$$

One has then to choose the electric field's amplitude and the injection energy, that is g_o and γ_o. The electric field amplitude is indeed given by

$$E_o\,[kV \cdot m^{-1}] = 9.420 \times 10^4 \cdot \alpha\, \gamma_o\, g_o \left(\frac{a}{1\,cm}\right)^{-1}. \tag{23}$$

Of course, γ_{max} has to be lower then γ_{im}, which is precised through

$$\left(\frac{\gamma_{\ell im}}{\gamma_{max}}\right)^4 = 2.816 \times 10^{10}\, g_o\, \gamma_o\, \alpha^3 \left(\frac{a}{1\,cm}\right)^{-3} \left(\frac{B_{max}}{1\,T}\right)^{-4}. \tag{24}$$

Time scaling to reach γ_{max} can be evaluated with the critical profile $b_{crit2}(t)$:

$$\Delta t_{crit}[\mu s] = 9.053 \times 10^{-6}\, (\gamma_{max} - \gamma_o) \cdot \frac{1}{\gamma_o}\, \frac{1}{\alpha}\, \frac{1}{g_o} \left(\frac{a}{1\,cm}\right). \tag{25}$$

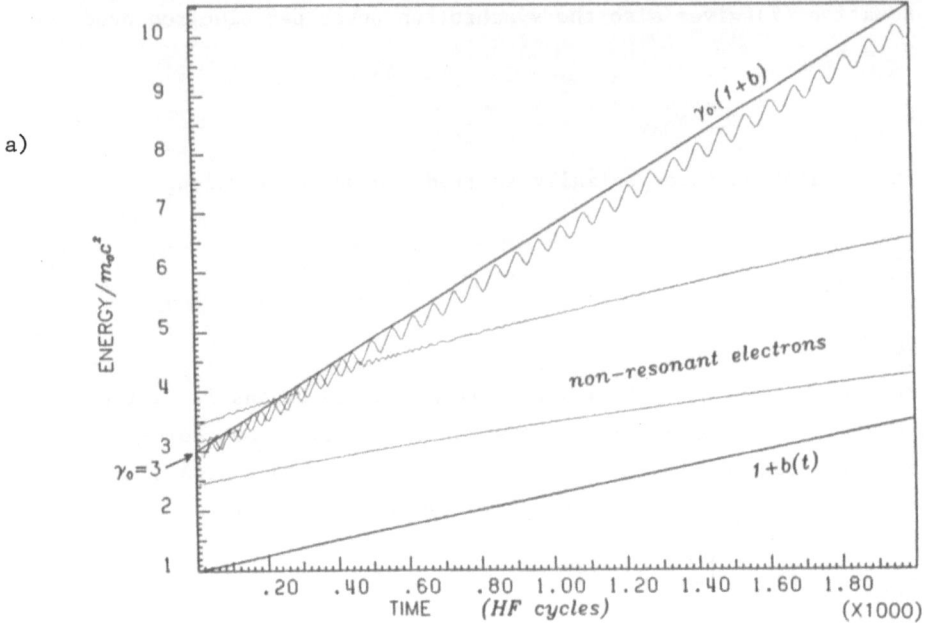

a)

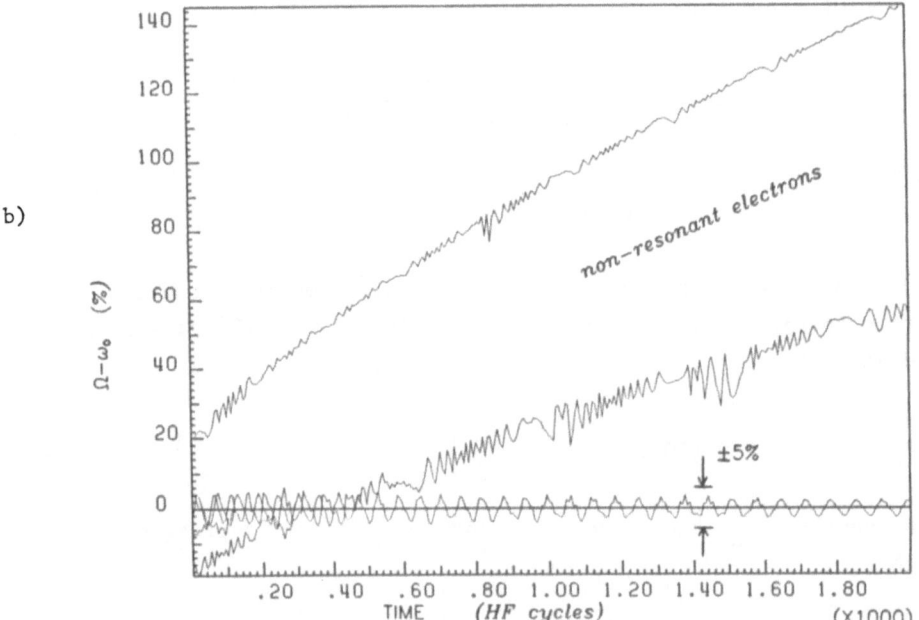

b)

Fig. 6. Autoresonance in realistic model. Injection is not studied. Five electrons are injected on the Larmor radius corresponding to γ_0 with different initial γ around γ_0 (calculated with data Eq. (26); $b(t) = 0.1 \times b_{crit2}$).

Numerical example

$a = 12\text{cm}$, $\alpha = 1.2$ → $1 = 31\text{cm}$, $\omega_0/2\pi = 0.88$ GHz

$$B_{max} = 5T \qquad \rightarrow \quad \gamma_{max} = 160 \text{ (80MeV)}, \ P_S = 10^{-8}\text{W/el}, \ \lambda_S = 850\text{\AA} \qquad (26)$$

$$\gamma_0 = 3 \text{ (1 MeV)} \qquad \rightarrow \quad B_0 = 940 \text{ G}$$

$$g_0 = 1.10^{-3} \qquad \rightarrow \quad E_0 = 28 \text{ kVm}^{-1}, \ \Delta t_{crit} = 5 \text{ }\mu s, \ \frac{\gamma_{\ell im}}{\gamma_{max}} = 3.4$$

If we consider 10^{11} resonant electrons, that is a 14 A current, the total synchrotron power over the whole trajectory is 1 kW.

RESULTS AND CONCLUSIONS

We have performed numerical integration of the 3D equations of motion in that more realistic model.

As one can see on Fig. 6, the results are really encouraging:

· non-homogeneous fields don't destroy the autoresonant mechanism;

· the movement in z-direction which induces variation of B_z seen by the electron is not fatal for the resonance;

· the results are close to those obtained with the simplified model.

But we have to mention fundamental flaws in this work:

· the problem of injection is not studied yet: electrons are simply put ex nihilo in the middle of the box! This question could be a real difficulty in the future;

· a serious study has to be done to know what happens to the fields in the cavity with the radiations due to the electrons (it will be necessary to precise the synchrotron spectrum and to consider a higher number of modes);

· one will have also to evaluate the space charge effects, which could be nevertheless weak enough (see the numerical example).

REFERENCES

Gal, O., 1985, Report PMI 1616 (Ecole Polytechnique).
Golovanivsky, K. S., 1980, Phys. Scripta, 22, 126.
Golovanivsky, K. S., 1983, IEEE Trans. on Pl. Sc., PS-11(1), 28.
Golovanivsky, K. S., 1984, Phys. Letters, 100A(7), 357.
Roberts, C. and Buchbaum, S., 1964, Phys. Rev. 135, 381.

THE SERC PLASMA BEAT WAVE EXPERIMENT

T. Garvey*, A. E. Dangor, A. K. L. Dymoke-Bradshaw,
A. Dyson, and I. Mitchell

Blackett Laboratory
Imperial College of Science and Technology
South Kensington, London, U.K.

A. J. Cole, C. N. Danson, C. B. Edwards, and R. G. Evans
Rutherford Appleton Laboratory

The possibility of using laser generated beat waves in an underdense plasma to create high accelerating fields was first proposed by Tajima and Dawson. This paper describes the aims of an IC/RAL collaborative experiment, funded by the SERC and materially supported by CERN, to demonstrate the excitation of such a longitudinal plasma wave and to measure the temporal evolution of the wave. Results of a recent experiment are presented and plans for a future experiment are briefly outlined.

INTRODUCTION

Over the last fifty years the accelerator energy available to the high energy physics (HEP) community has increased by eight orders of magnitude while the cost (taking account of inflation) has only increased by a factor of ten (Johnsen, 1985). However, this trend is now beginning to saturate and as a result of technological and economic limits conventional accelerator techniques may not be able to provide the required beam energies necessary for significant advances in HEP. This situation has resulted in a number of proposals for novel accelerator (Proceedings of the ECFA-RAL Meeting, 1982). Tajima and Dawson have proposed using the ponderomotive force of two co-propagating laser beams, at frequencies

---

* Present address: LEP Division
CERN
Geneva, Switzerland

ω_1 and and ω_2, to drive a large amplitude Langmuir wave (Tajima, 1979; Dawson, 1979). The wave is excited in a plasma which has a density such as to make it resonant with the beat frequency of the laser beams, i.e.,

$$\omega_p = \omega_1 - \omega_2$$

where ω_p is the angular electron plasma frequency. (Strictly speaking, this should be the angular Bohm and Gross frequency but for the high phase velocity plasma wave the difference is negligible.)

Experiments on beat wave generation have already taken place at UCLA (Clayton et al., 1985) and INRS (Ebrahim et al., 1985) using the CO_2 laser rotational bands at 9.6 µm and 10.6 µm. As the phase velocity of the beat wave, V_{ph}, is characterized by a Lorentz factor, γ, where

$$\gamma = (1 - V_{ph}^2/c^2)^{-1/2} \simeq \omega_1/\omega_p ,$$

the beat waves in these experiments had a γ factor of ten. Following the report of the Rutherford Appleton Laboratory (RAL) beat wave study group (Lawson, 1983) the SERC (Science and Engineering Research Council) approved funding for a joint Imperial College/RAL experiment. This experiment uses the RAL Nd glass laser (VULCAN) configured to produce radiation of 1.064 µm and 1.053 µm thus providing a $\gamma \simeq 100$. The higher γ is thus complementary to those of the UCLA and INRS experiments and is of greater interest to the accelerator community. A plasma beatwave experiment has already been performed at RAL and a second experiment is being planned.

EXPERIMENTAL PARAMETERS

The laser frequencies are determined by the crystalline oscillator materials YLF and YAG which produce wavelengths of 1.053 µm and 1.064 µm respectively. The resulting resonant density of $1.1 \times 10^{23} m^{-3}$ ($\omega_p = 1.8 \times 10^{13} s^{-1}$) is conveniently produced by a Z-pinch hydrogen plasma (2 kV, 5 kJ) constructed by the Imperial College group and diagnosed as being stable with a lifetime of a few microseconds. The plasma so pro-duced exhibits a uniform density, to within 5%, over a radial distance of 5 mm. The experiment requires laser energies of $\simeq 100$ J at each wave-length in a pulse width of approximately 300 ps. With a 3-metre focusing lens the spot size is limited by the irreducible divergence of the beams to $\simeq 300$ µm resulting in an irradiance of $\gtrsim 3 \times 10^{18}$ W m^{-2}. The laser beams are incident perpendicular to the axis of the pinch and have a Rayleigh length of approximately 3 mm in the focused waist of the beams.

Theory predicts that the linear growth rate of the plasma density perturbation ε (= $\delta n/n$) is given by (Rosenbluth, 1972; Lin, 1972)

$$\frac{d\varepsilon}{dt} = \frac{\omega_p \alpha_1 \alpha_2}{4}$$

where α_1 and α_2 are the ratio of the electron quiver velocities to the speed of light at the frequencies ω_1 and ω_2 respectively. Relativistic detuning, due to the mass increase of the accelerated plasma electrons, is predicted to limit the growth of ε to a saturation value, ε_s, given by

$$\varepsilon_s = \left(\frac{16}{3}\alpha_1\alpha_2\right)^{1/3}$$

For the parameters of the IC/RAL experiment the values of α_1 and α_2 should reach 0.016 and ε_s should be limited to 10% in a saturation time of 90 ps. The use of optical streak cameras with a resolution of 10 ps will allow the growth of the beam wave to be studied. It is of interest to note that even if ε is limited to 0.1 the resulting Langmuir wave amplitude, E_p, which is given by

$$E_p = \frac{\omega_p c \varepsilon}{\eta},$$

(c = speed of light, η = charge to mass ratio of the electron) will still be 3 GVm^{-1}.

EXPERIMENTAL MEASUREMENTS

The excitation of the plasma beat wave and its subsequent growth result in a forward Raman scattered spectrum of lines separated by the plasma frequency. These lines are easily resolved by the use of a spectrometer in the transmitted light path. The growth of this "cascade" spectrum will be observed with a streak camera optically coupled to the spectrometer and the ratio of successive cascade amplitudes should provide a means of calculating the beat wave amplitude.

The temporal development of the plasma density and temperature can be measured by Thomson scattering of a probe beam at the first harmonic of the YLF line (0.53 μm). Novel multi-angle collection optics employing the use of retro-reflecting mirrors allow the detection of light from two coupled spectrometer and streak camera as for the cascade channel. This allows the signals appropriate to a range of α values to be recorded [α is the normal Thomson scattering parameter (Evans, 1969; Katzenstein, 1969)]. Measurements on the rate of plasma heating allow one to determine the time for which the beat wave is useful for acceleration before the wave energy transfers to thermal energy via turbulence.

EXPERIMENTAL RESULTS

The 1986 annual report to the Central Laser Facility (CLF) committee contains a more detailed description of the beat wave experiment than is given here (RAL annual report to the Central Laser Facility Committee, 1986). In addition the required oscillator development and results are discussed. A brief summary of the experimental results will be presented here.

The first attempt, in January, 1986, to measure the cascade signal due to the plasma wave was foiled due to the unfortunate coincidence of the beat frequency of the pump waves (98.2 cm^{-1}) with the S(11) rotational line of nitrogen of 99 cm^{-1}. The simultaneous propagation of the two laser wavelengths through the long air path between the laser output and the plasma tube gave rise to strong Stokes and anti-Stokes sidebands due to Raman scattering in the atmosphere, thus making it impossible to detect any plasma beat wave.

However, encouraging results were obtained from the Thomson scattering channels during a brief experiment on multiphoton ionization. This process is of importance to the beat wave project as it offers the possibility of producing a longer uniform column of plasma than is available from a Z-pinch. Clearly the prospects of using plasma beat waves for acceleration to high energy depend of sufficiently long plasmas being available. In this experiment the infra-red beams were not used and the pinch tube was simply filled with hydrogen gas at various pressures between 0.5 torr and 4 torr. The green probe beam was then used to create the plasma (by multiphoton ionization) and as a probe for scattering. The results indicate a rapidly formed plasma whose density and temperature are constant during the laser pulse width. The dependence of the plasma density and temperature on fill pressure and laser irradiance is discussed in the report to the CLF committee. More recently a further study of multiphoton ionization involving spatially resolved measurements has illustrated that the plasma column is at least 12 mm in length.

CONCLUSION

Due to the problem of Raman scattering in nitrogen it is clear that a different experimental arrangement will be required for a second beat wave experiment. It is hoped that the generation of each laser line can be done in separate amplifier trains and the two waves then combined just prior to entering the plasma vessel. In addition the success with multiphoton ionization studies may obviate the need for the Z-pinch in future studies. All previous diagnostic channels will be retained and, in addition, a Raman shifted probe beam will be used to measure the beat wave amplitude by small angle Raman-Nath scattering.

REFERENCES

Clayton, C. E., et al., 1985, Phys. Rev. Lett., Vol. 54, p. 2343.

Ebrahim, N. A., et al., 1985, IEEE Trans. Nucl. Sci., Vol. NS-32, p. 3539.

Evans, D. E. and Katzenstein, J, 1969, Rep. Prog. Phys., Vol. 32, p. 207.

Johnsen, K., 1985, CERN Accelerator School, held in Paris, France, Sept. 84, CERN Report 85-19.

Lawson, J. D., 1983, Rutherford Appleton Laboratory Report, RAL-83-057.

Proceedings of the ECFA-RAL Meeting, held at New College, Oxford, Sept. 1982.

RAL annual report to the Central Laser Facility Committee (1986).

Rosenbluth, M. N. and Liu, C. S., 1972, Phys. Rev. Lett., Vol. 29, p. 701.

Tajima, T. and Dawson, J. M., 1979, Phys. Rev. Lett., Vol. 43, p. 267.

SCALING OF CURRENT DENSITY, TOTAL CURRENT, EMITTANCE, AND
BRIGHTNESS FOR HYDROGEN NEGATIVE ION SOURCES

J. R. Hiskes

Lawrence Livermore National Laboratory
University of California
Livermore, California 94550

ABSTRACT

The atomic and molecular processes that play a principal role in
negative ion formation in a hydrogen negative ion discharge are
discussed. The collisions of energetic electrons with gas molecules
within the discharge lead to vibrationally excited molecules. Thermal
electrons in turn attach to these excited molecules and generate negative
ions via the dissociative attachment process. A system geometry chosen
to optimize these collision processes is discussed that consists of a
high-power discharge in tandem with a low electron temperature bath, the
two regions separated by a magnetic filter. The current density
extracted from such a system is found to scale inversely with the system
scale length provided the gas density and electron density are also
increased inversely with scale length. If a system is scaled downward in
size to provide a new beamlet but one with increased current density, and
these beamlets are packed to fill the original dimension, the new total
extracted current will exceed the original total current by the scale
factor. The emittance, ε, of the new system remains unchanged. The
brightness, J/ε^2, of the new system will also be increased in proportion
to the scale factor.

INTRODUCTION

The development of volume-type negative hydrogen ion sources has
reached a sufficiently advanced stage that one can anticipate sources
producing large total currents with moderately large current densities.
It follows then that the scaling properties of the different beam
qualities, i.e., current density, total current, emittance, and
brightness, should be of some general interest. In this paper a summary

of the system geometry and source operation is presented together with a discussion of the scaling rules for these beam qualities.

SYSTEM CONCEPT

A schematic of a tandem two-chamber system is shown in the first figure. The present-day working hypothesis for negative ion formation adopts a two-step process of vibrational excitation of the background H_2 molecules, occurring in chamber 1, followed by low-energy electron attachment to these molecules to form negative ions in the second chamber (Hiskes and Karo, 1982). The first chamber is a high-power, high-density discharge with a relatively large fraction of energetic (E > 25 eV) electrons that generate $H_2(v")$ via singlet electronic excitation. The second chamber is a low-electron-temperature bath (kT ~ 1 eV) suited to maximize negative ion formation. The magnetic filter configuration separating these chambers functions so as to isolate the two electron energy distributions. In addition to the direct electron excitation process, the Auger neutralization of excited $H_2(^3\Pi_u)$ molecules or hydrogen molecular ions incident upon the chamber walls can be an important source of vibrational excitation. These molecular configurations are also generated by the high-energy electrons. Under certain conditions the $H_2(^3\Pi_u)$ source can dominate the $H_2(v")$ production in low-density discharges (Hiskes and Karo, 1982), and the H_2^+ source can dominate $H_2(v")$ production for H_2^+-rich discharges (Hiskes and Karo, 1983). For high-density high-current extraction sources however, we are concerned principally with the singlet excitations.

An equilibrium population distribution spanning all fourteen vibrational levels, v", of the ground electronic state vibrational spectrum is shown in Fig. 2. This vibrational distribution is calculated for a first chamber electron density of 3×10^{12} electrons cm^{-3}, and a gas density of 3×10^{14} molecules cm^{-3}. In the upper distribution the atomic density is set equal to zero, in the lower distribution the atomic density is set equal to the molecular density. The difference arises because of the induced vibrational relaxation caused by atom collisions.

In addition to the fore-mentioned excitation process, several other collision process contribute to vibrational relaxation. For the gas densities of interest in ion source work, typically less than 3×10^{16} molecules cm^{-3}, the V-V relaxation process can be ignored and the set of fourteen coupled rate equations that describe the equilibrium population distribution reduce to a set of linear coupled algebraic equations. These equations can be written in matrix form as

$$AN = -K ,$$

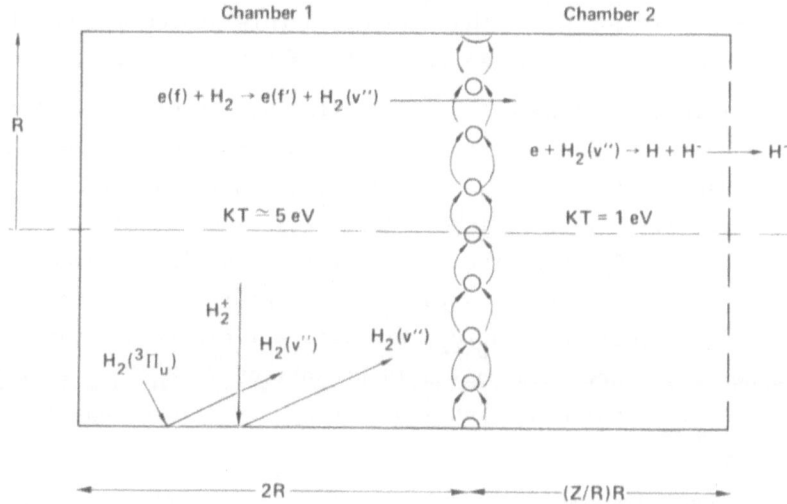

Fig. 1.   Schematic diagram for a tandem negative ion generator.

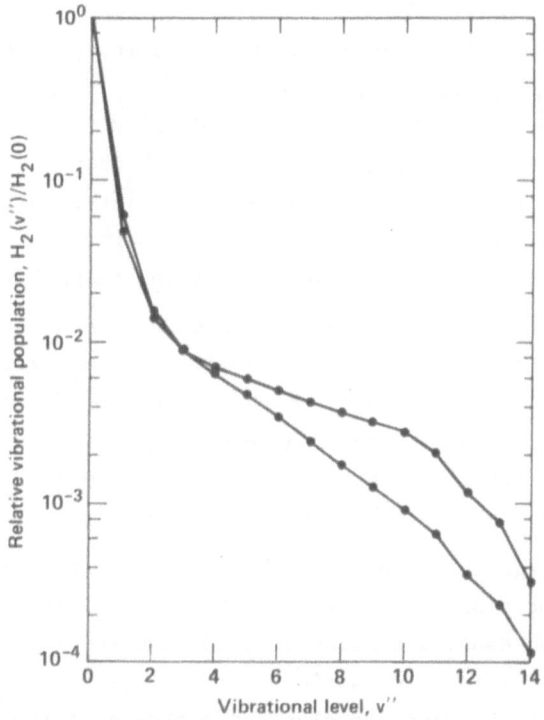

Fig. 2.   Vibrational population distributions for a high density
system: upper distribution, no atomic relaxation;
lower distribution, atomic concentration equal to
molecular concentration.

with solution $N = A^{-1} K$. The N is a column matrix whose elements, $N(v")$, are the populations of the respective vibrational levels, $v"$. The elements of the A matrix are the relaxation rate processes, and the K matrix is a column matrix whose elements are the excitation source terms.

If the vibrational population densities have been obtained for an ion source geometry with scale length R_0, then one can predict the population densities for a system with a different scale length, R_b, using a simple scaling law (Hiskes, Karo, Willmann, 1985). If for simplicity we take R_0 to be the radius of a cylindrical discharge, then for a new and smaller scale length R_b all elements of the product $A^{-1}K$ will scale in proportion to R_0/R_b, provided all electron, atom, and molecule densities are also scaled in proportion to R_0/R_b. It follows then that the population of each vibrational level is enhanced by the factor R_0/R_b.

In the second chamber the solution for the negative ion density, measured as a function of position along the chamber in units of the discharge radius R, has the form

$$N^-(Z/R) = N(v")F \left[n \left(\frac{Z}{R} \right) R \right],$$

where n represents either the electronic, atomic, or molecular density (Hiskes, Karo, Willmann, 1985). Under this scaling process the function F remains unchanged and the negative ion density scales in proportion to the vibrational population, $N(v")$. Hence both chambers scale in the same way, and the negative ion density at a particular aspect ratio Z/R scales in proportion to R_0/R_b. Since the extracted negative ion current density is proportional N^-, the current density scales as R_0/R_b.

Having obtained the scaling for the current density the scaling of the remaining beam qualities follows easily. Upon scaling a system of initial radius R_0 and with total extracted current J_0 to a smaller beamlet system R_b, and then packing several beamlets to fill the initial cylinder R_0, the new total current is increased to $J_b = J_0 f R_0/R_b$, where f is the packing fraction.

In a discharge the negative ion velocities are determined by the initial gas temperature and the dispersion of the dissociative attachment process. For the density scaling assumed here the gas temperature will not change in a first approximation. If the scaling and packing is accomplished while holding the electron energy distribution constant, then the resulting negative ion velocity, $v(-)$, and hence the emittance, $\varepsilon = R_0 v(-)$, are unchanged. The brightness, $B = J_b/\varepsilon^2$, derived from the above quantities, is seen to scale inversely with R_b.

CONCLUSIONS

The emittance remains invariant but the other three qualities, current density, total current, and brightness, are enhanced under the scaling process. A cluster of beamlet sources appears as a more attractive system than a single large cylindrical source.

ACKNOWLEDGMENT

This work was performed under the auspices of the U.S. Department of Energy by the Lawrence Livermore National Laboratory under contract number W-7405-ENG-48.

REFERENCES

Hiskes, J. R. and Karo, A. M., "Electron Energy Distributions, Vibrational Population Distributions, and Negative Ion Concentrations in Hydrogen Discharges," presented at the NATO Advanced Study Institute on Atomic and Molecular Processes, Palermo, Italy, July 19-30, 1982. Lawrence Livermore National Laboratory Report No. UCRL-87779, June, 1982.

Hiskes, J. R. and Karo, A. M., 1983, AIP Conference Series No. 111, p. 125.

Hiskes, J. R., Karo, A. M., and Willmann, P. A., 1985, J. Appl. Phys. 58(5):1759.

Hiskes, J. R., Karo, A. M., and Willmann, P. A., 1985, J. Vac. Sci. Technol. A3(3):1229.

CYCLOTRON RESONANCE LASER ACCELERATOR

S. P. Kuo

Polytechnic University
Route 110
Farmingdale, New York   11735   USA

ABSTRACT

It is shown that in the cyclotron resonance laser acceleration the
Doppler shifted cyclotron resonance conditions can self-maintain throughout the entire interaction period of interest. This leads the acceleration mechanism to be very effective. In addition, we find that there is
an upper bound, independent of the wave amplitude, on the transverse
velocity of the electron. Therefore, the produced electron beam has very
small transverse energy spread.

INTRODUCTION

With the availability of high power lasers, the possibility of
accelerating electrons of GeV energies in reasonable short distances
(i.e., few Rayleigh lengths) has been studied extensively (Tajima, 1979;
Dawson, 1979), (Joshi, 1981; Tajima, 1981; Dawson, 1981). A number of
interesting concepts for achieving high-gradient accelerators have been
reported (Channell, ed., 1982). Among them, a mechanism (Sprangle, 1981;
Tang, 1981), (Sprangle, 1983; Vlahos, 1983; Tang, 1983), (Kuo, 1986;
Schmidt, 1986) which utilizes the cyclotron resonance interaction between
an electron beam and a relativistically strong laser beam of right hand
(R-H) circular polarization propagating collinearly along a dc magnetic
field is studied in the present work. We elaborate this cyclotron resonance interaction process for electron acceleration. We will assume that
the laser intensity is constant throughout the entire interaction period
in the analysis. We will show that this acceleration process is indeed
effective and the produced energetic beam can have very small transverse
energy spread.

ANALYSIS

We consider the interaction between an electron beam and a large amplitude EM wave of R-H circular polarization proprogating collinearly along a uniform magnetic field $B_o z$. A vector potential is used to represent the total electromagnetic fields in the system. It is

$$\vec{A} = \vec{A}_o + \vec{A}_1 = (B_o/2)(-y\hat{x} + x\hat{y})$$

$$+ A_1[\hat{x} \sin(kz-\omega t) + \hat{y} \cos(kz-\omega t)] \tag{1}$$

where $\vec{A}_o$ is the vector potential of the background magnetic field and $\vec{A}_1$ is the vector potential of the wave fields. When the diffraction effect is neglected, we then have $A_1 = A_1(z)$ in general.

Using Eq. (1) to represent the fields, the transverse and the parallel component of the electron momentum equation are written by

$$\frac{d}{dt} P^- + i\Omega P^- = \frac{e}{c} \frac{d}{dt} A_1^- \tag{2}$$

$$\frac{d}{dt} P_z = -\frac{e}{c} R_e \left[\left(\frac{\partial}{\partial z} A_1^-\right) v^{-*} \right] \tag{3}$$

Where $P^- = P_x - iP_y$, $v^- = v_x - iv_y$, $A^- = A_x - iA_y = A_o^- + A_1^-$, $A_o^- = -i(B_o/2)(x - iy)$, $A_1^- = -iA_1 e^{i(kz-\omega t)}$, $\Omega_o = \Omega_o/\gamma$, $\Omega_o = eB_o/mc$, and R_e and $*$ stand for real part and complex conjugates respectively.

The energy equation is given by

$$mc^2 \frac{d}{dt} \gamma = \frac{e}{c} R_e \left[\left(\frac{\partial}{\partial t} A_1^-\right) v^{-*} \right] \tag{4}$$

To determine the electron dynamics in the wavefield Eq. (1), we have to solve Eqs. (2)-(4). It is aided by first using the single mode (forward propagation) wave equation

$$\left[\frac{\partial}{\partial t} + (kc^2/\omega)\frac{\partial}{\partial z} \right] A_1^- - i(kc^2/\omega)e^{i(kz-\omega t)}\frac{\partial}{\partial z} A_1 = -i(\omega_p^2/\omega)(mc/e)v^- \tag{5}$$

where $\omega_p = 4\pi n_o e^2/m$ and n_o is the beam density. Equations (3) and (4) can then be combined to be

$$(kc/\omega)\frac{d}{dt} P_z = mc \frac{d}{dt} \gamma + \frac{ke^2}{2mc\omega} \frac{1}{\gamma} \frac{\partial}{\partial z} A_1^2 \tag{6}$$

where $P^- \simeq eA_1^-/c$ is used for obtaining Eq. (6). This approximation is reasonable for $\Omega \ll \omega$ as will be the case.

We next use the energy conservation equation to obtain the relation

$$c \frac{\partial}{\partial z} A_1^2 = -(\omega_p^2/\omega^2)(mc^2/e)^2 \frac{d}{dt} \gamma,$$ (7)

substituting Eq. (7) into Eq. (6). A constant of motion is derived to be

$$\gamma[\omega - kv_z - \frac{1}{2} kc(\omega_p^2/\omega^2)\frac{1}{\gamma} \ell_n \gamma] = \text{const.} =$$

$$a = \gamma_0[\omega - kv_{zo} - \frac{1}{2} kc (\omega_p^2/\omega^2)\frac{1}{\gamma_0} \ell_n \gamma_0]$$ (8)

Assuming that the beam and wave are set at resonance initially, i.e., $\omega = \Omega_0/\gamma_0 + kv_{zo}$, so that $a = \Omega_0 - \frac{1}{2} kc (\omega_p^2/\omega^2)\ell_n \gamma_0$, it is shown by Eq. (8) that the resonant condition will prevail over the entire interaction length of interest if $|\frac{1}{2} kc(\omega_p^2/\omega^2)(\ell_n \gamma_0/\gamma_0)(\ell/c)| \ll \pi$. This condition is satisfied for beam density $n_0 \ll 10^{14} \text{cm}^{-3}$, which is indeed true in general, where ℓ = 15m and γ_0 = 100 are used in the calculation. Therefore, the constant of motion under the resonance can be approximated to be $\omega - \Omega_0/\gamma - kv_z \equiv 0 = \omega - \Omega_0/\gamma - kv_{zo}$, a relation indicates that the Doppler shifted cyclotron resonance will be maintained by itself through the entire interaction period of interest, i.e., a self-resonance effect. Using this resonance condition together with the definition of $\gamma = [1-(v_\perp^2 + v_z^2/c^2]^{-1/2}$, the relationship between v_z and $v_\perp$ are obtained as

$$v_z/c = [\alpha \pm \sqrt{1 - (1 + \alpha)v_\perp^2/c^2}] \Big/ (1 + \alpha)$$ (9)

where $\alpha = \omega^2/\Omega_0^2 = 1/\gamma_0^2(1-v_{zo}/c)^2$. Since v has to be real, one constraint is deduced from Eq. (9), namely,

$$v_\perp^2/c^2 \leq 1/(1 + \alpha)$$ (10)

which sets an upper bound on the transverse velocity of the electron beam obtained through the resonance interaction. It is noted that the upper bound is independent of the wave field amplitude. This is understandable because the wave magnetic field gives rise to a Lorentz force whose transverse component acts to reduce the effective transverse electric force on electrons. We not integrate Eq. (2) formally to

$$P^- = -\Omega_0(e/c)e^{i(kz-\omega t)} [\int_0^t dt'[A_1/\gamma(t')]]$$ (11)

This result is then substituted into Eq. (4) to yield

$$\gamma = \gamma_0 + \frac{1}{2} \, \mathfrak{L}_0 \omega (e/mc^2)^2 \left[\int_0^t dt' \, (A_1/\gamma(t'))\right]^2 \sim \left[\frac{3}{2} \sqrt{2\bar{v}_q^2 \mathfrak{L}_0 \omega} \; t \right]^{2/3} \qquad (12)$$

where $\bar{v}_q = eA_1(0)/mc^2$. Equation (12) shows that γ is an increasing function of time if the wave field amplitude A_1 is maintained. Since $v_\perp$ is bounded as given by Eq. (1), it simply means that v_z can be increased indefinitely to approach c. Therefore, only the negative sign in Eq. (9) holds initially. However, it should be changed to the positive sign after $(v_\perp/c)max = 1/(1+\alpha)^{1/2}$ is reached. Subsequently, $v_\perp$ decreases as v_z increases. This is because the increase of γ makes the electron become heavier.

A numerical example is illustrated in the following. We use a CO_2 laser pulse having power flux $1.2 \times 10^{14} W/cm^2$ to accelerate the electron beam, it gives $\bar{v}_q = 0.064$. In order to match the resonance condition, a background magnetic field is needed. However, this field has an upper limit from the technological consideration. It would be reasonable to limit it at the value of $B_0 = 50KG$. Since $\mathfrak{L}_0 = \gamma_0 \omega (1-v_{zo}/c) = \omega/2\gamma_0$, it then requires $\gamma_0 = 100$. This means that a 50MeV beam is needed initially for satisfying the resonance condition. It is fortunate to be technologically feasible. Presented in Fig. 1 is functional dependence of γ on t obtained from Eq. (12). We find that for a 15m interaction length the electron beam is already accelerated to $\gamma_f \sim 2 \times 10^3$, i.e., 1GeV beam.

CONCLUSION

We, therefore, have shown that the cyclotron resonance laser (CRL) acceleration mechanism is very effective. this is mainly because the resonance condition will self-maintain throughout the entire interaction period. In addition, we have shown by Eq. (10) that the transverse velocity of the electron beam has an upper bound independent of the wave field strength. This result can also be seen from Fig. 2, where the temporal dependence of the transverse velocity of the electron beam is presented. Therefore, the produced energetic beam will have very small transverse energy spread. We may then conclude that the cyclotron resonance laser (CRL) accelerator is a potential candidate of high gradient accelerator.

ACKNOWLEDGEMENT

This work is supported by the Air Force Office of Scientific Research, Air Force Systems Command, under Grant No. AFOSR-85-0316.

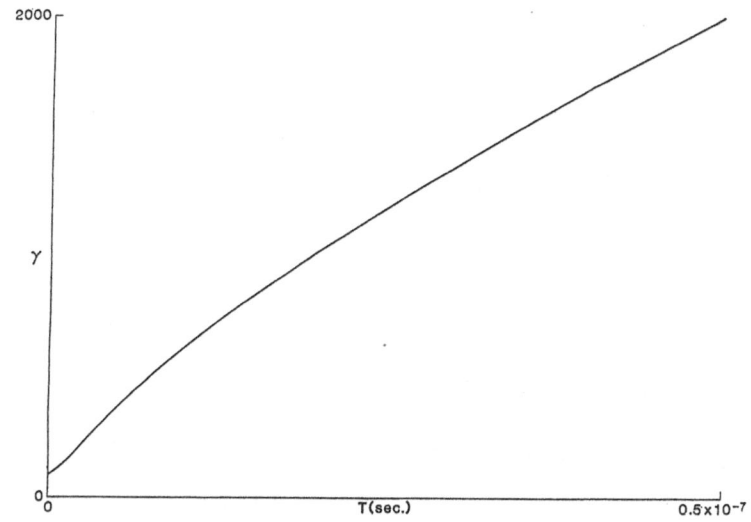

Fig. 1.   Functional dependence of electron beam γ on time t.

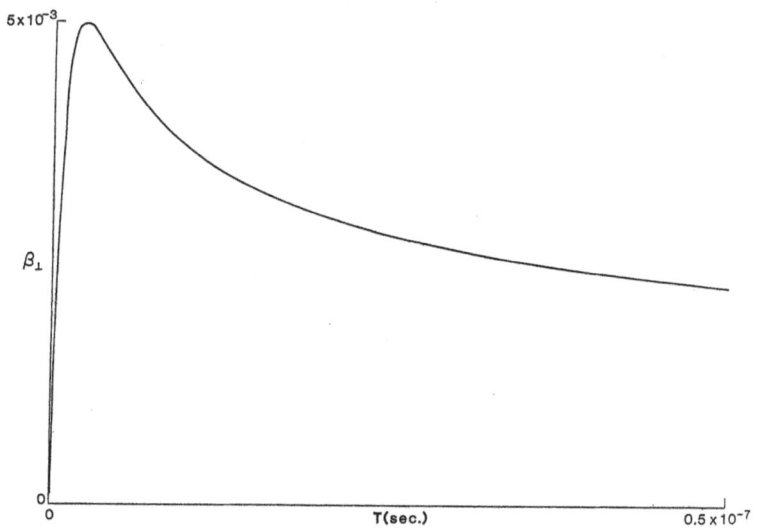

Fig. 2.   Temporal evolution of the transverse velocity β
of the electron beam.

REFERENCES

Channell, Paul J., ed., 1982, Laser Acceleration of Particles, AIP Conf.
    Proceedings No. 91, AIP, N.Y.
Joshi, C., Tajima, T., Dawson, J. M., Baldis, H. A., and Ebrahim, N. A.,
    (1981), Phys. Rev. Lett., 47, 1285.
Kuo, S. P. and Schmidt, G., (1985), J. Appl. Phys., 58, 3646; AIP Conf.
    Proceedings, July 1986.
Sprangle, P. and Tang, C. M., (1981), IEEE Trans. Nucl. Sci., NS-28,
    3346.
Sprangle, P., Vlahos, L., and Tang, C. M., (1983), IEEE Trans. Nucl.
    Sci., NS-30 3177.
Tajima, T. and Dawson, J. M., (1979), Phys. Rev Lett., 43, 267.

BEAM "SELF-TRAPPING" IN THE NRL MODIFIED BETATRON ACCELERATOR*

F. Mako, J. Golden, D. Dialetis**, L. Floyd, N. King,***
and C. A. Kapetanakos

Plasma Physics Division
Naval Research Laboratory
Washington, DC 20375 USA

ABSTRACT

The electron beam in the NRL Modified Betatron is observed to "self-trap" within a narrow range of parameters. It appears that the cause of the "self-trapping" is the localized magnetic field generated by the residual diode stalk current. In this paper, the experimental observations on the "self-trapping" of the beam are briefly summarized, and a model is presented that accurately predicts the experimental results.

INTRODUCTION

The modified betatron (Sprangle, 1978; Kapetanakos, 1978), (Rostoker, 1980), (Kapetanakos, 1983; Sprangle, 1983; Chernin, 1983; Marsh, 1983; Haber, 1983) is a toroidal induction accelerator that has the potential to generate high current beams. Its field configuration includes a strong toroidal magnetic field in addition to the time varying betatron magnetic field which is responsible for the acceleration. The toroidal magnetic field substantially improves the stability of the circulating electron ring. Preliminary results obtained, so far, from the NRL modified betatron have demonstrated some important aspects of the concept (Mako, 1985; Golden, 1985; Floyd, 1985; McDonald, 1985; Smith, 1985; Kapetanakos, 1985), (Golden, 1986; Mako, 1986; Floyd, 1986; McDonald, 1986; Smith, 1986; Marsh, 1986; Dialetis, 1986; Kapetanakos, 1986), including: (1) the beneficial effect of the toroidal magnetic

----
*   work supported by ONR.
**   SAIC, McLean, VA, USA
*** Sachs-Freeman Associates, Landover, MD, USA

field on the expansion of the ring's minor radius, (2) the pronounced effect of image forces on the ring equilibrium, (3) the drift (bounce) motion of the ring in the poloidal direction and (4) "self trapping" of the multi-kiloampere beam with efficiency as high as 80%. In this paper, we briefly summarize the experimental observation of "self trapping" of the beam and also present a model that explains its origins.

EXPERIMENTAL SET-UP AND OBSERVATION

A schematic plan view of the NRL Modified Betatron (Golden, 1983) is shown in Fig. 1. The support structure is an equilateral triangle. Twelve rectangular toroidal field (TF) coils are equally spaced azimuthally about the major axis (perpendicular to the page). Eighteen circular vertical (i.e., betatron) field (VF) coils are located within the TF coils and generate the betatron field that accelerates the electrons. A toroidal vacuum chamber (major radius r_o = 1 m, minor radius a = 15.3 cm) is contained within the coil system. The electron beam is generated in a diode located inside the vacuum chamber. Typically a 0.8 MeV, 1.5-4 kA electron beam is injected tangential to the toroidal field, on the midplane, at a radius of 109 cm.

As theoretically predicted (Kapetanakos, 1982; Sprangle, 1982; Marsh, 1982), when the betatron field (B_z) is a particular value B_{zm} ~ 48 G) the beam returns to the injection position after one revolution around the major axis. For lower values of B_z, the beam drifts poloidally on a nearly circular trajectory and returns to the injector after a poloidal period (typically 200-500 nsec).

The poloidal trajectory is observed by open shutter photography of the light produced when the beam passes through a thin polycarbonate film that spans the minor cross section of the vacuum chamber. The film target is 2 μm thick and is coated with 2-3 μm of carbon to reduce electrostatic charging.

"Self trapping" is observed for a narrow range of B_{zo} (~ 29 G). In this case, as the beam approaches the injector, it is deflected away from the diode and subsequently performs additional oscillations (see Fig. 2a). At higher betatron magnetic fields the deflection is insufficient, and the beam strikes the injector. At lower B_{zo}, the beam drifts into the wall during the first poloidal bounce.

The "self trapping" results from the localized magnetic field associated with the injection diode. Both electric and magnetic fields are associated with the diode because of the applied potential and the current flowing in the cathode stalk (see Fig. 3). However, during the main injection pulse, the fringing electric forces and the magnetic

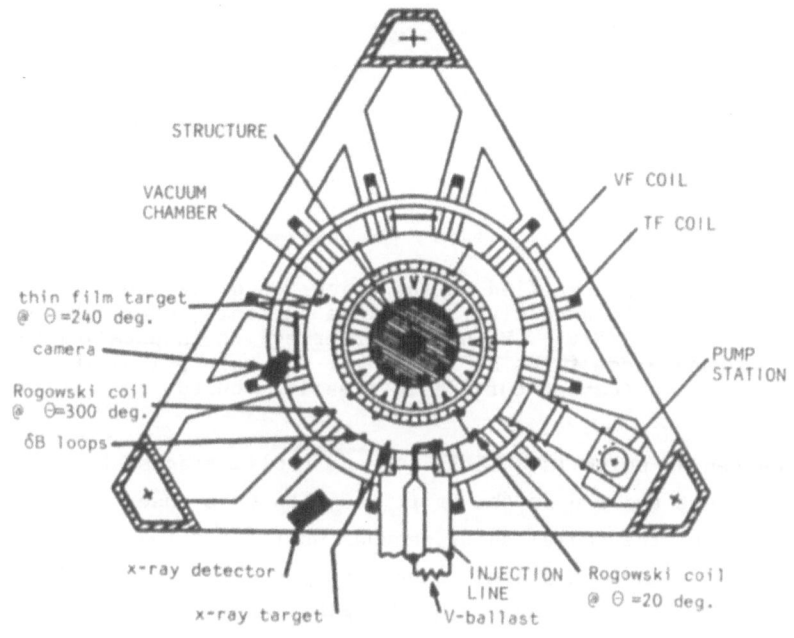

Fig. 1.   Schematic plan view of the NRL modified betatron.

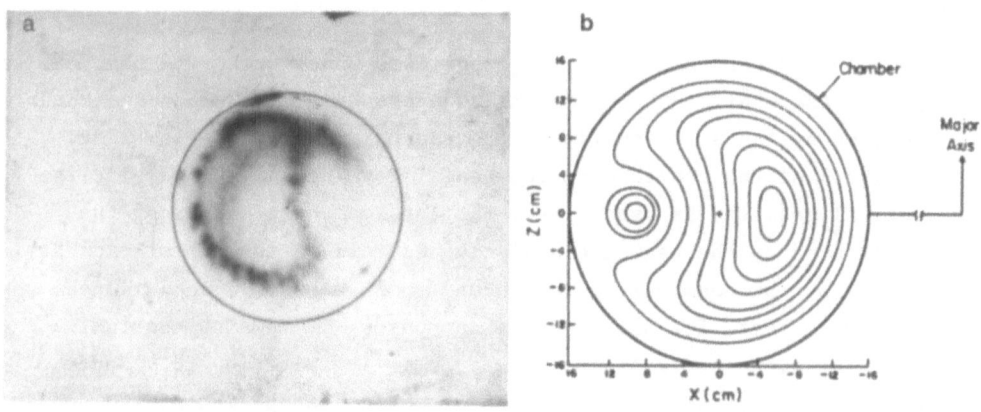

Fig. 2.   Experimentally observed (a) and theoretically predicted (b)
poloidal motion of the beam.

forces from the 15 cm long azimuthal segment of the cathode stalk nearly
cancel. A second pulse (dashed portion of the traces in Fig. 3) follows
the main applied voltage pulse by 200 nsec and has an ~ 150 kV peak and a
150 nsec duration. During the afterpulse, the diode impedance is reduced
by plasma closure and a 1-4 kA current flows in the stalk. In this case,
the magnetic forces are larger than the electric forces and the net
component produces a radial inward drift of the beam. This explanation
is supported by the observation that when a diverter sparkgap that is in
parallel with the diode fires at the end of the flat portion of the main

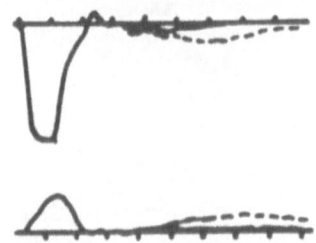

Fig. 3. Potential applied to the injection diode (upper, 850 kV
peak, 50 nsec/div), and diode current (lower, 6kA peak,
50 nsec/div). When a diverter switch in parallel with
the diode fires at the end of the main pulse, the second
pulse (dashed curves) is greatly reduced.

pulse, the second pulse is greatly reduced (solid traces, Fig. 3) and
only a small deformation of the drift trajectory is observed. Further
evidence is that the radial shift depends on the length of the cathode
stalk. A 2-3 cm cathode stalk produces only a slight distortion of the
poloidal orbit. A 30 cm long cathode stalk produces a gross distortion
of the poloidal orbit.

THE THEORETICAL MODEL

A simple understanding of the conditions under which the beam can be
trapped in the modified betatron in the presence of the diode stalk can
be provided by the slow equation (Kapetanakos, 1986; Dialetis, 1986;
Marsh, 1986) of the centroid of the beam. For the sake of simplicity,
cylindrical geometry is used, i.e., the toroidal corrections of the
fields induced by the beam are neglected as well as those of the diode
stalk. Also, the beam energy is assumed to have a small variation as the
beam moves so that, for all practical purposes, it remains constant.
finally, the diode stalk is assumed to be of infinite length. Under
these assumptions, the slow equations of motion can be integrated and an
expression for the orbits of the beam centroid can be obtained, namely,

$$
\left[\frac{\beta\Omega_{zo}r_o}{c}\right]\left[\frac{n}{2}\left(\frac{x}{r_o}\right)^2 - \frac{x}{r_o} - \frac{n}{2}\left(\frac{z}{r_o}\right)^2\right] + \left[\beta^2\gamma\right]\ell n\left(1 + \frac{x}{r_o}\right)
$$

$$
- \left[\frac{\nu}{\gamma^2}\right]\ell n\left(1 - \frac{x^2+z^2}{a^2}\right) - \frac{1}{4\pi}\frac{|e|}{mc^2}\left(V_{ds}-120\pi\beta I_{ds}\right)\left\{\ell n\left[\left(\frac{x-\Delta}{a}\right)^2 + \left(\frac{z}{a}\right)^2\right]\right.
$$

$$
\left. - \ell n\left[\left(\frac{x\Delta}{a^2} - 1\right)^2 + \left(\frac{z\Delta}{a^2}\right)^2\right]\right\} = K. \tag{1}
$$

Rationalized MKS units are used. Here, (x,z) is the position of the beam centroid, ν Budker's parameter, β the beam velocity normalized to the velocity of light c, and γ is the relativistic factor of the beam. Also, r_0 and a are the major and minor radii of the toroidal chamber while n and B_{zo} are the field index and the betatron magnetic field on the minor axis ($\Omega_{zo} = |e|B_{zo}/m$, where e and m are the electron charge and mass). Finally, Δ is the distance of the diode stalk from the minor axis, V_{ds} and I_{ds} are the stalk voltage and stalk current. The integration constant K is determined by the initial position of the beam. The various terms in Eq. (1) can be easily identified. Thus, the first and second terms are due to the betatron field and the centrifugal force, while the third term originates from the induced charge and currently by the beam on the chamber walls. The fourth term comes from the fields of the diode stalk and its image due to the conducting walls. When all four terms are comparable to each other, then the possibility arises for the beam to be trapped as Fig. (2b) demonstrates.

Excellent agreement is seen in the comparison of theory and experiment shown in Fig. 2. The experimental parameters have been used in the theoretical model to predict the poloidal orbits. In this case, $\gamma = 1.64$, $r_0 = 100$ cm, $I_{ds} = 1$ kA, $\Delta = 9$ cm, a = 16 cm, n = 0.5, the beam current is 1 kA, and the toroidal field is 2 kGauss. Because the model does not include toroidal effects, the model is evaluated with $B_{zo} = 22.16$ G which is equivalent to 29 G in the experiment.

REFERENCES

Golden, J., Mako, F., Floyd, L., McDonald, K., Smith, T., Marsh, S. J., Dialetis, D., and Kapetanakos, C. A., 1986, Beams '86, Proc. of Laser Society of Japan.

Golden, J. et al., 1983, IEEE Trans. Nucl. Sci., NS-30, 2114.

Kapetanakos, C. A., Dialetis, D., and Marsh, S. J., 1986, "Beam Trapping in a Modified Betatron with Torsatron Windings," NRL Memo Report 5619, Washington, DC (to be published in Particle Acceleration).

Kapetanakos, C. A., Sprangle, P., Chernin, D. P., Marsh, S. J., Haber, I., 1983, Phys. Fluids, 26, 1634.

Kapetanakos, C. A., Sprangle, P., and Marsh, S. J., 1982, Phys. Rev. Lett., 49, 741.

Mako, F., Golden, J., Floyd, L., McDonald, K., Smith, T., and Kapetanakos, C. A., 1985, IEEE Trans. Nucl. Sci., NS-32, 3027.

Rostoker, N., 1980, Comments Plasma Phys., 6, 91.

Sprangle, P. and Kapetanakos, C. A., 1978, J. Appl. Phys., 49, 1.

ELECTRON CYCLOTRON MASER USING A PULSED RELATIVISTIC ELECTRON BEAM

A. D. R. Phelps, A. Z. Maatug and S. N. Spark

Department of Physics and Applied Physics
University of Strathclyde
Glasgow G4 ONG, Scotland

ABSTRACT

Measurements of high power millimetre-wave radiation from a pulsed electron cyclotron maser operating in the 75-110 GHz frequency range are reported. A mildly relativistic electron beam excites the electron cyclotron maser. The annular electron beam is extracted from a cylindrical cold cathode and propagates through a pulsed magnetic field of up to 4 T. The beam current in these experiments has been varied from 5 A to 1.25 kA. The pulse duration of the millimetre-wave emission is typically a few hundred ns and corresponds to the time for which energetic electrons are present. Gap closure within the accelerating diode limits the pulse duration.

The millimetre-wave radiation is produced within a highly overmoded cavity. An echelette grating spectrometer has been used to measure the frequency spectra. Far-field radiation patterns have also been measured. The ability to tune the emission frequency by varying the magnetic field has been successfully demonstrated.

The electron cyclotron maser in its performance at the longer wavelength end of the electromagnetic spectrum complements the short and medium wavelength free electron laser. It appears that the electron beam requirements for successful electron cyclotron maser operation are less stringent than those of the free electron laser.

INTRODUCTION

The electron cyclotron maser instability can occur when a beam of relativistic electrons having a high transverse energy passes through a magnetic field region. In devices which exploit this instability it is usual for the electron beam/magnetic field interaction region to be located within a resonant cavity of some kind. This can be either a microwave cavity resonator or a quasi-optical maser cavity. The operating frequency of the device lies near to the electron cyclotron frequency or one of its harmonics. The electron beam cyclotron mode (Allen and Phelps, 1977) has the characteristic equation

$$\omega - k_z v_{//} - s\Omega_c = 0 \tag{1}$$

where $v_{//}$ is the axial velocity of the electron beam, s is the cyclotron harmonic number and Ω_c is the relativistically correct cyclotron frequency

$$\Omega_c = \frac{eB}{\gamma m} \tag{2}$$

There is coupling between the fast wave represented by Eq. (1) and the waveguide electromagnetic mode, which is also fast and is represented by

$$\omega^2 - k_z^2 c^2 - \omega_{c0}^2 = 0 \tag{3}$$

where ω_{c0} is the cutoff frequency for the particular electromagnetic mode.

In the case of a single cavity

$$k_z = \frac{\pi q}{\ell} \tag{4}$$

where q is the axial eigenmode number and so Eqs. (1) and (3) become

$$\omega = \frac{q \pi v_{//}}{\ell} + \frac{seB}{\gamma m} \tag{5}$$

and

$$\omega = \left(\frac{q^2 \pi^2 c^2}{\ell^2} + \omega_{c0}^2 \right)^{1/2} \tag{6}$$

Equation (6) is the resonance frequency of the cavity for the particular electromagnetic mode and determines to first order the frequency at which interaction occurs.

With a given cavity, as the frequency of the electron cyclotron mode is varied, by varying the magnitude of the magnetic field B, as described by Eq. (5), it is possible to excite in turn each particular electromagnetic mode of the cavity. Each mode has its characteristics value of ω_{c0} determined by the transverse dimensions of the cavity and by the characteristic structure of that particular mode.

Cyclotron resonance devices have been reviewed by Symons and Jory (Symons and Jory, 1981) and an interesting comparison of the different models used to describe gyrotrons has been published by Lindsay (Lindsay, 1981).

This research at Strathclyde University started in 1978, the first emissions being measured in 1981. The results from these early experiments were concentrated in the 8-12 GHz range (Garvey, 1983; Phelps and Garvey, 1983). A magnetically-shielded diode was used to accelerate the electron beam in the first experiments, whereas in a second series of experiments a magnetic-field-immersed diode was used (Phelps et al., 1984; Phelps et al., 1984). Several configurations were constructed and operated (Hasaani, 1986). The MkV electron cyclotron maser used a higher magnetic field of up to 1.6 T and successfully operated in the 26.5-40 GHz range. In the experiments reported here the magnetic field has been further increased to a maximum of 4 T to provide emission in the 75-110 GHz range.

EXPERIMENTS

A schematic diagram of the experiment is shown in Fig. 1. The Marx bank was designed and constructed as part of the experiment. It consists of ten 0.5 µF low inductance capacitors charged in positive and negative polarity pairs resulting in only five spark gaps being needed. The lowest two spark gaps are electrically triggered to enable the electron beam pulse to be precisely timed with respect to the peak of the much longer magnetic field pulse. In this way the electron beam interacts with a quasistatic magnetic field of known magnitude. In these experiments the capacitors are normally charged within the range 15-35 kV to obtain a mildly relativistic electron beam. The potential difference applied to the accelerator diode is monitored with a voltage divider and the beam current is measured using a Rogowski loop. The magnitude of the magnetic field has been calibrated using a standard magnetic coil probe.

The millimetre-wave cavity used in the these experiments is made using 23 mm internal diameter stainless steel tubing with a constriction which provides partial wave reflection at the cathode end. There is a small but calculable and measurable screening effect of the cavity on the relatively slowly varying magnetic field as it penetrates the wall of the cavity.

Three methods have been used to measure the millimetre-wave emission (i) direct reception of the emissions for power monitoring, (ii) spectral dispersion and measurement using a blazed echelette grating spectrometer,

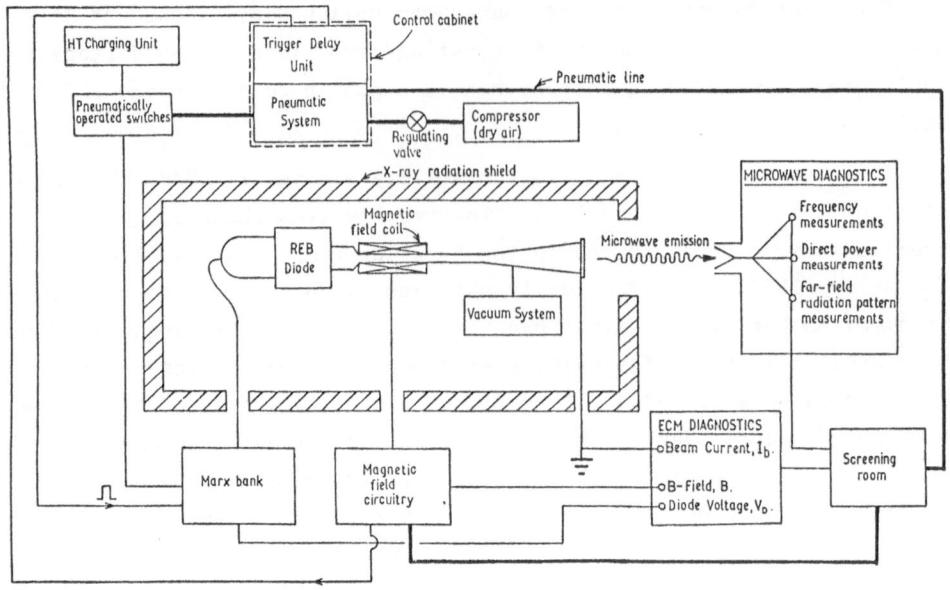

Fig. 1. Schematic diagram of the experiment.

(iii) far-field radiation pattern plots using a single scanned detector and relying on reproducibility to identify operating modes.

A pair of spectra are shown in Fig. 2, for a beam current of 1.25 kA. It is clearly demonstrated that as the magnetic field increases from 2.80 T to 3.33 T the emission spectrum shifts to higher frequencies. It is apparent from the mode scale at the top of these spectral plots that these are in high order modes and, as expected, the far-field pattern plots in these cases have a complicated appearance. The upshift in frequency is exactly what Eq. (5) predicts. Although the output levels shown in Fig. 2 are not yet absolutely calibrated, the known high insertion loss of the spectrometer and previous observations of megawatt power emissions in the MkV experiments indicate that these measurements correspond to quite high powers.

CONCLUSIONS

Using an overmoded cavity implies possible mode competition, but an apparent compensation, demonstrated by these experiments, is that a spectrum of available modes allows the electron cyclotron maser to tune in frequency, albeit discretely, as the magnetic field is varied. This could have useful implications in any situation where it is desirable to change the operating frequency between one short pulse and the next. The powers available and the reasonable efficiencies of electron cyclotron

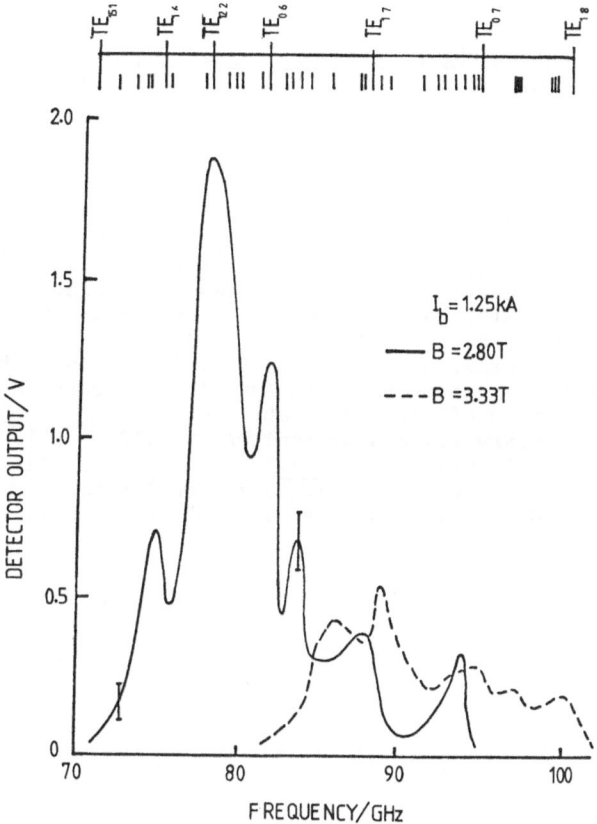

Fig. 2. Spectral analysis of the millimetre-wave emission.

masers tend to make them attractive alternatives to free electron lasers a the microwave and long millimetre-wave end of the spectrum.

It is anticipated that as the electron beam current increases there should be some spread in the particle velocities because of space charge. This lowering of the beam quality does not appear to prevent the electron cyclotron maser operating and leads to the suggestion that the electron cyclotron maser may operate with a lower electron beam quality than that required by a free electron laser. A quantitative investigation of the beam emittance and beam brightness is needed to confirm this.

ACKNOWLEDGMENTS

The support of the SERC and the earlier contributions of Dr. T. Garvey, Dr. A. S. Hasaani and Mr. P. S. Bansal are gratefully acknowledged.

REFERENCES

Allen, J. E. and Phelps, A. D. R., 1977, Waves and Microinstabilities in Plasmas - Linear Effects. Rep. Prog. Phys. 40:1305-68.

Garvey, T., 1983, Investigation of the Interaction of High-Current Relativistic Electron Beams with Electromagnetic Fields, Ph.D. Thesis, Strathclyde University.

Hasaani, A. S., 1986, Pulsed Electron Cyclotron Maser Experiments with Different Configurations, Ph.D. Thesis, Strathclyde University.

Lindsay, P. A., 1981, Gyrotrons (Electron Cyclotron Masers): Different Mathematical Models, IEEE J. Quant. Electron. 17:1327-33.

Phelps, A. D. R. and Garvey T., 1983, Gyroresonant Relativistic Electron Beam Emissions, Xth. Annual Conference on Plasma Physics, Institute of Physics, UCNW, Bangor, UK.

Phelps, A. D. R., Hasaani, A. S., and Bansal, P. S., 1984, Electron Cyclotron Maser Experiments, XIth Annual Conference on Plasma Physics, Institute of Physics, Cambridge, UK.

Phelps, A. D. R., Garvey T., and Hasaani, A. S., 1984, Pulsed Electron Cyclotron Maser Experiments. Int. J. Electron. 57:1141-50.

Symons, R. S., and Jory, H. R., 1981, Cyclotron Resonance Devices, Adv. Electronics and Electron Physics, Vol. 55, (New York: Academic Press), 1-75.

TRANSPORT OF HIGH-BRIGHTNESS ELECTRON BEAMS WITH ION FOCUSING

John R. Smith and Ralph F. Schneider

Naval Surface Weapons Center
White Oak, Silver Spring, Maryland   20903-5000

ABSTRACT

Transport of high-brightness relativistic electron beams via ion focusing in lieu of the more traditional magnetic focusing is of current interest. For example, the Advanced Test Accelerator at Lawrence Livermore National Laboratory has used ion focusing with considerable success. As a second example, ion focusing is being considered for beam transport in´a recirculating linac at Sandia National Laboratories. In order to achieve increased control over this method, and more completely understand its limitations, a parameter study of beam transport with a <u>well diagnosed plasma channel</u> is required. To obtain channels amenable to such a study we have employed channel creation by ionization of a low-pressure gas (less than a milliTorr) by means of a low-energy electron gun (400eV). This method is capable of producing dc channels with suitable levels of ionization in a wide range of gases. A low magnitude magnetic field (140 Gauss) confines the plasma to a channel. Channel ion density profiles are measured with Langmuir probe. Beam transport characteristics are measured by injection of a relativistic electron beam (100 ns, 2 kA, 700 kV) into the preformed channels. The preliminary results which are presented in this report are for channels created in hydrogen gas.

INTRODUCTION

Ion focusing regime (IFR) propagation is the electrostatic focusing of an intense relativistic electron beam using an ion channel. The channel's charge reduces the beam's space charge field which allows radial confinement of the beam by the beam's self magnetic field. The

degree to which the channel's charge compensates the beam's space charge
is expressed in terms of the space charge neutralization fraction, f,
which may be defined as the ratio of the channel ion line density to beam
electron line density. Initially a plasma channel is created by ionization of a gas. Subsequently, high electric fields associated with the
relativistic beam eject plasma electrons which results in an ion channel.
The body of the electron beam may then propagate in this channel. Plasma
channels for IFR propagation have been created by several different
methods. Some examples of these methods and their features are listed
below.

Ionization by relativistic beam electrons (Miller, 1972; Gerardo,
1972; Poukey, 1972), (Struve, 1983; Lauer, 1983; Chambers, 1983)

> (i) uses the relativistic electron beam's own electrons and elec-
> tric field to ionize the background gas,

> (ii) is the simplest method since no additional ionizer is
> required,

> (iii) may be used for ionization in any gas, and

> (iv) produces ionization that varies greatly both in time and
> space.

Laser produced IFR plasma channels (Prono, 1985; Caporaso, 1985),
(Shope, 1985; Frost, 1985; Leifeste, 1985; Crist, 1985, Kiekel, 1985;
Poukey, 1985; Godfrey, 1985)

> (i) can be produced only with heavy organic-type gases,

> (ii) can be created over long distances, and

> (iii) may possess rather uniform ionization over the entire channel
> length.

Low energy electron gun plasma channels (Shope, 1986; Frost, 1986;
Ekdahl, 1986; Freeman, 1986; Hasti, 1986; Leifeste, 1986; Mazarakis,
1986; Miller, 1986; Poukey, 1986; Tucker, 1986)

> (i) can provide adequate levels of ionization for IFR propagation
> in both light and heavy gases,

> (ii) can be produced on a steady state basis, but

> (iii) require a low level external magnetic guide field.

The choice of ionizer clearly depends on the particular goal of a given
experiment. Our primary objective in this work was to investigate propagation of a relativistic electron beam in a well diagnosed channel. To
this end we have used the method of channel creation which employs a low
energy electron gun. As previously mentioned the low energy electron gun

produces plasma channels on a dc basis which greatly simplifies channel diagnosis.

PLASMA CHANNELS

The plasma channel is produced in a 15 cm diameter, 2 m long stain less steel drift tube. The tube is evacuated to a base pressure on the 10^{-6} Torr scale, and then gas is flowed raising the pressure to the desired level. The electron gun is located at one end of the tube and is fabricated from an automobile tail-light filament. Approximately 2 A is passed through the filament in order to obtain a heated cathode and the filament is biased at -400 V. Wire is coiled around the drift tube to produce an external magnetic field which was 140 Gauss. Electrons emitted from the gun produce ionization of the background gas. The resultant plasma is confined to a channel by the external field. Current in this channel was of the order of 10 to 20 mA. The channel ion density was measured with a Langmuir probe. In the preliminary work reported here we have measured a density profile for the case of hydrogen gas at 2.7×10^{-4} Torr. The plasma channel has a peak ion density of 9×10^{10} cm^{-3} and a line density of 2.5×10^{11} cm^{-1} (Fig. 1).

BEAM TRANSPORT

Attached to the drift tube end opposite of the electron gun was the Transbeam accelerator which produces a 100 ns, 700 kV electron beam. The current in the diode was apertured so that the current injected into the drift tube was about 2 kA. The relativistic beam profile was measured and has a width similar to that of the channel. We emphasize that the channel is created on a dc basis and the pulsed relativistic beam is then fired into it. For this work the accelerator used a foilless diode. The anode plate is a carbon disk with a 2 cm diameter aperture on center, therefore the plasma channel extends into the diode as far as the cathode.

The drift tube has two Rogowski coils which monitor net current. Plasma currents (which are usually manifest as slowly decaying tails at the end of a beam pulse) do not appear significant in our results. Therefore, we take the Rogowski coil signals as being a fairly good representation of the relativistic electron beam current. Rogowski A is positioned 34 cm downstream of the anode foil and Rogowski B 160 cm downstream. Midway between the two Rogowski coils is an X-Y displacement monitor (@ z = 97 cm). Results from these diagnostics for a series of tests at different pressures are given in Fig. 2. Good current transport from location A to B is present for all the pressures. The lowest pres- sure shows a reduced upstream current (Rogowski A) as compared with

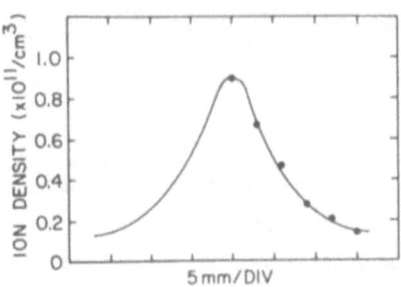

Fig. 1. Channel profile created in hydrogen gas at a pressure of 2.7×10^{-4} Torr.

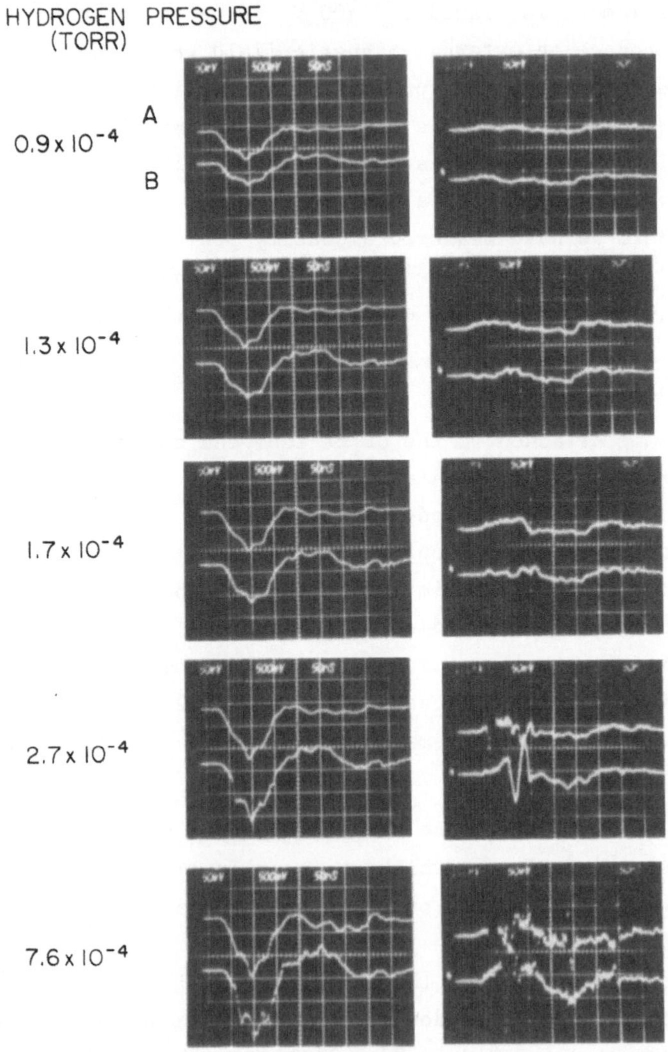

HYDROGEN PRESSURE
(TORR)

Fig. 2. Results from Rogowski coils and displacement monitors. Rogowski A was located at z=34 cm (measured downstream from the anode foil) and Rogowski B was located at z=160 cm. The scale factor for Rogowski A is 1 kA/div and for Rogowski B is 1.2 kA/div. The X and Y displacement monitors were located at z=97 cm.

0.9×10^{-4} A
                    B

1.3×10^{-4}

1.7×10^{-4}

2.7×10^{-4}

7.6×10^{-4}

the other pressures. Beam displacement is relatively small for the three lowest pressures. However, at 2.7×10^{-4} Torr the y displacement is strikingly different from the other shots in that a distinct beam oscillation is present. This oscillation occurs during the latter half of the

788

beam and has a frequency of 25 MHz. Similar oscillations were observed in many other shots at this pressure. The channel's line density at this pressure was 2.5×10^{11} cm^{-1} which is 1/2 the beam line density (@ maximum current). Therefore, charge neutralization of (f=1/2) was realized on this shot. The parameters in this shot are particularly favorable (i.e., light ion mass, high f) for a transverse beam oscillation known as the ion resonance instability (Uhm, 1980; Davidson, 1980). Moreover, the frequency of oscillation is comparable to that expected for this instability. At the highest pressure the displacement signals are very noisy and contain a frequency of oscillations so high that the signals are not recorded on the oscilloscope film. At this point we may have exceeded the pressure required for propagation in the ion focusing regime. At gas pressures such that the ion density exceeds the beam density, the plasma may be heated by the beam and streaming instabilities are likely.

DISCUSSION

The significance of this work lies in the investigation of beam propagation in an ion channel where the space charge neutralization fraction has been measured. Knowledge of the neutralization fraction is important for understanding the beam dynamics of IFR propagation and for analysis of beam oscillations such as the ion resonance instability. The preliminary work described here will be expanded to include channel measurements and beam transport measurements for other pressures and gas species.

ACKNOWLEDGMENTS

We gratefully acknowledge funding by the Independent Research Fund at NSWC. We would also like to express appreciation to Sandia National Laboratories for collaboration in building the low energy electron gun.

REFERENCES

Miller, P. A., Gerardo, J. B., and Poukey, J. W., 1972, J. Appl. Phys., 43, 3001.

Prono, D. S. and Caporaso, G. J., in Energy and Technology Review, (UCRL-52000-85-3, Lawrence Livermore National Laboratory, Livermore, CA, 1985), p. 1.

Shope, S. L., Frost, C. A., Leifeste, G. T., Crist, C. E., Kiekel, P. D., Poukey, J. W., and Godfrey, B. B., 1985, IEEE Trans. Nucl. Sci., NS-32, 3092.

Shope, S. L., Frost, C. A., Ekdahl, C. A., Freeman, J. R., Hasti, D. E., Leifeste, G. T., Mazarakis, M. G., Miller, R. B., Poukey, J. W., and Tucker, W. K., 1986, in Conference Record-Abstracts 1986 IEEE International Conference on Plasma Science, IEEE Cat. No. 86CH2317-6, 46.

Struve, K. W., Lauer, E. J., Chambers, F. W., in Proceedings of the Fifth International Conference on High-Power Particle Beams, edited by R. J. Briggs and A. J. Toepfer (CONF-830911, Lawrence Livermore National Laboratory, Livermore, CA, 1983), p. 408.

Uhm, H. S. and Davidson, R. C., 1980, Phys. Fluids 23, 813.

GENERATION OF ENERGETIC ION BEAMS FROM A PLASMA FOCUS

I. Ueno, M. Tanimoto*, A. Donaldson**, K. Koyama*,
Y. Matsumoto*, and K. Nakajima***
Department of EE
University of Tokyo
Tokyo 113, Japan

ABSTRACT

In certain applications, such as the breeding of fossil fuels, high current plasma focus devices can be used in place of traditional linear accelerators for the generation of 1 Gev ion beams. Several mechanisms have been examined experimentally and theoretically which may be responsible for the generation of the electric fields necessary to produce the high energy particles in a plasma focus. The three sources of electric fields discussed are in inductive field induced by the pinch plasma, resistive fields generated in a magnetically insulated pinch plasma, and the fields resulting from both macroscopic and microscopic plasma instabilities. A universal scaling law for the electric field as a function of current and the ion energy as a function of the charge state of the ions is also presented. In addition, expressions are given for the beam current, beam power, and pulse duration in order to more readily compare this means of ion beam production with more traditional linear accelerators. One sample calculation yields a beam current of 15 mA, beam power of 15 MW, and a pulse duration of 1 ns.

INTRODUCTION

Plasma focus machines, investigated so far principally as neutron sources, are now being considered for use as energetic ion and electron beam sources also. A wide range of experimental and theoretical work is being carried out, and already several theoretical models of energetic

---

*   Electrotechnical Laboratory, Tsukuba, 1-1-4 Umezono, Sakura-mura, Niihari-gun, Ibaraki 305
** Department of EE/CS, Texas Tech University, Lubbock, TX 79409
*** Department of EE, University of Maryland, College Park, MD 20742

beam generation have been proposed. In a previous paper (Tanimoto, 1984; Ueno et al., 1984), a comparative study was made of three models for the driving source of the electric field: (1) the rapidly-increasing inductance of the pinching current sheath; (2) the enhanced resistivity of a magnetically insulated plasma; and (3) plasma instabilities. A summary of those findings is given here along with calculations of typical beam parameters for the case of resistive ion acceleration.

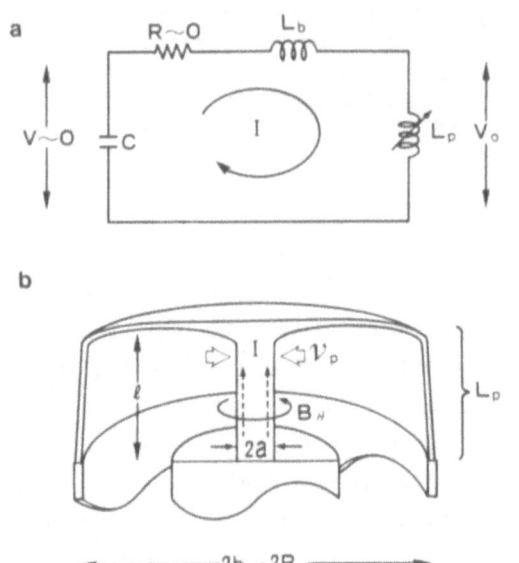

Fig. 1.   (a) Equivalent circuit of plasma focus.
(b) Pinch model.

ELECTRIC FIELD GENERATION

Tanimoto et al. (Tanimoto et al., 1984) derive an expression for the induced electric field resulting from the change in inductance accompanying the plasma pinch illustrated in Fig. 1.

$$E_z \simeq \mu_o I / 2\pi\tau_p \; , \tag{1}$$

where τ_p is the pinch time and I is the current in the plasma. Using typical values of $I = 1$ MA and $\tau_p = 30$ ns, we find that $E_z = 6.7$ MV/m and the maximum energy of the ions of charge state Z, after transit across plasma column length of 3 cm, is approximately

$$eV_{max} \simeq 200Z \; (keV) \; . \tag{2}$$

Making use of a relationship between the current and τ_p given by
Imshennik (D'yachenko and Imshennik), the maximum attainable energy obeys
the following scaling law:

$$eV_{max} \sim ZI^2/(Mn_o)^{1/2}R_o{}^2,\tag{3}$$

where Mn_o is the mass density of the initially filled gas, and R_o is the
radius of the external discharge electrode.

A detailed discussion of the generation of a resistive electric
field in a magnetically insulated pinch plasma is given (Tanimoto et al.,
1982), so only a summary of main points is mentioned here.

An intense azimuthal self-magnetic field B_θ approaching 1 MG can be
generated at maximum pinch, driven by an axial current in a plasma focus
device (Koyama, 1981; Tanimoto et al., 1981). The plasma electrons,
insulated in the axial direction by the magnetic field, drift with a
velocity $v_d = E_z B_\theta$ in the radial direction under the influence of an
axial electric field. Referring to Fig. 2, the axial conductivity will
be reduced by a factor of $h_e{}^2$ from that for the value without magnetic
field, σ_o, to a value of $\sigma_o/h_e{}^2$. The electron-hole parameter, h_e, is
defined as $h_e = \omega_{ce}\tau_e$ ($\gg 1$). The conventional notation is used for the
various parameters. Let a be the radius of the plasma at the maximum
pinch; the magnetic insulation will then be effective for a period of
time given approximately by $\tau_H = a/v_d$. Since the current will be
continuous in time due to the finite inductance of the discharge circuit,
the electric field E_z generated in the plasma at the start of the magnetic insulation and τ_H are related as follows:

$$E_z = \mu_o I/2\pi\tau_H,\tag{4}$$

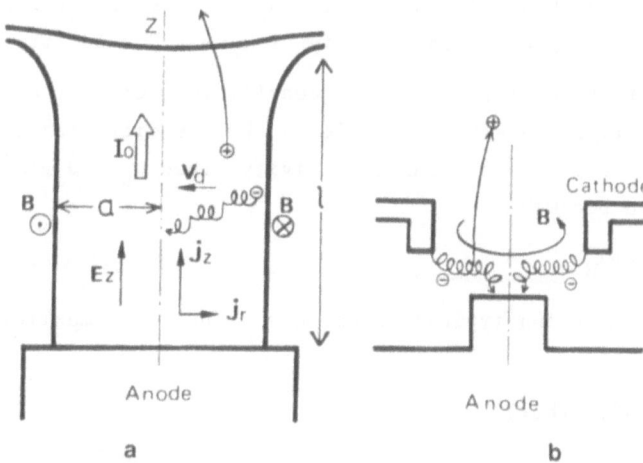

Fig. 2. (a) Comparison of magnetically-insulated focus with
(b) Self-pinch ion diode.

The maximum ion kinetic energy is calculated in (Tanimoto et al., 1984) to be given by the simple relation

$$eV_i \simeq 960ZI^2 \text{ (keV)}. \tag{5}$$

where the ion mass M_i is assumed to be given by $2ZM_p$ (M_p = proton mass).

A third source of electric fields in the plasma focus are the macroscopic and microscopic instabilities. An m=0-MHD macroscopic instability produced in a plasma column is depicted in Fig. 3 and has been observed experimentally in Fig. 4. Since the electric field is generated by the local inductance changes, the same analysis used to produce Eq. (1) yields, for this case,

$$E_z \simeq \mu_o I/2\pi\tau_i, \tag{6}$$

where $\tau_i = a/v_A$, and v_A is the Alfven velocity. Typical values for a deuterium plasma column yield $\tau_i = 4.1 \times 10^{-9}$ s, and $E_z = 4.9 \times 10^7$ V/m for I = 1 MA (Tanimoto et al., 1984). If we assume that the maximum effective acceleration distance is half of the wavelength λ (~ 0.5 cm), the maximum energy is of the order of

$$eV_{max} \simeq 120Z \text{ (keV)}. \tag{7}$$

The microscopic instabilities in a plasma focus are not presently well understood, and only the basic treatment given in (Tanimoto et al., 1984) of the generation of the electric field due to an increase in resistance enhanced by the instability is presented here.

From general measurements, Gary gives a tentative value of $\sigma\star = 2.4 \times 10^4/\Omega m$ for the effective conductivity at the point of maximum plasma constriction, assuming that the magnetic field diffuses/penetrates a distance of radius a = 1 mm in a time τ_R = 30 ns (Gary, 1974). Comparing this to the plasma conductivity of $\sigma_o = 2.8 \times 10^7/\Omega m$, expected in a plasma of temperature 1 keV and density 10^{19} cm^{-3}, the resistivity is about 1200 times larger. If we let τ_R be the diffusion time of the magnetic field due to anomalous resistivity, then $\tau_R = \mu_o\sigma\star a^2$ and the electric field becomes

$$E_z = \mu_o I/\pi\tau_R . \tag{8}$$

For I = 1 MA and using typical parameters values, the maximum ion energy is

$$eV_{max} = 400Z \text{ (keV)} . \tag{9}$$

We have considered the electric field generated in the plasma focus by various processes along with the energy attained by the ions

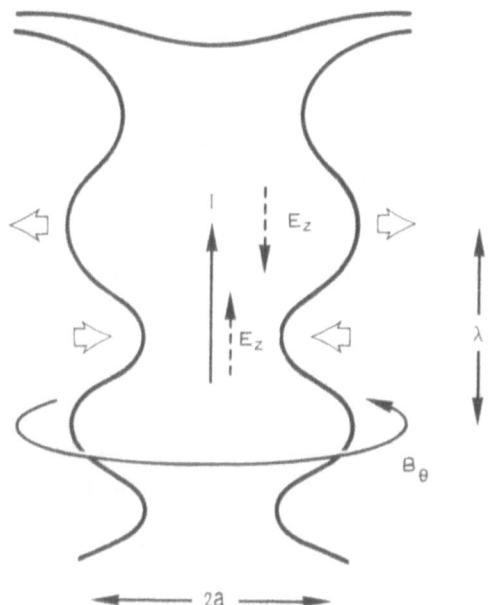

Fig. 3. "m=0" MHD instability.

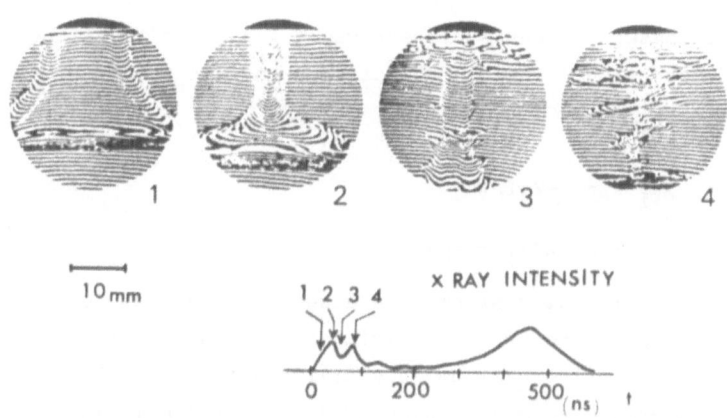

Fig. 4. Optical picture of the development of an "m=0" MHD instability.

accelerated by such fields. In spite of the differences between the various driving mechanisms, the induced electric field is given by the simple, universal formula

$$E = \alpha \, \mu_o I / \pi \tau \,.\tag{10}$$

The numerical factor α varies from 1/8 to 1 for the various cases. The characteristic time constant τ has been discussed for the respective cases, and corresponds more or less to the duration of the driving source.

The value of the electric field can be analytically determined using the following equation:

$$E \simeq 4\alpha\, I/\tau \; (MV/cm) \tag{11}$$

where I and τ are in units of MA and ns, respectively (the same units are used below.)

The distance that ions of mass $M = 2ZM_p$ travel in a time τ under the influence of an electric field E is given by

$$d = 0.96\alpha\tau I \; (cm). \tag{12}$$

The final energy W_i that the ions attain depends on whether d (or τ) is greater or smaller that the plasma length (or the transit time τ_t required to travel over the pinched plasma length l) and can be divided into two categories as follows:

[1] $\tau < \tau_t$ (or $d < l$):

$$W_i = (\alpha\,\mu_o eZI/\pi)^2/2M$$

$$\simeq 3.8\alpha^2 ZI^2 \; (MeV). \tag{13}$$

In this case, the energy is expressed very simply, is independent of τ, and is proportional to z-number of ions and the square of the current.

[2] $\tau > \tau_i$ (or $d > l$):

$$W_i = \alpha\mu_o eZlI/\pi\tau$$

$$\simeq 4.0\alpha ZlI/\tau \; (MeV), \tag{14}$$

Where l is given in units of cm. We assume for the case of a periodic electric field produced by the m=0-microscopic instability that we can use for l the effective maximum accelerated distance $\lambda/2$. In this case, also, the expression is a very simple one, but since it includes τ, it has a rather complicated parametric dependence.

In either of the above cases, if we use an appropriate discharge parameter constant, then the kinetic energy of the ions generated will be proportional to the charge state of ions Z. This scaling law is consistent with the following experimental Z-dependence:

$$W_i = ZW_o, \tag{15}$$

where W_o is a scaling coefficient around 1 MeV.

BEAM CHARACTERISTIC

Among the various ion-acceleration models discussed above, energetic ions are generated most effectively in the magnetically insulated pinch plasma, where the mobile electrons are frozen by the self-generated magnetic field. Furthermore the ions get the highest kinetic energy in this mode. Then we estimate some characteristics of the proton beam accelerated by this magnetically enhanced resistivity in the following.

Noting $M_i = M_p$ for protons instead of the relation $M_i = 2ZM_p$ assumed in Eq. (5), the maximum proton kinetic energy is given by

$$W_i \sim 1.9 \, I^2 \text{ (MeV)}, \tag{16}$$

which amounts to 1 GeV for the discharge current of 23MA. Provided that the electron current is completely prohibited from flowing, continuity of the current should be sustained by the ion beam because of presence of the finite inductance in the discharge circuit. Then we may roughly approximate the averaged ion beam current I_b by half the initial beam current I:

$$I_b \sim 0.5I. \tag{17}$$

The duration of the enhanced electric field τ_H is given in a general formula (Tanimoto et al., 1984):

$$\tau_H \sim 1.3 \times 10^{-10} I^2 / (kT)^{7/2} \quad \text{(sec)}, \tag{18}$$

where kT is the plasma temperature in keV. Noting the plasma current dependence of the temperature given by the simple MHD consideration (D'yachenko et al.), $kT \sim I^{4/7}$, τ_H becomes independent of the discharge current, ranging from several tens to a hundred ps. The field strength induced in the plasma is typically 6 GeV/m for I = 1 MA and kT = 1.5 keV. From Eqs. (16) and (17) one may expect the pulsed beam power,

$$P_b \sim 0.96 I^3 \text{ (TW)} \tag{19}$$

for the period of τ_H. Further experimental studies are important to examine these theoretical predictions.

REFERENCES

D'yachenko, V. F. and Imshennik, V. S., Reviews of Plasma Physics, ed. M. A. Leontovich, Vol. 5, p. 447.
Gary, S. P., 1974, Phys. Fluids 17, 2135.
Koyama, K., Tanimoto, M., et al., 1981, Jpn. J. Appl. Phys. 20, L95.
Tanimoto, M. and Koyama, K., 1982, Jpn. J. Appl. Phys. 21, L491.
Tanimoto, M., Ueno, I., et al., 1984, J. Appl. Phys. 23, 1470.

BEAM DYNAMICS ANALYSIS IN AN RFQ*

J. H. Whealton, R. J. Raridon, M. A. Bell, K. E. Rothe,
B. D. Murphy, and P. M. Ryan

Oak Ridge National Laboratory
Oak Ridge, Tennessee 37831 USA

ABSTRACT

Details of a fully 3-D time-dependent Vlasov-Poisson analysis are
presented including treatment of space charge, boundaries, convergence,
nonlinearities, plasma, accuracy, economy, validity, and limitations.
The potential applications to an RFQ at high beam current are indicated
with particular emphasis on beam halo or apparent emittance growth or
instabilities due to beam space charge and aberrations. A single kick
numerical runaway phenomenon is described which is important for intense
ion beams. The analysis embodies all orders of nonlinearities, all
orders of nonrelativistic image charges, and accuracies on the order of
10^{-8} radians in the ion velocity distribution function. Evolution in the
analysis has led to resource economy improvement of $\sim 10^{9}$ while simultaneously increasing accuracy by as much as 10^{6}. Such improvements permit
the consideration of complicated problems.

FORMULATION AND SIGNIFICANCE

An analysis of the time-dependent 6-D Vlasov-Poisson equations of
the form

$$\frac{\partial f(\vec{r},\vec{v},t)}{\partial t} + \vec{v} \cdot \nabla f(\vec{r},\vec{v},t) + \nabla \phi(\vec{r},t) \cdot \nabla_{\vec{v}} f(\vec{r},\vec{v},t) = 0 \; , \qquad (1)$$

* Research sponsored by IRAD funds of ORNL/Martin Marietta Energy Systems, Inc., and by the U.S. Department of Defense under Interagency
Agreement DOE No. 40-1442-84 and Army No. W31RPD-53-A180 with the Office
of Fusion Energy, U.S. Department of Energy, under Contract No. DE-AC05-
840R21400 with Martin Marietta Energy Systems, Inc.

$$\nabla^2 \phi(\vec{r},t) = \int f(\vec{r},\vec{v},t)d\vec{v} - \exp[-\phi(\vec{r},t)/T] \ , \tag{2}$$

$$\phi(\vec{s}_n,t) = \phi_n(t), \ n = 0, \ 1, \ 2, \ \ldots \ N \ , \tag{3}$$

is presented. The Vlasov equation, Eq. (1), represents the evolution of an ion distribution function, $f(\vec{r},\vec{v},t)$, subject to its inertia and the influence of time-dependent electric fields, $\nabla\phi(\vec{r},t)$. Equations (2) and (3) indicate the causes of these fields. A set of N prescribed surfaces, $\vec{s}_n$, $0 \le n \le N$, having a corresponding time-dependent Dirichlet boundary condition for the electric field potential, is one cause of the forces. In the limit of a zero-intensity beam, these boundary conditions would be the only cause of the forces, and the right-hand side of Eq. (2) would be zero, leaving just a Laplace equation. Equations (1) and (2) would only be coupled one way: $\phi \rightarrow f$. However, the presence of ion space charge causes the inhomogeneous term, the first term on the right hand side of Eq. (2), to fully couple the two Eqs. (1) and (2): $\phi \leftrightarrow f$.

Much of the physics of RFQs is contained in an accurate analysis of these equations. Besides the basic linear physics of the beam betatron oscillations, synchrotron oscillations, bunching, and acceleration, many other phenomena are in the ambit of Eqs. (1)-(3). For example, one can study beam emittance growth, or beam halo, due to (a) coherent oscillations from misalignment, vane voltage nonuniformity, and aberrant intervane phase; (b) optical aberrations due to van shape and vane-beam proximity; (c) nonlinear space-charge instabilities due to nonuniform beam (Hofmann et al., 1983; Wangler et al., 1985) and charge redistribution; and (d) all combinations of items (a) through (c) acting on the ion betatron, synchrotron, and bunching motion (McMichael, 1985). Since arbitrary tune depression can be considered, one can study accurately the compromise between space charge and emittance growth in an RFQ and perhaps minimize the emittance growth for any given beam space charge. One can study RFQ end effects, beam scrapers, novel geometries, and multiple beam funneling. Specifically the funneling scheme of Stokes (Stokes and Minerbo, 1985) or Krejcik (Krejcik, 1985) could be brought to a validated embodiment.

Finally, the last term on the right-hand side of Eq. (2) represents an equilibrium plasma of temperature T balancing out the space charge of the ions. This extremely nonlinear term produces sheaths around the plasma and allows plasma waves to propagate. Such a plasma term might be relevant to the junction between the ion source and the RFQ, and for low-frequency RFQs might be relevant near the nodes in the RF cycle. The possible importance of such phenomena in beam transport systems is indicated (Reiser, 1985), (Evans and Warner, 1971).

Two phenomena are not described by Eqs. (1)-(3). First, the vane voltages (for an RFQ) must be specified; they are not calculated. One could go a level deeper and solve the relevant Maxwell equation accounting for the 3-D cavity modes, in particular the dipole modes known to be present to some extent. An investigation along this line is underway (Whealton et al., 1986; Jaeger et al., 1986; Whealton, et al., June 1986).

Second, Eqs. (1)-(3) neglect all relativistic effects. The ion space charge waves propagate at infinite velocity, and the charges on the metallic surfaces (if assumed perfectly conducting) react to the ion beam instantly. No wake fields are produced by the finite transit time of E-M waves. These phenomena should not be confused with the term image charges frequently used (McMichael, 1985), (Schreiber, 1981); image charges in this context denotes a replacement of the boundary conditions, Eq. (3), by a series of image charges in the presence of the beam. These image charges interact with each other, and the correction charges are normally inserted in compensation. As the beam distribution changes, the series of image charges also must change to obtain Eq. (3). If one solves Eqs. (2) and (3) directly with rigorously applied boundary conditions, as presented herein, then image charges are not needed.

While space-charge waves propagate at infinite velocity in the absence of a plasma (the last term on the right-hand side of the Poisson equation), the presence of a plasma alters this. Equation (2) embodies ion acoustic eaves when, for example, the Bohm sheath criterion is not locally met. These ion acoustic waves move with the ion acoustic velocity (Whealton, Raridon, Bell, and Rothe, 1986) and cause apparently incoherent instabilities and emittance growth. The equilibrium species in the plasma [last term on right-hand side of Eq. (2)] of course move with infinite velocity in this formulation. Summarizing for the plasmaless Vlasov-Poisson system

$$|v_B| \ll |c| \ll |v_{SCW}| \approx \infty, \tag{4}$$

and for the Vlasov-Poisson with equilibrium plasma, Eqs. (1) and (2),

$$|v_B| \approx |v_{IAW}| \ll |c| \ll |v_{EPW}| \approx \infty, \tag{5}$$

where the subscripts B, SCW, IAW, EPW are beam, space charge wave, ion acoustic wave, and equilibrium plasma wave, respectively. The speed of light in vacuum is denoted as c. We will examine these points later because it is possible to get a significant variance from Eq. (4) when solving the plasmaless Vlasov-Poisson Eqs. (1) and (2).

ANALYSIS: PEDIGREE AND VALIDATION

An analysis solving time-independent Vlasov-Poisson Eqs. (1)-(3) in two dimensions is presented (Whealton et al., 1978), the first time the 2-D plasma sheath equation was solved. The difficulty is the extremely nonlinear exponential plasma term [in Eq. (2)] which causes numerical difficulties. A further resolution of this nonlinearity is provided (Whitson et al., 1978; Whealton and Whitson, 1980). Finally, the Vlasov solution was made more accurate and quicker (by orders of magnitude) (Whealton, 1981). These analyses have seen considerable experimental confirmation (Grisham et al., 1977; Kim et al., 1978; Whealton et al., 1978; Whealton, 1978; Menon, 1980; Meixner, 1981) and are used as design tools (Whealton and Whitson, 1979; Whealton and McGaffey, Stirling, 1981; Whealton, Wooten and McGaffey, 1982; Whealton and McGaffey, 1982). An extension from two dimensions to three dimensions of these steady-state equations was first made (Wooten et al., 1981) and applied to end effects in Tokamak Fusion Test Reactor injectors (Wooten, Whealton, and McCollough, 1981). This 3-D analysis was a purely finite-element calculation with 20 mode bricks of arbitrary shape (but six sides). A newer 3-D steady-state analysis was constructed which takes 1/400 of the computing resources (Whealton, McGaffey, and Meszaros, 1986). It has been applied to the design of a surface production negative ion source (McGaffey, Whealton, Raridon, Ohr, and Bell, 1984). This extra speediness is very helpful in doing complicated problems. Finally, the analysis has been made time dependent [a 3-D solution to Eqs. (1)-(3)] during the past year and a half. A comparison was made with the Los Alamos National Laboratory (LANL) RFQ code PARMTEQ (Crandall, Stokes, and Wangler, 1979) for an RFQ section where PARMTEQ is expected to be accurate (paraxial beam, hyperbola van shape, and no ion current). The two analyses agree to within 1% for the transverse emittance. Some confidence in the accuracy of the beam dynamics calculation is obtained from the validation presented in this section.

ATTRIBUTES

The attributes of the subject analysis can best be understood with reference to Fig. 1, which shows the path of the calculation. First, the Poisson equation is considered. For this first pass the source terms are set equal to zero and a Laplace equation is solved. It is solved by SOR, finite difference, and boundary interpolation within a cell (Hornsby, 1963) using a Gauss-Seidel implicit method (Smith, 1978). Considering the attributes A1, resource utilization, and A2, accuracy, SOR reduces memory requirements (A1), and boundary interpolation contributes to the accuracy per cell (A2). Generally, convergence of the solution is not

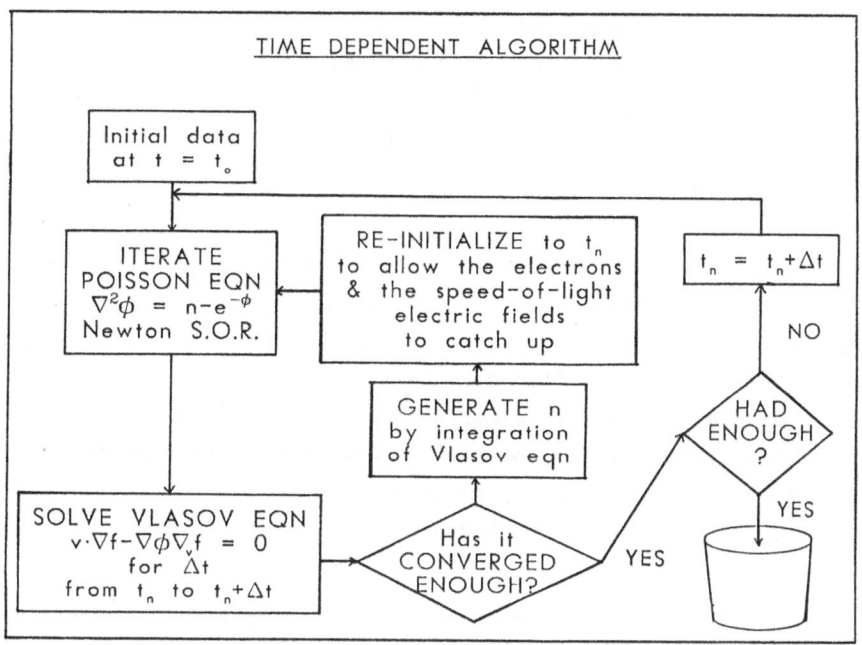

Fig. 1. Calculation procedure described in detail in the section, ATTRIBUTES.

warranted on each pass (contributes to A1), and the iteration procedure lends itself to deliberate nonconvergence. As noted before, finite difference has in our experience reduced A1 by a factor of 20 for the Poisson solution (ref. 18 vs rev. 20) for the same accuracy. Boundary conditions for arbitrarily shaped metal surfaces can be specified as time-dependent Dirichlet, or ramped Dirichlet, conditions (Hornsby, 1963) (contributes to A2). Neumann boundary conditions can also be specified.

Second, the Vlasov equation is solved for an arbitrary initial condition using the solution to the Laplace equation above for a time step Δt. The technique is described (Whealton, J. Compt. Phys., 1981; Whealton, Nucl. Instrum. Methods, 1981; Whealton, IEEE Trans. Nucl. Sci., 1981; Whealton, McGaffey and Meszaros, 1986) where significant advances in A1 and A2 are reported. The methods described by Whealton (1981, cited above) speeds up the Vlasov solver by a factor of 10 (contributes to A1) over work published by Whealton et al. (1978) or Whitson et al. (1978) while at the same time improving the accuracy by over a factor of 10 (contributes to A2). Whealton, McGaffey and Meszaros (1986) decreases resource utilization (A1) over Wooten et al., (1981) by a factor of 400 with the same accuracy (as mentioned in ANALYSIS: PEDIGREE AND VALIDATION). The trivial relationship between the coordinates inside an element and the global elements for the uniform Cartesian grid used in this algorithm allows a factor of 20 (of the 400) savings in the Vlasov

803

solver (A1) over that employed in the irregular elements of Wooten et al., (1981). As mentioned in Whealton (1981, cited above) the Vlasov solver is made self-regulating in accuracy whereupon trajectory refinement is undertaken only in those places that need it (A2).

Third, charge deposition is done in full 3-D by interpolation over the grid and is "exact" in the sense that as the 3-D grid is made more fine and the number of trajectories is increased, a result as accurate as described can be obtained (A2). Notice that nowhere is any paraxial-like assumption made, and the fields "to the orders" are directly calculated (attribute A3, nonlinear effects). Therefore aberrations (to all orders) are also directly computed. Other nonlinear optics effects (A3) computed include space charge "to all orders" due to nonuniform beam density and/or boundaries. (Boundaries cause nonlinear space charge forces also since they alter the delicate dependence of ϕ or r required to keep it linear.)

Fourth, the beam charge and the exponential plasma term (A3) are taken as inhomogeneous terms [as in Eq. (2)] to the Laplace equation solved in step 1 above. Now the two inhomogeneous terms are, in many cases, large, of opposite sign, extremely nonlinear, and three dimensional. This is the cause of numerical difficulties that were first surmounted (in 2-D steady state) in ref. 13. The technique used, accelerated under-relaxation, improved the prior art (Jaeger and Whitson, 1975) by a factor of 1000 (A1) in the beam perveance of interest, and by a greater factor for higher perveance. Another factor of 10 (A1) increase in speed was achieved while at the same time increasing the accuracy by more than a factor of 100 (A2) in Whitson et al. (1978) and Whealton and Whitson (1980). This technique was extended to three dimensions in Wooten et al (1981) and McGaffey et al. (1984). Essentially the best technique we have found is to use an unconverged Newton SOR outside of its established range of validity (Ortega and Rheinboldt, 1970).

Fifth, the time is moved back Δt, the ions are moved back to their phase space positions a time Δt ago, and the Vlasov equation is resolved with the new fields computed from the Poisson equation solution of step 4. The trajectories are different than those computed in step 2 because of the presence of the space charge terms (steps 3 and 4).

Sixth, since the trajectories of step 5 are different from those of step 2, the steps 3, 4, and 5 are repeated (Vlasov-Poisson iteration) until no change obtains. This completes the convergence procedure (A2), and it is time to proceed to the next time step. However, one should note the implications of the iteration consisting of steps 5 and 6. By deleting these steps one has the so-called "single kick" model which is almost universally used to analyze RFQs. In this model the ions can outrun the subsonic components of the space charge waves (SSSCW), or Eq. (4) is replaced by

$$|v_{SSSCW}| < |v_B| \ll |c| \ll |v_{SCW}|. \tag{6}$$

BEAM EVOLUTION IN AN RFQ

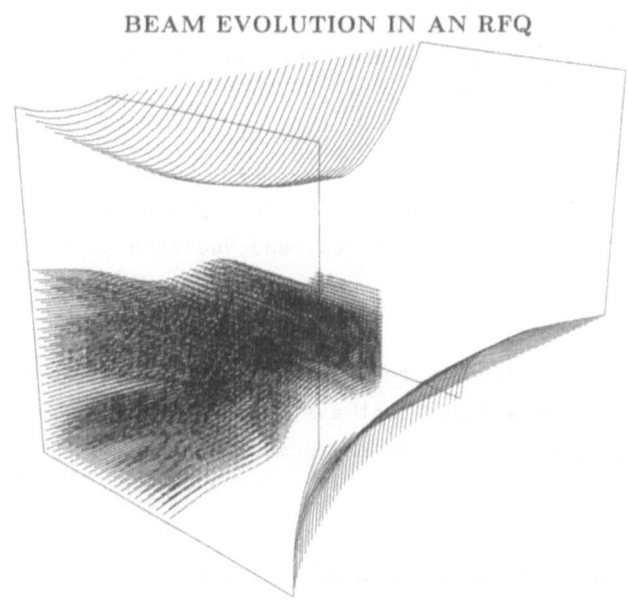

RADIAL MATCHING SECTION FROM 3D
TIME DEPENDENT VLASOV-POISSON ANALYSIS

Fig. 2. Example showing, at an instant of time, a continuous beam from an ion source, or low-energy beam transport system entering the radial matching section of an RFQ; the betatron oscillations of the beam are clearly evident.

These numerically artificial subsonic space charge waves can be shown to predominate at high beam currents (denoted as the single kick numerical runaway). The errors so introduced are likely to be cumulative.

Seventh, advance the time by Δt and do steps 2 through 6. This performs the beam evolution through the device under consideration. An example showing the transport of beam through the radial matching section of an RFQ is shown in Fig. 2.

The attributes A1 through A3 provide orbit accuracies of up to 10^{-8} radians in speedy calculations with significant nonlinearities. Six items contributing to a decrease in resource utilization (A1) total about 2×10^9 in the product of memory saved and CPU time (the accounting procedure leading to the figure of 2×10^9 is somewhat ambiguous). Five items contributing to increased accuracy (A2) make an improvement of about 10^6 for a significantly nonlinear problem.

EXTRAPOLATIONS

Magnetic fields, multiple species, and multiple plasmas can and have been included in the antecedents to the analysis described herein. Many relativistic effects can be trivially included; a full relativistic E-M analysis with all contributions to wake field effects is on the horizon as is a fully nonrelativistic E-M analysis. Some of the missing components of the latter already appear in the works of Whealton et al. (ORNL TM-7992, 1986), Jaeger et al. (1986), and Whealton et al. (June 1986).

CONCLUSIONS

The physics embodied in the subject analysis combined with its accuracy and economy suggest that a significant number of outstanding accelerator problems may now be addressed.

ACKNOWLEDGMENTS

For assistance in some computational issues we appreciate the assistance provided by G. L. Chen, D. E. Wooten, S. Y. Ohr, R. W. McGaffey, T. C. Jernigan, and D. E. Greenwood; for advice about the RFQ code PARMTEQ and other matters relating to RFQ beam dynamics theory, we appreciate the assistance of K. R. Crandall and T. P. Wangler of LANL; for their support, we appreciate the assistance of W. L. Stirling and H. H. Haselton.

REFERENCES

Crandall, K. R., Stokes, R. H., and Wangler, T. P., 1979 LINAC Conference, p. 205.
Evans, L. R. and Warner, D. J., 1971, CERN/MPS/LIN 71-2.
Grisham, L. R., Tsai, C. C., Whealton, J. H., and Stirling, W. L., 1977, Rev. Sci. Instrum. 48, 1037; Kim, J., Whealton, J. H., and Schilling, G., 1978, J. Appl. Phys. 49, 517; Whealton, J. H., Grisham, L. R., Tsai, C. C., and Stirling, W. L., 1978, J. Appl. Phys. 49, 3091; Whealton, J. H. et al., 1978, Appl. Phys. Lett. 33, 278; Menon, M. M. et al., 1980, Rev Sci. Instrum. 51, 1163; Meixner, C. N. et al., 1981, Rev. Sci. Instrum. 52, 1625.
Hofmann, I., Laslett, L. J., Smith, L., and Haber, I., 1983, Part. Accel. 13, 145.
Hornsby, J. S., 1963, CERN-63-7, A 3-D time-dependent extension was performed.

Jaeger, E. F., Batchelor, D. B., Weitzner, H., and Whealton, J. H., 1986, Comput. Phys. Commun. 40, 33.

Jaeger, E. F. and Whitson, J. C., 1975, ORNL/TM-4990 (unpublished).

Krejcik, P., in Proc. of the High Current, High Brightness, and High Duty Factor Ion Injectors (San Diego, 1985), ed. by G. H. Gillespie, Y. Y. Kuo, D. Keefe, and T. P. Wangler, AIP Conf. Proc. No. 139, p. 179.

McGaffey, R. W., Whealton, J. H., Raridon, R. J., Ohr, S. Y., and Bell, M. A., 1984, in Proc. of the 1983 Negative Ion Conference (Brookhaven National Laboratory), AIP Conf. Proc. No. 111.

McMichael, G. E., in Proc. of the High Current, High Brightness, and High Duty Factor Ion Injectors (San Diego, 1985), ed. by G. H. Gillespie, Y. Y. Kuo, D. Keefe, and T. P. Wangler, AIP Conf. Proc. No. 139, p. 153.

Ortega, J. M. and Rheinboldt, W. D., (Academic Press, N.Y., 1970), Interative Solution of Nonlinear Equations in Several Variables.

Reiser, M., in Proc. of the High Current, High Brightness, and High Duty Factor Ion Injectors (San Diego, 1985), ed. by G. H. Gillespie, Y. Y. Kuo, D. Keefe, and T. P. Wangler, AIP Conf. Proc. No. 139, p. 45.

Schreiber, S. O., 1981, presented at the High-Current Beam Dynamics Workshop, Chalk River, Ontario, Canada (unpublished).

Smith, G. D., (Oxford Univ. Press, Oxford, 1978), Numerical Solution of Partial Difference Equations: Finite Difference Methods, 2nd ed.

Stokes, R. H. and Minerbo, G. N., in Proc. of the High Current, High Brightness, and High Duty Factor Ion Injectors (San Diego, 1985), ed. by G. H. Gillespie, Y. Y. Kuo, D. Keefe, and T. P. Wangler, AIP Conf. Proc. No. 139, p. 79.

Wangler, T. P., Crandall, K. R., Mills, R. S., and Reiser, M., in Proc. of the High Current, High Brightness, and High Duty Factor Ion Injectors (San Diego, 1985), ed. by G. H. Gillespie, Y. Y. Kuo, D. Keefe, and T. P. Wangler, AIP Conf. Proc. No. 139, p. 133.

Whealton, J. H., 1981, J. Comput. Phys 40, 491; Whealton, J. H., 1981, Nucl. Instrum. Methods 189, 55; Whealton, J. H., 1981, IEEE Trans. Nucl. Sci. NS28, 1358.

Whealton, J. H., Chen, G. L., McGaffey, R. W., Raridon, R. J., Jaeger, E. F., Bell, M. A., and Hoffman, D. J., 1986, ORNL/TM-9792.

Whealton, J. H., Chen, G. L., Raridon, R. J., Bell, M. A., Rothe, K. E., 1986, Application to the dipole modes in RFQ ends of refs. 8-9 at the LINAC conference SLAC.

Whealton, J. H., Jaeger, E. J., Whitson, J. C., 1978, J. Comput. Phys. 27, 32.

Whealton, J. H., McGaffey, R. W., and Meszaros, 1986, J. Comput. Phys. 63, 20.

Whealton, J. H., Raridon, R. J., Bell, M. A., and Rothe, K. E., 1986, shown at the Charged Particle Optics Conference, Albuquerque, N. Mex. (movie).

Whealton, J. H. and Whitson, J. C., 1979, J. Appl. Phys. 50, 3964; Whealton, J. H., McGaffey, R. W., and Stirling, W. L., 1981, J. Appl. Phys. 52, 3787; Whealton, J. H., Wooten, J. W., and McGaffey, R. W., 1982, J. Appl. Phys., 53, 2806; Whealton, J. H. and McGaffey, R. W., 1982, Nucl. Instrum. Methods 203, 377.

Whitson, J. C., Smith, J., and Whealton, J. H., 1978, J. Comput. Phys., 28, 408; Whealton, J. H. and Whitson, J. C., 1980, Part. Accel. 10, 235.

Wooten, J. W., Whealton, J. H., and McCollough, D. A., 1981, J. Appl. Phys. 52, 6418.

Wooten, J. W., Whealton, J. H., McCollough, D. A., McGaffey, R. W., Akin, J. E., and Drooks, L. J., 1981, J. Comput. Phys. 43, 95.

APPENDIX B: ORGANIZING COMMITTEE,

LECTURERS, AND PARTICIPANTS

ORGANIZING COMMITTEE

Dr. H. Doucet
Laboratorie de Physique des
  Milieux Ionises
Ecole Polytechnique Plateau
  Palaiseau
91126 Palaiseau, France

Dr. T. Godlove
Department of Energy
Office of High Energy Nuclear
  Physics ER-16/GTN
Washington, DC 20545 USA

Dr. A. H. Guenther (Co-Director)
AFWL/CCN
Air Force Weapons Laboratory
Kirtland AFB, NM 87117-6008
USA

LtCol R. Gullickson
Office of the Secretary of Defense
The Pentagon
Washington, DC 20301 USA

Dr. A. K. Hyder (Co-Director)
Office of the Vice President
  for Reserch
202 Samford Hall
Auburn University
Auburn, AL 36849 USA

Dr. B. Miller
Department 1270
Sandia National Laboratories
Albuquerque, NM 87185 USA

Dr. J. Nation
224 Phillips Hall
Cornell University
Ithaca, NY 14851 USA

Dr. J. P. Rager
Comm Euro Community
DGX11 Programme Fusion
Rue de la Loi, 200
B-1049 Brussels, Belgium

Dr. C. Roberson
Physics Division
Office of Naval Research
800 North Quincy Street
Arlington, VA 22217 USA

Dr. M. F. Rose (Co-Director)
314 Nuclear Science Center
Auburn University
Auburn University, AL 36849 USA

Dr. W. Schmidt
Science Liaison Office
Karlsruhe Nuclear Research Center
One Farragut Square
Washington, DC 20006 USA

Dr. A. J. Toepfer
Science Applications Intl Co
1710 Goodridge Drive
McLean, VA 22102 USA

Mr. R. L. Verga
Office of the Secretary of Defense
The Pentagon
Washington, DC 20301 USA

Dr. J. M. Buzzi
Laboratorie de Physique
  des Milieux Ionises
Ecole Polytechnique Plateau
  Palaiseau
AS91128 Palaiseau, France

Dr. F. Cole
Fermi National Accelerator
  Laboratory
Batavia, IL 60510 USA

Dr. R. Cooper
MS H829
Los Alamos National Laboratory
Los Alamos, NM USA

Dr. M. Craddock
TRIUMF
University of British Columbia
Y6T 2A3 British Columbia
Canada

Dr. P. Elleaume
European Synchrotron Radiation
  Facility
BP 220
38043 Grenoble
France

Dr. B. Godfrey
Mission Research Corporation
1712 Randolph Road SE
Albuquerque, NM 87106 USA

Dr. T. Godlove
Department of Energy
Office of High Energy Nuclear
Physics ER-16/GTN
Washington, DC 20545 USA

Dr. T. Green
Culham Laboratory
Abingdon
Oxfordshire OX14 3DB
United Kingdom

Dr. S. Humphries
IAPBD
Farris Engineering Center
University of New Mexico
Albuquerque, NM 87131 USA

Dr. R. Jameson
AT-DD, MS, HB11
Los Alamos National Laboratory
Los Alamos, NM 87545 USA

Dr. D. Keefe
MS-47
Lawrence Berkeley Laboratory
University of California Berkeley
Berkeley, CA 94720 USA

Dr. R. Keller
Gesellschaft-fuer
Schwerionenforschung-mbH
Planckstrasse 1 D-6100
Darmstadt-Arheilgen
Federal Republic of Germany

Dr. H. Klein
Institut fuer Angewandte Physik
Johann Wolfgang Goethe
  Universitat
Robert-Mayer-Strasse 2-4
D-6000 Frankfurt 1
Federal Republic of Germany

Dr. J. L. Laclare
Laboratoire National Saturne
CEN de Saclay
91191 GIF-S/YVETTE France

Prof. J. Lawson
Rutherford Appleton Laboratory
Chilton Oxon OXII OQX
England

Dr. B. Miller
Department 1270
Sandia National Laboratories
Albuquerque, NM 87185 USA

Dr. G. Mueller
Fachbereich
  8-Naturwissenschafter I
Bergische Universitaet-
  Gesamthochschule Wuppertal
Postfach 100127
D-5600 Wuppertal 1
Federal Republic of Germany

Dr. K. Neil
1626 Lawrence Livermore Laboratory
P. O. Box 808
Livermore, CA   94550   USA

Mr. Y. Petroff
LURE, Bat 209D
Universite Paris-XI
91405 ORSAY CEDEX
France

Dr. K. Prestwich
Department 1240
Sandia National Laboratories
Albuquerque, NM   87185   USA

Dr. J. M. Reid
Kelvin Laboratory
University of Glasgow
East Kilbride   G12-8QQ
United Kingdom

Dr. C. Roberson
Physics Division
Office of Naval Research
800 North Quincy St.
Arlington, VA   22217   USA

Dr. N. Rostoker
Department of Physics
University of California
Irvine, CA   92717   USA

Dr. E. Sabia
ENEA-TIB-FIS
P. O. Box 65
00044, Frascati
Rome, Italy

Dr. T. Wangler
Los Alamos National Laboratory
At-1, MS H817
P. O. Box 1663
Los Alamos, NM   87545   USA

Dr. P. Wilson
Stanford Linear Accelerator Center
P. O. Box 4349
Stanford, CA   94305   USA

PARTICIPANTS

Dr. E. S. Ball
Naval Surface Weapons Center
Code F-12
Dahlgren, VA 22448 USA

Dr. F. Ballester
Inst di Fisica Corpuscular
Avda. Dr. Moliner, 50
Burjasot (Valencia)
Spain

Dr. C. Bameire
SGDN/AST
51, Bd de Latcur-Mauberg
Paris 75700
France

Dr. W. Bell
Department of Electrical and
Computer Engineering
University of Newcastle Upon Tyne
NEI 7RU
England

Mr. C. Burkhart
University of New Mexico
Department of Chemical Engineering
Albuquerque, NM 87131
USA

Dr. E. Casal
Inst de Fisica Corpuscular
Avda. Dr. Moliner, 50
Nurjasot (Valencia)
Spain

Mr. R. A. Charles
Culham Lab
Abingdon, Oxfordshire
OX14 3DB
United Kingdom

Dr. H. C. Chen
Naval Surface Weapons Center
R41, White Oak
Silver Spring, MD 20903 USA

Dr. D. Chernin
Science Applications International
Corporation, Division 157
P. O. Box 1303
McLean, VA 22102 USA

Dr. F. Ciocci
ENEA-TIB-FIS
P. O. Box 65
00044 Frascati
Rome, Italy

Dr. G. Clark
Department of Physics
University of St. Andrews
Fife KY16 9SS
Scotland

Dr. R. J. Commisso
Naval Research Lab
Code 4770
4555 Overlook Avenue, SW
Washington, DC 20575 USA

Dr. R. Cutler
National Bureau of Standards
Building 245
Room B116
Gaithersburg, MD 20899 USA

Dr. S. Darendelioglu
Agdulaziz Mah.
Sirin Hanim Sok
No =5/9
Turkey

Dr. R. Dewitt
Naval Surface Weapons Center
Code F-12
Dahlgren, VA 22448 USA

Dr. J. Diaz
Inst de Fisica Corpuscular
Avda. Dr. Moliner, 50
Burjasot (Valencia)
Spain

Mr. A. Donaldson
Texas Tech University
Box 4439, TTU
Lubbock, TX 79409 USA

Ms. S. Embry
U.S. Army Strat Def Comm
DASD-H-WD-P
P. O. Box 1500
Huntsville, AL 35807 USA

Dr. J. A. Farrell
Los Alamos National Laboratory
P. O. Box 1663
Los Alamos, NM 97545 USA

Dr. J. Ferrero
Inst de Fisica Corpuscular
Avda. Dr. Moliner, 50
Burjasot (Valencia)
Spain

Dr. T. R. Fisher
Dept 91-10
Building 203
Lockheed Missiles and Space
  Company
3251 Hanover Street
Palo Alto, CA 94304 USA

Mr. O. Gal
Laboratoire de Physique
  des Mileux Ionises
Ecole Polytechnique Plateau
  Palaiseau
91128 Palaiseau
France

Mr. F. G. Gallagher-Daggitt
SDI-PO, Room 350
Northumberland House
Northumberland Avenue
London WCAV 5BP
England

Dr. R. Gandy
Auburn University
Physics Department
Allison Laboratory
Auburn University, AL 36849
USA

Dr. T. Garvey
LEP Division
CERN CH-1211
Geneva, Switzerland

Dr. S. Gecim
Hacettepe University
Department of Electrical
  Engineering
Baytape-Ankara
Turkey

Dr. G. Gillespie
Physics Dynamics Inc.
P. O. Box 1883
La Jolla, CA 92038 USA

Dr. S. Gralnick
Grumman Corporation
MS-C47-05
Bethpage, NY 11714 USA

Dr. W. Graybeal
AFWL/AWY
Kirtland AFB, NM 87117-6008
USA

Dr. P. Guimbal
B.P.N. 12, C.E.A.
91680 Bruyeres-LeChatel
France

Dr. J. Head
AFWL/AWY
Kirtland AFB, NM 87117-6008
USA

Ms. M. Herranz
Dep. Fisica Nuclear
E.S.I. Industriales
Alda, de Urquijo, s/n
48013 Bibao
Spain

Dr. J. Hiskes
Lawrence Livermore National
Laboratory, L-630
P. O. Box 5511
Livermore, CA 94550 USA

Dr. A. J. T. Holmes
Culham Laboratory
Abingdon
Oxfordshire OX 14 3DB
United Kingdom

Dr. R. Kribel
Auburn University
Physics Department
Allison Laboratory
Auburn University, AL 36849
USA

Dr. S. Kuo
Polytechnic University
Route 110
Farmingdale, NY 11735 USA

Dr. M. Law
TRIUMF
4004 Wesbrook Mall
University of British Columbia
Vancouver, British Columbia
  V6T 2A3
Canada

Mr. R. Lucey
University of Michigan
121 Cooley Building
2355 Bonisteel
Ann Arbor, Michigan 48109 USA

Mr. L. H. Luessen
Naval Surface Warfare Center
Code F-12
Dahlgren, VA 22448 USA

Dr. F. Mako
Naval Research Laboratory
Code 4711
4555 Overlook Avenue
Washington, DC 20375 USA

Dr. J. Marilleau
CEL-V
94190 Villeneuve
St. Georges France

Dr. R. A. Meger
Naval Research Laboratory
Plasma Physics Division
Code 4750
Washington, DC 20375 USA

Dr. K. Mittag
IK/KFK
P. O. Box 3640
7500 Karlsruhe
Federal Republic of Germany

Dr. M. Molen
Old Dominion University
School of Engineering
Norfolk, VA 23508 USA

Dr. L. Orphanos
Institute of Nuclear & Particle
   Physics
Department of Physics
University of VA
Charlottesville, VA 22901 USA

Dr. A. D.R. Phelps
Strathclyde University
Physics Department
Glasgow G4 ONG
Scotland

Dr. C. Pidgeon
Department of Physics
Heriot-Watt University
Edinbourgh, England

Dr. C. Pirrie
English Elect Valve
Waterhouse Lane
Chemsford, Essex CMi 2QU
England

Dr. C. Planner
Rutherford Appleton Laboratory
Chilton Didcot
Oxon OX11 QX
England

Dr. D. Prono
Lawrence Livermore National
   Laboratory
P. O. Box 808
L-626
Livermore, CA 94550 USA

Dr. H. Pugh
AFSTC/NPT
Kirtland AFB, NM 87117-6008
USA

Dr. S. Putnam
Pulse Sciences, Inc.
14796 Wicks Boulevard
San Leandro, CA 94577 USA

Dr. J. Rawls
GA Technologies
P. O. Box 85608
San Diego, CA 92138 USA

Mr. I. Roth
Department of Electrical
Engineering
Cornell University
909 Mitchell Street
Ithaca, NY 14850 USA

Dr. J. M. Salome
Commission European Community
CBNM-EURATOM
Steenweg naar Retie
B-2440 Geel
Belgium

Dr. J. Shiloh
P. O. Box 2250
Department 23
Rafael, Haifa
Israel

Dr. B. Smith
Air Force Office of
   Scientific Research
Building 410
Bolling AFB, DC 20332
USA

Dr. T. Teich
Federal Inst of Technology
F. G. Hochspannung Physic
str. 3
Zuerich CH-8092
Switzerland

Dr. I. Ueno
Department of Electrical
Engineering
University of Tokyo
7-3-1 Hongo
Bunkyo-Ku, Tokyo   113
Japan

Dr. S. Webb
Space Department R14
Royal Aircraft Est
Franborough
Hants GU14 GTD
United Kingdom

Dr. J. Whealton
Martin Marietta Energy Systems
P. O. Box Y
Oak Ridge, TN   37831   USA

Dr. J. Williams
Auburn University
Physics Department
Allison Laboratory
Auburn University, AL   36849
USA

Dr. C. Wilson
Department of Physics
University of St. Andrews
Fife KY16 9SS
Scotland

Dr. M. Wilson
National Bureau of Standards
Building 245
B102
Gaithersburg, MD   20899   USA

Ms. P. Whited (Institute
  Secretary)
AFWL/CCN
Air Force Weapons Laboratory
Kirtland AFB, NM   87177-6008   USA

INDEX